Series: Environmental and Agricultural Nanotechnology

Series Editor: Kamel A. Abd-Elsalam

Volume

Green Nanotechnology for Stress-Resilient Crops and Sustainable Agroecosystems

Books in the Series

- Biopolymer Nanoparticles: A Novel Approach for Eco-conscious Environmental Applications
  Ayat F. Hashim, Rabab M.A. El-Sherif, Farah K. Ahmed and Sabu Thomas (eds.), 2026
- Green Nanotechnology for Stress-Resilient Crops and Sustainable Agroecosystems
  Jen-Tsung Chen (ed.), 2026
- Soil Nanotechnology
  Akhtar Hameed and Muhammad Zahid (eds.), 2026
- Smart Nanomaterials for Environmental Remediation
  Megha Barot and Azamal Husen (eds.), 2026

Series: Environmental and Agricultural Nanotechnology

Series Editor: Kamel A. Abd-Elsalam

Volume

Green Nanotechnology for Stress-Resilient Crops and Sustainable Agroecosystems

Editor

Jen-Tsung Chen

Department of Life Sciences
National University of Kaohsiung
Taiwan

CRC Press
Taylor & Francis Group
Boca Raton London New York

CRC Press is an imprint of the
Taylor & Francis Group, an **informa** business

A SCIENCE PUBLISHERS BOOK

First edition published 2026
by CRC Press
2385 NW Executive Center Drive, Suite 320, Boca Raton FL 33431

and by CRC Press
4 Park Square, Milton Park, Abingdon, Oxon, OX14 4RN

CRC Press is an imprint of Taylor & Francis Group, LLC

Library of Congress Cataloging-in-Publication Data (applied for)

ISBN: 978-1-032-87912-3 (hbk)
ISBN: 978-1-032-93791-5 (pbk)
ISBN: 978-1-003-56768-4 (ebk)

DOI: 10.1201/9781003567684

Typeset in Times New Roman
by Prime Publishing Services

Preface to the Series

Welcome to the "Environmental and Agricultural Nanotechnology" book series, a specialty forum revealing the breakthrough promise of nanotechnology in two of the most vital areas of human endeavor: environment and agriculture. As Kamel A. Abd-Elsalam, Series Editor from the Plant Pathology Research Institute, Agricultural Research Center (ARC), I'm pleased to present this series of publications by leading figures in the field.

Nanotechnology is a promise technology, with creative solutions to the majority of the globe's most contentious problems. It presents innovative alternatives in environmental applications to prevent pollution, protect resources, and effectively address environmental effects. Consider new nanomaterials with previously unavailable pollutant absorption properties from air, water, and soil, or new sensors with specific environmental detection and cleansing. These are no longer concepts on paper anymore; they are becoming things and hard facts in a rush to redefine our way of environmental stewardship.

Parallel to the concerns, nanotechnology is also going to revolutionize agriculture, ushering in new avenues of food production and sustainable land management. This includes higher yields from crops through improved delivery of nutrients, optimization of nutrient use efficiency, enhanced fertility of the soil, and development of climate change-resilient agro-systems. The combination of the nanotechnological advances is the door to a future where farming is not only productive but considerably more sustainable.

This book series is to be a comprehensive treatise, taking the reader through the fundamentals of and concepts in nanoscience, nanomaterials, and their unique properties in relation to environmental and agronomic applications. We cover the specifics of the application of nanotechnology in controlling environmental pollution, e.g., its worth in contaminant decontamination and environmental monitoring. Moreover, we detail how nanotechnology can be applied in sustainable agricultural use and resource management, such as the application of targeted agrochemical delivery as well as crop adaptation to climate change.

Through these volumes, we hope to impart a deep appreciation of recent advances and envision future breakthroughs. This volume is a must-read for researchers, academics, policymakers, and entrepreneurs interested in leveraging nanotechnology for a sustainable and green future. In our opinion, by bridging the gap between frontier nanoscience and practical use in the environment and agriculture, we can contribute importantly to global efforts for sustainable development and higher living standards.

Taylor & Francis
Taylor & Francis Group
http://taylorandfrancis.com

Preface

Food and nutrition security have faced serious threats globally due to the changing climate, the increasing human population, and pollution over the recent decades. Fortunately, scientists have developed various technologies that are expected to mitigate the damage, among which, nanotechnology is a highly promising approach for developing smart agriculture and organizing sustainable agroecosystems.

For years, a range of engineered nanomaterials/nanoparticles (NPs) have been proven to possess positive impacts on agricultural production and environments. To advance their practical usage, green synthesis of NPs, i.e., green nanotechnology, has set a new generation for producing smart, eco-friendly, cost-effective, and safety-engineered nanomaterials.

This book provides cutting-edge knowledge of green nanotechnology and its current applications in regulating plant stress responses and managing agricultural environments. It covers all types of plant stressors consisting of biotic and abiotic factors, such as temperature, salinity, drought, heavy metals, pests, and pathogens. This book proposes strategies to enhance plant stress resilience based on green nanotechnology and, additionally, discusses and shares opinions on limitations and future directions when applying green nanotechnology in agri-food systems.

This book summarizes the current achievements of green nanotechnology in developing smart and sustainable agriculture highlighting the applications of advanced approaches using biogenic nanoparticles, nanofertilizers, nanopesticides, and nanobiosensors, for reducing the negative effects of environmental stress and plant diseases in agricultural production, and additionally, advance the development of sustainable agroecosystems through bioremediation and the beneficial of eco-friendly, cost-effective, safety and smart engineered green nanoparticles, in which to support "SDGs" and the "Paris Agreement".

The book editor, Jen-Tsung Chen, thanks all contributors for their innovative and insightful chapters. The instruction and help from the publisher are greatly appreciated.

Taylor & Francis
Taylor & Francis Group
http://taylorandfrancis.com

Contents

Preface to the Series v

Preface vii

1. Green Nanotechnology for Advancing Agricultural and Environmental Management: An Updated Overview 1
Nidaa Harun, Sidra Anwar, Robina Kousar, Hummera Nawaz and *Khawaja Shafique Ahmad*

2. Green Nanotechnology for Improving Plant Abiotic Stress Resilience under Climate Change 26
Ratneswar Poddar, Susnigdha Sengupta, Arup Sen, Arindam Sarkar and *Akbar Hossain*

3. Green Nanotechnology for Alleviating Salinity Stress of Plants 41
Sureshkumar Tharani, Govindarajan Ramkumar, Krishnan Kutty Sunitha Kumari, Jaison Sarah, Siva Daniel Ajay Samuel and *Balasundaram Saraswathy Chithra Devi*

4. Enhancing Drought Stress Resilience in Plants Based on Green Nanotechnology 56
Muhammad Faheem Iqbal, Iqra Irshad, Huma Arshad, Muhammad Uzair, Murtaza Hasan and *Ghazala Mustafa*

5. Temperature Stress Resilience in Plants Induced by Green Nanotechnology 73
Kuchipudi Rakesh, Ankan Das, Rajashree Mitra Thakur and *Sandip Debnath*

6. Green Nanotechnology for Mitigating Heavy Metals Stress in Plants 97
Saiqa Nazir, Hajra Abid, Sajjad Hussain, Anar Mehmood and *Khawaja Shafique Ahmad*

7. Green Nanofertilizers for Sustainable Agriculture 111
Ankita Gogoi, Nikha Borah, Kabyashree Buragohain, Sukanya Sonowal, Ratul Nath, Bhaben Chowardhara, Debojyoti Moulick and *Akbar Hossain*

8. Nanobiosensors for Smart and Sustainable Agriculture 140
Soumya Ranjan Mahapatra, Anish Bebarta, Akash Kumar Jena, Amiya Ranjan Tripathy and *Jyoti Prakash Sahoo*

9. Green Nanotechnology for Plant Disease Management: An Overview 156
Mohammed Radi, Mohammed Taoussi, Khadija Benamar and *Rachid Lahlali*

10. Plant Viral Disease Management Based on Green Nanotechnology 195
Esha Kumari, Sneha Kumari, Sai Santosh Das and *Jyoti Prakash Sahoo*

11. Green Nanotechnology for Protecting Plants against Bacterial Diseases 205
Aldo Córdoba Guerrero, Adrian Salcedo Gómez, Daniel Mendoza Jiménez, Fidel Landeros Jaime, José Antonio Cervantes Chávez and *Karen Esquivel Escalante*

12. Plant Pest Control by Green Nanotechnology 236
Kabyashree Buragohain, Sukanya Sonowal, Dhruba Jyoti Bordoloi, Ankita Gogoi, Nikha Borah, Ratul Nath, Bhaben Chowardhara, Debojyoti Moulick and *Akbar Hossain*

13. Multi-Omics of Nanotechnology-Mediated Abiotic Stress Resilience in Crops 258
Saher Fatima, Shaharyar, Saira Zaheer, Huma Arshad, Sunbal Khalil Chaudhari, Murtaza Hasan and *Ghazala Mustafa*

14. Green-Engineered Nanoparticles for Smart and Sustainable Agriculture 270
Ahmed M. Galal, Thomas Mounir, Karim Ehab Lasheen, Hala Hisham, Farah Esmail Mohamed and *Sozan E. El-Abeid*

15. Green Nanotechnology for Supporting Global Nutrition and Food Security 288
Hanane Choubbane, Khadija El hazzam and *Fatima El Amerany*

16. Nanoparticle-Delivering Biostimulants for High-Yield, High-Quality and Stress-Resilient Crops 305
Mohsen Mohamed Elsharkawy

17. Plant-Derived Nanoparticles for Stress-Resilient and Sustainable Agriculture 322
Israt Jahan, Meskul Zannat, Arup Ratan Basak, Arpita Rani Roy, Srishti Chauhan and *Prodipto Bishnu Angon*

18. Microbial-Derived Nanoparticles for Stress-Resilient and Sustainable Agriculture 351
Swapna Battana, Srinivasan Kameswaran and *Mandala Ramakrishna*

19. Green Nanofertilizers: Types, the Mode of Action and Pollution Mitigation for Sustainable Agriculture and Agroecosystems 370
Kun Chen, Mozammel Hoque, Naser Khan, Philip Kwong, Mohammad Boshir Ahmed and *Volker Hessel*

Index 387
About the Editor 389

CHAPTER-1

Green Nanotechnology for Advancing Agricultural and Environmental Management

An Updated Overview

Nidaa Harun,[1] *Sidra Anwar,*[1] *Robina Kousar,*[1] *Hummera Nawaz*[2] and *Khawaja Shafique Ahmad*[3,*]

1. Introduction

Agriculture has made a significant contribution to the human ecological success (Smith, 2007) and most of the population in the world (more than 60%) trusts agriculture for survival (Bruinsma, 2017). Since agriculture produces both food for human consumption and raw resources for other sectors, it is one of the most important and secure industries (Tipu et al., 2021). It is an activity that has kept pastures and food available. It is important as a source of income, particularly in developing nations. In addition to farmers' disinterest in the sector, the agriculture industry is presently facing certain domestic and international challenges, such as climate change, low productivity, organic matter loss in soil, inadequate supply of micro/macronutrients, inefficient nutrient use, shortage of water and agricultural land, and other environmental problems like rising pesticide residue levels and a labor shortage (Chen and Yada, 2011). Additionally, the cost of materials is very high. According to an estimate nearly 70% of all freshwater reservoirs, 187 million tons of fertilizer, 4 million tons of insecticides, and greater than 2 quadrillion BTU of energy will be required for the 3 billion tons of annual crop production worldwide (Usman et al., 2020). 'Zero hunger' is one of the 17 SDGs (sustainable development goals) of the UN (United Nations) and sustainable agriculture is crucial to achieve that goal. Food and Agriculture Organization (Usman et al., 2020), stated that by 2050 population of the world is predicted to an extent 10 billion, which will boost the 50% world's food necessities, particularly in developing nations. This increased agricultural demand also puts pressure to use unsustainable and non-friendly agricultural approaches (like chemically synthesized pesticides, etc.) which ultimately contaminate our environment. In addition to rapid urbanization and industrialization, extensive use of metals and synthesized organic agents also plays a significant

[1] Department of Botany, Faculty of Life Sciences, University of Okara, 56130, Okara, Pakistan.
[2] Department of Botany, University of Poonch Rawalakot, Rawalakot, 12350, Pakistan.
[3] Department of Botany, Division of Science and Technology, University of Education, Lahore, 55210, Pakistan.
* Corresponding author: ahmadks@upr.edu.pk

DOI: 10.1201/9781003567684-1

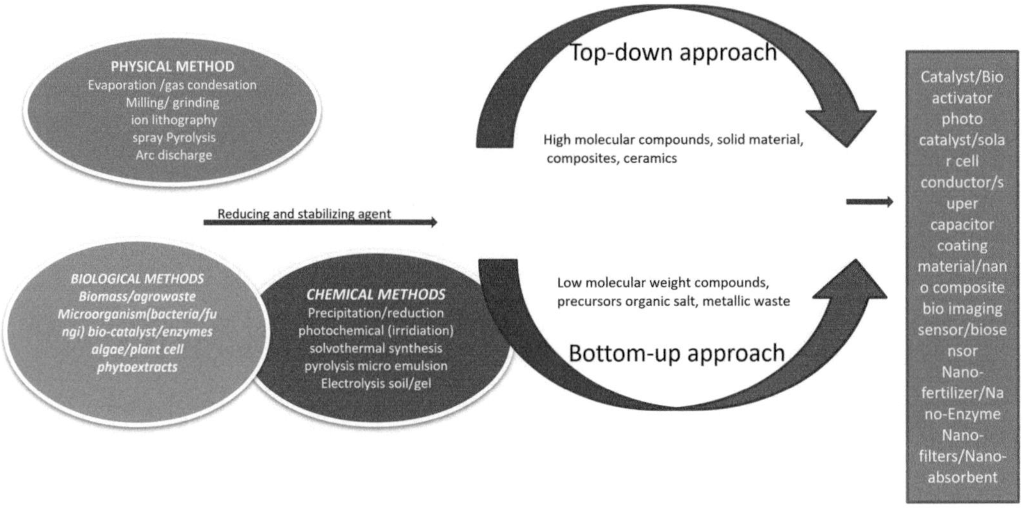

Figure 1. Multiple approaches for the nanoparticle synthesis.

role in leveling up environmental impurity (Benjamin et al., 2019). To meet these challenges alternate sustainable strategies must be incorporated. The sustainable agriculture notion became significant after the report of Brundtland Commission Report (1987). Department of Agriculture, United States (2012) has provided its definition as "agriculture is basically about livestock and production of various crops having an impact on the environment". The preservation of the balance between the need for food and the conservation of natural assets is the fundamental objective of sustainable agriculture. Other aims like reduction in the application of inorganic fertilizers, water assets protection, waste reduction, and biodiversity conservation of biodiversity are also associated (Velten et al., 2015; Mahmoud et al., 2018). Supporting farmers financially and using innovative methods while also providing improved facilities to improve their quality of life is another goal of the sustainable approach (Campbell et al., 2014). Currently, nanotechnology is considered as promising tool to upsurge the competence in the agriculture field by offering green solutions for better food yield and security.

Nanotechnology-across-domainsubjectencompassingthefollowing:engineering,biotechnology, chemistry, physics and physics and microbiology, focusing on the fervent development, application, and improvement of nanoscale objects-1 to 100 nanometers in diameter (Chaudhary et al., 2022; Nawaz et al., 2024). Two types of synthesis approaches are in practice, i.e., top-down and bottom-up (Fig. 1). Macroscopic components are reduced in size to form appropriate-sized aggregates, which is how nanocomposites are made (Palmero, 2015). Utilizing chemical, physical, and biological processes, bottom-up approaches create nanoscale structures from atomic and molecular materials (Bhandari Geeta et al., 2023). Stability, physical strength, optical properties, electrical conductivity, reactivity, and magnetic characteristics are some of the new attributes that make nanoparticles (NPs) distinct (Nejatzadeh, 2021). Because of this, NPs are utilized in many different fields, like pharmaceutical and skin care sectors, as well as in the fields of electrochemistry, electronics, food processing, textiles, sensors, optical devices, agriculture, sensors, and energy (Fig. 2). As these particles are vastly in use in almost every aspect of our life therefore it is necessary to develop them through the non-toxic approach. Several chemical techniques are in practice to synthesize nanoparticles (NPs) from various sources, but nowadays there has been a noticeable trend toward "green" technologies that use inexpensive, renewable, and environment-friendly ingredients (Patil and Chandrasekaran (2020). Numerous research conducted in past years also strengthen and support this argument (Singh et al., 2015; Bansal et al. 2022; Singh et al., 2021).

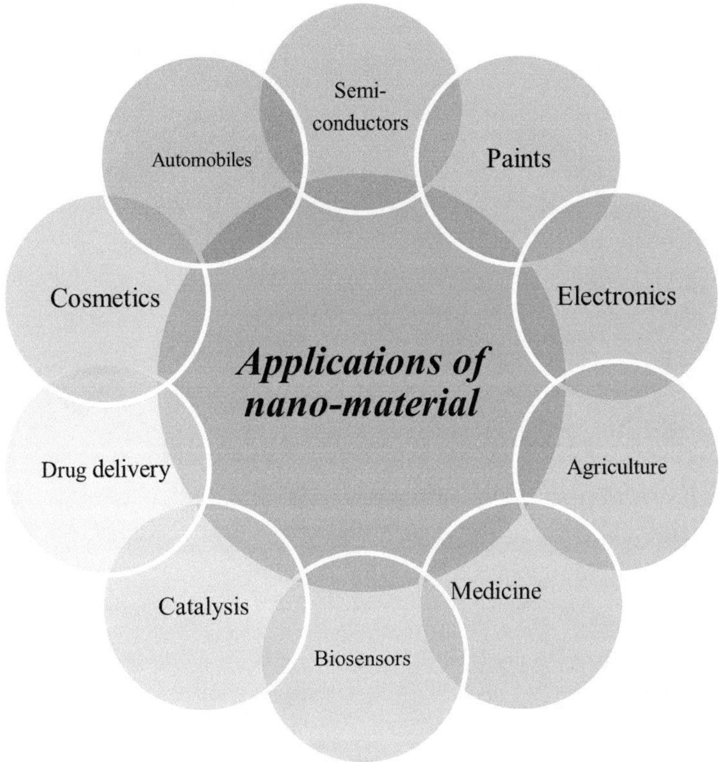

Figure 2. Applications of nanomaterial in different fields.

2. Application of green nanotechnology in agricultural

Adaptive drug delivery systems, plant growth promoters, genetically modified animals and plants, biosensors, nutraceuticals, nanoherbicides, nanopesticides, and nanofertilizers are just a few examples of how nanotechnology might change agriculture (Chaudhary et al., 2022). By employing nanotechnology for "sustainable intensification," it is feasible for plants' protection, can track plant development, identify animal as well as plant illnesses, increase food production worldwide, improve food value, and cut unwanted waste (Sekhon, 2014). The field of nanotechnology has emerged as one that has fundamentally changed agronomic techniques and given sustainable agriculture its true meaning. Nanomaterials in agriculture offer a wide variety of applications in the production of crops, water-soil management, and chemical usage in a controlled way, plant protection, and diagnostic testing because of their special characteristics, lesser size, and surface-to-volume ratio. One of the significant advancements in better farming has been the application of nanotechnology, which has led to the creation of nano-based insecticides, herbicides, and fertilizers, early pathogen diagnosis tools (Fig. 3) (Nazir et al., 2020). This nanotechnology has gained its place in conventional agricultural practices due to its numerous benefits like improved soil quality, smart monitoring system, improved enzymatic activity, and more efficient nutrient uptake (Aslam et al., 2022).

2.1 Nanofertilizers

There is a dearth of published research on nanofertilizers technology in scientific journals, despite its highly inventive nature. By using nanofertilizers in place of conventional fertilizer application techniques, nutrients can be released into the soil slowly and under control, avoiding eutrophication

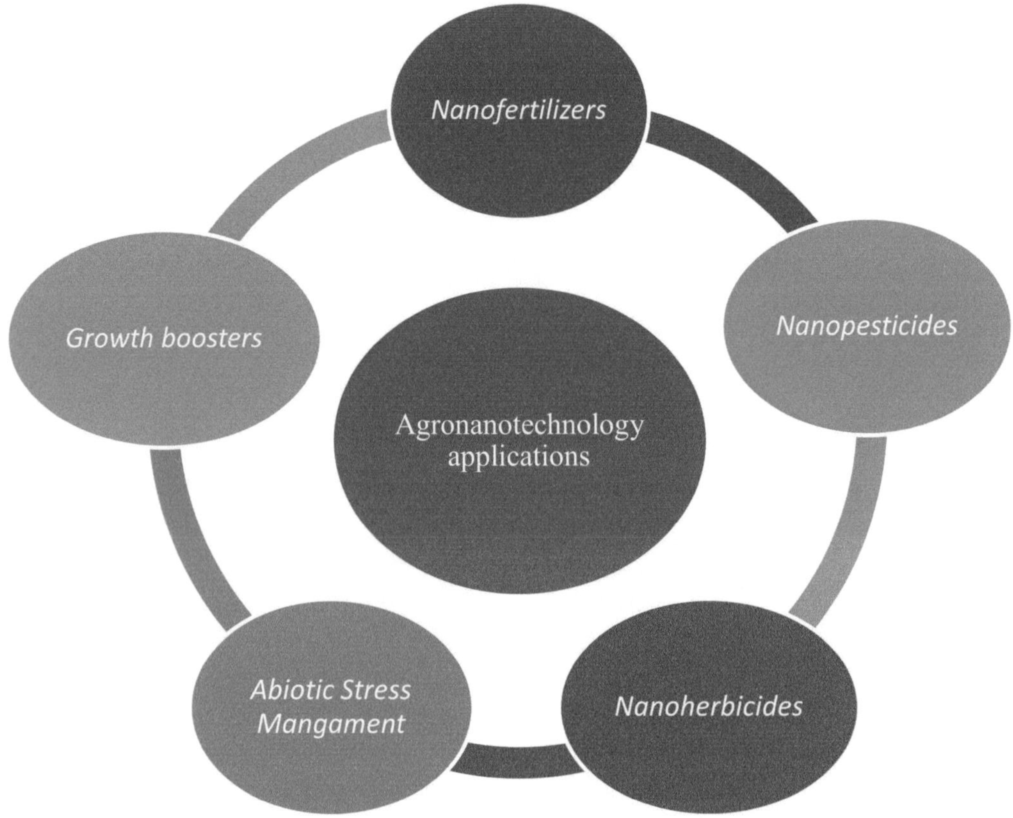

Figure 3. Diagrammatic representation of nanoparticles applications in the field of agriculture.

and pollution of water reservoirs (Naderi and Abedi, 2012). A study reported that maize treated with TiO_2 nanoparticles showed significant growth effects in comparison to bulk TiO_2 treatment. Phenomena like light absorption and photo-energy transfer were both enhanced by titanium nanoparticle treatment. Another study reported that a combination of SiO_2 and TiO_2 nanoparticles produces a profound effect on the ability of plants to absorb water and nutrients by increasing the activity of nitrate reductase in soybeans (Lu et al., 2002). An environmentally viable nano-organic iron-chelated fertilizer has been created by Iranian researchers. Ultrahigh absorption, improved crop production, increased photosynthesis, and a notable increase in the surface area of the leaves are some of the distinctive characteristics of nanofertilizers (Iran Nanotechnology Initiative Council, 2014).

Nanofertilizers are of three types, i.e., Macronutrient Nanofertilizer, Micronutrient Nanofertilizer and Nano-particulate Fertilizer. Macronutrient Nanofertilizers are those nutrients that are mainly necessary for the plants' development and growth, e.g., Nitrogen, Sulfur, Phosphorus, Magnesium, Calcium and Potassium. As the nanomaterials have a comparatively high volume/surface ratio, therefore, it works more efficiently rather customary macronutrient fertilizers (Sekhon, 2014). Applying macronutrient nanofertilizers significantly increased crop yield, according to research findings (Liu and Lal, 2014; Ditta et al., 2015).

The utilization of nanofertilizer increases element efficiency, reduces soil toxicity, mitigates the negative impacts of fertilizer overconsumption, and reduces fertilizer application frequency (Table 1) (Naderi and Danesh-Shahraki, 2013). The beneficial effects of zinc oxide nanoparticles on the plant tomato, such as their uptake and penetration into the leaves, pave the way for their possible

Table 1. The effects of different agricultural nanofertilizers on crop on multiple crops

Sr. no	Agricultural Crop	Particular Nano-fertilizer	Amount [PPM]	Major highlights	Citations
1	*Arachis hypogaea*	Calcium	-	Plant growth and nutrient substances	(Liu et al., 2005)
2	*Cucumis sativus*	Ceric oxide	400	Globuline content and quality of food	(Zhao et al., 2014)
3	*Cicer arietinum*	Molybdenum oxide	8	The increased mass of plant	(Taran et al., 2014)
	Glycine max	Iron oxide	30–60	Effects on chlorophyll	(Ghafariyan et al., 2013)
	Glycine max	Phosphorus	100	Increment in growth rate	(Liu and Lal., 2014)
	Lolium	Zinc	1–2000	Root length elongation	(Lin and Xing, 2008)
	Lactuca sativa	Copper	130–600	Shoot length, germination of seed and improvement in soil microbe community	(Shah and Belozerova, 2009)
	Oryza sativa	Manganese	-	Zn uptake was improved	(Yuvaraj and Subramanian, 2015)
	Pennisetum glaucum	Gold	50	Seedling growth and seed germination effect	(Parveen et al., 2016)
	Phoenix dactylifera	Carbon nanotube	0.05–0.1	Effect on Leaf number and shoot length	(Taha et al., 2016)
5	*Spinacia oleracea*	Titanium Dioxide	0.25–4	N_2 fixation was improved	(Zheng et al., 2005)
9	*Vigna unguiculata*	Magnesium	2.5	Increment in chlorophyll	(Delfani et al., 2014)
10	*Vigna radiata*	Manganese	0.05–1	Chlorophyll content was increased	(Pradhan et al., 2013)
11	*Vigna radiata*	Zinc oxide	1–2000	Growth rate increased	(Mahajan et al., 2011)
14	*Zea mays*	Copper oxide	10	Plant growth and roots	(Adhikari et al., 2016)

application as a nanofertilizer in the future. According to pot tests using foliar spray technology, plants sprayed with a 20 mg mL^{-1} zinc oxide nanoparticle solution exhibited better growth and biomass output than control plants

In place of conventional fertilizers, using nanofertilizers can help control the flow of nutrients in the soil and lessen losses brought on by chemical fertilizers (Tripathi et al., 2018). By using nanoparticles coated in a thin layer or administered as emulsions during nanofertilization, nutrients can be entrapped. Crop nutrient usage is enhanced by fertilizers coated with nanoparticles because of their gradual discharge of nutrients. Iqbal (2019) investigated the impact of nanofertilizers on agricultural sustainability. According to a number of other studies, nanofertilizers might be used as small-quantity, reasonably priced substitutes for conventional fertilization techniques (Fig. 4) (Kaila et al., 2021; Liu et al., 2021).

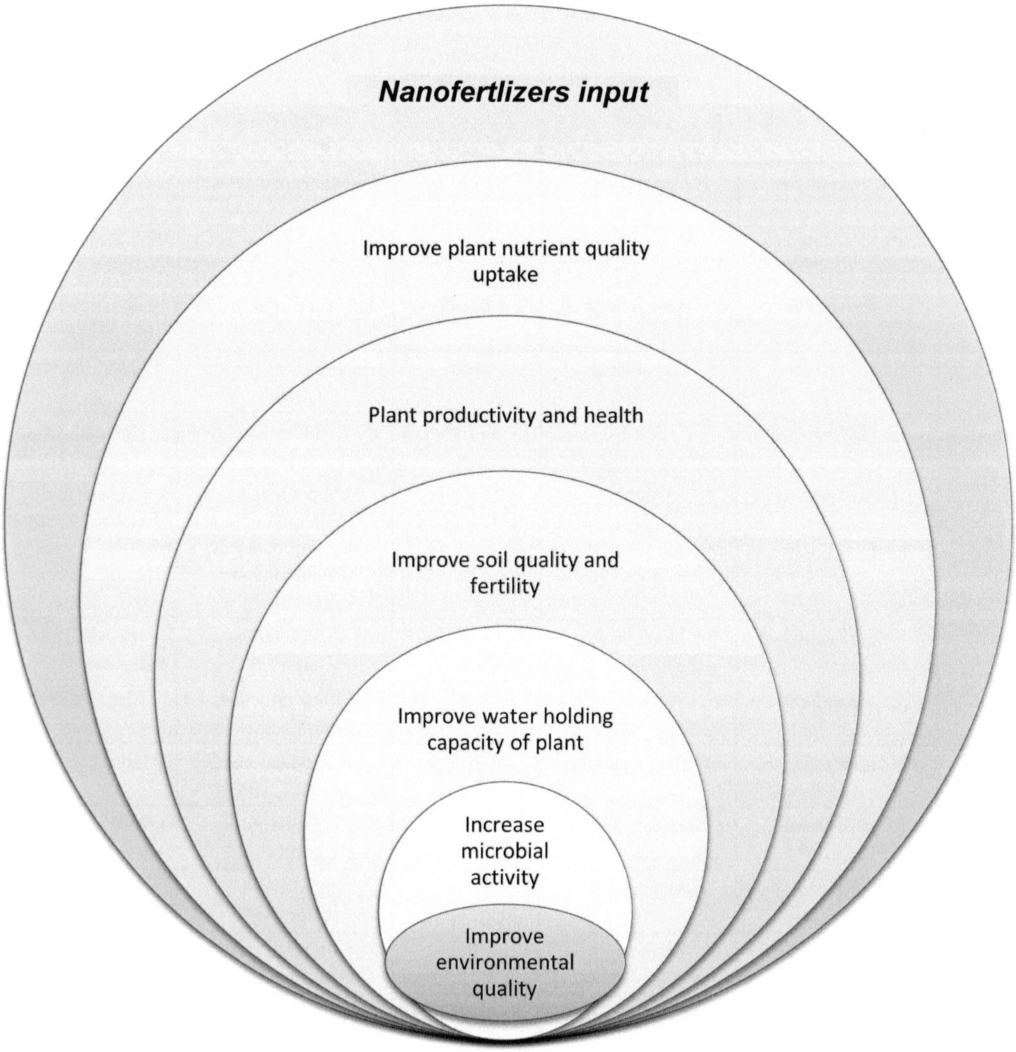

Figure 4. Diagrammatic representation of applications as nanofertilizers.

2.2 Growth boosters

Nanotechnology is incredibly promising for solving a number of agricultural problems. Nanoparticles, which bridge the gap between macroscopic materials and minute atomic or molecular structures, have attracted a lot of scientific interest because of their potential to completely transform agricultural methods (Shang et al., 2019). Conventional plant fertilizers have several drawbacks, including excessive application requirements, harmful environmental impacts, and toxicity to non-target species. Overreliance on nitrogen and phosphate-rich fertilizers can lead to water contamination through runoff and algal blooms, compromising freshwater resources. Furthermore, agriculturalists face substantial financial burdens in managing fertilizer demands annually, resulting in significant economic losses (Savci, 2012). Zinc, an indispensable plant nutrient, plays a crucial role in sustaining cellular metabolic processes in plants. Zinc ionic species activate various biocatalysts that regulate diverse physiological functions, including hydrocarbon metabolism, cell membrane integrity, leaf development, membrane physiology, and protein synthesis (Mori et al., 2016; Hafeez et al., 2013). A zinc shortage causes stunted growth, decreased water absorption, chlorosis, smaller leaves, dysfunctional flowers, and, eventually, decreased fertility. The study of nanostructure (NS) ranks as one of the most fascinating and active areas of biotechnological research nowadays. Because of their incredibly large size, finely adjusted structure, and physio-chemical qualities in the agriculture industries, NS encompasses an extensive variety of modern uses (Anwar et al., 2016; Nai et al., 2010). Among various methods for engineering application-targeted nanoparticles (NS), plants have emerged as the optimal choice due to their exceptionally low cytotoxicity and high biocompatibility (Sharma et al., 2009). *Solanum lycopersicum*, commonly known as the tomato plant, belongs to the Solanaceae family. As a diploid species, it possesses 24 chromosomes ($2n = 24$). Characterized as a perennial herbaceous plant, it typically reaches heights of 2–3 meters, featuring fragile, short-lived woody stems and yellow-hued flowers. The fruit size varies greatly, ranging from 1–10 cm in diameter, transitioning to a vibrant red color upon full ripeness (Mirnezhad et al., 2010) *Lycopersicum esculentum* stands out as a nutrient-dense crop worldwide, boasting an abundance of vitamin C, and vitamin A, and lycopene, a powerful antioxidant. This compound shields cellular environments from oxidative stress caused by free radicals, such as hydroxyl (OH^-) and superoxide ($O2^-$) ions (Ilahy et al., 2011).

2.3 Nanopesticides

Nanotechnology offered a promising solution for overcoming challenges associated with traditional chemical pesticides in insect pest management. Research has shown that nano-based materials exhibit enhanced efficacy due to their increased surface area, allowing greater contact with insect pests. Additionally, nanomaterials provide targeted, selective, and sustained release formulations, making them a more environmentally friendly (Zannat et al., 2021).

Insect pests are a primary factor impacting crop quality and can be managed by using synthetic insecticides. This has driven the search for alternative, eco-friendly solutions that are biologically based and sustainable (Subashini et al., 2004). Crop output is declining day by day due to significant pest invasions from weeds, insect pests, rodents, birds, mites, phytopathogens and mollusks such as fungus, bacteria, and viruses. The species listed above are all categorized as tissue consumers, photosynthetic rate reducers, and accelerators of leaf senescence. Crop yields are significantly reduced by chewing and sucking insects (Shahid Muhammad, 2020).

Pesticides utilized to manage pests globally are approximately two million tons per year, with Europe alone consuming 45%, 25% is consumed by the United States, and the rest of the world using 25% pesticides (De et al., 2014). The excessive dependence on chemical pesticides has spawned numerous environmental and health issues, including; Rising pesticide poisoning cases among farmers, a decline in soil biodiversity, Mammalian toxicity concerns, insecticide resistance development, biomagnification of pesticides in ecosystems, residual insecticides on produce, hazards to non-target organisms and pollinators. To address these challenges, nano pesticides offer a promising alternative. These innovative formulations can combine organic (e.g., active ingredients, polymers) and inorganic components (e.g., metal oxides) in various configurations (e.g., particles, micelles). Nanoformulations share key objectives with conventional pesticide formulations, specifically in enhancing the bioavailability of active ingredients with limited solubility and controlling active ingredient release for sustained efficacy and/or shielding against environmental degradation (Ragaei and Sabry, 2014).

Harmful substances, including genetically modified organisms, pathogens, and pesticide-resistant organisms, are permeating food chains, posing significant environmental and health risks (Wieczorek, 2003). Nanotechnology has transformed key sectors, including food, textiles, energy, and communication, addressing pressing global challenges. As the world population grows, food demand rises. To meet this need, researchers are exploring innovative solutions to mitigate pest infestations and boost agricultural productivity. Biologically synthesized nanoparticles have emerged as a game-changer in pesticide formulations, crop production and filtration of water (Barik et al., 2008; Prasad et al., 2014).

Nanotechnology has far-reaching applications across various sectors, enabling the development of innovative formulations and solutions in; sustainable biomass production, sensing and diagnostic

Table 2. Effects of nanoparticles on the Growth and Development of plants.

Nanoparticles	Pathogen	Plant disease	Effect	Supporting Ref.
Silver	*Xanthomonas campestris pv. campestris*	Bacterial blight	Reduction in bacterial growth	(Rajesh et al., 2012)
Copper	*Fusarium* sp.	Fungal disease	Antifungal	(Bramhanwade et al., 2016)
	Phytophthora infestans	Disease in tomato	Antifungal	(Giannousi et al., 2013)
	Xanthomonas axonopodis pv. punicae	Bacterial blight	Improved bacterial growth inhibition	(Mondal and Mani, 2012)
Silver-GO composite	*Xanthomonas performance*	Bacterial Spots	Improved bacterial growth inhibition	(Ocsoy et al., 2013)
Chitosan	*Colletotrichum capsici*	Disease in chili	Reduce mycelia growth rate	(Chookhongkha et al., 2012)
	Macrophomina phaseolina	Fungal disease	Stopped spore germination	(Saharan et al., 2013)
Iron Oxide	*F. oxysporum*	Fungal disease	Growth inhibition	(Chhipa and Kaushik., 2015)
Ag core-DHPAC shell	*Phytophthora nicotianae*	Fungal disease	Growth inhibition	(Ho et al., 2015)

tools, enhanced food and nutrition systems, advanced paint and coating technologies, paper and pulp industry innovations, sensing and diagnostic tools, agrochemical industries and fertilizer, and plant protection solutions (Sadowski, 2010) (Table 2). Nano-based pesticides, incorporating zinc oxide (ZnO), copper (Cu), silver (Ag), and silica (SiO2) nanoparticles, demonstrate exceptional efficacy and environmental benefits. These formulations offer broad-spectrum pest control, minimize water usage, and mitigate environmental pollution. As compared to traditional insecticides, nano-based pesticides give a more sustainable solution (Chhipa, 2017). Zinc is a vital nutrient that is helpful in the development and growth of plants and is poisonous to insect pests. Silver has a lot of potential for application in medicine, living things, pest control, and plant management. It can be utilized in a more sustainable manner and has better fungal, microbiological, pesticidal, larvicidal, antiviral and antibacterial properties (Chopra et al., 1994; Kharissova et al., 2009; Gao et al., 2014; Chhipa, 2017). Researchers investigated the antifungal properties of zinc oxide (ZnO) nanoparticles against a devastating *Fusarium graminearum,* a plant pathogen (Dimkpa et al., 2013). The properties of metal nanoparticles include stability, reduced particle sizes, and environmentally beneficial consequences (Iravani, 2011). Nanoformulations are the greatest feasible alternative management approaches for controlling insect pests (Fig. 5).

2.4 Nanoherbicides

Herbicides are toxic compounds used to eliminate undesired plants, such as weeds in crops. Herbicides are extremely important in agriculture, and their production makes it one of the most profitable businesses. Nanoherbicides are developed by combining nano-based particles used in herbicide manufacturing with an effective delivery method based on nano particles. The advantages of nanoherbicides include a reduction in the number of synthetic pesticides, an effective delivery mechanism, personal safety, and a productive reaction time. Nano herbicides are an advanced technological solution to every drawback of conventional herbicides. One of the main issues with traditional herbicides is the development of weed resistance, which requires careful planning in order to eradicate it completely. Properties such as high penetration capability and efficient delivery mechanism of nanoherbicides can be useful in resolving such issues (Patil-Patankar and Sanghvi, 2022).

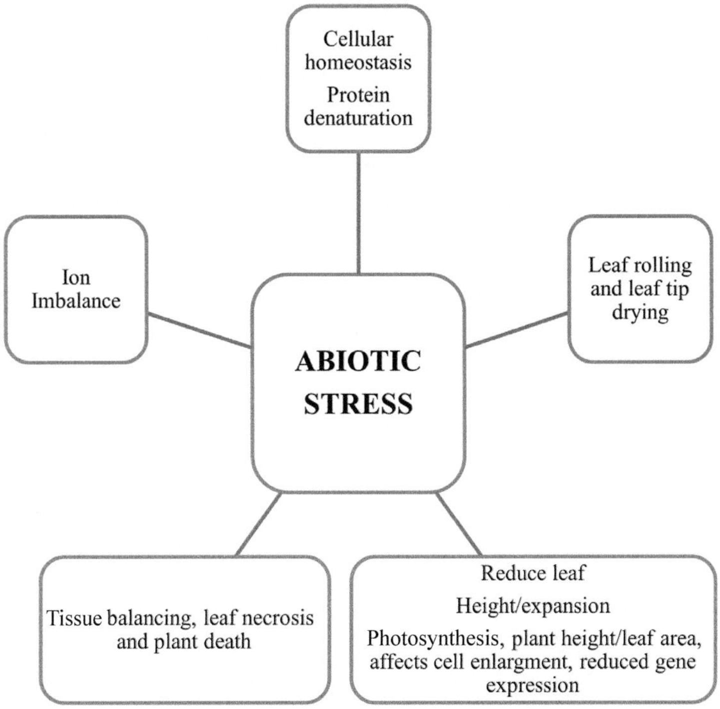

Figure 5. Types of Abiotic stress facing by plants.

By using novel techniques, nanoherbicides are able to gradually release herbicides while also efficiently diminishing weed seed banks and germination inhibitors. Innovative techniques for seed bank depletion include the use of carbon nanotubes and causing harm to weed pollen grains. Formulations with controlled release ensure effective and long-lasting weed suppression with minimal negative effects on the environment. Furthermore, adding metal ions promotes the breakdown of herbicide residue, reducing its lasting impact on the environment. Nanopesticide application promotes plant growth, emphasizing their potential as long-term weed-control options. However, further research is necessary to confirm their efficacy and security before widespread implementation in agriculture for commercial purposes, addressing possible concerns associated with their use (Adarsh et al., 2024). Nanoherbicides not only minimize the load of herbicides on the environment but also aid in weed eradication without leaving significant harmful residues in soils and the environment. These herbicide-loaded 'nano-bullets' can precisely target a particular plant organ or tissue through controlled herbicide release. This strategy to weed reduces expenses and ensures little toxicological implications, in addition to enhancing herbicide absorption into weeds (Hakeem and Pirzadah, 2020). Nanoherbicides offer a revolutionary approach to weed management, significantly reducing the environmental impact of traditional herbicides. By utilizing precision targeting, these innovative formulations eradicate weeds without leaving persistent toxic residues in soil and water. These herbicide-loaded 'nano-bullets' release herbicides in a controlled manner, specifically targeting the desired plant organ or tissue. As a result, nanoherbicides minimize herbicide load on the environment, ensuring a safer and more sustainable solution for weeds (Hakeem and Pirzadah, 2020).

2.5 Management of Abiotic Stress

Abiotic stress disrupts the delicate balance of the soil-plant-atmosphere continuum, severely impacting crop productivity and plant growth. This stress arises from both natural factors and human activities, constituting a primary contributor to global crop yield reductions. To address this pressing issue, there is a critical need for environmentally sustainable technologies that: enhance stress tolerance in plants, improve resource utilization efficiency and promote sustainable agricultural practices. A powerful tool for reducing the effects of biotic and abiotic stress is nanomaterials. Because of their extremely small size, nanoparticles can penetrate specific cellular regions and have gained some unique features that play an important role in protecting plants from various abiotic stressors. NMs protect plants from oxidative stress by mimicking the role of antioxidative enzymes such as superoxide dismutase (SOD), catalase (CAT), and peroxidase (POX) under stressful conditions such as drought, water logging, salinity, excessive temperature, UV radiation, etc. (Mahto et al., 2021).

The world's agricultural sector and green plants face numerous global challenges, including: sustainability concerns due to population growth and urbanization, climate change, habitat destruction and biodiversity loss, environmental degradation (Chen and Yada, 2011). Plants, a critical component of our world, are constantly subjected to various environmental changes and challenges throughout their lives. Plants, unlike animals, these organisms are unable to move when stressed, making them vulnerable to biotic and abiotic threats. Fortunately, they possess inherent defense mechanisms that enable them to thrive in hostile environments. Plants adopt many methods, including escapism, tolerance, and avoidance, to deal with harsh conditions, but responses can differ substantially among plants of the same species. To promote sustainable agriculture and crop production, identifying resilient plant species remains a top priority. Fortunately, nanotechnology presents a vast research landscape, unlocking diverse possibilities and benefits in agricultural biotechnology, including, disease prevention (Carmen et al., 2003). Nanotechnology offers innovative solutions for sustainable agriculture through: efficient nutrient management via nanofertilizers (Priester et al., 2012). Targeted pest control using nanoherbicides and nanopesticides (Prasad, 2014; Marchiol, 2012), mitigating abiotic stress factors (Saxena et al., 2016). Moreover, nanotechnology holds significant promise for addressing abiotic stress-related challenges and promoting agricultural sustainability.

Throughout their life cycle, plants face numerous stress factors. According to available data, abiotic stress factors significantly impact potential maximum yields, with reported decreases ranging from 54 to 82% (Bray, 2000). Abiotic stress comes in many forms for crops, and it is well known that the main factors that negatively impact plant growth and productivity globally are ion toxicity (heavy metals, salinity), availability of water (drought, flooding), and temperature extremes (freezing, heat and cold) (Gill et al., 2011; Tuteja, 2007; Mahajan and Tuteja, 2005; Khan and Singh, 2008). These abiotic challenges are associated with osmotic stress, which disrupts ion transport and cell homeostasis. Abiotic stress conditions adversely harm crop plants.

Nanotechnology serves as a transformative platform for developing innovative tools and technologies aimed at enhancing and optimizing biological systems (Fortina et al., 2005). Nanoparticles (NPs) are tiny aggregates of molecules, typically measuring between 1 and 100 nanometers in size (Roco, 2003). Studies have demonstrated the efficacy of NPs in improving plant development, productivity, and growth, while also enhancing crop resilience to abiotic and biotic stresses (Giraldo et al., 2014). Over the past ten years, nanotechnology research has emerged as a promising means of reducing the stresses related to the aforementioned concerns and securing the global agricultural sector. The discoveries made in this field have the potential to greatly expand the applications of biotechnology and agriculture.

Sustainable agriculture relies on technological advancements that optimize food production while minimizing crop losses. Abiotic stress poses a significant threat to agricultural productivity, manifesting in diverse forms such as; environmental stresses (submergence, salinity, waterlogging, drought,), heavy metal-induced stresses, and metal and mineral toxicity/deficiencies. Addressing these abiotic stress factors is crucial for ensuring crop resilience and sustainability (Tripathi et al., 2015; Tripathi et al., 2012; Upadhyaya et al., 2008). Abiotic stressors significantly impede plant productivity. Plants must develop complex defense systems to withstand various stressors throughout their life cycle. Nanoparticle (NP) research offers promising solutions, demonstrating plant growth and development are influenced by concentration-dependent responses (Dimpka et al., 2017; Mishra et al., 2017; Mishra and Singh, 2016). Studies have shown that important antioxidant enzymes, such as catalase (CAT), peroxidase (POD), and superoxide dismutase (SOD), can have their activity increased by nanoparticles (NPs) (Laware and Raskar, 2014). The interactions between nanoparticles and plants are multifaceted, resulting in changes to plant structure and function, which are dependent on the NPs' size, chemical composition, reactivity, and surface coating. These interactions significantly impact crop plants. While NPs can enhance plant growth and development, they also exhibit both beneficial and adverse effects (Ma et al., 2010). It has been shown that certain nanoparticles (NPs) have both positive and negative impacts on the growth and development of plants.

2.6 Comparative analysis of Traditional carry out and Nanotechnology

Nowadays, target framing is becoming more widespread in the field of nano agriculture, where nanoparticles are used to enhance crop growth and eminence (Batsmanova et al., 2013; Scott and Chen, 2013). Carbon nanotubes (CNTBs) have emerged as a revolutionary tool, surpassing traditional methods in enhancing seed development, germination of plants, and growth promotion (Khodakovskaya et al., 2013). Nanosized bacteriophages have replaced the use of Cu bactericides in the past. They are more effective than any other conventional methods because of their smaller size and higher surface-to-volume ratio. Figure 5 presents a comparative analysis of nanomaterials and traditional materials. Nanoparticles (NPs) penetrate the apoplast and subsequently accumulate within the epidermal and cortical layers through aggregative processes (Larue et al., 2012). Rico et al. (2011) investigated the symplastic pathway, revealing a more structured and controlled mechanism for nanoparticle penetration (Larue et al., 2012).

2.7 Demerits

The rapid proliferation of nanoparticle production and its applications across various industries is a widely acknowledged trend (Piccinno et al., 2012). Although current nanoparticle usage and mixing levels are considered non-hazardous, it is crucial to acknowledge the potential risk of increasing NP concentrations in the future (Johnson and Park, 2012; Nicolodi and Gianello, 2014). Nanotechnology presents both benefits and drawbacks (Kumar et al., 2019). Nanoparticles (NPs) can originate from natural sources or anthropogenic production, entering soil through rainfall or agricultural activities. Notably, soil contains higher concentrations of NPs as compared to air and water, due to their slow movement and prolonged persistence in soil environments (Gottschalk and Nowack, 2011). Nanofertilizers are a primary source of nanoparticle contamination, as plants absorb and accumulate NPs from the soil, facilitating their translocation to various plant parts. Upon human consumption of these contaminated plants, NPs can cause harmful effects, including increased reactive oxygen species production, chromosomal aberrations, genetic defects and organ damage. Nanoparticles can enter the human body through various routes, including inhalation, ingestion, and medical injections. However, the primary concern lies in their interactions and behavior within the body, specifically how they interact with tissues and organs (Kumar et al., 2019). Using large doses of nanoparticles increases their cytotoxicity. Large-scale production of nanoparticles during the last few decades has led to their harmful effects on the environment. According to research, Ag NPs are used to make socks because they reduce odor, but when they are released into the wash, they have negative effects (Khan et al., 2019). Silver nanoparticles (Ag NPs) exhibit bacteriostatic properties, potentially harming beneficial bacteria essential for organic matter degradation in wastewater treatment facilities. Recognizing these risks, regulatory agencies like the EPA and FDA have initiated preventive measures to mitigate NP-related hazards (Kumar et al., 2019).

3. Green nanotechnology for environmental management

Nanomaterials play a crucial role in environmental sustainability, offering effective solutions for remediation and green technologies. The application of nanomaterials is vital for the cleanup of hazardous waste sites and the treatment of pollutants. Nanoscale surface coatings with self-cleaning properties can reduce the need for chemical cleaning agents in maintenance protocols. The rapidly expanding applications of Fe NPs in water purification and soil remediation have generated considerable interest in these nanoparticles (Samuel et al., 2022).

3.1 Bioremediation

Agricultural pollution poses a significant global challenge, primarily due to unsustainable agrochemical practices. This leads to the accumulation of harmful substances, including, phenols, halogenated compounds, organic compounds, and heavy metals like insecticides and pesticides (Singh et al., 2020a). As a result, dangerous materials such as phenols, halogenated chemicals, organic compounds, and heavy metals like herbicides and insecticides build up. All these agricultural pollutants are particularly hazardous and could damage air, soil, and water (Khan et al., 2020; Ken and Sinha, 2020; Singh et al., 2020a). Organic pollutants are generally composed of phenols, phthalic esters, azo dyes, insecticides, Persistent Organic Pollutants (POPs), chlorinated and other compounds. Inorganic contaminants include heavy metals like nickel, arsenic, cadmium, lead, chromium, and mercury. These agro-pollutants exhibit high toxicity in living organisms and have a significant negative influence on soil and water pollution (Saxena et al., 2019).

The rapid integration of nanotechnology across various scientific disciplines and environmental applications can be attributed to its distinct properties, including exceptionally high surface area and nano size, which offer vast potential in diverse fields (Hidangmayum and Dwivedi, 2022). Although there are many remediation options available for both *ex* and *in situ* applications, nano bioremediation techniques have recently attracted a lot of attention because of their environmental friendliness and efficient removal of contaminants from the environment. As per the US-EPA (Environmental Protection Agency of US), bioremediation transforms dangerous materials into less harmful or innocuous molecules by consuming naturally present microbes (Singh et al., 2020b; Rajput et al., 2021).

Bioremediation principally involves the application of microorganisms to mitigate pollutants in aquatic and terrestrial environments (Saxena et al., 2019; Bhojiya et al., 2021; Halecký and Kozliak, 2020). Keep in mind that modern bioremediation techniques use a variety of cutting-edge methods, like the use of emulsifiers, biosurfactants, enzymes, genetically modified microbes and biopesticides (Halecký and Kozliak, 2020). Integrating nanotechnology with bioremediation enhances remediation efficiency, expedites the process, and minimizes environmental impact, thereby amplifying advantages (Ghani et al., 2022; El-Ramady et al., 2020; Mallikarjuna et al., 2011; Rajput et al., 2021).

The three main techniques in bioremediation are plants, microbes, and enzyme-mediated remediation. The process of nano-bioremediation involves using nanomaterials to break down pollutants into biodegradable forms. This is followed by biodegradation, which eliminates the contaminants. A key component of this process is nanoparticles, which are derived from natural sources like microbes or plant extracts. These nanoparticles play a crucial role in removing pollutants from contaminated land and water. Due to their exceptional efficacy, cost, and environmental friendliness, nanomaterials have become a competitive treatment alternative throughout the past 20 years (Dastjerdi and Montazer, 2010). According to Tratnyek and Johnson (2006), iron nanoparticles were the first to be applied to environmental rehabilitation. Several iron-based treatments offer promising solutions for cleaning up groundwater and treating contaminated soil. Notably, zero-valent iron nanoparticles have proven effective in remediating acidic water polluted with heavy metals (Iravani, 2011; Saif et al., 2016).

Carbon nanotubes (CNTs), noble metal nanomaterials magnetic nanoparticles, and quantum dots are considered efficient for water quality monitoring, especially for the detection of trace contaminants and pathogens (Kumari and Singh, 2016). Additionally, applying zero-valent or elemental metals in nanoparticle form, such as palladium, nickel, and iron, to contaminated areas containing a range of pollutants has demonstrated encouraging results (Li et al., 2016). These metals

Table 3. Efficacy of Nano formulations in Controlling Pest Populations.

Cam	Pesticides	Pathogen	Plant disease	Effect	References
Porous hallow silica	Validamycin	-	-	Increase the loading capacity with 36% efficacy	(Liu et al., 2006)
Calcium carbonate	Validamycin	*Rhizoctonia solani*	-	Controlled rate of release	(Qian et al., 2011)
Lignin	Diuron	-	-	Controlled rate of release	(Yearla and Padmasree, 2016)
Cellulose/silica nanocomposites	Tebuconazole	Molds and rust	Leaf deterioration	Enhanced rate of release	(Mattos and Magalhaes, 2016)
Titanium dioxide	With silver and Zinc	*Xanthomonas perforans*	Tomato infection	Reduction in bacterial growth	(Paret et al., 2013)

are primarily used to stabilize transition metals like arsenic and chromium and to dehalogenate obstinate organic molecules (Rizwan et al., 2014; Thomé et al., 2015).

Nanotechnology boosts phytoremediation productivity because nanoparticles are a viable way to purify polluted surroundings. Heavy metals, inorganic contaminants, and other dangerous materials can be successfully eliminated from soil and water using them (Table 3). Nanoparticles and phytoremediation can be combined in compound-based bioremediation. By combining nanoparticles with phytoremediation, complex organic pollutants can be broken down into simpler compounds. These simpler compounds are then rapidly degraded by the synergistic actions of the roots of plants and microorganisms (Pillai and Kottekottil, 2016).

Nano-remediation has rapidly become a key focus area for researchers and developers, offering vast possibilities for cleaning up contaminated sites and preventing future pollution (Dixit et al., 2015). It provides deep soil remediation capabilities, aligning with bioremediation (Huang et al., 2016). Exploring the potential of nanofertilizers (via bioaugmentation and bio-stimulation), nano minerals (through bio-stimulation), and green-synthesized nanooxidizers (for PAH oxidation) can unlock the full potential of nanoremediation in removing Polycyclic Aromatic Hydrocarbons (PAHs). Breakthroughs in genetic engineering and transgenic technologies have transformed the field of bioremediation. In order to carry out specific processes, Genetically Modified Microorganisms (GMMs) and plants combine advantageous enzymes and degradation pathways from several species into a single host, acting as "exclusive biocatalysts" (Nkansah et al., 2012; Jin et al., 2016; Lawal, 2017).

3.2 Biosensors

Advances in biosensor technology have revolutionized multiple sectors, encompassing food safety, environmental surveillance, bioprocess optimization, agricultural monitoring, medical diagnostics, and military operations (Rodríguez-Mozaz et al., 2004). Biosensors offer a rapid and targeted solution for environmental monitoring and control, providing real-time data on contaminated sites' biological, ecological, and chemical quality, including inorganic and organic priority pollutants (da Costa Silva et al., 2011). Conventional environmental monitoring techniques for pollutant detection, including surface-enhanced Raman scattering spectroscopy, capillary electrophoresis, enzyme-linked immunosorbent assay, and chromatography, provide high sensitivity and selectivity but are limited by the need for skilled technicians, costly equipment, time-consuming sample preparation, and high reagent costs (Lang et al., 2016).

Conventional methods are no longer effective in situations requiring immediate, on-site analysis, such as accidental pesticide releases or acute poisoning cases. Their limitations hinder effective *in situ* measurement and real-time monitoring. In contrast, environmental monitoring biosensors offer a promising solution, providing: rapid detection, miniaturized design, portability and on-site analysis capabilities (Bo et al., 2017). Globally, Harmful Algal Blooms (HABs) pose significant threats to human health, environmental sustainability, and economic stability. To combat this issue, biosensors have been developed to detect six majors marine biotoxin groups produced by algae and microorganisms in harvested fish and shellfish, promoting ocean food safety (McPartlin et al., 2016).

3.3 Wastewater Treatment

Global water contamination poses a significant threat to human and environmental well-being as access to potable water is crucial for human survival. Water contamination has severe environmental, health, and socioeconomic implications, necessitating effective mitigation strategies, particularly in the face of escalating pollution driven by technological advancements (Butnariu et al., 2019). Nanotechnology has emerged as a highly effective and advanced approach for wastewater treatment, leveraging the unique properties of nanoparticles (NPs), including enhanced reaction, adsorption, and interaction capabilities due to their nanoscale dimensions and high surface atom density (Abigail et al., 2015; Samuel et al., 2018a; Samuel et al., 2018; Chidambaram, 2016). Nanoparticles (NPs) can be dispersed in aqueous solutions, forming colloids that facilitate efficient water treatment. Due to their miniature size, NPs conserve energy, leading to cost savings. Moreover, NPs offer significant benefits for treating water at considerable depths and in remote or hard-to-reach locations where conventional technologies are ineffective (Samuel et al., 2018a; Needhidasan et al., 2017; Needhidasan et al., 2014). Green nanomaterials exhibit a broad spectrum of capabilities for remediating water polluted by harmful organic and inorganic contaminants, metal ions and pathogenic microorganisms. Numerous nanomaterials, such as Molecularly Imprinted Polymers (MIPs), bioactive nanoparticles, nano sorbents, nano catalysts, nanostructured catalytic membranes, and biomimetic membranes, are the subject of intensive research and commercialization efforts to eliminate harmful metal ions, pathogenic microorganisms, and organic and inorganic pollutants from water (Samuel et al., 2019).

3.4 Degradation of dyes

Dyes are organic compounds that impart color to materials through their unique structure and affinity for specific substrates, enabling permanent coloration in textiles, food, and various industries. Iron nanoparticles are synthesized using *Chlorophytum comosum* leaf extract, demonstrating effective decolorization of methyl orange and antimicrobial properties. In this procedure, a water dispersion of *Chlorophytum comosum* leaf extract serves as the sole reduction and stabilizing agent, added to the iron salt precursor. Methyl orange is employed as a model contaminant to assess the dye removal efficacy of H_2O_2-catalyzed iron nanoparticles in degrading organic pollutants. This study's results showed that the optimal degradation of methyl orange, reached 77%, was achieved after a 6-hour period (Aragaw et al., 2022). Iron oxide nanoparticles synthesized using Carica papaya leaf extract exhibited efficient photocatalytic degradation of yellow RR dye, achieving 77% removal in 6 hours using sunlight at 0.8 mg/L dosage (Bhuiyan et al., 2020). In a separate study, green-synthesized iron oxide NPs were found to have remarkable performance for methyl orange breakdown in the presence of the H_2O_2 aqueous solution. The data showed that the bulk of IONPs are spherical in shape, with diameters ranging from 4.6–30.6 nm. As a result, the environmentally friendly and ecologically safe technology for producing IONPs via plant-mediated synthesis can be used for a range of applications, such as a heterogeneous Fenton-like catalyst for the breakdown of azo dyes (Lohrasbi et al., 2019). According to research (Chen et al., 2017), MGO-laccase retains 59.8% of its original activity after ten cycles and has a reasonable magnetic response and reusability at ideal pH (3.0) and temperature (35°C). By eliminating 94.7% of Crystal Violet (CV), 95.6% of Malachite Green (MG), and 91.4% of Brilliant Green (BG) from aqueous solutions, the MGO-laccase system showed remarkable decolorization effectiveness. The findings of the experiment showed that MGO-laccase nanoparticles significantly increased the processing efficiency, increasing the enzymes' commercial usability. Additionally, Piper betle extract was used to successfully create iron oxide nanoparticles (FeONPs) in a natural setting. The phenolic compounds present in Piper betle likely served as dual-functional capping and reducing agents, facilitating the synthesis of iron oxide nanoparticles (FeONPs) that efficiently catalyzed the degradation of Methyl Orange (MO) and malachite green (MG) dyes (Badmapriya, et al., 2016). The degradation of malachite green (MG) by OT-FeNP followed pseudo-first-order kinetics, achieving 75.5% removal efficiency for an initial concentration of 50 mg/L, indicating the potential of OT-FeNP as a green nanomaterial for environmental remediation (Huang et al., 2017). Upon irradiation, photocatalytic dye degradation occurs through the generation of electron-hole pairs, which subsequently react on the nanoparticle surface to produce hydroxyl radicals. Upon exposure to visible or UV light, nanoparticles exhibit photocatalytic activity, inducing a redox reaction that generates electron-hole pairs through the creation of electrons in the conduction band and H^+ ions in the valence band, facilitating the degradation of toxic pigments via radical-mediated oxidation. A series of redox processes are started by the electron-hole pairs created on the surface of the photocatalyst, which also makes it easier for H+ ions to interact with water molecules to form hydroxyl (OH) free radicals during oxidation. On the nanoparticle surface, conduction band electrons react with oxygen to form reactive intermediates, including superoxide radicals, perhydroxyl radicals, and hydrogen peroxide, ultimately yielding hydroxyl radicals (•OH) through reduction, facilitating photocatalytic degradation of harmful dyes (Mehta et al., 2021; Alinezhad et al., 2022).

3.5 Removal of heavy metals

Heavy metals are present in the environment naturally or by anthropogenic means. Human activities such as agricultural practices, mining, and municipal sewage sludge disposal contribute significantly to heavy metal contamination in aquatic ecosystems. Conversely, natural processes like rock weathering, erosion, and volcanic eruptions also release heavy metals into water bodies. Heavy metal contamination poses a significant environmental threat. Therefore, it is essential to regulate and mitigate heavy metal levels in wastewater before their release. To address this concern, a novel approach involves the rapid, single-stage, and entirely eco-friendly production of iron oxide nanoparticles (Fe_3O_4NPs). A novel approach to mitigate heavy metal pollution involves biosynthesizing nanoparticles using plant extracts. Studies have demonstrated that Ramalina Sinensis extract, containing sulfate polysaccharides, can efficiently decrease lead and cadmium levels when combined with iron chloride solution. Thermodynamic analysis and adsorption studies revealed that lead removal follows the Langmuir adsorption model, while cadmium removal conforms to the Freundlich isotherm model (R2 = 0.999), demonstrating that spontaneous exothermic removal is an adsorption process. A second-order kinetic model produced a correlation value of 0.99 and properly reflected kinetic investigations on the elimination of lead and cadmium from aqueous solutions. For example, magnetic iron oxide nanoparticles demonstrated impressive removal efficiencies of 82% for lead and 77% for cadmium under ideal circumstances, namely an initial concentration of 50 mg/L and a pH range of 4–5 (Arjagh et al., 2021). A separate investigation involved the synthesis of EDTA-modified Fe_3O_4/SC nanoparticles and magnetic nanoparticles infused with sawdust carbon. Batch mode experiments demonstrated the effectiveness of these materials in adsorbing Cd (II) ions from aqueous solutions. The Langmuir isotherm revealed a maximum adsorption capacity of 63.3 mg/g for EDTA@Fe_3O_4/SC nanocomposites, surpassing the 51 mg/g capacity of Fe_3O_4/SC nanocomposites. Moreover, the pseudo-second-order model provided the best description of the adsorption kinetics of cadmium (II). Cadmium (II) adsorption onto EDTA@ Fe_3O_4/SC nanocomposites was studied thermodynamically, and the results showed that the process is endothermic, spontaneous, and feasible. However, the efficacy of cadmium (II) adsorption may be impacted by the coexistence of divalent cations (Ca^{2+} and Mg^{2+}). Furthermore, desorption and recycling studies demonstrated that EDTA@ Fe_3O_4//SC nanocomposites serve as a cost-effective and durable adsorbent for metal removal from aqueous solutions. A comparative analysis revealed that EDTA modification significantly enhances the Cd (II) adsorption potential of Fe_3O_4//SC nanocomposites. The results underscore the crucial role of surface-bound carboxyl groups on EDTA@ Fe_3O_4//SC nanocomposites in facilitating Cd (II) adsorption.

The adsorption process was governed by both electronic and chemical interactions. Our study highlights the potential of biogenic green synthesis as a viable method for producing EDTA@ Fe_3O_4//SC nanocomposites tailored for heavy metal remediation in wastewater (Kataria et al., 2018). The presence of carboxyl groups on the surface of EDTA@ Fe_3O_4/SC nanocomposites was found to be crucial for the adsorption of Cadmium (II). Specifically, the mechanism of adsorption involved a dual interaction, comprising chemical adsorption and electronic attraction. The findings of this study indicate that biogenic green synthesis can be effectively utilized to produce EDTA@ Fe_3O_4//SC nanocomposites, offering a promising solution for heavy metal remediation in wastewater treatment (García et al., 2019). Another investigation revealed that iron nanoparticles can be successfully synthesized using blueberry leaf extracts, exhibiting reduced agglomeration compared to nanoscale zero-valent iron (nZVI). Notably, this eco-friendly approach eliminates the need for dispersants due to the inherent capping mechanism occurring during synthesis. Studies on the decontamination efficacy of BB-Fe nanoparticles (NPs) have revealed a reduced capacity for arsenate removal due to decreased active surface areas resulting from residual organic matter adsorbed during synthesis. Nevertheless, leveraging these organic remnants enables the fabrication of affordable, environmentally friendly nanomaterials (Manquián-Cerda et al., 2017). The biosorption efficacy of Fe3O4-NPs alginate beads, synthesized using *Padina pavonica* and *Sargassum acinarium* plant extracts, was evaluated for Pb removal. The results revealed that *P. pavonica*-derived beads exhibited a superior bio removal capacity (91%) compared to *S. acinarium*-derived beads (78%) within 75 minutes (El-Kassas et al., 2016).

4. Conclusion

To address the escalating global food demand, boosting crop productivity is crucial. This requires innovative solutions that enhance yields while minimizing environmental impact. The final solution required for the agricultural sector's development has been made possible by the integration of nanotechnology. Since they target and control the release of safe chemicals that protect soil microbes and non-target species, including crops, nanoparticles typical alter agricultural particles. Ultimately, they are not natural. Nano fertilizers offer a cost-effective and environmentally friendly solution by releasing nutrients gradually and consistently, eliminating the need for frequent reapplication. Additionally, these innovative fertilizers: regulate soil environmental pressure and optimize active ingredient release as needed. Despite their promising potential, nanotechnology applications in agriculture remain in their early stages. The ecological implications of nano-based agricultural particles are not yet fully understood. Once a comprehensive understanding of nanoparticles' ecological impact and long-term consequences is achieved, their benefits can be fully utilized.

The development of nanoparticles has raised concerns about unintended consequences, preferring interest in greener, more sustainable remediation solutions. Green synthesis offers a dual benefit: eliminating secondary environmental contamination and lowering production costs. However, there are still research gaps that need to be filled in order to support the field's advancement. For instance, to achieve further anticipated results, an explanation of the specific mechanisms underlying green synthesis is crucial. Green synthesis research has made progress, but a deeper understanding of the fundamental mechanisms is necessary. Furthermore, novel methods are required to fabricate NPs with intricate composite structures and enhanced surface areas. Modulating the crystalline morphology of environmentally friendly NPs to access novel properties beyond their bulk material characteristics presents a promising research direction for green synthesis investigators. Nanotechnology, functioning at the nanometer scale, has the potential to transform various industries. By manipulating molecules, atoms, and macromolecules in the 1–100 nm range, researchers can develop materials with unique characteristics, unlocking novel applications.

By emulating nature's principles, scientists have discovered an innovative method for synthesizing metal oxide nanoparticles using microorganisms and extracts from plants. This biological approach provides a cost-effective and sustainable pathway to producing biocompatible particles, holding immense promise for advancements in healthcare.

The emergence of bio-green synthesis has revolutionized NP production, eliminating harmful chemicals and paving the way for innovative solutions in clinical research, energy harvesting, and environmental sustainability. Cancer treatment stands as a pressing concern, necessitating innovative solutions. Nanotechnology offers a promising avenue to enhance existing therapeutic approaches, addressing the limitations of conventional chemotherapeutics that often result in debilitating side effects. As a leading global cause of mortality and morbidity, cancer demands more effective and targeted treatments. A new approach is urgently needed to improve cancer treatment. Fortunately, recent breakthroughs in nanotechnology provide promising solutions. These advancements enable personalized treatments that can overcome the limitations of traditional medications, providing new hope in the fight against cancer. Research has demonstrated that nanomaterials can enhance food processing outcomes and nutritional content without altering the fundamental properties of food products, making them valuable additives. Nanomaterials have also demonstrated exceptional potential in bioremediation, particularly in wastewater treatment applications. The versatility of nanotechnology has led to its widespread adoption across various disciplines, positioning it as a pivotal strategy for addressing several challenges.

5. Future challenges and perspectives

What significant hurdles need to be overcome to enable the practical and effective utilization of nanomaterials in agricultural applications? Despite notable advancements in laboratory research on polymeric nanoparticles for agrochemical delivery, significant challenges hinder the practical application of nanomaterials in agriculture, contrasting with their more established uses in biomedicine and pharmacology. To advance research in this field, a few key problems need to be addressed;

1. It is essential to investigate the nanotoxicological profile of agrochemical-loaded polymeric nanoparticles, focusing on environmentally relevant concentrations and their acute and chronic effects on animals, plants, humans, and ecosystems (Sadeghi et al., 2017). Investigations are focused on elucidating the environmental behavior, plant toxicity, and soil biota effects of nanoparticles to inform guidelines for their safe and sustainable use in agriculture (Iavicoli et al., 2017; Tripathi et al., 2017).
2. The potential risks of nanoparticle-biological system interactions require comprehensive evaluation. Although significant advancements in understanding polymeric nanoparticle uptake, distribution, and bioavailability in plants, further research is needed to address existing gaps (Pe´rez-de-Luque, 2017).
3. Despite being primarily made from biocompatible and biodegradable materials, polymeric nanoparticles (such as pectin and chitosan), further research is needed to understand how the nanoparticles accumulate and internalize in plant cells. Plant cell compartments can internalize, transport, and accumulate nanoparticles, triggering a range of unpredictable effects. When nanoparticles accumulate in plant organs, they can either settle there or be transported to other unintended areas, a phenomenon also seen in animals.
4. The chemical and physical characteristics of engineered polymeric nanoparticles are highly influenced by various factors. These include: surface chemistry, size distribution, morphology, surface charge, reactivity, concentration, target organism, hydrophobicity, aggregation state and composition of test media. Moreover, the behavior of these nanoparticles is strongly dependent on the specific environment in which they are applied. The behavior of polymeric nanoparticles loaded with agrochemicals can be influenced by various chemical and physical parameters of soil and water, including pH, ionic strength, and temperature.

5. Scalability and manufacturing cost are important factors in nanoparticle development; chitosan-based polymeric nanoparticles provide a workable option because of their cheap cost, biocompatibility, and biodegradability.
6. More study is required on important aspects including manufacturing schedules, long-term stability, and possible impacts in order to properly transition nanoparticles from laboratory testing to actual agricultural application.
7. Critical research on production scalability, long-term stability, and environmental effects is crucial to facilitating the transition of nanoparticles from the laboratory to real-world agricultural uses.
8. Finally, consumers are still uncertain about nanotechnology and are wary of nanoparticles containing agrochemicals, echoing concerns surrounding transgenic crops. Addressing these knowledge gaps is crucial for polymer nanoparticle technology to jump from laboratory to real-world agricultural implementation (Pe´rezde- Luque 2017).

References

Abigail, M. E. A., Samuel, S. M. and Ramalingam, C. (2015). Addressing the environmental impacts of butachlor and the available remediation strategies: A systematic review. *International Journal of Environmental Science and Technology*, 12(11): 4025–4036.

Adarsh, S., Shilpa, S., Ameena, M., Susha, V. S., Sreelekshmi, K., Renjan, B. et al. (2024). A review on nano herbicides: The future of weed management. *Journal of Advances in Biology & Biotechnology*, 27(7): 1244–1253.

Adhikari, T., Sarkar, D., Mashayekhi, H., and Xing, B. (2016). Growth and enzymatic activity of maize (*Zea mays* L.) plant: Solution culture test for copper dioxide nanoparticles. *Journal of Plant Nutrition*, 39(1): 99–115.

Alinezhad, H., Hajiabbas Tabar Amiri, P., Mohseni Tavakkoli, S., Muhiebes, R., and Fakri Mustafa, Y. (2022). Progressive types of Fe_3O_4 nanoparticles and their hybrids as catalysts. *Journal of Chemical Reviews*, 4(1): 288–312.

Alsharef, J.M., Taha, M.R., Firoozi, A.A., and Govindasamy, P. (2016). Potential of using nanocarbons to stabilize weak soils. *Applied and Environmental Soil Science* 2016, 5060531.

Anwaar, S., Maqbool, Q., Jabeen, N., Nazar, M., Abbas, F., Nawaz, B. et al. (2016). The effect of green synthesized CuO nanoparticles on callogenesis and regeneration of *Oryza sativa* L. *Frontiers in Plant Science*, 7: 1330.

Aragaw, T. A., Bogale, F. M., and Aragaw, B. A. (2021). Iron-based nanoparticles in wastewater treatment: A review on synthesis methods, applications, and removal mechanisms. *Journal of Saudi Chemical Society*, 25(6): 101280.

Arjaghi, S. K., Alasl, M. K., Sajjadi, N., Fataei, E., and Rajaei, G. E. (2021). Green synthesis of iron oxide nanoparticles by RS Lichen extract and its application in removing heavy metals of lead and cadmium. *Biological Trace Element Research*, 199(2): 763–768.

Aslam, A. A., Aslam, A. A., Aslam, M. S., and Quazi, S. (2022). An overview of green synthesis of nanomaterials and their advanced applications in sustainable agriculture. *Preprints* 2022, 2022020315.

Aslam, A. A., Aslam, M. S., and Aslam, A. A. (2022). An overview on green synthesis of nanoparticles and their advanced applications in sustainable agriculture. *International Journal of Applied Chemical and Biological Sciences,* 3(2): 70–99.

Badmapriya, D., and Asharani, I. V. (2016). Dye degradation studies catalysed by green synthesized iron oxide nanoparticles. *International Journal of ChemTech Research*, 9(3): 409–416.

Bansal, K., Hooda, V., Verma, N., Kharewal, T., Tehri, N., Dhull, V. et al. (2022). Stress alleviation and crop improvement using silicon nanoparticles in agriculture: a review. *Silicon*: 1–14.

Barik, T. K., Sahu, B., and Swain, V. (2008). Nanosilica from medicine to pest control. *Parasitology Research*, 103(2): 253–258.

Batsmanova, L. M., Gonchar, L. M., Taran, N. Y., and Okanenko, A. A. (2013). Using a colloidal solution of metal nanoparticles as micronutrient fertiliser for cereals (Doctoral dissertation, Sumy State University).

Benjamin, S. R., Lima, F. D., Florean, E. O. P. T., and Guedes, M. I. F. (2019). Current Trends in Nanotechnology for Bioremediation. *International Journal of Environment and Pollution*, 66(1-3): 19–40.

Bhandari, G., Dhasmana, A., Chaudhary, P., Gupta, S., Gangola, S., Gupta, A. et al. (2023). A perspective review on green nanotechnology in agro-ecosystems: opportunities for sustainable agricultural practices & environmental remediation. *Agriculture*, 13(3): 668.

Bhojiya, A. A., Joshi, H., Upadhyay, S. K., Vaishnav, A., Choudhary, K. K., and Sharma, R. (2021). Screening and optimization of zinc removal potential in *Pseudomonas aeruginosa*-HMR1 and its plant growth-promoting attributes. *Bulletin of Environmental Contamination and Toxicology*, 107(6): 1184–1193.

Bhojiya, A. A., Joshi, H., Upadhyay, S. K., Srivastava, A. K., Pathak, V. V., Pandey, V. C. et al. (2022). Screening and optimization of zinc removal potential in *Pseudomonas aeruginosa*-HMR1 and its plant growth-promoting attributes. *Bulletin of Environmental Contamination and Toxicology*, 108(3), 468–477.

Bhuiyan, M. S. H., Miah, M. Y., Paul, S. C., Das Aka, T., Saha, O., Rahaman, M. M. et al. (2020). Green synthesis of iron oxide nanoparticle using Carica papaya leaf extract: Application for photocatalytic degradation of remazol yellow RR dye and antibacterial activity. Heliyon, 6(9): e04603.

Bo, X., Zhou, M., & Guo, L. (2017). Electrochemical sensors and biosensors based on less aggregated graphene. *Biosensors and Bioelectronics*, 89: 167–186.

Bramhanwade, K., Shende, S., Bonde, S., Gade, A., and Rai, M. (2016). Fungicidal activity of Cu nanoparticles against *Fusarium* causing crop diseases. *Environmental Chemistry Letters*, 14(1): 229–235.

Bray, E. A. (2000). Responses to abiotic stresses. *Biochemistry and Molecular Biology of Plants*, 1158–1203.

Bruinsma, J. (2017). World agriculture: Towards 2015/2030: An FAO study. Routledge.

Butnariu, I. C., Stoian, O., Voicu, Ş., Iovu, H., and Paraschiv, G. (2019). Nanomaterials used in treatment of wastewater: A review. *Annals of Faculty Engineering Hunedoara*, 17(2): 175–179.

Cai, L., Liu, M., Liu, Z., Yang, H., Sun, X., Chen, J. et al. (2018). MgONPs can boost plant growth: evidence from increased seedling growth, morpho-physiological activities, and Mg uptake in tobacco (*Nicotiana Tabacum* L.). *Molecules*, 23(3375): 1–12.

Campbell, B. M., Thornton, P., Zougmoré, R., Van Asten, P., and Lipper, L. (2014). Sustainable intensification: what is its role in climate smart agriculture? *Current Opinion in Environmental Sustainability*, 8: 39–43.

Carmen, I. U., Chithra, P., and Huang, Q. (2003). Nanotechnology: A new frontier in food science. *Food Technology*, 57, 24–29.

Chaudhary, P., Chaudhary, A., Bhatt, P., Kumar, G., Khatoon, H., Rani, A. et al. (2022). Assessment of soil health indicators under the influence of nanocompounds and *Bacillus* spp. in field condition. *Frontiers in Environmental Science*, 9: 769871.

Chen, H., and Yada, R. (2011). Nanotechnologies in agriculture: New tools for sustainable development. *Trends in Food Science and Technology*, 22: 585–594.

Chen, J., Leng, J., Yang, X., Liao, L., Liu, L., and Xiao, A. (2017). Enhanced performance of magnetic graphene oxide-immobilized laccase and its application for the decolorization of dyes. *Molecules*, 22(2): 221.

Chhipa, H. 2017. Nanofertilizers and nanopesticides for agriculture. *Environmental Chemistry Letters*, 15(1): 15–22.

Chhipa, H., and Kaushik, N. (2015). Development of nano-bio-pesticide using iron and *Eucalyptus* plant extract and their application in pest management. In *Conference Proceedings of Symposium on Recent Advances in Biotechnology for Food and Fuel* (pp. 19–20). TERI, New Delhi.

Chidambaram, R. (2016). Application of rice husk nanosorbents containing 2,4-dichlorophenoxyacetic acid herbicide to control weeds and reduce leaching from soil. *Journal of the Taiwan Institute of Chemical Engineers*, 63: 318–326.

Chookhongkha, N., Sopondilok, T., and Photchanachai, S. (2012, February). Effect of chitosan and chitosan nanoparticles on fungal growth and chilli seed quality. In *I International Conference on Postharvest Pest and Disease Management in Exporting Horticultural Crops (PPDM2012)*, 973: 231–237.

Chopra, R. N., Badhwar, R., and Ghosh, S. (1994). *Poisonous Plants of India*. Indian Council of Agricultural Research, New Delhi, India.

da Costa Silva, L. M., Melo, A. F., and Salgado, A. M. (2011). Biosensors for environmental applications. *Environmental Biosensors*, 3–16.

Dastjerdi, R., and Montazer, M. (2010). A review on the application of inorganic nano-structured materials in the modification of textiles: Focus on anti-microbial properties. *Colloids and Surfaces B: Biointerfaces*, 79: 5.

de la Rosa, G., Lopez-Moreno, M. L., De Haro, D., Botez, C. E., Peralta-Videa, J. R., and Gardea-Torresdey, J. (2013). Effects of ZnO Nanoparticles in Alfalfa, Tomato, and Cucumber at the Germination Stage: Root Development and X-ray Absorption Spectroscopy Studies. *Pure and Applied Chemistry*, 85(12): 2161–2174.

De, A., Bose, R., Kumar, A., and Mozumdar, S. (2014). Worldwide pesticide use. In *Targeted Delivery of Pesticides Using Biodegradable Polymeric Nanoparticles*, SpringerBriefs in Molecular Science, 29–45.

Delfani, M., Baradarn Firouzabadi, M., Farrokhi, N., and Makarian, H. (2014). Some physiological responses of black-eyed pea to iron and magnesium nanofertilizers. *Communications in Soil Science and Plant Analysis*, 45(4): 530–540.

Dimkpa, C. O., Bindraban, P. S., Fugice, J., Agyin-Birikorang, S., Singh, U. et al. (2017). Composite micronutrient nanoparticles and salts decrease drought stress in soybean. *Agronomy for Sustainable Development*, 37, 5.
Dimkpa, C. O., McLean, J. E., Britt, D. W. and Anderson, A. J. 2013. Antifungal activity of ZnO nanoparticles and their interactive effect with a biocontrol bacterium on growth antagonism of the plant pathogen *Fusarium graminearum*. *Biometals*, 26(6): 913–921.

Ditta, A., Arshad, M., and Ibrahim, M. (2015). Nanoparticles in sustainable agricultural crop production: applications and perspectives. In *Nanotechnology and Plant Sciences*, 55–75.

Dixit, R., Wasiullah, X., Malaviya, D., Pandiyan, K., Singh, U. B., Sahu, A. et al. (2015). Bioremediation of heavy metals from soil and aquatic environment: An overview of principles and criteria of fundamental processes. *Sustainability,* 7(2): 2189–2212.

El-Kassas, H. Y., Aly-Eldeen, M. A., and Gharib, S. M. (2016). Green synthesis of iron oxide (Fe_3O_4) nanoparticles using two selected brown seaweeds: Characterization and application for lead bioremediation. *Acta Oceanologica Sinica*, 35(8): 89–98.

El-Ramady, H., El-Henawy, A., Amer, M., Omara, A. E. D., Elsakhawy, T., Salama, A. M. et al. (2020). Agro-pollutants and their nano-remediation from soil and water: A mini-review. *Environment, Biodiversity and Soil Security*, 4: 361–375.

Fortina, P., Kricka, L. J., Surrey, S., and Grodzinski, P. (2005). Nanobiotechnology: The promise and reality of new approaches to molecular recognition. *Trends in Biotechnology*, 23: 168.

Gao, Y., Huang, Q., Su, Q. and Liu, R. (2014). Green synthesis of silver nanoparticles at room temperature using kiwifruit juice. *Spectroscopy Letters* 47(10): 790–795.

García, F. E., Senn, A. M., Meichtry, J. M., Scott, T. B., Pullin, H., Leyva, A. G. et al. (2019). Iron-based nanoparticles prepared from yerba mate extract: Synthesis, characterization and use on chromium removal. *Journal of Environmental Management*, 235: 1–8.

Ghafariyan, M. H., Malakouti, M. J., Dadpour, M. R., Stroeve, P., and Mahmoudi, M. (2013). Effects of magnetite nanoparticles on soybean chlorophyll. *Environmental Science & Technology*, 47(18): 10645–10652.

Ghani, M. I., Saleem, S., Rather, S. A., Rehmani, M. S., Alamri, S., Rajput, V. D. et al. (2022). Foliar application of zinc oxide nanoparticles: An effective strategy to mitigate drought stress in cucumber seedling by modulating antioxidant defense system and osmolytes accumulation. *Chemosphere*, 289: 133202.

Giannousi, K., Avramidis, I., and Dendrinou-Samara, C. (2013). Synthesis, characterization, and evaluation of copper-based nanoparticles as agrochemicals against *Phytophthora infestans*. *RSC Advances*, 3(44): 21743–21752.

Gill, S. S., Khan, N. A., Anjum, N. A., and Tuteja, N. (2011). Amelioration of cadmium stress in crop plants by nutrients management: morphological, physiological and biochemical aspects. *Plant Stress*, 5(1): 1–23.

Giraldo, J. P., Landry, M. P., Faltermeier, S. M., McNicholas, T. P., Iverson, N. M., Boghossian, A. A. et al. (2014). Plant nanobionics approach to augment photosynthesis and biochemical sensing. *Nature Materials*, 13(4): 400–408.

Gottschalk, F., and Nowack, B. 2011. The release of engineered nanomaterials to the environment. *Journal of Environmental Monitoring*, 13(5): 1145–1155.

Hafeez, B., Khanif, Y.M., and Saleem, M. (2013). Role of zinc in plant nutrition – a review. *American Journal of Experimental Agriculture* 3: 374–380.

Hakeem, K. R., and Pirzadah, T. B. (2020). *Nanobiotechnology in agriculture*. Springer International Publishing.

Halecký, M., and Kozliak, E. (2020). Modern bioremediation approaches: use of biosurfactants, emulsifiers, enzymes, biopesticides, GMOs. *Advanced Nano-Bio Technologies for Water and Soil Treatment*, 495–526.

Hidangmayum, A., and Dwivedi, P. (2022). Chitosan based nanoformulation for sustainable agriculture with special reference to abiotic stress: A review. *Journal of Polymers and the Environment*, 30(4): 1264–1283.

Ho, V. A., Le, P. T., Nguyen, T. P., Nguyen, C. K., Nguyen, V. T., and Tran, N. Q. (2015). Silver core-shell nanoclusters exhibiting strong growth inhibition of plant-pathogenic fungi. *Journal of Nanomaterials*, 2015(1): 241614.

Huang, L., Weng, X., Chen, Z., Megharaj, M., and Naidu, R. (2014). Synthesis of iron-based nanoparticles using oolong tea extract for the degradation of malachite green. Spectrochimica Acta Part A: *Molecular and Biomolecular Spectroscopy*, 117: 801–804.

Huang, Y., Fulton, A. N., and Keller, A. A. (2016). Simultaneous removal of PAHs and metal contaminants from water using magnetic nanoparticle adsorbents. *Science of the Total Environment*, 571: 1029–1036.

Iavicoli, I., Leso, V., Beezhold, D. H., and Shvedova, A. A. (2017). Nanotechnology in agriculture: Opportunities, toxicological implications, and occupational risks. *Toxicology and Applied Pharmacology*, 329: 96–111.

Igiebor, F. A., Ikhajiagbe, B., and Asia, M. (2023). Green Nanotechnology: A modern tool for sustainable agriculture in nigeria – a review. *International Journal of Horticultural Science and Technology*, 10(3): 269–286.

Ilahy, R., Hdider, C., Lenucci, M.S., et al. (2011). Phytochemical composition and antioxidant activity of high-lycopene tomato (*Solanum lycopersicum* L.) Cultivars Grown in Southern Italy. *Scientia Horticulturae*, 127(3): 255–261.

Iqbal, M. A. (2019). Nano-fertilizers for sustainable crop production under changing climate: a global perspective. In *Sustainable Crop Production*, edited by Hasanuzzaman, M., M. C. M. T. Filho, M. Fujita, and T. A. R. Nogueira, 325–338. London: IntechOpen.

Iran Nanotechnology Initiative Council. (2009). First nano-organic iron chelated fertilizer invented in Iran. *Iran Nanotechnology Initiative Council.* Available at: http://www.iranreview.org/content/Documents/Iranians_Researchers_Produce_Nano_Organic_Fertilizer.htm. Accessed April 11, 2014.

Iran Nanotechnology Initiative Council. (2009). First Nano-Organic Iron Chelated Fertilizer Invented in Iran." Webpage on the Internet. Tehran, Iran: Iran Nanotechnology Initiative Council.http://www.iranreview.org/content/Documents/Iranians_Researchers_Produce_Nano_Organic_Fertilizer.htm. Accessed April 11, 2014.

Iravani, S. (2011). Green synthesis of metal nanoparticles using plants. *Green Chemistry*, 13(10): 2638–2650.

Jin, X., Yu, B., Lin, J., and Chen, Z. (2016). Integration of biodegradation and nano-oxidation for removal of PAHs from aqueous solution. *ACS Sustainable Chemistry & Engineering*, 4(8): 4717–4723. https://doi.org/10.1021/acssuschemeng.6b00933.

Johnson, A. C., and Park, B. 2012. Predicting contamination by the fuel additive cerium oxide engineered nanoparticles within the united kingdom and the associated risks. *Environmental Toxicology and Chemistry*, 31(11): 2582–2587.

Kaila, T. M., Wahiduzzaman, M. D., and Vamsi, G. (2021). Nanofertilizers in agriculture. *Acta Scientific Agriculture*, 5(3): 35–46.

Kataria, N., and Garg, V. K. (2018). Green synthesis of Fe_3O_4 nanoparticles loaded sawdust carbon for cadmium (II) removal from water: Regeneration and mechanism. *Chemosphere*, 208: 818–828.

Ken, D. S., and Sinha, A. (2020). Recent developments in surface modification of nano zero-valent iron (nZVI): Remediation, toxicity and environmental impacts. *Environmental Nanotechnology, Monitoring & Management*, 14, 100344.

Khan, I., Saeed, K., and Khan, I. 2019. Nanoparticles: properties, applications and toxicities. *Arabian Journal of Chemistry* 12(7): 908–931.

Khan, M. I., Cheema, S. A., Anum, S., Niazi, N. K., Azam, M., Bashir, S. et al. (2020). Phytoremediation of agricultural pollutants. *Phytoremediation: In-situ Applications*, 27–81.

Khan, N. A., and Singh, S. (2008). Abiotic stress and plant responses.

Kharissova, O. V., Kharisov, B. I., García, T. H. and Méndez, U. O. (2009). A review on less-common nanostructures. *Synthesis and Reactivity in Inorganic, Metal-Organic, and Nano-Metal Chemistry*, 39(10): 662–684.

Khodakovskaya, M. V., de Silva, S., Nedosekin, P., Dervishi, E., Biris, A., Shashkov, E. et al. (2013). Carbon nanotubes as plant growth regulators: effects on tomato growth, reproductive system, and soil microbial community. *Small*, 9(1): 115–123.

Kumar, A., Gupta, K., Dixit, S., Mishra, K., and Srivastava, S. 2019. A review on positive and negative impacts of nanotechnology in agriculture. *International Journal of Environmental Science and Technology*, 16(4): 2175–2184.

Kumari, B., and Singh, D. P. (2016). A review on multifaceted application of nanoparticles in the field of bioremediation of petroleum hydrocarbons. *Ecological Engineering*, 97: 98–105.

Lang, Q., Han, L., Hou, C., Wang, F., and Liu, A. (2016). A sensitive acetylcholinesterase biosensor based on gold nanorods modified electrode for detection of organophosphate pesticide. *Talanta*, 156: 34–41.

Larue, C., Laurette, J., Herlin-Boime, N., Khodja, H., Fayard, B., Flank, A. M. et al. 2012. Comparative Uptake and Impact of TiO2 Nanoparticles in Wheat and Rapeseed. *Journal of Toxicology and Environmental Health, Part A*, 75(13–15): 722–734.

Larue, C., Laurette, S., Herlin-Boime, H., Khodja, D., Fayard, A., Flank, G. et al. (2012). Accumulation, translocation and impact of TiO2 nanoparticles in wheat (*Triticum aestivum* spp.): Influence of Diameter and Crystal Phase. *Chemosphere*, 431: 197–208.

Lawal, A. T. (2017). Polycyclic aromatic hydrocarbons: A review. *Cogent Environmental Science*, 3: 1–89. https://doi.org/10.1080/23311843.2017.1339841.

Laware, S. L., and Raskar, S. (2014). Effect of titanium dioxide nanoparticles on hydrolytic and antioxidant enzymes during seed germination in onion. *International Journal of Current Microbiology and Applied Sciences*, 3(7): 749–760.

Li, J., Li, Q., Ma, X., Tian, B., Li, T., Yu, J. et al. (2016). Biosynthesis of gold nanoparticles by the extreme bacterium Deinococcus radiodurans and an evaluation of their antibacterial properties. *International Journal of Nanomedicine*, 5931–5944.

Lin, D., and Xing, B. (2008). Root uptake and phytotoxicity of ZnO nanoparticles. *Environmental Science & Technology*, 42(15): 5580–5585.

Liu, C., Zhou, H., and Zhou, J. (2021). The applications of nanotechnology in crop production. *Molecules*, 26(23): 7070.

Liu, G. D., Dong, S. T., and Zhang, J. W. (2005). Effects of calcium on peanut growth and nutrient uptake. *Plant Nutrition and Fertilizer Science*, 11(1): 77–82.

Liu, R., and Lal, R. (2014). Synthetic apatite nanoparticles as a phosphorus fertilizer for soybean (*Glycine max*). *Scientific Reports*, 4(1): 5686.

Liu, R., and Lal, R. (2015). Potentials of engineered nanoparticles as fertilizers for increasing agronomic productions. *Science of the Total Environment*, 514: 131–139.

Liu, X. M., Feng, Z. B., Zhang, F. D., Zhang, S. Q., and He, X. S. (2006). Preparation and testing of cementing and coating nano-subnanocomposites of slow/controlled-release fertilizer. *Agricultural Sciences in China*, 5(9): 700–706.

Liu, X., Zhang, F., Zhang, S., He, X., Wang, R., Feng, Z., and Wang, Y. (2005). Response of peanut to nano-calcium carbonate, 385–389.

Lohrasbi, S., Kouhbanani, M. A. J., Beheshtkhoo, N., Ghasemi, Y., Amani, A. M., and Taghizadeh, S. (2019). Green synthesis of iron nanoparticles using Plantago major leaf extract and their application as a catalyst for the decolorization of azo dye. *BioNanoScience*, 9(2): 317–322.

Lu, C. M., Zhang, C. Y., Wen, J. Q., Wu, G. R., and Tao, M. X. (2002). Research of the effect of nanometer materials on germination and growth enhancement of *Glycine Max* and its mechanism. *Soybean Science*, 21(3): 168–171.

Ma, X., Geiser-Lee, J., Deng, Y., and Kolmakov, A. (2010). Interactions between engineered nanoparticles (ENPs) and plants: Phytotoxicity, uptake and accumulation. *Science of the Total Environment*, 408(16): 3053–3061.

Mahajan, P., Dhoke, S. K., and Khanna, A. S. (2011). Effect of nano-ZnO particle suspension on growth of mung (*Vigna radiata*) and gram (*Cicer arietinum*) seedlings using plant agar method. *Journal of Nanotechnology*, 2011, Article ID 696535.

Mahajan, S., and Tuteja, N. (2005). Cold, salinity and drought stress: An overview. *Archives of Biochemistry and Biophysics*, 444: 139–158.

Mahmoud, A. W. M., and Taha, S. S. (2018). Main sulphur content in essential oil of *Eruca Sativa* as affected by nano iron and nano zinc mixed with organic manure. *Agriculture*, 64(2): 65.

Mahto, R., Rani, P., Bhardwaj, R., Singh, R. K., Prasad, S. K., and Rakshit, A. (2021). Nanotechnology: A potential approach for abiotic stress management. In *Advances in Nano-Fertilizers and Nano-Pesticides in Agriculture* (pp. 249–264). Woodhead Publishing.

Mallikarjuna, K., Narasimha, G., Dillip, G. R., Praveen, B., Shreedhar, B., Lakshmi, C. S. et al. (2011). Green synthesis of silver nanoparticles using *Ocimum* leaf extract and their characterization. *Digest Journal of Nanomaterials and Biostructures*, 6(1): 181–186.

Manquián-Cerda, K., Cruces, E., Rubio, M. A., Reyes, C., and Arancibia-Miranda, N. (2017). Preparation of nanoscale iron (oxide, oxyhydroxides and zero-valent) particles derived from blueberries: Reactivity, characterization and removal mechanism of arsenate. *Ecotoxicology and Environmental Safety*, 145: 69–77.

Maqbool, Q., Iftikhar, S., Nazar, M., Abdallah, H. H., Parveen, T., Hussain, T et al. (2017). Green fabricated CuO nanobullets via *Olea Europaea* leaf extract shows auspicious antimicrobial potential. *IET Nanobiotechnology*, 11(4): 463–468.

Marchiol, L. (2012). Synthesis of metal nanoparticles in living plants. *Italian Journal of Agronomy*, 7: 37.

Mattos, B. D., and Magalhães, W. L. (2016). Biogenic nanosilica blended by nanofibrillated cellulose as support for slow-release of tebuconazole. *Journal of Nanoparticle Research*, 18(1): 1–10.

McPartlin, D. A., Loftus, J. H., Crawley, A. S., Silke, J., Murphy, C. S., and O'Kennedy, R. J. (2017). Biosensors for the monitoring of harmful algal blooms. *Current Opinion in Biotechnology*, 45: 164–169.

Mehta, M., Sharma, M., Pathania, K., Jena, P. K., and Bhushan, I. (2021). Degradation of synthetic dyes using nanoparticles: A mini-review. *Environmental Science and Pollution Research*, 28(39): 49434–49446.

Mirnezhad, M., Romero-González, R. R., Leiss, K. A., Choi, Y. H., Verpoorte, R., and Klinkhamer, P. G. L. (2010). Metabolomic Analysis of Host Plant Resistance to Thrips in Wild and Cultivated Tomatoes." *Phytochemical Analysis*, 21(1): 110–117.

Mishra, S., and Singh, H. B. (2016). Preparation of biomediated metal nanoparticles. *Indian Patent Filed*, 201611003248.

Mishra, S., Keswani, C., Abhilash, P. C., Fraceto, L. F., and Singh, H. B. (2017). Integrated approach of agrinanotechnology: Challenges and future trends. *Frontiers in Plant Science*, 8: 471. https://doi.org/10.3389/fpls.2017.00471.

Moaveni, P., and Kheiri, T. (2011). TiO2 nanoparticles affected on maize (*Zea Mays* L). In *Proceedings of the 2nd International Conference on Agricultural and Animal Science*, Maldives. Singapore: IACSIT Press, 22: 160–163. http:// jm.birjand.ac.ir/archive_jou/escs/139261/4P-ESCS61-A91216.pdf.

Mondal, K. K., and Mani, C. (2012). Investigation of the antibacterial properties of nanocopper against *Xanthomonas axonopodis* pv. *punicae*, the incitant of pomegranate bacterial blight. *Annals of Microbiology*, 62: 889–893.

Mori, A., Kirk, G. J. D., Lee, J. S., Morete, M. J., Nanda, A. K., Johnson-Beebout, S. E. et al. (2016). Rice genotype differences in tolerance of zinc-deficient soils: evidence for the importance of root-induced changes in the rhizosphere. *Frontiers in Plant Science*, 6: 1160.

Nawaz, H., Ahmad, K. S., Aslam, I., Mahmood, A., Mehmood, A., Khan, A. et al. (2024). Modulations in Physiological and Biochemical attributes of Citrus Limon by selenium nanoparticles (SeNPs) under salinity stress. *Biocatalysis and Agricultural Biotechnology*, 62: 103438.

Naderi, M. R., and Abedi, A. (2012). Application of nanotechnology in agriculture and refinement of environmental pollutants. *Journal of Nanotechnology*, 11(1): 18–26.

Naderi, M. R., and Danesh-Shahraki, A. (2013). Nanofertilizers and their roles in sustainable agriculture. *International Journal of Agriculture and Crop Sciences*, 5(19): 2229–2232.

Nair, R., Varghese, S. H., Nair, B. G., Maekawa, T., Yoshida, Y., and Kumar, D. S. (2010). Nanoparticulate material delivery to plants. *Plant Science*, 179(3): 154–163.

Nazir, R., Ayub, Y., and Tahir, L. (2020). Green-nanotechnology for precision and sustainable agriculture. In *Biogenic Nano-Particles and Their Use in Agro-Ecosystems*, 317–357.

Needhidasan, S., Samuel, M., and Chidambaram, R. (2014). Electronic waste—An emerging threat to the environment of urban India. *Journal of Environmental Health Science and Engineering*, 12: 36.

Needhidasan, S., and Ramalingam, C. (2017). Stratagems employed for 2,4-dichlorophenoxyacetic acid removal from polluted water sources. *Clean Technologies and Environmental Policy*, 19(6): 1607–1620.

Nejatzadeh, F. (2021). Effect of silver nanoparticles on salt tolerance of *Satureja hortensis* L. During *in vitro* and *in vivo* germination tests. *Heliyon*, 7: e05981.

Nicolodi, M., and Gianello, C. (2014). Understanding soil as an open system and fertility as an emergent property of the soil system. *Sustainable Agriculture Research*, 4: 526–201637873.

Nie, Z., Petukhova, A., and Kumacheva, E. (2010). Properties and emerging applications of self-assembled structures made from inorganic nanoparticles. *Nature Nanotechnology*, 5(1): 15–25.

Nkansah, M. A., Christy, A. A., and Barth, T. (2012). Catalytic oxidation and reduction of polycyclic aromatic hydrocarbons (PAHs) present as mixtures in hydrothermal media. *Polycyclic Aromatic Compounds*, 32: 408–422. https://doi.org/10.1080/10406638.2012.663451.

Ocsoy, I., Paret, M. L., Ocsoy, M. A., Kunwar, S., Chen, T., You, M. et al. (2013). Nanotechnology in plant disease management: DNA-directed silver nanoparticles on graphene oxide as an antibacterial against *Xanthomonas perforans*. *ACS Nano*, 7(10): 8972–8980.

Palmero, P. (2015). Structural ceramic nanocomposites: a review of properties and powders' Synthesis Methods. *Nanomaterials*, 5(2): 656–696.

Panwar, J., Jain, N., Bhargaya, A., Akhtar, M. S., and Yun, Y. S. (2012). Positive effect of zinc oxide nanoparticles on tomato plants: a step towards developing 'Nano-Fertilizers'. In *Proceedings of 3rd International Conference on Environmental Research and Technology (ICERT)*, May 30–June 1, 2012, Penang, Malaysia.

Paret, M. L., Vallad, G. E., Averett, D. R., Jones, J. B., and Olson, S. M. (2013). Photocatalysis: Effect of light-activated nanoscale formulations of TiO2_22 on *Xanthomonas perforans* and control of bacterial spot of tomato. *Phytopathology*, 103(3): 228–236.

Parveen, A., Mazhari, B. B. Z., and Rao, S. (2016). Impact of bio-nanogold on seed germination and seedling growth in *Pennisetum glaucum*. *Enzyme and Microbial Technology*, 95: 107–111.

Patil, S., and Chandrasekaran, R. (2020). Biogenic nanoparticles: a comprehensive perspective in synthesis, characterization, application and its challenges. *Journal of Genetic Engineering & Biotechnology*, 18: 67.

Patil-Patankar, V., and Sanghvi, G. (2022). Nanoherbicides: A sustainable option for field applications. In *Relationship Between Microbes and the Environment for Sustainable Ecosystem Services, Volume 1* (pp. 335–355). Elsevier.

Pérez-de-Luque, A. (2017). Interaction of nanomaterials with plants: what do we need for real applications in agriculture? *Frontiers in Environmental Science*, 5: 12.

Piccinno, F. Gottschalk, F., Seeger, S., and Nowack, B.. (2012). Industrial production quantities and uses of ten engineered nanomaterials in europe and the world. *Journal of Nanoparticle Research*, 14(9): 1109.

Pillai, H. P., and Kottekottil, J. (2016). Nano-phytotechnological remediation of endosulfan using zero valent iron nanoparticles. *Journal of Environmental Protection*, 7(5): 734.

Pradhan, S., Patra, P., Das, S., Chandra, S., Mitra, S., Dey, K. K. et al. (2013). Photochemical modulation of biosafe manganese nanoparticles on *Vigna radiata*: A detailed molecular, biochemical, and biophysical study. *Environmental Science & Technology*, 47(22): 13122–13131.

Prasad, R. (2014). Synthesis of silver nanoparticles in photosynthetic plants. *Journal of Nanoparticles*, 963961: 1–8.

Prasad, R., Kumar, V., and Prasad, K. S. (2014). Nanotechnology in sustainable agriculture: present concerns and future aspects. *African Journal of Biotechnology*, 13(6): 705–711.

Priester, J. H., Ge, Y., Mielke, R. E., Horst, A. M., Moritz, S. C., Espinosa, K. et al. (2012). Soybean susceptibility to manufactured nanomaterials with evidence for food quality and soil fertility interruption. *Proceedings of the National Academy of Sciences*, *109*(37), E2451-E2456.

Qian, K., Shi, T., Tang, T., Zhang, S., Liu, X., & Cao, Y. (2011). Preparation and characterization of nano-sized calcium carbonate as controlled release pesticide carrier for validamycin against *Rhizoctonia solani*. *Microchimica Acta*, 173: 51–57.

Ragaei, M., and Sabry, A. K. H. (2014). Nanotechnology for insect pest control. *International Journal of Science, Environment and Technology*, 3(2): 528–545.

Rajesh, S., Raja, D. P., Rathi, J. M., and Sahayaraj, K. (2012). Biosynthesis of silver nanoparticles using *Ulva fasciata* (Delile) ethyl acetate extract and its activity against *Xanthomonas campestris* pv. *malvacearum*. *Journal of Biopesticides*, 5: 119–124.

Rajput, V. D., Singh, A., Minkina, T., Rawat, S., Mandzhieva, S., Sushkova, S. et al. (2021). Nano-enabled products: Challenges and opportunities for sustainable agriculture. *Plants*, 10(12): 2727.

Rico, C. M., Majumdar, S., Duarte-Gardea, M., Peralta-Videa, J. R., and Gardea-Torresdey, J. L. 2011. Interaction of nanoparticles with edible plants and their possible implications in the food Chain. *Environmental International*, 59(8): 3485–3498.

Rizwan, M. D., Singh, M., Mitra, C. K., and Morve, R. K. (2014). Ecofriendly application of nanomaterials: nanobioremediation. *Journal of Nanoparticles*, 2014(1): 431787.

Roco, M. C. (2003). Broader societal issue of nanotechnology. *Journal of Nanoparticle Research*, 5: 181–189.

Rodriguez-Mozaz, S., Marco, M. P., De Alda, M. L., and Barceló, D. (2004). Biosensors for environmental applications: Future development trends. *Pure and Applied Chemistry*, 76(4): 723–752.

Sadeghi, R., Rodriguez, R. J., Yao, Y., and Kokini, J. L. (2017). Advances in nanotechnology as they pertain to food and agriculture: Benefits and risks. *Annual Review of Food Science and Technology*, 8(1): 467–492.

Sadowski, Z. (2010). Biosynthesis and Application of Silver and Gold Nanoparticles. In *Silver Nanoparticles*, 257–276. Dordrecht: Springer.

Saharan, V., Mehrotra, A., Khatik, R., Rawal, P., Sharma, S. S., and Pal, A. (2013). Synthesis of chitosan-based nanoparticles and their *in vitro* evaluation against phytopathogenic fungi. *International Journal of Biological Macromolecules*, 62: 677–683.

Saif, S., Tahir, A., and Chen, Y. (2016). Green synthesis of iron nanoparticles and their environmental applications and implications. *Nanomaterials*, 6(11): 209. https://doi.org/10.3390/nano6110209.

Samuel, M. S., Bhattacharya, J., Parthiban, C., Viswanathan, G., and Singh, N. P. (2018a). Ultrasound-assisted synthesis of metal organic framework for the photocatalytic reduction of 4-nitrophenol under direct sunlight. *Ultrasonics Sonochemistry*, 49: 215–221.

Samuel, M. S., Shah, S. S., Subramaniyan, V., Qureshi, T., Bhattacharya, J., and Singh, N. P. (2018b). Preparation of graphene oxide/chitosan/ferrite nanocomposite for Chromium(VI) removal from aqueous solution. *International Journal of Biological Macromolecules*, 119: 540–547.

Samuel, M. S., Bhattacharya, J., Raj, S., Santhanam, N., Singh, H., and Singh, N. P. (2019). Efficient removal of Chromium(VI) from aqueous solution using chitosan grafted graphene oxide (CS-GO) nanocomposite. *International Journal of Biological Macromolecules*, 121: 285–292.

Samuel, M. S., Ravikumar, M., John, J. A., Selvarajan, E., Patel, H., Chander, P. S. et al. (2022). A review on green synthesis of nanoparticles and their diverse biomedical and environmental applications. *Catalysts*, 12(5): 459.

Savci, S. (2012). An agricultural pollutant: chemical fertilizer. *International Journal of Environmental Science and Development* 3: 73.

Saxena, G., Purchase, D., Mulla, S. I., and others. (2019). Phytoremediation of heavy metal-contaminated sites: Eco-environmental concerns, field studies, sustainability issues, and future prospects, pp. 71–131.

Saxena, G., Purchase, D., Mulla, S. I., Saratale, G. D., and Bharagava, R. N. (2020). Phytoremediation of heavy metal-contaminated sites: eco-environmental concerns, field studies, sustainability issues, and future prospects. *Reviews of Environmental Contamination and Toxicology Volume*, 249: 71–131.

Saxena, R., Tomar, R. S., and Kumar, M. (2016). Exploring nanobiotechnology to mitigate abiotic stress in crop plants. *Journal of Pharmaceutical Sciences and Research*, 8(9): 974.

Scott, N., and Chen, H. (2013). Nanoscale science and engineering for agriculture and food systems. *Industrial Biotechnology*, 9(1): 17–18.

Sekhon, B. S. (2014). Nanotechnology in Agri-Food Production: An Overview. *Nanotechnology, Science and Applications*, 7: 31–53.

Shah, V., and Belozerova, I. (2009). Influence of metal nanoparticles on the soil microbial community and germination of lettuce seeds. *Water, Air, and Soil Pollution*, 197: 143–148.

Shahid, M., Naeem-Ullah, U., Khan, W., Saeed, D. S., and Razzaq, K. (2021). Application of nanotechnology for insect pests management: a review. *Journal of Innovative Science*, 7: 28–39.

Shang, Y., Hasan, M. K., Ahammed, G. J., Li, M., Yin, H., and Zhou, J. (2019). Applications of nanotechnology in plant growth and crop protection: a review. *Molecules*, 24(14): 2558. https://doi.org/10.3390/molecules24142558.

Sharma, V. K., Yngard, R. A., and Lin, Y. (2009). Silver nanoparticles: Green synthesis and their antimicrobial activities. *Advances in Colloid and Interface Science*, 145(1–2): 83–96.

Singh, A., Singh, N. B., Hussain, I., Singh, H., and Singh, S. C. (2015). Plant-nanoparticle interaction: an approach to improve agricultural practices and plant productivity. *International Journal of Pharm Sci Invent*, 4(8): 25–40.

Singh, B. N., Hidangmayum, A., Singh, A., Guru, A., Yashu, B. R., and Singh, G. S. (2020a). Effect of emerging contaminants on crops and mechanism of toxicity. *Sustainable Agriculture Reviews*, 40: 217–241.

Singh, R., Behera, M., and Kumar, S. (2020b). Nano-bioremediation: an innovative remediation technology for treatment and management of contaminated sites. *Bioremediation of Industrial Waste for Environmental Safety: Volume II: Biological Agents and Methods for Industrial Waste Management*, 165–182.

Singh, R. P., Handa, R., and Manchanda, G. (2021). Nanoparticles in sustainable agriculture: an emerging opportunity. *Journal of Controlled Release*, 329: 1234–1248.

Smith, B. D. (2007). *Rivers of change: Essays on early agriculture in eastern North America*. University of Alabama Press.

Subashini, H. D., Malarvannan, S., and Pillai, R. R. (2004). Dodonaea Angustifolia – A Potential Biopesticide against Helicoverpa Armigera. *Current Science*, 86(1): 26–28.

Taha, R. A., Hassan, M. M., Ibrahim, E. A., Abou Baker, N. H., and Shaaban, E. A. (2016). Carbon nanotubes impact on date palm in vitro cultures. *Plant Cell, Tissue and Organ Culture (PCTOC)*, 127: 525–534.

Taran, N. Y., Gonchar, O. M., Lopatko, K. G., Batsmanova, L. M., Patyka, M. V., and Volkogon, M. V. (2014). The effect of colloidal solution of molybdenum nanoparticles on the microbial composition in the rhizosphere of *Cicer arietinum* L. *Nanoscale Research Letters*, 9(1): 1–8.

Thome, A., Reddy, K. R., Reginatto, C., and Cecchin, I. (2015). Water, air, and soil pollution. *Water, Air, & Soil Pollution*, 226: 1–20.

Tipu, M. M. H., Baroi, A., Rana, J., Islam, S., Jahan, R., Miah, M. S. et al. (2021). Potential applications of nanotechnology in agriculture: a smart tool for sustainable agriculture. *In*: Asaduzzaman, M. and Afroz, M. (eds.). *Agricultural Development in Asia - Potential Use of Nano-Materials and Nano-Technology*. IntechOpen.

Tratnyek, P. G., and Johnson, R. L. (2006). Nanotechnologies for environmental cleanup. *Nano Today*, 1: 44.

Tripathi, D. K., Singh, V. P., Kumar, D., and Chauhan, D. K. (2012). Impact of exogenous silicon addition on chromium uptake, growth, mineral elements, oxidative stress, antioxidant capacity, and leaf and root structures in rice seedlings exposed to hexavalent chromium. *Acta Physiologiae Plantarum*, 34: 279–289. https://doi.org/10.1007/s11738-011-0826-5.

Tripathi, D. K., Singh, S., Singh, S., Mishra, S., Chauhan, D. K., and Dubey, N. K. (2015). Micronutrients and their diverse role in agricultural crops: Advances and future prospective. *Acta Physiologiae Plantarum*, 37(7): 1–14.

Tripathi, D. K., Singh, S., Singh, S., Pandey, R., Singh, V. P., Sharma, N. C. et al. (2017). An overview on manufactured nanoparticles in plants: Uptake, translocation, accumulation and phytotoxicity. *Plant Physiology and Biochemistry*, 110: 2–12.

Tripathi, M., Kumar, S., Kumar, A., Tripathi, P., and Kumar, S. (2018). Agro-nanotechnology: a future technology for sustainable agriculture. *International Journal of Current Microbiology and Applied Sciences*, 7: 196–200.

Tuteja, N. (2007). Mechanisms of high salinity tolerance in plants. *Methods in Enzymology*, 428; 419–438.

Upadhyaya, H., Panda, S. K., and Dutta, B. K. (2008). Variation of physiological and antioxidative responses in tea cultivars subjected to elevated water stress followed by rehydration recovery. *Acta Physiologiae Plantarum*, 30(4): 457–463.

Usman, M., Farooq, M., Wakeel, A., Nawaz, A., Cheema, S. A., ur Rehman, H. et al. (2020). Nanotechnology in agriculture: current status, challenges and future opportunities. *Science of the Total Environment*, 721: 137778.

Velten, S., Leventon, J., Jager, N., and Newig, J. (2015). What is sustainable agriculture? a systematic review. *Sustainability*, 7(6): 7833–7865.

Wieczorek, A. (2003). Use of biotechnology in agriculture—benefits and risks. *Biotechnology*, 11(3): 45–53.

World Commission on Environment and Development (WCED). (1987). Our Common Future: Report of the World Commission on Environment and Development. United Nations.

Yearla, S. R., and Padmasree, K. (2016). Exploitation of subabul stem lignin as a matrix in controlled-release agrochemical nanoformulations: A case study with herbicide diuron. *Environmental Science and Pollution Research*, 23: 18085–18098.

Yuvaraj, M., and Subramanian, K. S. (2015). Controlled-release fertilizer of zinc encapsulated by a manganese hollow core shell. *Soil Science and Plant Nutrition*, 61(2): 319–326.

Zannat, R., Rahman, M. M., and Afroz, M. (2021). Application of nanotechnology in insect pest management: a review. *SAARC Journal of Agriculture*, 19(2): 1–11.

Zhao, L., Peralta-Videa, J. R., Rico, C. M., Hernandez-Viezcas, J. A., Sun, Y., Niu, G. et al. (2014). CeO_2 and ZnO nanoparticles change the nutritional qualities of cucumber (*Cucumis sativus*). *Journal of Agricultural and Food Chemistry*, 62(13): 2752–2759.

Zheng, L., Hong, F., Lu, S., and Liu, C. (2005). Effect of nano-TiO2_22 on the strength of naturally aged seeds and growth of spinach. *Biological Trace Element Research*, 104(1): 83–91.

CHAPTER-2

Green Nanotechnology for Improving Plant Abiotic Stress Resilience under Climate Change

Ratneswar Poddar,[1,*] *Susnigdha Sengupta,*[1] *Arup Sen,*[2] *Arindam Sarkar*[2] and *Akbar Hossain*[3,*]

1. Introduction

Most people in many nations rely on agriculture as their primary source of income. Thus, modern technology interventions in the agricultural sector are crucial in developing nations to feed the burgeoning populations (Shang et al., 2019). Currently, chemical fertilizers and pesticides are used largely in intensive agriculture to ensure higher crop yields and profits (Meena et al., 2016). As a result, unscientific and excessive use of agrochemicals has resulted in degraded soil health and reduced productivity of crops (Meena et al., 2020).

On the contrary, various abiotic stresses, including heat, cold, salinity, drought, and nutrient toxicity, significantly limit agricultural production (Bashir et al., 2019). With the rapidly increasing global population, these challenges are expected to become even more critical. The world population has recently surpassed 8 billion, and projections by the United Nations indicate that it could be to the tune of 8.5 billion by 2030 and 9.7 billion by 2050 (Sexton, 2025). This escalating demand for food, combined with environmental changes, presents a substantial challenge for agriculture. Unfortunately, a significant portion of arable land is subjected to stressful conditions (Cramer et al., 2011). Compared with potential record yields, abiotic stresses can lead to yield reductions of more than 60% (Kopecka et al., 2023). A recent study revealed that extreme climatic conditions could account for a global yield reduction of up to half of the present condition (Vogel et al., 2019). This study further suggested that under regular irrigation conditions, heat stress is more associated with yield variation than moisture factors. In this context, we discuss the various crucial abiotic stressors affecting crop plants, including drought, salinity, heat, cold, and heavy metals. Various earlier studies revealed that nanotechnology is a promising green technology for the sustainability

[1] Department of Agricultural Chemistry and Soil Science, Bidhan Chandra Krishi Viswavidyalaya, Mohanpur, Nadia, West Bengal, India-741252.
[2] Department of Agronomy, Bidhan Chandra Krishi Viswavidyalaya, Mohanpur, Nadia, West Bengal, India-741252.
[3] Soil Science Division, Bangladesh Wheat and Maize Research Institute, Dinajpur 5200, Bangladesh.
* Corresponding authors: rpoddar.bckv@rediffmail.com; akbarhossainwrc@gmail.com

DOI: 10.1201/9781003567684-2

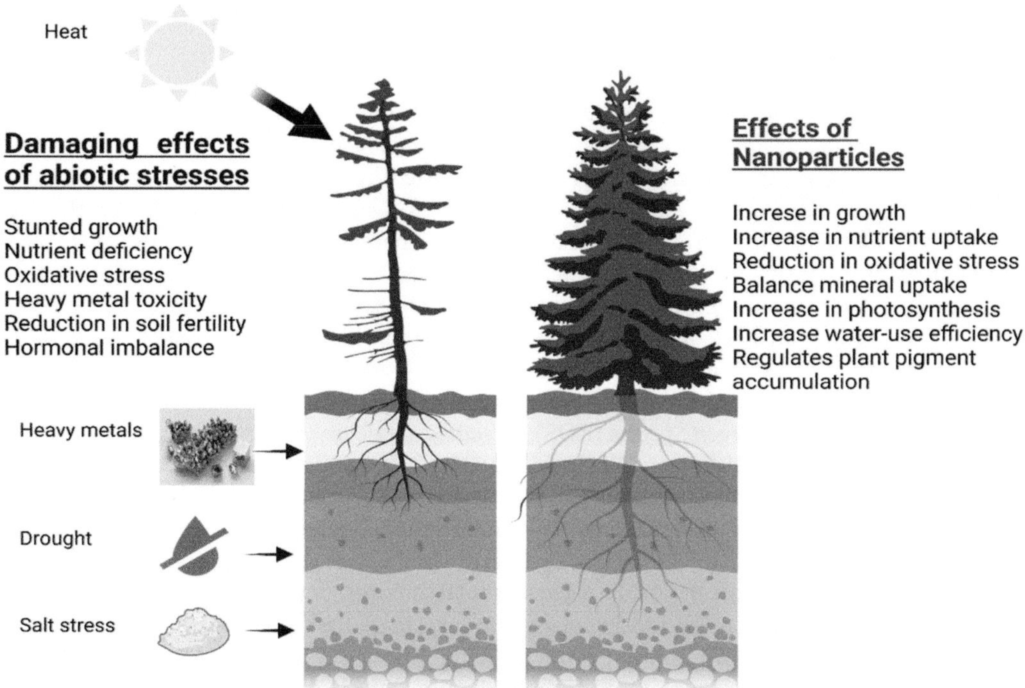

Figure 1. Various roles of nanoparticles (NPs) during abiotic stress in plants. Adapted modified from Azameti and Imoro (2023).

of crop production by improving the tolerance of plants due to abiotic stress. The various roles of nanoparticles (NPs) in plants under abiotic stress conditions are presented in Fig. 1 (Azameti and Imoro 2023).

Nanotechnology is an interdisciplinary field of study that emphasizes mainly the production and application of nanoparticles having 1 to 100 nm in size. A large surface area endows them with unique features and has the opportunity to apply in various fields (Khan et al., 2019). Owing to their unique physicochemical characteristics, nanomaterials offer solutions to contemporary issues that affect modern society and agriculture. Several nations rely heavily on the use of nanotechnology to increase agricultural output and improve food security (Marchiol, 2018; Prasad et al., 2017). The advancement of nanoscience in the field of agriculture has led to the development of GMOs, precision farming, and the application of nanosensors for disease detection, food quality maintenance, and agricultural waste management.

The agricultural sector is expected to face numerous obstacles in the near future, including population explosions, shrinking land areas, increased costs of pesticides and fertilizers, and environmental concerns (Wield et al., 2013; Pingali et al., 2019). To address these problems and increase crop output and nutritional value, nanotechnology is a better choice (Manjunatha et al., 2016). From the storage of water to the precise distribution of nutrients and fertilizers, nanomaterials are essential to agriculture (Usman et al., 2020). With the use of nanosensors, diseases can be easily identified earlier. Nanofungicides, nanobactericides, and nano pesticides are applied to protect crops against several types of biological stress resulting from microorganisms. Nanosensors are also helpful in minimizing postharvest losses by improving packing and transportation and monitoring food quality. The current chapter highlights the prospects of nanotechnology for abiotic stress tolerance in plants under a changing climate. A detailed discussion regarding the use of nanotechnology for agricultural sustainability is presented below.

2. Abiotic Stresses Significantly Limit Agricultural Production

2.1 High Temperature

The increase in global temperature has led to frequent occurrence of dangerous heat events, which pose significant abiotic stress for growth of the plants and enhance crop productivity (Zhao et al., 2020). According to the Intergovernmental Panel on Climate Change, if current trends persist, global temperatures may attain 1.5°C more between 2030–2052 (Masson-Delmotte et al., 2019). Heat stress can cause a great deal of damage to crop plants by permanently reducing their growth and development (Kopecka et al., 2023). This stress usually reduces photosynthetic efficiency, shortens the plant life cycle, and decreases overall productivity ((Xalxo et al., 2020). One of the primary physiological effects of heat stress is membrane dysfunction. Elevated temperatures increase the kinetic energy and mobility of biomolecules within membranes, weakening chemical bonds and causing lipid disintegration, which in turn increases membrane fluidity. Plants under such stress tend to accumulate Reactive Oxygen Species (ROS), including singlet oxygen ($^1O_2$), superoxide radicals (O_2^-), hydrogen peroxide (H_2O_2), and hydroxyl radicals (OH^-), leading to oxidative stress (Nosaka and Nosaka, 2017). These ROS, such as O_2- and H_2O_2, can disrupt membrane integrity, degrade proteins, and deactivate enzymes, ultimately compromising cell viability (Zhao et al., 2020). Higher temperatures also adversely affect various stages of plant development, especially reproductive processes, which can result in significant yield losses (Khan et al., 2022).

In an aeroponic study on salad rocket plants (*Eruca sativa*), the root-zone temperature from by 5°C both in the upper and lower ground part, along with a decline in net photosynthesis (He et al., 2016). Another study on tomato plants revealed that more than six days of continuous exposure of the plant to 35°C unexpectedly increased the photosynthesis rate; however, if plants are exposed to 42°C there was a decline in photosynthesis and stomatal conductance (Giri et al., 2017). Similarly, Guo et al. (2022) showed that exposure to 45±2°C for 52 days reduced the photosynthetic rate and greenness index in Ahmar plants (*Solanum lycopersicum* L.), with higher temperatures significantly affecting reproductive development. For Brassica species, the number of flowers and pods, seed weight, and overall yield were reduced if the crop was in flowering stages and faced temperatures above 29.5°C (Morrison and Stewart, 2002). Nutrient concentrations vary depending on whether the entire plant or only the roots are exposed to heat stress. Both chronic and acute heat stress in tomato plants decreases the concentration of iron (Fe), Nitrogen (N), and total Carbon (C) in the roots and N% in the shoot (Giri et al., 2017).

2.2 Drought Stress

High temperatures are closely linked to drought stress, a condition characterized by prolonged periods of insufficient or absent rainfall, leading to significant hydrological imbalances. Drought is a recurring challenge for farmers, with serious implications for agriculture each year. It is a natural occurrence that lasts for a long time because there is not enough freshwater available to meet ecosystems' and humans' demands (Ahluwalia et al., 2021). Current climate projections and anticipated global warming are expected to significantly affect moisture patterns, thereby increasing the frequency and severity of droughts globally. Plants encounter various stresses (drought, low light, temperature variations, salt stress, heat waves, and exposure to heavy metals) throughout their growth and development. Among all these stresses drought significantly affects plant morphology at every stage of development (Mansoor et al., 2020). This stress specifically disrupts plant metabolism, potentially leading to a 30% reduction in productivity worldwide by 2025 (Bashir et al., 2021). Agriculture currently consumes 70% of the world's clean water supply, a demand projected to rise significantly in the coming years (Boretti and Rosa, 2019). The effect of drought on crop yield is largely dependent on crop severity; extreme drought can lead to total yield loss in the absence of irrigation, as observed in field studies with barley (Samarah et al., 2019).

A significant consequence of drought is the decline in photosynthesis, primarily caused by reduced leaf development, damage to the photosynthetic pigments, early leaf senescence, and a subsequent decrease in yield (Seleiman et al., 2021). Drought stress leads to changes in photosynthetic pigments and components, disrupts the photosynthetic apparatus, and reduces the activity of Calvin cycle enzymes, all of which play a crucial role in lowering crop yields (Ahluwalia et al., 2021). A study on tomato plants found that combined drought and heat stress led to an 80% reduction in chlorophyll levels and a 57% decline in carotenoid content (Raja et al., 2020). Drought and heat stress can also negatively impact pistil and pollen growth, ovule function, and grain mass (Prasad et al., 2011). One of the earliest effects of drought is reduced germination, resulting in improper development of seedlings. For instance, an experiment on peas found that drought stress inhibited germination and quick development of seedlings in each of the five cultivars that were tested (Okcu et al., 2005). Soil water deficits decrease crop production mainly through three mechanisms: (1) decreased canopy absorption of photosynthetically active radiation, (2) reduced light use efficiency, and (3) a lower harvest index (Farooq et al., 2009). An analysis of field experiment data revealed that a 40% reduction in water supply resulted in yield losses ranging from approximately 21% in wheat to 39% in maize (Daryanto et al., 2016). Moreover, drought negatively impacts crop quality; for example, rapeseed has a lower seed oil content under water-deficient conditions (Hussain et al., 2018). The developmental stage at which plants experience drought is also a critical factor in yield loss. Plants exposed to drought during the reproductive phase are particularly vulnerable, making the timing of plant development a crucial consideration for improving yields (Kopecka et al., 2023). Prolonged drought stress during the reproductive stage can reduce grain filling, flower production, seed composition, and longevity (Farooq et al., 2016).

2.3 Salt Stress

Soil salinity has been one of the most significant challenges in agriculture throughout history (Nandal and Hooda, 2013), as it severely restricts agricultural productivity, particularly by reducing crop yields (Munns and Tester, 2008). Globally, 1125 million ha of cultivated land are affected by salinity, with 76 million ha suffering from salinity and sodicity caused by human activities (Wicke et al., 2011). In India alone, approximately 6.73 million ha of land is impacted by salinity (Choudhary et al., 2023). Currently, salinity affects one in five irrigated farms, and each year, 1.5 mha of agricultural land becomes unsuitable for farming. If these trends continue, it is predicted that by 2050, 50% of the world's cultivated land will be on the verge of degradation (Yildiz et al., 2021).

Excess soluble salt concentration is an indication of soil salinity, and soil is called saline when the electrical conductivity of the soil extract is 4 dS/m or higher. The primary contributors to soil salinity are salts such as NaCl and Na_2SO_4 (Pessarakli et al., 2010). Salinity induces stress in plants primarily in two ways: high concentrations of salt in soil hinder water uptake by roots, and high salt concentrations within plants can lead to toxicity. Salts outside the plant roots directly inhibit cell expansion and growth, whereas toxic levels of salts within the plant take time to accumulate before their harmful effects manifest (Yildiz et al., 2021). Salinity disrupts ionic balances within crops, leading to osmotic stress, which reduces water availability to the plant, thereby causing physiological drought. Following this, ionic stress occurs as the plant's ion balance deteriorates. Owing to ionic stress, the increase in Na+ and Cl^- ions compete with essential mineral nutrients such as K^+, Ca^{2+}, and Mg^{2+}, leading to their deficiencies. While osmotic and ionic stresses are the direct effects of salinity, structural damage and the synthesis of toxic compounds are secondary consequences (Botella et al., 2007).

The effect of increasing salinity on crop yield varies not only by plant species but also by cultivar, as demonstrated in recent laboratory studies on olives (*Olea europaea*) and maize and in field studies on wheat, barley, and potatoes (Kopecká et al., 2023). Among the crops, rice responds differently to salinity stress. The crop is very susceptible to the effect of salt stress, with significant reductions in growth observed during its early developmental stages. High salt concentrations lead to reduced seed growth and a decrease in plant fresh weight (Kazemi and Eskandari, 2011). The first organ affected by salinity is the root, where excessive Na^+ accumulation results in lowering the growth of the root and shoot of the plant (Hussain et al., 2017). In wheat, osmotic stress caused by water unavailability due to salt stress lowers germination rates (Motos et al., 2017). In addition to germination, salinity also leads to reductions in shoot and root growth, leaf development, and cell function. Salt stress at the beginning of the crop period is most vulnerable in wheat (Hussain et al., 2019). In cotton, salinity negatively affects the leaf area due to osmotic stress, hinders overall plant growth, and reduces root and shoot development. Additionally, it decreases photosynthetic activity, fibre quality, and metabolic processes (Muhammad et al., 2018). Maize, although generally less susceptible to salt stress than C_4 plants from the Poaceae family (Chinnusamy et al., 2005), also suffers from disrupted shoot growth due to osmotic stress. Excessive Na^+ ion concentrations in maize lead to ion imbalances, making it difficult for the plant to absorb essential ions such as K^+ and Ca^{2+}, resulting in water loss and necrosis (Farooq et al., 2015).

2.4 Cold Stress

Plants are subjected to cold stress when exposed to temperatures ranging from 0 to 15°C (Manasa et al. 2022). Each plant species has an optimal temperature range for its growth and development, and what may be optimal for one crop could be stressful for others (Kopecka et al., 2023). Plants native to warmer climates are generally sensitive to low, non-freezing temperatures and often exhibit signs of stress or damage when exposed to such conditions. For example, species like maize (*Zea mays*), soybean (*Glycine max*), cotton (*Gossypium hirsutum*), tomato (*Lycopersicon esculentum*), and banana (*Musa* sp.) may exhibit wound symptoms when exposed to low-temperature (Devi et al., 2023). However, the severity of these symptoms varies among species, depending on their sensitivity to colds (Bhat et al., 2022).

Cold stress-induced injuries typically manifest after 48 to 72 hours of exposure. Affected plants present different symptoms, such as a decrease in leaf elongation, wilting, chlorosis (leaf yellowing), and potentially necrosis (tissue death) (Kopecka et al., 2023). In addition to these vegetative symptoms, cold stress can severely impact reproductive development (Suzuki et al., 2008). For example, in rice, exposure to cold during the anthesis stage (flower opening) can result in flower sterility (Jiang et al., 2002). Generally, cold stress leads to reduced seed germination, low seedling growth, yellowing, wilting, and decreased tillering (Ding et al., 2019). During the reproductive phase, it can delay heading and cause pollen sterility infertility, a key factor contributing to reduced grain yield (Suzuki et al., 2008). One of the primary detrimental effects of cold stress is damage to the plasmalemma, often due to dehydration (Aslam et al., 2022). Even when temperatures are not lethal, low temperatures can still adversely affect both the vegetative and reproductive growth stages of plants. The impact extends beyond yield reduction and affects seed and overall plant quality (Kopecka et al., 2023). For example, cold conditions have been shown to significantly decrease the size of chickpea (*Cicer arietinum*) seeds by 24% and reduce starch, protein, and fat reserves by 34, 33, and 43%, respectively, while increasing sugar levels (Nayyar et al., 2005). In wheat, low temperatures lower the concentrations of amylose, amylopectin, and total starch in grains, with more severe effects observed when cold occurs during the booting stage rather than the jointing stage (Zhang et al., 2021).

2.5 Heavy Metal Stress

Heavy metal stress refers to a class of nonbiodegradable inorganic chemicals characterized by an atomic weight greater than twenty and a density of more than 5 g cm^{-3} (Yadav et al., 2020). These metals are known for their lethal effects on cells, leading to a mutagenic footprint on plants by polluting water, soil, and the environment (Emamverdian et al., 2015). Metals found in soils can be categorized into two groups: essential micronutrients, which are necessary for normal plant growth (e.g., iron, magnesium, molybdenum, zinc, manganese, copper, and nickel), and nonessential elements, which have no known physiological or biological role (e.g., chromium, cadmium, cobalt, arsenic, antimony, lead, selenium, and mercury) (Yadav et al., 2020).

In many parts of the world, agricultural soils are mild to moderately contaminated with heavy metals. This contamination mainly stems from prolonged use of phosphate fertilizers, sewage sludge application, industrial emissions, waste disposal, and improper irrigation practices (Yadav et al., 2010). When exposed to high concentrations of heavy metals, plants typically respond by generating Reactive Oxygen Species (ROS) (Mishra et al., 2017). Heavy metals can be absorbed by plant surfaces both underwater and above ground (Ali and Gill, 2022). Essential metals are vital for enzymes and protein structures and are required in trace amounts for plant growth, and development. However, maintaining balanced concentrations of both essential and nonessential metals is crucial, as their excessive presence can inhibit plant growth and reduce crop yield (Yadav et al., 2020). At toxic concentrations, heavy metals disrupt normal plant functions and interfere with various metabolic processes (Tiwari and Lata, 2018). This interference can occur through the displacement or alteration of protein structural components, often through the formation of bonds between heavy metals and sulphydryl groups or through interactions with functional groups of vital cellular molecules (Ghori et al., 2019).

3. Types of Nanoparticles and their use in Agriculture

Nano-particles pose a wide range of features because of their smaller size and shape and higher surface area. Different types of nanomaterials are found in the market which can be classified as nanocarriers, nanotubes, nanoclays, nanocomposites, nanosensors, nanofibres, and nanocapsules (Fig. 2).

3.1 Nanocarriers

Nanocarriers act as carriers for the transportation of different types of chemicals (Patra et al., 2018). Polymers, liposomes, and micelles are examples of different types of nanocarriers (Yan et al., 2018). Encapsulated NPs mostly administered in the form of growth regulators, chemicals, and micronutrients (Shang et al., 2019). Nanocarriers release encapsulated material gradually in a slow and sustained manner (Huang et al., 2018). Thus, nanocarriers ensure the optimal concentration of these elements by regulating their solubility and preventing their loss from the system (Wang and Liu 2019).

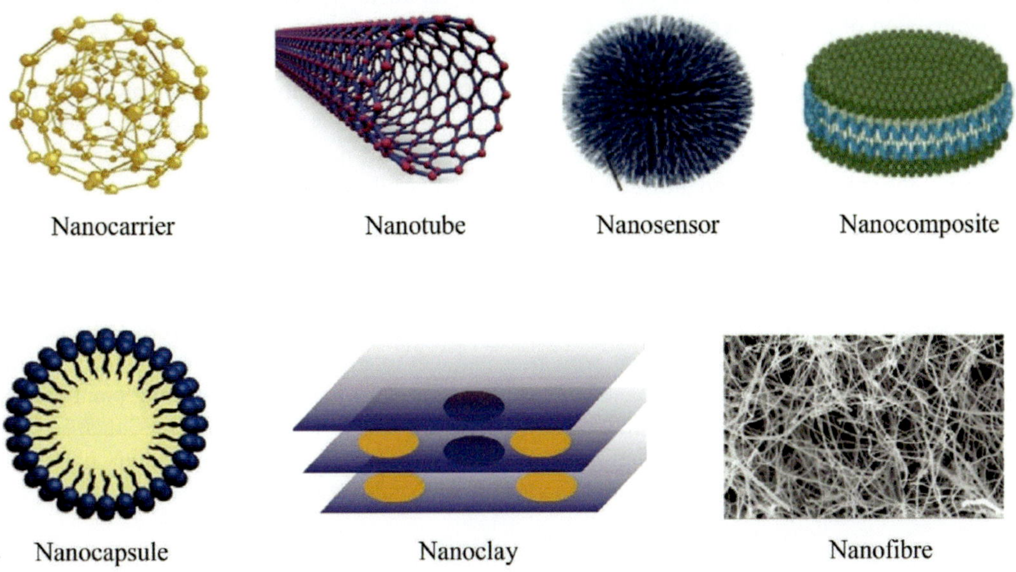

Figure 2. Different types of nanoparticles used for agricultural sustainability.

3.2 Nanotubes

Nanotubes bear carbon molecules and are cylindrical in shape. Carbon nanotubes (CNTs) have a diameter of 1–3 nm and a length of thousands of nm (Saifuddin et al., 2013). They present higher surface-to-volume ratios, strengths, water solubilities, and better immobilization efficiencies (He et al., 2013). CNTs are used in agriculture to improve seed germination and increase plant biomass by controlling plant growth. In addition to serving as carriers of chemicals, carbon nanotubes also shield seeds from disease.

3.3 Nanosensors

Bioreceptor molecules used as nanosensors are employed to target analyte molecules (Nikhil et al., 2016). For monitoring crop health, deciding harvesting time, and determining crop contamination by microbes or chemicals, nanosensors are commonly used (Misra et al., 2013). It can also be used to track soil conditions such as pH and moisture content. In addition, these biosensors can detect contaminants for environmental monitoring purposes (Sekhon 2014).

3.4 Nanocapsules

The most preferred delivery method for bioactive and functional compounds is the use of nanocapsules. A liquid or polymeric matrix makes up the interior of nanocapsules, whereas an exterior polymeric membrane surrounds the entire structure (Deng et al., 2020). The bioactive materials comprise of polymeric inner matrix, and it selectively targets specific plant parts to release their contents (Martinez-Ballesta et al., 2018). Ionic pregelation/coacervation, polymerization, and dispersion of preformed polymers are the most common techniques for nano capsule synthesis. Chitosan, zein, casein, alginate, and sodium tripolyphosphate are the main polymer ingredients that are used to prepare nanocapsules (Grillo et al., 2014). Micronutrient fertilizers and various plant protection compounds can be easily applied in the form of nano capsules to improve crop growth and yield.

3.5 Nanoclays

Natural mineral particles generally make nanoclays. They are composed of several layers of silicate layers and other elements (Guo et al., 2018). Owing to the absence of sufficient bonding between layers, nanoclays are able to absorb enormous amounts of water, resulting in the expansion or contraction of these materials. Mica, talc, kaolin seprhite, montmorillonite, and serpentine are common nanoclays. They can store large amounts of water and release the plant nutrients slowly over time (Cecilia et al., 2018). Owing to their metal binding affinity, nanoclays are often used to sequester metals in contaminated soil (Sen et al., 2015) and restore the ecological health of the system. When sprayed directly into dry sandy soil, liquid nanoclay binds sand particles and improves the soil structure and moisture-holding capacity of the soil.

3.6 Nanocomposites

Nanocomposites are nanoscale materials in a solid matrix. The polymers in these nanocomposites are biocompatible and biodegradable and are ideal for delivering nutrients to crops (Guha et al., 2020). They often exhibit antibacterial properties and can interact with proteins and DNA (Wang et al., 2017). The potential uses of nanocomposites in agriculture include promoting plant growth and modifying the microbial counts in the soil (Usman et al., 2020). Different fillers, such as clay/nanoclay and nanocellulose, are often reinforced in the polymeric matrix of various polymeric nanocomposites and used for nutrient management in crop cultivation.

3.7 Nanofibres

Nanofibres are one-dimensional nanomaterials with diameters ranging from tens to hundreds of nanometres. A wide range of natural, synthetic, and hybrid polymers are used to create nanofibres (Xue et al., 2019). These fibre-like nanomaterials possess several unique properties, such as a large surface area relative to their volume, interconnected nano porosity, as well as efficient mass transfer and transport capabilities (Lim 2017). In addition, the carbon nanofibres can infiltrate the seed coating. They serve as carriers for the transport of micronutrients and slow-release metal nanoparticles in plants (Ashfaq et al., 2017). They increase seed germination and act as growth stimulators. Stojanov and Berlec (2020), reported their antifungal activity against Fusarium and Alternaria when they were treated with Trichoderma. Carbon nanofibres are considered the next-generation fertilizers that enhance nutrient absorption and provide protection to plants against pests, insects, and harmful microorganisms (Mohamed et al., 2018).

4. Effect of Nanoparticles in the Mitigation of Abiotic Stress

At the nanoscale, physical and chemical properties such as strength, conductivity, and reactivity differ significantly from those of bulk ingredients. The use of nanoparticles (NPs) has gained significant traction in crop production, with increasing recognition of their potential (Misra et al., 2013; Staroń et al., 2020; Verma et al., 2019). The exclusive characteristics of NPs, including high reactivity, more surface area, adjustable pore width, and varied shapes, have opened new opportunities for agricultural research. In recent years, nanoparticle-based treatments have become increasingly popular for mitigating abiotic stress while promoting robust plant growth. Different types of nanoparticles play a key role in enhancing crop productivity by alleviating abiotic stress, thus improving agricultural output. Common methods for applying NPs include seed coating, foliar spraying, and soil drenching, all of which are designed to optimize nanoparticle uptake by plants.

4.1 Mitigation of heat stress

Some examples in which nanoparticles have been used to alleviate heat stress in crop plants are listed in Table 1.

Table 1. Nanoparticle-based mitigation of heat stress in plants.

Name	Example	References
Silver	Under heat stress conditions the application of AgNPs extracted from moringa significantly enhanced dry matter production and LAI of wheat.	Iqbal et al., 2019
Titanium oxide	Titanium oxide ($TiO2$) NPs improved the heat tolerance ability by adjusting the oscillation of the stomata of tomato plants.	Qi et al., 2013
Cerium oxide	The application of cerium oxide ($CeO2$) nanoparticles enhanced maize growth under heat stress by upregulating HSP70.	Zhao et al., 2012

4.2 Mitigation of Drought Stress

The following nanoparticles may be used to counter the devastating effects of drought on crop performance in terms of their growth and yield. The results of the mitigation effects of nanoparticle-based methods on drought stress in plants are presented in Table 2.

4.3 Mitigation of Salt Stress

In major salt stress areas, osmotic stress generated through the excessive presence of salt creates ion toxicity and disturbs the balance of minerals which ultimately affects plant growth. Under such conditions, the application of certain nanoparticles helps reduce the effect of salinity on plants

Table 2. Nanoparticle-based mitigation of drought stress in plants.

Name	Example	References
Silver	Under drought stress conditions in the lentil plant, the application of NPs of Ag enhanced the germination of seed and dry matter production.	Hojjat and Ganjali, 2016
Nanosilicon dioxide	Treatment with nanosilicon dioxide ($nSiO_2$) NPs resulted in increased potential of seed germination, and fresh and dry weight of seedlings; in tomato plants.	Siddiqui et al., 2014
Iron oxide	Iron oxide nanoparticles effectively enhanced antioxidant activities (CAT, SOD, POX), osmolyte content, and elevated biochemical levels (iron, H_2O_2, and proline), aiding in drought stress tolerance.	Sreelakshmi et al., 2021

Table 3. Nanoparticle-based mitigation of salt stress in plants.

Name	Example	References
Titanium oxide	*Dracocephalum moldavica* can counter salt stress through the activation defence mechanism by producing antioxidants when there is an application of TiO2-NPs.	Gohari et al., 2020
Silica	ABA is released through MesoporousSilica Nanoparticles (MSNs) to increase the Arabidopsis resistance against.	Sun et al., 2018
Selenium and Zinc oxide	*Brassica napus* impacted positively on various parameters of germination when they are treated with NPs ((Se-NPs and ZnO NPs) under salt stress conditions at the early growth stage.	El-Badri et al., 2021

through different mechanisms. A detailed discussion of nanotechnology-based salt stress tolerance in plants is presented in Table 3.

4.4 Mitigation of Heavy Metal Stress

NPs can be utilized to effectively alleviate heavy metal stress. After the application of NPs, they make an attachment with heavy metals and make a complex that prevents microorganisms to access them by creating a barrier in adsorption in plants and hindering the movement throughout the plants their movement throughout the plant and minimizing their biological activity (Wang et al., 2021). The effects of nanoparticle-based mitigation of heavy metal stress in plants are discussed in detail in Table 4.

Table 4. Nanoparticle-based mitigation of heavy metal stress in plants.

Name	Example	References
Cerium dioxide	A foliar spray of CeO2 NPs (200 mg/L) improved the antioxidant defence mechanism through antioxidant production and prevented Cd buildup in rice plants grown in hydroponic.	Wang et al., 2019
Titanium dioxides	The rate of photosynthesis and plant physiological parameters in soybean were not only enhanced by the application of TiO2 NPs but also reduced the Cd toxicity.	Singh & Lee, 2016
Zinc oxide	Seed germination along with root and embryo development were enhanced by accelerating the formation of cell walls and glycolytic metabolism through the application of ZnO NPs (50 mg/L) during seed treatment.	Luying et al., 2020
Selenium	Se treated rice plant was found to uptake less Cd and lower the toxicity level.	Hu et al., 2014; Lin et al., 2012
Silicon	The toxicity level of Al in maize is decreased by the application of Si NPs (4 mg/kg) due to the activation of the deference mechanism by antioxidants.	de Sousa et al., 2019

4.5 Mitigation of cold stress

Several plant species can effectively induce cold tolerance by the use of nanomaterials. Some examples in which such nanoparticles have been used to mitigate cold stress in plants are summarized in Table 5.

Table 5. Nanoparticle-based mitigation of cold stress in plants.

Name	Example	References
Selenium	Application of NPs with Se in tomato plants under cold stress improved the level of photosynthetic pigments, boosted the antioxidant defence system in plants, and ultimately improved plant growth.	Haghighi et al., 2014
Titanium oxide	Nanoparticle like titanium oxide (TiO2) makes chickpea plants withstand cold stress condition	Amini et al., 2017; Mohammadi et al., 2014
Chitosan	Applying chitosan nanoparticles (CH-NPs) at concentrations of 100–400 mg L^{-1} in water mitigated the negative impacts of chilling on banana plants.	Wang et al., 2021

5. Conclusion

The detrimental effects of abiotic stress on crop production under a changing climate have become a top priority in recent eras. In the current era, nanotechnology has been recognized as a potential technology to mitigate the negative impacts of abiotic stresses. For example, many studies have revealed that when fertilizers are used in their nano-based form (nanoparticles; NPs), crop production has improved because of a greater capacity for nutrient retention and a reduction in nutrient losses. Nanofertilizers can also enhance the ability of plants to survive against abiotic stress by improving their growth and stand establishment, nutrient use efficiency, the biosynthesis of growth regulators, antioxidant activity, and the reduction of ROS. We believe that this information will be helpful in the development of efficient nano-based agricultural approaches to meet the food security requirements of increasing populations by addressing various abiotic stresses.

References

Ahluwalia, O., Singh, P. C., and Bhatia, R. (2021). A review on drought stress in plants: Implications, mitigation and the role of plant growth promoting rhizobacteria. *Resources, Environment and Sustainability*, 5:100032. https://doi.org/10.1016/j.resenv.2021.100032.

Ali, B., and Gill, R. A. (2022). Editorial: Heavy metal toxicity in plants: Recent insights on physiological and molecular aspects, volume II. *Frontier in Plant Science*, 2022 21(13): 1016257. https://doi.org/10.3389/fpls.2022.1016257.

Amini, S., Amiri, M. R., Mohammadi, R., and Shahandashti, S. S. K. (2017). cDNA-AFLP analysis of transcripts induced in chickpea plants by TiO2 nanoparticles during cold stress. *Plant Physiology and Biochemistry*, 111: 39–49.

Ashfaq, M., Talreja, N., Chuahan, D., and Srituravanich, W. (2019). Polymeric nanocomposite-based agriculture delivery system: emerging technology for agriculture. *In*: Jamal, F. (Eds.). Genetic Engineering-A Glimpse of Techniques and Applications. *IntechOpen*. https://doi.org/10.5772/intechopen.89702.

Ashfaq, M., Verma, N., and Khan, S. (2017). Carbon nanofibers as a micronutrient carrier in plants: efficient translocation and controlled release of Cu nanoparticles. *Environmental Science: Nano*, 4(1): 138–148.

Aslam, M., Fakher, B., Ashraf, M. A., Cheng, Y., Wang, B., and Qin, Y. (2022.) Plant low-temperature stress: signalling and response. *Agronomy*, 12: 702. https://doi.org/10.3390/agronomy12030702.

Awasthi, A., Jadhao, P., and Kumari, K. (2019). Clay nanoadsorbent: structures, applications and mechanism for water treatment. *SN Applied Science*, 1(9): 1076.

Azameti, M. K., and Imoro, A. W. M. (2023). Nanotechnology: A promising field in enhancing abiotic stress tolerance in plants. *Crop Design*, 2(5): 100037. http://dx.doi.org/10.1016/j.cropd.2023.100037.

Bashir, K., Matsui, A., Rasheed, S., and Seki, M. (2019). Recent advances in the characterization of plant transcriptomes in response to drought, salinity, heat, and cold stress. *F1000Research*, 8: 658.

Bashir, S. S., Hussain, A., Hussain, S. J., Wani, O. A., ZahidNabi, S., Dar, N. A. et al. (2021). Plant drought stress tolerance: understanding its physiological, biochemical and molecular mechanisms. *Biotechnology and Biotechnological Equipment*, 35(1): 1912–1925. https://doi.org/10.1080/13102818.2021.2020161.

Bhat, K. A., Mahajan, R., Pakhtoon, M. M., Urwat, U., Bashir, Z., Shah, A. A. et al. (2022). Low temperature stress tolerance: an insight into the omics approaches for legume crops. *Frontier in Plant Science*, 13: 888710. https://doi.org/10.3389/fpls.2022.888710.

Boretti, A., and Rosa, L. 2019. Reassessing the projections of the world water development report.*NPJ Clean Water Nature*, 2: 15.

Botella, M. A., Rosado, A., Bressan, R., and Hasegawa, P. M. (2007). Plant adaptive responses to salinity stress. pp. 37–70. *In*: Jenks, M. A. and Hasegawa, P.M. (Eds.). Plant Abiotic Stress. USA, Iowa, Blackwell Publishing. https://doi.org/10.1002/9780470988503.ch3.

Boyer, J. S. (1982). Plant Productivity and Environment. *Science*, 218: 443–448.

Cecilia, J. A., García-Sancho, C., Vilarrasa-García, E., Jiménez-Jiménez, J., and Rodriguez-Castellón, E. (2018). Synthesis, characterization, uses and applications of porous clays heterostructures: a review. *Chemical Record*, 18(7-8): 1085–1104.

Chinnusamy, V., Jagendorf, A., and Zhu, J. (2005). Understanding and improving salt tolerance in plants. *Crop Science*, 45(2): 437–448. https://doi.org/10.2135/cropsci2005.0437.

Choudhary, M., Jat, H. S., Mukhopadhyay, R., Kakraliya, M., Poonia, T., Phogat, A. et al. (2023). Functional diversity and behavioral changes of microbial communities under salt affected soils. *Applied Soil Ecology*, 190: 105017. https://doi.org/10.1016/j.apsoil.2023.105017.

Cramer, G. R., Urano, K., Delrot, S., Pezzotti, M., and Shinozaki, K. (2011). Effects of abiotic stress on plants: A systems biology perspective. *BMC Plant Biology*, 11: 1–14.

Daryanto, S., Wang, L., and Jacinthe, P. A. (2016). Global synthesis of drought effects on maize and wheat production. *PLOS ONE*, 11: e0156362.

de Sousa, A., Saleh, A. M., Habeeb, T. H., Hassan, Y. M., Zrieq, R., Wadaan, M. A. M. et al. (2019). Silicon dioxide nanoparticles ameliorate the phytotoxic hazards of aluminium in maize grown on acidic soil. *Science of the Total Environment*, 693: 133636. https://doi.org/10.1016/j.scitotenv.2019.133636.

Deng, S., Gigliobianco, M. R., Censi, R., and Di Martino, P. (2020). Polymeric nanocapsules as nanotechnological alternatives for drug delivery systems: Current status, challenges, and opportunities. *Nanomaterials*, 10(5): 847.

Devi, V., Kaur, A., Sethi, M., and Avinash, G. (2023). Perspective Chapter: Effect of Low-Temperature Stress on Plant Performance and Adaptation to Temperature Change. *IntechOpen*. https://doi.org/10.5772/intechopen.110168.

Ding, Y., Shi, Y., and Yang, S. (2019). Advances and challenges in uncovering cold tolerance regulatory mechanisms in plants. *New Phytologist*, 222: 1690–1704. https://doi.org/10.1111/nph.15696.

El-Badri, A. M., Batool, M., Wang, C., Hashem, A. M., Tabl, K. M., Nishawy, E. et al. (2021). Selenium and zinc oxide nanoparticles modulate the molecular and morpho-physiological processes during seed germination of *Brassica napus* under salt stress. *Ecotoxicology and Environmental Safety*, 225: 112695.

Emamverdian, A., Ding, Y., Mokhberdoran, F., and Xie, Y. (2015). Heavy metal stress and some mechanisms of plant defense response. *The Scientific World Journal*, 756120. https://doi.org/10.1155/2015/756120.

Farooq, M., Wahid, A., Kobayashi, N., Fujita, D., and Basra, S. M. A. (2009). Plant drought stress: Effects, mechanisms, and management. *Agronomy for Sustainable Development*, 29: 185–212. https://doi.org/10.1051/agro:2008021.

Farooq, M., Hussain, M., Wakeel, A., and Siddique, K. H. M. (2015). Salt stress in maize: Effects, resistance mechanisms, and management—a review. *Agronomy for Sustainable Development*, 35(2): 461–481. https://doi.org/10.1007/s13593-015-0287-0.

Farooq, M., Gogoi, N., Barthakur, S., Baroowa, B., Bharadwaj, N., Alghamdi, S. S. et al. (2016). Drought stress in grain legumes during reproduction and grain filling. *Journal of Agronomy and Crop Science*, 203(2): 81–102. https://doi.org/10.1111/jac.12169.

Ghori, N. H., Ghori, T., Hayat, M. Q., Imadi, S. R., Gul, A., Altay, V. et al. (2019). Heavy metal stress and responses in plants. *International Journal of Environmental Science and Technology*, 16: 1807–1828. https://doi.org/10.1007/s13762-019-02215-8.

Giri, A., Heckathorn, S., Mishra, S., and Krause, C. (2017). Heat stress decreases the levels of nutrient-uptake and assimilation proteins in tomato roots. *Plants*, 6(6): 6.

Gohari, G., Mohammadi, A., Akbari, A., Panahirad, S., Dadpour, M. R., Fotopoulos, V. et al. (2020). Titanium dioxide nanoparticles (TiO_2 NPs) promote growth and ameliorate the effects of salinity stress on the essential oil profile and biochemical attributes of *Dracocephalum moldavica. Scientific Reports*, 10(1): 1–14.

Grillo, R., Pereira, A. E., Nishisaka, C. S., De Lima, R., Oehlke, K., Greiner, R. et al. (2014). Chitosan/tripolyphosphate nanoparticles loaded with paraquat herbicide: An environmentally safer alternative for weed control. *Journal of Hazardous Materials*, 278: 163–171.

Guha, T., Gopal, G., Kundu, R., and Mukherjee, A. (2020). Nanocomposites for delivering agrochemicals: A comprehensive review. *Journal of Agricultural and Food Chemistry*, 68(12): 3691–3702.

Guo, F., Aryana, S., Han, Y., and Jiao, Y. 2018. A review of the synthesis and applications of polymer–nanoclay composites. *Applied Sciences*, 8(9): 1696.

Guo, T., Gull, S., Ali, M. M., Yousef, A. F., Ercisli, S., Kalaji, H. M. et al. (2022). Heat stress mitigation in tomato (*Solanum lycopersicum* L.) through foliar application of gibberellic acid. *Scientific Reports*, 12(1): 1–13.

Haghighi, M., Abolghasemi, R., and da Silva, J. A. T. (2014). Low and high temperature stress affects the growth characteristics of tomato in hydroponic culture with Se and nano-Se amendment. *Scientia Horticulturae*, 178: 231–240.

He, H., Pham-Huy, L. A., Dramou, P., Xiao, D., Zuo, P., and Pham-Huy, C. (2013). Carbon nanotubes: Applications in pharmacy and medicine. *Biomed Research International*, 2013: Article 734518.

He, J., See, X. E., Qin, L., and Choong, T. W. (2016). Effects of root-zone temperature on photosynthesis, productivity and nutritional quality of aeroponically grown salad rocket (Eruca sativa) vegetable. *American Journal of Plant Sciences*, 7: 1993–2005.

He, X., Deng, H., and Hwang, H. M. (2019). The current application of nanotechnology in food and agriculture. *Journal of Food and Drug Analysis, 27(*1): 1–21.

Hojjat, S. S., and Ganjali, A. (2016). The effect of silver nanoparticle on lentil seed germination under drought stress. *International Journal of Farm Allied Sciences*, 5(3): 208–212.

Hu, Y., Norton, G. J., Duan, G., Huang, Y., and Liu, Y. (2014). Effect of selenium fertilization on the accumulation of cadmium and lead in rice plants. *Plant and Soil*, 384(1–2): 131–140.

Huang, B., Chen, F., Shen, Y., Qian, K., Wang, Y., Sun, C. et al. (2018). Advances in targeted pesticides with environmentally responsive controlled release by nanotechnology. *Nanomaterials*, 8(2): 102.

Hussain, M., Farooq, S., Hasan, W., Ul-Allah, S., Tanveer, M., Farooq, M. et al. (2018). Drought stress in sunflower: Physiological effects and its management through breeding and agronomic alternatives. *Agricultural Water Management*, 201: 152–166.

Hussain, S., Bai, Z., Huang, T., Cao, X., Zhu, L., Zhu, C. et al. (2019). 1-Methylcyclopropene modulates physiological, biochemical and antioxidant responses of rice to different salt stress levels. *Journal of Integrative Agriculture*, 10(124): 1–18. https://doi.org/10.3389/fpls.2019.00124.

Hussain, S., Zhong, C., Bohr, J. A., Hu, J. J., Jin, Q., Cao, X. C. et al. (2017). Effects of salt stress on rice growth and development characteristics and the regulatory mechanisms: A review. *Journal of Integrative Agriculture*, 16(11): 2357–2374. https://doi.org/10.1016/S2095-3119(16)61608-8.

Ibrahim, K. S. (2013). Properties and applications of carbon nanotubes: A review. *Carbon Letters*, 14(3): 131–144.

Iqbal, M., Raja, N. I., Mashwani, Zia-ur-Rehman, Hussain, M., Ejaz, M. and Yasmeen, F. (2019). Effect of silver nanoparticles on the growth of wheat under heat stress. *Iranian Journal of Science and Technology, Transactions A: Science*, 43(2): 387–395.

Jiang, Y., and Zhang, Q. (2002). Effect of low temperature on fertility and agronomic traits of rice. *Plant Science*, 163(3): 525–532. https://doi.org/10.1016/S0168-9452(02)00173-9.

Kah, M., and Kookana, R. (2020). Emerging investigator series: Nanotechnology to develop novel agrochemicals: Critical issues to consider in the global agricultural context. *Environmental Science: Nano*, 7: 1867–1873.

Kah, M., Tufenkji, N. and White, J. C. (2019). Nanoenabled strategies to enhance crop nutrition and protection. *Nature Nanotechnology*, 14(6): 532–540.

Kazemi, K. and Eskandari, H. 2011. Effects of salt stress on germination and early seedling growth of rice (*Oryza sativa* L.) cultivars in Iran. *African Journal of Biotechnology*, 10(77): 17789–17792. https://doi.org/10.5897/AJB11.2219.

Khan, A. H., Min, L., Ma, Y., Zeeshan, M., Jin, S., and Zhang, X. (2022). High-temperature stress in crops: Male sterility, yield loss, and potential remedy approaches. *Plant Biotechnology Journal*, 21: 680–697.

Khan, I., Saeed, K., and Khan, I. (2019). Nanoparticles: Properties, applications, and toxicities. *Arabian Journal of Chemistry*, 12(7): 908–931.

Khan, M. R., and Rizvi, T. F. (2014). Nanotechnology: Scope and application in plant disease management. *Plant Pathology Journal*, 13(3): 214–231.

Kopecká, R., Kameniarová, M., Černý, M., Brzobohatý, B., and Novák, J. (2023). Abiotic stress in crop production. *International Journal of Molecular Sciences*, 24: 6603. https://doi.org/10.3390/ijms24076603.

Lim, C. T. (2017). Nanofiber technology: Current status and emerging developments. *Progress in Polymer Science*, 70: 1–17.

Lin, L., Zhou, W., Dai, H., Cao, F., Zhang, G., and Wu, F. (2012). Selenium reduces cadmium uptake and mitigates cadmium toxicity in rice. *Journal of Hazardous Materials*, 235–236: 343–351.

Luying, S., Fengbin, S., Xiangnan, L., Xiancan, Z., Shengqun, L., Yang, W. et al. (2020). Effects of ZnO nanoparticles on seed germination and root carbon metabolism in maize (*Zea mays* L.). Soils and Crops, 9: 40–49.

Manasa, S. L., Panigrahy, M., Panigrahi, K. C. S., and Rout, G. R. (2022). Overview of cold stress regulation in plants. Botanical Review, 88: 359–387. https://doi.org/10.1007/s12229-021-09267-x.

Manjunatha, S. B., Biradar, D. P., and Aladakatti, Y. R. 2016. Nanotechnology and its applications in agriculture: A review. *Journal of Farm Sciences*, 29(1): 1–13.

Mansoor, S., Kour, N., Manhas, S., Zahid, S., Wani, O. A., Sharma, V.,et al. (2020). Biochar as a tool for effective management of drought and heavy metal toxicity. *Chemosphere*, 271: 129458.

Marchiol, L. (2018). Nanotechnology in agriculture: New opportunities and perspectives. *New Visions in Plant Science*, 9: 121.

Martinez-Ballesta, M., Gil-Izquierdo, Á., García-Viguera, C., and Domínguez-Perles, R. (2018). Nanoparticles and controlled delivery for bioactive compounds: Outlining challenges for new "smart foods" for health. *Foods*, 7(5): 72.

Masson-Delmotte, V., Zhai, P., Pörtner, H.-O., Roberts, D., Skea, J., Shukla, P. R. et al. (2019). *IPCC, 2018: Global warming of 1.5°C. An IPCC specializes in the impacts of global warming of 1.5°C above preindustrial levels and related global greenhouse gas emission pathways in the context of strengthening the global response to the threat of climate change, sustainable development, and efforts to eradicate poverty.*

Meena, R. S., Gogoi, N., and Kumar, S. (2016). Alarming issues on agricultural crop production and environmental stresses. *Journal of Cleaner Production*, 142: 3357–3359.

Meena, R. S., Kumar, S., Datta, R., Lal, R., Vijayakumar, V., Brtnicky, M. et al. (2020). Impact of agrochemicals on soil microbiota and management: A review. *Land*, 9(2): 34. https://doi.org/10.3390/land9020034.

Mishra, J., Singh, R., and Arora, N. K. (2017). Alleviation of heavy metal stress in plants and remediation of soil by rhizosphere microorganisms. *Frontiers in Microbiology*, 8: 1706. https://doi.org/10.3389/fmicb.2017.01706.

Misra, A. N., Misra, M., and Singh, R. (2013). Nanotechnology in agriculture and food industry. *International Journal of Pure and Applied Sciences and Technology*, 16(2): 1.

Mohamed, M. A., Hashim, A. F., Alghuthaymi, M. A., and Abd-Elsalam, K. A. (2018). Nanocarbon: Plant growth promotion and protection. In *Nanobiotechnology applications in plant protection* (pp. 155–188). Springer, Cham.

Mohammadi, R., Maali-Amiri, R., and Mantri, N. (2014). Effect of TiO_2 nanoparticles on oxidative damage and antioxidant defense systems in chickpea seedlings during cold stress. *Russian Journal of Plant Physiology*, 61(6): 768–775.

Morrison, M. J., and Stewart, D. W. (2002). Heat stress during flowering in summer brassica. *Crop Science*, 42(3): 797–803.

Motos, J. A., Ortuño, M. F., Vivancos, P. D., Vicente, A. B., Hernández, J. A., and Sánchez, B. M. (2017). Plant responses to salt stress: Adaptive mechanisms. *Agronomy*, 7(1): 18. https://doi.org/10.3390/agronomy7010018.

Muhammad, S., Ditta, A., Iqbal, M. S., Hussain, S. B., Khan, M. I., Ramzan, M. et al. (2018). Effect of salinity stress on cotton growth and role of marker-assisted breeding and agronomic practices for salinity tolerance. *Scholars Report*, 4(1): 1–13.

Munns, R., and Tester, M. (2008). Mechanisms of salinity tolerance. *The Annual Review of Plant Biology*, 59(1): 651–681.

Nandal, M., and Hooda, R. (2013). Salt tolerance and physiological response of plants to salinity: A review. *International Journal of Scientific & Engineering Research*, 4(1): 44–67.

Nayyar, H., Bains, T. S., Kumar, S., and Kaur, G. (2005). Chilling effects during seed filling on accumulation of seed reserves and yield of chickpea. *Journal of Science of Food and Agriculture*, 85(12): 1925–1930.

Nikhil, B., Pawan, J., Nello, F., and Pedro, E. (2016). Introduction to biosensors. *Essays in Biochemistry*, 60(1): 1–8.

Nile, S. H., Baskar, V., Selvaraj, D., Nile, A., Xiao, J., and Kai, G. (2020). Nanotechnologies in food science: Applications, recent trends, and future perspectives. *Nano-Micro Letters*, 12(1): 45.

Nosaka, Y., and Nosaka, A. Y. (2017). Generation and detection of reactive oxygen species in photocatalysis. *Chemical Reviews*, 117(17): 11302–11336. https://doi.org/10.1021/acs.chemrev.7b00161.

Okcu, G., Kaya, M. D., and Atak, M. (2005). Effects of salt and drought stresses on germination and seedling growth of pea (*Pisum sativum* L.). *Turkish Journal of Agriculture and Forestry*, 29(4): 237–242.

Patra, J. K., Das, G., Fraceto, L. F., Campos, E. V. R., del Pilar Rodriguez-Torres, M., Acosta-Torres, L. S. et al. (2018). Nanobased drug delivery systems: Recent developments and future prospects. *Journal of Nanobiotechnology*, 16(1): 71.

Pessarakli, M., and Szabolcs, I. (2010). Soil salinity and sodicity as particular plant/crop stress factors. pp. 1–19. *In*: Pessarakli, M. (Ed.). *Handbook of plant and crop stress*. Routledge Handbooks Online. https://doi.org/10.1201/B10329-3.

Pessarakli, M., Szabolcs, I., and Pardue, J. H. (2010). Soil salinity: Historical perspectives and future prospects. *In*: Pessarakli, M. (Ed.). *Handbook of Plant and Crop Stress* (3rd ed., pp. 3–21). CRC Press, Boca Raton.

Pingali, P., Aiyar, A., Abraham, M. and Rahman, A. 2019. Agricultural technology for increasing competitiveness of smallholders. In Pingali, P., et al. (Eds.), *Transforming food systems for a rising India* (pp. 215–240). Palgrave Macmillan, Cham.

Prasad, P. V. V., Boote, K. J., Allen, L. H., Sheehy, J. E., and Thomas, J. M. G. 2011. Species, ecotype and cultivar differences in crop sensitivity to heat stress. *Plant, Cell and Environment*, 34(8): 1396–1415. https://doi.org/10.1111/j.1365-3040.2011.02378.x.

Prasad, R., Kumar, M., and Kumar, V. (2017). Nanotechnology: An agriculture paradigm. Springer Nature Singapore Pte Ltd. https://doi.org/10.1007/978-981-10-4573-8.

Qi, M., Liu, Y., and Li, T. (2013). Nano-TiO_2 improve the photosynthesis of tomato leaves under mild heat stress. *Biological Trace Element Research*, 156(1): 323–328.

Raja, V., Majeed, U., Kang, H., Andrabi, K. I., and John, R. (2020). Impact of drought and heat stress individually and in combination on physio-biochemical parameters, antioxidant responses, and gene expression in *Solanum lycopersicum*, *PLOS ONE*, 15(11): e0239697. https://doi.org/10.1371/journal.pone.0239697.

Regni, L., del Pino, A. M., Mousavi, S., Palmerini, C. A., Baldoni, L., Mariotti, R. et al. (2019). Behavior of four olive cultivars during salt stress. *Frontiers in Plant Science*, 10: 867. https://doi.org/10.3389/fpls.2019.00867.

Robles-Garcia, M. A., Rodriguez-Felix, F., Marquez-Rios, E., Aguilar, J. A., Barrera-Rodriguez, A., and Aguilar, J. (2016). Applications of nanotechnology in the agriculture, food, and pharmaceuticals. *Journal of Nanoscience and Nanotechnology*, 16(8): 8188–8207.

Saifuddin, N., Raziah, A. Z., and Junizah, A. R. (2013). Carbon nanotubes: A review on structure and their interaction with proteins. *Journal of Chemistry*, 2013: 1–18.

Samarah, N. H., Alqudah, A. M., Amayreh, J. A., and McAndrews, G. M. (2009). The effect of late-terminal drought stress on yield components of four barley cultivars. *Journal of Agronomy and Crop Science*, 195(6): 427–441.

Sekhon, B. S. (2014). Nanotechnology in agri-food production: An overview. *Nanotechnology, Science and Applications*, 7: 31–53.

Seleiman, M. F., Al-Suhaibani, N., Ali, N., Akmal, M., Alotaibi, M., Refay, Y. et al. (2021). Drought stress impacts on plants and different approaches to alleviate its adverse effects. *Plants*, 10(2): 259. https://doi.org/10.3390/plants10020259.

Sen, J., Prakash, P., and De, N. (2015). Nanoclay composite and phyto-nanotechnology: A new horizon to food security issue in Indian agriculture. *Journal of Global Biosciences*, 4(5): 2187–2198.

Sexton, R.J. (2025). Food Fight: Misguided Policies, Supply Challenges, and the Impending Struggle to Feed a Hungry World. *Univ of California Press*, 246 p.

Shang, Y., Hasan, M., Ahammed, G. J., Li, M., Yin, H., and Zhou, J. (2019). Applications of nanotechnology in plant growth and crop protection: A review. *Molecules*, 24(14): 2558. https://doi.org/10.3390/molecules24142558.

Siddiqui, M. H., and Al-Whaibi, M. H. (2014). Role of nano-SiO_2 in germination of tomato (*Lycopersicon esculentum* Mill.) seeds. *Saudi Journal of Biological Sciences*, 21(1): 13–17. https://doi.org/10.1016/j.sjbs.2013.04.005.

Singh, J., and Lee, B. (2016). Influence of nano-TiO2 particles on the bioaccumulation of Cd in soybean plants (*Glycine max*): A possible mechanism for the removal of Cd from the contaminated soil. *Journal of Environmental Management*, 170: 88–96. https://doi.org/10.1016/j.jenvman.2016.01.010.

Singh, S., Singh, B. K., Yadav, S. M., and Gupta, A. K. (2015). Applications of nanotechnology in agriculture and their role in disease management. *Research Journal of Nanoscience and Nanotechnology*, 5(1): 1–5.

Sreelakshmi, B., Induja, S., Adarsh, P. P., Rahul, H. L., Arya, S. M., Aswana, S. et al. (2021). Drought stress amelioration in plants using green synthesized iron oxide nanoparticles. *Materials Today: Proceedings*, 41: 723–727. https://doi.org/10.1016/j.matpr.2020.06.162.

Staroń, A., Długosz, O., Pulit-Prociak, J., and Banach, M. 2020. Analysis of the exposure of organisms to the action of nanomaterials. *Materials*, 13(2): 349. https://doi.org/10.3390/ma13020349.

Stojanov, S., and Berlec, A. (2020). Electrospun nanofibers as carriers of microorganisms, stem cells, proteins, and nucleic acids in therapeutic and other applications. *Frontiers in Bioengineering and Biotechnology*, 8: 130. https://doi.org/10.3389/fbioe.2020.00130.

Sun, D., Hussain, H. I., Yi, Z., Rookes, J. E., Kong, L., and Cahill, D. M. (2018). Delivery of abscisic acid to plants using glutathione-responsive mesoporous silica nanoparticles. *Journal of Nanoscience and Nanotechnology*, 18(3): 1615–1625.

Suzuki, K., Miki, S., Shiki, S., Wang, Z., and Ohkubo, M. (2008). Time resolution improvement of superconducting NbN stripline detectors for time-of-flight mass spectrometry. *Applied Physics Express*, 1(3): 031702. https://doi.org/10.1143/APEX.1.031702.

Tiwari, S., and Lata, C. (2018). Heavy metal stress, signalling, and tolerance due to plant-associated microbes: An overview. *Frontiers in Plant Science*, 9: 452. https://doi.org/10.3389/fpls.2018.00452.

Usman, M., Farooq, M., Wakeel, A., Nawaz, A., and Cheema, S. A. (2020). Nanotechnology in agriculture: Current status, challenges, and future opportunities. *Science of the Total Environment*, 721: 137778. https://doi.org/10.1016/j.scitotenv.2020.137778.

Vega-Vásquez, P., Mosier, N. S., and Irudayaraj, J. (2020). Nanoscale drug delivery systems: From medicine to agriculture. *Frontiers in Bioengineering and Biotechnology*, 8:79 https://doi.org/10.3389/fbioe.2020.00079.

Verma, S. K., Das, A. K., Gantait, S., Kumar, V., and Gurel, E. (2019). Applications of carbon nanomaterials in the plant system: A perspective view on the pros and cons. *Science of the Total Environment*, 667: 485–499. https://doi.org/10.1016/j.scitotenv.2019.02.390.

Vogel, E., Donat, M. G., Alexander, L. V., Meinshausen, M., Ray, D. K., Karoly, D. et al. (2019). The effects of climate extremes on global agricultural yields. *Environmental Research Letters*, 14: 054010. https://doi.org/10.1088/1748-9326/ab154b.

Wang, A., Li, J., Al-Huqail, A. A., Al-Harbi, M. S., Ali, E. F., Wang, J., Ding, Z. et al. (2021). Mechanisms of chitosan nanoparticles in the regulation of cold stress resistance in banana plants. *Nanomaterials*, 11(10): 2670. https://doi.org/10.3390/nano11102670.

Wang, K., Wang, Y., Wan, Y., Mi, Z., Wang, Q., Wang, Q. et al. (2021). The fate of arsenic in rice plants (*Oryza sativa* L.): Influence of different forms of selenium. *Chemosphere*, 264: 128417. https://doi.org/10.1016/j.chemosphere.2020.128417.

Wang, L., and Liu, X. (2019). Sustained release technology and its application in environmental remediation: A review. *International Journal of Environmental Research and Public Health*, 16(12): 2153. https://doi.org/10.3390/ijerph16122153.

Wang, L., Hu, C., and Shao, L. (2017). The antimicrobial activity of nanoparticles: Present situation and prospects for the future. *International Journal of Nanomedicine*, 12: 1227–1239.

Wang, Y., Wang, L., Ma, C., Wang, K., Hao, Y., Chen, Q. et al. (2019). Effects of cerium oxide on rice seedlings as affected by coexposure of cadmium and salt. *Environmental Pollution*, 252: 1087–1096. https://doi.org/10.1016/j.envpol.2019.06.007.

Wicke, B., Smeets, E., Dornburg, V., Vashev, B., Gaiser, T., Turkenburg, W. C. et al. (2011). The global technical and economic potential of bioenergy from salt-affected soils. *Energy & Environmental Science*, 4(8): 2669–2681. https://doi.org/10.1039/C1EE01029H.

Wield, D., Hanlin, R., Mittra, J., and Smith, J. (2013). Twenty-first century bioeconomy: Global challenges of biological knowledge for health and agriculture. *Science and Public Policy*, 40(1): 17–24.

Xalxo, R., Yadu, B., Chandra, J., Chandrakar, V., and Keshavkant, S. (2020). Alteration in carbohydrate metabolism modulates thermotolerance of plant under heat stress. pp. 77–115. *In*: Wani, S. H., and Kumar, V. (Eds.). *Heat Stress Tolerance in Plants: Physiological, Molecular and Genetic Perspectives*. John Wiley & Sons Ltd.

Xue, J., Wu, T., Dai, Y., and Xia, Y. (2019). Electrospinning and electrospun nanofibers: Methods, materials, and applications. *Chemical Reviews*, 119(8): 5298–5415.

Yadav, S. (2010). Heavy metals toxicity in plants: An overview on the role of glutathione and phytochelatins in heavy metal stress tolerance of plants. *South African Journal of Botany*, 76: 167–179. https://doi.org/10.1016/j.sajb.2009.10.007.

Yadav, S., Modi, P., Dave, A., Vijapura, A., Patel, D., and Patel, M. (2020). Effect of abiotic stress on crops. In *Abiotic Stress in Plants.* IntechOpen. https://doi.org/10.5772/intechopen.88434.

Yan, G., Li, A., Zhang, A., Sun, Y., and Liu, J. (2018). Polymer-based nanocarriers for codelivery and combination of diverse therapies against cancers. *Nanomaterials*, 8(2): 85. https://doi.org/10.3390/nano8020085.

Yildiz, M., Poyraz, İ., Çavdar, A., Özgen, Y., and Beyaz, R. (2021). Plant responses to salt stress. In *Abiotic Stress in Plants.* IntechOpen. https://doi.org/10.5772/intechopen.93920.

Zhang, C., Gu, K., Gu, D., Zhang, S., and Wu, J. (2021). Quantifying the effect of low-temperature events on the grain quality formation of wheat. *Journal of Cereal Science*, 100: 103257. https://doi.org/10.1016/j.jcs.2021.103257.

Zhao, J., Lu, Z., Wang, L., and Jin, B. (2020). Plant responses to heat stress: Physiology, transcription, noncoding RNAs, and epigenetics. *International Journal of Molecular Sciences*, 22(1): 117. https://doi.org/10.3390/ijms22010117.

Zhao, L., Peng, B., and Yin, Y. (2012). Stress response and tolerance of *Zea mays* to CeO_2 nanoparticles: Cross talk among H_2O_2, heat shock protein, and lipid peroxidation. *ACS Nano*, 6(11): 9615–9622. https://doi.org/10.1021/nn303477k.

CHAPTER-3

Green Nanotechnology for Alleviating Salinity Stress of Plants

Sureshkumar Tharani,[1] *Govindarajan Ramkumar,*[1] *Krishnan Kutty Sunitha Kumari,*[2] *Jaison Sarah,*[2] *Siva Daniel Ajay Samuel*[1] and *Balasundaram Saraswathy Chithra Devi*[2,*]

1. Introduction

Salinity stress is a major threat to global agriculture, negatively impacting plant growth, development, and productivity (Balasubramaniam et al., 2023; Raza et al., 2023). Plants respond to salinity through various mechanisms, including ion homeostasis modulation, antioxidant defense system induction, and osmoprotectant biosynthesis (Raza et al., 2023). Recent advances in genomic and proteomic technologies have enabled the identification of salt-stress-responsive genes and proteins involved in these tolerance mechanisms (Balasubramaniam et al., 2023). However, many questions remain unanswered, such as the exact mechanisms of salinity sensing and the relative importance of symplastic versus apoplastic pathways for ion uptake (Isayenkov and Maathuis, 2019). Additionally, the role of ion toxicity in salinity stress may be less significant in the cytosol than previously thought, with potentially greater impact in the vacuole (Isayenkov and Maathuis, 2019). Modern biotechnological tools, including omics approaches and genome editing, offer promising avenues for improving crop salt tolerance and addressing global food security challenges (Raza et al., 2023).

Salinity stress poses a significant threat to global agriculture and food security, affecting over 20% of agricultural soils worldwide (Singh et al., 2024). It reduces crop yields through ion toxicity and osmotic imbalances (Sultan et al., 2023), potentially causing annual losses equivalent to feeding 170 million people (Russ et al., 2020). Climate change exacerbates this issue, potentially spreading salinity to currently unaffected regions (Mukhopadhyay et al., 2021). Various mitigation strategies have been proposed, including traditional methods like soil amendments and irrigation management (Mukhopadhyay et al., 2021), as well as innovative approaches such as climate-smart agriculture and microbial-assisted phytoremediation (Sultan et al., 2023). Emerging technologies like nanoparticles show promise in enhancing salt tolerance in crops by influencing plant hormone signaling and antioxidant defense systems (Singh et al., 2024). Addressing soil salinity is crucial for achieving sustainable agriculture, alleviating poverty, and meeting the UN Sustainable Development Goals (Sultan et al., 2023).

[1] BIRAC E-YUVA Centre, PSGR Krishnammal College for Women, Coimbatore -641004, Tamil Nadu, India- 641004.
[2] Department of Botany, PSGR Krishnammal College for Women, Coimbatore -641004, Tamil Nadu, India- 641004.
* Corresponding author: bchitradevi@psgrkcw.ac.in

DOI: 10.1201/9781003567684-3

Advanced technologies play a crucial role in managing salinity stress in agriculture. Biotechnological tools are being employed to develop salt-tolerant plants, offering efficient and cost-effective solutions (Mbinda and Mukami, 2021). Genomics, metabolomics, transcriptomics, and proteomics are used to understand plant-bacterial interactions for combating salt stress (Slama et al., 2023). Sustainable strategies include phytoremediation, phyto desalination, and the use of halophytes, combined with beneficial microorganisms like plant growth-promoting bacteria and arbuscular mycorrhizal fungi (Navarro-Torre et al., 2023). Other promising approaches involve crop residue management, biochar application, and agroforestry systems, along with plant-oriented techniques such as plant growth-promoting microbes and nanotechnology (Mishra et al., 2023). Remote sensing and geographical information systems are essential for monitoring and assessing salt-affected soils, aiding in long-term policy development to address threats to crop productivity and sustainability (Mishra et al., 2023).

Green nanotechnology provides promising solutions for alleviating salinity stress in plants. Engineered nanoparticles (ENPs) can enhance plant growth under salt stress by improving nutrient uptake, water balance, and antioxidant defense mechanisms. Various nanoparticles, including silica and chitosan, have shown effectiveness in increasing chlorophyll content, photosynthetic efficiency, and water-use efficiency in salt-stressed crops (Junedi et al., 2023b). Green-synthesized silver nanoparticles have demonstrated the ability to mitigate salinity stress by activating antioxidant enzymes, increasing osmoprotectants, and enhancing flavonoid and anthocyanin production (Sadat-Hosseini et al., 2022b). Nanoparticles can also protect cell membranes, optimize hormone levels, and regulate gene expression to improve plant resilience (Rehman et al., 2024). While nano-management approaches show great potential for sustainable agriculture, concerns about environmental impacts and safety necessitate further research to optimize their efficacy and address potential nanotoxicity (Junedi et al., 2023b; Sári et al., 2023a).

2. Salinity Stress in Plants

2.1 Causes of Soil Salinity

Soil salinity, a major challenge in drylands and irrigated agriculture, arises from both natural and anthropogenic causes (Stavi et al., 2021). Natural factors include climate, lithology, topography, and pedology, while human activities, particularly irrigated agriculture, contribute significantly (Pedrotti et al., 2015; Stavi et al., 2021). Historical evidence shows that salinization has affected civilizations for millennia, with notable examples in Mesopotamia (Zaman et al., 2018). The global extent of salt-affected soils is substantial, with estimates ranging from 932 to 954 million hectares (Zaman et al., 2018). Salinization impacts agricultural productivity, environmental health, and food security (Stavi et al., 2021). Mitigation strategies include advanced irrigation techniques, chemical amendments, and biological approaches such as using halophytes and beneficial microorganisms (Sarita et al., 2024). Integrating physical, chemical, and biological processes provides comprehensive solutions for sustainable soil salinity management and improved agricultural productivity (Sarita et al., 2024).

2.2 Physiological and Biochemical Effects of Salinity on Plants

Salinity is a major abiotic stress that significantly impacts plant growth and productivity worldwide (Okon, 2019). It induces osmotic, ionic, and oxidative stresses in plants, leading to various physiological and biochemical changes (Arif 2020; Polash et al., 2019). Salinity negatively affects photosynthesis, water relations, nutrient uptake, and membrane integrity (Rasool et al., 2022). Plants respond to salt stress through multiple mechanisms, including ion homeostasis, antioxidant defense systems, and osmoprotectant production (Arif et al., 2020). Hormonal regulation plays a crucial role in salt stress tolerance, with various phytohormones involved in mitigating salinity effects (Arif et al., 2020). Plants also exhibit morphological adaptations, such as changes in root-to-shoot ratio and leaf anatomy (Okon, 2019). Understanding these complex physiological and biochemical responses is essential for developing salt-tolerant crop varieties and ensuring food security in salt-affected areas (Polash et al., 2019; Rasool et al., 2022).

2.3 Mechanisms of Salinity Tolerance in Plants

Plants employ various mechanisms to cope with salinity stress, which poses a significant threat to agricultural productivity. Key strategies include osmotic adjustment through the production of osmolytes (Mansour and Salama, 2019; Nie et al., 2020), and maintaining ion homeostasis by regulating Na+/K+ balance and compartmentalization (Balasubramaniam et al., 2023; Hussain et al., 2021). Plants also activate antioxidant defense systems to combat reactive oxygen species generated under salt stress (Mansour and Salama, 2019; Nie et al., 2020). Hormonal regulation and modifications in membrane composition play crucial roles in salt tolerance (Nie et al., 2020). Some plants have developed specialized salt glands for salt excretion (Nie et al., 2020). Recent advances in genomics and proteomics have identified numerous genes and proteins involved in these salt-tolerance mechanisms (Balasubramaniam et al., 2023; Hussain et al., 2021). Understanding these complex mechanisms is essential for developing salt-tolerant crops and ensuring food security in the face of increasing soil salinization (Balasubramaniam et al., 2023; Mansour and Salama, 2019) (Fig. 1).

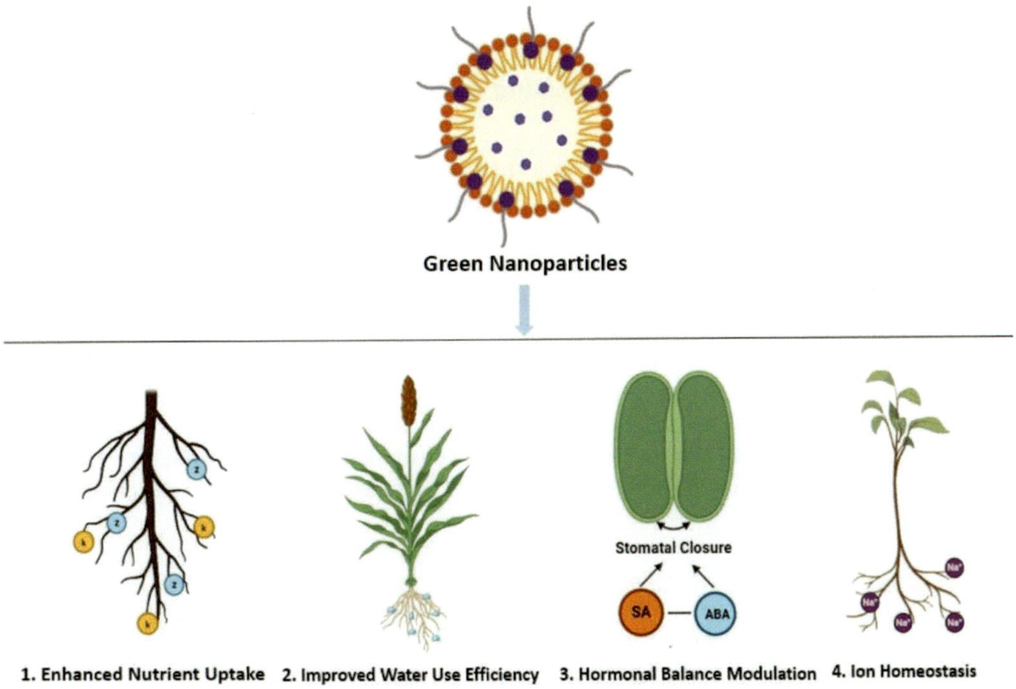

Figure 1. Mechanisms of Nanoparticles in Alleviating Salinity Stress.

2.4 Conventional Methods to Mitigate Salinity Stress: Limitations and Challenges

Conventional methods to mitigate salinity stress in plants face limitations and challenges. While agro-hydrotechnical, biological, chemical, and bio-ecological approaches can be effective (Ondrasek et al., 2022), they may not be sufficient to meet future crop production demands. Arbuscular Mycorrhizal Fungi (AMF) offer a promising biotechnological solution, enhancing nutrient uptake, water use efficiency, and antioxidant metabolism in plants (Evelin et al., 2019). Similarly, Plant Growth-Promoting Bacteria (PGPB) assist plants in coping with salinity stress by producing compounds like ACC deaminase, IAA, and antioxidants (Kumar et al., 2020). These microbial strategies improve plants' physiological, biochemical, and molecular responses to salinity. However, challenges remain in applying these biological tools effectively in field conditions. Advances in computer science and artificial intelligence may aid in predicting salinization processes and developing integrated strategies to combat salt stress (Ondrasek et al., 2022).

3. Principles of Green Nanotechnology

Green nanotechnology is an innovative field that applies nanotechnology principles to enhance environmental sustainability (Khan, 2020). It focuses on developing clean technologies and sustainable solutions to address global issues like energy shortages and water scarcity (Nanotechnology et al., 2022). By integrating green chemistry and engineering principles, this approach aims to produce safe and eco-friendly nanoparticles without using toxic chemicals (Md Ashif Ikbal and Dutta Choudhury, 2023). Green nanotechnology emphasizes the use of plant products, renewable inputs, and energy-efficient processes to minimize health and environmental risks (Jaiswal, 2022). Its applications span various sectors, including environmental remediation, clean energy production, and the development of sustainable materials (Khan, 2020). The field promotes the miniaturization of products, reducing material usage and transportation costs, thereby saving energy and fuel (Nanotechnology et al., 2022). Overall, green nanotechnology offers a promising pathway towards a more sustainable future by addressing environmental concerns and promoting cleaner technologies.

3.1 Fabrication of Nanomaterials Using Eco-Friendly Methods

Green synthesis of nanomaterials offers an eco-friendly alternative to conventional chemical and physical methods, which often involve toxic reagents and high energy consumption (Gomez-Muñoz et al., 2023; Merugu, 2020). Biological approaches utilizing plants, microorganisms, and agricultural wastes provide economical and sustainable routes for nanoparticle production (Singh et al., 2021). These methods exploit natural reducing and capping agents, eliminating the need for additional stabilizers (Merugu, 2020). Green nanomaterials (GNMs) have gained attention in various applications, particularly in wastewater treatment, due to their superior removal capabilities and intrinsic characteristics (Paradva et al., 2023). Techniques such as hydrothermal, sol-gel, and sonochemical methods can be employed for GNM synthesis (Paradva et al., 2023). While green fabrication offers numerous advantages, further research is needed to address potential environmental toxicity concerns associated with nanomaterial release (Merugu, 2020). Overall, green synthesis approaches present promising solutions for sustainable nanomaterial production and environmental remediation.

3.2 Unique Properties of Nanomaterials for Plant Applications

Nanomaterials possess unique properties that make them valuable for plant applications. Their small size (1–100 nm) and high surface area enhance nutrient absorption and facilitate easier penetration into plant tissues (Awad-Allah, 2023; Faraz et al., 2019; Omar 2019). The increased reactivity of nanoparticles compared to bulk materials allows for more efficient reactions and targeted delivery of substances (Faraz et al., 2019; Omar et al., 2019). These properties enable nanomaterials to function as effective fertilizers, pesticides, and stress-mitigating agents (Awad-Allah, 2023; Omar et al., 2019). Silica nanoparticles, for example, demonstrate greater solubility and absorption by plants due to their unique physicochemical characteristics (Awad-Allah, 2023). In plant pathology, nanoparticles serve as bactericides, fungicides, and biosensors for disease diagnostics (Elmer and White, 2018). The integration of nanotechnology in agriculture offers promising solutions for sustainable crop production, improved plant health, and enhanced food security (Awad-Allah, 2023; Elmer and White, 2018).

3.3 Advantages of Green Nanotechnology Over Traditional Nanotechnology

Conventional methods to mitigate salinity stress in plants face limitations and challenges. While agro-hydrotechnical, biological, chemical, and bio-ecological approaches can be effective (Ondrasek et al., 2022), they may not be sufficient to meet future crop production demands. Arbuscular Mycorrhizal Fungi (AMF) offer a promising biotechnological solution, enhancing nutrient uptake, water use efficiency, and antioxidant metabolism in plants (Evelin et al., 2019). Similarly, Plant Growth-Promoting Bacteria (PGPB) assist plants in coping with salinity stress by producing compounds like ACC deaminase, IAA, and antioxidants (Kumar et al., 2020). These microbial strategies improve plants' physiological, biochemical, and molecular responses to salinity. However, challenges remain in applying these biological tools effectively in field conditions. Advances in computer science and artificial intelligence may aid in predicting salinization processes and developing integrated strategies to combat salt stress (Ondrasek et al., 2022) (Table 1).

Table 1. Comparison of Conventional Methods vs. Green Nanotechnology in Salinity Stress Management.

Aspect	Conventional Methods	Green Nanotechnology	References
Nutrient Management	Conventional fertilizers are used to supply essential nutrients but may not be efficient under salinity stress	Nano-fertilizers enhance nutrient uptake and improve nutrient management under salinity stress1	(Adil et al., 2022; Nasrallah et al., 2022; Shoukat et al., 2024a)
Yield Improvement	Limited improvement in yield under salinity stress conditions	Significant yield improvement observed with the use of nanoparticles	(Adil et al., 2022; Nasrallah et al., 2022)
Stress Mitigation	Conventional methods may not effectively mitigate salinity stress	Nanoparticles improve stress tolerance by enhancing physiological and biochemical responses	(Etesami et al., 2021; Nasrallah et al., 2022; Shoukat et al., 2024a)
Environmental Impact	Potential for environmental pollution and inefficiency in nutrient delivery	Eco-friendly and efficient, with reduced environmental impact	(Adeleye et al., 2016; Manzoor et al., 2022; Sári et al., 2023b)
Cost and Economic Viability	Generally lower initial cost but may require more frequent applications	Potentially higher initial cost but more cost-effective in the long term due to efficiency	(Adeleye et al., 2016; Malik et al., 2022)

4. Nanomaterials for Salinity Stress Management

4.1 Types of Green Nanomaterials

Green nanomaterials have gained significant attention for environmental remediation and biomedical applications due to their enhanced properties and eco-friendliness. These materials can be categorized into three main types: inorganic, carbon-based, and polymeric-based (Guerra et al., 2018). Carbon-based nanocomposites, including carbon fibers, graphene, and carbon nanotubes, are extensively studied for heavy metal removal from water due to their high surface area and stability (Mandal et al., 2020). Plant-based green synthesis of metallic nanomaterials, such as silver, gold, and copper, provides advantages in terms of cost-effectiveness, safety, and environmental compatibility (Ullah and Lim, 2022). Green nanofillers derived from animal, plant, and natural sources are being explored as alternatives to synthetic fillers in polymeric systems, improving thermo-mechanical and physical properties (Mohan and Kanny, 2020). While these nanomaterials show promise in various applications, including antimicrobial, anticancer, and environmental remediation, safety concerns and regulations surrounding their use are also being addressed.

4.2 Role of Nanoparticles in Alleviating Salinity Stress

Nanoparticles (NPs) have emerged as a promising solution for mitigating salinity stress in crops, which significantly affects agricultural productivity (Soni et al., 2024b). Various types of NPs, including metal oxides like zinc oxide and silicon dioxide, have shown potential in alleviating salt stress by influencing cellular processes, counteracting reactive oxygen species, improving water-use efficiency, nutrient uptake, and antioxidant activities (Shoukat et al., 2024b; Singh et al., 2024; Soni et al., 2024b). Nano-fertilizers improve plants' salt tolerance by optimizing the potassium-to-sodium ratio, crucial for maintaining ionic homeostasis (Shoukat et al., 2024b). They also boost sugar metabolism and limit sodium accumulation under saline conditions (Shoukat et al., 2024b). NPs enhance crop resilience by modulating osmotic potential, facilitating water movement, and improving overall nutrient availability and uptake (Shoukat et al., 2024b). However, long-term field trials, are needed to confirm the efficacy and economic viability of nano-fertilizers in sustainable agriculture (Shoukat et al., 2024b; Singh et al., 2024).

NPs protect cell membranes, enhance photosynthetic efficiency, and regulate stress-related hormones like abscisic acid (ABA) and Jasmonic Acid (JA) (Rehman et al., 2024; Singh et al., 2024). For instance, silica NPs increase chlorophyll content and photosynthetic efficiency in salt-stressed tomato plants, while chitosan NPs improve water-use efficiency in maize (Junedi et al., 2023b). NPs modulate hormone signaling networks, including auxin, gibberellins, and abscisic acid, which are crucial for stress responses (Singh et al., 2024; Tripathi et al., 2022). The effectiveness of NPs in stress mitigation depends on factors such as size, composition, and plant species (Soni et al., 2024b). Despite their promising applications, concerns about environmental impacts and safety necessitate further research to optimize NP efficacy and safety in agriculture (Junedi et al., 2023b; Singh et al., 2024) (Table 2).

Table 2. Types of Nanoparticles and Their Roles in Alleviating Salinity Stress.

Type of Nanoparticle	Role in Alleviating Salinity Stress	Example Crops	References
Titanium Dioxide (TiO2)	Enhances growth, photosynthetic performance, and antioxidant enzyme activity; reduces H2O2, MDA, and electrolyte leakage.	Stevia	(Sheikhalipour et al., 2021)
Selenium Functionalized with Chitosan (Cs–Se)	Increases growth, photosynthetic performance, and essential oil content; enhances stevioside and rebaudioside A concentration.	Stevia	(Sheikhalipour et al., 2021)
Zinc, Boron, Silicon, and Zeolite	Improves plant growth, physiological parameters, and nutrient concentration; reduces leaf abscisic acid and transpiration rate.	Potato	(Mahmoud et al., 2019)
Cerium Oxide (CeO2)	Ameliorates chlorophyll damage and reduces proline, MDA, and electrolyte leakage; does not affect H2O2 content.	Grapevine	(Gohari et al., 2021)
Chitosan-Magnesium Oxide (CMgO)	Enhances root length, biomass, and antioxidative enzyme activities; maintains higher K+/Na+ ratio.	Rice	(Song et al., 2023)
Zinc Oxide (ZnO)	Improves biomass, photosynthesis, and antioxidant enzyme activities; more effective in tolerant genotypes.	Barley	(Ali et al., 2022)
Iron Oxide (FeO)	Stimulates growth, increases nutrient concentration, and reduces Na+ and Cl-; restricts cadmium uptake.	Wheat	(Manzoor et al., 2021)
$B_2\,O_3$	Enhances photosynthesis, maintains mineral and redox status, and reduces oxidative stress.	Maize	(El-Shafey et al., 2023)
Chitosan Nanoparticles (CsNPs)	Improves yield, pigment fractions, and the antioxidant system; reduces hydrogen peroxide and lipid peroxidation.	Bean	(Alenazi et al., 2024)

4.3 Mechanistic Insights into Nanoparticle-Plant Interactions Under Salinity Stress

Nanoparticles (NPs) have emerged as a promising solution to mitigate salinity stress in plants, which affects over 20% of agricultural soils globally (Singh et al., 2024). Various types of NPs, including metal, metal oxide, and carbon-based, can alleviate salinity stress through multiple mechanisms (Soni et al., 2024b). These include regulating ion homeostasis, osmotic balance, and antioxidant defense systems (Singh et al., 2024). NPs influence cellular processes by impacting signal transduction, gene expression, and plant hormone signaling networks (Singh et al., 2024; Soni et al., 2024b). They also enhance photosynthesis, nutrient uptake, and overall plant growth under saline conditions (Soni et al., 2024b). However, the effectiveness of NPs depends on factors such as size, composition, and environmental conditions (Soni et al., 2024a). While NPs show great potential in improving plant resilience to salinity stress, concerns about their potential toxicity and environmental implications necessitate further research (Etesami et al., 2021; Thabet and Alqudah, 2024).

5. Synthesis of Green Nanomaterials

5.1 Biological Synthesis Using Microorganisms, Plant Extracts, and Biopolymers

Biological synthesis of nanoparticles using microorganisms and plants has emerged as a green, efficient, and cost-effective approach in nanotechnology (Jeevanandam et al., 2022; Singh et al., 2016). These biogenic nanoparticles exhibit enhanced pharmacological activity and stability compared to physicochemically synthesized counterparts (Singh et al., 2016). Recent studies have explored the potential of biogenic nanoparticles in mitigating abiotic stresses, particularly salinity stress, which significantly impacts agricultural productivity (Chakraborty et al., 2022; Etesami et al., 2021). Nanoparticles can alleviate salt stress-induced damage in plants by modulating various physiological, biochemical, and molecular mechanisms (Etesami et al., 2021). The synthesis of nanoparticles using plant extracts and microorganisms involves biomolecules and phytochemicals that act as reducing and stabilizing agents (Jeevanandam et al., 2022). This eco-friendly approach provides reduced toxicity, lower energy consumption, and potential for commercial-scale production (Jeevanandam et al., 2022). Ongoing research aims to elucidate the interactions between nanoparticles and plants under salinity stress to develop sustainable agricultural solutions (Chakraborty et al., 2022; Etesami et al., 2021).

5.2 Environmental and Economic Benefits of Green Synthesis

Green synthesis of nanoparticles, particularly silver nanoparticles (AgNPs), using plant extracts provides an eco-friendly and economical approach to nanotechnology (Rai et al., 2018). These green-synthesized AgNPs have shown promise in mitigating the adverse effects of salinity stress on plants by enhancing antioxidant enzyme activity, increasing osmoprotectants, and improving flavonoid and anthocyanin content (Sadat-Hosseini et al., 2002a). AgNPs can boost antioxidant defense mechanisms and methylglyoxal detoxification, thereby developing tolerance to abiotic stress-induced oxidative damage in various plant species (Alabdallah and Hasan, 2021). The interaction between nanoparticles and plants under salinity stress occurs at physiological, biochemical, and molecular levels, helping plants recover from salt stress (Etesami et al., 2021). While nanotechnology shows potential in addressing salinity stress in agriculture, concerns remain about the fate of nanoparticles in the environment and their potential health impacts due to nanotoxicity (Rai et al., 2018).

5.3 Case Studies on Green Nanomaterial Synthesis for Agricultural Applications

Green nanotechnology offers promising solutions for sustainable agriculture by utilizing eco-friendly synthesis methods and materials. Biomass-derived nanoparticles from plants, microorganisms, and agro-industrial waste can reduce energy consumption and costs while increasing production output compared to conventional methods (Osman et al., 2024). These green nanomaterials have diverse applications in agriculture, including nanofertilizers, nanopesticides, and nano(bio)sensors (Bartolucci et al., 2020). They can supply micronutrients, balance plant hormones, improve soil quality, stimulate plant growth, and minimize the use of toxic chemicals (Malini et al., 2022). Biosynthesized nanoparticles using plants, bacteria, and fungi provide a cost-effective and environmentally friendly approach to nanomaterial production (Chaudhry et al., 2018). While green nanotechnology shows great potential for revolutionizing agriculture, further research is needed to understand complex behaviors, ensure targeted delivery, and establish regulatory norms for effective implementation in sustainable farming practices (Malini et al., 2022).

6. Applications of Green Nanotechnology in Salinity Stress Alleviation

6.1 Nanofertilizers and their Impact on Salinity Tolerance

Recent studies have demonstrated the potential of nanotechnology in mitigating salinity stress in plants. Green nanosilica has been shown to enhance salt tolerance in Williams banana, improving fruit yield and quality while reducing Na+ accumulation in leaves (Ding et al., 2022). Various nanoparticles and nanofertilizers have exhibited promising results in salt stress management by regulating adaptive mechanisms in plants (Zulfiqar and Ashraf, 2021). Nanofertilizers, such as ZnO and Fe3O4 nanoparticles, have been found to alleviate salt stress in Moringa peregrina by enhancing growth parameters, chlorophyll content, and antioxidant activity (Sh et al., 2015). These nanoparticles can improve nutrient utilization efficiency through slow and sustainable nutrient release (Mariyam et al., 2023). However, factors such as nanoparticle size, shape, and charge, as well as plant physiology, influence their effectiveness. Further research is needed to fully understand the potential of nanotechnology in sustainable agriculture and to address implementation challenges (Mariyam et al., 2023).

6.2 Nano-bio-stimulants for Stress Resistance

Nanotechnology offers promising solutions for mitigating salinity and drought stress in crops, addressing global food security challenges. Engineered nanoparticles (ENPs) can enhance plant growth, stress tolerance, and yield by improving nutrient uptake, water balance, and antioxidant defense mechanisms (Junedi et al., 2023a). Various nanoparticles, including silica and chitosan, have shown effectiveness in alleviating salt stress in plants (Junedi et al., 2023a). Nanobiotechnology combines with genome editing techniques like CRISPR/Cas to develop stress-smart crops (Raza et al., 2023). Nanoparticles protect cell membranes, enhance photosynthetic efficiency, optimize hormone levels, and regulate gene expression, strengthening plant resilience to stress (Rehman et al., 2024). However, concerns about environmental impacts and safety necessitate further research (Junedi et al., 2023a). Green synthesis of nanoparticles is emphasized for its potential benefits in agriculture (Saxena et al., 2016).

6.3 Role of Nano-based Delivery Systems for Plant Growth Regulators

NPs can act as carriers for plant growth regulators like salicylic acid, enabling controlled release and maintaining optimal concentrations (Polyakov et al., 2023). The effectiveness of NPs depends on factors such as size, composition, and application method (Zhao et al., 2020). Studies have demonstrated that NPs can enhance chlorophyll content, photosynthetic efficiency, and water-use efficiency in salt-stressed plants (Junedi et al., 2023a). However, concerns about environmental impacts and safety necessitate further research to optimize their efficacy and safety in agricultural applications (Junedi et al., 2023a; Zhao et al., 2020).

6.4 Field-level Studies and Success Stories

Recent field studies have demonstrated the potential of green nanotechnology in alleviating salinity stress in crops. Foliar application of green nanosilica significantly improved yield and quality of Williams banana under saline irrigation, enhancing salt tolerance mechanisms (Ding et al., 2022). Various nano-management approaches, including biological nanomaterials and nano-enabled nutrients, have shown promise in mitigating salinity stress effects on plants both in field and *in vitro* conditions (Sári et al., 2023b). Engineered nanoparticles (ENPs) made of biocompatible materials have been effective in enhancing plant growth and reducing salt stress-induced damage by improving nutrient uptake and antioxidative defense mechanisms (Junedi et al., 2023b). In a hydroponic study, green-synthesized selenium nanoparticles significantly alleviated salt stress complications in mustard plants by enhancing antioxidant enzyme activities, phenolic content, and improving various growth parameters (Sarkar and Kalita, 2022). These findings highlight the potential of green nanotechnology in sustainable agriculture under saline conditions.

7. Environmental and Socioeconomic Implications

7.1 Potential Risks of Nanomaterials in the Environment

Engineered nanomaterials (ENMs) offer significant potential in various applications, but their environmental and health risks require careful assessment (Keshavan et al., 2024). The entire lifecycle of nanomaterials, from synthesis to disposal, presents potential hazards that need comprehensive evaluation (Dos Santos et al., 2022). To address these challenges, integrated science-based methodologies have been proposed to assess ENM risks across societal, environmental, and economic domains (Tolaymat et al,., 2015). Key factors influencing nanomaterial behavior include size, shape, and capping agents, which complicate risk comparisons (Dos Santos et al., 2022). Understanding nanomaterial transport, transformation, and fate in the environment is crucial for accurate risk assessment (Lowry and Casman, 2009). Researchers advocate for a balanced approach incorporating early hazard identification, safe-by-design principles, and life cycle assessments to ensure responsible nanotechnology development (Keshavan et al., 2024). Addressing knowledge gaps and adapting existing risk analysis methods are essential steps in effectively evaluating and mitigating the environmental implications of nanomaterials (Lowry and Casman, 2009).

7.2 Regulatory Challenges and Safety Concerns

The integration of Socio-Economic Considerations (SECs) into biosafety regulatory frameworks for genetically modified organisms presents significant challenges and implications (Falck-Zepeda et al., 2016; Falck-Zepeda and Zambrano, 2011). These challenges extend to emerging technologies like synthetic biology and engineered nanomaterials, which pose new regulatory hurdles for environmental, health, and safety assessments (Carter et al., 2014; Schierow, 2008). The inclusion of SECs in decision-making processes requires careful consideration of scope, timing, methods, and integration modalities to avoid negatively impacting technology flows and creating unworkable biosafety systems (Falck-Zepeda and Zambrano, 2011). As the number and diversity of engineered organisms increase, regulatory agencies may face resource and expertise constraints in effectively overseeing these products (Carter et al., 2014). Policymakers must navigate trade-offs between competing considerations when addressing these challenges, potentially necessitating new legislative approaches to ensure adequate regulation of novel biotechnologies (Carter et al., 2014; Schierow, 2008).

7.3 Cost-benefit Analysis of Implementing Green Nanotechnology in Agriculture

Nanotechnology in agriculture offers promising solutions for sustainable practices, including nano fertilizers and nano pesticides that can increase crop yields and reduce environmental impacts (Bhandari et al., 2023; Su et al., 2022). These nanoformulations provide targeted delivery, slow release, and improved efficiency, potentially increasing net revenue and lowering greenhouse gas emissions (Arora et al., 2022; Su et al., 2022). Green nanotechnology, utilizing biogenic nanoparticles from plants, bacteria, or fungi, presents a more eco-friendly and economical approach (Bhandari et al., 2023). However, challenges remain, such as the need to improve nanoformulation efficiency for widespread adoption and address potential risks associated with nanoparticle accumulation in soil and water bodies (Arora et al., 2022; Su et al., 2022). Long-term life cycle analyses are crucial to assess the impacts of nanoparticles on plants and the environment (Kumar et al., 2018). Overall, nanotechnology in agriculture requires careful consideration of both benefits and risks for sustainable implementation.

7.4 Ethical Considerations and Public Acceptance

The ethical implications and public acceptance of environmental and socioeconomic issues related to new technologies are complex and multifaceted. Social acceptance studies alone are insufficient to address the broader ethical impacts of risky technologies, necessitating a combination of empirical and conceptual approaches (Taebi, 2018). Socioeconomic impacts of technologies like GMOs can be far-reaching and potentially irreversible, affecting social institutions and relations (Daño, 2008). Environmental remediation projects, such as those addressing radionuclide contamination, can have significant social and economic consequences beyond health risks, requiring consideration of socio-ethical aspects in their implementation (Oughton, 2013). Biotechnology in agriculture raises ethical issues that span medical, environmental, and social domains, demanding a multidisciplinary approach to decision-making that incorporates scientific data, religious values, and philosophical perspectives (Macer, 1998). These studies emphasize the importance of comprehensive assessments that consider both ethical acceptability and social acceptance in the governance of new technologies.

8. Future Perspectives and Challenges

Nanotechnology offers promising solutions for sustainable agriculture, particularly in addressing salinity stress challenges. Nanoparticles can enhance crop resilience, improve nutrient use efficiency, and mitigate environmental impacts (Raliya et al., 2017; Singh et al., 2024). Green nanotechnology integrates with conventional breeding and biotechnology to develop stress-tolerant crop varieties (Prasad et al., 2017). However, scaling up from the laboratory to field applications faces challenges in cost-effective production and regulatory approval (Mishra et al., 2017). Research gaps include understanding long-term environmental impacts, standardizing application protocols, and assessing toxicity risks (Mishra et al., 2017; Singh et al., 2024). Future directions involve developing multifunctional nanoparticles, exploring interactions with plant microbiomes, and establishing regulatory frameworks (Mishra et al., 2017; Raliya et al., 2017). Advancing the field requires interdisciplinary collaboration to build resilient agricultural systems against salinity stress while ensuring environmental safety and sustainable practices (Prasad et al., 2017; Singh et al., 2024).

9. Conclusion

Green nanotechnology offers a promising and sustainable approach to alleviating salinity stress in plants, addressing a critical challenge in global agriculture. By enhancing nutrient uptake, improving water use efficiency, and modulating hormonal balance and signaling pathways, nanoparticles have demonstrated their potential to improve plant resilience and productivity under saline conditions.

Their unique properties enable targeted and efficient solutions while reducing the environmental footprint of conventional agricultural practices. However, the successful implementation of green nanotechnology requires overcoming challenges related to large-scale production, environmental safety, and regulatory frameworks. Integrating nanotechnology with traditional breeding and modern biotechnological tools presents a transformative opportunity to develop stress-resilient crops and ensure agricultural sustainability in salt-affected regions. Continued research, field trials, and interdisciplinary collaboration will be essential to harness the full potential of green nanotechnology for mitigating salinity stress and securing global food security.

References

Adeleye, A., Conway, J., Garner, K., Huang, Y., Su, Y., and Keller, A. (2016). Engineered nanomaterials for water treatment and remediation: Costs, benefits, and applicability. *Chemical Engineering Journal*, 286: 640–662. doi:10.1016/J.CEJ.2015.10.105.

Adil, M., Bashir, S., Bashir, S., Aslam, Z., Ahmad, N., Younas, T. et al. (2022). Zinc oxide nanoparticles improved chlorophyll contents, physical parameters, and wheat yield under salt stress. *Frontiers in Plant Science*, 13. doi:10.3389/fpls.2022.932861.

Alabdallah, N. M., and Hasan, M. (2021). Plant-based green synthesis of silver nanoparticles and its effective role in abiotic stress tolerance in crop plants. *Saudi Journal of Biological Sciences*, 28: 5631–5639.

Alenazi, M., El-Ebidy, A., El-Shehaby, O., Seleiman, M., Aldhuwaib, K., and Abdel-Aziz, H. (2024). Chitosan and Chitosan Nanoparticles Differentially Alleviate Salinity Stress in Phaseolus vulgaris L. Plants. *Plants*, 13. doi:10.3390/plants13030398.

Ali, B., Saleem, M., Ali, S., Shahid, M., Sagir, M., Tahir, M. et al. (2022). Mitigation of salinity stress in barley genotypes with variable salt tolerance by application of zinc oxide nanoparticles. *Frontiers in Plant Science*, 13. doi:10.3389/fpls.2022.973782.

Arif, Y., Singh, P., Siddiqui, H., Bajguz, A., and Hayat, S. (2020). Salinity induced physiological and biochemical changes in plants: An omic approach towards salt stress tolerance. *Plant Physiology and Biochemistry*, 156: 64–77.

Arora, S., Murmu, G., Mukherjee, K., Saha, S. K., and Maity, D. (2022). A Comprehensive Overview of Nanotechnology in Sustainable Agriculture. *Journal of Biotechnology*.

Awad-Allah, E. F. (2023). Effectiveness of silica nanoparticle application as plant nano-nutrition: a review. *Journal of Plant Nutrition*, 46(11): 2763–2776.

Balasubramaniam, T., Shen, G., Esmaeili, N., and Zhang, H. (2023). Plants' response mechanisms to salinity stress. *Plants*, 12(12), 2253.

Bartolucci, C., Antonacci, A., Arduini, F., Moscone, D., Fraceto, L. F., Campos, E. V. R. et al. (2020). Green nanomaterials fostering agrifood sustainability. *TrAC Trends in Analytical Chemistry*.

Bhandari, G., Dhasmana, A., Chaudhary, P., Gupta, S., Gangola, S., Gupta, A. et al. (2023). A Perspective Review on Green Nanotechnology in Agro-Ecosystems: Opportunities for Sustainable Agricultural Practices & Environmental Remediation. *Agriculture*.

Carter, S. R., Rodemeyer, M., Garfinkel, M., and Friedman, R. M. (2014). *Synthetic Biology and the U.S. Biotechnology Regulatory System: Challenges and Options*.

Chakraborty, S., Singh, A. K., and Roychoudhury, A. (2022). Biogenic nanoparticles and generation of abiotic stress-resilient plants: A new approach for sustainable agriculture. *Plant Stress*.

Chaudhry, N., Dwivedi, S., Chaudhry, V., Singh, A., Saquib, Q., Azam, A. et al. (2018). Bio-inspired nanomaterials in agriculture and food: Current status, foreseen applications and challenges. *Microbial Pathogenesis*, 123: 196–200.

Daño, E. C. (2008). *Potential Socio-Economic, Cultural and Ethical Impacts of GMOs: Prospects for Socio- Economic Impact Assessment*.

Ding, Z., Zhao, F., Zhu, Z., Ali, E. F., Shaheen, S. M., Rinklebe, J. et al. (2022). Green nanosilica enhanced the salt-tolerance defenses and yield of Williams banana: A field trial for using saline water in low fertile arid soil. *Environmental and Experimental Botany*.

Dos Santos, O. A. L., Pizzorno Backx, B., Abumousa, R. A., and Bououdina, M. (2022). Environmental Implications Associated with the Development of Nanotechnology: From Synthesis to Disposal. *Nanomaterials*, 12.

El-Shafey, N., Avramova, V., Beemster, G., Korany, S., and AbdElgawad, H. (2023). B2 O3 nanoparticles alleviate salt stress in maize leaf growth zones by enhancing photosynthesis and maintaining mineral and redox status. *Physiologia Plantarum*, 175 5. doi:10.1111/ppl.14033

Elmer, W., and White, J. C. (2018). The future of nanotechnology in plant pathology. *Annual Review of Phytopathology,* 56(1): 111-133.

Etesami, H., Fatemi, H., and Rizwan, M. (2021). Interactions of nanoparticles and salinity stress at physiological, biochemical and molecular levels in plants: A review. *Ecotoxicology and Environmental Safety*, 225: 112769.

Evelin, H., Devi, T. S., Gupta, S., and Kapoor, R. (2019). Mitigation of salinity stress in plants by arbuscular mycorrhizal symbiosis: current understanding and new challenges. *Frontiers in Plant Science*, 10: 470.

Falck-Zepeda, J. B., Smyth, S. J., and Ludlow, K. (2016). Zen and the Art of Attaining Conceptual and Implementation Clarity: Socio-economic Considerations, Biosafety and Decision-making. *The Estey Centre Journal of International Law and Trade Policy*, 17: 117–136.

Falck-Zepeda, J. B., and Zambrano, P. (2011). Socio-economic Considerations in Biosafety and Biotechnology Decision Making: The Cartagena Protocol and National Biosafety Frameworks. *Review of Policy Research*, 28: 171–195.

Faraz, A., Faizan, M., Sami, F., Siddiqui, H., Pichtel, J., and Hayat, S. (2019). Nanoparticles: biosynthesis, translocation and role in plant metabolism. *IET Nanobiotechnology*, 13(4): 345–352.

Gohari, G., Zareei, E., Rostami, H., Panahirad, S., Kulak, M., Farhadi, H. et al. (2021). Protective effects of cerium oxide nanoparticles in grapevine (Vitis vinifera L.) cv. Flame Seedless under salt stress conditions. *Ecotoxicology and Environmental Safety*, 220: 112402. doi:10.1016/j.ecoenv.2021.112402.

Gomez-Muñoz, C. L., Zaca-Morán, P., and Zaca-Moran, O. (2023). Los nanomateriales y su importancia en la síntesis amigable con el ambiente. *RD-ICUAP*, 162–170.

Guerra, F. D., Attia, M. F., Whitehead, D. C., and Alexis, F. (2018). Nanotechnology for Environmental Remediation: Materials and Applications. *Molecules : A Journal of Synthetic Chemistry and Natural Product Chemistry*, 23.

Hussain, S., Hussain, S., Ali, B., Ren, X., Chen, X., Li, Q. et al. (2021). Recent progress in understanding salinity tolerance in plants: Story of Na+/K+ balance and beyond. *Plant Physiology and Biochemistry*, 160: 239–256.

Isayenkov, S. V., and Maathuis, F. J. (2019). Plant salinity stress: many unanswered questions remain. *Frontiers in Plant Science*, 10: 80.

Jaiswal, S. (2022). Applications of nanoparticles: a step towards a green environment. *International Journal of Biological Innovations*, 4(01): 64–70.

Jeevanandam, J., Kiew, S. F., Boakye-Ansah, S., Lau, S. Y., Barhoum, A., Danquah, M. K. et al. (2022). Green approaches for the synthesis of metal and metal oxide nanoparticles using microbial and plant extracts. *Nanoscale*, 14(7): 2534–2571.

Junedi, M. A., Mukhopadhyay, R., and Manjari, K. S. (2023a). Alleviating salinity stress in crop plants using new engineered nanoparticles (ENPs). *Plant Stress*.

Junedi, M. A., Mukhopadhyay, R., and Manjari, K. S. (2023b). Alleviating salinity stress in crop plants using new engineered nanoparticles (ENPs). *Plant Stress*, 100184.

Keshavan, S., Petri-Fink, A., and Rothen-Rutishauser, B. (2024). Understanding The Benefits and Risks of Sustainable Nanomaterials in a Research Environment. *Chimia*, 78 6: 397–402.

Khan, S. H. (2020). Green nanotechnology for the environment and sustainable development. *Green materials for wastewater treatment*, 13–46.

Kumar, A., Gupta, K., Dixit, S., Mishra, K. K., and Srivastava, S. (2018). A review on positive and negative impacts of nanotechnology in agriculture. *International Journal of Environmental Science And Technology*, 16: 2175–2184.

Kumar, A., Singh, S., Gaurav, A. K., Srivastava, S., and Verma, J. P. (2020). Plant growth-promoting bacteria: biological tools for the mitigation of salinity stress in plants. *Frontiers in Microbiology*, 11: 1216.

Lowry, G. V., and Casman, E. A. (2009). *Nanomaterial Transport, Transformation, and Fate in the Environment A Risk-Based Perspective on Research Needs.*

Macer, D. R. J. (1998). *Biotechnology in Agriculture: Ethical Aspects and Public Acceptance.*

Mahmoud, A., Abdeldaym, E., Abdelaziz, S., El-Sawy, M., and Mottaleb, S. (2019). Synergetic effects of zinc, boron, silicon, and zeolite nanoparticles on confer tolerance in potato plants subjected to salinity. *Agronomy.* doi:10.3390/agronomy10010019.

Malik, S., Dhasmana, A., Preetam, S., Mishra, Y., Chaudhary, V., Bera, S. P. et al. (2022). Exploring Microbial-Based Green Nanobiotechnology for Wastewater Remediation: A Sustainable Strategy. *Nanomaterials*, 12. doi:10.3390/nano12234187.

Malini, S., Raj, K., Madhumathy, S. M., El-Hady, K. M., Islam, S. M. F., and Dutta, M. (2022). Bioinspired Advances in Nanomaterials for Sustainable Agriculture. *Journal of Nanomaterials*.

Mandal, S., Adhikari, S., Shengyan, P., Mishra, A. K., and Patel, R. K. (2020). *Functional Green Carbon Nanocomposites for Heavy Metal Treatment in Water.*

Mansour, M. M. F., and Salama, K. H. A. (2019). Cellular mechanisms of plant salt tolerance. *Microorganisms in Saline Environments: Strategies and Functions*, 169–210.

Manzoor, N., Ahmed, T., Noman, M., Shahid, M., Nazir, M. M., Ali, L. et al. (2021). Iron oxide nanoparticles ameliorated the cadmium and salinity stresses in wheat plants, facilitating photosynthetic pigments and restricting cadmium uptake. *The Science of the Total Environment*, 769: 145221. doi:10.1016/j.scitotenv.2021.145221.

Manzoor, N., Ali, L., Ahmed, T., Noman, M., Adrees, M., Shahid, M. et al. (2022). Recent Advancements and Development in Nano-Enabled Agriculture for Improving Abiotic Stress Tolerance in Plants. *Frontiers in Plant Science*, 13. doi:10.3389/fpls.2022.951752.

Mariyam, S., Upadhyay, S. K., Chakraborty, K., Verma, K. K., Duhan, J. S., Muneer, S. et al. (2023). Nanotechnology, a frontier in agricultural science, a novel approach in abiotic stress management and convergence with new age medicine-A review. *The Science of the Total Environment*, 169097.

Mbinda, W., and Mukami, A. (2021). A review of recent advances and future directions in the management of salinity stress in finger millet. *Frontiers in Plant Science*, 12: 734798.

Md Ashif Ikbal, A., and Dutta Choudhury, P. (2023). Green nanotechnology- an innovative pathway for sustainability. *Paripex Indian Journal of Research*.

Merugu, R. (2020). Green synthesis approach to fabricate nanomaterials. *Nanomaterials in Biofuels Research*, 75–113.

Mishra, A. K., Das, R., George Kerry, R., Biswal, B., Sinha, T., Sharma, S. et al. (2023). Promising management strategies to improve crop sustainability and to amend soil salinity. *Frontiers in Environmental Science*, 10: 962581.

Mishra, S., Keswani, C., Abhilash, P. C., Fraceto, L. F., and Singh, H. B. (2017). Integrated Approach of Agri-nanotechnology: Challenges and Future Trends. *Frontiers in Plant Science*, 8.

Mohan, T., and Kanny, K. (2020). *Green Nanofillers for Polymeric Materials*.

Mukhopadhyay, R., Sarkar, B., Jat, H. S., Sharma, P. C., and Bolan, N. S. (2021). Soil salinity under climate change: Challenges for sustainable agriculture and food security. *Journal of Environmental Management*, 280: 111736.

Nanotechnology, G., Goel, A., and Bhatnagar, S. (2022). *Green Nanotechnology*.

Nasrallah, A., Kheder, A., Kord, M., Fouad, A., El-Mogy, M., and Atia, M. (2022). Mitigation of Salinity Stress Effects on Broad Bean Productivity Using Calcium Phosphate Nanoparticles Application. *Horticulturae*. doi:10.3390/horticulturae8010075.

Navarro-Torre, S., Garcia-Caparrós, P., Nogales, A., Abreu, M. M., Santos, E., Cortinhas, A. L. et al. (2023). Sustainable agricultural management of saline soils in arid and semi-arid Mediterranean regions through halophytes, microbial and soil-based technologies. *Environmental and Experimental Botany*, 212, 105397.

Nie, H., Xu, L., Zhao, Q., Wang, N., You, C., Zhang, F. et al. (2020). Mechanisms of plant response to salinity stress: current understanding and recent progress. *Pak. J. Bot*, 52(5): 1879–1883.

Okon, O. G. (2019). Effect of salinity on physiological processes in plants. *Microorganisms in Saline Environments: Strategies and Functions*, 237–262.

Omar, R. A., Afreen, S., Talreja, N., Chauhan, D., and Ashfaq, M. (2019). Impact of nanomaterials in plant systems. *Plant nanobionics: Volume 1, Advances in the Understanding of Nanomaterials Research and Applications*, 117–140.

Ondrasek, G., Rathod, S., Manohara, K. K., Gireesh, C., Anantha, M. S., Sakhare, A. S. et al. (2022). Salt stress in plants and mitigation approaches. *Plants*, 11(6): 717.

Osman, A. I., Zhang, Y., Farghali, M., Rashwan, A. K., Eltaweil, A. S., Abd El-Monaem, E. M. et al. (2024). Synthesis of green nanoparticles for energy, biomedical, environmental, agricultural, and food applications: A review. *Environmental Chemistry Letters*.

Oughton, D. H. (2013). Social and ethical issues in environmental remediation projects. *Journal of Environmental Radioactivity*, 119: 21–25.

Paradva, K. C., Jangir, R., and Kalla, S. (2023). Green nanomaterials: Synthesis and applications in wastewater treatment. *Inorganic Chemistry Communications*, 111584.

Pedrotti, A., Chagas, R. M., Ramos, V. C., do Nascimento Prata, A. P., Tadeu Lucas, A. A., and dos Santos, P. B. (2015). Causes and consequences of the process of soil salinization. *Revista Eletronica Em Gestao Educacao E Tecnologia Ambiental*, 19(2): 1308–1324.

Polash, M. A. S., Sakil, M. A., and Hossain, M. A. (2019). Plants responses and their physiological and biochemical defense mechanisms against salinity: A review. *Trop. Plant Res*, 6(2): 250–274.

Polyakov, V., Bauer, T., Butova, V., Minkina, T., and Rajput, V. D. (2023). Nanoparticles-Based delivery systems for salicylic acid as plant growth stimulator and stress alleviation. *Plants*, 12(8): 1637.

Prasad, R., Bhattacharyya, A., and Nguyen, Q. D. (2017). Nanotechnology in Sustainable Agriculture: Recent Developments, Challenges, and Perspectives. *Frontiers in Microbiology*, 8.

Rai, P. K., Kumar, V., Lee, S.-S., Raza, N., Kim, K. H., Ok, Y. S. et al. (2018). Nanoparticle-plant interaction: Implications in energy, environment, and agriculture. *Environment International*, 119: 1–19.

Raliya, R., Saharaṇ, V., Dimkpa, C. O., and Biswas, P. (2017). Nanofertilizer for precision and sustainable agriculture: current state and future perspectives. *Journal of Agricultural and Food Chemistry*, 66 26: 6487–503.

Rasool, A., Shah, W. H., Mushtaq, N. U., Saleem, S., Hakeem, K. R., and ul Rehman, R. (2022). Salinity-induced changes on different physiological and biochemical features of plants. In *Plant Abiotic Stress Physiology* (pp. 201–224): Apple Academic Press.

Raza, A., Charagh, S., Salehi, H., Abbas, S., Saeed, F., Poinern, G. E. J. et al. (2023). Nano-enabled stress-smart agriculture: Can nanotechnology deliver drought and salinity-smart crops? *Journal of Sustainable Agriculture and Environment*.

Raza, A., Tabassum, J., Fakhar, A. Z., Sharif, R., Chen, H., Zhang, C. et al. (2023). Smart reprograming of plants against salinity stress using modern biotechnological tools. *Critical Reviews in Biotechnology*, 43(7): 1035–1062.

Rehman, A., Khan, S., Sun, F., Peng, Z., Feng, K., Wang, N. et al. (2024). Exploring the nano-wonders: unveiling the role of Nanoparticles in enhancing salinity and drought tolerance in plants. *Frontiers in Plant Science*, 14: 1324176.

Rehman, A., Khan, S., Sun, F., Peng, Z., Feng, K., Wang, N. et al. (2024). Exploring the nano-wonders: unveiling the role of Nanoparticles in enhancing salinity and drought tolerance in plants. *Frontiers in Plant Science*, 14.

Russ, J. D., Zaveri, E. D., Damania, R., Desbureaux, S. G., Escurra, J. J., and Rodella, A.-S. (2020). Salt of the earth: quantifying the impact of water salinity on global agricultural productivity. *World Bank Policy Research Working Paper* (9144).

Sadat-Hosseini, M., Naeimi, A., Boroomand, N., Aalifar, M., and Farajpour, M. (2022a). Alleviating the adverse effects of salinity on Roselle plants by green synthesized nanoparticles. *Scientific Reports*, 12.

Sadat-Hosseini, M., Naeimi, A., Boroomand, N., Aalifar, M., and Farajpour, M. (2022b). Alleviating the adverse effects of salinity on Roselle plants by green synthesized nanoparticles. *Scientific Reports*, 12(1): 18165.

Sári, D., Ferroudj, A., Abdalla, N., El-Ramady, H., Dobránszki, J., and Prokisch, J. (2023a). Nano-management approaches for salt tolerance in plants under field and *in vitro* conditions. *Agronomy*, 13(11): 2695.

Sári, D., Ferroudj, A., Abdalla, N., El-Ramady, H., Dobránszki, J., and Prokisch, J. (2023b). Nano-management approaches for salt tolerance in plants under field and *in vitro* Conditions. *Agronomy*.

Sarita, Bhupnesh, and Goyal, V. (2024). Promoting Sustainable Agriculture: Approaches for Mitigating Soil Salinity Challenges: A ReviewPromoting Sustainable Agriculture: Approaches for Mitigating Soil Salinity Challenges: A Review. *Arcc Journals*. doi:10.18805/ag.R-2648.

Sarkar, R. D., and Kalita, M. C. (2022). Green Synthesized Se Nanoparticle-mediated Alleviation of Salt Stress in Field Mustard TS-36 Variety. *Journal of Biotechnology*.

Saxena, R. S., Tomar, R. S., and Kumar, M. (2016). *Exploring Nanobiotechnology to Mitigate Abiotic Stress in Crop Plants*.

Schierow, L.-J. (2008). *Engineered Nanoscale Materials and Derivative Products: Regulatory Challenges*.

Sh, A., Elfeky, S. A., and Darwish, E. (2015). Alleviation of salt stress on Moringa peregrina using foliar application of nanofertilizers. *Journal of Horticulture and Forestry*, 7: 36–47.

Sheikhalipour, M., Esmaielpour, B., Gohari, G., Haghighi, M., Jafari, H., Farhadi, H. et al. (2021). Salt stress mitigation via the foliar application of chitosan-functionalized selenium and anatase titanium dioxide nanoparticles in stevia (Stevia rebaudiana Bertoni). *Molecules*, 26. doi:10.3390/molecules26134090.

Shoukat, A., Saqib, Z., Akhtar, J., Aslam, Z., Pitann, B., Hossain, M. S., and Mühling, K. (2024a). Zinc and Silicon Nano-Fertilizers Influence Ionomic and Metabolite Profiles in Maize to Overcome Salt Stress. *Plants*, 13. doi:10.3390/plants13091224.

Shoukat, A., Saqib, Z. A., Akhtar, J., Aslam, Z., Pitann, B., Hossain, M. S., and Mühling, K. H. (2024b). Zinc and Silicon Nano-Fertilizers Influence Ionomic and Metabolite Profiles in Maize to Overcome Salt Stress. *Plants*, 13(9): 1224.

Singh, A., Agrawal, S., Rajput, V. D., Ghazaryan, K., Yesayan, A., Minkina, T. et al. (2024). Nanoparticles in revolutionizing crop production and agriculture to address salinity stress challenges for a sustainable future. *Discover Applied Sciences*, 6(6): 317.

Singh, A., Agrawal, S., Rajput, V. D., Ghazaryan, K. A., Yesayan, A., Minkina, T. M. et al. (2024). Nanoparticles in revolutionizing crop production and agriculture to address salinity stress challenges for a sustainable future. *Discover Applied Sciences*.

Singh, P., Kim, Y.-J., Zhang, D., and Yang, D.-C. (2016). Biological Synthesis of Nanoparticles from Plants and Microorganisms. *Trends in Biotechnology*, 34 7: 588–599.

Singh, T., Srivastava, N., Mishra, P., and Bhatiya, A. (2021). Analysis of various green methods to synthesize nanomaterials: an eco-friendly approach. *Bioenergy Research: Revisiting Latest Development*, 181–205.

Slama, H. B., Chenari Bouket, A., Alenezi, F. N., Luptakova, L., Baranov, O., Ahadi, R. et al. (2023). Impacts of salt stress on the rhizosphere and endophytic bacterial role in plant salt alleviation. *International Journal of Plant Biology*, 14(2): 361–376.

Song, Y., Zheng, C., Li, S., Chen, J., and Jiang, M. (2023). Chitosan-magnesium oxide nanoparticles improve salinity tolerance in rice (*Oryza sativa* L.). *ACS Applied Materials & Interfaces*. doi:10.1021/acsami.3c00043.

Soni, S., Jha, A. B., Dubey, R. S., and Sharma, P. (2024a). Nanowonders in agriculture: Unveiling the potential of nanoparticles to boost crop resilience to salinity stress. *The Science of the Total Environment*, 171433.

Soni, S., Jha, A. B., Dubey, R. S., and Sharma, P. (2024b). Nanowonders in agriculture: Unveiling the potential of nanoparticles to boost crop resilience to salinity stress. *Science of the Total Environment*, 171433.

Stavi, I., Thevs, N., and Priori, S. (2021). Soil salinity and sodicity in drylands: A review of causes, effects, monitoring, and restoration measures. *Frontiers in Environmental Science*, 9: 712831.

Su, Y., Zhou, X., Meng, H., Xia, T., Liu, H., Rolshausen, P. E. et al. (2022). Cost-benefit analysis of nanofertilizers and nanopesticides emphasizes the need to improve the efficiency of nanoformulations for widescale adoption. *Nature food*, 3 12: 1020–1030.

Sultan, M. T., Mahmud, U., and Khan, M. Z. (2023). Addressing soil salinity for sustainable agriculture and food security: Innovations and challenges in coastal regions of Bangladesh. In (pp. 100260): Elsevier.

Taebi, B. (2018). How to bridge the gap between social acceptance and ethical acceptability. *Environmental Health Risks*.

Thabet, S. G., and Alqudah, A. M. (2024). Unraveling the role of nanoparticles in improving plant resilience under environmental stress condition. *Plant and Soil*, 1–18.

Tolaymat, T., El Badawy, A. M., Sequeira, R., and Genaidy, A. M. (2015). An integrated science-based methodology to assess potential risks and implications of engineered nanomaterials. *Journal of Hazardous Materials*, 298: 270–281.

Tripathi, D., Singh, M., and Pandey-Rai, S. (2022). Crosstalk of nanoparticles and phytohormones regulate plant growth and metabolism under abiotic and biotic stress. *Plant Stress*, 6: 100107.

Ullah, A., and Lim, S. I. (2022). Plant extract-based synthesis of metallic nanomaterials, their applications, and safety concerns. *Biotechnology and Bioengineering*, 119: 2273–2304.

Zaman, M., Shahid, S. A., Heng, L., Shahid, S. A., Zaman, M., and Heng, L. (2018). Soil salinity: Historical perspectives and a world overview of the problem. *Guideline for Salinity Assessment, Mitigation and Adaptation Using Nuclear and Related Techniques*, 43–53.

Zhao, L., Lu, L., Wang, A., Zhang, H., Huang, M., Wu, H. et al. (2020). Nanobiotechnology in Agriculture: Use of Nanomaterials To Promote Plant Growth and Stress Tolerance. *Journal of Agricultural and Food Chemistry*.

Zulfiqar, F., and Ashraf, M. A. (2021). Nanoparticles potentially mediate salt stress tolerance in plants. *Plant Physiology And Biochemistry : PPB*, 160: 257–268.

CHAPTER-4

Enhancing Drought Stress Resilience in Plants Based on Green Nanotechnology

Muhammad Faheem Iqbal,[1,2] *Iqra Irshad,*[1,2] *Huma Arshad,*[1] *Muhammad Uzair,*[2] *Murtaza Hasan*[3,*] and *Ghazala Mustafa*[1,*]

1. Introduction

Global warming and its related problems are the major focus of national and international organizations worldwide (Janni et al., 2024; Zandalinas et al., 2021). Crop yields are predicted to vary across different regions of the world due to unexpected temperature increases and changes in rainfall patterns caused by various human and other environmental activities (Md Gulzar and Mazumder, 2024; Qiao et al., 2024). In future this will further increase the frequency and intensity of droughts. The non-motile nature of plants has led them to be victims of the ever-changing global changes (Grange et al., 2024). Plants, like all other creatures, need water for their normal functioning. Environmental stressors like heat, salt, and drought have a great impact on the overall survival of the plant's yield and quality (Chaffai et al., 2024b; Li et al., 2024). Plants undergo different morphological, physiological, biochemical, and molecular changes (Abdelaal et al., 2021; Hasanuzzaman et al., 2020; Yang et al., 2023) due to the increasing severity of drought, which is experimentally confirmed.

Water deficit is the limiting factor for the growth and yield of plants by disruption in cellular regulation, an increase of ROS production, and oxidative damage (Cotrina et al., 2023; Siddique and Helen, 2020). It prevents the plant from reaching its full genetic potential (Tian et al., 2023). On the demand of ever-increasing global warming and the severity of the environment, these creatures developed a strategic pattern of changes to compete with these stresses (Dastborhan et al., 2021; Melelli et al., 2022). According to a report by FAO (2024), water is a crucial factor for agriculture, food security, and the ecosystem, and inadequate agricultural systems along with environmental changes are proving to be a major hindrance in achieving SDG 2, zero hunger by 2030.

[1] Department of Plant Sciences, Faculty of Biological Sciences, Quaid-i-Azam University, Islamabad 45320, Pakistan.
[2] National Institute for Genomics and Advanced Biotechnology (NIGAB), National Agricultural Research Centre (NARC), Park Road, Islamabad-45500, Pakistan.
[3] Department of Biotechnology, Faculty of Chemical and Biological Sciences, The Islamia University of Bahawalpur, 63100, Pakistan.
* Corresponding authors: murtaza@iub.edu.pk; mghazala@qau.edu.pk

DOI: 10.1201/9781003567684-4

The global food demands can be reached using a novel resilient crop production system. Conventional farming proved ineffective due to population overflow and extreme climatic changes (Haris et al., 2023; Pramanik et al., 2020; Yadav et al., 2023). The emergence of promising approaches like nanotechnology can ripe into sustainable crop production (Gondal and Tayyiba, 2022; Ijaz et al., 2023; Wang et al., 2021). Nanobiotechnology can enhance nutrient delivery and absorption, pest management, genome editing efficiency, and integration of smart sensors for agriculture thus improving overall health and addressing global food security concerns.

The decrease in crop traits caused by stressors like drought is a result of complex gene interactions in plants (Chandrashekar et al., 2023; Mishra et al., 2018) that can be mitigated using new approaches like nanobiotechnology, especially green nanobiotechnology (Wahab et al., 2023). Nanoparticles can improve crops through genetic modifications (Ali et al., 2021), water use efficiency (Qureshi et al., 2018), nutrient transfer, and other routes, complementing conventional methods. Some nanoparticles have been reported to have no side effects on the physiological functioning of the plants till now but have successfully triggered their drought stress resilience (Amjad et al., 2021). A recent approach includes an eco-friendly approach to synthesizing nanoparticles that are believed to be non-toxic (Jiang et al., 2022; Constantinescu and Sicuia, 2020). Green nanobiotechnology, especially phyto-synthesized nanoparticles, holds an ecological and economic edge over traditional methods. Despite these nanoparticles' ability to boost photosynthesis, seed germination, water absorption, phytohormonal responses, and antioxidant defense system, the risk of its impact on non-targeted organisms persists (Wahab et al., 2023). Therefore, further research can uncover these novel methods' potential advantages and limitations.

2. Plant Drought Stress: their Causes and Consequences

Drought is the most prominent environmental stressor generally caused by declining rainfall (Tannehill, 1947), salinity, extreme temperatures, high light intensity, flooding, dry wind, and increased evaporation or low precipitation rate (Vicente-Serrano et al., 2020). A drought may be defined as the normal recurring force of nature occurring in nearly all areas (Zandalinas et al., 2024). Drought is a natural phenomenon that occurs due to a prolonged lack of rainfall, especially in arid and semiarid regions (Iqbal and Munir, 2024). The decrease in soil moisture negatively affects the plant's ability to absorb minerals and transpire water for cooling effects, leading to a decline in overall yield (Qiao et al., 2024). The severity, duration, and timing of drought, as well as the plant's ability to recover, are critical factors in determining its impact on growth despite different definitions of drought (Abu Arra et al., 2024; Athukoralalage et al., 2024). Drought stress reduces water availability to the plant, which leads to inhibition of photosynthesis, altered respiration rate, reduced ATP production (Seleiman et al., 2021), biomass, and plant growth.

This disruption also affects hormone balance and protein synthesis, causes membrane damage, and increases ROS production. Which in turn causes cellular damage, further leading to metabolic changes and plant stress responses (Salehi-Lisar and Bakhshayeshan-Agdam, 2016). Drought severely affects energy production, food, water, and a range of socio-economic activities (Mishra and Singh, 2010; Mondal et al., 2023; Chiang et al., 2021). Numerous studies have shown that droughts are becoming more frequent, prolonged, and severe in the background of global warming. Elevated anthropogenic activities in recent and present eras have contributed to global warming and subsequently water scarcity (Agha Kouchak et al., 2021). The responsive mechanism of plants includes morpho-physiological, biochemical molecular, and cellular responses. These result in an improved root system, leaf structure, RWC, osmotic balance, and stomatal conductance. Along with Ca^{2+} and phytohormones, these are the key factors in coping with drought (Ilyas et al., 2021). Some microorganisms, genes, and transgenic approaches, along with exogenous methods, are also being considered to maintain a positive impact on reducing the effects of drought stress.

The plant's structure, leaf characteristics, vascular system, and flowering time significantly impact photosynthetic efficiency and crop productivity. Optimizing these features has the potential to increase biomass and yield (Mathan et al., 2016). From an agricultural point of view, the overall yield is dependent on yield which in turn is dependent on the production machinery of the plant. Stomatal closure induced by ABA signaling reduces photosynthesis, leading to decreased activity of key photosynthetic enzymes and increased ROS production (Abdelaal et al., 2021). Downregulation of the non-cyclic electron transport chain further disrupts ATP production and the overall photosynthesis process. However, limited soil water availability can lead to reduced photosynthesis, potentially resulting in a negative carbon balance unless plant growth and carbon use are simultaneously reduced to maintain equilibrium (Malik and Bouskill, 2022; Ilyas et al., 2021). Therefore, drought tolerance involves tightly regulated processes, including energy conservation and metabolic adjustments, to maintain plant function under drought stress.

The severity of drought stress directly correlates with the production of reactive oxygen species, leading to increased peroxidation of membrane lipids and degradation of nucleic acids and proteins (Challabathula et al., 2022). Chloroplasts, mitochondria, and peroxisomes serve as primary targets for these reactive oxygen species during drought stress (Farooq et al., 2009). Drought stress delays germination and inhibits photosynthesis, leading to abnormal seed establishment and increased susceptibility to pathogenic attacks. This stress also causes early senescence, low accumulation of secondary compounds, and low fruit production due to insufficient pollination and maturation failure. Drought stress significantly affects all stages of plant growth, from germination to fruit and seed development (Dietz et al., 2021). Seedling germination is prone to drought stress (Liu et al., 2020). Lack of water during the vegetative stage delays panicle development and maturation, directly reducing yield (Abid et al., 2016; Lilley and Fukai, 1994; Moulick et al., 2024; Seleiman et al., 2021; Singh et al., 2012). Drought furthermore takes part in reducing the quality of the seeds and the overall agricultural sustainability.

3. Mechanisms of Plant Resilience and Drought Tolerance

When faced with drought stress, plants adapt by reducing leaf size, developing deeper roots, closing stomata, increasing water efficiency, and producing osmolytes, organic acids, polyols, and antioxidants (Ali et al., 2022). Farooq et al. (2009) have reported the expression of drought-responsive genes such as aquaporins, LEA proteins, and dehydrins to aid in stress tolerance. In response to drought stress, plants exhibit noticeable morphological changes such as reduced height, leaf wilting, and decreased leaf size or number (Bashir et al., 2021). These alterations stem from reduced cell expansion and heightened senescence. Drought resistance in plants is a multifaceted strategy that combines various mechanisms to survive and reproduce despite drought stress (Martignago et al., 2020). Drought tolerance involves changes in physiological and metabolic processes (Hoover et al., 2021; Thomas, 2022), enabling plants to withstand drought stress even at low water potential through osmotic adjustments, solute accumulation, photosynthesis maintenance, antioxidant production, and photosynthesis adjustments.

In contrast, drought avoidance focuses on maintaining high water potential by enhancing water uptake via deep and extensive root systems and conserving water through small leaves and stomatal closure along leafroll and efficient transpiration (Pamungkas and Farid, 2022; Yang et al., 2021). Drought escape is a mechanism where plants complete their life cycle before the onset of drought, typically seen in species (Raymundo et al., 2021) with shorter life cycles and having early maturity, early flowering, and high photosynthesis. Together, these mechanisms—tolerance, avoidance, and escape—constitute the overall drought resistance strategy, ensuring plant survival and reproduction under drought conditions (Fang and Xiong, 2015; Khan et al., 2018; Pamungkas and Farid, 2022). Drought-resistant species develop thicker leaves with higher tissue density to minimize water loss, while their root systems become deeper and more intricate, thereby enhancing water absorption (Yang et al., 2021). These adaptations serve to improve drought tolerance and ensure survival in water-scarce environments.

Plants have evolved physiological adaptations to combat drought stress. These include stomatal closure to reduce water loss, osmotic adjustments to maintain cell turgor, and the production of antioxidants to counteract damage from reactive oxygen species. Additionally, plants may undergo reduced photosynthesis, develop deeper roots, and alter leaf morphology to alleviate water loss. Hormones like abscisic acid play a crucial role in regulating these responses, allowing plants to withstand prolonged drought conditions (Oguz et al., 2022). In the face of oxidative stress induced by reactive oxygen species, plants mobilize enzymatic (e.g., SOD, CAT, APX) and non-enzymatic (e.g., ascorbate, glutathione) antioxidants to modulate ROS levels (Bashir et al., 2021). This process serves to uphold cellular integrity, bolster stress tolerance, and fortify plant growth, thus fortifying their defense against environmental adversities.

In response to drought stress, plants employ two primary molecular pathways: the ABA-dependent and ABA-independent mechanisms. The ABA-dependent pathway relies on abscisic acid (ABA) to promote stomatal closure and activate stress-responsive genes like RD29B, while also engaging ROS and calcium signaling. The ABA-independent pathway involves intriguing transcription factors such as DREB, WRKY, and bZIP, which interact with drought-responsive genes independently of ABA (Oguz et al., 2022). Both pathways work synergistically to maintain osmotic balance, scavenge ROS, and facilitate the synthesis of protective molecules, ultimately enabling plants to adeptly survive drought-induced stress.

Nanoparticles (NPs) offer a promising solution to alleviate drought stress in plants. They enhance plant growth by stabilizing membranes, improving nutrient and water uptake, and protecting the photosynthetic system under drought (Yasmeen et al., 2023). NPs also boost water relations, stomatal conductance, and photosynthetic efficiency (Rasheed et al., 2022). Additionally, they promote the accumulation of osmolytes, antioxidants, and stress-related hormones, and activate stress-responsive genes, fortifying plant defenses against drought stress at the molecular level.

4. Innovations and Concepts in Green Nanotechnology

Green nanotechnology combines green chemistry with nano-engineering to create sustainable products, reducing environmental and health risks. It utilizes the 12 Green Chemistry Principles published in 1998, making nanotechnology a valuable tool for responsible innovation (Soni et al., 2022). In recent years, green technology has gained traction as a sustainable approach to producing nanoparticles. Researchers are exploring environmentally friendly methods to synthesize nanoparticles, aiming to minimize the use of hazardous materials and reduce the environmental impact. By utilizing various bio-sources such as plants, bacteria, fungi, yeasts, and algae, a diverse array of nanoparticles in different shapes and sizes can be prepared. This highlights the potential of green nanotechnology in nanoparticle synthesis (Alqarni et al., 2022). Metal nanoparticles synthesized through plant-based methods offer a more economical, user-friendly (Zaheer et al., 2023), and advanced alternatives to those manufactured using physical or chemical processes (Mustafa et al., 2023). This ground-breaking approach demonstrates the potential for sustainable and efficient production of high-quality nanoparticles, with significant implications for various industrial applications (Alagesan and Venugopal, 2019; Iravani, 2011). By harnessing the phytochemicals found in various plant components such as fruits, leaves, stems, and roots, this process yields green nanoparticles, presenting a promising avenue for sustainable and eco-friendly nanoparticle synthesis.

Commonly synthesized metallic nanoparticles involve using eco-friendly preparations harnessing harmless plant parts as stabilizing or reducing/capping agents in the synthesis of nanoparticles. The current research community focuses on using minimal materials to achieve maximum output without compromising health and the environment. Green nanotechnology aims to address these gaps and promote a green economy for a sustainable future. These green methods can provide sustainable solutions to our current demanding problems, such as efficiency, availability for later use, and reliability of energy-producing systems (Soni et al., 2022). Nano-sized

Metal Oxide Fertilizers (NMOFs), including zinc, copper, and iron nanoparticles, facilitate precise nutrient delivery and improve plant uptake efficiency (Zafar et al., 2024). They play a critical role in enhancing plant metabolism, supporting crop growth, and increasing resilience to drought by optimizing soil water retention (Thangavelu et al., 2024). Innovative strategies are essential to address the urgent issue of global hunger. One promising solution is the sustainable application of carbon-based or graphene-based nanomaterials (GNMs) (Zhang et al., 2025). These materials have significant potential to enhance crop resilience, optimize the use of agrochemicals, improve fertilizer retention, enable controlled release, and improve soil quality and antimicrobial activities for pest and disease control. It also aids in stresses like drought as it promotes water retention by forming a layer over the surface of a medium, plant drought resilience, carbon sequestration, enhanced photosynthesis, and bio-remedial sensors and monitoring agents. These characteristics can greatly increase plant productivity and contribute to food security (Chacko et al., 2021; Ahmad et al., 2020; Patel et al., 2020). Similar properties are exhibited in carbon nanotubes synthesized through environmentally friendly methods.

Polymeric nanoparticles stand out from other types due to their suitability for sensing technologies and biotechnological applications (Balusamy et al., 2023). These nanoparticles can act as a direct source of micronutrients and enhance the delivery of bioactive agrochemicals. Their adaptable physicochemical properties allow various agricultural applications. Promising results have been observed in seed priming and foliar applications

Along with these, other nanoparticles taking part in the agriculture sector like Fullerenes significantly boost plant growth, enhance pathogen resistance, enrich soil quality, and improve nutrient absorption. Nanodiamonds and layered double hydroxides are pivotal in optimizing the delivery of agrochemicals and nutrients. Carbon nano-horns facilitate the controlled release of fertilizers and pesticides. Additionally, carbon quantum dots have been shown to improve seed germination and plant growth (Ayanda et al., 2024; Baig et al., 2021). Further studies are needed to maximize the benefits of each type of NPs and address any potential health risks.

5. Green Nanotechnology in Enhancing Plant Tolerance to Drought

Leveraging nanotechnology to enhance agricultural inputs by precisely targeting distribution, regulating release, and improving solubility and adhesion, while minimizing damage, significantly boosts crop yield and stress tolerance. Direct application of nanoparticles to plants optimizes physiological, biochemical, and molecular systems under stress (Ullah et al., 2023). Specifically, nano-priming, and foliar exposure strengthen plant defense mechanisms by elevating the expression of genes, proteins, and metabolites associated with oxidative stress, thereby enhancing seed germination, seedling growth, and plant maturation (Raza et al., 2023).

Nutritional imbalances resulting from drought conditions considerably hinder the ability of plants to absorb, transport, and utilize essential nutrients, particularly nitrogen. This limitation not only stunts plant growth but also reduces overall productivity. Even when nutrient levels in the soil appear adequate, extreme drought can severely restrict the plants' capacity to take these nutrients up. As a result, efforts to apply additional nutrients often yield minimal benefits for plant growth under these adverse circumstances. This intricate relationship between water scarcity and nutrient availability highlights the challenges faced in achieving sustainable agriculture in regions prone to drought (Rouphael et al., 2012). Nutrient deficiency during drought significantly impacts plant health and productivity. Drought reduces nitrogen content in biomass, leading to changes in nitrite levels that alter root pH and trigger the release of abscisic acid, which affects stomatal regulation. Research shows that nitrogen supply during drought can enhance stress resilience in plants, highlighting the need for targeted nutrient management strategies like in the roots and aerial parts of *Vicia faba, Nicotiana tabacum*, and *Pisum sativum* (Gloser et al., 2020). Nano-fertilizers provide a pivotal solution to mitigating agricultural drought by ensuring the controlled and consistent delivery of nutrients, thus reducing nutrient leaching and enhancing soil nutrient retention

(Ain et al., 2024; Chaffai et al., 2024a; Chandrashekar et al., 2023; Rasheed et al., 2022). They improve water retention and facilitate better nutrient uptake, aiding plants in maintaining optimal hydration and nutrition. Additionally, nano-fertilizers bolster antioxidant defenses against reactive oxygen species produced during drought stress, support the maintenance of stomatal closure, and help balance phytohormones, further enhancing plant resilience under drought conditions.

The upregulation of nutrient uptake, photosynthetic efficacy, and antioxidant defense system using nano-fertilizers is evident in many studies such as (Ahmed et al., 2021; Hafez et al., 2021) and (Silva et al., 2022) the latter also explained the increase of phytohormones and osmolytes during the water-scarce condition for titanium and zinc-based nanoparticles. in another study, SiO_2 NPs significantly decreased the devastating impacts of water stress by increasing transpiration, photosynthesis, water uptake, and Relative Water Content (RWC) (Sutulienė et al., 2021; Djanaguiraman et al., 2018). The stomatal closure can also be controlled using metal-based nanoparticles like cerium oxide nano-fertilizers.

Nano-pesticides can now target specific parts of plants, minimizing excessive pesticide use, reducing environmental impact, and conserving water, crucial during drought stress. With engineered slow-release mechanisms, they cut down the need for frequent applications and protect active ingredients from degradation by UV rays and high temperatures. This technology also reduces toxicity to non-target organisms like beneficial insects and soil microorganisms, essential for crop and soil health. Nanotechnology enhances pesticide efficacy by improving their solubility and bioavailability, even during drought conditions. Different nanocarriers like that of nano chitosan, magnetic, LayeredDouble Hydroxides (LDHs), and zein nanoparticles are extensively used for study purposes (Pan et al., 2023). Nano-tin showing anti-UV activities can be harvested to formulate more sustainable pesticides (Jalali et al., 2020). α-Amylase-Responsive Avermectin (AVM) Carrier is a nano-pesticide that targets only effectively against *Plutella xylostella* larvae (a type of pest in the Brassica family) (Kaziem et al., 2018), minimizing the harm to beneficial organisms and provides a sustainable release thus reducing environmental hazards.

Nanoparticles play a vital role in enhancing drought tolerance. For instance, oxides of zinc and titanium can boost photosynthesis by increasing light absorption and decreasing photorespiration, thereby raising the levels of useful energy molecules needed to cope with drought stress (Rehman et al., 2024). Additionally, certain nanoparticles like cerium oxide act as ROS scavengers, safeguarding organelles from cellular damage. Furthermore, some nanoparticles trigger drought-responsive networks of genes, enhancing the plant's ability to withstand drought conditions.

6. Soil and nutrient management with green nanotechnology

Fertilizer consumption has increased globally in recent decades with soil nutrient loss (Chugh et al., 2021) due to its cumulative addition to enhance crop productivity (Lin et al., 2019; Savci, 2012; Verma et al., 2022) with loss in soil health and rising environmental issues.

Nano fertilizers could lessen the negative effects of synthetic fertilizers (Cicek and Nadaroglu, 2015; Rajput et al., 2021) in addition to the benefit of making up for nutritional shortages so that the plants may grow properly (Baloch et al., 2008; Bernal et al., 2007). By encapsulating nutrients in nanoparticles (NPs), which enable targeted and controlled release, nanotechnology might greatly improve the efficiency of nutrient delivery to crops (Kottegoda et al., 2011; Yadav et al., 2023). By reducing crop nutritional components, the soil nutrient deficit might endanger the soil profile (Khan et al., 2021; Zulfiqar et al, 2019). By reducing crop dietary components, the soil nutrient deficit might endanger the soil profile.

Accordingly, most plants utilize around half of the fertilizers provided for optimal results (Fellet et al., 2021; Mittal et al., 2020), negatively impacting the agroecosystem. However, excessive use of chemical fertilizers can harm the microflora and soil profile, upend underground food webs, and cause genetic mutations that alter natural ecosystems and biodiversity (Mandal, 2021; Solanki et al., 2015). Therefore, sustainable substitutes might be investigated to enhance fertilizer's functional applications in plants using phytoremediation (Adisa et al., 2019; Pradhan and Mailapalli, 2017).

Micronutrients and macronutrients control how well plants withstand hazardous conditions. By boosting productivity, stress tolerance, and disease resistance, NPs may boost the dietary requirements of plants (Verma et al., 2022; Zhao et al., 2020). Research has indicated that the utilization of NFs can lead to substantial enhancements in the efficiency of plant absorbance of nutrients. In one case, research found that applying ZnO NP-coated urea boosted wheat grain production by 51% when compared with the control group (Dimkpa et al., 2020). Similarly, it has been shown that zinc oxide nanoparticles (NPs) promote plant growth and grain zinc content by increasing the efficiency of zinc absorption in rice plants (Ali et al., 2019). Nanotechnology-based PA can help establish more particular and sustainable nutrition management strategies in addition to optimizing the efficiency of nutrient use (Khan and Fatima, 2024) For example, farmers may now adjust fertilizer application rates and timings based on real-time data provided by nano-sensors that have been designed to monitor soil nutrient levels. The nanoscale physicochemical characteristics of NFs offer a large surface area for nutrients and aid in the reduction of biotic and abiotic stresses (Zulfiqar et al., 2019) enhance photosynthesis (Janmohammadi et al., 2016), improve water usage efficiency (Rico et al., 2011), and mitigate abiotic stress.

The framework for innovative and sustainable nutrient delivery systems that make use of the nanoporous surfaces of plant components is therefore provided by this technology. Plant health and soil fertility are restored, environmental pollution and agroecological degradation are decreased, and the effectiveness of applied fertilizer is increased using encapsulated nanoparticles, nano-clays, and zeolites (Janmohammadi et al., 2016). Silica, iron, titanium dioxide, zinc oxide nanoparticles (ZnONPs), ZnS/ZnCdSe core shell quantum dots (QDs), InP/ZnS core-shell QDs, Mn/ZnSe QDs, gold nanorods, Al_2O_3, TiO_2, CeO_2, and FeO are some examples of the constituents of nano fertilizers (Pradhan and Mailapalli, 2017). Additionally, nanofertilizers might be designed to help plants that are lacking in certain nutrients. The ability to organize the atoms on the surfaces of nanomaterials to achieve distinctively varied characteristics makes this conceivable.

Li et al. (2016) have shown that metals and anionic nanoparticles are strongly absorbed by the soil or porous surfaces, making them excessively available as nutrients for food or even pollutants when unwanted. Additionally, some scientists have created and patented "Nano-Leucite Fertilizer," an environmentally friendly nanofertilizer that may decrease food nutrient loss and boost crop and food output overall (Kamran et al., 2016). In a nutshell, nanofertilizers can improve soil fertility in nutrient-deficient soil, making them perhaps the greatest development in the agricultural revolution. Nonetheless, it might be viewed as "an additional tool in the toolbox."

7. Improving plant and soil water retention using green nanotechnology

NPs treatments enhance water control, stomatal conductance, photosynthesis, and nitrogen absorption from NP-based fertilizers, they can help plants that are experiencing drought stress (Adrees et al., 2020; Aghdam et al., 2016). Studies demonstrate the benefits of carbon-based nanomaterials, especially multi-walled carbon nanotubes (MWNTs).

NPs improve growth metrics and biomass to increase plant drought resilience (Ali et al., 2020), as well as maximize the absorption of nutrients and water (Chandrashekar et al., 2023). Several studies have demonstrated that treating drought-stressed agricultural plants with nanoparticles may greatly enhance their vigor, morphological traits, and physio-biochemical parameters (Zaheer et al., 2023). Nanoparticles have been reported to bolster biological defenses by boosting the efficiency of antioxidants which include both enzymatic and non-enzymatic (Ali et al., 2021) such as catalase (CAT), glutathione (GSH), ascorbate peroxidase (APX), peroxidase (POD), and superoxide dismutase (SOD) (Djanaguiraman et al., 2018; Taran et al., 2017). As demonstrated by recent research, Reactive Oxygen Species (ROS) involving superoxide (O^{2-}) and hydrogen peroxide (H_2O_2) produce oxidative damage that nanoparticles (NPs) serve to ameliorate. Additionally, NPs may elevate prolines, glycine betaine, soluble sugars, amino acids, other osmolytes, and osmo protectants. Under drought stress, these substances aid promote osmotic adjustments (Mustafa et al.,

2021; Van Nguyen et al., 2022). Exposure to nanoparticles under drought stress increases signaling pathways for the development of phytohormones and osmo-protectants.

Upon NP treatment, drought-stressed plants showed higher amounts of proline, glycine betaine, total proteins, and soluble carbohydrates. In a study by (Ghani et al., 2022), application of zinc oxide nanoparticles (ZnO NPs) to drought-stressed cucumber plants preserves the amount of soluble sugar while increasing proline, glycine betaine, and amino acid levels (Mustafa et al., 2024). By controlling the expression of its biosynthetic genes, ZnO facilitates the effective production of proline, a crucial osmolyte for maintaining osmotic equilibrium. Iron-based nanoparticles, especially Fe_3O_4 (Mozafari et al., 2018), help plants develop and absorb iron better in their roots, which helps them withstand drought even at low concentrations.

Alidoust and Isoda (Alidoust and Isoda, 2013) discovered that after delivering citric acid-coated Fe_2O_3 NPs directly to Glycine max, Fe_3O_4 NPs boosted root length and photosynthesis (Dhoke et al., 2013). Furthermore, plantlets of Fragaria × ananassa exposed to Fe_3O_4 NPs exhibited enhanced drought tolerance relative with the varieties that were not treated. Oryza sativa plants treated with Fe3O4 NPs showed accelerated nutrient uptake, photosynthesis, biomass, and antioxidant activity during drought stress (Ahmed et al., 2021). Silver nanoparticles (Ag-NPs) in chickpeas improve nutritional absorption, antioxidant activity, and photosynthesis. Similarly, through boosting nutrient availability, sunlight absorption, and oxidative damage elimination, Ag-NPs extracted from *Carum copticum* leaves enhanced plant development (Danish et al., 2022; Mohamed et al., 2017; Zulfiqar et al., 2019). Copper nanoparticles made from extract via basil leaves have stimulated photosynthesis and antioxidant effectiveness in drought-stricken tomato plants, preventing oxidative damage (Jiang et al., 2021; Tripathi et al., 2022).

8. Case Studies: Green Nanotechnology in Crop Management

In agriculture, phyto-synthesized NPs are crucial because they increase the productivity of crops and drought resilience on their own (Zaheer et al., 2024). Their unique features, such as elevated surface-area-to-volume ratios and efficiency, stimulate better nutrient absorption and have a favorable impact on plant development. These nanoparticles can also interact with plants at the molecular level and pass through biological barriers (Ebrahimi et al., 2022; Pattanayak et al., 2022). According to research by Janmohammadi et al. (2016), wheat growth may be successfully enhanced by low concentrations of titanium dioxide nanoparticles (TiO_2NPs) in drought-prone environments. TiO_2NP treatment increased a variety of metrics, such as plant height, biomass, harvest index, total yield, ear weight, number of ears and seeds, and so on.

The use of green-produced silver NPs improved *Tephrosia apollinea's* antibacterial activity against Staphylococcus aureus and Escherichia coli under drought stress (Ali et al., 2019). When lentils were subjected to drought stress, the better response showed that Ag-synthesized green nanoparticles significantly increased seed germination (89.5%), rate of germination (6.88), and seedling biomass. The improvement related to the retention of tissue water balance. Dhoke et al. (2013) demonstrated that ZnONP application enhanced both root biomass and above-ground tissues, according to research on the effect of ZnONPs on *Vigna radiata* seedling development. Under drought stress, foliar application of ZnONPs has been shown to reduce transpiration rate, raise stomatal conductance, promote photosynthetic activity, increase chlorophyll content, and improve water usage efficiency (Sun et al., 2021). Additionally, they helped to stabilize the ultrastructure of the chloroplasts and mitochondria in *Zea mays* during water stress.

Dimkpa et al. (2020) showed that ZnONPs can improve nitrogen absorption under drought stress conditions, promote yield, enhance edible grains with essential elements like zinc, and accelerate the evolution of *Sorghum bicolor*. Foroutan et al. (2018) demonstrated that applying zinc oxide nanoparticles (ZnONPs) to *M. Peregrina* greatly increases its resistance to drought in water-deficient environments. The foliar treatment of ZnONPs largely reduces oxidative stress and elevates the activities of POD and superoxide dismutase (SOD) in the leaves as compared to the control group.

Some plant extracts, such as lemon peel, grape seeds, and green tea, effectively generate gold nanoparticles (Au-NPs). Throughout dry seasons, these plants may boost yield and growth, particularly when green tea extract is applied (Ying et al., 2022). Wahab et al. (2023) discovered that Au-NPs maximize plants' food supplies by improving nutrient uptake and photosynthetic pigments (Alam et al., 2013; Wahab et al., 2023). By modulating gene expression and stimulating pathways, Au-NPs optimize tissue growth, seed germination, and plant development (Khan et al., 2021). Increased drought resilience and improved average yield are more likely to be observed among crops treated with copper nanoparticles (Cu-NPs) (Imtiaz et al., 2023). This affordable and environmentally conscious green synthesis strategy has proven beneficial for growing plants, particularly in photosynthesis and under challenging conditions.

A local strain of Bacillus (RNT1) was applied to generate iron oxide nanoparticles (IONPs). Rice plants treated with IONP revealed enhanced biomass, enhanced activity of antioxidant enzymes, and improved photosynthesis (Zafar et al., 2024). Furthermore, compared to controlled plants, the treated plants exhibited reduced cadmium (Cd) translocation and lower reactive oxygen species levels under both normal and drought conditions (Ahmed et al., 2021). *Catharanthus roseus*'s oxidative stress caused by dehydration is well mitigated by chitosan nanoparticles (CSNPs) (Liu et al., 2020). They improve membrane integrity and lower hydrogen peroxide (H_2O_2) and malondialdehyde (MDA) levels by boosting non-enzymatic (proline) and enzymatic (catalase and ascorbate peroxidase) antioxidants and developing plant growth, relative water content, stomatal conductance, and chlorophyll content (Hossain et al., 2015). Applying cerium nanoparticles (CeNPs) (Mohammadi et al., 2021) through efficient scavenging of Reactive Oxygen Species (ROS) and reduced levels of hydrogen peroxide and lipid peroxidation, 50 ppm incorporated into drought-stressed wheat could boost relative water content, antioxidant enzyme activity, and photosynthetic parameters (Hossain et al., 2015).

9. Challenges and prospects

Scalability and reproducibility are the primary challenges of green nanobiotechnology (Saleh and Yousaf, 2018). The green synthesis of nanoparticles includes various methods, such as extracting materials from plants, fungi, or bacteria. Variations in nanoparticle characteristics, including size, morphology, and surface properties, can arise from different sources. These aspects are essential for effective applications (Raliya et al., 2013). Optimization of conditions such as pH, temperature, and extract concentration are crucial. Each biological material has distinct properties and reacts differently to synthesis conditions, necessitating a customized approach for each source (Ying et al., 2022). Producing nanoparticles on a large scale using biological methods is challenging. The potential lifetime effects of NMs on worker health must also be carefully considered while developing green nanotechnology (Schulte et al., 2013). Workers doing discovery research in laboratories that create, synthesize, and test NMs for several uses as well as assess their toxicological and environmental effects may be exposed to NMs for the first time.

A key limitation in green nanobiotechnology is the lack of understanding of the biological mechanisms involved in nanoparticle synthesis. This lack of understanding restricts the ability to enhance these processes for applications. Iravani et al. (2014) noted that although biosynthesis provides an eco-friendly alternative, much about the biochemical pathways and molecular interactions involved is still unknown, which impedes precise control and predictability.

Green synthesis offers an environment-friendly approach, yet we must acknowledge that there are still significant uncertainties about the long-term health and ecological impacts of nanoparticles. It is crucial to address these concerns to ensure a safe and sustainable future (Santhosh et al., 2022). This technology aims to be cost-effective by utilizing biological waste and natural resources; however, it faces economic challenges. The processing of biomolecules for green synthesis is often extensive, and the availability of biological materials can vary seasonally (Brar et al., 2022). The absence of regulatory standards for green nanobiotechnology limits commercialization, as few frameworks currently address the production, handling, and disposal of green-synthesized nanoparticles.

Biosynthesized nano fertilizers can enhance plant resilience to abiotic stress, but their widespread use raises concerns about potential ecological impacts. More research is needed to understand the harmful effects of residual nanoparticles on environmental systems, particularly their interactions with plants (Rai et al., 2023; Younas et al., 2022). Scientists can use advanced techniques like genomics, proteomics, and metabolomics to explore the molecular processes of enhancing plants with nanoparticles (NPs). With the increasing use of phyto-synthesized NPs in agriculture, it is crucial to evaluate their potential environmental impacts and ensure they are safe and sustainable for broader applications (Dappula et al., 2023). One of the major ecological hazards is the accumulation of phyto-synthesized nanoparticles (NPs) in the soil, which crucially undermines soil health and disrupts microbial communities (Noman et al., 2023; Prasad et al., 2021). Furthermore, the runoff of NPs into water bodies poses a serious threat, causing aquatic toxicity and significantly disturbing marine ecosystems.

10. Conclusion

Addressing abiotic stress effects on crop production is crucial for researchers. This chapter discusses how green nanotechnology improves plants' resilience to drought stress. Numerous research studies have shown that phyto-synthesized nanoparticles (NPs) can significantly enhance seed germination and seedling growth.

They help regulate water relations, promote photosynthesis, increase chlorophyll content, and activate antioxidant defense mechanisms in plants experiencing drought stress. Moreover, these nanoparticles can regulate plant hormones, providing a valuable defense against drought stress and enhancing agricultural productivity (Mustafa, 2016). However, it is important to consider the potential dangers and challenges that may arise from using phyto-synthesized nanoparticles in agriculture.

The toxicity of nanoparticles to non-target organisms and their effects on soil health must be considered. Understanding their environmental fate and movement is crucial to prevent adverse ecosystem impacts. Future research should enhance synthesis processes to improve nanoparticle stability and bioavailability for safe use in agriculture. Developing clear policies and regulations for the ethical use of phyto-synthesized nanoparticles (NPs) in agriculture is crucial. It is important to optimize the positive effects of these nanoparticles while minimizing their negative impacts on the environment, non-target species, and human health. Incorporating phyto-synthesized nanoparticles into agricultural practices, especially under drought-stressed conditions, holds significant promise for sustainable crop production. To fully realize the potential of these advanced nanomaterials for a resilient and secure food future, further research, risk analysis, and effective policy implementation are needed.

References

Abdelaal, K., AlKahtani, M., Attia, K., Hafez, Y., Király, L., and Künstler, A. (2021). The role of plant growth-promoting bacteria in alleviating the adverse effects of drought on plants. *Biology*, 10(6): 520.

Abid, M., Tian, Z., Ata-Ul-Karim, S. T., Liu, Y., Cui, Y., Zahoor, R. et al. (2016). Improved tolerance to post-anthesis drought stress by pre-drought priming at vegetative stages in drought-tolerant and-sensitive wheat cultivars. *Plant Physiology and Biochemistry*, 106: 218–227.

Abu Arra, A., Birpınar, M. E., Gazioğlu, Ş. A., and Şişman, E. (2024). Critical drought characteristics: A new concept based on dynamic time period scenarios. *Atmosphere*, 15(7): 768.

Adisa, I. O., Pullagurala, V. L. R., Peralta-Videa, J. R., Dimkpa, C. O., Elmer, W. H., Gardea-Torresdey, J. L. et al. (2019). Recent advances in nano-enabled fertilizers and pesticides: a critical review of mechanisms of action. *Environmental Science: Nano*, 6(7): 2002–2030.

Adrees, M., Khan, Z. S., Ali, S., Hafeez, M., Khalid, S., ur Rehman, M. Z. et al. (2020). Simultaneous mitigation of cadmium and drought stress in wheat by soil application of iron nanoparticles. *Chemosphere*, 238: 124681.

Agha Kouchak, A., Mirchi, A., Madani, K., Di Baldassarre, G., Nazemi, A., Alborzi, A. et al. (2021). Anthropogenic drought: Definition, challenges, and opportunities. In: *Wiley Online Library*.

Aghdam, M. T. B., Mohammadi, H., and Ghorbanpour, M. (2016). Effects of nanoparticulate anatase titanium dioxide on physiological and biochemical performance of Linum usitatissimum (Linaceae) under well-watered and drought stress conditions. *Brazilian Journal of Botany*, 39: 139–146.

Ahmad, J., Qamar, S., Kausar, N., and Qureshi, M. I. (2020). Nanoparticles: the magic bullets in mitigating drought stress in plants. *Nanobiotechnology in Agriculture: An Approach Towards Sustainability*, 145–161.

Ahmed, T., Noman, M., Manzoor, N., Shahid, M., Abdullah, M., Ali, L. et al. (2021). Nanoparticle-based amelioration of drought stress and cadmium toxicity in rice via triggering the stress responsive genetic mechanisms and nutrient acquisition. *Ecotoxicology and Environmental Safety*, 209: 111829.

Ain, Q. u., Hussain, H. A., Zhang, Q., Kamal, F., Charagh, S., Imran, A. et al. (2024). Deciphering the Role of Nanoparticles in Stimulating Drought and Salinity Tolerance in Plants: Recent Insights and Perspective. *Journal of Plant Growth Regulation*, 43(5): 1605–1630.

Alagesan, V., and Venugopal, S. (2019). Green synthesis of selenium nanoparticle using leaves extract of withania somnifera and its biological applications and photocatalytic activities. *Bionanoscience*, 9: 105–116.

Alam, M. N., Roy, N., Mandal, D., and Begum, N. A. (2013). Green chemistry for nanochemistry: exploring medicinal plants for the biogenic synthesis of metal NPs with fine-tuned properties. *Rsc Advances*, 3(30): 11935–11956.

Ali, M. A., Mosa, K. A., El-Keblawy, A., and Alawadhi, H. (2019). Exogenous production of silver nanoparticles by Tephrosia apollinea living plants under drought stress and their antimicrobial activities. *Nanomaterials*, 9(12): 1716.

Ali, M. H., Sobze, J.-M., Pham, T. H., Nadeem, M., Liu, C., Galagedara, L. et al. (2020). Carbon nanotubes improved the germination and vigor of plant species from peatland ecosystem via remodeling the membrane lipidome. *Nanomaterials*, 10(9): 1852.

Ali, S., Mehmood, A., and Khan, N. (2021). Uptake, translocation, and consequences of nanomaterials on plant growth and stress adaptation. *Journal of Nanomaterials*, 2021(1): 6677616.

Ali, S., Rizwan, M., Noureen, S., Anwar, S., Ali, B., Naveed, M. et al. (2019). Combined use of biochar and zinc oxide nanoparticle foliar spray improved the plant growth and decreased the cadmium accumulation in rice (*Oryza sativa* L.) plant. Environmental Science and Pollution Research, 26: 11288–11299.

Ali, S., Tyagi, A., Park, S., Mir, R. A., Mushtaq, M., Bhat, B. et al. (2022). Deciphering the plant microbiome to improve drought tolerance: mechanisms and perspectives. Environmental and Experimental Botany, 201: 104933.

Alidoust, D., and Isoda, A. (2013). Effect of $\gamma Fe2O3$ nanoparticles on photosynthetic characteristic of soybean (*Glycine max* (L.) Merr.): foliar spray versus soil amendment. *Acta Physiologiae Plantarum*, 35: 3365–3375.

Alqarni, L. S., Alghamdi, M. D., Alshahrani, A. A., and Nassar, A. M. (2022). Green nanotechnology: recent research on bioresource-based nanoparticle synthesis and applications. *Journal of Chemistry*, 2022.

Amjad, S., Tiwari, S., and Serajuddin, M. (2021). Applications of Nanobiotechnology in overcoming drought stress in crops. *Nanobiotechnology: Mitigation of Abiotic Stress in Plants*, 399–415.

Athukoralalage, D., Brookes, J., McDowell, R. W., and Mosley, L. M. (2024). Impact of hydrological drought occurrence, duration, and severity on Murray-Darling basin water quality. *Water Research*, 252: 121201.

Ayanda, O. S., Mmuoegbulam, A. O., Okezie, O., Durumin Iya, N. I., Mohammed, S. a. E., James, P. H. et al. (2024). Recent progress in carbon-based nanomaterials: critical review. Journal of Nanoparticle Research, 26(5): 1–42.

Baig, N., Kammakakam, I., and Falath, W. (2021). Nanomaterials: A review of synthesis methods, properties, recent progress, and challenges. *Materials Advances*, 2(6): 1821–1871.

Baloch, Q., Chachar, Q., and Tareen, M. (2008). Effect of foliar application of macro and micro-nutrients on production of green chilies (*Capsicum annuum* L.). *Journal of Agricultural Technology*, 4(2): 177–184.

Balusamy, S. R., Joshi, A. S., Perumalsamy, H., Mijakovic, I., and Singh, P. (2023). Advancing sustainable agriculture: a critical review of smart and eco-friendly nanomaterial applications. *Journal of Nanobiotechnology*, 21(1): 372.

Bashir, S. S., Hussain, A., Hussain, S. J., Wani, O. A., Zahid Nabi, S., Dar, N. A. et al. (2021). Plant drought stress tolerance: Understanding its physiological, biochemical and molecular mechanisms. *Biotechnology & Biotechnological Equipment*, 35(1): 1912–1925.

Bernal, M., Cases, R., Picorel, R., and Yruela, I. (2007). Foliar and root Cu supply affect differently Fe-and Zn-uptake and photosynthetic activity in soybean plants. *Environmental and Experimental Botany*, 60(2): 145–150.

Brar, K. K., Magdouli, S., Othmani, A., Ghanei, J., Narisetty, V., Sindhu, R. et al. (2022). Green route for recycling of low-cost waste resources for the biosynthesis of nanoparticles (NPs) and nanomaterials (NMs)-A review. *Environmental Research*, 207: 112202.

Chacko, A. R., Plathanam, N. J., Korah, B. K., Abraham, T., and Mathew, B. (2021). Green synthesized carbon-based nanomaterials: applications and future developments. In *Renewable Materials and Green Technology Products* (pp. 135–165): Apple Academic Press.

Chaffai, R., Ganesan, M., and Cherif, A. (2024a). Nanoparticles for Sustainable Agriculture and Abiotic Stress Resistance. In *Plant Adaptation to Abiotic Stress: From Signaling Pathways and Microbiomes to Molecular Mechanisms* (pp. 195–208): Springer.

Chaffai, R., Ganesan, M., and Cherif, A. (2024b). Plant response and tolerance to environmental stresses. In *Plant Adaptation to Abiotic Stress: From Signaling Pathways and Microbiomes to Molecular Mechanisms* (pp. 31–47): Springer.

Challabathula, D., Analin, B., Mohanan, A., and Bakka, K. (2022). Differential modulation of photosynthesis, ROS and antioxidant enzyme activities in stress-sensitive and-tolerant rice cultivars during salinity and drought upon restriction of COX and AOX pathways of mitochondrial oxidative electron transport. *Journal of Plant Physiology*, 268: 153583.

Chandrashekar, H. K., Singh, G., Kaniyassery, A., Thorat, S. A., Nayak, R., Murali, T. S. et al. (2023). Nanoparticle-mediated amelioration of drought stress in plants: a systematic review. *3 Biotech*, 13(10): 336.

Chiang, F., Mazdiyasni, O., and AghaKouchak, A. (2021). Evidence of anthropogenic impacts on global drought frequency, duration, and intensity. *Nature Communications*, 12(1): 2754.

Chugh, G., Siddique, K. H., and Solaiman, Z. M. (2021). Nanobiotechnology for agriculture: Smart technology for combating nutrient deficiencies with nanotoxicity challenges. *Sustainability*, 13(4): 1781.

Cicek, S., and Nadaroglu, H. (2015). The use of nanotechnology in the agriculture. *Advances in Nano Research*, 3(4): 207.

Constantinescu, F., and Sicuia, O. A. B. (2020). Phytonanotechnology and plant protection. In *Phytonanotechnology* (pp. 245–287): Elsevier.

Cotrina Cabello, G. G., Ruiz Rodriguez, A., Husnain Gondal, A., Areche, F. O., Flores, D. D. C., Astete, J. A. Q. et al. (2023). Plant adaptability to climate change and drought stress for crop growth and production. *CABI Reviews*.

Danish, M., Shahid, M., Zeyad, M. T., Bukhari, N. A., Al-Khattaf, F. S., Hatamleh, A. A. et al. (2022). Bacillus mojavensis, a metal-tolerant plant growth-promoting bacterium, improves growth, photosynthetic attributes, gas exchange parameters, and Alkalo-Polyphenol Contents in Silver Nanoparticle (Ag-NP)-Treated *Withania somnifera* L. (Ashwagandha). ACS omega, 7(16): 13878–13893.

Dappula, S. S., Kandrakonda, Y. R., Shaik, J. B., Mothukuru, S. L., Lebaka, V. R., Mannarapu, M. et al. (2023). Biosynthesis of zinc oxide nanoparticles using aqueous extract of Andrographis alata: Characterization, optimization and assessment of their antibacterial, antioxidant, antidiabetic and anti-Alzheimer's properties. *Journal of Molecular Structure*, 1273: 134264.

Dastborhan, S., Kalisz, A., Kordi, S., Lajayer, B. A., and Pessarakli, M. (2021). Morphological, physiological, and biochemical responses of plants to drought and oxidative stresses. In *Handbook of plant and crop physiology* (pp. 517–541): CRC Press.

Dhoke, S. K., Mahajan, P., Kamble, R., and Khanna, A. (2013). Effect of nanoparticles suspension on the growth of mung (*Vigna radiata*) seedlings by foliar spray method. *Nanotechnology Development*, 3(1): e1–e1.

Dietz, K. J., Zörb, C., and Geilfus, C. M. (2021). Drought and crop yield. *Plant Biology*, 23(6): 881–893.

Dimkpa, C. O., Andrews, J., Fugice, J., Singh, U., Bindraban, P. S., Elmer, W. H. et al. (2020). Facile coating of urea with low-dose ZnO nanoparticles promotes wheat performance and enhances Zn uptake under drought stress. *Frontiers in Plant Science*, 11: 168.

Djanaguiraman, M., Nair, R., Giraldo, J. P., and Prasad, P. V. V. (2018). Cerium oxide nanoparticles decrease drought-induced oxidative damage in sorghum leading to higher photosynthesis and grain yield. *ACS omega*, 3(10): 14406–14416.

Ebrahimi, A., Samari, F., Eftekhar, E., and Yousefinejad, S. (2022). Rapid and efficient colorimetric sensing of clindamycin and Fe^{3+} using controllable phyto-synthesized silver/silver chloride nanoparticles by *Syzygium cumini* fruit extract. *Journal of Analytical Science and Technology*, 13(1): 10.

Fang, Y., and Xiong, L. (2015). General mechanisms of drought response and their application in drought resistance improvement in plants. *Cellular and Molecular Life Sciences*, 72: 673–689.

FAO. (2024). Water Scarcity – One of the greatest challenges of our time. FAO Newsroom. Retrieved from https://www.fao.org/newsroom/story/Water-Scarcity-One-of-the-greatest-challenges-of-our-time/en.

Farooq, M., Wahid, A., Kobayashi, N., Fujita, D., and Basra, S. M. (2009). Plant drought stress: effects, mechanisms and management. Sustainable agriculture, 153–188.

Fellet, G., Pilotto, L., Marchiol, L., and Braidot, E. (2021). Tools for nano-enabled agriculture: Fertilizers based on calcium phosphate, silicon, and chitosan nanostructures. *Agronomy*, 11(6): 1239.

Foroutan, L., Solouki, M., Abdossi, V., and Fakheri, B. A. (2018). The effects of zinc oxide nanoparticles on enzymatic and osmoprotectant alternations in different Moringa peregrina populations under drought stress. International *Journal of Basic Science in Medicine*, 3(4): 178–187.

Ghani, M. I., Saleem, S., Rather, S. A., Rehmani, M. S., Alamri, S., Rajput, V. D. et al. (2022). Foliar application of zinc oxide nanoparticles: An effective strategy to mitigate drought stress in cucumber seedling by modulating antioxidant defense system and osmolytes accumulation. *Chemosphere*, 289: 133202.

Gloser, V., Dvorackova, M., Mota, D. H., Petrovic, B., Gonzalez, P., and Geilfus, C. M. (2020). Early changes in nitrate uptake and assimilation under drought in relation to transpiration. *Frontiers in Plant Science*, 11: 602065.

Gondal, A. H., and Tayyiba, L. (2022). Prospects of using nanotechnology in agricultural growth, environment and industrial food products. *Reviews in Agricultural Science*, 10: 68–81.

Grange, G., Brophy, C., Vishwakarma, R., and Finn, J. A. (2024). Effects of experimental drought and plant diversity on multifunctionality of a model system for crop rotation. *Scientific Reports*, 14(1): 10265.

Hafez, E. M., Osman, H. S., Gowayed, S. M., Okasha, S. A., Omara, A. E.-D., Sami, R. et al. (2021). Minimizing the adversely impacts of water deficit and soil salinity on maize growth and productivity in response to the application of plant growth-promoting rhizobacteria and silica nanoparticles. *Agronomy*, 11(4): 676.

Haris, M., Hussain, T., Mohamed, H. I., Khan, A., Ansari, M. S., Tauseef, A. et al. (2023). Nanotechnology–A new frontier of nano-farming in agricultural and food production and its development. *Science of the Total Environment*, 857: 159639.

Hasanuzzaman, M., Bhuyan, M. B., Parvin, K., Bhuiyan, T. F., Anee, T. I., Nahar, K. et al. (2020). Regulation of ROS metabolism in plants under environmental stress: A review of recent experimental evidence. *International Journal of Molecular Sciences*, 21(22): 8695.

Hoover, D. L., Pfennigwerth, A. A., and Duniway, M. C. (2021). Drought resistance and resilience: The role of soil moisture–plant interactions and legacies in a dryland ecosystem. *Journal of Ecology*, 109(9): 3280–3294.

Hossain, M. A., Bhattacharjee, S., Armin, S. M., Qian, P., Xin, W., Li, H. Y. et al. (2015). Hydrogen peroxide priming modulates abiotic oxidative stress tolerance: insights from ROS detoxification and scavenging. *Frontiers in Plant Science*, 6: 420.

Iavicoli, I., Leso, V., Ricciardi, W., Hodson, L. L., and Hoover, M. D. (2014). Opportunities and challenges of nanotechnology in the green economy. Environmental Health, 13: 1–11.

Ijaz, M., Khan, F., Ahmed, T., Noman, M., Zulfiqar, F., Rizwan, M. et al. (2023). Nanobiotechnology to advance stress resilience in plants: Current opportunities and challenges. *Materials Today Bio*, 100759.

Ilyas, M., Nisar, M., Khan, N., Hazrat, A., Khan, A. H., Hayat, K. et al. (2021). Drought tolerance strategies in plants: a mechanistic approach. *Journal of Plant Growth Regulation*, 40: 926–944.

Imtiaz, H., Shiraz, M., Mir, A. R., Siddiqui, H., and Hayat, S. (2023). Nano-priming techniques for plant physio-biochemistry and stress tolerance. *Journal of Plant Growth Regulation*, 42(11): 6870–6890.

Iqbal, Z., and Munir, M. (2024). Multifaceted natural drought response mechanisms in three elite date palm cultivars uncovered by expressed sequence tags analysis. *Scientific Reports*, 14(1): 23186.

Iravani, S. (2011). Green synthesis of metal nanoparticles using plants. *Green Chemistry*, 13(10): 2638–2650.

Jalali, E., Maghsoudi, S., and Noroozian, E. (2020). A novel method for biosynthesis of different polymorphs of TiO2 nanoparticles as a protector for Bacillus thuringiensis from Ultra Violet. *Scientific Reports*, 10(1): 426.

Janmohammadi, M., Amanzadeh, T., Sabaghnia, N., and Dashti, S. (2016). Impact of foliar application of nano micronutrient fertilizers and titanium dioxide nanoparticles on the growth and yield components of barley under supplemental irrigation. *Acta Agriculturae Slovenica*, 107(2): 265–276.

Janni, M., Maestri, E., Gullì, M., Marmiroli, M., and Marmiroli, N. (2024). Plant responses to climate change, how global warming may impact on food security: a critical review. *Frontiers in Plant Science*, 14: 1297569.

Jiang, M., Song, Y., Kanwar, M. K., Ahammed, G. J., Shao, S., and Zhou, J. (2021). Phytonanotechnology applications in modern agriculture. Journal of Nanobiotechnology, 19: 1–20.

Jiang, Y., Zhou, P., Zhang, P., Adeel, M., Shakoor, N., Li, Y. et al. (2022). Green synthesis of metal-based nanoparticles for sustainable agriculture. *Environmental Pollution*, 309: 119755.

Kamran, A., Haroon, Z., Muhammad, Z., Imdad, H., and Zeeshan, A. (2016). Nano-zinc oxide as a future fertilizer. *Weekly Technology Times*, 27.

Kaziem, A. E., Gao, Y., Zhang, Y., Qin, X., Xiao, Y., Zhang, Y. et al. (2018). α-Amylase triggered carriers based on cyclodextrin anchored hollow mesoporous silica for enhancing insecticidal activity of avermectin against Plutella xylostella. *Journal of Hazardous Materials*, 359: 213–221.

Khan, A., Pan, X., Najeeb, U., Tan, D. K. Y., Fahad, S., Zahoor, R. et al. (2018). Coping with drought: stress and adaptive mechanisms, and management through cultural and molecular alternatives in cotton as vital constituents for plant stress resilience and fitness. *Biological Research*, 51.

Khan, A., and Fatima, F. (2024). Agro-Nanotechnology: A Way Towards Sustainable Agriculture. *In Recent Advancements in Multidimensional Applications of Nanotechnology*: Volume 1 (pp. 104–123). Bentham Science Publishers.

Khan, M. K., Pandey, A., Hamurcu, M., Gezgin, S., Athar, T., Rajput, V. D. et al. (2021). Insight into the prospects for nanotechnology in wheat biofortification. *Biology*, 10(11): 1123.

Kottegoda, N., Munaweera, I., Madusanka, N., and Karunaratne, V. (2011). A green slow-release fertilizer composition based on urea-modified hydroxyapatite nanoparticles encapsulated wood. *Current Science*, 73–78.

Li, H., Shan, C., Zhang, Y., Cai, J., Zhang, W., and Pan, B. (2016). Arsenate adsorption by hydrous ferric oxide nanoparticles embedded in cross-linked anion exchanger: effect of the host pore structure. *ACS Applied Materials and Interfaces*, 8(5): 3012–3020.

Li, Y., Narayanan, M., Shi, X., Chen, X., Li, Z., and Ma, Y. (2024). Biofilms formation in plant growth-promoting bacteria for alleviating agro-environmental stress. *Science of the Total Environment*, 907: 167774.

Lilley, J., and Fukai, S. (1994). Effect of timing and severity of water deficit on four diverse rice cultivars I. Rooting pattern and soil water extraction. *Field Crops Research*, 37(3): 205–213.

Lin, W., Lin, M., Zhou, H., Wu, H., Li, Z., and Lin, W. (2019). The effects of chemical and organic fertilizer usage on rhizosphere soil in tea orchards. *PloS one*, 14(5): e0217018.

Liu, Z., Li, H., Gou, Z., Zhang, Y., Wang, X., Ren, H. et al. (2020). Genome-wide association study of soybean seed germination under drought stress. *Molecular Genetics and Genomics*, 295: 661–673.

Malik, A. A., and Bouskill, N. J. (2022). Drought impacts on microbial trait distribution and feedback to soil carbon cycling. *Functional Ecology*, 36(6): 1442–1456.

Mandal, D. (2021). Nanofertilizer and its application in horticulture. *Journal of Applied Horticulture*, 23(1).

Martignago, D., Rico-Medina, A., Blasco-Escámez, D., Fontanet-Manzaneque, J. B., and Caño-Delgado, A. I. (2020). Drought resistance by engineering plant tissue-specific responses. *Frontiers in Plant Science*, 10: 1676.

Mathan, J., Bhattacharya, J., and Ranjan, A. (2016). Enhancing crop yield by optimizing plant developmental features. *Development*, 143(18): 3283–3294.

Md Gulzar, A. B., and Mazumder, P. B. (2024). Seed priming and subsequent inoculation of drought-tolerant plant growth-promoting rhizobacteria Bacillus tropicus KH90 improves rice (*Oryza sativa* L.) plant health under drought and non-drought conditions. *Cereal Research Communications*, 1–21.

Melelli, A., Durand, S., Alvarado, C., Kervoëlen, A., Foucat, L., Grégoire, M. et al. (2022). Anticipating global warming effects: A comprehensive study of drought impact of both flax plants and fibres. 184: 115011.

Mishra, A. K., and Singh, V. P. (2010). A review of drought concepts. *Journal of Hydrology*, 391(1-2): 202–216.

Mishra, P., Singh, N., Jain, A., Jain, N., Mishra, V., Pushplatha, G. et al. (2018). Identification of cis-regulatory elements associated with salinity and drought stress tolerance in rice from co-expressed gene interaction networks. *Bioinformation*, 14(3): 123.

Mittal, D., Kaur, G., Singh, P., Yadav, K., and Ali, S. A. (2020). Nanoparticle-based sustainable agriculture and food science: Recent advances and future outlook. *Frontiers in Nanotechnology*, 2: 579954.

Mohamed, A. K. S., Qayyum, M. F., Abdel-Hadi, A. M., Rehman, R. A., Ali, S., and Rizwan, M. (2017). Interactive effect of salinity and silver nanoparticles on photosynthetic and biochemical parameters of wheat. *Archives of Agronomy and Soil Science*, 63(12): 1736–1747.

Mohammadi, M. H. Z., Panahirad, S., Navai, A., Bahrami, M. K., Kulak, M., and Gohari, G. (2021). Cerium oxide nanoparticles (CeO2-NPs) improve growth parameters and antioxidant defense system in Moldavian Balm (*Dracocephalum moldavica* L.) under salinity stress. Plant Stress, 1: 100006.

Mondal, S., K. Mishra, A., Leung, R., and Cook, B. (2023). Global droughts connected by linkages between drought hubs. *Nature communications*, 14(1): 144.

Moulick, D., Choudhury, S., Brestic, M., and Hossain, A. (2024). The impact of abiotic stresses on agriculture: mitigation through climate smart strategies. In (Vol. 15, pp. 1391051): *Frontiers Media SA*.

Mozafari, A. a., Havas, F., and Ghaderi, N. (2018). Application of iron nanoparticles and salicylic acid in in vitro culture of strawberries (*Fragaria× ananassa* Duch.) to cope with drought stress. *Plant Cell, Tissue and Organ Culture* (PCTOC), 132: 511–523.

Mustafa, G., Miyagawa, S., Hasan, M. et al. (2023). Bio-Synthesized Nanoflowers And Chemically Synthesized Nanowires Zinc-Oxide Induced Changes In The Redox And Protein Folding In Soybean Seedlings: A proteomic analysis. *J. Plant Growth Regul.*, 42: 2570–2584. https://doi.org/10.1007/s00344-022-10728-9.

Mustafa, H., Ilyas, N., Akhtar, N., Raja, N. I., Zainab, T., Shah, T. et al. (2021). Biosynthesis and characterization of titanium dioxide nanoparticles and its effects along with calcium phosphate on physicochemical attributes of wheat under drought stress. *Ecotoxicology and Environmental Safety*, 223: 112519.

Mustafa, G. (2016). Proteomic Analysis of Nanoparticles-Mediated Soybean Response under Flooding Stress.

Mustafa, G., Chaudhari, S. K., Manzoor, M., Batool, S., Hatami, M., and Hasan, M. (2024). Zinc oxide nanoparticles mediated salinity stress mitigation

Noman, M., Ahmed, T., Ijaz, U., Hameed, A., Shahid, M., Azizullah et al. (2023). Microbe-oriented nanoparticles as phytomedicines for plant health management: an emerging paradigm to achieve global food security. *Critical Reviews in Food Science and Nutrition*, 63(25): 7489–7509.

Oguz, M. C., Aycan, M., Oguz, E., Poyraz, I., and Yildiz, M. (2022). Drought stress tolerance in plants: Interplay of molecular, biochemical and physiological responses in important development stages. *Physiologia*, 2(4): 180–197.

Pamungkas, S. S. T., and Farid, N. (2022). Drought stress: responses and mechanism in plants. *Reviews in Agricultural Science*, 10: 168–185.

Pan, X., Guo, X., Zhai, T., Zhang, D., Rao, W., Cao, F. et al. (2023). Nanobiopesticides in sustainable agriculture: developments, challenges, and perspectives. *Environmental Science: Nano*, 10(1): 41–61.

Patel, D. K., Kim, H.-B., Dutta, S. D., Ganguly, K., and Lim, K.-T. (2020). Carbon nanotubes-based nanomaterials and their agricultural and biotechnological applications. *Materials*, 13(7): 1679.

Pattanayak, D. S., Pal, D., Thakur, C., Raut, P., and Wasewar, K. (2022). Catalytic potential of phyto-synthesized silver nanoparticles for the degradation of pollutants. *Sustainable Engineering, Energy, and the Environment*, 465–481.

Pradhan, S., and Mailapalli, D. R. (2017). Interaction of engineered nanoparticles with the agri-environment. *Journal of Agricultural and Food Chemistry*, 65(38): 8279–8294.

Pramanik, P., Krishnan, P., Maity, A., Mridha, N., Mukherjee, A., and Rai, V. (2020). Application of nanotechnology in agriculture. *Environmental Nanotechnology Volume* 4: 317–348.

Prasad, A. R., Williams, L., Garvasis, J., Shamsheera, K., Basheer, S. M., Kuruvilla, M. et al. (2021). Applications of phytogenic ZnO nanoparticles: A review on recent advancements. *Journal of Molecular Liquids*, 331: 115805.

Qiao, M., Hong, C., Jiao, Y., Hou, S., and Gao, H. (2024). Impacts of drought on photosynthesis in major food crops and the related mechanisms of plant responses to drought. *Plants*, 13(13): 1808.

Qingying Zhan, Ashfaq Ahmad, Huma Arshad, Bingxian Yang, Sunbal Khalil Chaudhari, Sana Batool et al. (2024). The role of reduced graphene oxide on mitigation of lead phytotoxicity in *Triticum aestivum* L. plants at morphological and physiological levels. *Plant Physiology and Biochemistry*, 211: 108719.

Qureshi, A., Singh, D., and Dwivedi, S. (2018). Nano-fertilizers: a novel way for enhancing nutrient use efficiency and crop productivity. *International Journal of Current Microbiology and Applied Sciences*, 7(2): 3325–3335.

Rai, M., Shende, S. S., and Paralikar, P. (2023). *In vivo* antifungal efficacy of copper and sulfur nanoparticles on soft rot of ginger (*Zingiber officinale Rosc.*). *BioNanoScience*, 13(4): 1988–2005.

Rajamuthuramalingam, T., da Silva, W., Zuverza-Mena, N., Dimkpa, C., and White, J. C. (2024). Nano-sized metal oxide fertilizers for sustainable agriculture: balancing benefits, risks, and risk management strategies. *Nanoscale*.

Rajput, V. D., Singh, A., Minkina, T., Rawat, S., Mandzhieva, S., Sushkova, S. et al. (2021). Nano-enabled products: challenges and opportunities for sustainable agriculture. *Plants*, 10(12): 2727.

Raliya, R., Saran, R., Choudhary, K., and Tarafdar, J. (2013). Biosynthesis and characterization of nanoparticles. *Current Trends in Advancement of Scientific Research and Opinion in Applied Research and Opinion in Applied Microbiology and Biotechnology*, 6.

Rasheed, A., Li, H., Tahir, M. M., Mahmood, A., Nawaz, M., Shah, A. N. et al. (2022). The role of nanoparticles in plant biochemical, physiological, and molecular responses under drought stress: A review. *Frontiers in Plant Science*, 13: 976179.

Rassam, G. (2020). Evaluation of water deficit on morphological and physiological traits of hyssop (*Hyssopus officinalis* L.). *Agricultural Knowledge*, 5(10): 1–12.

Raymundo, R., Sexton-Bowser, S., Ciampitti, I. A., and Morris, G. P. (2021). Crop modeling defines opportunities and challenges for drought escape, water capture, and yield increase using chilling-tolerant sorghum. *Plant Direct*, 5(9): e349.

Raza, A., Charagh, S., Salehi, H., Abbas, S., Saeed, F., Poinern, G. E. et al. (2023). Nano-enabled stress-smart agriculture: Can nanotechnology deliver drought and salinity-smart crops? *Journal of Sustainable Agriculture and Environment*, 2(3): 189–214.

Rehman, A., Khan, S., Sun, F., Peng, Z., Feng, K., Wang, N. et al. (2024). Exploring the nano-wonders: unveiling the role of Nanoparticles in enhancing salinity and drought tolerance in plants. *Frontiers in Plant Science*, 14: 1324176.

Rico, C. M., Majumdar, S., Duarte-Gardea, M., Peralta-Videa, J. R., and Gardea-Torresdey, J. L. (2011). Interaction of nanoparticles with edible plants and their possible implications in the food chain. *Journal of Agricultural and Food Chemistry*, 59(8): 3485–3498.

Rouphael, Y., Cardarelli, M., Schwarz, D., Franken, P., and Colla, G. (2012). Effects of drought on nutrient uptake and assimilation in vegetable crops. *Plant Responses to Drought Stress: From Morphological to Molecular Features*, 171–195.

Saleh, N., and Yousaf, Z. (2018). Tools and techniques for the optimized synthesis, reproducibility and scale up of desired nanoparticles from plant derived material and their role in pharmaceutical properties. In *Nanoscale Fabrication, Optimization, Scale-Up and Biological Aspects of Pharmaceutical Nanotechnology* (pp. 85–131): Elsevier.

Salehi-Lisar, S. Y., and Bakhshayeshan-Agdam, H. (2016). Drought stress in plants: causes, consequences, and tolerance. *Drought stress tolerance in plants, Vol 1: Physiology And Biochemistry*, 1–16.

Santhosh, P. B., Genova, J., and Chamati, H. (2022). Green synthesis of gold nanoparticles: An eco-friendly approach. *Chemistry*, 4(2): 345–369.

Savci, S. (2012). An agricultural pollutant: chemical fertilizer. *International Journal of Environmental Science and Development*, 3(1): 73.

Schulte, P. A., Geraci, C. L., Hodson, L., Zumwalde, R., Kuempel, E., Murashov, V. et al. (2013). Overview of risk management for engineered nanomaterials. *Paper presented at the Journal of Physics: Conference Series.*

Seleiman, M. F., Al-Suhaibani, N., Ali, N., Akmal, M., Alotaibi, M., Refay, Y. et al. (2021). Drought stress impacts on plants and different approaches to alleviate its adverse effects. *Plants*, 10(2): 259.

Siddique, K. H., and Helen, B. (2020). Water deficits: development. In *Fresh Water and Watersheds* (pp. 215–219): CRC Press.

Silva, S., Dias, M. C., and Silva, A. M. (2022). Titanium and zinc based nanomaterials in agriculture: A promising approach to deal with (a) biotic stresses? *Toxics*, 10(4): 172.

Singh, C. M., Binod, K., Suhel, M., and Kunj, C. (2012). Effect of drought stress in rice: a review on morphological and physiological characteristics. Trends in Biosciences, 5(4): 261–265.

Solanki, P., Bhargava, A., Chhipa, H., Jain, N., and Panwar, J. (2015). Nano-fertilizers and their smart delivery system. *Nanotechnologies in Food and Agriculture*, 81–101.

Soni, R. A., Rizwan, M. A., and Singh, S. (2022). Opportunities and potential of green chemistry in nanotechnology. *Nanotechnology for Environmental Engineering*, 7(3): 661–673.

Sun, L., Song, F., Zhu, X., Liu, S., Liu, F., Wang, Y., and Li, X. (2021). Nano-ZnO alleviates drought stress via modulating the plant water use and carbohydrate metabolism in maize. *Archives of Agronomy and Soil Science*, 67(2): 245–259.

Sutulienė, R., Ragelienė, L., Samuolienė, G., Brazaitytė, A., Urbutis, M., and Miliauskienė, J. (2021). The response of antioxidant system of drought-stressed green pea (*Pisum sativum* L.) affected by watering and foliar spray with silica nanoparticles. *Horticulturae*, 8(1): 35.

Tannehill, I. R. (1947). *Drought, its causes and effects* (Vol. 64): LWW.

Taran, N., Storozhenko, V., Svietlova, N., Batsmanova, L., Shvartau, V., and Kovalenko, M. (2017). Effect of zinc and copper nanoparticles on drought resistance of wheat seedlings. *Nanoscale Research Letters*, 12: 1–6.

Thangavelu, R. M., da Silva, W. L., Zuverza-Mena, N., Dimkpa, C. O., and White, J. C. (2024). Nano-sized metal oxide fertilizers for sustainable agriculture: balancing benefits, risks, and risk management strategies. *Nanoscale*, 16(43): 19998–20026.

Thomas, H. (2022). Drought resistance in plants. In *Mechanisms of environmental stress resistance in plants* (pp. 1–42): Routledge.

Tian, T., Wang, S., Yang, S., Yang, Z., Liu, S., Wang, Y. et al. (2023). Genome assembly and genetic dissection of a prominent drought-resistant maize germplasm. *Nature Genetics*, 55(3): 496–506.

Tripathi, D., Singh, M., and Pandey-Rai, S. (2022). Crosstalk of nanoparticles and phytohormones regulate plant growth and metabolism under abiotic and biotic stress. *Plant Stress*, 6: 100107.

Ullah, H., Khan, I., Mustafa, G. et al. (2023). Molecular characterization and mitigative role of silicon dioxide nanoparticles in *Ocimum Basilicum* Under Lead (Pb) Stress. *Silicon* 15: 2551–2565. https://doi.org/10.1007/s12633-022-02178-5.

Van Nguyen, D., Nguyen, H. M., Le, N. T., Nguyen, K. H., Nguyen, H. T., Le, H. M. et al. (2022). Copper nanoparticle application enhances plant growth and grain yield in maize under drought stress conditions. *Journal of Plant Growth Regulation*, 41(1): 364–375.

Verma, K. K., Song, X.-P., Joshi, A., Tian, D.-D., Rajput, V. D., Singh, M. et al. (2022). Recent trends in nano-fertilizers for sustainable agriculture under climate change for global food security. *Nanomaterials*, 12(1): 173.

Vicente-Serrano, S. M., Quiring, S. M., Peña-Gallardo, M., Yuan, S., and Domínguez-Castro, F. (2020). A review of environmental droughts: Increased risk under global warming? *Earth-Science Reviews*, 201: 102953.

Wahab, A., Batool, F., Muhammad, M., Zaman, W., Mikhlef, R. M., and Naeem, M. (2023). Current knowledge, research progress, and future prospects of phyto-synthesized nanoparticles interactions with food crops under induced drought stress. *Sustainability*, 15(20): 14792.

Wahab, A., Munir, A., Saleem, M. H., AbdulRaheem, M. I., Aziz, H., Mfarrej, M. F. B. et al. (2023). Interactions of metal-based engineered nanoparticles with plants: an overview of the state of current knowledge, research progress, and prospects. *Journal of Plant Growth Regulation*, 42(9): 5396–5416.

Wang, Y., Deng, C., Rawat, S., Cota-Ruiz, K., Medina-Velo, I., and Gardea-Torresdey, J. L. (2021). Evaluation of the effects of nanomaterials on rice (*Oryza sativa* L.) responses: Underlining the benefits of nanotechnology for agricultural applications. *ACS Agricultural Science and Technology*, 1(2): 44–54.

Yadav, A., Yadav, K., Ahmad, R., and Abd-Elsalam, K. A. (2023). Emerging frontiers in nanotechnology for precision agriculture: advancements, hurdles and prospects. *Agrochemicals*, 2(2): 220–256.

Yadav, N., Garg, V. K., Chhillar, A. K., and Rana, J. S. (2023). Recent advances in nanotechnology for the improvement of conventional agricultural systems: A review. *Plant Nano Biology*, 4: 100032.

Yang, X., Lu, M., Wang, Y., Wang, Y., Liu, Z., and Chen, S. (2021). Response mechanism of plants to drought stress. *Horticulturae*, 7(3): 50.

Yang, Z., Cao, Y., Shi, Y., Qin, F., Jiang, C., and Yang, S. (2023). Genetic and molecular exploration of maize environmental stress resilience: Toward sustainable agriculture. *Molecular Plant*, 16(10): 1496–1517.

Yasmeen, F., Mustafa, G., Jhanzab, H. M. and Komatsu, S. (2023). Current scenario and future trends of plant nano-interaction to mitigate abiotic stresses: a review. *In*: Sahu, S.C. (Ed.). *Impact of Engineered Nanomaterials in Genomics and Epigenomics*. https://doi.org/10.1002/9781119896258.ch16.

Ying, S., Guan, Z., Ofoegbu, P. C., Clubb, P., Rico, C., He, F., and Hong, J. (2022). Green synthesis of nanoparticles: Current developments and limitations. *Environmental Technology and Innovation*, 26: 102336.

Younas, Z., Naseer, S., Kazmi, A., Ali, A., Wahab, A., Sultana, T. et al. (2022). Assessment of diversity among important Brinjal (*Solanum melongena*) cultivars using morphological markers. *Journal of Food Quality*, 2022(1): 4255554.

Zafar, S., Farooq, A., Batool, S., Tariq, T., Hasan, M., and Mustafa, G. (2024). Green synthesis of iron oxide nanoparticles for mitigation of chromium stress and anti-oxidative potential in *Triticum aestivum. Hybrid Advances*, 5: 100156.

Zaheer, S., Shehzad, J., Chaudhari, S. K., Hasan, M., and Mustafa, G. (2023). Morphological and biochemical responses of Vigna radiata L. seedlings towards green synthesized SiO2 NPs. *Silicon,* 15(14): 5925–5936.

Zaheer, S., Shehzad, J., Chaudhari, S. K., and Mustafa, G. (2024). The Nanotechnology Application to Accelerate Plant Adaptation to Abiotic Stresses. *Advanced Nanotechnology in Plants: Methods and Applications*, 13.

Zandalinas, S. I., Fritschi, F. B., and Mittler, R. (2021). Global warming, climate change, and environmental pollution: recipe for a multifactorial stress combination disaster. *Trends in Plant Science*, 26(6): 588–599.

Zandalinas, S. I., Peláez-Vico, M. Á., Sinha, R., Pascual, L. S., and Mittler, R. (2024). The impact of multifactorial stress combination on plants, crops, and ecosystems: how should we prepare for what comes next? *The Plant Journal*, 117(6): 1800–1814.

Zhang, C., Zheng, Y., Lin, X., and Weng, S. (2025). Research progress on the preparation, performance, mechanisms, and applications of carbon nanomaterials against Gram-negative bacteria. *RSC Advances*, 15(28): 22180–22201.

Zhao, L., Lu, L., Wang, A., Zhang, H., Huang, M., Wu, H. et al. (2020). Nano-biotechnology in agriculture: use of nanomaterials to promote plant growth and stress tolerance. *Journal of Agricultural and Food Chemistry*, 68(7): 1935–1947.

Zulfiqar, F., Navarro, M., Ashraf, M., Akram, N. A., and Munné-Bosch, S. (2019). Nanofertilizer use for sustainable agriculture: Advantages and limitations. *Plant Science*, 289: 110270.

CHAPTER-5

Temperature Stress Resilience in Plants Induced by Green Nanotechnology

Kuchipudi Rakesh,[1] *Ankan Das,*[2] *Rajashree Mitra Thakur*[2] and *Sandip Debnath*[1,*]

1. Introduction

1.1 Overview of Temperature Stress in Plants

Among the abiotic environmental factors, temperature is the most important factor which significantly effects the life process of all organisms (Bacelar et al., 2024). Temperature stress experienced by plants are usually 3 types: (i) chilling stress (0 – 15°C) (occurring at temperature below freezing), (ii) freezing stress (< 0°C) (occurring at low temperature below freezing), and (iii) high temperature stress. Normally the growth permitting range of temperature is 10°C – 45°C (Hamdan and Salam, 2024).

The consequences of chilling and freezing or heat stress are: (a) reduction in photosynthesis, (b) reduction in plant growth, (c) effect on reproductive development, (d) reduction in yield, (e) anatomical changes and (f) reduction in produce quality (Sher et al., 2024).

1.2 Importance of Enhancing Plant Resilience

Enhancing plant resistance to various temperature stress is of paramount importance due to several interconnected reason: 1. Food Security: Temperature extremes, such as heatwaves and frost can severely reduce the crop yields and quality. Improving plant resilience ensures a stable food supply, which is critical for addressing global hunger and ensuring nutritional security for growing population. 2. Climate change Adaptation: Temperature variations are predicted to become more frequent and intense as climate change advances. Developing crops that can tolerate these stresses is essential for maintaining productivity of agriculture despite erratic weather conditions.

[1] Department of Genetics and Plant Breeding, Palli Siksha Bhavana (Institute of Agriculture), Visva-Bharati, Sriniketan, West Bengal-731236, India.
Email: rakeshkuchipudi909@gmail.com, ORCiD: https://orcid.org/0009-0002-8690-7908
[2] Palli Siksha Bhavana (Institute of Agriculture), Visva-Bharati, Sriniketan, West Bengal-731236, India.
Email: rickdas.official@gmail.com, ORCiD: https://orcid.org/0009-0008-1410-8158
Email: rajashreemitrathakur4@gmail.com, ORCiD: https://orcid.org/0009-0004-3940-8983
* Corresponding author:sandip.debnath@visva-bharati.ac.in, ORCiD: https://orcid.org/0000-0002-0234-6633

DOI: 10.1201/9781003567684-5

3. Economic Stability: Agriculture plays a crucial role in the economies of many countries. Crop failures due to temperature stress can lead to significant economic losses for farmers and disrupt food supply chains, affecting prices and availability. Enhancing crop resilience helps stabilize incomes and supports economic stability. 4. Biodiversity and Ecosystem Health: Temperature stress can threaten biodiversity by affecting the survival of various plant species. Resilient plants help maintain ecosystem stability and biodiversity, which are vital for the health of ecosystems and the services they provide, such as pollination and soil fertility (Diyaolu and Folarin, 2024). 5. Resource Efficiency: Temperature stresses often increase the demand for water and nutrients as plants struggle to survive. Resilient plants use resources more efficiently, reducing the need for additional inputs like irrigation and fertilizers, and promoting more sustainable agricultural practices. 6. Technological and Scientific Advancement: Efforts to enhance plant resilience drive research and innovation in fields such as biotechnology, genetics, and agronomy. These advancements lead to the development of new techniques and technologies that can improve crop performance under stress conditions.

1.3 Role of Nanotechnology in Agriculture

It has been determined that nanotechnology is among the numerous technological advances that might improve the lives of the poor and revive the food and agriculture industries (Rajput, 2024). Norio Taniguchi coined the term "nanotechnology" in 1974 (Milani et al., 2023). According to the US EPA, nanotechnology is the scientific field that investigates stuff with nanometre dimensions (1–100) (Ramadevi and Ushasri, 2023). It transforms matter into new forms with completely different or different qualities from those of individual molecules, atoms, and molecular clusters. It encompasses a wide range of applications, including the construction, manipulation, and organization of materials and devices at the nanoscale that are chemical, physical, and biological in nature. The National Nanotechnology Initiative (NNI) was launched by the US Federal Government in 2001 with the goal of advancing nanotechnology. According to the National Nanotechnology Initiative (NNI), nanotechnology involves manipulating matter at the nanoscale, specifically within the size range of 1 to 100 nanometres. This emerging field holds significant potential to address critical global challenges, including those related to agriculture, healthcare, water resources, energy, and biodiversity, as identified by the 2002 United Nations Johannesburg Summit on Sustainable Development (Karnwal et al., 2024). A United Nations survey underscores the importance of nanotechnology in achieving the Millennium Development Goals, highlighting that increasing agricultural productivity in developing countries through nanotechnology applications is a crucial factor in their successful implementation (Lekshmi et al., 2024). The first two critical areas of nanotechnology applications are energy conversion and storage, and the third is water treatment. Nanotechnology has a lot of promise for use in food and agriculture systems, however there are obstacles in the way of its broad acceptance. A key factor in deciding these apps' eventual viability on a worldwide basis will be stakeholder acceptability (Jones et al., 2024). Establishing suitable governance and regulatory processes and systems that involve all stakeholders is essential for the application of nanotechnology in agriculture. Consequently, nanotechnology has the potential to usher in the second green revolution that the global agriculture industry urgently needs, with a focus on sustainable production.

2. Understanding Temperature Stress in Plants

2.1 Impact of Temperature Stress on Plant Growth and Development

Temperature affects plant growth and development, and each species has a range of ideal, maximum, and lowest temperatures. Extreme temperature events can last a few days and exceed typical temperatures by more than 5 degrees Celsius. Kumudini et al. (2014) discovered that extreme events throughout the summer season have a significant effect on plant productivity. Nevertheless, not much study has been done on this subject. In their 2015 analysis of temperature extremes in wheat

(*Triticum aestivum* L.), Barlow et al. (2015) discovered that high heat decreased grain number and filling period, whereas frost induced sterility and abortion of produced grains. Grain yield may suffer due to these modifications. Developing adaptation strategies requires an understanding of how temperature changes will affect plant growth and development over the next 30 years.

The timing, intensity, and duration of stress are the primary factors that determine the impact of prolonged exposure to high temperatures. However, high temperature stress has a greater influence during the reproductive stage (Tenorio et al., 2013). Southeast Asia's mean annual temperature increase is expected to rise by 0.7–0.9°C every decade, translating into 4.8° (IPCC 2014; Asian Development Bank, 2009; Burhan et al., 2017). According to Fahad et al. (2016) rice grain production may drop by 10% for every 1°C increase in temperature. Many Asian nations frequently experience cold stress (Zhang et al., 2014a,b). Crops in other areas, like Europe, South America, East and West Africa, and the United States, need to be able to withstand cold temperatures because frequent cold stress threatens crop growth and yield in these areas. More than 60% of South America's rice is produced in Rio Grande do Sul, the southernmost state of Brazil, which is also known for its cold climate. Depending on whether the plant is in the vegetative or reproductive stage, the physiological factors that cause yield losses in crops due to cold stress can vary (Arshad et al., 2017).

2.2 Mechanism of Plant Response to Temperature Stress

2.2.1 High Temperature Stress

A major environmental stressor on plant productivity that usually coexists with drought conditions is high temperature stress. Plants cope with high temperature stress in many ways:

2.2.1.1 Morphological Adaptations

Similar to drought resistance, plants evolve morphological adaptations to prevent extreme sun radiation and warming, including, (i) increased vertical leaf orientation, (ii) like grasses, they roll their leaves along the axis, (iii) leaf hair that reflects, (iv) leaf surfaces with wax and (v) smaller, deeply cut leaves to increase heat loss through convection and conduction by reducing the thickness of the boundary layer. Some desert plants have dimorphic leaves viz. small nearly hairless leaves found during winter and white pubescent leaves developed in summer. This dimorphism of leaves protect plants against excessive heating during summer (Liu et al., 2020).

2.2.1.2 Stabilization of Membrane

Stabilization of O, evolving complex of PSII in thylakoid membranes is carried out by glycine betaine, glycerol and sucrose, Carotenoid Zeaxanthin increases heat tolerance. Increased temperature decreases the lumen pH which in turn favours dissociates Violaxanthin into Zeaxanthin by de-epoxidation.

- Violaxanthin $^{\text{xanthophyll deep-oxide}}$ Zeaxanthin

Zeaxanthin decreases membrane fluidity and stabilizes the membrane.

2.2.1.3 Synthesis of GABA under Heat Stress

Heat stress significantly elevates GABA levels in plants, with increases observed to be six to ten times higher than under normal conditions (Ramos-Ruiz et al., 2019). This accumulation is driven by the enzyme glutamate decarboxylase (GAD), which converts glutamate to GABA. GAD activity is tightly regulated by calcium-calmodulin signalling. Heat stress triggers an increase in cytosolic calcium, activating calmodulin and subsequently stimulating GAD activity (Ficociello, 2017). The GABA accumulation seen during heat stress is mostly due to this calcium signalling mechanism. Furthermore, heat stress disrupts cellular pH homeostasis by inhibiting proton pumps, leading to cytosolic acidification. The conversion of glutamate to GABA consumes protons, effectively counteracting this pH drop and potentially contributing to cellular pH regulation.

2.2.1.4 Heat Shock Proteins (HSP)

In response to sudden temperature spikes of around 5 to 10°C, plants create a unique family of proteins called heat shock proteins (Kan et al., 2023). The majority of HSPs function as molecular chaperones to assist cells in withstanding heat stress. HSPs prevent the high temperature induced denaturing, unfolding and misfolding of normal enzymatic and structural proteins and thus facilitates normal functioning of the cell at elevated and stressful temperatures. When temperature rises, plants produce HSPs of varying sizes. With molecular weights ranging from 15 to 104 kDa, the HSPs may be categorized into five groups based on their sizes. The nucleus is home to many HSPs. cytoplasm, endoplasmic reticulum, mitochondria, and chloroplasts. While members of the HSP60, HSP70, HSP90, and HSP100 families act as molecular chaperones, stabilizing and folding proteins in ATP-dependent ways and building oligomeric proteins, other HSPs help move polypeptides across membranes into cellular compartments. HSP90 is associated with hormone receptors in animal cells and may be required for their activation, while their role in plants is uncertain. Induced HSP-producing cells have enhanced heat tolerance and can withstand exposure to otherwise fatal temperatures (Ern et al., 2023). Stress at high temperatures is not the only condition that induces HSP expression. Abiotic stressors such as wounds and water scarcity can also cause some HSPs. The effects of oxidative stress are salinity, temperature, and heavy metals. For this reason, having already experienced one stressor offers cross-protection against another. For instance, tomato fruits subjected to a 48-hour heat shock at 38°C produce HSPs, which in turn shield the cells from 21-day freezing stress at 2°C (Angon et al., 2024).

Under stressful situations, the synthesis of Heat Shock Proteins (HSPs) significantly rises whereas that of other proteins considerably decreases (Mondal et al., 2023). To mediate the heat shock response, a Heat Shock Factor (HSF) regulates the transcription of HSP mRNAs. HSF is found in a monomeric form in non-stressed cells, usually linked to HSP70 proteins. HSP70 separates from HSF during heat stress, which causes the monomers to trimerize (Puustinen, 2023). By attaching to certain DNA sequences in the promoters of HSP genes is known as Heat Shock Elements (HSEs), these active trimers trigger the transcription of HSP mRNAs and result in the translation of specific HSPs, including HSP70. It is simpler for newly produced HSP70 to bind to the phosphorylated trimers when the trimeric HSFs connected to HSEs are phosphorylated. The HSP70-HSF trimer complex separates from HSEs as a result of this interaction, then disassembles and dephosphorylates into monomeric HSF forms. The resting HSP70/HSF complex is then formed when these monomers reassociate with HSP70. This regulatory cycle makes sure that HSPs build up to an appropriate amount during stress, and once their abundance is sufficient, it stops the creation of further HSP mRNA.

2.2.1.5 Antioxidative Enzyme System

It protects against the active oxygen species, neutralizing them before they can damage cellular components (Hodges, 2023).

2.2.2 Low Temperature Stress

In order to respond to low temperatures, plants have evolved complex processes that allow them to endure and flourish in such environments (Cao et al., 2024). Limiting ice formation is a crucial tactic that greatly enhances freezing tolerance. This process is greatly aided by specialized plant proteins called antifreeze proteins, which are produced in response to low temperatures. By binding to ice crystal surfaces, these proteins use a non-colligative mechanism to stop or reduce the formation of ice crystals (Mimi et al., 2024). The cells that round the intercellular spaces and the epidermal layer of rye leaves contain antifreeze proteins, which effectively stop extracellular ice from forming. Additionally, sugars and cold-induced proteins act as cryoprotectants by stabilizing proteins and membranes during dehydration caused by low temperatures. For instance, greater sucrose content in winter wheat correlates with enhanced freezing tolerance, with sucrose acting as a cryoprotectant. Similarly, raffinose, fructans, sorbitols, or mannitols serve comparable protective functions in other species. Additionally, a cryoprotective glycoprotein has been extracted from cold-acclimated cabbage leaves, demonstrating the variety of metabolic responses that plants use.

Certain woody plants demonstrate remarkable acclimation to extremely low temperatures. While some plants are freeze-sensitive and sustain injury from short-term exposure to freezing temperatures, woody plants from temperate zones exhibit significant freeze tolerance. These plants can endure prolonged exposures to temperatures as low as –20°C to –30°C, or even lower, without visible injury. However, freezing tolerance varies across different plant species, organs, and developmental stages. Slow dehydration and supercooling are common methods of freezing resistance. Supercooling, or cooling a liquid below its freezing point without crystallizing, is an essential adaptation of plants like oak, elm, maple, beech, ash, walnut, rose, rhododendron, apple, pear, peach, and plum. Deep supercooling enables the inhibition of ice crystal formation in several species found in hardwood forests at temperatures far lower than the freezing point. This resistance is temporary, though, since plant growth picks back up in the spring. Subalpine fir stem tissues, for instance, that in May could supercool and survive at temperatures below –35°C, lose this capacity by June and are vulnerable to freezing damage at –10°C.

During deep supercooling, protoplasts also inhibit ice nucleation. A strong vapour pressure gradient drives water loss from the protoplast to extracellular ice, and cell walls act as barriers to ice growth from intercellular voids into the wall. The living cells in the bark and the vegetative buds of temperate fruit trees, and the flower buds of apples and pears, on the other hand, do not supercool but rather withstand dehydration during the creation of extracellular ice. Woody species from northern Canada, Alaska, and northern Europe, where yearly minimum temperatures can drop as low as –40°C, have acquired this tolerance to a great degree. In intercellular spaces, ice usually forms between –3°C and –5°C. As water is gradually removed from the protoplast, which is not frozen, crystals build. To survive freezing temperatures, extracellular spaces must be able to hold expanding ice crystals, and protoplasts must be able to withstand dehydration (Luo et al., 2023).

Advancements in biotechnology have further bolstered plant tolerance to low-temperature stress. Genetically modified bacterial strains, specifically non-ice nucleators, are commercially deployed to outcompete native ice-nucleating bacterial strains on plant surfaces, thereby reducing potential ice nucleation points under low-temperature conditions. Additionally, abscisic acid (ABA) and protein synthesis play integral roles in acclimation to cold stress, as ABA mediates various physiological adjustments necessary for survival.

Antifreeze protein expression is markedly elevated during cold stress, and several genes are induced during cold acclimation (Zhang et al., 2024). These proteins, also known as Thermal Hysteresis Proteins (THPs), are members of a class of pathogen-related proteins that are also triggered by pathogen infection, including endoglucanase and endo-chitinase. Transcription activators known as C-repeat Binding Factors (CBFs) include CBF1, CBF2, and CBF3 (also known as DREB1b, DREB1c, and DREB1a, respectively), which coordinate the regulation of cold-induced gene expression. In order for plants to adapt to low temperatures, these transcription factors are necessary for the activation of genes that are induced by cold stress.

2.3 Consequence of Temperature Stress on Crop Yield

Temperature extremes, both high and low, are becoming increasingly frequent and severe due to climate change (Zhou et al., 2023). These extremes, commonly referred to as heat stress and cold stress, significantly constrain agricultural production worldwide. The primary calorie sources for humans—wheat, rice, maize, and soybeans—are particularly vulnerable to these changes. Maintaining food security requires assessing how these essential crops are affected by global warming, but study results are frequently inconsistent. For example, short periods of intense heat during flowering can severely reduce yields in annual crops like wheat and groundnut. Post-flowering heat stress in wheat alters grain quality by decreasing the starch-to-gliadin ratio while increasing the glutelin ratio, negatively affecting both grain yield and flour quality (Ashraf, 2014). Even a 1°C rise in temperature could lead to cereal yield reductions ranging from 4.1 to 10%.

High temperatures not only reduce yield quantity but also affect quality by shrinking fruits, grains, and tubers. Elevated temperatures can impair photosynthesis by damaging critical enzymes and cellular components. Conversely, cold stress, such as frost, hinders photosynthesis by slowing the metabolic processes needed for growth, thereby reducing energy availability. This slows down grain filling and decreases yield quality and marketability. For instance, cold stress at the late filling stage in chickpea (12°C/3.4°Cday/night) was found to decrease pod filling, leading to smaller seeds and a 43% yield loss per plant due to pod abortion (Kaur et al., 2008). Similarly, cold stress reduces photosynthesis and carbon assimilation rates, which disrupt the flow of photo-assimilates from source tissues to sink tissues, as observed in rice (Xu et al., 2023), wheat (Li et al., 2017), tomato, and maize (Leipner and Stamp, 2009). These disruptions lead to insufficient carbon resources for grain development, resulting in poor grain filling and yield formation. Reduced grain filling due to inadequate source-sink interactions further limits yield potential (Mitchell and Madore, 1992; Thakur et al., 2010).

3. Nanotechnology Applications in Agriculture

3.1 Introduction to Nanotechnology

Technology implemented at the nanoscale with practical uses is known as nanotechnology, "TECHNOLOGY ON NANOSCALE." Nanomaterials (NMs) have special chemical and physical properties that make them useful for society. A megatrend that has developed into a multipurpose technology is nanotechnology. The control of materials at the nanoscale is a key component of nanotechnology, a revolutionary method to technological advancement (Malik et al., 2023). The essence and potential of nanoscience and nanotechnology are best illustrated by the fact that materials' chemical, physical, and biological properties can vary greatly at the nanoscale compared to their bulk counterparts. When a material's dimensions are decreased to less than 100 nm, its characteristics may change dramatically. Thus, in addition to changes pertaining to size and structure, materials can be nanostructured to impart new properties or to provide a particular performance. These macromolecules and particles have specific physicochemical properties and are made up of a small number of molecules and range in size from 1 to 50 nm. When applied to comparable applications, nanoparticles exhibit attributes of improved performance in comparison to bulk materials. The National Nanotechnology Initiative was established by presidential action in 2000 and was formally institutionalized by the 21st Century Nanotechnology Research and Development Act of 2003. In FY 2015, the federal government invested around US $20 billion in R&D, whereas the private sector invested approximately US $1.5 billion per year, more than double the federal contribution. In FY 2012, nano-enabled product revenues exceeded US $200 billion in the US and US $700 billion globally. This represents a significant return on investment. Basic research and development, as well as investigations into potential nanotechnology safety hazards, workforce development, and education and curriculum, should continue.

Biological systems can be quite tiny. A lot of the cells are really small, yet those are extremely mobile; they produce numerous substances, walk around, wriggle, and perform a variety of amazing tasks—all on a microscale. They also keep data. Contemplate the notion that we too can create something incredibly little that does what we want—an object that can move at that level.

3.2 Nanotechnology's Advantages in Agriculture

Agriculture has continually benefitted from a number of technological innovations, including hybrid crops, artificial fertilizers, and insecticides. To address the global concerns of food security and climate change, agricultural scientists are now realizing that smart innovation, such as nanotechnology, is essential for agricultural expansion (Ridhi et al., 2024). Although the study was started almost 50 years ago, the significance of the importance of using nanotechnology in the field of agriculture just recently became apparent (Mukhopadhyay, 2014). Nanomaterials enable smart fertilizer and pesticide delivery systems, smart food packaging and processing systems, quick and early detection of pathogens and toxic chemicals in food products, agricultural food security regulation, and increased fertilization (or fertilizer) use efficiency and yields. Agricultural production can be raised through site-specific medication and gene delivery of molecules at the cellular or molecular level in animals and plants, genetic modification of animals and plants under stress utilizing nanoarrays, and genetically modified animals and plants caused by nanomaterials. Pollutants can be reduced and degraded using nano-filters or nano-catalysts, which will ultimately reduce pollution (Ghosh et al., 2024). Using nanotechnology in agriculture, one can create slow-releasing nano-fertilizers that plants can use; controlled and demand-release pesticides encapsulated in nanoparticles; site-specific drug and nutrient delivery in livestock and fisheries; soil treatment and water using nanoparticles, nano-brushes, and nanomembranes; maintenance and cleaning of fishponds and nano-sensors that evaluate plant health and soil quality. Nano-remediation has demonstrated significant potential in reducing the costs and time required for decontaminating large polluted areas by eliminating the necessity to remediate and dispose of contaminants in the soil. It can bring contamination levels down to nearly zero at the original site, providing an efficient solution for environmental cleanup. However, careful consideration is essential to mitigate any possible adverse environmental impacts, especially in large-scale studies involving entire ecosystems. Nanotechnology also holds promise for expanding the application of existing scientific knowledge into new domains. For example, nano-zeolites could potentially be utilized to neutralize soil acidity, presenting innovative opportunities for soil management.

3.3 Green Nanotechnology and Sustainable Agriculture

The massive demand for agricultural production, that the population growth and global climate change have created an unprecedented challenge for the modern agricultural system. There has been a lot of attention lately in using nanotechnology in agriculture to develop environmentally remediation solutions and sustainable farming technologies. Nanotechnology is the application of nanoparticles and offers the possibility of creating new materials and products with higher quality. Among the many applications for nanomaterials in contemporary agriculture are nano-sensors, nanocides, nanofertilizers, nano-barcodes, and nano-remediators (Alrabie et al., 2024). However, the physical and chemical processes needed to create nanoparticles are neither economical nor environmentally beneficial. NPs are created by a variety of chemical and natural processes, including pyrolysis, arc discharge, sonication, laser blasting, and radiation. Toxic byproducts are produced by the chemical procedures, which employ sodium borohydrate and hydrazine, two dangerous reducing chemicals. These technologies' combined energy and financial needs, as well as the usage of toxic substances, nonpolar solvents, artificial additives, and coating agents, restrict their applicability in the fields of medicine, biology, agriculture, and food. Thus, "green" chemistry and bioprocesses have become the centre of attention for researchers in their pursuit of a stable, safe, environmentally safe, and nontoxic method to produce NPs. Researchers have examined using plant parts (seed, flower, leaf, bark, peel), bacteria, fungus, actinomycetes, yeast, viruses, and enzymes to create homogeneous, stable NP synthesizing agents at physiological pH, temperature, and pressure in an eco-friendly way. These organisms are naturally able to produce NPs through biosynthesis in either an extracellular or intracellular context, making them potential bio-factories for NP synthesis (Fig. 1). In biological synthesis, amino acids, peptides, protein, polysaccharides, tannins, flavonoids, phenols, and vitamins play a crucial role as reducing and capping agents and are essential for regulating the size and form of nanoparticles. High dimensions, stability, and polydispersity characterize the biogenic NP. There are many reports on the use of biogenic NP in the biological and therapeutic domains, but there are not many that discuss how they might be applied in the environmental and agricultural domains (Bahadur et al., 2023). This review offers a thorough understanding of the various approaches and characterisation techniques for biogenic NP synthesis from various sources. Furthermore discussed are the potential uses of these biogenic NPs for improving agricultural yield, reducing abiotic stress, and cleaning up environmental pollutants.

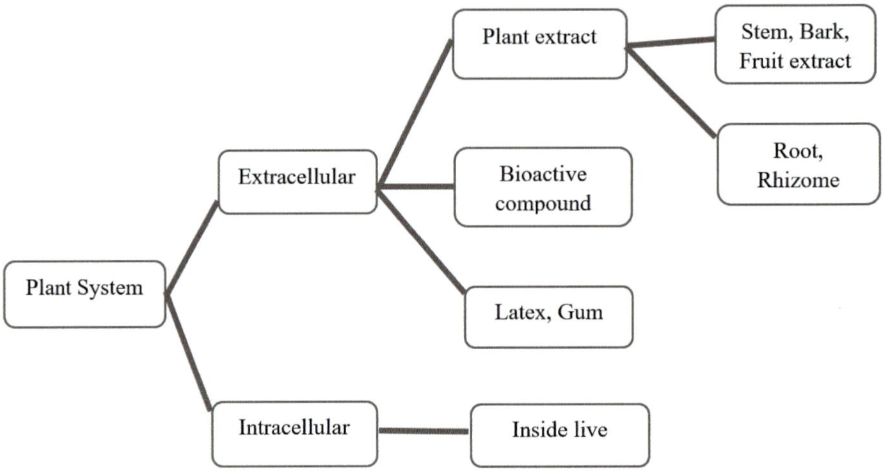

Figure 1. Various techniques for creating metal nanoparticles using plant systems.

3.3.1 Green synthesis

The utilization of plants and plant extracts in green synthesis has become more common due to its rapid advancement, one-step method, cost-effective protocol, non-pathogenicity, and environmentally benign nature (Fig. 2). Plant-based green synthesis usually happens faster than other microbes like bacteria and fungi (Alsaiari et al., 2023). Therefore, to date, a number of studies and inquiries have been prompted by using plant extract in green synthesis.

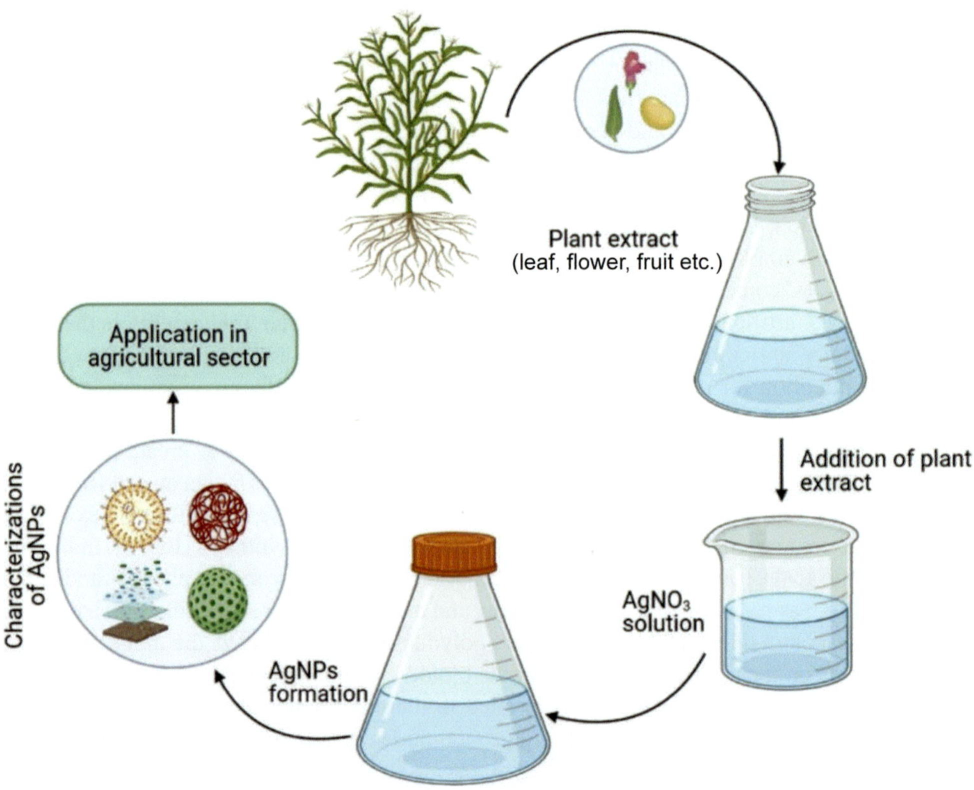

Figure 2. Green synthesis of nanoparticles.

4. Green Nanotechnology for Enhancing Temperature Stress Resilience

4.1 Green Nanotechnology for Stress Mitigation

The area of green nanotechnology holds great promise since it focuses on creating sustainable and eco-friendly nanomaterials and processes. In the context of stress mitigation, green nanotechnology offers several potential applications:

- **Biosensors for Stress Detection:** Highly sensitive and selective biosensors for identifying stress biomarkers in biological samples can be created using green nanomaterials (Pechyen et al., 2024). These sensors can provide early warning signs of stress and enable timely interventions.
- **Stress-Relieving Agents:** Green nanoparticles can be designed to directly transport stress-relieving substances to specific cells, including hormones or neurotransmitters (Verma et al., 2024). This targeted delivery can enhance the effectiveness of these agents and minimize side effects.
- **Biocompatible Nanomaterials for Stress Management:** Green nanomaterials can be used to create biocompatible implants or coatings that promote tissue regeneration and reduce inflammation, both of which are associated with stress (Saikia, 2024).
- **Environmental Remediation for Stress Reduction:** Green nanomaterials can be used to reduce pollutants and toxins in the environment, reducing exposure to stressors that can contribute to stress and health problems.
- **Sustainable Agriculture for Stress Mitigation:** Green nanotechnology can be used to create sustainable farming methods that require fewer fertilizers and pesticides, reducing environmental stress and encouraging the production of nutritious food.

Overall, green nanotechnology holds significant potential for developing innovative solutions for stress mitigation. By utilizing the special qualities of green nanomaterials, we can create new tools and technologies that promote well-being and reduce the impact of stress on individuals and society.

4.2 Nanoparticles' Function in Controlling How Plants React to Temperature Stress

Both high and low temperatures have a major effect on growth and productivity. The plant's capacity to grow, develop, and produce is eventually impacted by these stresses, which impair physiological, biological, biochemical, and molecular processes. A complicated and expanding field of study is how plants interact with nanoparticles while they are under temperature stress. Because of their distinct physicochemical characteristics, nanoparticles can affect how plants react to temperature stress. Plant growth, development, and general performance can be significantly impacted by temperature stress, whether it comes from heat or cold. Many investigations have found that when plants are exposed to heat stress, nanoparticles can have both beneficial and detrimental effects. Positively, certain nanoparticles have been demonstrated to act as stress reducers, enhancing the plant's capacity to withstand temperature fluctuations (Abasi et al., 2024). These nanoparticles may improve the uptake of water and nutrients, boost photosynthetic efficiency, and strengthen antioxidant defence, all of which may help the plant adapt to problems brought on by changes in temperature. On the other hand, certain nanoparticles might make temperature stress on plants worse. They could cause oxidative stress, interfere with cellular functions, or compromise membrane integrity, which would hinder growth and lessen resistance to temperature fluctuations.

Comprehending the complex relationship between nanoparticles and plant reactions to temperature fluctuations is essential for creating environmentally friendly farming methods and maximizing the utilization of nanoparticles in a range of settings. More research is needed to identify the characteristics of nanoparticles that can be modified for advantageous interactions with plants under temperature stress and help develop resilient and adaptable types of crops as well as to better understand the underlying mechanisms.

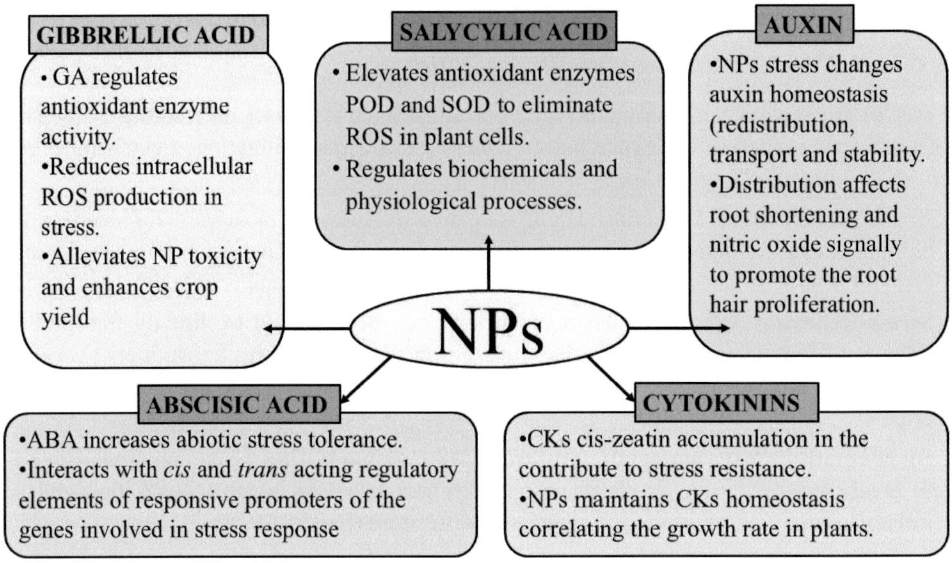

Figure 3. NPs' impact on phytohormone regulation.

The significance of phytohormones, which control plant growth and development, in NPs-induced abiotic stress is unclear (Wahab et al., 2023). Plant abiotic stress is indicated by phytohormone levels. The hormonal balance and metabolic activities of plants are altered by abiotic stress. Nanoparticles have a beneficial or detrimental impact on the generation of phytohormones, depending on their concentration as shown in Fig. 3. Examples of plant antioxidant systems include ABA, methyl jasmonate, salicylic acid, jasmonate, and other plant hormones (Santos-Espinoza et al., 2021; Yuan et al., 2013). Under stress, plants respond to nanoparticles (NPs) by producing more ABA and SA, two stress hormones. Because auxin and cytokinins (CKs) are reduced when NPs accumulate in the apices, growth is inhibited. Gibbrellic Acid (GA) enhances the antioxidant response, which lowers NP buildup.

Abiotic stress causes a range of physiological, morphological, and anatomical changes in plants. Numerous physiological processes, such as photosynthesis, respiration, transpiration, osmotic regulation, and reproductive capacity, are adversely affected by heat stress. One of the physiological functions of plants that is most susceptible to heat is photosynthesis. According to Farooq et al., (2024), heat stress affects the thylakoid membrane and inhibits important photosynthetic enzymes including Rubisco and 1, 5-bisphosphate carboxylase. Because the PSII complex is heat-intolerant, photosystem II (PSII) activity drastically reduces or ceases at high temperatures (Song et al., 2014; Zhao et al., 2021a,b). Plant physiology has been demonstrated to be impacted by nanoparticles under temperature stress. Because they increase biomass output by amplifying metabolic, photocatalytic, and light energy conversion processes, NPs are a valuable source of nutrients. NPs present a number of difficulties and factors to take into account, including toxicity and environmental effects, cost and scalability, and safety and regulatory concerns. NPs have the potential to improve plants' resistance to temperature stress in a number of ways. To completely comprehend their impacts and create safe and affordable applications, more research is necessary.

4.3 Mechanism of Action of Nanoparticles in Enhancing Plant Resilience

Increased agricultural crop production is required due to the population's rapid rise. However, it confronts pressure from climate change, dwindling arable lands, and unforeseen threats. Several biotic and abiotic variables, such as temperature, heavy metals, osmotic pressure, salt, and drought, can induce plant stress (Zhang et al., 2023). Reactive oxygen species production, reduced photosynthetic efficiency, membrane damage, and other issues can be made worse by plant stressors.

These issues can change crop growth and development by influencing physiological, molecular, and biochemical processes, which can lead to large productivity losses. Advanced nanotechnology applications in agriculture offer suitable ways to increase food production and confer stress resistance (Ullah et al., 2023). Because of their small size and large surface area, nanomaterials are being used more and more as a new tool to initiate special biochemical events that are connected to the physiological output of plants. During stress reactions, they can alter genes related to energy transfer, electron transport, and biosynthesis and organization. By increasing anti-stress chemicals, enhancing antioxidant enzyme activity, modifying systemic signals, or increasing photosynthetic rates, nanoparticles help people become more resilient to adverse environmental conditions. To control different agricultural factors, they also serve as soil enhancers, nano pesticides, fertilizers, growth promoters, and sensors. Beyond the effectiveness of nanoparticles in helping plants tolerate stress, more scientific study is required to comprehend how and why nanoparticles affect crops under stress.

4.3.1 Interaction between Nanoparticles and Plants

Nanoparticles may interfere with normal plant biological processes, often leading to oxidative stress. This disruption typically occurs when NPs interact with the chloroplasts' and mitochondria's electron transport chains resulting in the causing an oxidative burst to start and the generation of Reactive Oxygen Species (ROS). However, too many ROS can harm cells by causing lipid peroxidation, DNA damage, and protein changes, it also plays a key role in the processes of plant tolerance and stress responses. Fortunately, plants possess antioxidant defence systems to mitigate the negative consequences of ROS. These systems include both enzymatic defences (such as superoxide dismutase, guaiacol peroxidase and catalase) and inorganic molecules (like ascorbate, carotenoids, and glutathione).

In addition to oxidative stress, NP-plant interactions can disrupt hormonal balance, affecting key metabolic processes and stress responses. Nanoparticles can either activate or inhibit hormonal pathways, influencing plant growth, development, and stress resistance. For example, a study on *Capsicum annuum* showed an increase in levels of cytokinin in reaction to silver nanoparticles (AgNPs), whereas in *Gossypium* sp., exposure to copper oxide nanoparticles (CuO-NPs) led to reduced levels of abscisic acid (ABA) and auxin. These changes indicate that NP-plant interactions can significantly alter plant hormonal balance and metabolism.

Furthermore, the expression of several genes is impacted by nanoparticles, which includes microRNAs (miRNAs), which regulate critical processes such as seed germination, root growth, and free radical scavenging. Studies have shown that nanoparticles affect miR-398 and miR-408, may contribute to the stress reaction and seedling development. Additionally, the downregulation of miR-164, a miRNA critical for auxin signalling, has been linked to the increased root growth seed in nanoparticle-treated plants. These findings suggest that the impact of nanoparticles on gene expression plays an important role in modulating plant responses to environmental stress.

4.3.2 Nanoparticle Production and its Role in Ecological Stress Mitigation

As the phrase suggests, nanoparticles have a variety of physicochemical properties and vary in size between 1 to 100 nm. Nanoparticles are commonly created using a combination of top-down and bottom-up processes. The top-down technique includes converting bigger molecules into tiny particles. Electro-explosion, mechanical milling, chemical etching, laser ablation, and sputtering are examples of top-down techniques. The majority are metal oxide nanoparticles. Pyrolysis, condensation, the sol-gel process, and spinning are examples of bottom-up procedures. The bottom-up process involves the synthesis of green nanoparticles, which are usually produced from biological components. Additionally, NPs can be produced via the green synthesis method, which entails reacting processed plant, fungal, and microbe extracts with salt solutions of metals, capping and reduction to produce metal nanoparticles coated with desired biomolecules. Because of their small size, green-synthesized nanoparticles contribute to better results and more efficient distribution to the target site. The following characteristics are necessary for the creation of nanoparticles to be successful:

- Defined morphology via imaging (transmission electron microscopy and scanning).
- Structural characterisation by Raman, Zeta-size analysers, XPS and XRD.
- Characterization of photoluminescence, visible light, and ultraviolet optical properties.

4.3.3 Nanoparticles that are Utilized Commonly for Abiotic Stress

The green production of nanoparticles (NPs) has made use of a number of metals and their oxides, including copper (Cu), gold (Au), silver (Ag), iron (Fe), zinc (Zn) and titanium (Ti), as well as plant extracts, microbes and membranes, and viral DNA, including diatoms.

According to recent reports, the application of cerium oxide (CeO) NPs preserved the subatomic yield of photosystem II and carbon dioxide absorption by removing reactive oxygen species, especially (H_2O_2) hydrogen peroxide, which is produced by abiotic stress. In Duckweed (Lemna minor) plants, the application of titania (TiO_2) NPs decreased oxidative stress and enhanced the expression of glutathione peroxidase (GPOX), catalase (CAT) and superoxide dismutase (SOD). Applications of NPs in a variety of crops have been shown to produce a number of positive and stress-relieving properties, such as increased growth in *Capsicum annuum, Oryza sativa,* and *Allium cepa.* Thus, it can be said that crop plants can be protected against a variety of unwanted consequences brought on by abiotic stresses by utilizing the stress-relieving capacity of nanoparticles.

4.3.4 Delivery Methods of Nanoparticles in Crop Plants

Nanoparticles can be delivered to crop plants utilizing several techniques (Fig. 4) to improve growth, nutrient uptake, and reduce environmental stress. The various application methods are given below.

4.3.4.1 Seed Coating

The surface of seeds is coated with nanoparticles. Before planting, seeds are dipped in or covered with a nanoparticle-containing solution. This technique promotes early engagement with the plant by guaranteeing direct contact between nanoparticles and growing roots.

4.3.4.2 Soil Drenching

The soil surrounding the plant root zone is treated immediately with a solution of nanoparticles. The soil surface is coated with nanoparticle solutions, which are then sprayed or poured on, allowing the roots to take up the particles efficient in supplying the root system with nanoparticles, encouraging the uptake of nutrients and the growth of roots.

4.3.4.3 Foliar Spray

Nanoparticles are sprayed directly on the shoot of plants. This procedure involves spraying a nanoparticle solution over the vegetative portions. Plants assimilate foliar nanoparticles by stomatal openings, potassium channels, amino acid transporters, fissures or water gaps, endocytosis, stigma, wounds, and trichomes (Huang et al., 2024). This technique improves foliar fertilizer absorption and assimilation by allowing for quick uptake and regulated release of NPs at specific places.

4.3.4.4 Hydroponic System

In hydroponic cultivation techniques, nanoparticles are added to fertilizer solutions. Nanoparticles are dispersed in nutrient solution and absorbed by plant roots in a soilless growth medium. This enables precise nutrient delivery and absorption.

4.3.4.5 Root Injection

The root system of the plant is directly injected with nanoparticles. Injecting nanoparticle solutions inside the soil surrounding the root zone or straight in the roots provides a focused and effective delivery mechanism.

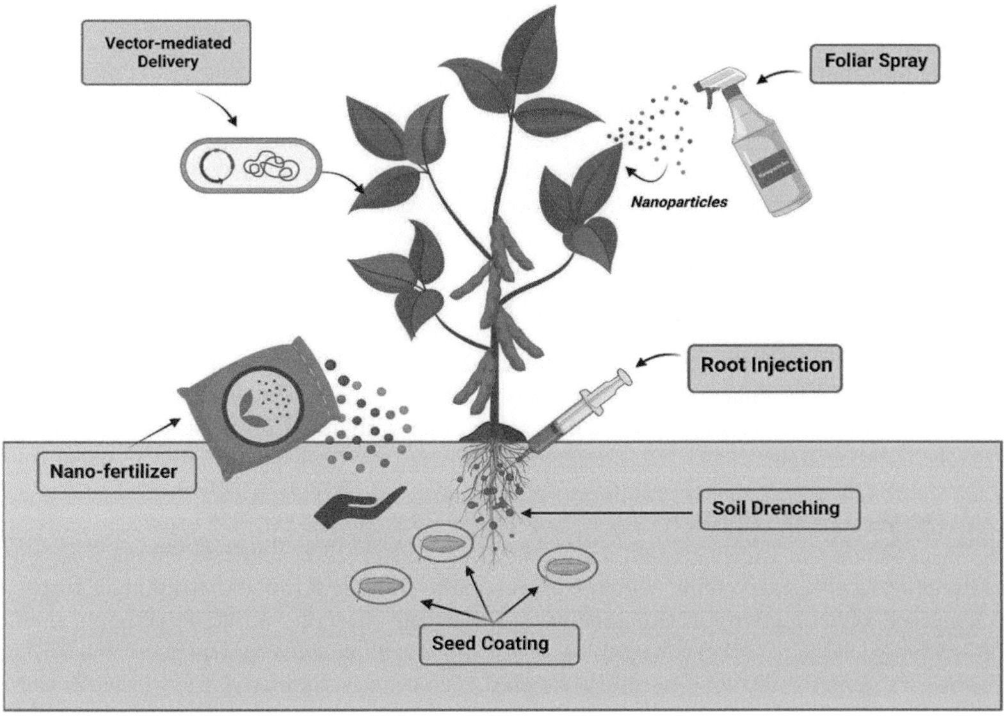

Figure 4. Diagrammatic illustration of several delivery mechanisms utilized for application of nanoparticles in agricultural plants.

4.3.4.6 Vector-based Delivery

Nanoparticles are transported to plants via biological transporters that include bacteria/ fungi. Synthetic microbes carrying nanoparticles are inserted into the plant, enhancing uptake while utilizing a natural biological method for nanoparticle delivery.

4.3.4.7 Fertilizers Ingrained with Nanoparticles

Nanoparticles (NPs) have been incorporated into conventional fertilizers, enabling their application alongside traditional fertilization methods. This combined approach allows for the simultaneous administration of fertilizers and nanoparticles throughout agricultural practices. After being sprayed, NPs travel through apoplastic and symplastic pathways after being absorbed by plant roots or leaves. Upon crossing the initial barrier of the cell walls, NPs are distributed between the plasma membrane and the cell wall, with their movement influenced by capillary forces and osmotic pressure. NPs can penetrate plant cells through mechanisms such as endocytosis, cell membrane disruption, transport, or carrier proteins like aquaporins, and ion channels. Despite these insights, a detailed mechanistic framework explaining the precise roles and locations of nanoparticles within plant cells remains absent. Further studies are necessary to explore their accumulation, translocation, and absorption rates, alongside the associated biochemical, molecular, and physiological processes in plants.

4.3.5 Plants' Biochemical Responses to Nanoparticles Amid Abiotic Stress

When plants encounter environmental stresses, they undergo a series of complex chemical reactions that influence their electro-physiology and signalling pathways. These reactions activate stress-related phytohormones, trigger the production of reactive oxygen species (ROS), initiate kinase cascades and stimulate ion channels. This leads to the activation of a species- and stress-specific gene network that regulates defence and repair mechanisms to counteract the damage caused by stress. Key stress-induced proteins play crucial roles in signalling pathways, such as the Salt Overly

Sensitive (SOS) kinase pathway, mitogen-activated protein (MAP) kinase pathway, and ROS neutralization mechanisms.

According to recent research, when applied to plants experiencing abiotic stress, nanoparticles can successfully trigger these chemical reactions. Several types of nanoparticles, including aluminum oxide (Al2O3), silicon, multi-walled carbon nanotubes, zinc oxide (ZnO), and titanium dioxide (TiO2), have been identified as effective in enhancing plant stress resilience. For example, Zhao et al. (2020), found that nanomaterials like silver (AgNPs) and copper oxide (CuO-NPs) can stimulate stress responses and enhance the plant's defence mechanisms, improving its ability to withstand environmental stresses.

The molecular impacts of NPs on different plant species have been examined using proteomic and transcriptomic methods. These studies have demonstrated that the morphological and physiological responses of plants to nanoparticles influenced by factors such as the NP dosage, type, size, and shape. By modulating molecular pathways and enhancing stress resilience, nanoparticles have shown promise as a tool to enhance how plants react to abiotic stressors such as salt, drought, and severe temperatures.

4.3.6 Nanoparticles Mitigate the Effects of High Temperatures on Plants

The plant's ability to withstand cold and heat stress is governed by intricate interplay of genes, transcription factors, and proteins, with their expression levels directly influencing the plant's capacity to withstand adverse environmental conditions. For exemplification, in rice, foliar application of ZnO nanoparticles (ZnO-NPs) enhanced the production of antioxidative system genes and chilling-response transcription factors in seedlings exposed to cold stress. Similarly, in soybeans, ZnO-NPs increased the production of photosynthetic pigments, proline content, antioxidant enzyme activity, and overall plant yield by upregulating a number of genes, including EREB, R2R3MYB, HSF-34, WRKY1, MAPK1, HDA3, and CAT.

Copper-based nanoparticles (Cu-NPs), particularly those measuring 50 nm, were shown to influence genes related to oxidative stress, brassinosteroid synthesis, and root development. In wheat, 40 nm-sized Cu nanoparticles promoted the accumulation of secondary metabolites involved in cell signalling and defence responses. Silicon treatment in wheat enhanced the expression of aquaporin genes under heat stress, resulting in improved water retention. In Arabidopsis thaliana, ZnO-NPs mitigated gene repression induced by heat stress, while in maize, by increasing the Rubisco activase content and activity, nano-anatase nanoparticles improved oxidative CO_2 fixation.

Additionally, lanthanum oxide (La_2O_3) treatment in maize seedlings modulated the Aquaporin gene expression in the tips of the roots, further influencing plant water management. These diverse impacts of various nanoparticles include the modulation of several transcription factors and genes involved in the response to heat and cold stress (Table 1), ultimately enhancing plant resilience through a combination of physiological, molecular, and biochemical modifications.

5. Case Studies and Research Findings

5.1 Examples of Nanoparticles Applications in Different Crop Species

Nanoparticles are emerging as innovative tools in agriculture, offering a range of benefits that enhance crop resilience, growth, and yield across diverse species. For wheat, zinc oxide (ZnO) nanoparticles improve growth and grain yield by facilitating zinc uptake and mitigating abiotic stresses like drought and salinity. Additionally, titanium dioxide (TiO_2) nanoparticles boost heat tolerance by promoting photosynthesis and reducing oxidative damage, helping maintain chlorophyll content and overall plant vigour under stress.

In rice, silicon dioxide (SiO_2) nanoparticles enhance resilience to drought, heavy metal, and salinity contamination by improving water retention and nutrient intake (Faisal et al., 2024). These nanoparticles also maintain cellular integrity and enhance antioxidant enzyme activity, mitigating high-temperature stress. Similarly, in tomatoes, silver (Ag) nanoparticles demonstrate antimicrobial

Table 1. Genetic influence in plant utilization of nanoparticles during temperature stress.

Stress	**NPs**	**Plant**	**Genes**	**Major regulated genes**
Cold stress	TiO_2	*Cicer arietinum* L.	Rubisco	Enhanced expression of large subunit Rubisco (LRubisco), small subunit Rubisco (SRubisco), chlorophyll a/b-binding protein and phosphoenolpyruvate carboxylase (PEPC), which plays an important role in improving photosynthetic efficiency and plant stress tolerance
			APX, CHN, RLK	Genes associated with cellular defence, cell signalling, transcriptional control, and chromatin modification are expressed differently and exhibit upregulation
Heat stress	CeO_2	*Zea mays*	Hsp70	Heat shock proteins (Hsp70) are upregulated, resulting in increased H2O2 production
	MMWCNTs	*Solanum lycopersicum*	HSP90	Upregulation of HSP90

properties, protecting plants from bacterial and fungal infections. Iron oxide (Fe_3O_4) nanoparticles improve growth and fruit quality by enhancing iron availability, while ZnO nanoparticles bolster heat stress resilience by strengthening the antioxidant defence system and reducing oxidative stress.

In maize, TiO_2 nanoparticles increase photosynthesis and growth rates, while copper oxide (CuO) nanoparticles protect against fungal pathogens. Cerium oxide (CeO_2) nanoparticles enhance heat tolerance by reducing oxidative damage and improving antioxidant enzyme activity, maintaining growth and productivity under high temperatures. Soybean benefits from molybdenum (Mo) nanoparticles, which promote nitrogen fixation by stimulating nitrogen-fixing bacteria, improving growth and yield. Additionally, Ag nanoparticles regulate heat shock proteins and stress-related genes, enhancing heat tolerance.

Potatoes show improved resistance to late blight, caused by *Phytophthora infestans*, with the application of chitosan nanoparticles. In cotton, carbon nanotubes (CNTs) facilitate water and nutrient uptake, boosting seed germination, growth, and resistance to abiotic stresses. Barley, on the other hand, benefits from silicon nanoparticles, which enhance heat stress tolerance by maintaining water balance and reducing cellular damage caused by high temperatures. These examples highlight the diverse applications of nanoparticles in agriculture, offering promising solutions for improved nutrient absorption, stress resilience, disease resistance, and total plant health. Their potential to overcome the problems provided by changing climate conditions and resource limitations underscores their role in sustainable agricultural practices.

6. Future Directions and Potential Applications

Given the rising global need for food grains, as well as the numerous concerns linked with climate shifts and agrochemicals, there was a need for novel and environmentally friendly agricultural approaches to improve crop yield and quality. In recent years, numerous options based on nanotechnology have been extensively researched to overcome all these challenges. In this sector, green nanotechnology is projected to play an important role in the near future. As a result, green nanotechnology contributes enticing breakthroughs such as innovative pesticides, herbicides, and fertilizers, rapid disease detection with nano-based kits, and better plant nutrient uptake (Ciambelli et al., 2019). Green nanomaterials can help to minimize nutrient expenditures, increase crop yields, and mitigate the harmful impact of agricultural operations on human health and the environment.

6.1 Nanofertilizers

Nutrient fertilization is a requisite for enhancing crop productivity and maintaining soil fertility. However, traditional fertilizers pose environmental and health risks and are often cost-inefficient in production. In this context, green nanotechnology offers an eco-friendly alternative by providing fertilizers with high nutrient efficiency. Fertilizers in nanoform can offer multiple benefits in nutrient management, such as significantly improving nutrient use efficiency, thereby boosting crop growth and productivity under temperature stress conditions. For example, nano-adsorbents combined with nutrients release them at a much slower rate compared to traditional fertilizers. Nanomaterials have been paired with essential macronutrients like N, P, K, Mg, S and Ca to deliver precise nutrient levels tailored to crop needs (Zulfiqar et al., 2019). Urea, widely used as a fertilizer due to its high solubility, rapid nitrogen availability, and plant uptake, suffers from inefficiencies, with only about 25% fertilizing efficacy due to losses from volatilization and leaching. These losses contribute to nitrous oxide emissions, contaminating both air and water. To address these issues, slow-release urea-coated nanoparticles, such as those made from hydroxyapatite, have been developed as a sustainable solution.

6.2 Nanopesticides

Chemical pesticides have revolutionized agriculture, but they also pose new issues such as pest resistance, insects, and environmental concerns. Nanopesticides are vital for pest management and can lessen the environmental impact of synthetic insecticides. According to Kah et al., (2019), nanopesticides differ from traditional pesticides in order to improve efficiency. Nano-range particles can be transported in both colloidal and dissolved forms, with a unique response mechanism compared to traditional solutes. Nanopesticides can provide precise delivery to pest-affected areas, reducing overall chemical use and mitigating the impact of temperature on pest proliferation. Enhanced stability and efficiency of pesticides at varying temperatures can be achieved, ensuring better protection for crops.

6.3 Nanoherbicides

Weeds are a major issue in agriculture, and traditional methods for eradicating them are time-consuming. Herbicides are commonly used to eradicate weeds in fields, but they can also harm crops, pollute soil, and reduce fertility. Nanoherbicides can effectively eliminate weeds from crops without harming the environment or soil. Using the same chemicals consistently over time can lead to weed resistance. Nanoherbicides use environmentally benign polymeric components to increase herbicide efficacy. Encapsulating herbicides in polymeric substances has been shown to safeguard the environment (Kumar et al., 2015). Atrazine was encapsulated using poly (ε-caprolactone) due to its excellent physicochemical properties and high biocompatibility (Abigail and Chidambaram, 2017). Compared to free atrazine, polymeric nanoparticles containing atrazine exhibit enhanced herbicidal effectiveness, primarily attributed to their improved stability, and reduced mobility within the soil. These features ensure that the herbicide remains more localized and active at the target site, thereby increasing its efficiency in weed control while minimizing environmental dispersion (Pereira et al., 2014). Another study (Grillo et al., 2012, 2014) found that polymer-encapsulated herbicides such as paraquat, ametryn, simazine, and atrazine followed similar tendencies. Herbicides containing nanoparticles have been developed to target weed roots. According to (Ali et al., 2014), these chemicals infiltrate the roots of weeds and restrict glycolysis, causing plant harm. Nanocapsules containing atrazine were discovered to be more efficient than commercial products in reducing hairy beggar ticks and slender amaranth (Sousa et al., 2018).

6.4 Nanobactericides and Nanofungicides

Crop losses caused by bacterial and fungal pathogens pose a significant challenge for farmers and horticulturists, as these microorganisms inflict severe damage on crops, fresh produce, and stored plant products. While fungicides and bacteriocides are traditionally employed to manage such infections, the use of green nanoparticles has emerged as a promising alternative. These nanoparticles can effectively address a range of bacterial and fungal infections through spraying or soaking applications (Jabeen and Anum, 2020). Among the metals used, silver nanoparticles stand out due to their superior antifungal properties. Silver ions disrupt thiol groups in fungal cell walls, interfering with the electron transport chain, energy metabolism, and transmembrane processes, thereby enhancing antifungal efficacy. The size and structure of silver nanoparticles have a major impact on their efficacy.

It has been demonstrated that silver nanoparticles can fight off a variety of plant diseases, including *Colletotrichum gloeosporioides, Sclerotinia sclerotiorum, Fusarium culmorum, Rhizoctonia solani, Phoma, Magnaporthe grisea, Pythium ultimum, Botrytis cinerea,* and *Bipolaris sorokiniana* (Chhipa and Joshi, 2016). Additionally, copper nanoparticles exhibit potent antibacterial characteristics against both Gram-positive and Gram-negative bacteria, as well as fungi, making them ideal for agricultural practices (Giannousi et al., 2013). Furthermore, Campos et al. (2015) developed polymeric nano capsules and solid lipid nanoparticles to enhance the fungicidal efficacy of carbendazim and tebuconazole, providing a more efficient approach to controlling fungal infections in agriculture.

7. Challenges and Considerations

7.1 Environmental Impacts of Nanoparticles Use

The chemical and physical properties of nanomaterials, along with their interactions with other environmental contaminants, dictate their fate in the environment (Maiti et al., 2016). These materials can originate from natural events such as volcanic eruptions, soil erosion, weathering of minerals, dust storms, and forest fires. Additionally, human activities, both deliberate and accidental, contribute significantly to their release. Examples include mining operations, demolition processes, fossil fuel combustion, emissions from vehicles, and the manufacturing or disposal of nanomaterials (NMs). Once introduced into the environment, nanomaterials disperse and accumulate across different medium such as air, water, soil, and sediments (Iavicoli et al., 2014). This section delves into the pathways and environmental behaviour of engineered nanomaterials (ENMs), examining how they interact with various environmental matrices after their release.

7.1.1 Air

There are significant lacks of understanding about the environmental impact of nanomaterials, particularly after release into the atmosphere. An ENM's whole life cycle, involving generation, interpreting, transportation, handling, and disposal, may result in environmental discharge (Caballero-Guzman and Nowack, 2016). Automobile discharge, burning, explosions, and gas decomposition all contribute to ambient NM formation (John et al., 2017).

Nanotechnology-engineered materials (NMs) released into the atmosphere are subject to greater exposure to sunlight and ultraviolet radiation compared to other environmental compartments (Mitrano et al., 2015). This heightened exposure significantly increases the likelihood of photochemical transformations in these materials. Atmospheric conditions can induce various alterations in nanomaterials, such as the condensation of low-volatility compounds or a reduction in particle size due to the evaporation of adsorbed water or volatile substances. Consequently, particle size distributions may shift, even as the overall numerical concentration of particles remains unchanged (Soni et al., 2015). The deposition of NMs from the atmosphere is influenced by gravitational settling velocities, which are directly proportional to particle size, further determining their atmospheric residence time and transport dynamics (Hartmann et al., 2014). Moreover, emissions of engineered nanomaterials (ENMs) from waste incineration facilities show significant variability depending on the composition of the waste. Carbon-based and biologically derived nanomaterials are typically incinerated completely, leaving minimal residues. In contrast, nanoclays often persist in combustion residues due to their higher thermal stability (Ounoughene et al., 2015).

7.1.2 Soil

Engineered nanomaterials have the potential to infiltrate soils via several routes and sources, such as the application of plant protection chemicals and fertilizers, sewage water, floodplains and biosolids

(Batley et al., 2013). A complex and multilayered interface between various materials (such as gases, solids, water, and organic and inorganic components) and living things is what makes up soil. The increased surface area of nanomaterials allows them to cling to soil particles and pass-through soil pores (Mukhopadhyay, 2014). Moreover, Mukhopadhyay (2014) states that large NM aggregates can be rendered immobile through the processes of precipitation, purification, or straining in microscopic pores. The physical-chemical features of the nanoparticles, the environment and soil, and the interactions between the NMs and naturally occurring colloidal material are some of the factors that affect the movement of nanomaterials in soil (Jafar and Hamzeh, 2013). Earlier studies have demonstrated that silver nanomaterials (Ag NMs) can function as a soil reservoir for silver ions ($Ag+$), with their aggregation greatly impacted by the amount of Ag NMs in the soil (Klitzke et al., 2015). Additionally, research indicates that plants have the capacity to uptake and translocate nanomaterials from the soil, a process that can potentially impact seed germination as well as overall plant growth and development (Hong et al., 2014).

7.1.3 Water

Nanoparticles can infiltrate aquatic ecosystems through various pathways, including surface runoff from soils, industrial emissions, and wastewater discharge from treatment facilities (Batley et al., 2013). Once introduced, their fate in water systems is influenced by several processes such as accumulation, degradation, and interactions with other elements or aquatic organisms. Both biological pathways, including aerobic and anaerobic decomposition, and abiotic mechanisms, like hydrolysis and photolysis, contribute to their breakdown (Vale et al., 2015). Nanomaterials often aggregate due to interactions with colloids, suspended particulates, and organic matter, leading to sedimentation rather than dispersing uniformly in water suspensions (Rocha et al., 2015). Aspects including size of particles, shape, surface chemistry, pH, ionic strength, and dissolved organic carbon content of the aquatic environment significantly affect their agglomeration behaviour (Baker et al., 2014). For example, a study showed that the type of coating on silver nanoparticles, including citrate (Cit-AgNPs) and polyvinylpyrrolidone (PVP-AgNPs), along with particle size, plays a crucial role in assessing their fate in water (Jang et al., 2014). When nanoparticles interact with water, they often associate with solids rather than remaining suspended (Grillo et al., 2015). Hendren et al., (2013) developed a mass-balance model utilizing a parsimony method to describe the behaviour of engineered nanomaterials (ENMs), highlighting that variations in coatings can significantly affect the removal efficiency of silver nanoparticles. Their findings showed that over 95% of nano-Ag concentrations in effluent and sludge fractions were below 0.12 $\mu g\ L^{-1}$ and 0.35 $\mu g\ L^{-1}$, respectively.

7.2 Legislative System for Nanotechnology in Agriculture

The existing legislative framework for nanotechnology primarily relies on applying current rules and laws to address issues related to this emerging field. An analysis of the regulatory landscape indicates that, given the current level of technical expertise, these existing frameworks may be adequate for managing the present state of nanotechnology. Agencies such as the Food and Drug Administration (FDA), Environmental Protection Agency (EPA), National Institute for Occupational Safety and Health (NIOSH), and the Consumer Product Safety Commission (CPSC) can enhance their regulatory efforts by leveraging the expertise of technical and industrial professionals through advisory committees.

7.2.1 Agriculture Risk Protection Act of 2000

Through Section 221 of the Carbon Cycle Research effort under the Agriculture Risk Protection Act of 2000, a US $15 million grant was awarded to an alliance of scientists from several institutions called the Consortium for Agricultural Soils Mitigation of Greenhouse Gases. The consortium's goal is to advance carbon cycle research by designing, assessing, and applying strategies at various levels—national, regional, and local. The funding is dedicated to exploring land management approaches that foster soil carbon sequestration, emphasizing the application of innovative technologies like biotechnology and nanotechnology.

7.2.2 21st Century Nanotechnology Research and Development Act of 2003

Members of the National Nanotechnology Initiative (NNI) started meeting on a regular basis in November 1996 to talk about the goals and activities of their different agencies in the field of nanotechnology. The continued interagency collaboration was formalized in September 1998 when the informal group established the Interagency Working Group on Nanotechnology (IWGN) under the National Science and Technology Council (NSTC). The Clinton administration responded to the working group's conclusions by establishing the NNI in January 2000. Since Congress authorized financing for the NNI in 2001, the White House and Congress have supported it in a bipartisan manner. The Nanoscale Science, Engineering, and Technology (NSET) Subcommittee was created by the NSTC to take the position of IWGN when the NNI got underway.

7.2.3 National Institute of Food and Agriculture, US Department of Agriculture

With the goal of providing expertise in food, farming, natural resources, development in rural areas, and wellness based on good government policies, research, and effective administration, President Abraham Lincoln founded the United States Department of Agriculture (USDA) in 1862. At that time, most Americans were employed and living on agriculture, but the USDA has evolved alongside the transformation of the U.S. economy. Recognizing advancements in technology, including nanotechnology, the USDA has integrated these innovations into its agricultural leadership strategies.

The National Institute of Food and Agriculture (NIFA) was created under the Farm Bill of 2008 to address critical agricultural issues through federal grants managed by the Agriculture and Food Research Initiative (AFRI). Collaborating with scientists, policymakers, educators, and specialists, NIFA has made significant contributions to addressing both indigenous and global challenges. Eleven institutions received US $5.2 million from AFRI in March 2016 to fund studies on nanotechnology's applications in enhancing food safety, increasing crop yields, improving renewable fuel efficiency, and managing agricultural pests. By July 2017, AFRI had provided an additional US $4.6 million through 13 grants to projects focused on using nanotechnology for food production, safety, nutrition, and sustainable agriculture.

Despite its engagement in promoting nanotechnology-based solutions, USDA NIFA does not hold regulatory authority over nanotechnology. Consequently, it is categorized within the "low power, low interest" quadrant regarding nanotechnology regulation.

7.3 Ethical Considerations in the use of Nanotechnology for Crop Improvement

The incorporation of nanotechnology into agriculture began with the recognition that traditional farming techniques had reached their limits in boosting productivity and could not fully restore ecosystems degraded by these methods. Concerns arose over the foreseeable impacts of reliance on "miracle seeds" alongside fertilizers, irrigation systems, and chemical pesticides, prompting scrutiny involving technical and social perspectives. In response, nanotechnology has emerged as a promising alternative, supported by substantial public funding in recent years. However, despite agriculture's diverse domains, progress in applying nanotechnology has been incremental. Moreover, while the potential for crop enhancement through nanotechnology is significant, it also raises a number of ethical concerns that warrant careful consideration.

- **Environmental Impact:** The introduction of nanoparticles into the environment could potentially disrupt ecosystems and biodiversity. Nanoparticles can collect in soil and water, impacting non-target animals including beneficial soil bacteria, insects, and aquatic life. The long-term environmental impacts of nanoparticles are still largely unknown. Continuous exposure to nanoparticles might lead to unforeseen consequences, such as soil and water contamination or the development of resistant pests and diseases

- **Human Health:** Nanoparticles may enter the food chain, generating worries regarding the influence on human health. Comprehensive research is required to assure that plants treated with nanotechnology are healthy for human consumption. Nanoparticles used in agriculture can pass through the human body in a number of methods, include respiration, intake of unclean food and water, or dermal contact.
- **Regulation and Oversight:** Current regulations could fail to be adequate to solve the specific issues presented by the nanotechnology in agriculture. Adequate regulatory frameworks are mandatory to ensure safe development and use. There is a need for transparency in the development and deployment of nanotechnology in agriculture. Engaging with the public and stakeholders, including farmers, consumers and environmental groups is essential to build trust and ensure that their concerns are addressed.
- **Equity and Access:** There is a risk that the benefits of nanotechnology in agriculture could be unequally distributed, favouring large-scale farmers or wealthier nations, while smallholder farmers and developing countries may be left behind.
- **Societal Acceptance:** The public's understanding and acceptability of nanotechnology in the farming industry play an important impact. There may be resistance due to fears or ethical objections, and open discussion regarding the advantages and drawbacks is necessary.
- **Unintended Consequences:** The potential for unforeseen consequences must be considered. Introducing novel technologies can lead to unexpected interactions and effects, necessitating careful and ongoing evaluation.
- **Intellectual property rights:** The patenting of nanotechnology innovations can lead to ethical dilemmas regarding ownership and access. Intellectual property rights should be balanced to encourage innovation while ensuring that small-scale farmers and developing countries can access and benefit from these technologies. There is a risk of biopiracy, where multinational corporations exploit indigenous knowledge and biological resources without fair compensation to local communities. Ethical considerations should include respecting the liberties and contributions of indigenous and regional groups.
- **Ethical Research Practices:** Researchers and companies should ensure that communities and individuals are fully informed about the possible risks and advantages of utilizing nanotechnology in farming. Informed consent should be obtained before deploying nanotechnology-based solutions in any community. Nanotechnology's advantages in agriculture should be shared equitably among all stakeholders, including farmers, consumers, and local communities. Mechanisms should be put in place to ensure that the benefits do not disproportionately favour one group over others.
- Addressing these ethical considerations requires a collaborative approach involving scientists, policy makers, farmers, and the public to guarantee that the application of nanotechnology in crop improvement is safe, equitable and sustainable.

8. Conclusion

8.1 Summary

Green nanotechnology, an emerging field that merges nanotechnology with green chemistry principles, offers innovative approaches to enhance temperature stress resilience in plants. This strategy leverages sustainable and eco-friendly nanomaterials to bolster plant defence against extreme temperatures, which is crucial for maintaining agricultural productivity in the face of climate change.

Temperature stress, encompassing both extreme heat and cold, is a major challenge for crop productivity worldwide. Traditional methods of improving crop resilience to temperature extremes, such as selective breeding and genetic modification, have shown limited success and often require significant time and resources. Green nanotechnology has become a viable substitute in recent years to increase crops' resistance to temperature stress. Green nanotechnology involves the use of nanoparticles synthesized through environmentally friendly methods, often using plant extracts or other biological materials, to enhance the plant health and stress resilience. Nanoparticles can enhance crop resilience to temperature stress in several ways. They can improve the uptake and efficiency of nutrients, thereby boosting plant growth and development. For instance, nanoparticles of essential nutrients like zinc and iron can be more readily absorbed by plants, leading to improved photosynthetic efficiency and stress tolerance. Additionally, nanoparticles can facilitate the controlled release of fertilizers and other agrochemicals, ensuring that plants receive the necessary inputs at optimal times, which is crucial during periods of temperature stress.

Nanoparticles can enhance crop resilience to temperature stress in several ways. They can improve the uptake and efficiency of nutrients, thereby boosting plant growth and development. For instance, nanoparticles of essential nutrients like zinc and iron can be more readily absorbed by plants, leading to improved photosynthetic efficiency and stress tolerance. Additionally, nanoparticles can facilitate the controlled release of fertilizers and other agrochemicals, ensuring that plants receive the necessary inputs at optimal times, which is crucial during periods of temperature stress. The creation of nano-enabled sensors and gadgets for real-time environmental and plant health monitoring is another use for green nanotechnology. These advanced tools can provide valuable data on soil moisture, nutrient levels and temperature fluctuations, allowing for timely interventions to protect crops from temperature stress. Farmers can increase crop resilience to climatic unpredictability and maximize resource use by integrating these technologies with precision farming techniques.

In summary, green nanotechnology offers a multifaceted approach to enhancing crop resilience to temperature stress. Through improved nutrient delivery, modulation of stress-related physiological processes, and advanced monitoring tools, nanoparticles can significantly bolster the ability of crops to withstand extreme temperatures. This innovative approach holds great promise for maintaining nutritional stability in the midst of climate change.

8.2 Implications of Green Nanotechnology for Enhancing Plant Resilience to Temperature Stress

Green nanotechnology provides promising implications for enhancing plant resilience to temperature stress through various innovative approaches. One of the primary implications is the ability to develop nanoparticles from natural source such as plant extracts, which can be used to fortify plants against extreme temperatures. Nanoparticle-based delivery systems enable targeted delivery of nutrients, growth regulators, and stress-tolerance enhancing compounds to plants, helping them cope with temperature stress. Nanomaterials, including carbon nanotubes and various nanoparticles, have the capacity to boost the efficacy of photosynthesis in plants. By optimizing light absorption and energy transfer, these materials can support plants in sustaining critical metabolic processes, even under challenging stress conditions such as drought, salinity, or extreme temperatures. By enhancing photosynthetic efficiency and reducing oxidative stress through scavenging of reactive oxygen species, nanoparticles contribute to better plant health and growth. Furthermore, green nanotechnology can help in the controlled and sustained release of growth regulators and stress ameliorants, providing a steady supply of essential compounds to plants during periods of temperature fluctuations. This technology also opens avenues for precision agriculture, where nanoparticle-based sensors can monitor environmental conditions and plant health, allowing for timely interventions.

Nanobiosensors can monitor plant responses to temperature stress in real-time, facilitating timely interventions. Nanoparticles can serve as carriers for gene editing tools like CRISPR-

Cas9 to introduce temperature stress-tolerance genes into plants. Smart materials incorporating nanotechnology can regulate plant microenvironments, providing optimal growth conditions even under temperature stress. The application of nanoparticles as soil amendments can lead to significant improvements in soil structure by promoting better aggregation and porosity. This enhancement supports increased water retention, ensuring that the soil retains moisture for longer periods, which is beneficial for plant growth. Additionally, nanoparticles can enhance nutrient availability by facilitating the release and absorption of essential nutrients, thereby supporting plant growth under challenging conditions. By leveraging these implications, green nanotechnology can support sustainable agriculture and food security. Ultimately, integrating green nanotechnology in agricultural practices promotes eco-friendly farming, reduces the reliance on chemical inputs, and improves crop yield and quality under challenging climatic conditions.

References

Abasi, F., Raja, N. I., Mashwani, Z. U. R., Ali, H., Mehmood, A., and Rahimi, M. (2024). Nanomaterials for plant resilience: a green approach to combating biotic stress. *BioNanoScience*, 1–16.

Alabdallah, M. N., and Hasan, Md. (2021). Plant-based green synthesis of silver nanoparticles and its effective role in abiotic stress tolerance in crop plants. *Saudi Journal of Biological Sciences*, 28(10): 5631–5639.

Alrabie, Hissah, Alsamadany, Hameed, and El-Zohri, Manal. (2024). Green Synthesis of Phytogenic nanoparticles using Plant Extracts and their Promising Roles in Sustainable Agriculture: A Review. *Bioscience Research*. 21: 738–749.

Alsaiari, N. S., Alzahrani, F. M., Amari, A., Osman, H., Harharah, H. N., Elboughdiri, N. et al. (2023). Plant and microbial approaches as green methods for the synthesis of nanomaterials: synthesis, applications, and future perspectives. *Molecules*, 28(1): 463.

Angon, P. B., Das, A., Roy, A. R., Khan, J. J., Ahmad, I., Biswas, A. et al. (2024). *Discover Plants*.

Arshad, M. S., Farooq, M., Asch, F., Krishna, J. S., Prasad, P. V., and Siddique, K. H. (2017). Thermal stress impacts reproductive development and grain yield in rice. *Plant Physiology and Biochemistry*, 115: 57–72.

Bacelar, E., Pinto, T., Anjos, R., Morais, M. C., Oliveira, I., Vilela, A. et al. (2024). Impacts of climate change and mitigation strategies for some abiotic and biotic constraints influencing fruit growth and quality. *Plants*, 13(14): 1942.

Bahadur, R., Kumar, A., Trivedi, A., Nandeha, N., Agrawal, S., Ahirwar, S. B. et al. (2023). An overview of green nanotechnology as a contemporary instrument for sustainable agriculture. *International Journal of Environment and Climate Change*, 13(12): 704–715.

Barlow, K. M., Christy, B. P., O'Leary, G. J., Riffkin, P. A., and Nuttall, J. G. (2015). Simulating the impact of extreme heat and frost events on wheat crop production: A review. *Field Crops Research*, 171: 109–119.

Bhandari, G., Dhasmana, A., Chaudhary, P., Gupta, S., Gangola, S., Gupta, A. et al. (2023). A Perspective Review on Green Nanotechnology in Agro-Ecosystems: Opportunities for Sustainable Agricultural Practices & Environmental Remediation. Agriculture, 13: 668. https://doi.org/10.3390/agriculture13030668.

Bhandari, K., and Nayyar, H. (2014). Low Temperature Stress in Plants: An Overview of Roles of Cryoprotectants in Defence. Physiological Mechanisms and Adaptation Strategies 193 in Plants Under Changing Environment: Volume 1, DOI 10.1007/978-14614-8591-9_9, © Springer Science + Business Media New York 2014.

Bharti, B., Kumar, R., Kumar, H., Li, H., Zha, X., and Ouyang. (2022). Advanced applications and current status of green nanotechnology in the environmental industry. *Green Functionalized Nanomaterials for Environmental Applications*. https://doi.org/10.1016/B978-0-12-823137-1.00012-9.

Bhushan, B. (2017). Introduction to Nanotechnology. Springer handbook of Nanotechnology, 1–19, 2017.

Cao, J., Bao, J., Lan, S., Qin, X., Ma, S., and Li, S. (2024). Research progress on low-temperature stress response mechanisms and mitigation strategies in plants. *Plant Growth Regulation*, 1–22.

Dhingra, M. (2015). Physiological responses and tolerance mechanisms of low temperature stress in plants. *International Journal of Advanced Research*, 3(2): 637–646.

Diyaolu, C. O., and Folarin, I. O. (2024). The role of biodiversity in agricultural resilience: Protecting ecosystem services for sustainable food production. *Int. J. Res. Publ. Rev.*, 5(10): 1560–1573.

Essemine, J., Ammar, S., and Bouzid, S. (2010). Impact of het stress on germination and growth in higher plants: physiological, biochemical and molecular repercussion and mechanisms of defence. *Journal of Biological Sciences*, 10(6): 565–572, 2010 ISSN 1727-3048.

Ern, R., Andreassen, A. H., and Jutfelt, F. (2023). Physiological mechanisms of acute upper thermal tolerance in fish. *Physiology*, 38(3): 141–158.

Fahad, S., Hussain, S., Saud, S., Hassan, S., Ihsan, Z., Shah, A. N. et al. (2016). Exogenously applied plant growth regulators enhance the morpho-physiological growth and yield of rice under high temperature. *Frontiers in Plant Science*, 7: 1250.

Faisal, M., Faizan, M., Soysal, S., and Alatar, A. A. (2024). Synergistic application of melatonin and silicon oxide nanoparticles modulates reactive oxygen species generation and the antioxidant defense system: a strategy for cadmium tolerance in rice. *Frontiers in Plant Science*, 15: 1484600.

Farooq, M., Rehman, A., Wahid, A., and Siddique, K. H. (2024). Physiology of Photosynthesis in Plants 31 under High-Temperature Stress.

Ficociello, G. (2017). Analysis of the role of the secretory pathway Ca2+/Mn2+-ATPase, PMR1/ATP2C1, in calcium homeostasis: from oxidative stress to skin disease.

Ghosh, P., Konar, A., Halder, S., Shivaranjani, D. S., and Kaur, J. (2024). Impact of nanotechnology on the environment: a review. *Environ Sci. Arch*, *3*(1), 152–163.

Hamdan, H. Z., and Salam, D. A. (2023). Sediment microbial fuel cells for bioremediation of pollutants and power generation: a review. *Environmental Chemistry Letters*, 21(5): 276–2787.

Hatfield, J. L., and Prueger, J. H. (2015). Temperature Extremes: Effects on Plant Growth and Development. http://dx.doi.org/10.1016/j.wace.2015.08.001 2212-0947.

Hodges, D. M. (2023). Chilling effects on active oxygen species and their scavenging systems in plants. In *Crop Responses and Adaptations to Temperature Stress* (pp. 53–76). CRC Press.

Huang, M., Huang, J., Yu, H., Fangzhang, Q., and Liu, M. (2024). Assessment of the phytotoxicity of HNTs on two plants, Cynodon dactylon and Brassica rapa L. *Environmental Science: Nano*.

Igiebor, F, A., Asia, M., and Ikhajiagbe, B. (2022). Green Nanotechnology: A modern tool for sustainable agriculture. *International Journal of Horticultural Science and Technology*. Journal homepage: http://ijhst.ut.ac.ir.

Jones, J. L., Berube, D. M., Cuchiara, M., Grieger, K., Hubal, E. A. C., Karikó, S. J. et al. (2024). Positioning nanotechnology to address climate change. *Environment Systems and Decisions*, 1–15.

Kan, Y., Mu, X. R., Gao, J., Lin, H. X., and Lin, Y. (2023). The molecular basis of heat stress responses in plants. *Molecular Plant*, 16(10): 1612–1634.

Karnwal, A., Jassim, A. Y., Mohammed, A. A., Sharma, V., Al-Tawaha, A. R. M. S., and Sivanesan, I. (2024). Nanotechnology for healthcare: plant-derived nanoparticles in disease treatment and regenerative medicine. *Pharmaceuticals*, 17(12): 1711.

Kumudini, S., Andrade, F. H., Boote, K. J., Brown, G. A., Dzotsi, K. A., Edmeades, G. O. et al. (2014). Predicting Maize Phenology: Intercomparison of functions for developmental response to temperature. *Agronomy Journal*, 106(6): 2087–2097.

Lekshmi, A. A., Lekshmi, S., Hema, S., Gopal, K., Sambhudevan, S., and Chandran, S. S. (2024). Nanofertilizers for the sustainable management of agricultural resources: A pathway to meet UN Millennium Sustainable Development Goals (MDGs). In *Nanofertilizer Delivery, Effects and Application Methods* (pp. 129–147). Elsevier.

Liu, J., Zhang, R., Xu, X., Fowler, J. C., Miller, T. E., and Dong, T. (2020). Effect of summer warming on growth, photosynthesis and water status in female and male Populus cathayana: Implications for sex-specific drought and heat tolerances. *Tree Physiology*, 40(9): 1178–1191.

Luo, J., Huang, S., Chang, Y., Li, H., and Guo, G. (2023). Physiological and transcriptomic analyses reveal tea plant (*Camellia sinensis* L.) adapts to extreme freezing stress during winter by regulating cell wall structure. *BMC genomics*, 24(1): 558.

Malik, S., Muhammad, K., and Waheed, Y. (2023). Nanotechnology: A revolution in modern industry. *Molecules*, 28(2): 661.

Milani, M., Curia, R., Shevlyagina, N. V., and Tatti, F. (2023). Nanoworld. In *Bacterial Degradation of Organic and Inorganic Materials: Staphylococcus aureus Meets the Nanoworld* (pp. 39–54). Cham: Springer International Publishing.

Mimi, A., Amin, M. R., Haque, M. A., Islam, M. W., and Tareq, M. Z. (2024). Characterization, structure-function relationship and mechanism of antifreeze proteins. *Bangladesh J*: 46: 61–72.

Mondal, S., Karmakar, S., Panda, D., Pramanik, K., Bose, B., and Singhal, R. K. (2023). Crucial plant processes under heat stress and tolerance through heat shock proteins. *Plant Stress*, 100227.

Nair, K. M., Mohanan, D., Ramya, K., and Francis, K. (2024). Nanotechnology in Agriculture and Food. *Applications of Nanotechnology*.

Pechyen, C., Tangnorawich, B., Toommee, S., Marks, R., and Parcharoen, Y. (2024). Green synthesis of metal nanoparticles, characterization, and biosensing applications. *Sensors International*, 100287.

Pramanik, P., Krishnan, p., Mity, A., Mridha, N., Mukherjee, A., and Rai, V. (2020). © Springer Nature Switzerland AG 2020. N. Dasgupta et al. (eds.). Environmental Nanotechnology Volume 4, Environmental Chemistry for a Sustainable World 32, https://doi.org/10.1007/978-3-030-26668-4_9.

Prasad, R., Bhattacharyya, A., and Nguyen, Q. D. (2017). Nanotechnology in sustainable agriculture: recent developments, challenges, and perspectives. *Front. Microbiol.*, 8: 1014. doi: 10.3389/fmicb.2017.01014.

Puustinen, M. C. (2023). HSF1 and HSF2 in Stress and EMT-Associated Transcriptional Networks.

Rajput, V. D. (Ed.). (2024). *Nanotechnology Applications and Innovations for Improved Soil Health.* IGI Global.

Ramadevi, D., and Ushasri, K. (2023). Nanotechnology Applications in the Environment. *24th & 25th Feb,* 2023, 17.

Ramos-Ruiz, R., Martinez, F., and Knauf-Beiter, G. (2019). The effects of GABA in plants. *Cogent Food & Agriculture*, 5(1): 1670553.

Raza, A., Wang, D., Zou, X., and Prakash, C.S. (2023). Developing temperature-resilient plants: a matter of present and future concern for sustainable agriculture. *Agronomy*, 13(4): 1006; https://doi.org/10.3390/agronomy13041006.

Ridhi, R., Saini, G. S. S., and Tripathi, S. K. (2024). Nanotechnology as a sustainable solution for proliferating agriculture sector. *Materials Science and Engineering: B*, 304: 117383.

Saikia, N. (2024). Inorganic-based nanoparticles and biomaterials as biocompatible scaffolds for regenerative medicine and tissue engineering: Current advances and trends of development. *Inorganics*, 12(11): 292.

Santos-Espinoza, A. M., González-Mendoza, D., Ruiz-Valdiviezo, V. M., Luján-Hidalgo, M. C., Jonapa-Hernández, F., Valdez-Salas, B. et al. (2021). Changes in the physiological and biochemical state of peanut plants (*Arachis hypogaea* L.) induced by exposure to green metallic nanoparticles. *International Journal of Phytoremediation*, 23(7): 747–754.

Saxena, A., Jain, A., Upadhyay, P., and Gauba, P. (2018). Applications of Nanotechnology in Agriculture.

Schlenker, W., and Roberts, J. M. (2009). Nonlinear effects indicate severe damage to US crop yields under climate change. *Proceedings of the National Academy of Sciences*., 106 (37), 15594–15598.

Sher, A., Noor, M. A., Li, H. X., Nasir, B., Manzoor, M. A., Hussain, S. et al. (2024). Heat stress effects on legumes: challenges, management strategies and future insights. *Plant Stress*, 100537.

Siddhartha, S. Mukhopadhyay. (2014). Nanotechnology in agriculture: Prospects and constraints. *Nanotechnology, Science and Applications*, 63–71, DOI: 10.2147/NSA.S39409. https://doi.org/10.2147/NSA.S39409.

Sidhu, A. K., Sharma, M., Agarwal, S. B., Bhavsar, P., and Samota, M. K. (2024). Nanomaterial Strategies for Enhancing Plant Resilience in the Face of Temperature Stress. *CABI Agriculture and Bioscience*, 5: 60 https://doi.org/10.1186/s43170-024-00255-w.

Singh, S., Sangwan, S., Sharma, P., Devi, p., and Moond, M. (2021). Nanotechnology for sustainable agriculture: an emerging perspective. *Journal of Nanoscience and Nanotechnology*, 21: 3453 – 3465, www.aspbs.com/jnn.

Soualiou, S., Duan, F., Li, X., and Zhou, W. (2022). Crop production under cold stress: an understanding of plant responses. *Acclimation process and Management Strategies*. https://doi.org/10.1016/j.plaphy.2022.08.024.

Usman, M., Farooq, M., Wakeed, A., Nawaz, A., Cheema, S. A., Rehman, H. et al. (2020). Nanotechnology in Agriculture: Current Status, Challenges and Future Opportunities https://doi.org/10.1016/j.scitotenv.2020.1377780048-9697/© 2020 Elsevier B.V.

Ullah, I., Toor, M. D., Basit, A., Mohamed, H. I., Gamal, M., Tanveer, N. A. et al. (2023). Nanotechnology: an Integrated Approach Towards Agriculture Production and Environmental Stress Tolerance in Plants. *Water, Air, & Soil Pollution*, 234(11): 666.

Verma, A., Inslicht, S. S., and Bhargava, A. (2024). Gut-brain axis: role of microbiome, metabolomics, hormones, and stress in mental health disorders. *Cells*, 13(17): 1436.

Wahab, A., Batool, F., Muhammad, M., Zaman, W., Mikhlef, R. M., and Naeem, M. (2023). Current knowledge, research progress, and future prospects of phyto-synthesized nanoparticles interactions with food crops under induced drought stress. *Sustainability*, 15(20): 14792.

www.pnas.org/lookup/suppl/doi:10.1073/pnas.1701762114/-/DCSupplemental.

Xu, Z., Zhang, J., Wang, X., Essemine, J., Jin, J., Qu, M. et al. (2023). Cold-induced inhibition of photosynthesis-related genes integrated by a TOP6 complex in rice mesophyll cells. *Nucleic Acids Research*, 51(4): 1823–1842.

Zhang, T., Jia, L., Niu, Z., Li, X., Men, S., Jiang, L. et al. (2024). Comparative transcriptomic analysis delineates adaptation strategies of Rana kukunoris toward cold stress on the Qinghai-Tibet Plateau. *BMC Genomics*, 25(1): 363.

Zhang, Y., Xu, J., Li, R., Ge, Y., Li, Y., and Li, R. (2023). Plants' response to abiotic stress: Mechanisms and strategies. *International Journal of Molecular Sciences*, 24(13): 10915.

Zhao, L., Lu, L., Wang, A., Zhang, H., Huang, M., Wu, H. et al. (2020). Nano-biotechnology in agriculture: use of nanomaterials to promote plant growth and stress tolerance. *Journal of Agricultural and Food Chemistry*, 68(7): 1935–1947.

Zhou, S., Yu, B., and Zhang, Y. (2023). Global concurrent climate extremes exacerbated by anthropogenic climate change. *Science Advances*, 9(10): eabo1638.

Zulfiqur, H., Afroze, B., Shakoor, S., Bhutta, M. S., Ahmed, M., Hassan, S. et al. (2024). Nanoparticles in Agriculture: Enhancing Crop Resilience and Productivity against Abiotic Stresses. DOI: 10.5772/intechopen.114843.

CHAPTER-6

Green Nanotechnology for Mitigating Heavy Metals Stress in Plants

Saiqa Nazir, Hajra Abid, Sajjad Hussain, Anar Mehmood and *Khawaja Shafique Ahmad**

1. Introduction

Globally, heavy metal contamination has become a serious issue due to growing industrialization and rapid urbanization (Khan et al., 2020; Arshad et al., 2023). Heavy metal-induced stress is one of the primary abiotic stresses that adversely affect plant growth and crop productivity worldwide (Ojuederie et al., 2019). Heavy metals are defined as elements with a density exceeding 5 g/cm^3 (Gao et al., 2024) and an atomic number greater than 20, encompassing various semi-metals, transition metals, lanthanides, and actinides (Inobeme et al., 2023). According to the World Health Organization (WHO), pollution caused by heavy metals from industrial and human activities is a critical environmental and public health concern (Ranjbar et al., 2023). It is estimated that over 20 million hectares of land globally are contaminated with metals such as cadmium, arsenic, mercury, nickel, zinc, lead, copper, and chromium (Liu et al., 2018). The Agency for Toxic Substances and Disease Registry (ATSDR) categorizes cadmium, arsenic, lead, and mercury as extremely hazardous heavy metals, posing significant risks to both plant and human health (Mansoor et al., 2021). Although these elements occur naturally in trace amounts due to processes such as volcanic explosions and weathering of rocks (Konate et al., 2017), anthropogenic activities have significantly elevated their concentrations in the environment, exacerbating their toxicity and persistence (Sheng et al., 2012; Shahid et al., 2015). They accumulate through leaching or release during oxidation in soil, which promotes their absorption by plants and ultimately threatens human health through bioaccumulation in the food chain (Hossain et al., 2008). Moreover, with the global population projected to reach 8.54 billion by 2030 and 9.73 billion by 2050, food production must increase by 70–100% to satisfy the demand of the growing population (UN, 2020; Mueller et al., 2012). Thus, enhancing plant tolerance to heavy metal stress is a critical objective to mitigate the negative impact on agricultural productivity and ensure sufficient food supply for the expanding global population (Shahy and Wu, 2019).

Department of Botany, University of Poonch Rawalakot, Rawalakot 12350, Pakistan.
* Corresponding author: ahmadks@upr.edu.pk

DOI: 10.1201/9781003567684-6

Soil contaminated with heavy metals adversely impacts numerous biological processes in plants and soil microbiota, altering their diversity, population density, and genetic structure (Rajput et al., 2022). This contamination diminishes enzymatic activities, reduces biomass, and lowers crop productivity while increasing oxidative stress due to the overproduction of Reactive Oxygen Species (ROS) (Huang et al., 2019). After infiltrating plant tissues, heavy metals disrupt the plant's life cycle, resulting in inhibited growth and decreased yields (Rizwan et al., 2019a). The severity of damage caused by heavy metals depends on factors such as plant species, the specific type of metal, and the physicochemical properties of the soil (Chaplygin et al., 2021). Metal accumulation triggers both direct and indirect detrimental effects, leading to morpho-physiological irregularities such as reduced photosynthesis, impaired growth, and decreased biomass and yield. Additionally, it promotes ROS generation, disrupting redox balance, damaging cellular components such as membranes, proteins, lipids, and nucleic acids, and interfering with metabolic functions, ultimately resulting in lower agricultural productivity (Hassan et al., 2017; Khan et al., 2019). Physiologically, plants often respond to heavy metal stress by accumulating elevated levels of phenolic compounds as a defense mechanism, which plays a role in mitigating toxicity by chelating metal ions or alleviating oxidative stress. The development and widespread adoption of effective technologies for the removal or immobilization of heavy metals in contaminated farmland are crucial for mitigating these adverse effects (Nouman et al., 2020).

Nanotechnology has emerged as a highly innovative and effective strategy during recent decades for managing environmentally induced abiotic stresses, improving agricultural yield, addressing nutrient scarcities, and revolutionizing biological systems (El-Saadony et al., 2022). Nanoparticles (NPs), defined as very small-sized particles with at least two dimensions between 1–100 nm in size (Afzal and Singh, 2022), have gained significant attention in this context. Although concerns remain regarding their synthesis methods, cellular uptake pathways, and potential ecotoxicological effects, the unique properties of NPs such as their robust structure, nanoscale size, large surface area, and exceptional adsorption capacity make them promising agents for heavy metal remediation in plants (Fatemi et al., 2021). NPs can enter plants primarily through stomatal openings, which have an average diameter of about 10 nm initiate various physiological and biochemical changes, including the reduction of ROS levels within cells, thereby mitigating oxidative damage and enhancing plant stress tolerance (Rubio et al., 2019; Singh et al., 2021; Emamverdian et al., 2022; Emamverdian et al., 2023).

Numerous studies have demonstrated the potential of NPs to enhance plant growth, development, and productivity (Faraz et al., 2018; Faizan et al., 2018). NPs have been shown to improve plant resistance to various toxic metals such as chromium (Cr), cadmium (Cd), iron (Fe), aluminum (Al), and manganese (Mn) (Shen et al., 2009; Skiba and Wolf, 2019; Zadeh et al., 2019). The application of NPs has been reported to mitigate the uptake and translocation of heavy metals like lead (Pb) (Cai et al., 2017), Cd (Faraji et al., 2018), and Mn (Ragab and Saad-Allah, 2020) in plant tissues. Furthermore, they play a significant role in protecting plants from environmental stresses by enhancing antioxidant enzyme activities, accumulating osmolytes, and facilitating the synthesis of free amino acids, as well as improving nutrient availability (Farhangi-Abriz and Torabian, 2018). In response to abiotic stresses, predominantly heavy metal toxicity, nanomaterials have been shown to increase plant adaptability and agricultural productivity (Yan et al., 2021; Sardar et al., 2022). This is achieved by supplying essential nutrients, minimizing nutrient losses due to their high surface area and nutrient-retention ability (Ahmed et al., 2021), maintaining phytohormone levels, boosting antioxidant defenses, enhancing photosynthesis, and reducing the accumulation of heavy metals in plant tissues. Owing to their unique properties, NPs not only provide superior remediation effects compared to traditional materials but also introduce novel functionalities (Moharem et al., 2019). In addition to soil application, foliar spraying is another effective method for enhancing plant resistance to heavy metal stress (Usman et al., 2020; Toumey, 2020). For instance, foliar applications of selenium and silicon NPs have been shown to alleviate Cd and Pb toxicity in rice

(Hussain et al., 2020). Similarly, the foliar application of TiO_2 NPs has proven more effective than soil application in mitigating Cd toxicity in maize (Lian et al., 2020). This chapter provides an overview of different types of nanoparticles and their role and mechanism in the mitigation of heavy metals stress tolerance in plants.

2. Green Synthesis of Nanoparticles

Biological methods utilizing natural stabilizing and reducing agents enable the production of NPs in an eco-friendly and cost-effective manner. Green nanotechnology is rapidly emerging as an innovative approach for the synthesis of NPs with minimal environmental impact. Biological methods utilizing natural stabilizing and reducing agents enable the production of NPs in an eco-friendly and cost-effective manner. Unlike traditional chemical and physical techniques, these methods require less energy and avoid the use of hazardous materials. Green synthesis, is the process that employs biological entities such as bacteria, fungi, algae, and plants for NPs production (Patil and Chandrasekaran, 2020; Andrade-Zavaleta et al., 2022), has gained significant attention due to its eco-friendly, non-toxic nature, and adaptability to biological systems (Mohd Yusof et al., 2019). Bacteria are widely utilized in the synthesis of NPs due to their inherent ability to reduce metal ions (Agarwal et al., 2017). Proteins present in the bacterial cytoplasm and cell wall, containing amino acids like tryptophan and tyrosine, contribute to the stabilization and reduction of bacterial NPs. Moreover, certain sugars, such as aldose and ketose, exhibit stabilizing and reducing properties. These bioactive compounds, commonly found in different bacterial species, interact with metal ions to reduce their size and promote the formation of complex structures, such as spherical NPs. The ease of handling and culturing microorganisms has led to the growing popularity of bacterial NP synthesis (Adelere and Lateef, 2016). The final size of the synthesized NPs depends on their interactions with biologically active substances and their localization, whether extracellular or intracellular (Thakkar et al., 2010). Fungi have been utilized to synthesize various NPs, including those of gold, silver, zinc oxide, and titanium dioxide, due to their high capacity for bioaccumulation and ease of cultivation. Enzyme reductases, which facilitate enzymatic reduction, along with proteins in fungal cell walls and cells, are pivotal in the biological synthesis of metallic NPs (Mohanpuria et al., 2008). Algae also play a significant role in nanomaterial synthesis. For instance, certain species like *Sargassum tenerrimum* contain a diverse range of bioactive compounds, including phenolic compounds, alkaloids, flavonoids, carbohydrates, amino acids, sterols, tannins, and saponins. After purification, these compounds are critical in modulating the size, shape, and functionality of nanomaterials. Characterizing algae is a fundamental step toward advancing future research and exploring their potential applications in NP synthesis (Saleh, 2020).

Plant-based synthesis among other biological approaches is particularly advantageous as it is more economical, efficient, and simpler to scale up compared to microorganism-based methods. The use of plant-derived extracts eliminates the challenges associated with microbial isolation, reduces biohazard risks, and minimizes the likelihood of contamination, making it a more reliable and sustainable alternative. Plants offer a more efficient, straightforward, cost-effective, and sustainable alternative to traditional NP synthesis methods (Akhtar et al., 2013). They are rich in bioactive compounds such as sugars, vitamins, and polyphenols, which facilitate the reduction of metal ions to atoms and promote the synthesis of metal and metal oxide NPs (Purnomo et al., 2023). Key factors influencing the green synthesis of NPs include the concentration of the plant extract, the metal salt, the pH, and the temperature (Mittal et al., 2013). The process of green NPs synthesis can be summarized in the following steps (Singh et al., 2018):

(a) Preparation of microbe or plant-derived extracts and mixing them with a metal salt solution.

(b) Biochemical reduction and fabrication of the metal salt solution, where monovalent or divalent metal ions are reduced to their zero-valent state, followed by their nucleation and growth into NPs.

(c) Analysis of the synthesized NPs using UV-Vis spectroscopy to confirm their formation.

(d) Characterization of NPs using advanced techniques such as Transmission Electron Microscopy (TEM), Scanning Electron Microscopy (SEM), and X-Ray Diffraction (XRD), followed by their application in various fields.

3. Nano-remediation of Heavy Metal Contaminated Soil

Nano-remediation refers to the removal or neutralization of environmental contaminants using nanomaterials or NPs synthesized through green methods (Amin et al., 2021; Baragano et al., 2020). This eco-friendly and inexpensive strategy employs mechanisms such as adsorption, reduction of toxic ions to stable metallic states, and acceleration of degradation processes, proving highly effective in remediating heavy metals (Mar Gil-Diaz et al., 2014; Baragano et al., 2020). Nanomaterials having high surface area and the capability to adjust chemical composition and physical properties gained significant attraction for soil remediation in recent years. This process breaks down pollutants in the soil and may be capable of eliminating, retaining, or reducing the concentration of pollutants present in the soil (Akhtar et al., 2022). Nano-remediation uses various technical approaches to address heavy metal contamination, including adsorption, heterogeneous catalysis, photodegradation, electro-nanoremediation (using electrical fields), and microbial involvement (nano-bioremediation) to remove or immobilize heavy metals (Mukhopadhyay et al., 2022). Within bioremediation strategies, phytoremediation is considered one of the most efficient, sustainable, and economical methods. Derived from the Greek word "phyton" (plants) and the Latin "remediare" (to remediate), phytoremediation involves plants and soil microbes working together to transform toxic pollutants into harmless or even beneficial substances. This process is particularly effective for removing various soil impurities, leveraging the interaction between plants and nanoparticles to enhance pollutant availability to plants, stimulate plant growth, and eliminate toxic contaminants from polluted soils. Key mechanisms of phytoremediation include phytostabilization, phytoextraction, phytovolatilization, phytodegradation, and rhizofiltration which enable plants to play a crucial role in detoxifying contaminated environments (Singh and Lee, 2016).

Nanotechnology has emerged as a promising tool in agriculture, significantly enhancing plant growth and development due to the unique properties of NPs as numerous studies during recent years have demonstrated that the adverse effects of heavy metals on plants can be mitigated through the application of NPs (Rizwan et al., 2019b; Lobzenko et al., 2022). Research indicates that NPs effectively reduce the toxicity and accumulation of metals in plants by limiting metal uptake, enhancing mineral nutrition, improving photosynthesis, and boosting the activity of antioxidant enzymes (Adrees et al., 2020; Irshad et al., 2021). Recent findings suggest that the external application of NPs can improve plant adaptation to heavy metals stress by regulating morphological, physiological, and molecular mechanisms (Bidi et al., 2021; Nazir et al., 2022a; Nazir et al., 2022b). Nanomaterials have been shown to increase plant resilience and improve agricultural productivity under HM toxicity by supplying essential nutrients, minimizing nutrient losses through their large surface area and high nutrient-binding capacity, maintaining phytohormone levels, enhancing photosynthetic efficiency, strengthening antioxidant system, and reducing metal uptake (Yan et al. 2021) with minimal environmental impact. Biological methods utilizing natural stabilizing and reducing agents enable the production of NPs in an eco-friendly and cost-effective manner. Sardar et al. (2022). The nutrient retention capability of NPs can be attributed to their distinct physical and chemical properties. For instance, carbon-based NPs with nanoscale pores effectively absorb and retain micronutrients, while multi-nutrient NPs, such as those composed of nitrogen, phosphorus, and potassium (NPK-NPs), provide an integrated nutrient delivery system to support plant growth and mitigate stress.

The application of NPs not only positively influences plant growth and development but also plays a significant role in removing heavy metals from contaminated environments, particularly through phytoremediation approaches (Wu, 1994). Various types of NPs have been developed depending on their intended applications, including inorganic nonmetallic NPs, metallic NPs, carbon-based NPs, and organic polymeric materials. Carbon nanotubes (CNTs), for example, have shown promise in environmental remediation. Camilli et al. (2014) reported the ability of CNTs to remove hazardous organic solvents, such as dichlorobenzene, as well as the heavy metal Cd from contaminated soils. Similarly, iron- and sulfur-based nanosponges have been found to enhance the absorption efficiency of cadmium and other pollutants in soil. Similarly, CNTs immobilize heavy metal ions, including nickel (Ni^{2+}), zinc (Zn^{2+}), lead (Pb^{2+}), and copper (Cu^{2+}), making them effective tools for soil remediation and environmental cleanup (Matos et al., 2017).

4. Types of Nanoparticles

4.1 Types and effects of metal based NPs on heavy metals stress in plants

4.1.1 Silver Nanoparticles (Ag NPs)

AgNPs exhibit remarkable properties compared to other NPs due to their unique physicochemical properties, including antioxidant and antimicrobial activities (Mohamed et al., 2017). AgNPs play a pivotal role in mitigating heavy metal toxicity in plants, as evidenced by various studies. For instance, Chen et al. (2022) demonstrated that AgNPs at concentrations of 10–50 mg/L effectively alleviated Pb stress in plants. They enhanced plant growth, biomass, productivity, chlorophyll content, photosynthetic efficiency, antioxidants activities, water-use efficiency, and ionic homeostasis while simultaneously reducing Pb accumulation in roots, ROS production, and malondialdehyde (MDA) levels in *Vigna radiata*. Faiz et al. (2022) observed that AgNPs stimulated the activity of antioxidant enzymes such as peroxidase (POX), phenylalanine ammonia-lyase (PAL), and polyphenol oxidase (PPO), along with improving chlorophyll content and growth-related parameters in *Daucus carota* under Cd stress. Moreover, AgNPs significantly reduced Cd uptake, ROS generation, and MDA content, thereby highlighting their potential to mitigate heavy metal stress in plants. AgNPs were also reported to reduce the Cd toxicity in *Datura stramonium* callus culture by enhancing fresh weight, dry weight, relative water content, chlorophyll contents, antioxidant activities, and decreasing H_2O_2 (Jdayea et al., 2023).

4.1.2 Iron Nanoparticles (Fe NPs)

FeNPs particularly Fe_2O_3, are integral to plant metabolic processes such as photosynthesis, respiration, production of photosynthetic pigments, and synthesis of DNA. They also possess exceptional adsorptive capabilities due to their unique structural and electronic properties. Magnetic NPs (γ-Fe_2O_3 and Fe_3O^4) are particularly advantageous as they can be efficiently separated from the adsorption medium using magnetic fields (Lin et al., 2019). The mechanisms by which FeNPs alleviate heavy metal stress include metal adsorption, their role as an essential element, the formation of iron films on root surfaces, oxidative defense system stimulation, and ROS scavenging (Konate et al., 2017). Numerous studies highlight the significant potential of FeNPs in agricultural applications, particularly as fertilizers. For instance, nanoscale Zero-Valent Iron (nZVI) has been found to decrease the accumulation of heavy metals in sunflowers, enhance superoxide dismutase (SOD) and peroxidase (POD) activity in leaves as well as plant growth (Michalkova et al., 2017). FeO nanoparticles coated with polyvinylpyrrolidone (PVP) have been utilized effectively by Gram-negative bacteria (*Halomonas* sp.) for bioremediation of Pb and Cd-contaminated soils. This approach achieved almost 100% Pb removal within 24 hours and approximately 100% Cd removal within 48 hours, overtaking either bacteria or NPs alone (Alabresm et al., 2018). Hussain et al. (2019) investigated the role of FeO NPs in mitigating Cd stress. Application of FeNPs at 20 ppm eliminates Cd toxicity by enhancing Fe uptake, improving plant growth, boosting antioxidant enzyme activity, and increasing photosynthetic efficiency, thus lowering Cd concentrations in plants. The application of Fe_2O_3 NPs (5, 10, 15, and 20 ppm) through soil and leaves in wheat exposed to Cd stress reduced leaf electrolyte escape rate and Cd content in grains, while increasing antioxidant enzyme activity and dry biomass. Foliar application was found to be more effective than soil application, as soil absorption of Fe is influenced by factors like pH and interactions with other minerals during uptake.

Moreover, combining FeNPs foliar spraying with biochar soil application further enhanced Cd stress alleviation in rice (Rizwan et al., 2019).

4.1.3 Copper Nanoparticles (Cu NPs)

CuNPs exhibit unique characteristics compared to other metallic NPs (Grigore et al., 2016) and have been demonstrated to mitigate Cd stress in *Solanum melongena* by improving antioxidant enzyme activity, photosynthetic pigments, osmoregulators, and plant growth parameters (Alomrani et al., 2024). Noman et al. (2020) recognized that CuNPs promoted plant growth ,and nutrient accumulation while lowering Cd transportation from soil to *Triticum aestivum*. They have also been shown to play a fundamental role in enhancing food safety and nutritional quality improvement, particularly in rice grown in As-contaminated areas, where they significantly reduced As levels in grains (Liu et al., 2018). Similarly, Wang et al. (2022) demonstrated the efficacy of CuNPs in alleviating Cd toxicity in *Brassica* species by enhancing the activity of antioxidant enzymes such as SODD, POD, and CAT along with increasing plant biomass. These findings underscore the potential of CuNPs as effective tools for enhancing plant resilience to heavy metal stress while improving crop quality.

4.1.4 Zinc oxide Nanoparticles (ZnO NPs)

ZnO NPs have long been utilized to provide essential nutrients to crops grown in zinc-deficient areas. Moreover, ZnO NPs contribute to improving physiological functions and promoting plant growth by mitigating the accumulation and toxicity of heavy metals. The application of ZnO NPs with arsenite (AsIII) or arsenate (AsV) significantly reduced total As levels in rice plants (Wang et al., 2018). Hussain et al. (2018) investigated the interaction of ZnO-NPs with *Triticum aestivum* and reported that ZnO NPs enhanced SOD and POD activities while reducing Cd concentrations in the plants. Likewise, these were found to eliminate phytotoxicity and considerably decrease As and Cd accumulation in *Oryza sativa* grains (Ma et al., 2020). Wu et al. (2020) further demonstrated that ZnO NPs promoted the growth and biomass of *O. sativa* under As stress by reducing As accumulation and increasing the uptake of zinc. Moreover, the combined application of Fe and ZnO NPs has been shown to increase Cd accumulation in *Boehmeria nivea* (Gong et al., 2017). In another study, ZnO NPs applied to two varieties of chickpea plants decreased Cr absorption in roots and its translocation within the plants, while improving photosynthetic efficiency and the activity of antioxidant enzymes under elevated Cr stress (Singh et al., 2023). These findings highlight the multifunctional role of ZnO NPs in alleviating HM toxicity and enhancing crop productivity.

4.1.5 Titanium oxide Nanoparticles (TiO_2 NPs)

TiO_2 NPs are mainly recognized for their significant potential in environmental applications due to their remarkable physical and chemical stability, less toxicity, corrosion resistance, and cost-effectiveness (Samanta and Roychoudhury, 2021). Globally, TiO_2 NPs are among the most widely manufactured NPs (Zhou et al., 2020). Their strong oxidizing properties, exceptional resistance to photodegradation, and ability to mediate selective redox reactions make them highly suitable for environmental remediation (Wu et al., 2021). The application of nano-TiO_2 has been shown to enhance photosynthesis and growth parameters while mitigating the adverse effects of Cd toxicity in *Glycine max* (Singh and Lee, 2016; Usman et al., 2020). Nano-TiO_2 reduces oxidative damage and scavenges free radicals induced by Cd stress, thereby decreasing Cd toxicity while increasing Cd uptake in soybean plants. Similar outcomes were observed by Sardar et al., (2022), who demonstrated that nano-TiO_2 effectively remediated Cd in coriander by reducing Cd content, minimizing oxidative damage caused by Cd stress, and improving agronomic performance. Singh et al. (2016) further reported improved photosynthetic rates and growth parameters in soybean with the application of TiO_2 NPs. The exogenous application of TiO_2 NPs on *Helianthus annuus* under Cr toxicity revealed that these NPs significantly reduced Cr accumulation in both the roots and leaves, further highlighting their potential to enhance plant resilience to heavy metal stress (Kumar et al., 2023).

4.2 Types and Effects of Non-metal Based NPs on Heavy Metals Stress in Plants

Non-metal based nanoparticles have shown promising effects in mitigating heavy metal stress in plants. Non-metal based nanoparticles are tiny particles made of non-metallic materials, such as ceramics, carbon, or silicon. These nanoparticles can be used to reduce the bioavailability of heavy metals in soil, thereby minimizing their uptake by plants (Zhou et al., 2021).

4.2.1 Carbon-based nanoparticles: These nanoparticles have been shown to reduce the toxicity of heavy metals such as cadmium and lead in plants. These nanoparticles have different types such as

Carbon nanotubes (CNTs): CNTs have been shown to reduce heavy metal uptake and improve plant growth.

Graphene oxide (GO): GO has been shown to reduce oxidative stress and improve plant tolerance to heavy metals.

Fullerenes: Fullerenes have been shown to reduce heavy metal toxicity and improve plant growth.

Carbon dots (CDs): CDs have been shown to reduce oxidative stress and improve plant tolerance to heavy metals.

4.3 Silica Nanoparticles

These nanoparticles can increase the antioxidant activity of plants, helping them to cope with heavy metal stress. Silica nanoparticles (SiNPs) are nanoparticles composed of silicon dioxide (SiO_2). They have gained significant attention in recent years due to their unique properties such as high surface area, stability, biocompatibility, and optical properties and potential applications such as (Selvarajan et al., 2020).

1. Biomedical applications: SiNPs are being explored for their potential in biomedical applications, such as drug delivery, imaging, and diagnostics.
2. Catalysis: SiNPs can be used as catalysts to facilitate chemical reactions, such as hydrogenation and oxidation.
3. Environmental remediation: SiNPs can be used to remove pollutants and contaminants from water and soil.
4. Agricultural applications: SiNPs are being explored for their potential in agricultural applications, such as plant protection and nutrition.
5. Improved plant growth: SiNPs have been shown to improve plant growth and development by increasing nutrient uptake and reducing stress.
6. Enhanced antioxidant activity: SiNPs can enhance antioxidant activity in plants, helping to protect them from oxidative stress.
7. Increased drought tolerance: SiNPs can help plants tolerate drought by improving water retention and reducing transpiration.
8. Reduced heavy metal toxicity: SiNPs can reduce the toxicity of heavy metals in plants by binding to them and reducing their bioavailability.

4.4 Polymeric Nanoparticles

These nanoparticles can be used to deliver nutrients and other beneficial substances to plants, helping them to mitigate heavy metal stress (Altammar, 2023). Polymeric nanoparticles

(PNPs) are nanoparticles composed of polymers, which are long chains of repeating molecules. Polymeric nanoparticles are nanoscale-sized objects constructed from polymers such as polyethylene glycol (PEG) or polylactide-co-glycolide (PLGA), with applications in the in vitro delivery of therapeutic and diagnostic agents, alongside a number of additional biomedical functions such as (Zielinska et al., 2020).

1. Drug delivery: PNPs can be used to deliver drugs and other therapeutic agents to specific sites in the body (Xiao et al., 2022).
2. Imaging: PNPs can be used as contrast agents for imaging techniques, such as MRI and CT scans.
3. Diagnostics: PNPs can be used to detect biomarkers and other diagnostic targets.
4. Agricultural applications: PNPs can be used to deliver pesticides, fertilizers, and other agricultural chemicals (Vauthier, 2019).
5. Enhanced antioxidant activity: PNPs can enhance antioxidant activity in plants, helping to protect them from oxidative stress.
6. Increased drought tolerance: PNPs can help plants tolerate drought by improving water retention and reducing transpiration.
7. Reduced heavy metal toxicity: PNPs can reduce the toxicity of heavy metals in plants by binding to them and reducing their bioavailability.

5. The Mechanism of Action of Nanoparticles (NPs)

The mechanism of action of nanoparticles (NPs) can vary depending on their composition, size, shape, and surface properties. Here are some general mechanisms of action of NPs (Chen et al., 2023):

5.1 Physical Mechanisms

5.1.1 Adsorption: NPs can adsorb heavy metals, pollutants, or other substances, reducing their bioavailability and toxicity.

5.1.2 Aggregation: NPs can aggregate with other particles or substances, altering their physical and chemical properties.

5.1.3 Penetration: NPs can penetrate plant cell walls and membranes, delivering nutrients, hormones, or other substances.

5.2 Chemical Mechanisms

5.2.1 Redox reactions: NPs can participate in redox reactions, altering the oxidative state of surrounding molecules.

5.2.2 Catalysis: NPs can catalyze chemical reactions, enhancing or inhibiting specific processes.

5.2.3 Complexation: NPs can form complexes with ions, molecules, or other substances, altering their chemical properties.

5.3 Biological Mechanisms

5.3.1 Cell signaling: NPs can interact with plant cells, influencing signaling pathways and gene expression, helping to regulate plant growth and development.

5.3.2 Antioxidant activity: NPs can scavenge Reactive Oxygen Species (ROS), reducing oxidative stress and promoting plant growth.

5.3.3 Microbial interactions: NPs can interact with microorganisms, influencing their growth, metabolism, and interactions with plants.

5.4 Plant-Specific Mechanisms

5.4.1 Nutrient uptake: NPs can enhance or inhibit nutrient uptake by plants, by improving soil fertility and reducing nutrient leaching, depending on their composition and properties. PNPs can deliver nutrients and other beneficial substances to plants, improving growth and development (Vasile, 2018).

5.4.2 Hormone regulation: NPs can influence plant hormone regulation, affecting growth, development, and stress responses.

5.4.3 Defense mechanisms: NPs can trigger or enhance plant defense mechanisms, such as Systemic Acquired Resistance (SAR) or Induced Systemic Resistance (ISR).

5.4.4 Heavy metal binding: SiNPs can bind to heavy metals and reduce their bioavailability, helping to protect plants from heavy metal toxicity.

These mechanisms can vary depending on the specific type of NP, plant species, and environmental conditions. Further research is needed to fully understand the mechanisms of action of NPs in plants.

6. Challenges and Future Directions of Non-Metal-Based Nanoparticles

6.1 Scalability and cost-effectiveness: Large-scale production of non-metal-based nanoparticles for phytoremediation can be expensive and challenging.

6.2 Toxicity and environmental impact: The long-term toxicity and environmental impact of non-metal-based nanoparticles need to be thoroughly assessed.

6.3 Optimization and standardization: Further research is needed to optimize and standardize the use of non-metal-based nanoparticles for phytoremediation.

7. Conclusion

Green nanotechnology has emerged as a promising tool for mitigating heavy metal stress in plants. The use of eco-friendly nanoparticles, derived from natural sources such as plants, microorganisms, and biomolecules, offers a sustainable and environmentally benign approach to reducing heavy metal toxicity in plants. The application of green nanotechnology has shown significant potential in reducing heavy metal uptake and translocation in plants, enhancing antioxidant activity and stress tolerance in plants, promoting plant growth and development under heavy metal stress, and providing a cost-effective and scalable solution for heavy metal remediation. However, further research is needed to explore the mechanisms of action of green nanoparticles in plants, investigate the potential toxicity and environmental impact of green nanoparticles, and develop scalable and cost-effective methods for green nanoparticle synthesis and application. Overall, green nanotechnology has the potential to revolutionize the field of heavy metal remediation in plants, offering a sustainable and eco-friendly solution for mitigating heavy metal stress and promoting plant growth and development.

References

Adelere, I. A., and Lateef, A. (2016). A novel approach to the green synthesis of metallic nanoparticles: the use of agro-wastes, enzymes, and pigments. *Nanotechnology Revision*, 5: 567–587.

Adrees, M., Khan, Z. S., Ali, S., Hafeez, M., Khalid, S., ur Rehman, M. Z. et al. (2020). Simultaneous mitigation of cadmium and drought stress in wheat by soil application of iron nanoparticles. *Chemosphere*, 238: 124681.

Afzal, S., and Singh, N. K. (2022). Effect of zinc and iron oxide nanoparticles on plant physiology, seed quality and microbial community structure in a rice-soil-microbial ecosystem. *Environ. Pollut.*, 314: 120224.

Agarwal, H., Kumar V. S., and Rajeshkumar, S. (2017). A review on green synthesis of zinc oxide nanoparticles – An eco-friendly approach. *Resource Technology*, 3: 406–413.

Ahmed, T., Noman, M., Manzoor, N., Shahid, M., Hussaini, K. M., Rizwan, M. et al. (2021). Green magnesium oxide nanoparticles-based modulation of cellular oxidative repair mechanisms to reduce arsenic uptake and translocation in rice (*Oryza sativa* L.) plants. *Environmental Pollution*, 288: 117785.

Akhtar, M. S., Panwar, J., and Yun, Y. S. (2013). Biogenic synthesis of metallic nanoparticles by plant extracts. *ACS Sustainable Chemical Engineering*, 1: 591–602.

Akhtar, N., Ilyas, N., Meraj, T. A., Aboughadareh, A. P., Sayyed, R. Z., and Mashwani, Z. (2022). Improvement of plant responses by nanobiofertilizer: A step towards sustainable agriculture. *Nanomaterials*, 12: 965.

Alabresm, A., Chen, Y. P., Decho, A. W., and Lead, J. (2018). A novel method for the synergistic remediation of oil-water mixtures using nanoparticles and oil-degrading bacteria. *Science Total Environment*, 630: 1292–1297.

Alomrani, S. O., Kaleem, M., Aslam, M., Habib, F., Jamal, A., Waseem, M. et al. (2024). Copper nanoparticles alleviate cadmium stress in *Solanum melongena* through endogenous melatonin and regulation of some physiochemical attributes. *Sci. Horticult.*, 323: 112546.

Altammar, k. A. (2023). A review on nanoparticles: characteristics, synthesis, applications, and challenges. *Frontier Microbiology and Secondary Microbiotechnology*, 14. 28. https://doi.org/10.3389/fmicb.2023.1155622.

Amin, I., Nazir, R., and Rather, M. A. (2021). Nano-bioremediation: an innovative approach for remedying heavy metals using fungi. *Jornal of Bioremediation and Biodegradation*, 12: 2.

Andrade-Zavaleta, K., Chacon-Laiza, Y., Asmat-Campos, D., and Raquel-Checca, N. (2022). Green synthesis of superparamagnetic iron oxide nanoparticles with eucalyptus globulus extract and their application in the removal of heavy metals from agricultural soil. *Molecules*, 27: 1367.

Arora , N., Thangavelu, K., and Karanikolos, G. N. (2020). Bimetallic nanoparticles for antimicrobial applications. *Frontier Chemistry and Secondary Nanoscience*, 8: 223. https://doi.org/10.3389/fchem.2020.00412.

Arshad, K., Aqeel, M., Noman, A., Nazir, A., Mahmood, A., Rizvi, Z. F. et al. (2023). Ecological health risk assessment of microplastics and heavy metals in sediments, water, hydrophytes (Alternanthera philoxeroides, Typha latifolia, and Ipomoea carnea), and fish (*Labeo rohita*) in Marala wetlands in Sialkot, Pakistan. *Environmental Science and Pollution Research*, 30(14): 41272–41285.

Baragaoo, D., Forjan, R., Welte, L., and Gallego J. L. R. (2020) Nanoremediation of as and metals polluted soils by means of graphene oxide nanoparticles. *Science Reports*, 101(10): 1–10.

Bidi, H., Fallah, H., Niknejad, Y., and Tari, D. B. (2021). Iron oxide nanoparticles alleviate arsenic phytotoxicity in rice by improving iron uptake, oxidative stress tolerance and diminishing arsenic accumulation. *Plant Physiology and Biochemistry*, 163: 348–357.

Cai, F., Wu, X., Zhang, H., Shen, X., Zhang, M., Chen, W. et al. (2017). Impact of TiO_2 nanoparticles on lead uptake and bioaccumulation in rice (*Oryza sativa* L.). *Nanomaterials Impact*, 5: 101–108.

Camilli, L., Pisani, C., and Gautron, E. (2014). A three-dimensional carbon nanotube network for water treatment. *Nanotechnology*, 25: 065701.

Chaplygin, V. A., Minkina, T. M.,Mandzhieva, S. S., Nazarenko, O. G., Zimulina, I. V., and Bauer, T. V. (2021). Heavymetals in agricultural crops of Rostov region through the example of soft wheat (*Triticum aestivum*). In *IOP Conference Series: Earth and Environmental Science*, 624: 012204.

Chen, F., Aqeel, M., Maqsood, M. F., Khalid, N., Irshad, M. K., Ibrahim, M. et al. 2022. Mitigation of lead toxicity in *Vigna radiata* genotypes by silver nanoparticles. *Environmental Pollution*, 308: 119606.

Chen, J., Guo,Y., Zhang, X., Liu, J., Gong, P., and Su, Z. (2023). Emerging nanoparticles in food: sources, application, and safety. *Journal of Agriculture and Food Chemistry*, 71: 3564–3582.

El-Saadony, M. T., Saad, A. M., Soliman, S. M., Salem, H. M., Desoky, E. S. M., Babalghith, A. O. et al. (2022). Role of nanoparticles in enhancing crop tolerance to abiotic stress: A comprehensive review. *Front Plant Sci.*, 13: 946717.

Emamverdian, A., Ding, Y., Hasanuzzaman, M., Barker, J., Liu, G., Li, Y. et al. (2023). Insight into the biochemical and physiological mechanisms of nanoparticles-induced arsenic tolerance in bamboo. *FrontierPlant Science*, 14: 1121886.

Emamverdian, A., Hasanuzzaman, M., Ding, Y., Barker, J., Mokhberdoran, F., and Liu, G. (2022). Zinc oxide nanoparticles improve pleioblastus pygmaeus plant tolerance to arsenic and mercury by stimulating antioxidant defense and reducing the metal accumulation and translocation. *FrontierPlant Science*, 13: 841501.

Faiz, S., Shah, A. A., Naveed, N. H., Nijabat, A., Yasin, N. A. Batool, A. I. et al. (2022). Synergistic application of silver nanoparticles and indole acetic acid alleviate cadmium induced stress and improve growth of *Daucus carota* L. *Chemosphere*, 290: 133200.

Faizan, M., Faraz, A., Yusuf, M., Khan, S. T., and Hayat, S. (2018). Zinc oxide nanoparticle-mediated changes in photosynthetic efficiency and antioxidant system of tomato plants. *Photosynthetica*, 56: 678–686.

Faraji, J., Sepehri, A., and Salcedo-Reyes, J. C. (2018). Titanium dioxide nanoparticles and sodium nitroprusside alleviate the adverse effects of cadmium stress on germination and seedling growth of wheat (*Triticum aestivum* L.). *Universal Science*, 23: 61–87.

Faraz, A., Faizan, M., and Hayat, S. (2018). Effects of copper oxide nanoparticles on then photosynthesis and antioxidant levels of mustard plants (*Brassica juncea*). *Journal of Biological Chemistry Research*, 35: 418–426.

Farhangi-Abriz, S., and Torabian, S. (2018). Nano-silicon alters antioxidant activities of soybean seedlings under salt toxicity. *Protoplasma*, 255: 953–962.

Fatemi, H., Pour, B. E., and Rizwan, M. (2021). Foliar application of silicon nanoparticles affected the growth, vitamin C, flavonoid and antioxidant enzyme activities of coriander (*Coriandrumsativum* L.) plants grown in lead (Pb)-spiked soil. *Environmental Science Pollution Research*, 28: 1417–1425.

Gagliardi, A., Giuliano, E., Venkateswararao, E., Fresta, M., Bulotta, S., Awasthi, V. et al. (2021). Biodegradable Polymeric Nanoparticles for Drug Delivery to Solid Tumors. *Secondary Experimental Pharmacology and Drug Discovery*, 12. https://doi.org/10.3389/fphar.2021.601626.

Gao, X., Wei, M., Zhang, X., Xun, Y., Duan, M., Yang, Z. et al. (2024). Copper removal from aqueous solutions by white rot fungus Pleurotus ostreatus GEMB-PO1 and its potential in co-remediation of copper and organic pollutants. *Bioresource Technology*, 395: 130337.

Gong, X., Huang, D., Liu, Y., Zeng, G., Wag, R., Wan, J. et al. (2017). Stabilized nanoscale zerovalent iron mediated cadmium accumulation and oxidative damage of *Boehmeria nivea* (L.) gaudich cultivated in cadmium contaminated sediments. *Environmental Science Technology*, 51: 11308–11316.

Grigore, M. E., Biscu, E. R., Holban, A. M., Gestal, M. C., and Grumezescu, A. M. (2016). Methods of synthesis, properties and biomedical applications of CuO nanoparticles. *Pharm.*, 9(4): 75.

Hassan, T.U., Bano, A., and Naz, I. (2017). Alleviation of heavy metals toxicity by the application of plant growth promoting rhizobacteria and effects on wheat grown in saline sodic field. *International Journal of Phytoremediation*, 19: 522–529.

Hossain, M. B., Jahiruddin, M., Panaullah, G. M., Loeppert, R. H., Islam, M. R., and Duxbury, J. M. (2008). Spatial variability of arsenic concentration in soils and plants and its relationship with iron manganese and phosphorus. *Environmental Pollution*, 156: 739–744.

Huang, H., Ullah, F., Zhou, D. X., Yi, M., and Zhao, Y. (2019). Mechanisms of ROS regulation of plant development and stress responses. *Frontier Plant Science*, 10: 800.

Hussain, A., Ali, S., Rizwan, M., Rehman, M. Z. U., Qayyum, M. F., Wang, H., Rinklebe, J. (2019). Responses of wheat (*Triticum aestivum*) plants grown in a Cd contaminated soil to the application of iron oxide nanoparticles. *Ecotoxicolical Environmental Safety*, 173: 156–164.

Hussain, A., Ali, S., Rizwan, M., Zia ur Rehman, M., Javed, M. R., Imran, M. et al. (2018). Zinc oxide nanoparticles alter the wheat physiological response and reduce the cadmium uptake by plants. *Environmental Pollution*, 242: 1518–1526.

Hussain, B., Lin, Q., Hamid, Y., Sanaullah, M., Di, L., Hashmi, M. L. U. R. et al. (2020). Foliage application of selenium and silicon nanoparticles alleviates Cd and Pb toxicity in rice (*Oryza sativa* L.). *The Science Total Environment*, 712: 136497.

Inobeme, A, Mathew, J. T., Adetunji, C. O., Ajai, A. I., Inobeme, J., Maliki, M. et al. (2023). Recent advances in nanotechnology for remediation of heavy metals. *Environmental Monitoring Assess*, 195: 111.

Irshad, M. A., Zia-ur Rehman, M. Z., Anwar-ul-Haq, M., Rizwan, M., Nawaz, R., Shakoor, M. B. et al. (2021). Effect of green and chemically synthesized titanium dioxide nanoparticles on cadmium accumulation in wheat grains and potential dietary health risk: A field investigation. *Journal of Hazardous Materials*, 415: 125585.

Jdayea, N. A., Neamah, S. I., and Alalousi, M. A. (2023). Silver nanoparticles reduce the toxic effects of cadmium on *Datura stramonium* callus culture. *International Journal of Agronomy*, (1).

Khan, A., Khan, A. L., Muneer, S., Kim, Y. H., Al-Rawahi, A., and Al-Harrasi, A. (2019). Silicon and Salinity, Crosstalk in crop-mediated stress tolerance mechanisms. *Frontier Plant Science*, 10: 1429.

Khan, I. U., Rono, J. K., Liu, X. S., Feng, S. J., Li, H., Chen, X. et al. (2020). Functional Characterization of a New Metallochaperone for Reducing Cadmium Concentration in Rice Crop. *Journal of Clean Products*, 272: 123–152.

Konate, A., He, X., Zhang, Z., Ma, Y., Zhang, P., Alugongo, G. M. et al. (2017). Magnetic (Fe_3O_4) nanoparticles reduce heavy metals uptake and mitigate their toxicity in wheat seedling. *Sustainability*, 9: 790.

Kumar, D., Dhankher, O. P., Tripathi, R. D., and Seth, C. S. (2023). Titanium dioxide nanoparticles potentially regulate the mechanism(s) for photosynthetic attributes, genotoxicity, antioxidants defense machinery, and phytochelatins synthesis in relation to hexavalent chromium toxicity in *Helianthus annuus* L. *Journal of Hazardous Materials*, 454: 131418.

Lian, J., Zhao, L., Wu, J., Xiong, H., Bao, Y., Zeb, A. et al. (2020). Foliar spray of TiO2 nanoparticles prevails over root application in reducing Cd accumulation and mitigating Cd-induced phytotoxicity in maize (*Zea mays* L.). *Chemosphere*, 239: 124794.

Lin, J., He, F., Su, B., Sun, M., Owens, G., and Chen, Z. (2019). The stabilizing mechanism of cadmium in contaminated soil using green synthesized iron oxide nanoparticles under long-term incubation. *Journal of Hazardous Material*, 379: 120832.

Liu, J., Simms, M. and Song, S. (2018). Physiological effects of copper oxide nanoparticles and arsenic on the growth and life cycle of rice (Oryza sativa japonica 'Koshihikari'). *Environmental Science Technology*, 52: 13728–13737.

Lobzenko, I., Burachevskaya, M., Zamulina, I., Barakhov, A., Bauer, T., Mandzhieva, S. et al. (2022). Development of a unique technology for the pyrolysis of rice husk biochar for promising heavy metal remediation. *Agriculture*, 12: 1689.

Ma, X., Sharifan, H., Dou, F., and Sun, W. (2020). Simultaneous reduction of arsenic (As) and cadmium (Cd) accumulation in rice by zinc oxide nanoparticles. *Chemical Engineering Journal*, 384: 123802.

Mansoor, S., Kour, N., and Manhas, S. (2021). Biochar as a tool for effective management of drought and heavy metal toxicity. *Chemosphere*, 271: 29458.

Mar Gil-Daaz, M., Perez-Sanz, A., Angeles, M., and Vicente, M. (2014). Immobilisation of Pb and Zn in Soils Using Stabilised Zero-valent Iron Nanoparticles: Effects on Soil Properties. Clean-Soil Air Water, 42: 1776–1784.

Matos, M. P. S. R., Correia, A. A. S., and Rasteiro, M. G. (2017). Application of carbon nanotubes to immobilize heavy metals in contaminated soils. *Journal of Nanoparticles Research*, 19: 1–11.

Michalkova, Z., Fernandez, D. M., and Komarek, M. (2017). Interactions of two novel stabilizing amendments with sunflower plants grown in a contaminated soil. *Chemosphere*, 2: 123.

Mittal, A. K., Chisti, Y., and Banerjee, U. C. 2013. *Synthesis of metallic nanoparticles using plant extracts. Biotechnology Advertisement*, 31: 346–356.

Mohamed, A. K. S., Qayyum, M. F., Abdetl-Hadi, A. M., Rehman, R. A., Ali, S., and Rizwan, M. (2017). Interactive effect of salinity and silver nanoparticles on photosynthetic and biochemical parameters of wheat. *Archeology and Agronmy Soil Science*, 63: 1736–1747.

Mohanpuria, P., Rana, N. K., and Yadav, S. K. (2008). Biosynthesis of nanoparticles: Technological concepts and future applications. *Journal of Nanoparticles Research*, 10: 507–517.

Moharem, M., Elkhatib, E., and Mesalem, M. (2019). Remediation of chromium and mercury polluted calcareous soils using nanoparticles: Sorption-desorption kinetics, speciation and fractionation. *Environmental Research*, 170: 366–373.

Mohd Yusof, H., Mohamad, R., Zaidan, U. H., and Abdul Rahman, N. A. (2019). Microbial synthesis of zinc oxide nanoparticles and their potential application as an antimicrobial agent and a feed supplement in animal industry: A review. *Journal of Animal Science and Biotechnology*, 10: 1–22.

Mueller, N. D., Gerber, J. S., Johnston, M., Ray, D. K., Ramankutty, N., and Foley, J. A. (2012). Closing yield gaps through nutrient and water management. *Nature*, 490, 254–257.

Mukhopadhyay, D., Jana, S., and Koley, S. (2022). Is objective structured practical examination a cutting-edge assessment tool over traditional practical examination to prepare future "Indian Medical Graduate?" A study from Department of Pathology at a Government Medical College in West Bengal. *National Journal of Physiology Pharmacy and Pharmacology*, 102: 1.

Nazir, M. M., Li, Q., Noman, M., Ulhassan, Z., Ali, S., Ahmed, T. et al. (2022). Calcium oxide nanoparticles have the role of alleviating arsenic toxicity of barley. *Frontiers in Plant Sciences*, 13: 843795.

Nazir, M. M., Noman, M., Ahmed, T., Ali, S., Ulhassan, Z., Zeng, F. et al. (2022). Exogenous calcium oxide nanoparticles alleviate cadmium toxicity by reducing cd uptake and enhancing antioxidative capacity in barley seedlings. *Journal of Hazardous Materials*, 438: 129498.

Noman, M., Ahmed, T., Hussain, S., Niazi, M. B. K., Shahid, M., and Song, F. (2020). Biogenic copper nanoparticles synthesized by using a copper-resistant strain Shigella flexneri SNT22 reduced the translocation of cadmium from soil to wheat plants. *Journal of Hazardous Materials*, 398: 123175.

Nouman, W., Bashir, T., Gul, R., Gull, T., Olson, M. E., Shaheen, M. et al. (2020). Metalliferous conditions induce regulation in antioxidant activities, polyphenolics and nutritional quality of *Moringa oleifera* L. *International Journal of Phytoremediation*, 22: 1348–1361.

Ojuederie, O. B., Olanrewaju, O. S., and Babalola, O. O. (2019). Plant Growth Promoting Rhizobacterial Mitigation of Drought Stress in Crop Plants, Implications for Sustainable Agriculture. *Agronomy*, 9: 712.

Purnomo-Sari, K. R., Ikawati, Z., Danarti, R., and Hertiani, T. (2023). Micro-titer plate assay for measurement of total phenolic and total flavonoid contents in medicinal plant extracts. *Arabian Journal of Chemistry*, 16: 105003.

Ragab, G. A., and Saad-Allah, K. M. (2020). Green synthesis of sulfur nanoparticles using *Ocimum basilicum* leaves and its prospective effect on manganese-stressed *Helianthus annuus* (L.) seedlings. *Ecotoxicological Environmental Safety*, 191: 110242.

Rajput, V. D., Minkina, T., Upadhyay, S. K., Kumari, A., Ranjan, A., and Mandzhieva, S. (2022). Nanotechnology in the restoration of polluted soil. *Nanomaterials*, 12: 769.

Ranjbar, Z., Pourhadadi, D., Montazeri, S., and Modaberi, M. R. (2023). Lead compounds in paint and coatings: A review of regulations and latest updates. *Progress in Organic Coatings*, 174: 107247.

Rizwan, M., Ali, S., Ur Rehman, M. Z., Malik, S., Adrees, M., Qayyum, M. F. et al. (2019). Effect of foliar applications of silicon and titanium dioxide nanoparticles on growth oxidative stress and cadmium accumulation by rice (*Oryza sativa*). *Acta Physiology Plant*, 41: 35.

Rizwan, M., Noureen, S., Ali, S., Anwar, S., Rehman, M. Z. U., Qayyum, M. F. et al. (2019). Influence of biochar amendment and foliar application of iron oxide nanoparticles on growth, photosynthesis, and cadmium accumulation in rice biomass. *Journal of Soils Sediments*, 19: 3749–3759

Rubio, L., Pyrgiotakis, G., Beltran-Huarac, Z. Y., Gaurav, J., Deloid, G., and Demokritou, P. (2019). Safer-by-design flame-sprayed silicon dioxide nanoparticles: The role of silanol content on ROS generation, surface activity and cytotoxicity. *Particle Fibre Toxicology*, 16: 40.

Saleh, T. A. (2020). Nanomaterials: classification, properties, and environmental toxicities. *Environmental Technology Innovation*, 20: 101067.

Samanta, S., and Roychoudhury, A. (2021). Recent trend in nanoparticle research in regulating arsenic bioaccumulation and mitigating arsenic toxicity in plant species. *Journal of Plant Biochemistry and Biotechnology*, 30: 793–812.

Sardar, R., Ahmed, S., and Yasin, N. A. (2022). Titanium dioxide nanoparticles mitigate cadmium toxicity in *Coriandrum sativum* L. through modulating antioxidant system, stress markers and reducing cadmium uptake. *Environmental Pollution*, 292: 118373.

Selvarajan, V., Obuobi, S., and Lai. P. (2020). Silica Nanoparticles—A Versatile Tool for the Treatment of Bacterial Infections. *Frontier Chemist and Secondary Nanoscience*, 8. https://doi.org/10.3389/fchem.2020.00602.

Shahid, M., Khalid, S., Abbas, G., Shahid, N., Nadeem, M., and Sabir, M. (2015). Heavy metal stress and crop productivity. *In*: Hakeem, K. R. (Ed.). *Crop Production and Global Environmental Issues*; *Springer International Publishing: Cham, Switzerland*, 2: 1–25.

Shahv, F., and Wu, W. (2019). Soil and crop management strategies to ensure higher crop productivity within sustainable environments. *Sustainability*, 11: 1485.

Shen, Y., Tang, J., Nie, Z., Wang, Y., Ren, Y., and Zuo, L. (2009). Preparation and application of magnetic Fe3O4 nanoparticles for wastewater purification. *Seperation and Purification Technology*, 68: 312–319.

Sheng, J. J., Wang, X. P., Gong, P., Tian, L.D., and Yao, T. D. (2012). Heavy metals of the Tibetan top soils Level source spatial distribution temporal variation and risk assessment. *Environmental Science Pollution Research*, 19: 3362–3370.

Singh, D., Sharma, N. L., Singh, D., Siddiqui, M. H., Sarkar, S. K., Rathore, A. et al. (2023). Zinc oxide nanoparticles alleviate chromium-induced oxidative stress by modulating physio-biochemical aspects and organic acids in chickpea (*Cicer arietinum* L). *Plant Physiology and Biochemistry*, 206: 108166.

Singh, J., and Lee, B. K. (2016). Influence of nano-TiO2 particles on the bioaccumulation of Cd in soybean plants (Glycine max): A possible mechanism for the removal of Cd from the contaminated soil. *Journal of Environmental Management*, 170: 88–96.

Singh, J., Dutta, T., Kim, K. H., Rawat, M., Samddar, P., and Kumar, P. (2018). 'Green'synthesis of metals and their oxide nanoparticles: applications for environmental remediation. *Journal of Nanobiotechnology*, 16: 1–24.

Singh, R. P., Handa, R., and Manchanda, G. (2021). Nanoparticles in sustain-able agriculture: An emerging opportunity. *Journal of Control and Release*, 329: 1234–1248.

Skiba, E., and Wolf, W. M. (2019). Cerium Oxide Nanoparticles Affect Heavy Metals Uptake by Pea in a Divergent Way than Their Ionic and Bulk Counterparts. *Water Air Soil Pollution*, 230: 248.

Thakkar, K. N., Mhatre, S. S., and Parikh, R. Y. (2010). Biological synthesis of metallic nanoparticles. Nanomedicin Nanotechnology Biological Medication, 6: 257–262.

Toumey, C. (2020). Notes on environmental nanoscience. *Natural Nanotechnology*, 15: 250–251.

United Nations. 2020. World Population Prospects: The 2019 Revision, Key Findings and Advance Tables. Available online: https: //population.un.org/wpp/Download/Standard/Population/(accessed on 1 October 2020).

Usman, M., Farooq, M., Wakeel, A., Nawaz, A., Cheema, S.A., Rehman, H. U., et al. (2020). Nanotechnology in agriculture: Current status, challenges and future opportunities. *Science Total Environment*, 721: 137778.

Vasile, C. (2018). Polymeric Nanocomposites and Nanocoatings for Food Packaging: A Review. *Materials*, *11*(10), 1834. doi.org/10.3390/ma11101834.

Vauthier, C. (2019). *A journey through the emergence of nanomedicines with poly (alkylcyanoacrylate) based nanoparticles.* HAL Open Science. Available at: https://hal.science/hal-02359395/document.

Wang, S., Fu, Y., Zheng, S., Xu, Y., and Sun, Y. (2022). Phytotoxicity and Accumulation of Copper-Based Nanoparticles in *Brassica* under Cadmium Stress. *Nanomaterials*, 12: 1497.

Wang, X., Sun, W., Zhang, S., Sharifan, H., and Ma, X. (2018). Elucidating the effects of cerium oxide nanoparticles and zinc oxide nanoparticles on arsenic uptake and speciation in rice (*Oryza sativa*) in a hydroponic system. *Environmental Science Technolnology*, 52(17): 10040–7.

Wu, F., Fang, Q., Yan, S., Pan, L., Tang, X., and Ye, W. (2020). Effects of zinc oxide nanoparticles on arsenic stress in rice (*Oryza sativa* L.): Germination, early growth, and arsenic uptake. *Environmental Science Pollution Research*, 27: 26974–26981.

Wu, S. (1994). Effect of manganese excess on the soybean plant cultivated under various growth conditions. *Journal of Plant Nutre*, 17: 991–1003.

Wu, X., Hu, J., Wu, F., Zhang, X., Wang, B., Yang, Y. et al. (2021). Application of TiO2 nanoparticles to reduce bioaccumulation of arsenic in rice seedlings (*Oryza sativa* L.): a mechanistic study. *Jouranl of Hazardous Material*, 405: 124047.

Xiao, X., Teng, F., Shi, C., Chen, J., Wu, S., Wang, B. et al. (2022). Polymeric nanoparticles—Promising carriers for cancer therapy. Frontier *Bioengineering, Biotechnology and Secondary Biomaterials*, 10, https://doi.org/10.3389/fbioe.2022.1024143.

Yan, S., Wu, F., Zhou, S., Yang, J., Tang, X., and Ye, W. (2021). Zinc oxide nanoparticles alleviate the arsenic toxicity and decrease the accumulation of arsenic in rice (*Oryza sativa* L). *BMC Plant Biology*, 21(1): 150.

Zadeh, R. R., Arvin, S. M. J., Jamei, R., Mozaffari, H., and Nejhad, F. R. (2019). Response of tomato plants to interaction effects of magnetic (Fe3O4) nanoparticles and cadmium stress. *Journal of Plant Interaction*, 14: 474–481.

Zhai, X., Li, Z., Huang, B., Luo, N., Huang, M., Zhang, Q., and G. Zeng. (2018). Remediation of multiple heavy metal-contaminated soil through the combination of soil washing and in situ immobilization. *Science Total Environment*, 635: 92–99.

Zhou, P., Adeel, M., Guo, N., Hao, Y., and Azeem, I. (2020). Application of nanoparticles alleviates heavy metals stress and promotes plant growth: An overview. *Nanomaterials*, 11(1): 16.

Zhou, P., Adeel, M., Shakoor, N., Guo, M., Azeem, Y. H. I., and Mingshuli, M. (2021). Application of nanoparticles alleviates heavy metals stress and promotes plant growth: an overview. *Nanomaterials*, 11(1): 26. https://doi.org/10.3390/nano11101002.

Zielinska, A., Carreiro, F., Oliveira, A. M., Neves, A., Pires, B., Venkatesh, D. N. et al. (2020). Polymeric Nanoparticles: Production, Characterization, Toxicology and Ecotoxicology. *Molecules*, 25(16): 3731. doi.org/10.3390/molecules25163731.

CHAPTER-7

Green Nanofertilizers for Sustainable Agriculture

Ankita Gogoi,[1] *Nikha Borah,*[1] *Kabyashree Buragohain,*[1] *Sukanya Sonowal,*[1] *Ratul Nath,*[1] *Bhaben Chowardhara*[2] *Debojyoti Moulick*[3] and *Akbar Hossain*[4,*]

1. Introduction

The growing global population has become a significant obstacle to achieving sustainable food production that can fulfil the rising needs of humanity. By 2050, the global population is anticipated to reach 9.6 billion, requiring a substantial 70–100% increase in food production to address the demands of this growth (Rehman et al., 2024). Rapid environmental changes demand alternative strategies to increase crop productivity and ensure food security for the growing population (Harsh et al., 2024; Ghosh et al., 2023). Environmental stresses such as extreme temperatures, salinity, drought, flooding, and Heavy Metals (HMs) significantly reduce crop yields, which vary with crop type and stress duration (Chowardhara et al., 2024; Moulick et al., 2024a; Haggag et al., 2015). These stresses in cultivated plants can be managed through the use of antistress formulations (s), such as nitric oxide, pyraclostrobin, PGPR (Plant Growth-Promoting Rhizobacteria), melatonin, strigolactones, phytohormones, and nutrients such as silicon and selenium, including their nano formulations (Sonowal et al., 2025; Majumdar et al., 2025; Moulick et al., 2024b; Mondal et al., 2023; Shalaby et al., 2022). The limited nutrient use efficiency of chemical fertilizers, coupled with their environmental impact, presents significant obstacles to achieving agricultural sustainability. Overapplication increases costs for growers and leads to nutrient losses due to rapid release rates that exceed plant absorption capacity or the conversion of nutrients into nonbioavailable forms (Deb et al., 2023; Roy et al., 2024). To address these challenges, interest in innovative approaches to improve fertilizer use efficiency is increasing (Dey et al., 2025; Moulick et al., 2025a, b). The proposed strategies include targeted fertilization, staged or site-specific nutrient applications, fertigation, and the development of advanced fertilizers such as nanofertilizers (Zulfiqar et al., 2019).

Nanofertilizers (NFs) are advanced forms of traditional fertilizers created via nanotechnology to increase their efficiency. With particle sizes between 1–100 nm and a high surface area-to-volume

[1] Microbiology Laboratory, Department of Life Sciences, Dibrugarh University, Dibrugarh- 786004, Assam, India.
[2] Department of Botany, Arunachal University of Studies, Namsai-792103, Arunachal Pradesh, India.
[3] Department of Environmental Science, University of Kalyani, Nadia, West Bengal-741235, India.
[4] Soil Science Division, Bangladesh Wheat and Maize Research Institute (BWMRI), Nashipur, Dinajpur-5200, Bangladesh.
* Corresponding author: akbarhossainwrc@gmail.com

DOI: 10.1201/9781003567684-7

ratio, they offer superior availability and absorption. This facilitates better uptake through soil or leaves, increasing photosynthate production and biomass, which are crucial for healthier crop growth (Pattnaik et al., 2023; Kumar et al., 2021). NFs are created by combining nanoparticles (NPs) with biofertilizers, increasing nutrient use efficiency and improving the physical, chemical, and biological properties of soil. The process involves encapsulating biofertilizer cells in biodegradable materials such as calcium alginate and starch, ensuring bacterial protection and slow nutrient release. Synthesis includes growing biofertilizer cultures, encapsulating them with NPs, and assessing product quality. Microcapsules are formed via Plant Growth-Promoting Rhizobacteria (PGPR), sodium alginate, starch, and bentonite or through combinations such as ZnONPs and salicylic acid. Additionally, organic wastes such as flowers, cattle manure, and compostable waste can be decomposed or pyrolyzed and mixed with NPs to create effective soil-enriched nano biofertilizers (Akhtar et al., 2022). Nanotechnology has recently revolutionized the agricultural sector by introducing green nanomaterials, NFs, nanoagrochemicals, nano pesticides, nanoherbicides, nanobiosensors, and nanobased approaches for agricultural waste management. These innovations enhance food productivity through foliar and soil applications while promoting sustainable agricultural practices (Al-Mamun et al., 2021). The use of nanoderived products in agriculture is becoming increasingly important for increasing plant growth and increasing crop yields. Compared with traditional chemicals and methods, farmers opt for these products because they are compact, portable, easy to handle, have a long shelf life, are highly effective, and are nontoxic. As a result, the commercialization of nano based solutions is experiencing rapid growth worldwide (Singh et al., 2021). Different plant parts have been found to assimilate a wide range of NPs. Plants such as cucumbers, spinach, tobacco, common dry beans, and soybeans exhibit both beneficial and detrimental effects from NPs, including a decrease in photosynthesis, an increase in chlorophyll production, and stimulation of root growth (Hazarika et al., 2022). NPs can be synthesized via chemical, physical, or biological methods, with the biosynthesis approach relying on natural sources such as plants, fungi, and bacteria. With this method, the particle size is more effectively controlled, and the degree of toxicity is reduced. To meet large-scale application demands, synthesis techniques must ensure uniform physicochemical properties for precise and targeted nanoformulations. The bottom-up approach remains the most effective method for producing NFs because of its ability to generate well-defined NPs (Zulfiqar et al., 2019). NPs with distinct surface structures and reactivities enable the development of NFs through processes such as Reactive Oxygen Species (ROS) generation, redox reactions, free radical formation, and dissolution. Unlike bulk particles, these NPs may have unique ecological and toxicological effects. By improving mineral bioavailability, NFs enhance crop yields while also reducing pathogens due to their antimicrobial properties, making them promising tools for sustainable agriculture (Moulick et al., 2024d; Pattnaik et al., 2024; Mahapatra et al., 2022).

To ensure environmentally safe NPe production, green synthesis methods have been developed using biological systems such as yeast, fungi, bacteria, and plant extracts. These methods are cost-effective, scalable, and reproducible. Among these methods, plant-based synthesis stands out for its simplicity and adaptability. NPs produced in this way exhibit improved biochemical, physical, and biological properties due to their high surface area-to-volume ratio, making nanofertilizers more efficient at delivering nutrients and increasing crop yields. The synthesis of NPs from plants is becoming increasingly popular because of their simplicity and accessibility. During the synthesis process, plant-derived phytoconstituents function as stabilizers, reducers, and capping agents. This eco-friendly approach involves combining metal salt solutions with extracts from various plant parts, including leaves, roots, flowers, seeds, and fruits (Sohrabi et al., 2022). Eco-friendly and cost-effective, green-synthesized NFs play crucial roles in cleaning contaminated agricultural soils. They eliminate harmful pollutants such as pesticides, industrial dyes, and HMs, helping restore soil fertility and increasing crop yields. Soil contamination disrupts nutrient availability, impairs plant growth, and threatens human health through polluted groundwater and toxin-laden crops. This sustainable method holds great promise for environmental and agricultural recovery (Sundararajan et al., 2024). In addition to plant-mediated synthesis, various microorganisms and viruses have been found to be capable of producing metal NPs. The interaction between metals and microorganisms has long been utilized in processes such as biomineralization, bioremediation, bioleaching, and biocorrosion. Recently, microbial NP synthesis has gained significant research attention because of its unique advantages over conventional methods. NP synthesis can occur within the cell or outside the cell depending on the microbial species involved (Dikshit et al., 2021).

Compared with plant cells, bacterial cells achieve greater green synthesis efficiency. This method is preferred for its cost-effectiveness and high yield, with NP synthesis occurring either intracellularly or extracellularly, depending on the microorganism involved, owing to their distinctive surface structures and biomolecules such as exopolysaccharides, proteins, and enzymes that aid in NP formation. Frequently used microbes include *Pseudomonas*, *Bacillus*, *Aspergillus*, *Fusarium*, and *Candida* species, which reduce metal ions through their biomolecular activity (Dikshit et al., 2021; Saberi Riseh and Gholizadeh Vazvani, 2024; Selvakumar et al., 2014; Wadhwani et al., 2016). Metallic NPs can be synthesized by bacteria through intracellular and extracellular methods, with extracellular synthesis being more efficient and oxidation resistant. Additionally, studies have revealed that NP biosynthesis can occur via both living and dead bacterial cells (Alloosh et al., 2021). In intracellular synthesis, microorganisms take up metal ions from their surroundings and enzymatically reduce them into elemental forms. In extracellular synthesis, enzymes and proteins secreted by microorganisms facilitate the reduction of metal ions and stabilize the resulting particles (Jiang et al., 2022).

2. Types of Green NPs

Concerns regarding the hazards and drawbacks of conventional NPs have led to a recent surge in the interest in green NPs. As reducing and capping agents, these NPs are made via biological processes that use plants, microbes, or other natural materials. This is in line with the principles of green chemistry, which emphasize the reduction or elimination of hazardous substances in the design, manufacture, and application of chemical products (Silva et al., 2015; Huston et al., 2021; Singh et al., 2023). The need for green NPs is increasing as more industries, such as agriculture, medicine, and environmental remediation, seek to provide environmentally friendly substitutes (Khan et al., 2022; Singh et al., 2023). Depending on where they originate and the biological organisms that are employed to create them, green NPs can be divided into several general categories. The use of hazardous chemical additions is avoided by using biological techniques, which use renewable resources such as microbes and plants as reducing agents to stabilize and cap nanomaterials (Ijaz et al., 2020; Khan et al., 2022). Among the most common types of green NPs are the following:

***i. Plant-based NPs*:**

One of the primary biosynthesis methods being studied is the use of plant extracts, as most plants are readily available, reasonably priced, and safe for human consumption (Huston et al., 2021). Additionally, a variety of phytochemicals are found in plant extracts, such as flavonoids, terpenoids, phenols, proteins, glucosides, and polysaccharides, act as reductants and stabilizers in the synthesis of plant derived NPs (Saravanakumar et al., 2018; Fardsadegh and Jafarizadeh Malmiri, 2019). The biosynthesis process typically involves extracting bioactive compounds from plant materials or food scraps through simple processes such as drying, grinding, and immersion in hot water, followed by filtration and storage at 4°C. These methods eliminate the need for harsh chemicals and reduce energy consumption, making them highly sustainable (Huston et al., 2021). The mechanisms of phytosynthesis for green NPs are not yet fully understood due to the diversity of bioactive compounds involved. Identifying specific bioreductants and stabilizers responsible for NP synthesis is challenging because numerous phytochemicals act synergistically (Ayaz et al., 2017). For example, phenolic acids, which belong to the polyphenol family, have been extensively studied for their role in NP synthesis. Their structure, consisting of a phenolic ring and a functional carboxylic acid group, provides nucleophilic aromatic rings that support antioxidant enzyme activity and the chelation of metals. Acids such as caffeic, gallic, and ellagic acids are known to function as reductants in the biosynthesis of metallic NPs (MNPs) (Jha and Prasad, 2018; Ovais et al., 2018). Raja et al., demonstrated that gallic acid lowers silver ions (Ag^+) to form AgNPs by forming a transitional complex with phenolic hydroxyl groups, which oxidize into quinones (Raja et al., 2017). Additionally, phenolic chemicals contribute to the reduction of other metal ions. *Melaleuca nesophila*, *Rosmarinus officinalis*, and *Eucalyptus tereticornis* leaf extracts were used by Wang et al., to create iron-polyphenol NPs. NPs with organic surfaces are produced by the combination of polyphenols with charged iron ions (Wang et al., 2015). Similarly, tannins found in leaf extracts from *Gloriosa superba* transformed gold and silver ions into spherical nanoscale particles (Gopinath et al., 2016). Protocatechuic acid, catechol, and ascorbic acid in *Hydrilla* sp. aid in silver ion bioreduction by generating ions, which are also reduced when benzoquinone derivatives in *Cyperus* sp. undergo tautomerization, resulting in the formation of coalescent clusters (Jha et al., 2009).

Flavonoids, another major class of phytochemicals, are plant pigments with reducing capabilities linked to their ability to donate hydrogen atoms or electrons. This capacity has been identified as a primary bioreducing component in plant extracts. For example, luteolin and rosmarinic acid undergo keto-enol transformations, releasing H^+ ions that reduce silver ions to silver NPs (Ahmad et al., 2010; Dubey et al., 2010). Polyhydroxylated flavonoids not only reduce Ag^+ ions but also act as capping agents, enhancing the antibacterial properties of silver NPs (Sahu et al., 2016). Zhou et al. reported that reducing sugars and flavonoids in plant extracts interact with tetrachloroaurate ions to form AuNPs, with proteins acting as capping agents but not reductants (Zhou et al., 2010). Copper NPs have also been synthesized from plant polyphenols. Nasrollahzadeh et al., used quercetin from *Ginkgo biloba* leaf extracts to reduce copper ions, producing NPs stabilized by carbonyl groups formed during the redox process (Nasrollahzadeh and Mohammad Sajadi, 2015). Terpenoids, derived from plant essential oils, are a group of bioactive compounds involved in nanosynthesis. These oxygenated derivatives of terpenes act as surface-active chemicals that stabilize and reduce metal ions, although the precise mechanism is not yet fully understood (Mashwani et al., 2016). For example, eugenol, a key terpenoid in *Cinnamomum zeylanicum* and *Syzygium aromaticum*, has been used to synthesize gold and silver NPs (Shankar et al., 2003; Chandran et al., 2006). Brahmachari et al. devised a straightforward approach using *Ocimum sanctum* leaf extracts under sunlight to synthesize AgNPs. They postulated that sunlight exposure breaks the phenolic O–H bond in eugenol, producing H^+ radicals that reduce Ag^+ ions to metallic AgNPs (Brahmachari et al., 2014).

Proteins in plant extracts also contribute to NP biosynthesis by serving as capping ligands and stabilizers. Amino acids with exposed disulphide bonds and thiol groups act as nonenzymatic reducers (Ahmad et al., 2013). For example, Bhattacharjee et al. synthesized gold NPs via a tripeptide sequence containing tyrosine residues, which reduced the number of Au ions while stabilizing the NPs (Bhattacharjee et al., 2005). Similarly, Li et al. demonstrated that proteins in *Capsicum annuum* extracts trap silver ions through electrostatic interactions, reducing them to form silver nuclei, which further aggregate into stable NPs (Li et al., 2007). Proteins provide advantages over synthetic surfactants because of their lower cost, eco-friendliness, and ability to act as carriers for drugs or genes in biomedical applications (Rodriguez et al., 2013). Various NPs, including silver, gold, copper, and selenium, have been synthesized from plant-derived compounds. For example, tea leaves rich in polyphenols have been used to create colloidal silver NPs with biocompatible properties (Moulton et al., 2010). Aloe vera has been employed to synthesize selenium NPs with antibacterial and antifungal activities (Fardsadegh and Jafarizadeh-Malmiri, 2019). In addition to extracts, living plants such as *Brassica juncea* and *Festuca rubra* can reduce metal ions within their tissues, forming NPs through processes mediated by sugars, phenols, and ascorbic acid (Marchiol et al., 2014). The reducing, stabilizing, and capping capabilities of plant-derived bioactive compounds provide an environmentally friendly, cost-effective approach for NP synthesis. This method has great potential for use in medicine, catalysis, and environmental remediation (Singh et al., 2023).

***ii. Microbe-based NPs*:**

Microorganisms, including bacteria, fungi, yeast, algae, and viruses, have emerged as efficient and eco-friendly platforms for synthesizing NPs, offering sustainable alternatives to conventional chemical and physical methods (Huston et al., 2021). Among these, bacteria are particularly favoured because of their rapid growth rates, adaptability, and ease of cultivation under artificial conditions. Bacteria exhibit an exceptional ability to tolerate high metal concentrations and reduce inorganic ions into NPs via extracellular or intracellular mechanisms. In the extracellular synthesis process, microbial cultures produce enzyme-rich supernatants that interact with metal ions, leading to their bioreduction and NP formation. On the other hand, intracellular synthesis occurs when metal ions are absorbed by microbial cells and reduced enzymatically to form nanoclusters, which eventually diffuse out into the solution. Bacteria have been extensively studied for their ability to synthesize a variety of metal and metal oxide NPs, including Ag, Au, Cu, Se, Fe, and ZnO, with diverse applications in biosensing, diagnostics, antimicrobial therapy, and cancer treatment (Koul et al., 2021). AgNPs are particularly notable due to their wide-ranging applications, with bacterial strains such as *Escherichia coli*, *Lactobacillus* sp., and *Pseudomonas* sp. producing AgNPs through enzymatic bioreduction (Al-asbahi et al., 2024). Enzymes such as nitrate reductase play pivotal roles in AgNP green synthesis, and factors such as pH, temperature, and ionic concentration significantly influence NP size and morphology (Anil Kumar et al., 2007). Advances in genetic engineering have further optimized bacterial systems, such as *E. coli* strains expressing metal ion efflux systems, to produce homogenous NPs (Koul et al., 2021). Similarly, AuNPs are synthesized by bacteria such as *Bacillus subtilis* and *Shewanella* algae, where pH modulation influences whether synthesis occurs intracellularly or extracellularly (Beveridge and Murray, 1980; Konishi et al., 2007).

Fungi and yeasts have also demonstrated remarkable potential in NP biosynthesis because of their high tolerance for metal ions and robust cellular machinery. Fungi, in particular, produce NPs in larger quantities than bacteria do and simplify downstream processing by enabling extracellular synthesis, which avoids issues related to intracellular impurities (Tyagi et al., 2023; Mahanta et al., 2023; Singh et al., 2016). For example, *Hypocrea lixii* and *Trichoderma viride* have been employed for AuNP synthesis, with reaction conditions such as temperature and pH impacting the size and activity of the resulting NPs (Mishra et al., 2014). Similarly, *Fusarium oxysporum* has been shown to produce monodispersed AgNPs under optimized conditions (Birla et al., 2013). *Schizophyllum radiatum* generates AgNPs with potent antibacterial activity, whereas fungal-mediated zinc oxide (ZnO) NPs synthesized from *Aspergillus terreus* and *Fusarium keratoplasticum* display strong antimicrobial and cytotoxic effects (Metuku et al., 2014; Fouda et al., 2018; Mohamed et al., 2019). Copper oxide (CuO) and titanium dioxide (TiO_2) NPs derived from fungi have demonstrated biopesticidal effects, while iron oxide NPs synthesized by *Aspergillus japonicus* have shown promise for bioremediation (Bhargava et al., 2013; Koul et al., 2021). Yeasts such as *Yarrowia lipolytica* have been utilized for the biosynthesis of NPs via the use of melanin as a reducing agent (Apte et al., 2013). Genetically engineered yeasts, such as *Pichia pastoris*, increase NP production by generating uniform and stable particles, making them suitable for biomedical and industrial applications (Elahian et al., 2017).

Algae, as autotrophic organisms, have emerged as unique "nanofactories" owing to their minimal growth requirements and diverse secondary metabolites that act as natural capping agents during NP synthesis (Fawcett et al., 2017). The extracellular biosynthesis of silver chloride NPs (AgCl NPs) has been achieved via *Sargassum plagiophyllum* and *Chlorella vulgaris*, with the resulting particles demonstrating strong antimicrobial activities against pathogens such as *E. coli* and *S. aureus* (Koul et al., 2021). Similarly, the marine algae *Caulerpa racemosa* and *Laminaria japonica* have been employed for AgNP biosynthesis, yielding NPs that exhibit excellent catalytic and antimicrobial properties (Sharma et al., 2016). Recent studies using *Portieria hornemannii* and

Padina sp. have confirmed the production of uniform AgNPs with potent antibacterial effects against marine pathogens such as *Vibrio parahaemolyticus* and *P. aeruginosa* (Koul et al., 2021; Barciela et al., 2023).

AuNPs have also been synthesized using algae. Intracellular biosynthesis by *Tetraselmis kochinensis* and extracellular production via *Padina gymnospora* have yielded spherical AuNPs with sizes ranging from 5 to 67 nm (Khan et al., 2019). Algae such as *Euglena gracilis* and *Sargassum tenerrimum* have demonstrated faster growth kinetics and enhanced colloidal stability during AuNP synthesis, with the resulting particles showing diverse applications in dye degradation and cancer cell cytotoxicity (Koul et al., 2021). Palladium NPs (PdNPs), synthesized from *Chlorella vulgaris* and *Spirulina platensis*, have proven effective for bioremediation and organic pollutant degradation (Arsiya et al., 2017; Sayadi et al., 2018).

Algae-mediated biosynthesis extends to metal oxide NPs. *Turbinaria conoides* and *Sargassum muticum* have been utilized to produce crystalline ZnONPs, which have significant antimicrobial and anticancer properties against pathogens and human liver cancer cells (Taghizadeh et al., 2020; Koul et al., 2021). Moreover, *Arthrospira platensis* has yielded spherical ZnONPs with strong activity against multi resistant bacterial strains, whereas the green alga *Chlorella vulgaris* has facilitated the production of ZnO nanorods for UV protection and antibacterial applications (Taghizadeh et al., 2020; El-Belely et al., 2021). TiO_2NPs, which are fabricated from *Chlorella pyrenoidosa*, have been incorporated into graphene oxide composites, enhancing the photocatalytic degradation of organic dyes (Sharma et al., 2018). Fe_3O_4 NPs synthesized from *Sargassum acinarium* and *Padina pavonica* exhibit promising bioremediation potential, effectively removing HMs such as lead from contaminated water systems (El-Kassas et al., 2016). These findings reinforce the versatility of algae as cost-effective and eco-friendly platforms for synthesizing a wide range of NPs with applications spanning biomedicine, environmental remediation, and material science.

Owing to their precise nanoscale capsid structures (20–500 nm), viruses provide unique scaffolds for NP synthesis through capsid protein interactions with metal ions (Kobayashi et al., 2012). Genetic engineering allows the modification of viral capsid proteins to synthesize nanocomposites and nanoconjugates for applications such as targeted drug delivery and cancer therapy (Gahlawat and Choudhury, 2019). Studies have demonstrated the use of plant viruses such as Tobacco Mosaic Virus (TMV) and cowpea chlorotic mottle virus for fabricating gold NPs, which produce stable and uniform particles (Slocik et al., 2005; Kobayashi et al., 2012). Additionally, viruses such as the M13 bacteriophage have been employed as templates for semiconductor NPs such as cadmium sulphide (CdS) and zinc sulphide (ZnS) (Mao et al., 2003). Viral NPs have also been explored for drug delivery, as demonstrated by RedClover Necrotic Mosaic Virus (RCNMV)-based nanocarriers for the anticancer drug doxorubicin and TMV-derived hollow nanotubes for targeted delivery of chemotherapeutics. Virus-mediated NPs, such as hepatitis E virus NPs functionalized with fluorescent dyes and ligands, have proven effective for tumour imaging and tissue-specific targeting (Koul et al., 2021).

Despite advancements in microbial NP synthesis, challenges such as scalability, biocompatibility, and cost-effectiveness remain. Innovative biotechnological approaches, including genetic engineering and optimization of culture conditions, are crucial for addressing these limitations and unlocking the full potential of microbe-mediated NPs in industrial, environmental, and biomedical applications (Kang et al., 2008).

***iii. Biopolymer-based NPs*:**

Biopolymer-based NPs, comprising polysaccharides, polypeptides, and proteins, have gained significant attention as versatile platforms for drug delivery because of their biocompatibility, biodegradability, and tunable properties. These naturally derived materials exhibit structural and functional diversity, enabling tailored NP formulations for targeted therapies, gene delivery, and controllable release of therapeutics. Protein-based NPs offer unique advantages such as low toxicity, high stability, and functional flexibility. Proteins such as collagen, silk, elastin, gelatin, and albumin have been widely employed for NP fabrication because of their structural integrity and biological compatibility (Nitta and Numata, 2013). Recent advances in genetic engineering have further optimized protein-based carriers, such as Elastin-Like Polypeptides (ELPs), which exhibit precise self-assembly through hydrophobic domains and can be tailored for stimuli-responsive drug delivery. ELPs exhibit thermosensitive micelle formation and pH sensitivity, enabling phase behaviour modulation and tumour-specific delivery (Santos et al., 2019). Similarly, recombinant silk proteins, known for their mechanical strength and controlled biodegradability, have been employed to prepare stable silk fibroin NPs (150–170 nm), which enhance drug encapsulation and sustained release (Kundu et al., 2010). These NPs can also be conjugated with bioactive agents such as insulin to improve their *in vitro* stability and therapeutic half-life. Complexes such as silk–chitosan blends further extend their biomedical applications by achieving enhanced curcumin delivery for cancer therapy (Yan et al., 2009).

NPs made from albumin have been widely studied for Drug Delivery Systems (DDSs) since the 1970s. Techniques such as coacervation, desolvation, thermal gelation, and self-assembly enable their preparation. Human Serum Albumin (HSA) NPs are stabilized by crosslinking agents such as glutaraldehyde or heat denaturation, facilitating drug conjugation through sulphhydryl or amino groups. Alternative strategies using polysaccharides such as gum arabic have improved biocompatibility by avoiding toxic crosslinkers (Weber et al., 2000; Hedayati et al., 2012). Self-assembled HSA NPs encapsulating hydrophobic drugs, such as paclitaxel, exhibit tumour-targeting ability through albumin-binding proteins such as gp60 and SPARC. The clinical success of NP albumin-bound (nab) technology, exemplified by nab-paclitaxel, underscores its potential in metastatic cancer treatment by enhancing intratumoral drug accumulation. Thermal gelation further enables the fabrication of core-shell NPs, as seen in ibuprofen-loaded BSA-dextran conjugates, which improve stability and controlled release (Nitta and Numata, 2013).

Polypeptide-based NPs, such as those derived from ELPs and Silk-Elastin-Like Proteins (SELPs), have expanded the scope of protein-like nanocarriers. ELP NPs exhibit tunable properties, such as temperature-triggered micelle assembly and pH-responsive drug release, enhancing their cellular uptake and tumour specificity (Osborne et al., 2008; Macewan and Chilkoti, 2012). SELPs, which combine silk structural strength with elastin flexibility, form NPs through hydrophobic interactions at specific transition temperatures, making them suitable for tissue engineering and drug delivery (Xia et al., 2011). Additionally, zwitterionic polypeptides, such as poly(l-glutamic acid)-b-poly(l-lysine), have been designed to form pH-responsive vesicles for controlled release in tumour environments (Rodríguez-Hernández and Lecommandoux, 2005).

Polysaccharides, as natural polymers, offer excellent stability, bioadhesion, and chemical versatility for NP preparation. Commonly used polysaccharides include chitosan, alginate, heparin, hyaluronic acid, dextran, and pullulan (Chen et al., 2016). NPs can be formed through crosslinking, polyion complexation, and self-assembly. Chemically crosslinked chitosan NPs stabilized with biocompatible agents such as genipin have been developed for sustained drug delivery (Liu et al., 2007; Kaihara et al., 2011). Ionically crosslinked NPs, such as those in chitosan–tripolyphosphate (TPP) systems, effectively encapsulate proteins, plasmid DNA, and small molecules for ocular and mucosal delivery applications (De Campos et al., 2001; Xu and Du, 2003). Polysaccharide polyion complexes, such as those in the chitosan-alginate and chitosan-CMC systems, exhibit efficient gene delivery and drug loading capabilities (Cui and Mumper, 2001; Sarmento et al., 2007). Alginate NPs, which are crosslinked with calcium ions, provide pH-responsive release of therapeutic agents such as insulin and nucleic acids (You and Peng, 2005; Douglas et al., 2006; Reis et al., 2007). Self-assembled polysaccharide NPs are formed by the grafting of hydrophobic moieties, such as cholesterol or bile acids, onto hydrophilic backbones. Chitosan derivatives modified with deoxycholic acid have been used to successfully deliver chemotherapeutics such as doxorubicin and paclitaxel. Dextran-based NPs, which are functionalized with hydrophobic chains or bile acids, enhance antigen presentation and targeted drug release. Hyaluronic Acid (HA) NPs conjugated with cholanic acid target CD44-overexpressing tumour cells, combining passive and active targeting mechanisms (Akiyoshi et al., 1998; Nichifor et al., 1999; Choi et al., 2010). Similarly, pullulan NPs grafted with cholesterol achieve monodisperse sizes and enable sustained paclitaxel release. Starch derivatives and heparin-based NPs further extend the polysaccharide NP family, offering controlled drug release and specific targeting capabilities (Akiyoshi et al., 1998; Nitta and Numata, 2013).

3. Properties of Green Synthesized NPs

Green-synthesized NPs have emerged as an eco-friendly and sustainable alternative to chemically synthesized NPs, offering unique physicochemical and biological properties that make them suitable for diverse applications in medicine, agriculture, and environmental remediation. These properties arise from their small size, high surface-to-volume ratio, functionalized surfaces, and biocompatibility, as well as the biomolecular coating provided by natural reducing and stabilizing agents derived from plants, microorganisms, and other biological sources (Singh et al., 2023; Vijayaram et al., 2023).

One of the most significant advantages of green-synthesized NPs is their biocompatibility and reduced toxicity. The synthesis process employs biological extracts, such as plant-based polyphenols, flavonoids, and proteins, which act as reducing and capping agents, resulting in NPs that are inherently biocompatible (Hano and Abbasi, 2021; Singh et al., 2023). These natural biomolecules not only reduce the toxicity associated with chemically synthesized NPs but also provide a protective layer that enhances their interaction with biological systems (Parveen et al., 2016; Vijayaram et al., 2023). Unlike conventional NPs, which often involve hazardous chemicals such as sodium borohydride or toxic surfactants for stabilization, green-synthesized NPs are safer for human health and the environment (Ying et al., 2022). For example, silver NPs synthesized from *Azadirachta indica* (neem) extract exhibited excellent antibacterial activity with significantly reduced cytotoxicity, making them suitable for biomedical applications (Alharbi and Alsubhi, 2023). The presence of bioactive compounds from the synthesis process further enhances their compatibility with cells, tissues, and ecosystems.

The characteristics and uses of green-synthesized NPs are strongly influenced by their size and shape. Exact control over the size and form of the NPs is made possible by using biological reducing agents in NP synthesis. By using natural biological systems, such as microbes and plants, this technique provides an effective and environmentally benign way to reduce metal ions to create NPs. Compared with traditional chemical methods, these biological agents help with the reduction process and keep the NPs stable, making the synthesis more controlled (Mourdikoudis, 2021; Lavande et al., 2022; Kulkarni et al., 2023). To manage the development phases and nucleation of NPs and produce monodispersed NPs with distinct forms, such as spheres, rods, and cubes or triangles, bioactive substances such as proteins and polyphenols are essential. These substances strongly affect the size and shape of the NPs by functioning as both stabilizing and reducing agents. For example, according to Kulkarni et al. (2023), some biomolecules can form protective coatings on NPs, stabilizing them and establishing their unique sizes and shapes during synthesis. The production of extremely stable spherical NPs with sizes ranging from 5.0–20.0 nm has been demonstrated by studies on gold NPs made with extracts from *Magnolia kobus* plants. Yong Song et al. reported that AuNPs made from leaf extracts of *M. kobus* and *Diopyros kaki* exhibited spherical shapes, with the size of these particles being impacted by variables such as reaction temperature and leaf extract concentration (Song et al., 2009; Teimuri-mofrad et al., 2017). In applications such as drug delivery, catalysis, and sensing, where particle size and shape directly affect functional performance, the ability to modify the morphology of NPs by green synthesis is very beneficial (Azad et al., 2023; Radulescu et al., 2023).

As natural capping agents during synthesis, biomolecules improve the stability and dispersibility of green-synthesized NPs. By being adsorbed onto the NP surface, these biomolecules—which include proteins, polysaccharides, and flavonoids—create a barrier that prevents aggregation (Hano and Abbasi, 2021; Liaqat et al., 2022; Kulkarni et al., 2023). Since NP aggregation decreases their surface area and functional efficacy, it is a serious problem. As per Sidhu et al. (2022) and Vijayaram et al. (2023), green-synthesized NPs are guaranteed to be very stable and dispersible in a variety of solvents or biological conditions due to the stabilizing impact of natural capping agents. For example, silver NPs made with extracts from *Eucalyptus globulus* exhibit exceptional durability over extended periods because they contain phenolic chemicals that function as efficient capping agents (Ali et al., 2015). Proteins, flavonoids, ketones, aldehydes, tannins, carboxylic acids, and phenolics are among the biomolecules present in plant extracts that convert Ag^+ to Ag^0, which produces AgNPs (Liaqat et al., 2022). Applications in biological systems require NPs to retain their colloidal characteristics to interact with target molecules effectively; hence, this increased stability is crucial.

The biomolecular coating and surface functionalization of green-synthesized NPs, which result organically from the synthesis process, are other noteworthy characteristics. The NP surfaces are bound by functional groups such as hydroxyl, carboxyl, and amine groups found in the biological extracts employed in green synthesis. These functional groups enable additional changes, such as conjugation with medications, ligands, or other biomolecules for certain uses, in addition to stabilizing the NPs. Surface functionalization enhances the versatility of NPs, particularly in biomedical applications (Singh et al., 2023). According to Murugan et al. (2015), gold NPs created from extracts from *Cymbopogon citratus* (lemongrass) were functionalized with bioactive chemicals, increasing their potential as carriers for targeted drug delivery. In addition to increasing the bioactivity of the NPs, this biomolecular covering makes it possible for them to interact with pathogens or cellular receptors more successfully.

Owing to their exceptional antibacterial and antioxidant qualities, green-synthesized NPs are becoming increasingly valued in a variety of industries, including agriculture, medicine, and food preservation. Both the bioactive compounds generated from the natural materials utilized in their synthesis and the nanoscale characteristics of the NPs, such as their high surface-to-volume ratio, are substantially responsible for these capabilities (Mohanta et al., 2017; Salari et al., 2019; Ansar et al., 2020; Rizwana et al., 2023). Because they can interact directly with microbial cells and interfere with vital processes, green-synthesized NPs have antibacterial effects. The bioactive compounds that coat the NPs during production, such as polyphenols and alkaloids, significantly contribute to this action. These compounds frequently improve how well the NPs interact with biological constituents and microbial membranes (Nguyen et al., 2022). In the case of *E. coli* and *Staphylococcus aureus*, for example, AgNPs produced from *Moringa oleifera* leaf extract have demonstrated strong antibacterial properties (Mohammed and Hawar, 2022). reactive oxygen species (ROS) are generated as part of this process, and they oxidatively damage microbial DNA, proteins, and lipids, ultimately resulting in cell death. Furthermore, the NPs physically cling to and damage bacterial membranes, jeopardizing their structural stability. According to Mohammed and Hawar's (2022) study, these NPs are a possible substitute for traditional antibiotics because of their dual roles in antimicrobial efficacy—ROS generation and membrane disruption. Similarly, bioactive substances, including polyphenols, flavonoids, and tannins—all of which are renowned for their capacity to scavenge free radicals—are the main reason why green-synthesized NPs have strong antioxidant qualities (Moulick et al., 2024e). These molecules aggressively neutralize free radicals, lowering oxidative stress and preventing damage to biological constituents such as lipids, proteins, and DNA. They coat the NPs during their formation. AuNPs, which are made from the peel extract of the pomegranate (Punica granatum), are a striking example. Here, the extract, which is rich in polyphenols, not only makes it easier to synthesize the NPs but also gives them strong antioxidant properties. The ability of these NPs to efficiently neutralize free radicals was shown by Bawazeer et al. (2021), indicating that they may be useful in reducing oxidative stress-related illnesses in biological systems.

The adaptability of green-synthesized NPs is further demonstrated by their catalytic qualities. The functionalized surfaces, large surface area, and small size of these NPs make them effective catalysts for environmental cleanup and redox processes. For example, chitosan/PVA-supported silver NPs, which were made with extracts from *Ocimum basilicum*, could be used to remediate wastewater by catalyzing the breakdown of harmful organic dyes such as methyl orange and methylene blue (Meydan et al., 2024). Similarly, the removal of HM ions from contaminated water sources has been successfully accomplished via the use of AuNPs made with polysaccharides as capping agents.

Effective removal was made possible by the bioactive functional groups that the polysaccharides produced, which improved the binding affinity of the NPs for HMs (Qian et al., 2013). Another significant example is ZnO NPs made from algal extracts, which were tested for their ability to combat waterborne infections. The bioactive components of the algal extracts greatly enhanced the ability of the NPs to interact with bacterial cell walls, which increased their antibacterial activity (Alprol et al., 2023). Green-synthesized NPs are useful for addressing environmental issues because bioactive capping agents foster interactions between NP surfaces and target pollutants, increasing their catalytic activity. For green-synthesized NPs to be produced and function, bioactive capping agents such as proteins, polysaccharides, and secondary metabolites are essential.

These biomolecules coat the surface of NPs during or after their synthesis for two purposes: stabilizing the NPs to prevent aggregation and enhancing their functional properties (Sidhu et al., 2022). By forming a bioactive coating, these agents increase the surface area available for catalytic reactions and introduce functional groups such as hydroxyl, carboxyl, or amine groups. These functional groups enable specific interactions between the NP surfaces and target pollutants, facilitating processes such as binding, degradation, or transformation of contaminants. This significantly enhances the catalytic activity of the NPs (Huq et al., 2022; Kordy et al., 2022). The combination of these enhanced capabilities and the eco-friendly synthesis process makes green-synthesized NPs highly effective for addressing environmental challenges. These NPs provide an effective and sustainable solution for a number of cleaning tasks, including the removal of organic contaminants and HMs and the breakdown of persistent poisons (Moulick et al., 2024e). Their improved interaction with pollutants not only lessens the influence on the environment but also aids in the creation of greener and cleaner pollution control methods (Tavan et al., 2023; Villagrán et al., 2024).

The distinctive optical and electrical characteristics of green-synthesized NPs are attributed to their nanoscale size and Surface Plasmon Resonance (SPR) phenomenon (Kumar et al., 2020). When conduction electrons on the surface of metal NPs oscillate collectively in resonance with incident light at certain wavelengths, SPR takes place. The distinctive SPR bands that result from this interaction are strongly influenced by the size, shape, composition, and surrounding medium of the NP and cause noticeable light absorption and scattering (Bhangale et al., 2019). The strong SPR bands of metal NPs, especially those of gold and silver, are well known. According to Doria et al. (2012), these characteristics make them ideal for use in biosensing, imaging, and diagnostic applications. By tracking variations in the local refractive index caused by biomolecular interactions on the NP surface, SPR-based optical responses are employed, for instance, in biosensing, to identify certain biomolecules. This enables fast detection techniques with excellent sensitivity (Mahmudin et al., 2015). One noteworthy work used aloe vera extract to generate AuNPs. In addition to stabilizing the NPs, the phytochemicals present in aloe vera serve as capping and reducing agents and impart biological properties. The regulated size and SPR characteristics of the resulting gold NPs allow for tunable optical properties (Nalini and Vijayraghavan, 2020). In colorimetric biosensors, where the NPs act as indicators to identify certain biomolecules, these characteristics have been successfully utilized. NPs experience alterations in aggregation or surface chemistry upon binding to the target molecule, which results in a discernible colour shift that may be easily identified without the use of complex equipment (Malik et al., 2023; Nalini and Vijayaraghavan, 2020).

Because green-synthesized NPs are produced via energy-efficient techniques and clear of dangerous chemicals, they are well known for being environmentally benign. Plant extracts, microorganisms, fungi, and algae are examples of renewable biological materials that are used to create these NPs. These materials serve as natural capping, stabilizing, and reducing agents.

Owing to the requirement for hazardous chemical reagents that are frequently employed in traditional synthesis techniques, this technology greatly lowers the production of hazardous byproducts and guarantees less environmental damage. Green synthesis represents a more environmentally friendly approach to nanotechnology because it is in line with the concepts of sustainability and green chemistry (Abuzeid et al., 2023; Singh et al., 2023; Rathod et al., 2024). Green-synthesized NPs have several important benefits, including lower toxicity and biodegradability, which makes them perfect for environmental and agricultural applications where ecological balance is essential (Dikshit et al., 2021). In contrast to chemically produced NPs, which may leave harmful and enduring residues in the environment, green-synthesized NPs are covered with bioactive compounds from natural sources. These coatings reduce long-term ecological hazards by allowing the NPs to break down innocuously. Because NP residues should not damage nontarget organisms or disturb soil and water ecosystems, this characteristic is very beneficial in agriculture (Abuzeid et al., 2023). For example, neem leaf extract from *Azadirachta indica* was used to create silver NPs, which showed potent antibacterial efficacy against plant diseases. Neem extract phytochemicals function as capping and reducing agents, guaranteeing the effectiveness of the NPs while preserving soil health by preventing damage to beneficial soil microorganisms. The equilibrium of the ecosystem was also maintained because the biodegradable nature of these NPs ensures that they do not remain in the soil (Barik et al., 2024). Environmentally friendly NPs have also been created using algae. One study synthesized ZnONPs mediated by *Spirulina* for the purpose of purifying water. These NPs reduce the risk of secondary pollution by efficiently removing HMs and organic pollutants while being completely biodegradable (Alprol et al., 2023). Additionally, biopolymers have been used in green NP synthesis to improve their compatibility with the environment. For example, to create gold NPs for use as NFs, chitosan, a biopolymer made from shrimp shells, was utilized. These chitosan-capped NPs demonstrated the long-term advantages of green-synthesized NPs by increasing crop growth and nutrient uptake while removing harmful residues from the soil and water. Khalifa and Hasaneen (2018) examined how chitosan–PMAA–NPK NFs affect *Pisum sativum* plants, with an emphasis on how well they work as a slow-release fertilizer system. Owing to the combination of sustainable biological resources and eco-friendly synthesis techniques, green-synthesized NPs are now viable substitutes for their chemically synthesized counterparts. They are ideal for applications where ecological balance and environmental health are crucial because of their biodegradability, low toxicity, and natural degradation capabilities. With the least possible negative ecological impact, these developments show how green nanotechnology can be used to solve global issues in environmental remediation, water purification, and agriculture (Khan, 2020).

4. The Role of Green NFs in Sustainable Agriculture

Sustainable farming methods are now more important than ever due to the world's growing population, climate change, and environmental deterioration. Providing food security while protecting ecosystems has emerged as a major worldwide issue. Crop yields have increased over time owing in large part to conventional farming practices, especially the application of artificial fertilizers (Hossain et al., 2021; Choudhury and Moulick, 2022; Sarkar et al., 2023). They are becoming less sustainable, nonetheless, as a result of serious environmental problems caused by their extensive and excessive use (Aamir Iqbal, 2020). Chemical fertilizers, which are mostly made up of potassium, phosphorus, and nitrogen, are a contributing factor to a number of environmental issues. Over time, they impair fertility by upsetting the normal microbial balance of the soil, which deteriorates its quality (Majumdar et al., 2024). Aquatic ecosystems are disturbed, and water bodies become contaminated by nutrients that seep from these fertilizers, leading to eutrophication. The manufacture and use of these fertilizers also contribute significantly to greenhouse gas emissions, which exacerbates climate change. Overuse frequently causes nutrient imbalances in the soil, which harm nearby ecosystems and plant health (Pahalvi et al., 2021). Innovative and environmentally friendly solutions are desperately needed to address these issues. An innovative advancement in agricultural technology, green NFs represent a viable substitute. The inefficiencies and negative effects of conventional fertilizers in the environment are addressed by these fertilizers, which use nanotechnology to distribute nutrients in an effective and sustainable manner. Green nanofluids are created at the nanoscale via biocompatible and biodegradable components derived from natural sources such as microbial or plant extracts. According to Al Zoubi et al. (2024), their distinct characteristics and nanoscale size allow accurate nutrient delivery to plants, minimizing losses due to volatilization or leaching and guaranteeing maximum plant absorption. Green NFs use controlled-release systems to release nutrients, in contrast to traditional fertilizers, which do so in large quantities. By gradually releasing nutrients, this approach minimizes waste and pollution of the environment while meeting the needs of plant growth. By efficiently penetrating plant tissues and delivering nutrients to specific locations, such as roots or leaves, the high surface-area-to-volume ratio of NPs improves nutrient uptake efficiency. Restoring soil fertility is another way that many green NFs support soil health by encouraging beneficial microbial activity (Reis et al., 2007; Yadav et al., 2023; Haydar et al., 2024).

The advantages of green NFs are numerous. They lessen the ecological footprint of agriculture, reduce greenhouse gas emissions, and decrease nutrient runoff, all of which promote environmental sustainability. The effective distribution of nutrients increases agricultural productivity, increasing yield and quality. Because the application is targeted and managed, less fertilizer is needed, which lowers farmers' input expenditures. Furthermore, the utilization of sustainable raw materials and their contributions to emission reduction make green NFs important instruments for climate change mitigation (Nongbet et al., 2022; Sundararajan et al., 2023; Tavan et al., 2023). By utilizing nanoscale technology to maximize nitrogen delivery to plants and solve the environmental issues related to traditional fertilizers, green NFs are a game-changing invention in agriculture. According to Arora et al. (2024), these nanoscale nutrient transporters, which are usually 1–100 nanometres in size, are intended to improve fertilizer use efficiency, decrease waste, and lessen the negative environmental effects of fertilization. The interaction of NPs with plant tissues is greatly improved by their tiny size and high surface-area-to-volume ratio in green NFs. Because of these characteristics, fertilizers can more easily enter plant cells and provide nutrients to certain tissues, such as roots or leaves. Even with lower fertilizer dosages, this focused strategy increases crop output by increasing nutrient uptake and use. Additionally, green NFs help farmers save money and reduce environmental damage by encouraging precision agriculture, which reduces the overapplication of nutrients (Jakhar et al., 2022; Singh et al., 2024).

The synthesis of green NFs relies on eco-friendly processes that avoid the use of toxic chemicals. The use of environmentally safe methods instead of harmful compounds is essential for the production of green NFs. By using phytochemicals such as flavonoids and polyphenols in plant extracts as stabilizing and reducing agents, plant-based synthesis creates ecologically safe and biocompatible NPs. For example, ZnONPs have been demonstrated to increase plant growth and stress tolerance when they are made with *Azadirachta indica* (neem) extract (Bhuyan et al., 2015). The metabolic powers of bacteria, fungi, and algae are also utilized in microbial-based synthesis to produce NPs via bioreduction processes. *Aspergillus niger* and *Bacillus subtilis* are two examples of organisms that may create NPs such as gold or chitosan, which promote soil health and nutrient uptake. By utilizing natural polymers, including chitosan, alginate, and cellulose, biopolymer-based synthesis further improves the nutrient release effectiveness and environmental safety of green NFs (Alzamily et al., 2023; Sharma et al., 2023).

Compared with traditional fertilizers, green NFs may provide nutrients at lower dosages, which is one of their main advantages (Zulfiqar et al., 2019). In addition to lowering farmers' input costs, this efficiency also lessens the overuse of fertilizers, which can damage ecosystems. By matching nutrient availability to plant growth needs, the controlled release of nutrients reduces environmental contamination. For example, FeNPs improve photosynthetic efficiency and chlorophyll production by increasing iron bioavailability in plants (Ahmad et al., 2024). Furthermore, foliar sprays of NFs ensure faster absorption and better plant growth by avoiding soil-related issues such as nitrogen fixation. Iron and zinc NPs applied topically have been demonstrated to dramatically increase photosynthetic activity, resulting in healthier and more fruitful crops (Farnoosh et al., 2024).

Early plant development also benefits greatly from seed treatments containing green NFs. During germination, the application of biopolymer-based NPs, such as chitosan, to seeds supplies vital nutrients that support strong seedling establishment and early root development (Zaim et al., 2023). Chitosan NPs, for example, promote root elongation and accelerate germination rates in rice seeds, increasing crop output (Chouhan and Mandal, 2021). This technique lessens the need for conventional fertilizers in the early phases of growth while simultaneously increasing seedling vigour. Additionally, green NFs help crops become more resilient to stress. For example, crops such as maize can endure drought conditions because SiNPs strengthen plant cell walls and increase water-use efficiency (Ali and Singh, 2024; Wahi et al., 2024). This improves resistance to abiotic stressors, which is essential for maintaining productivity in regions where water scarcity is a problem. Owing to phosphorus-based NFs such as nanohydroxyapatite, legumes can better absorb and use phosphorus, which release phosphorus gradually and steadily. This enhances the overall production of biomass, nitrogen fixation, and root nodulation in crops such as beans and peas (Gupta et al., 2024).

Because of these characteristics, green NFs are a sustainable substitute for traditional fertilizers that can correct nutrient deficits and lessen their negative effects on the environment. However, high production costs, legal restrictions, and the requirement for lengthy field testing are still obstacles. Their widespread implementation in agriculture will depend on overcoming these obstacles with the help of research and policy (Kah et al., 2019). Farmers may increase yields, lessen their impact on the environment, and help ensure that food production worldwide is more sustainable in the future by incorporating green NFs into their farming operations (Sundararajan et al., 2023; Arora et al., 2024).

5. Synergistic role of PGPR and Green NFs

The overuse of agrochemicals poses serious environmental risks, while the effectiveness of PGPRs as bioinoculants is limited by factors such as slow action, environmental stress sensitivity, and reduced field performance. A promising solution is combining PGPRs with NMs, creating an eco-friendly alternative that combines the advantages of biofertilizers and NFs. This combination ensures better nutrient availability, minimizes chemical fertilizer use and enhances crop productivity through the slow-release properties of nanobiofertilizers (Upadhayay et al., 2023) (Table 1). PGPR are considered advantageous soil microbes that increase plant development through both direct and indirect mechanisms. They help mitigate the adverse effects of chemical fertilizers and toxic substances released into the soil due to industrial activities. Adding PGPR to the soil can serve as an effective, eco-friendly strategy to increase agricultural productivity (Basit et al., 2023). PGPR play crucial roles in agriculture by enhancing crop health through the fixation of N_2, solubilization of phosphate, HM reduction, phytohormone production, and suppression of phytopathogens. They contribute to agricultural biotechnology by providing new genes and insights into biochemical pathways for enzymes and bioproducts. PGPR potency depends on factors such as soil ecology, plant species, and growth stages, which aid in adjusting plant–soil chemistry and overall crop well-being (Mohanty et al., 2021). The integrated PGPR and NP applications positively influence soil properties, including physicochemical and biological aspects. Compared with the use of PGPR or NPs alone, this approach typically enhances the composition of mineral nutrients in rhizospheric soils. Notably, soil nitrate levels increase more effectively with their combined use than with a single nitrogen application. Moreover, the availability of organic carbon, phosphorus, and nitrogen is greater than that of untreated plants (Liu et al., 2022; Verma et al., 2024). PGPR combined with NPs increase the activity of soil enzymes, including phosphomonoesterase, protease, urease, sucrase, and invertase, in both acidic and alkaline environments. These enzymes facilitate nutrient transformation processes, increasing soil fertility by activating vital biochemical functions within

Table 1. Use of green synthesized metallic NPs in sustainable agriculture.

Nano fertilizers	**Biological component (source)**	**Targeted plants**	**Key aspects**	**References**
ZnNPs, Cu	*Ocimum basilicam* L.	*Ocimum basilicum* L	Improved basil plant growth, enhancing morphological traits, chlorophyll content, carotenoids, phenolic and flavonoid content, and antioxidant activity. Overall, Zn and Cu NPs application boosted basil's yield and quality.	(Abbasifar, Shahrabadi et al., 2020)
CuONPs	*Morus alba*	*Brassica oleracea*, *Solanum lycopersicum*	Exposure to CuONPs induced dose-dependent increases in lipid peroxidation, electrolyte leakage, and antioxidant enzyme activity in plants. Additionally, lignin deposition was observed in the roots of plants treated with the highest CuO NP concentration.	(Singh et al., 2017)
Mo NPs	vermicompost extract	*Spinacia oleracea* L.	Green-synthesized Mo NPs at 1 mg/L significantly increased Nitrate Reductase (NR) activity, reducing NO_3^- accumulation. This treatment also resulted in the highest chlorophyll a, b, and total chlorophyll content.	(Abbasifar, ValizadehKaji et al., 2020)
Fe NPs	*Green tea and Eucalyptus leaf extract*	-	Green-synthesized GT-Fe and EL-Fe NPs (NPs) showed strong stability and maintained nitrate removal efficiency even after two months of air exposure, unlike nZVI, which lost 50% efficiency.	(Wang et al., 2014)
FeO, MnO	*Paenibacillus polymyxa* DEBB B-358	*Zea mays*	It showed positive effects on maize seedling growth, especially using the bimetallic NPs, increasing seed germination, root development, and fresh weight.	(de França Bettencourt et al., 2020)
Zero valent Fe NPs	*Emblica officinalis*		ZV FeNPs were found to be very efficient for the remediation of Pb and other HMs.	(Kumar et al., 2015)
ZnO	*Azadirachta indica*	Green gram	Higher seed yield, protein content.	(Sahoo et al., 2021)
ZnO	*Bacillus subtilis*	*Zea mays* L.	Increased protein content of the plant, and leaf area, and all growth parameters	(Sabir et al., 2020)
Magnetite NPs	*Hevea brasiliensis* bark extract	*Oryza sativa*	decreased the accumulation of Cd and Na in rice plants.	(Sebastian et al., 2019)
AgNPs	aqueous leaf extract of *Hypericum perforatum*	*Solanum tuberosum*	The biosynthesized AgNPs demonstrated potent antibacterial activity against *Ralstonia solanacearum*.	(Tortella et al., 2019)
CuO	*Klebsiella pneumonia* SN35	*Triticum aestivum*	enhanced plant growth, biomass, and cellular antioxidant levels, while reducing ROS and the translocation of Cr from soil to roots and shoots, compared to control plants.	(Noman et al., 2020)

Table 1 contd. ...

...Table 1 contd.

Nano fertilizers	**Biological component (source)**	**Targeted plants**	**Key aspects**	**References**
Se	*Bacillus megaterium*	*Vicia faba*	Se-NPs effectively reduced both pre- and postemergence damping-off caused by *R. solani* and minimized the severity of root rot disease. Additionally, Se-NPs enhanced both morphological and metabolic indicators, leading to a significant boost in yield compared to the infected control.	(Hashem et al., 2021)
Ag	*Triticum aestivum* leaf extract	*Triticum aestivum*	Alleviated the negative effects of salt stress by lowering H_2O_2 levels, decreasing Na^+ and Cl^- content in roots and leaves, and enhancing the activity of SOD, APX, GR, and GPX.	(Wahid et al., 2020)
ZnO	Leaves of *Olea europaea,* flower of *Matricaria chamomilla L.,* and fruit of *Lycopersicon esculentum M.*	*Xanthomonas oryzae* pv. *oryzae* (Xoo) strain GZ 0003	ZnONPs exhibited the highest antibacterial activity at 16.0 µg/ml, with *Olea europaea*-synthesized ZnONPs forming the largest inhibition zone (2.2 cm), surpassing those synthesized from *Matricaria chamomilla* and *Lycopersicon esculentum.*	(Ogunyemi et al., 2019)
SNPs (Sulfur NPs)	*Moringa oleifera* Leaf extract	*Vicia faba*	SNPs treatment under salinity stress elevated the levels of osmolytes and nonenzymatic antioxidants, while decreasing oxidative stress biomarkers (MDA and H_2O_2). Additionally, SNPs treatment enhanced the activity of antioxidant enzymes such as APX and POD.	(Khalifa et al., 2024)
CuO	*Daucus carota* leaves	*Vigna unguiculata*	There was an increase in morphological attributes, chlorophyll content, and antioxidant enzyme activities.	(Mustafa et al., 2024)
AgNPS	*Melia azedarach*	*Solanum melongena* L.	The application of AgNPs at a concentration of 60 ppm significantly suppressed the growth *of Verticillium dahliae*, demonstrating their potential as an effective antifungal agent against this pathogen.	(Jebril et al., 2020)

the ecosystem (Verma et al., 2024). Evaluating the effects of NMs on microbial growth before they are combined with PGPR is important. NMs should support, rather than inhibit, microbial activity. Research has shown that NMs such as nano gypsum and nanozeolite enhance PGPR growth in soils affected by NMs. The synergistic interaction of NMs and PGPR stimulates plant physiological and functional advancement, enhances photosynthetic translocation in aerial plant parts, and results in better quality and increased yields (Swetha et al., 2024). Green-synthesized ZnO NPs from papaya fruit extract combined with stress-resistant PGPR (*Pseudomonas* sp.) improved the defence of wheat against drought, heat, and combined stresses. The treated plants presented improved growth, increased antioxidant activity, and reduced expression of stress markers such as MDA and H_2O_2. The combination effectively enhanced stress tolerance by increasing the contents of proline, antioxidant enzymes, and abscisic acid, confirming its potential for mitigating multiple environmental stresses in wheat (Azmat et al., 2022). AgNPs were synthesized in green via extracts from the leaves and flowers of *Tagetes erecta* (marigold), while PGPR were isolated from the spinach rhizosphere. The PGPR isolates produced IAA, ammonia, and HCN; solubilized tricalcium phosphate; and inhibited the growth of *Macrophomina phaseolina*. Among them, *Bacillus cereus* LPR2 was the most effective, significantly increasing maize seed germination and root and shoot growth, particularly when it was combined with AgNPs. *Bacillus cereus* and AgNPs were found to be potential bioinoculants and growth enhancers for maize (Kumar et al., 2020). $nSiO_2$ was synthesized from rice husk and evaluated alongside four selected PGPR strains: *Bacillus megaterium*, *Bacillus brevis*, *Pseudomonas fluorescens*, and *Azotobacter vinelandii*. This study aimed to assess the effects of nano silica and different silicon sources on soil properties. $nSiO_2$ triggered PGPR growth, increased the total soil bacterial population, maintained the soil pH, improved the soil nutrient levels, and enhanced maize seed germination. These findings suggested that $nSiO_2$ could be incorporated into fertilizer formulations to increase soil fertility and support beneficial microbial communities, ultimately improving crop yields. These results broaden the research scope for $nSiO_2$ applications in agriculture (Karunakaran et al., 2013). A study investigated the impact of biogenic AuNPs, which were analysed via FT-IR, UV–Vis spectroscopy, Transmission Electron Microscopy (TEM), and Energy Dispersive Spectroscopy (EDS), on the IAA-producing ability of *Pseudomonas monteilii* at various concentrations synthesized by *Bacillus subtilis*. Supplementing the culture medium with 50 µg/mL AuNPs significantly increased IAA production. This improvement translated into better growth of *Vigna unguiculata*, as confirmed by statistical

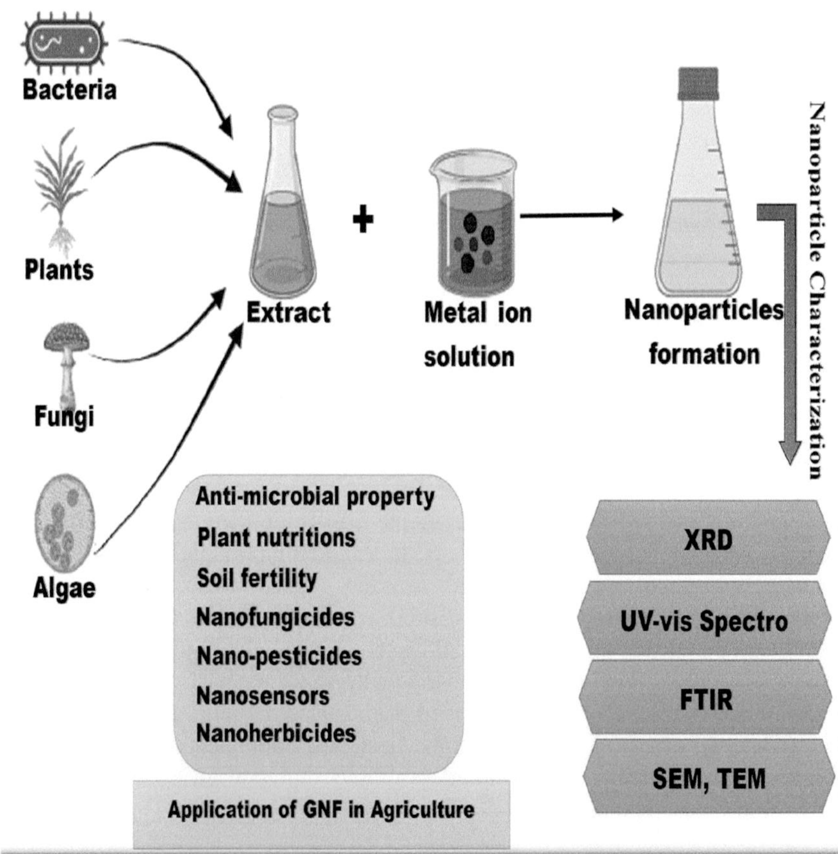

Figure 1. Synthesis of green NFs (GNFs) and their applications in agriculture (created in biorenders).

analysis of growth parameters. These results highlight the potential of AuNPs as nano stimulators for biofertilizers in agricultural practices (Panichikkal et al., 2019). Green-synthesized Iron Oxide NPs (IONPs) function as efficient nanofertilizers, significantly enhancing the morphophysiological traits, biochemical properties, and yield of *R. pusense*-inoculated green gram plants. The IONPs, synthesized from *S. cumini* leaf extract with a crystalline size of 33.5 nm, had no toxic effects on nodule bacteria or bacterial cell morphology, as confirmed by FTIR, SEM, and CLSM analyses. Additionally, IONP treatment sharpened the FTIR peaks without altering the functional groups. These results suggested that IONPs could be effectively used as nano biofertilizers for increasing crop production and for various nanobiotechnological applications, including iron fortification and improving iron fertilizer efficiency (Saleem and Khan, 2023).

6. Future Directions and Conclusions

Adverse agroclimatic conditions can disrupt plant growth and reduce crop yields. Nano fertilizers offer an efficient alternative to traditional fertilizers by improving nutrient use efficiency, reducing soil toxicity, minimizing overuse effects, and lowering application frequency. Slow release nanofertilizers combined with organic manure and PGPR can increase enzyme activity during germination and increase seed vigour. PGPR also support better NPK fertilizer utilization, leading to increased crop yields (Nayana et al., 2020). The synergistic approach of PGPR and NPs has shown promise in increasing crop productivity, increasing soil health, and increasing resilience to environmental stresses. The incorporation of these technologies into sustainable agricultural systems could minimize the reliance on synthetic fertilizers while improving crop performance, yield quality, and soil characteristics. However, few field trials have explored their combined effects, emphasizing the need for further research and extended field studies to better understand the interactive benefits of PGPR and NPs in agriculture (Verma et al., 2024). The focus on the synergetic interaction between NMs and PGPR in agriculture has increased, with the aim of increasing crop yields and maintaining soil health. This strategy addresses several objectives: plant disease control, increased productivity, food security, environmental stress management, soil health promotion, reduced dependence on chemical fertilizers, and sustainable bioeconomic development. Advances in nanotechnology support the controlled release of NFs and PGPR, although it remains vital to establish safe dosages of nanomaterials for environmental safety. Ultimately, the combined use of NFs and PGPR represents a powerful approach for improving crop quality and yield, with continued research being crucial for meeting global agricultural demands sustainably (Swetha et al., 2024).

Nanoenabled soil remediation provides a sustainable way to address soil degradation and increase agricultural productivity. Advancing nanotechnology plays a crucial role in addressing soil contamination, highlighting the need for innovative materials, enhanced phytoremediation techniques, and comprehensive ecological impact assessments. Effective soil remediation depends on interdisciplinary efforts among researchers and industry leaders. By integrating sustainability with cutting-edge nanotechnological solutions, we can develop precise methodologies to restore soil health and ensure its preservation for future generations. Despite these advancements, challenges remain in scaling up green nanotechnology for soil remediation. Critical areas for future research include evaluating the long-term ecological impacts of nanomaterials, improving their cost-effectiveness, and tailoring them for diverse soil types and contamination profiles. The development of comprehensive safety protocols and regulatory frameworks is also essential to ensure their responsible use. Future efforts must focus on interdisciplinary collaboration, comprehensive impact assessments, and the integration of nanotechnology with sustainable practices. Advancing research and innovation in this field will facilitate the creation of specialized, efficient, and eco-friendly soil remediation strategies, ensuring soil health and environmental sustainability for future generations.

References

Aamir Iqbal, M. (2020). Nano-Fertilizers for Sustainable Crop Production under Changing Climate: A Global Perspective. *In*: Hasanuzzaman, M., Carvalho Minhoto Teixeira Filho, M., Fujita, M., and Assis Rodrigues Nogueira, T. (Eds.). *Sustainable Crop Production*. IntechOpen. https://doi.org/10.5772/intechopen.89089.

Abbasifar, A., ValizadehKaji, B., and Iravani, M. A. (2020). Effect of green synthesized molybdenum nanoparticles on nitrate accumulation and nitrate reductase activity in spinach. *Journal of Plant Nutrition*, 43(1): 13–27. https://doi.org/10.1080/01904167.2019.1659340.

Abuzeid, H. M., Julien, C. M., Zhu, L., and Hashem, A. M. (2023). Green synthesis of nanoparticles and their energy storage, environmental, and biomedical applications. *Crystals*, 13(11): 1576. https://doi.org/10.3390/cryst13111576.

Ahmad, N., Sharma, S., Alam, Md. K., Singh, V. N., Shamsi, S. F., Mehta, B. R. et al. (2010). Rapid synthesis of silver Nanoparticles using dried medicinal plant of basil. *Colloids and Surfaces B: Biointerfaces*, 81(1): 81–86. https://doi.org/10.1016/j.colsurfb.2010.06.029.

Ahmad, S., Ahmad, N., Islam, M. S., Ahmad, M. A., Ercisli, S., Ullah, R. et al. (2024). Rice seeds biofortification using biogenic ıron oxide nanoparticles synthesized by using *Glycyrrhiza glabra*: A study on growth and yield ımprovement. *Scientific Reports*, 14(1): 12368. https://doi.org/10.1038/s41598-024-62907-1.

Ahmad, T., Wani, I. A., Manzoor, N., Ahmed, J., and Asiri, A. M. (2013). Biosynthesis, structural characterization and antimicrobial activity of gold and silver nanoparticles. *Colloids and Surfaces B: Biointerfaces*, 107: 227–234. https://doi.org/10.1016/j.colsurfb.2013.02.004.

Akhtar, N., Ilyas, N., Meraj, T. A., Pour-Aboughadareh, A., Sayyed, R. Z., Mashwani, Z.-R. et al. (2022). Improvement of plant responses by nanobiofertilizer: a step towards sustainable agriculture. *Nanomaterials*, 12(6): Article 6. https://doi.org/10.3390/nano12060965.

Akiyoshi, K., Kobayashi, S., Shichibe, S., Mix, D., Baudys, M., Wan Kim, S. et al. (1998). Self-assembled hydrogel nanoparticles of cholesterol-bearing pullulan as a carrier of protein drugs: Complexation and stabilization of insulin. *Journal of Controlled Release*, 54(3): 313–320. https://doi.org/10.1016/S0168-3659(98)00017-0.

Al Zoubi, O. M., Singh, A., Rajput, V. D., Varshney, A., Agrawal, S., Ghazaryan, K. et al. (2024). Green nanofertilizers: the need for modern agriculture, intelligent, and environmentally ly-friendly approaches. *Ecological Engineering & Environmental Technology*, 25(1): 1–21. https://doi.org/10.12912/27197050/172946.

Al-asbahi, M. G. S. S., Al-Ofiry, B. A., Saad, F. A. A., Alnehia, A., and Al-Gunaid, M. Q. A. (2024). Silver Nanoparticles biosynthesis using mixture of *Lactobacillus* sp. In addition, and *Bacillus* sp. Growth and their antibacterial activity. *Scientific Reports*, 14(1): 10224. https://doi.org/10.1038/s41598-024-59936-1.

Alharbi, N. S., and Alsubhi, N. S. (2023). Silver nanoparticles biosynthesized using *Azadirachta indica* fruit and leaf extracts: optimization, characterization, and anticancer activity. *Journal of Nanomaterials*, 2023(1): 9916777. https://doi.org/10.1155/2023/9916777.

Ali, A. M., and Bijay-Singh. (2024). Silicon: A crucial element for enhancing plant resilience in challenging environments. *Journal of Plant Nutrition*, 1–36. https://doi.org/10.1080/01904167.2024.2406479.

Ali, K., Ahmed, B., Dwivedi, S., Saquib, Q., Al-Khedhairy, A. A., and Musarrat, J. (2015). Microwave accelerated green synthesis of stable silver nanoparticles with *Eucalyptus globulus* leaf extract and their antibacterial and antibiofilm activity on clinical isolates. *PLOS ONE*, 10(7): e0131178. https://doi.org/10.1371/journal.pone.0131178.

Alloosh, M. T., Saleh, M. M., Alnaddaf, L. M., Almuhammady, A. K., Salem, K. F. M., and Al-Khayri, J. M. (2021). Biosynthesis and characterization of microorganisms-derived nanomaterials. *In*: Al-Khayri, J. M., Ansari, M. I., and Singh, A. K. (Eds.). Nanobiotechnology: Mitigation of Abiotic Stress in Plants (pp. 239–260). Springer International Publishing. https://doi.org/10.1007/978-3-030-73606-4_10.

Al-Mamun, Md. R., Hasan, Md. R., Ahommed, Md. S., Bacchu, Md. S., Ali, Md. R., and Khan, Md. Z. H. (2021). Nanofertilizers towards sustainable agriculture and environment. Environmental Technology & Innovation, 23: 101658. https://doi.org/10.1016/j.eti.2021.101658.

Alprol, A. E., Mansour, A. T., El-Beltagi, H. S., and Ashour, M. (2023). Algal extracts for green synthesis of zinc oxide nanoparticles: promising approach for algae bioremediation. *Materials*, 16(7): 2819. https://doi.org/10.3390/ma16072819.

Alzamily, I. A., Al-Shibly, M. K., and Hussein Ali, Z. A. (2023). Biosynthesis of gold nanoparticles by *Aspergillus Niger* and their role against pathogenic bacteria. *Clinical Medicine And Health Research Journal*, 3(4): 513–518. https://doi.org/10.18535/cmhrj.v3i4.224.

Anil Kumar, S., Abyaneh, M. K., Gosavi, S. W., Kulkarni, S. K., Pasricha, R., Ahmad, A. et al. (2007). Nitrate reductase-mediated synthesis of silver nanoparticles from AgNO3. *Biotechnology Letters*, 29(3): 439–445. https://doi.org/10.1007/s10529-006-9256-7.

Ansar, S., Tabassum, H., Aladwan, N. S. M., Naiman Ali, M., Almaarik, B., AlMahrouqi, S. et al. (2020). Eco friendly silver Nanoparticles synthesis by *Brassica oleracea* and its antibacterial, anticancer and antioxidant properties. *Scientific Reports*, 10(1): 18564. https://doi.org/10.1038/s41598-020-74371-8.

Apte, M., Sambre, D., Gaikawad, S., Joshi, S., Bankar, A., Kumar, A. R. et al. (2013). Psychrotrophic yeast *Yarrowia lipolytica* NCYC 789 mediates the synthesis of antimicrobial silver nanoparticles via cell-associated melanin. *AMB Express*, 3(1): 32. https://doi.org/10.1186/2191-0855-3-32.

Arora, P. K., Tripathi, S., Omar, R. A., Chauhan, P., Sinhal, V. K., Singh, A. et al. (2024). Next-generation fertilizers: The impact of bionanofertilizers on sustainable agriculture. *Microbial Cell Factories*, 23(1): 254. https://doi.org/10.1186/s12934-024-02528-5.

Arsiya, F., Sayadi, M. H., and Sobhani, S. (2017). Green synthesis of palladium nanoparticles using *Chlorella vulgaris*. *Materials Letters*, 186: 113–115. https://doi.org/10.1016/j.matlet.2016.09.101.

Ayaz, M., Junaid, M., Ullah, F., Subhan, F., Sadiq, A., Ali, G. et al. (2017). Anti-Alzheimer's Studies on β-Sitosterol Isolated from *Polygonum hydropiper* L. *Frontiers in Pharmacology*, 8: 697. https://doi.org/10.3389/fphar.2017.00697.

Azad, A., Zafar, H., Raza, F., and Sulaiman, M. (2023). Factors influencing the green synthesis of metallic nanoparticles using plant extracts: a comprehensive review. *Pharmaceutical Fronts*, 05(03): e117–e131. https://doi.org/10.1055/s-0043-1774289.

Azmat, A., Tanveer, Y., Yasmin, H., Hassan, M. N., Shahzad, A., Reddy, M. et al. (2022). Coactive role of zinc oxide nanoparticles and plant growth promoting rhizobacteria for mitigation of synchronized effects of heat and drought stress in wheat plants. *Chemosphere*, 297: 133982. https://doi.org/10.1016/j.chemosphere.2022.133982.

Barciela, P., Carpena, M., Li, N.-Y., Liu, C., Jafari, S. M., Simal-Gandara, J. et al. (2023). Macroalgae as biofactories of metal nanoparticles; biosynthesis and food applications. *Advances in Colloid and Interface Science*, 311: 102829. https://doi.org/10.1016/j.cis.2022.102829.

Barik, B., Satapathy, B. S., Pattnaik, G., Bhavrao, D. V., and Shetty, K. P. (2024). Sustainable synthesis of silver nanoparticles from *Azadirachta indica*: Antimicrobial, antioxidant and *in silico* analysis for periodontal treatment. *Frontiers in Chemistry*, 12. https://doi.org/10.3389/fchem.2024.1489253.

Basit, F., Bhat, J. A., and Guan, Y. (2023). Elucidation of synergistic interaction among metal oxide nanoparticles and PGPR on the plant growth and development. pp. 311–325. *In*: Husen, A. (Ed.). Nanomaterials and Nanocomposites Exposures to Plants: Response, Interaction, Phytotoxicity and Defense Mechanisms. *Springer Nature*. https://doi.org/10.1007/978-981-99-2419-6_16.

Bawazeer, S., Rauf, A., Nawaz, T., Khalil, A. A., Javed, M. S., Muhammad, N. et al. (2021). *Punica granatum* peel extracts mediated the green synthesis of gold nanoparticles and their detailed *in vivo* biological activities. *Green Processing and Synthesis*, 10(1): 882–892. https://doi.org/10.1515/gps-2021-0080.

Behravan, M., Hossein Panahi, A., Naghizadeh, A., Ziaee, M., Mahdavi, R., and Mirzapour, A. (2019). Facile green synthesis of silver nanoparticles using *Berberis vulgaris* leaf and root aqueous extract and its antibacterial activity. *International Journal of Biological Macromolecules*, 124: 148–154. https://doi.org/10.1016/j.ijbiomac.2018.11.101.

Beveridge, T. J., and Murray, R. G. (1980). Sites of metal deposition in the cell wall of *Bacillus subtilis*. *Journal of Bacteriology*, 141(2): 876–887. https://doi.org/10.1128/jb.141.2.876-887.1980.

Bhangale, H., Bachhav, S., Nerkar, D., Sarode, K. M., and Patil, D. (2019). Study on optical properties of green synthesized silver nanoparticles for surface plasmon resonance. *Journal of Nanoscience and Technology*, 5: 658–661. https://doi.org/10.30799/jnst.230.19050203.

Bhargava, A., Jain, N., Barathi L., M., Akhtar, Mohd. S., Yun, Y.-S., and Panwar, J. (2013). Synthesis, characterization and mechanistic insights of mycogenic iron oxide nanoparticles. *Journal of Nanoparticle Research*, 15(11): 2031. https://doi.org/10.1007/s11051-013-2031-5.

Bhattacharjee, R. R., Das, A. K., Haldar, D., Si, S., Banerjee, A., and Mandal, T. K. (2005). Peptide-assisted synthesis of gold nanoparticles and their self-assembly. *Journal of Nanoscience and Nanotechnology*, 5(7): 1141–1147. https://doi.org/10.1166/jnn.2005.166.

Bhuyan, T., Mishra, K., Khanuja, M., Prasad, R., and Varma, A. (2015). Biosynthesis of zinc oxide nanoparticles from *Azadirachta indica* for antibacterial and photocatalytic applications. Materials Science in Semiconductor Processing, 32: 55–61. https://doi.org/10.1016/j.mssp.2014.12.053.

Birla, S. S., Gaikwad, S. C., Gade, A. K., and Rai, M. K. (2013). Rapid synthesis of silver nanoparticles from *Fusarium oxysporum* by optimizing physicocultural conditions. *The Scientific World Journal*, 2013(1): 796018. https://doi.org/10.1155/2013/796018.

Brahmachari, G., Sarkar, S., Ghosh, R., Barman, S., Mandal, N. C., Jash, S. K. et al. (2014). Sunlight-induced rapid and efficient biogenic synthesis of silver nanoparticles using aqueous leaf extract of *Ocimum sanctum* Linn. With enhanced antibacterial activity. *Organic and Medicinal Chemistry Letters*, 4(1): 18. https://doi.org/10.1186/s13588-014-0018-6.

Chandran, S. P., Chaudhary, M., Pasricha, R., Ahmad, A., and Sastry, M. (2006). Synthesis of gold nanotriangles and silver nanoparticles using aloe vera plant extract. *Biotechnology Progress*, 22(2): 577–583. https://doi.org/10.1021/bp0501423.

Chen, L., Liu, X., and Wong, K.-H. (2016). Novel Nanoparticle materials for drug/food delivery-polysaccharides. *Physical Sciences Reviews*, 1(8). https://doi.org/10.1515/psr-2016-0053.

Choi, K. Y., Chung, H., Min, K. H., Yoon, H. Y., Kim, K., Park, J. H. et al. (2010). Self-assembled hyaluronic acid nanoparticles for active tumor targeting. *Biomaterials*, 31(1): 106–114. https://doi.org/10.1016/j.biomaterials.2009.09.030.

Choudhury, S., and Moulick, D. (2022). Response of Field Crops to Abiotic Stress: Current Status and Future Prospects (1st ed.). CRC Press. https://doi.org/10.1201/9781003258063.

Chouhan, D., and Mandal, P. (2021). Applications of chitosan and chitosan based metallic nanoparticles in agrosciences-A review. International *Journal of Biological Macromolecules*, 166: 1554–1569. https://doi.org/10.1016/j.ijbiomac.2020.11.035.

Chowardhara, B., Saha, B., Awasthi, J. P., Deori, B. B., Nath, R., Roy, S. et al.. (2024). An assessment of nanotechnology-based interventions for cleaning up toxic heavy metal/metalloid-contaminated agroecosystems: Potentials and Issues. *Chemosphere*, 359: 142178. https://doi.org/10.1016/j.chemosphere.2024.142178.

Cui, Z., and Mumper, R. J. (2001). Chitosan-based Nanoparticles for topical genetic immunization. *Journal of Controlled Release: Official Journal of the Controlled Release Society*, 75(3): 409–419. https://doi.org/10.1016/s0168-3659(01)00407-2.

De Campos, A. M., Sánchez, A., and Alonso, M. J. (2001). Chitosan Nanoparticles: A new vehicle for the improvement of the delivery of drugs to the ocular surface. Application to cyclosporin A. *International Journal of Pharmaceutics*, 224(1): 159–168. https://doi.org/10.1016/S0378-5173(01)00760-8.

de França Bettencourt, G. M., Degenhardt, J., Zevallos Torres, L. A., de Andrade Tanobe, V. O., and Soccol, C. R. (2020). Green biosynthesis of single and bimetallic nanoparticles of iron and manganese using bacterial auxin complex to act as plant biofbio-fertilizer. *Biocatalysis and Agricultural Biotechnology*, 30: 101822. https://doi.org/10.1016/j.bcab.2020.101822.

Deb, P., Mandal, A., Harendra, Choudhury, S. R., Das, A., Hazra, S. et al. (2023). Improving plant nutrient use efficiency for climate-resilient agriculture. pp. 209–243. *In*: Hasanuzzaman, M. (Eds.). Climate-resilient agriculture, vol 2: Agro-biotechnological advancement for crop production. Cham: Springer International Publishing. https://doi.org/10.1007/978-3-031-37428-9_10.

Dey, S., Majumdar, A., Dubey, P. K., Roychowdhury, T., Majumdar, J., Santra, S. C. et al. Involvement of soil parameters and rhizosphere microbiome in sustainable crop productivity. pp. 189–228 *In*: Panwar, J. S., and Sharma, V. (Eds.). Rhizomicrobiome in sustainable agriculture and environment 2025). Academic Press. https://doi.org/10.1016/B978-0-443-23691-4.00009-9.

Dikshit, P. K., Kumar, J., Das, A. K., Sadhu, S., Sharma, S., Singh, S. et al. (2021). Green Synthesis of Metallic Nanoparticles: Applications and Limitations. *Catalysts*, 11(8): Article 8. https://doi.org/10.3390/catal11080902.

Doria, G., Conde, J., Veigas, B., Giestas, L., Almeida, C., Assunção, M. et al. (2012). Noble metal nanoparticles for biosensing applications. *Sensors*, 12(2): Article 2. https://doi.org/10.3390/s120201657.

Douglas, K. L., Piccirillo, C. A., and Tabrizian, M. (2006). Effects of alginate inclusion on the vector properties of chitosan-based nanoparticles. *Journal of Controlled Release*, 115(3): 354–361. https://doi.org/10.1016/j.jconrel.2006.08.021.

Dubey, S. P., Lahtinen, M., Särkkä, H., and Sillanpää, M. (2010). Bioprospective of *Sorbus aucuparia* leaf extract in development of silver and gold nanocolloids. *Colloids and Surfaces B: Biointerfaces*, 80(1): 26–33. https://doi.org/10.1016/j.colsurfb.2010.05.024.

Elahian, F., Reiisi, S., Shahidi, A., and Mirzaei, S. A. (2017). High-throughput bioaccumulation, biotransformation, and production of silver and selenium using genetically engineered *Pichia pastoris*. *Nanomedicine: Nanotechnology, Biology and Medicine*, 13(3): 853–861. https://doi.org/10.1016/j.nano.2016.10.009.

El-Belely, E. F., Farag, M. M. S., Said, H. A., Amin, A. S., Azab, E., Gobouri, A. A. et al. (2021). Green synthesis of zinc oxide nanoparticle (ZnO-NPs) Using *Arthrospira platensis* (Class: Cyanophyceae) and Evaluation of their Biomedical Activities. *Nanomaterials*, 11(1): Article 1. https://doi.org/10.3390/nano11010095.

El-Kassas, H. Y., Aly ly-Eldeen, M. A., and Gharib, S. M. (2016). Green synthesis of iron oxide (Fe_3O_4) nanoparticles using two selected brown seaweeds: Characterization and application for lead bioremediation. Acta Oceanologica Sinica, 35(8): 89–98. https://doi.org/10.1007/s13131-016-0880-3.

El-Seedi, H. R., El-Shabasy, R. M., Khalifa, S. A. M., Saeed, A., Shah, A., Shah, R. et al. (2019). Metal nanoparticles fabricated by green chemistry using natural extracts: Biosynthesis, mechanisms, and applications. *RSC Advances*, 9(42): 24539–24559. https://doi.org/10.1039/C9RA02225B.

Fardsadegh, B., and Jafarizadeh-Malmiri, H. (2019). Aloe vera leaf extract mediated green synthesis of selenium nanoparticles and assessment of their *In vitro* antimicrobial activity against spoilage fungi and pathogenic bacteriala strains. *Green Processing and Synthesis*, 8(1): 399–407. https://doi.org/10.1515/gps-2019-0007.

Farnoosh, S., Masoudian, N., Safipour Afshar, A., Nematpour, F. S., and Roudi, B. (2024). Foliar-applied iron and zinc nanoparticles improved plant growth, phenolic compounds, essential oil yield, and rosmarinic acid production of lemon balm (*Melissa officinalis* L.). *Environmental Science and Pollution Research*, 31(25): 36882–36893. https://doi.org/10.1007/s11356-024-33680-5.

Fawcett, D., Verduin, J. J., Shah, M., Sharma, S. B., and Poinern, G. E. J. (2017). A review of current research into the biogenic synthesis of metal and metal oxide nanoparticles via marine algae and Seagrasses. *Journal of Nanoscience*, 2017, 1–15. https://doi.org/10.1155/2017/8013850.

Fouda, A., EL-Din Hassan, S., Salem, S. S., and Shaheen, T. I. (2018). *In-Vitro* cytotoxicity, antibacterial, and UV protection properties of the biosynthesized Zinc oxide Nanoparticles for medical textile applications. *Microbial Pathogenesis*, 125: 252–261. https://doi.org/10.1016/j.micpath.2018.09.030.

Gahlawat, G., and Choudhury, A. R. (2019). A review on the biosynthesis of metal and metal salt nanoparticles by microbes. *RSC Advances*, 9(23): 12944–12967. https://doi.org/10.1039/C8RA10483B.

Gericke, M., and Pinches, A. (2006). Microbial production of gold nanoparticles. Gold Bulletin, 39(1): 22–28. https://doi.org/10.1007/BF03215529.

Ghosh, S., Sarkar, S., Garai, S., Roy, A., Saha, S., Dey, S. et al. (2023). Climate change and global crop production: an inclusive insight. pp. 1–34. *In*: Hasanuzzaman, M. (Eds.). Climate-Resilient Agriculture, Vol 2: Agro-Biotechnological Advancement for Crop Production. *Cham: Springer International Publishing*. https://doi.org/10.1007/978-3-031-37428-9_1.

Gopinath, K., Kumaraguru, S., Bhakyaraj, K., Mohan, S., Venkatesh, K. S., Esakkirajan, M. et al. (2016). Green synthesis of silver, gold and silver/gold bimetallic nanoparticles using the Gloriosa superba leaf extract and their antibacterial and antibiofilm activities. *Microbial Pathogenesis*, 101: 1–11. https://doi.org/10.1016/j.micpath.2016.10.011.

Gupta, A., Tripathy, D. B., Kumar, G., Agarwal, P., and Ghosal, A. (Eds.). (2024). Nanopesticides, nanoherbicides, and Nanofertilizers: Formulations and applications (First edition). CRC Press.

Haggag, W. M., Abouziena, H. F., Abd-El-Kreem, F., and El Habbasha, S. (2015). Agriculture biotechnology for management of multiple biotic and abiotic environmental stress in crops. *J. Chem. Pharm. Res.*, 7(10): 882–889.

Hano, C., and Abbasi, B. H. (2021). Plant-based green synthesis of nanoparticles: production, characterization and applications. *Biomolecules*, 12(1): 31. https://doi.org/10.3390/biom12010031.

Harsh, K., Kumar, G., Kumari, P., Kundu, M., Choudhury, S. R., Rakshit, R. et al. (2024). The altered crop pattern and livelihoods in changing climate. pp. 345–369. *In*: Meena, V. S. et al. (Eds.). Waste Management for Sustainable and Restored Agricultural Soil. Academic Press. https://doi.org/10.1016/B978-0-443-18486-4.00018-X.

Hashem, A. H., Abdelaziz, A. M., Askar, A. A., Fouda, H. M., Khalil, A. M. A., Abd-Elsalam, K. A. et al. (2021). *Bacillus megaterium*-mediated synthesis of selenium nanoparticles and their antifungal activity against *Rhizoctonia solani* in faba bean plants. *Journal of Fungi*, 7(3): Article 3. https://doi.org/10.3390/jof7030195.

Haydar, M., Ghosh, D., and Roy, S. (2024). Slow and controlled release nanofertilizers as an efficient tool for sustainable agriculture: Recent understanding and concerns. *Plant Nano Biology*, 7: 100058. https://doi.org/10.1016/j.plana.2024.100058.

Hazarika, A., Yadav, M., Yadav, D. K., and Yadav, H. S. (2022). An overview of the role of nanoparticles in sustainable agriculture. Biocatalysis and Agricultural Biotechnology, 43: 102399. https://doi.org/10.1016/j.bcab.2022.102399.

Hedayati, R., Jahanshahi, M., and Attar, H. (2012). Fabrication and characterization of albumin-acacia nanoparticles based on complex coacervation as potent nanocarrier. *Journal of Chemical Technology & Biotechnology*, 87(10): 1401–1408. https://doi.org/10.1002/jctb.3758.

Hossain, A., Pramanick, B., Bhutia, K. L., Ahmad, Z., Moulick, D., Maitra, S. et al. (2021). Emerging roles of osmoprotectantglycinebetaineagainstsalt-inducedoxidativestressinplants:Amajoroutlookofmaize(*Zeamays*L.). pp. 567–587. *In*: Aftab, T., and Hakeem, K. R. (Eds.). Frontiers in plant–soilplant-soil interaction. Academic Press. https://doi.org/10.1016/B978-0-323-90943-3.00015-8.

Huq, Md. A., Ashrafudoulla, Md., Rahman, M. M., Balusamy, S. R., and Akter, S. (2022). Green Synthesis and Potential Antibacterial Applications of Bioactive Silver Nanoparticles: A Review. *Polymers*, 14(4): 742. https://doi.org/10.3390/polym14040742.

Huston, M., DeBella, M., DiBella, M., and Gupta, A. (2021). Green synthesis of nanomaterials. *Nanomaterials*, 11(8): 2130. https://doi.org/10.3390/nano11082130.

Ijaz, I., Gilani, E., Nazir, A., and Bukhari, A. (2020). Detail review on chemical, physical and green synthesis, classification, characterizations and applications of nanoparticles. *Green Chemistry Letters and Reviews*, 13(3): 223–245. https://doi.org/10.1080/17518253.2020.1802517.

Indian Farmers FertilizerFertiliser Cooperative Limited, IFFCO Sadan, C-1, District Centre, Saket Place, New Delhi-110017, and Kumar, Y. (2021). Nanofertilizers and their role in sustainable agriculture. Annals of Plant and Soil Research, 23(3): 238–255. https://doi.org/10.47815/apsr.2021.10067.

Jakhar, A. M., Aziz, I., Kaleri, A. R., Hasnain, M., Haider, G., Ma, J., and Abideen, Z. (2022). NanofNano-fertilizers: A sustainable technology for improving crop nutrition and food security. *NanoImpact*, 27: 100411. https://doi.org/10.1016/j.impact.2022.100411.

Jebril, S., Khanfir Ben Jenana, R., and Dridi, C. (2020). Green synthesis of silver nanoparticles using *Melia azedarach* leaf extract and their antifungal activities: *In vitro* and *in vivo*. *Materials Chemistry and Physics*, 248: 122898. https://doi.org/10.1016/j.matchemphys.2020.122898.

Jha, A. K., Prasad, K., Prasad, K., and Kulkarni, A. R. (2009). Plant system: Nature's nanofactory. *Colloids and Surfaces B: Biointerfaces*, 73(2): 219–223. https://doi.org/10.1016/j.colsurfb.2009.05.018.

Jha, A. K., and Prasad, K. (2018). Mechanistic Plethora of Biogenetic Nanosynthesis: An Evaluation. pp. 1–24. *In*: Prasad, R., Jha, A. K., and Prasad, K. (Eds.). Exploring the Realms of Nature for Nanosynthesis. Springer International Publishing. https://doi.org/10.1007/978-3-319-99570-0_1.

Jiang, Y., Zhou, P., Zhang, P., Adeel, M., Shakoor, N., Li, Y. et al. (2022). Green synthesis of metal-based nanoparticles for sustainable agriculture. *Environmental Pollution*, 309: 119755. https://doi.org/10.1016/j.envpol.2022.119755.

Kah, M., Tufenkji, N., and White, J. C. (2019). Nano-enabled strategies to enhance crop nutrition and protection. *Nature Nanotechnology*, 14(6): 532–540.

Kaihara, S., Suzuki, Y., and Fujimoto, K. (2011). *In situ* synthesis of polysaccharide nanoparticles via polyion complex of carboxymethyl cellulose and chitosan. *Colloids and Surfaces B: Biointerfaces*, 85(2): 343–348. https://doi.org/10.1016/j.colsurfb.2011.03.008.

Kang, S. H., Bozhilov, K. N., Myung, N. V., Mulchandani, A., and Chen, W. (2008). Microbial synthesis of CdS nanocrystals in genetically engineered *E. coli*. *Angewandte Chemie*, 120(28): 5264–5267. https://doi.org/10.1002/ange.200705806.

Kanwal, U., Irfan Bukhari, N., Ovais, M., Abass, N., Hussain, K., and Raza, A. (2018). Advances in nanodnano-delivery systems for doxorubicin: An updated insight. *Journal of Drug Targeting*, 26(4): 296–310. https://doi.org/10.1080/1061186X.2017.1380655.

Karunakaran, G., Suriyaprabha, R., Manivasakan, P., Yuvakkumar, R., Rajendran, V., Prabu, P. et al. (2013). Effect of nanosilica and silicon sources on plant growth promoting rhizobacteria, soil nutrients and maize seed germination. *IET Nanobiotechnology*, 7(3): 70–77. https://doi.org/10.1049/iet-nbt.2012.0048.

Khalifa, A. M., Safhi, F. A., and Elsherif, D. E. (2024). Green synthesis of a dual-functional sulfur nanofertilizer to promote growth and enhance salt stress resilience in faba bean. *BMC Plant Biology*, 24(1): 607. https://doi.org/10.1186/s12870-024-05270-7.

Khalifa, N. S., and Hasaneen, M. N. (2018). The effect of chitosan–PMAA–NPK Nanofertilizer on *Pisum sativum* plants. *3 Biotech*, 8(4): 193. https://doi.org/10.1007/s13205-018-1221-3.

Khan, A. U., Khan, M., Malik, N., Cho, M. H., and Khan, M. M. (2019). Recent progress of algae and blue–green algae-assisted synthesis of gold nanoparticles for various applications. *Bioprocess and Biosystems Engineering*, 42(1): 1–15. https://doi.org/10.1007/s00449-018-2012-2.

Khan, F., Shariq, M., Asif, M., Siddiqui, M. A., Malan, P., and Ahmad, F. (2022). Green nanotechnology: plant-mediated nanoparticle synthesis and application. *Nanomaterials*, 12(4): 673. https://doi.org/10.3390/nano12040673.

Khan, S. H. (2020). Green nanotechnology for the environment and sustainable development. pp. 13–46. *In*: Mu. Naushad, and Lichtfouse, E. (Eds.). Green Materials for Wastewater Treatment. Springer International Publishing. https://doi.org/10.1007/978-3-030-17724-9_2.

Kobayashi, M., Tomita, S., Sawada, K., Shiba, K., Yanagi, H., Yamashita, I. et al. (2012). Chiral meta-molecules consisting of gold nanoparticles and genetically engineered tobacco mosaic virus. *Optics Express*, 20(22): 24856. https://doi.org/10.1364/OE.20.024856.

Konishi, Y., Tsukiyama, T., Tachimi, T., Saitoh, N., Nomura, T., and Nagamine, S. (2007). Microbial deposition of gold nanoparticles by the metal-reducing bacterium *Shewanella algae*. *Electrochimica Acta*, 53(1): 186–192. https://doi.org/10.1016/j.electacta.2007.02.073.

Kordy, M. G. M., Abdel-Gabbar, M., Soliman, H. A., Aljohani, G., BinSabt, M., Ahmed, I. A. et al. (2022). Phyto-Capped Ag nanoparticles: Green synthesis, characterization, and catalytic and antioxidant activities. *Nanomaterials*, 12(3): 373. https://doi.org/10.3390/nano12030373.

Koul, B., Poonia, A. K., Yadav, D., and Jin, J.-O. (2021). Microbe-mediated biosynthesis of nanoparticles: applications and future prospects. *Biomolecules*, 11(6): 886. https://doi.org/10.3390/biom11060886.

Kulkarni, D., Sherkar, R., Shirsathe, C., Sonwane, R., Varpe, N., Shelke, S. et al. (2023). Biofabrication of nanoparticles: Sources, synthesis, and biomedical applications. *Frontiers in Bioengineering and Biotechnology*, 11: 1159193. https://doi.org/10.3389/fbioe.2023.1159193.

Kumar, A. S., Madhu, G., John, E., Kuttinarayanan, S. V., and Nair, S. K. (2020). Optical and antimicrobial properties of silver nanoparticles synthesized via green route using honey. *Green Processing and Synthesis*, 9(1): 268–274. https://doi.org/10.1515/gps-2020-0029.

Kumar, P., Pahal, V., Gupta, A., Vadhan, R., Chandra, H., and Dubey, R. C. (2020). Effect of silver nanoparticles and *Bacillus cereus*Bacillus cereus LPR2 on the growth of *Zea mays*Zea mays. *Scientific Reports*, 10(1): 20409. https://doi.org/10.1038/s41598-020-77460-w.

Kumar, R., Singh, N., and Pandey, S. N. (2015). Potential of green synthesized zero-valent iron nanoparticles for remediation of lead-contaminated water. *International Journal of Environmental Science and Technology*, 12(12): 3943–3950. https://doi.org/10.1007/s13762-015-0751-z.

Kundu, J., Chung, Y.-I., Kim, Y. H., Tae, G., and Kundu, S. C. (2010). Silk fibroin nanoparticles for cellular uptake and control release. *International Journal of Pharmaceutics*, 388(1): 242–250. https://doi.org/10.1016/j.ijpharm.2009.12.052.

Lavande, S., Jaiswal, S., Deore, R., Pawar, J., and Tale, V. (2022). Metal nanoparticle synthesis using fruit extracts as reducing agents and comparative studies with a chemical reducing agent. *Biosciences Biotechnology Research Asia*, 19(2): 487–496. https://doi.org/10.13005/bbra/3002.

Lee, S. J., Hong, G.-Y., Jeong, Y.-I., Kang, M.-S., Oh, J.-S., Song, C.-E. et al. (2012). Paclitaxel-incorporated nanoparticles of hydrophobized polysaccharide and their antitumourantitumor activity. *International Journal of Pharmaceutics*, 433(1): 121–128. https://doi.org/10.1016/j.ijpharm.2012.04.048.

Li, S., Shen, Y., Xie, A., Yu, X., Qiu, L., Zhang, L. et al. (2007). Green synthesis of silver nanoparticles using *Capsicum annuum L.* extract. *Green Chemistry*, 9(8): 852. https://doi.org/10.1039/b615357g.

Liaqat, N., Jahan, N., Khalil-ur-Rahman, Anwar, T., and Qureshi, H. (2022). Green synthesized silver nanoparticles: Optimization, characterization, antimicrobial activity, and cytotoxicity study by hemolysis assay. *Frontiers in Chemistry*, 10: 952006. https://doi.org/10.3389/fchem.2022.952006.

Liu, H., Chen, B., Mao, Z., and Gao, C. (2007). Chitosan nanoparticles for loading of toothpaste actives and adhesion on tooth analoguesgs. *Journal of Applied Polymer Science*, 106(6): 4248–4256. https://doi.org/10.1002/app.27078.

Liu, L., Nian, H., and Lian, T. (2022). Plants and rhizospheric environment: Affected by zinc oxide nanoparticles (ZnO NPs). A review. *Plant Physiology and Biochemistry*, 185: 91–100. https://doi.org/10.1016/j.plaphy.2022.05.032.

Londhe, S., Haque, S., and Patra, C. R. (2023). Silver and gold nanoparticles: Potential cancer theranostic applications, recent development, challenges, and future perspectives. *In Gold and Silver Nanoparticles* (pp. 247–290). Elsevier. https://doi.org/10.1016/B978-0-323-99454-5.00006-8.

Macewan, S. R., and Chilkoti, A. (2012). Digital switching of local arginine density in a genetically encoded self-assembled polypeptide nanoparticle controls cellular uptake. *Nano Letters*, 12(6): 3322–3328. https://doi.org/10.1021/nl301529p.

Mahanta, S., Shree, J., Santra, S. C., Moulick, D., and Hossain, A. (2023). Deciphering of mycogenic nanoparticles by spectroscopic methods. pp. 93–117. *In*: Kuddus, M. (Eds.). Myconanotechnology and Application of Nanoparticles in Biology. Academic Press. https://doi.org/10.1016/B978-0-443-15262-7.00003-6.

Mahapatra, D. M., Satapathy, K. C., and Panda, B. (2022). Biofertilizers and Nanofertilizers for sustainable agriculture: Phycoprospects and challenges. *Science of the Total Environment*, 803: 149990. https://doi.org/10.1016/j.scitotenv.2021.149990.

Mahmudin, L., Suharyadi, E., Utomo, A. B. S., and Abraha, K. (2015). Optical properties of silver nanoparticles for surface plasmon resonance (SPR)-based biosensor applications. *Journal of Modern Physics*, 6(8): Article 8. https://doi.org/10.4236/jmp.2015.68111.

Majumdar, A., Moulick, D., and Srivastava, S. (2024). 'Save Soil' by managing soil nutrient losses, agronomic practices and crop-microbial interaction: World Soil Day 2022. *Frontiers in Sustainable Food Systems*. 8: 1360937.

Majumdar, J., Biswas, J. K., Majumdar, A., Roychowdhury, T., Santra, S. C., Hossain, A. et al. (2025). Rhizomicrobiome: Role in management of heavy metal stress in plants. pp. 315-332. *In*: Panwar, J. S., and Sharma, V. (Eds.). Rhizomicrobiome in sustainable agriculture and environment. *Academic Press*. https://doi.org/10.1016/B978-0-443-23691-4.00013-0.

Malik, S., Niazi, M., Khan, M., Rauff, B., Anwar, S., Amin, F. et al. (2023). Cytotoxicity study of gold nanoparticle synthesis using aloe vera, honey, and gymnema sylvestre leaf extract. *ACS Omega*, 8(7): 6325–6336. https://doi.org/10.1021/acsomega.2c06491.

Mao, C., Flynn, C. E., Hayhurst, A., Sweeney, R., Qi, J., Georgiou, G. et al. (2003). Viral assembly of oriented quantum dot nanowires. *Proceedings of the National Academy of Sciences*, 100(12): 6946–6951. https://doi.org/10.1073/pnas.0832310100.

Marchiol, L., Mattiello, A., Pošćić, F., Giordano, C., and Musetti, R. (2014). *In vivo* synthesis of nanomaterials in plants: Location of silver nanoparticles and plant metabolism. *Nanoscale Research Letters*, 9(1): 101. https://doi.org/10.1186/1556-276X-9-101.

Mashwani, Z.-R., Khan, M. A., Khan, T., and Nadhman, A. (2016). Applications of plant terpenoids in the synthesis of colloidal silver nanoparticles. *Advances in Colloid and Interface Science*, 234: 132–141. https://doi.org/10.1016/j.cis.2016.04.008.

Metuku, R. P., Pabba, S., Burra, S., Hima Bindu, N., S. V. S. S. S. L., Gudikandula, K. et al. (2014). Biosynthesis of silver nanoparticles from *Schizophyllum radiatum* HE 863742.1: Their characterization and antimicrobial activity. *3 Biotech*, 4(3): 227–234. https://doi.org/10.1007/s13205-013-0138-0.

Meydan, I., Aygun, A., Elhouda Tiri, R. N., Gur, T., Kocak, Y., Seckin, H. et al. (2024). Chitosan/PVA-supported silver nanoparticles for azo dyes removal: Fabrication, characterization, and assessment of antioxidant activity. Environmental Science: *Advances*, 3(1): 28–35. https://doi.org/10.1039/D3VA00224A.

Mishra, A., Kumari, M., Pandey, S., Chaudhry, V., Gupta, K. C., and Nautiyal, C. S. (2014). Biocatalytic and antimicrobial activities of gold nanoparticles synthesized by *Trichoderma* sp. *Bioresource Technology*, 166: 235–242. https://doi.org/10.1016/j.biortech.2014.04.085.

Mohamed, A. A., Fouda, A., Abdel-Rahman, M. A., Hassan, S. E.-D., El-Gamal, M. S., Salem, S. S. et al. (2019). Fungal strain impacts the shape, bioactivity and multifunctional properties of green synthesized zinc oxide nanoparticles. *Biocatalysis and Agricultural Biotechnology*, 19: 101103. https://doi.org/10.1016/j.bcab.2019.101103.

Mohammed, G. M., and Hawar, S. N. (2022). Green Biosynthesis of silver nanoparticles from *Moringa oleifera* leaves and its antimicrobial and cytotoxicity activities. *International Journal of Biomaterials*, 2022, 1–10. https://doi.org/10.1155/2022/4136641.

Mohanta, Y. K., Panda, S. K., Jayabalan, R., Sharma, N., Bastia, A. K., and Mohanta, T. K. (2017). Antimicrobial, antioxidant and cytotoxic activity of silver nanoparticles synthesized by leaf extract of *Erythrina suberosa* (Roxb.). *Frontiers in Molecular Biosciences*, 4. https://doi.org/10.3389/fmolb.2017.00014.

Mohanty, P., Singh, P. K., Chakraborty, D., Mishra, S., and Pattnaik, R. (2021). Insight into the role of PGPR in sustainable agriculture and environment. *Frontiers in Sustainable Food Systems*, 5. https://doi.org/10.3389/fsufs.2021.667150.

Mondal, T., Sahoo, U., Sairam, M., Shankar, T., Lalichetti, S., Gaikwad, D. J. et al. (2023). Melatonin-induced plant tolerance against temperature stress. pp. 237–253. *In*: Roychoudhury, A. (Eds.). *Advancement of Melatonin Research in Plants*. CRC Press. https://doi.org/10.1201/9781003343752.

Moulick, D., Choudhury, S., Brestic, M., and Hossain, A. (2024a). The impact of abiotic stresses on agriculture: mitigation through climate smart strategies. *Frontiers in Plant Science.*, 15: 1391051. https://doi.org/10.3389/fpls.2024.1391051.

Moulick, D., Majumdar, A., Choudhury, A., Das, A., Chowardhara, B., Pattnaik, B. K. et al. (2024b). Emerging Concern of Nano-pollution in Agro-Ecosystem: Flip Side of Nanotechnology. *Plant Physiol. Biochem.*, 211: 108704. https://doi.org/10.1016/j.plaphy.2024.108704.

Moulick, D., Mukherjee, A., Das, A., Roy, A., Majumdar, A., Dhar, A. et al. (2024c). Selenium–An environmentally friendly micronutrient in agroecosystem in the modern era: An overview of 50-year findings. *Ecotoxicol. Environ. Safe.*, 270: 115832. https://doi.org/10.1016/j.ecoenv.2023.115832.

Moulick, D., Khan, F. R., Siddiqui, N., Pradhan, J., Khamari, P., and Mahanta, S. (2024d). Applications of nanotechnology in abiotic stress management and biofortification. pp. 50–66. *In*: Roy, S., and Hossain, A. (Eds.). The Nanotechnology Driven Agriculture. CRC Press.

Moulick, D., Hossain, A., Barek, V., Chowardhara, B., Mukherjee, A., Pattnaik, B. K. et al. (2024e). Heavy metal stress in the agro-environment: consequences, adaptations and remediation. *International Journal of Environmental Science and Technology.*, 21: 927–9340. https://doi.org/10.1007/s13762-024-05657-x.

Moulick, D., Santra, S. C., Majumdar, A., Das, A., Chowardhara, B., Saha, B. et al. (2025a). Efficacy of seed priming technology in ameliorating metals and metalloids toxicity in crops: prospective and issues. *Rev. Environ. Contam. Toxicol.*, 263(1): 1–41. https://doi.org/10.1007/s44169-024-00074-0.

Moulick, D., Mukherjee, A., Mahanta, S., Chowdhara, B., Pattnaik, B. K., Mukhopadhyay, A. K. et al. (2025b). Nano-composites and nanobnano-biochar: a novel solution to alleviate heavy metal stress in crops. In Nanomaterials and Nano-Biochar in Reducing Soil Stress (pp. 183–224). Apple Academic Press.

Moulton, M. C., Braydich-Stolle, L. K., Nadagouda, M. N., Kunzelman, S., Hussain, S. M., and Varma, R. S. (2010). Synthesis, characterization and biocompatibility of "green" synthesized silver nanoparticles using tea polyphenols. *Nanoscale*, 2(5): 763. https://doi.org/10.1039/c0nr00046a.

Mourdikoudis, S. (Ed.). (2021). Reducing Agents in Colloidal Nanoparticle Synthesis. The Royal Society of Chemistry. https://doi.org/10.1039/9781839163623.

Murugan, K., Benelli, G., Panneerselvam, C., Subramaniam, J., Jeyalalitha, T., Dinesh, D. et al. (2015). Cymbopogon citratus-synthesized gold nanoparticles boost the predation efficiency of copepod *Mesocyclops aspericornis* against malaria and dengue mosquitoes. *Experimental Parasitology*, 153: 129–138. https://doi.org/10.1016/j.exppara.2015.03.017.

Mustafa, M., Azam, M., Nawaz Bhatti, H., Khan, A., Zafar, L., and Rehan Abbasi, A. M. (2024). Green fabrication of copper nanofnano-fertilizer for enhanced crop yield in cowpea cultivar: A sustainable approach. *Biocatalysis and Agricultural Biotechnology*, 56: 102994. https://doi.org/10.1016/j.bcab.2023.102994.

Nalini, S. K., and Vijayaraghavan, K. (2020). Green synthesis of silver and gold nanoparticles using aloe vera gel and determining its antimicrobial properties on nanoparticle impregnated cotton fabric. *Journal of Nanotechnology Research*, 02(03). https://doi.org/10.2502/jnr.2688-85210015.

Nasrollahzadeh, M., and Mohammad Sajadi, S. (2015). Green synthesis of copper nanoparticles using *Ginkgo biloba* L. leaf extract and their catalytic activity for the Huisgen [3 + 2] cycloaddition of azides and alkynes at room temperature. *Journal of Colloid and Interface Science*, 457: 141–147. https://doi.org/10.1016/j.jcis.2015.07.004.

Nayana, A. R., Joseph, B. J., Jose, A., and Radhakrishnan, E. K. (2020). Nanotechnological Advances with PGPR Applications. *In*: Hayat, S., Pichtel, J., Faizan, M., and Fariduddin, Q. (Eds.). Sustainable Agriculture Reviews 41 (Vol. 41, pp. 163–180). Springer International Publishing. https://doi.org/10.1007/978-3-030-33996-8_9.

Nguyen, N. T. T., Nguyen, L. M., Nguyen, T. T. T., Nguyen, T. T., Nguyen, D. T. C., and Tran, T. V. (2022). Formation, antimicrobial activity, and biomedical performance of plant-based nanoparticles: A review. Environmental Chemistry Letters, 20(4): 2531–2571. https://doi.org/10.1007/s10311-022-01425-w.

Nichifor, M., Lopes, A., Carpov, A., and Melo, E. (1999). Aggregation in water of dextran hydrophobically modified with bile acids. *Macromolecules*, 32(21): 7078–7085. https://doi.org/10.1021/ma990408k.

Nitta, S., and Numata, K. (2013). Biopolymer-based nanoparticles for drug/gene delivery and tissue engineering. *International Journal of Molecular Sciences*, 14(1): 1629–1654. https://doi.org/10.3390/ijms14011629.

Noman, M., Shahid, M., Ahmed, T., Tahir, M., Naqqash, T., Muhammad, S. et al. (2020). Green copper nanoparticles from a native *Klebsiella pneumoniae* strain alleviated oxidative stress impairment of wheat plants by reducing the chromium bioavailability and increasing the growth. *Ecotoxicology and Environmental Safety*, 192: 110303. https://doi.org/10.1016/j.ecoenv.2020.110303.

Nongbet, A., Mishra, A. K., Mohanta, Y. K., Mahanta, S., Ray, M. K., Khan, M. et al. (2022). Nanofertilizers: A Smart and Sustainable Attribute to Modern Agriculture. *Plants*, 11(19): 2587. https://doi.org/10.3390/plants11192587.

Ogunyemi, S. O., Abdallah, Y., Zhang, M., Fouad, H., Hong, X., Ibrahim, E. et al. (2019). Green synthesis of zinc oxide nanoparticles using different plant extracts and their antibacterial activity against *Xanthomonas oryzae* pv. Oryzae. *Artificial Cells, Nanomedicine, and Biotechnology*, 47(1): 341–352. https://doi.org/10.1080/21691401.2018.1557671

Osborne, J. L., Farmer, R., and Woodhouse, K. A. (2008). Self-assembled elastin-like polypeptide particles. *Acta Biomaterialia*, 4(1): 49–57. https://doi.org/10.1016/j.actbio.2007.07.007.

Ovais, M., Khalil, A. T., Islam, N. U., Ahmad, I., Ayaz, M., Saravanan, M. et al. (2018). Role of plant phytochemicals and microbial enzymes in biosynthesis of metallic nanoparticles. *Applied Microbiology and Biotechnology*, 102(16): 6799–6814. https://doi.org/10.1007/s00253-018-9146-7.

Pahalvi, H. N., Rafiya, L., Rashid, S., Nisar, B., and Kamili, A. N. (2021). Chemical fertilizers and their impact on soil health. pp. 1–20. *In*: Dar, G. H., Bhat, R. A., Mehmood, M. A., and Hakeem, K. R. (Eds.). Microbiota and Biofertilizers, Vol 2: Ecofriendly tools for reclamation of degraded soil environs. *Springer International Publishing*. https://doi.org/10.1007/978-3-030-61010-4_1.

Panichikkal, J., Thomas, R., John, J. C., and Radhakrishnan, E. K. (2019). Biogenic gold nanoparticle supplementation to plant beneficial *Pseudomonas monteilii* was found to enhance its plant probiotic effect. *Current Microbiology*, 76(4): 503–509. https://doi.org/10.1007/s00284-019-01649-0.

Parveen, K., Banse, V., and Ledwani, L. (2016). Green synthesis of Nanoparticles: Their advantages and disadvantages. 020048. https://doi.org/10.1063/1.4945168.

Pattnaik, B. K., Sahu, C., Choudhury, S., Santra, S. C., and Moulick, D. (2023). Importance of soil management in sustainable agriculture. pp. 487–511. *In*: Hasanuzzaman M. (Eds.). Climate-Resilient Agriculture, Vol 1: Crop Responses and Agroecological Perspectives. Cham: Springer International Publishing.

Pattnaik, B. K., Sahu, C., and Moulick, D. (2024). Nanoparticles in soil–water–plant inference. pp. 85–105. *In*: Roy, S., and Hossain, A. (Eds.). The nanotechnology driven agriculture. CRC Press.

Qian, H., Pretzer, L. A., Velazquez, J. C., Zhao, Z., and Wong, M. S. (2013). Gold nanoparticles for cleaning contaminated water. *Journal of Chemical Technology & Biotechnology*, 88(5): 735–741. https://doi.org/10.1002/jctb.4030.

Radulescu, D.-M., Surdu, V.-A., Ficai, A., Ficai, D., Grumezescu, A.-M., and Andronescu, E. (2023). Green synthesis of metal and metal oxide nanoparticles: a review of the principles and biomedical applications. *International Journal of Molecular Sciences*, 24(20): 15397. https://doi.org/10.3390/ijms242015397

Raja, S., Ramesh, V., and Thivaharan, V. (2017). Green biosynthesis of silver nanoparticles using *Calliandra haematocephala* leaf extract, their antibacterial activity and hydrogen peroxide sensing capability. *Arabian Journal of Chemistry*, 10(2): 253–261. https://doi.org/10.1016/j.arabjc.2015.06.023.

Rajput, S., Werezuk, R., Lange, R. M., and McDermott, M. T. (2016). Fungal Isolate optimized for biogenesis of silver nanoparticles with enhanced colloidal stability. *Langmuir,* 32(34): 8688–8697. https://doi.org/10.1021/acs.langmuir.6b01813.

Rathod, S., Preetam, S., Pandey, C., and Bera, S. P. (2024). Exploring synthesis and applications of green nanoparticles and the role of nanotechnology in wastewater treatment. *Biotechnology Reports*, 41: e00830. https://doi.org/10.1016/j.btre.2024.e00830.

Rehman, A., Khan, S., Sun, F., Peng, Z., Feng, K., Wang, N. et al. (2024). Exploring the nanownano-wonders: Unveiling the role of nanoparticles in enhancing salinity and drought tolerance in plants. *Frontiers in Plant Science*, 14: 1324176. https://doi.org/10.3389/fpls.2023.1324176.

Reis, C. P., Ribeiro, A. J., Houng, S., Veiga, F., and Neufeld, R. J. (2007). Nanoparticulate delivery system for insulin: Design, characterization and *in vitro/in vivo* bioactivity. *European Journal of Pharmaceutical Sciences*, 30(5): 392–397. https://doi.org/10.1016/j.ejps.2006.12.007.

Rizwana, H., Aljowaie, R. M., Al Otibi, F., Alwahibi, M. S., Alharbi, S. A., Al asmari, S. A. et al. (2023). Antimicrobial and antioxidant potential of the silver nanoparticles synthesized using aqueous extracts of coconut meat (*Cocos nucifera* L). *Scientific Reports*, 13(1): 16270. https://doi.org/10.1038/s41598-023-43384-4.

Rodriguez, P. L., Harada, T., Christian, D. A., Pantano, D. A., Tsai, R. K., and Discher, D. E. (2013). Minimal "Self" peptides that inhibit phagocytic clearance and enhance delivery of nanoparticles. *Science*, 339(6122): 971–975. https://doi.org/10.1126/science.1229568.

Rodríguez-Hernández, J., and Lecommandoux, S. (2005). Reversible inside-out micellization of pH-responsive and water-soluble vesicles based on polypeptide diblock copolymers. Journal of the American Chemical Society, 127(7): 2026–2027. https://doi.org/10.1021/ja043920g.

Roy, D., Ghosh, D., Chander, S., Saha, B., and Moulick, D. (2024). Herbicide contamination in ground water and their probable remediation. pp. 111–140. *In*: Ganguly et al. (Eds.). Environmental Contaminants. *Apple Academic Press*. https://doi.org/10.1201/9781003412236.

Roy, S., and Hossain, A. (Eds.). (2024). The Nanotechnology Driven Agriculture: The Future Ahead. CRC Press.

Saberi Riseh, R., and Gholizadeh Vazvani, M. (2024). Green synthesis of metal nanoparticles using plant growth promoting rhizobacteria and application in agriculture. *Plant Nano Biology*, 10: 100111. https://doi.org/10.1016/j.plana.2024.100111.

Sabir, S., Zahoor, M. A., Waseem, M., Siddique, M. H., Shafique, M., Imran, M. et al. (2020). Biosynthesis of ZnO nanoparticles using *Bacillus Subtilis*: characterization and nutritive significance for promoting plant growth in *Zea mays* L. *Dose–Response*, 18(3): 1559325820958911. https://doi.org/10.1177/1559325820958911.

Sahoo, S. K., Dwivedi, G. K., Dey, P., and Praharaj, S. (2021). Green synthesized ZnO nanoparticles for sustainable production and nutritional biofortification of green gram. Environmental Technology & Innovation, 24: 101957. https://doi.org/10.1016/j.eti.2021.101957.

Sahu, N., Soni, D., Chandrashekhar, B., Satpute, D. B., Saravanadevi, S., Sarangi, B. K. et al. (2016). Synthesis of silver nanoparticles using flavonoids: Hesperidin, naringin and diosmin, and their antibacterial effects and cytotoxicity. *International Nano Letters*, 6(3): 173–181. https://doi.org/10.1007/s40089-016-0184-9.

Salama, A., Rashad, E., Elgarahy, A., and Elwakeel, K. (2023). Effect of green synthesized iron oxide nanoparticles on bacterial microbiome for clean up the crude oil. *Aswan University Journal of Environmental Studies*, 4(1): 49–81. https://doi.org/10.21608/aujes.2023.178805.1110.

Salari, S., Esmaeilzadeh Bahabadi, S., Samzadeh-Kermani, A., and Yosefzaei, F. (2019). *In vitro* evaluation of antioxidant and antibacterial potential of green synthesized silver nanoparticles using *Prosopis farcta* Fruit Extract. Iranian Journal of Pharmaceutical Research : IJPR, 18(1): 430–455.

Saleem, S., and Khan, Mohd. S. (2023). Phyto-interactive impact of green synthesized iron oxide nanoparticles and *Rhizobium pusense* on morpho-physiological and yield components of greengram. *Plant Physiology and Biochemistry*, 194: 146–160. https://doi.org/10.1016/j.plaphy.2022.11.013.

Santos, M., Serrano-Dúcar, S., González-Valdivieso, J., Vallejo, R., Girotti, A., Cuadrado, P. et al. (2019). Genetically engineered elastin-based biomaterials for biomedical applications. *Current Medicinal Chemistry*, 26(40): 7117–7146. https://doi.org/10.2174/0929867325666180508094637.

Saravanakumar, K., Chelliah, R., Shanmugam, S., Varukattu, N. B., Oh, D.-H., Kathiresan, K. et al. (2018). Green synthesis and characterization of biologically active nanosilver from seed extract of *Gardenia jasminoides* Ellis. *Journal of Photochemistry and Photobiology B: Biology*, 185: 126–135. https://doi.org/10.1016/j.jphotobiol.2018.05.032.

Sarkar, S., Saha, S., Ghosh, S., Paul, S. K., Dey, S., Moulick, D. et al. (2023). Abiotic stress sensitivity and adaptation in field crops. pp. 319–362. *In*: Hasanuzzaman, M. (Eds.). Climate-Resilient Agriculture, Vol 2:

agro-biotechnological advancement for crop production. Cham: Springer International Publishing. https://doi.org/10.1007/978-3-031-37428-9_15.

Sarmento, B., Ribeiro, A. J., Veiga, F., Ferreira, D. C., and Neufeld, R. J. (2007). Insulin-loaded nanoparticles are prepared by alginate ionotropic prepre-gelation followed by chitosan polyelectrolyte complexation. *Journal of Nanoscience and Nanotechnology*, 7(8): 2833–2841. https://doi.org/10.1166/jnn.2007.609.

Sayadi, M. H., Salmani, N., Heidari, A., and Rezaei, M. R. (2018). BiosBio-synthesis of palladium nanoparticle using Spirulina platensis alga extract and its application as adsorbent. *Surfaces and Interfaces*, 10: 136–143. https://doi.org/10.1016/j.surfin.2018.01.002.

Sebastian, A., Nangia, A., and Prasad, M. N. V. (2019). Cadmium and sodium adsorption properties of magnetite nanoparticles synthesized from *Hevea brasiliensis* Muell. Arg. bark: Relevance in amelioration of metal stress in rice. *Journal of Hazardous Materials*, 371, 261–272. https://doi.org/10.1016/j.jhazmat.2019.03.021.

Selvakumar, R., Seethalakshmi, N., Thavamani, P., Naidu, R., and Megharaj, M. (2014). Recent advances in the synthesis of inorganic nano/microstructures using microbial biotemplates and their applications. *RSC Advances*, 4(94): 52156–52169. https://doi.org/10.1039/C4RA07903E.

Shalaby, T. A., Bayoumi, Y., Eid, Y., Elbasiouny, H., Elbehiry, F., Prokisch, J. et al. (2022). Can nanofertilizers mitigate multiple environmental stresses for higher crop productivity? *Sustainability*, 14(6): Article 6. https://doi.org/10.3390/su14063480.

Shankar, S. S., Ahmad, A., Pasricha, R., and Sastry, M. (2003). Bioreduction of chloroaurate ions by geranium leaves and its endophytic fungus yields gold nanoparticles of different shapes. *Journal of Materials Chemistry*, 13(7): 1822. https://doi.org/10.1039/b303808b.

Sharma, A., Sharma, S., Sharma, K., Chetri, S. P. K., Vashishtha, A., Singh, P. et al. (2016). Algae as crucial organisms in advancing nanotechnology: A systematic review. *Journal of Applied Phycology*, 28(3): 1759–1774. https://doi.org/10.1007/s10811-015-0715-1.

Sharma, B., Tiwari, S., Kumawat, K. C., and Cardinale, M. (2023). Nano-biofertilizers as bioebio-emerging strategies for sustainable agriculture development: Potentiality and their limitations. *Science of The Total Environment*, 860: 160476. https://doi.org/10.1016/j.scitotenv.2022.160476.

Sharma, M., Behl, K., Nigam, S., and Joshi, M. (2018). TiO2-GO nanocomposite for photocatalysis and environmental applications: A green synthesis approach. *Vacuum*, 156: 434–439. https://doi.org/10.1016/j.vacuum.2018.08.009.

Sharma, V., and Janmeda, P. (2017). Extraction, isolation and identification of flavonoid from *Euphorbia neriifolia* leaves. *Arabian Journal of Chemistry*, 10(4): 509–514. https://doi.org/10.1016/j.arabjc.2014.08.019.

Sidhu, A. K., Verma, N., and Kaushal, P. (2022). Role of biogenic capping agents in the synthesis of metallic nanoparticles and evaluation of their therapeutic potential. *Frontiers in Nanotechnology*, 3: 801620. https://doi.org/10.3389/fnano.2021.801620.

Silva, L. P., Reis, I. G., and Bonatto, C. C. (2015). Green synthesis of metal nanoparticles by plants: current trends and challenges. pp. 259–275. *In*: Basiuk, V. A., and Basiuk, E. V. (Eds.). Green Processes for Nanotechnology. Springer International Publishing. https://doi.org/10.1007/978-3-319-15461-9_9.

Singh, A., Sharma, R., Rajput, V. D., Ghazaryan, K., Minkina, T., Al Tawaha, A. R. M. et al. (2024). Green synthesis of nanofertilizers and their application for crop production. pp. 205–231. *In*: Abd-Elsalam, K. A., and Alghuthaymi, M. A. (Eds.). Nanofertilizers for Sustainable Agroecosystems: Recent Advances and Future Trends. Springer Nature Switzerland. https://doi.org/10.1007/978-3-031-41329-2_8.

Singh, A., Singh, N. B., Hussain, I., and Singh, H. (2017). Effect of biologically synthesized copper oxide nanoparticles on metabolism and antioxidant activity to the crop plants *Solanum lycopersicum* and *Brassica oleracea* var. Botrytis. *Journal of Biotechnology*, 262: 11–27. https://doi.org/10.1016/j.jbiotec.2017.09.016.

Singh, H., Desimone, M. F., Pandya, S., Jasani, S., George, N., Adnan, M. et al. (2023). Revisiting the green synthesis of nanoparticles: uncovering influences of plant extracts as reducing agents for enhanced synthesis efficiency and its biomedical applications. *International Journal of Nanomedicine*, Volume 18, 4727–4750. https://doi.org/10.2147/IJN.S419369.

Singh, P., Kim, Y.-J., Zhang, D., and Yang, D.-C. (2016). Biological synthesis of nanoparticles from plants and microorganisms. *Trends in Biotechnology*, 34(7): 588–599. https://doi.org/10.1016/j.tibtech.2016.02.006.

Singh, R. P., Handa, R., and Manchanda, G. (2021). Nanoparticles in sustainable agriculture: An emerging opportunity. *Journal of Controlled Release*, 329: 1234–1248. https://doi.org/10.1016/j.jconrel.2020.10.051.

Slocik, J. M., Naik, R. R., Stone, M. O., and Wright, D. W. (2005). Viral templates for gold nanoparticle synthesis. *Journal of Materials Chemistry*, 15(7): 749. https://doi.org/10.1039/b413074j.

Sohrabi, Y., Sharifi Kalyani, F., Heydari, M., Yazdani, M., Omer, K. M., and Yousefi, A. R. (2022). Plant-based nanofnano-fertilizer prepared from Paulownia Tomentosa: Fabrication, characterization, and application on *Ocimum basilicum*. *Chemical and Biological Technologies in Agriculture*, 9(1): 82. https://doi.org/10.1186/s40538-022-00352-w.

Song, J. Y., Jang, H.-K., and Kim, B. S. (2009). Biological synthesis of gold nanoparticles using *Magnolia kobus* and *Diopyros kaki* leaf extracts. *Process Biochemistry*, 44(10): 1133–1138. https://doi.org/10.1016/j.procbio.2009.06.005.

Sonowal, S., Buragohain, K., Borah, N., Gogoi, A., Nath, R., Choudhury, S. et al. (2025). Rhizomicrobiome as a potential reservoir of heavy metal resistant microorganisms. pp. 275–314. *In*: Panwar, J. S., and Sharma, V. (Eds.). Rhizomicrobiome in Sustainable Agriculture and Environment. Academic Press. https://doi.org/10.1016/B978-0-443-23691-4.00012-9

Stozhko, N. Yu., Bukharinova, M. A., Khamzina, E. I., Tarasov, A. V., Vidrevich, M. B., and Brainina, K. Z. (2019). The effect of the antioxidant activity of plant extracts on the properties of gold nanoparticles. *Nanomaterials*, 9(12): 1655. https://doi.org/10.3390/nano9121655.

Sundararajan, N., Habeebsheriff, H. S., Dhanabalan, K., Cong, V. H., Wong, L. S., Rajamani, R. et al. (2023). Mitigating global challenges: harnessing green synthesized nanomaterials for sustainable crop production systems. *Global Challenges*, 8(1): 2300187. https://doi.org/10.1002/gch2.202300187.

Sundararajan, N., Habeebsheriff, H. S., Dhanabalan, K., Cong, V. H., Wong, L. S., Rajamani, R. et al. (2024). Mitigating global challenges: harnessing green synthesized nanomaterials for sustainable crop production systems. *Global Challenges*, 8(1): 2300187. https://doi.org/10.1002/gch2.202300187.

Swetha, S., Jeeva, M., and Minchitha, K. U. (2024). Integrating nanomaterials and plant growth-promoting rhizobacteria for enhanced plant disease management: An overview. *Journal of Plant Diseases and Protection*, 132(1): 26. https://doi.org/10.1007/s41348-024-01023-8.

Taghizadeh, S.-M., Lal, N., Ebrahiminezhad, A., Moeini, F., Seifan, M., Ghasemi, Y. et al. (2020). Green and economic fabrication of zinc oxide (ZnO) nanorods as a broadband UV Blocker and antimicrobial agent. *Nanomaterials*, 10(3): Article 3. https://doi.org/10.3390/nano10030530.

Tavan, M., Hanachi, P., Mirjalili, M. H., and Dashtbani-Roozbehani, A. (2023). Comparative assessment of the biological activity of the green synthesized silver nanoparticles and aqueous leaf extract of *Perilla frutescens* (L.). *Scientific Reports*, 13(1): 6391. https://doi.org/10.1038/s41598-023-33625-x.

Teimuri-mofrad, R., Hadi, R., Tahmasebi, B., Farhoudian, S., Mehravar, M., and Nasiri, R. (2017). Green synthesis of gold nanoparticles using plant extract: Mini-review. Nanochemistry Research, 2(1). https://doi.org/10.22036/ncr.2017.01.002.

Tortella, G., Navas, M., Parada, M., Durán, N., Seabra, A. B., Hoffmann, N. et al. (2019). Synthesis of silver nanoparticles using extract of weeds and optimized by response surface methodology to the control of soil pathogenic bacteria *Ralstonia solanacearum*. *Journal of Soil Science and Plant Nutrition*, 19(1): 148–156. https://doi.org/10.1007/s42729-019-00021-2.

Tyagi, S., Yadav, P., Chakraborty, A., Majumdar, A., Moulick, D., Santra, S. C. et al. (2023). Myconanotechnologies: an approach towards sustainable agriculture. pp. 139–157. *In*: Kuddus, M. (Eds.) Myconanotechnology and Application of Nanoparticles in Biology. Academic Press. https://doi.org/10.1016/B978-0-443-15262-7.00005-X.

Upadhayay, V. K., Chitara, M. K., Mishra, D., Jha, M. N., Jaiswal, A., Kumari, G. et al. (2023). Synergistic impact of nanomaterials and plant probiotics in agriculture: A tale of two-way strategy for long-term sustainability. Frontiers in Microbiology, 14. https://doi.org/10.3389/fmicb.2023.1133968.

Verma, K. K., Joshi, A., Song, X.-P., Singh, S., Kumari, A., Arora, J. et al. (2024). Synergistic interactions of nanoparticles and plant growth promoting rhizobacteria enhancing soil–plantsoil-plant systems: A multigenerational perspective. *Frontiers in Plant Science*, 15. https://doi.org/10.3389/fpls.2024.1376214.

Vijayaram, S., Razafindralambo, H., Sun, Y.-Z., Vasantharaj, S., Ghafarifarsani, H., Hoseinifar, S. H. et al. (2023). Applications of green synthesized metal nanoparticles—a review. *Biological Trace Element Research*, 1–27. https://doi.org/10.1007/s12011-023-03645-9.

Villagrán, Z., Anaya-Esparza, L. M., Velázquez-Carriles, C. A., Silva-Jara, J. M., Ruvalcaba-Gómez, J. M., Aurora-Vigo, E. F. et al. (2024). Plant-based extracts as reducing, capping, and stabilizing agents for the green synthesis of inorganic nanoparticles. Resources, 13(6): 70. https://doi.org/10.3390/resources13060070.

Wadhwani, S. A., Shedbalkar, U. U., Singh, R., Vashisth, P., Pruthi, V., and Chopade, B. A. (2016). Kinetics of synthesis of gold nanoparticles by *Acinetobacter* sp. SW30 Isolated from environment. *Indian Journal of Microbiology*, 56(4): 439–444. https://doi.org/10.1007/s12088-016-0598-0.

Wahi, D., Bisht, K., Gautam, S., Salvi, P., and Lohani, P. (2024). Green synthesized nano silica: Foliar and soil application provides drought endurance in *Eleucine coracana. Environmental Science: Nano*, 11(8): 3412–3429. https://doi.org/10.1039/D4EN00275J.

Wahid, I., Kumari, S., Ahmad, R., Hussain, S. J., Alamri, S., Siddiqui, M. H. et al. (2020). Silver nanoparticle regulates salt tolerance in wheat through changes in ABA Concentration, ion homeostasis, and defense systems. *Biomolecules*, 10(11): Article 11. https://doi.org/10.3390/biom10111506.

Wang, T., Lin, J., Chen, Z., Megharaj, M., and Naidu, R. (2014). Green synthesized iron nanoparticles by green tea and eucalyptus leaves extracts used for removal of nitrate in aqueous solution. *Journal of Cleaner Production*, 83: 413–419. https://doi.org/10.1016/j.jclepro.2014.07.006.

Wang, Z., Fang, C., and Mallavarapu, M. (2015). Characterization of iron–polyphenol complex nanoparticles synthesized by Sage (*Salvia officinalis*) leaves. *Environmental Technology & Innovation*, 4: 92–97. https://doi.org/10.1016/j.eti.2015.05.004.

Weber, C., Coester, C., Kreuter, J., and Langer, K. (2000). Desolvation process and surface characterization of protein nanoparticles. *International Journal of Pharmaceutics*, 194(1): 91–102. https://doi.org/10.1016/s0378-5173(99)00370-1.

Xia, X.-X., Xu, Q., Hu, X., Qin, G., and Kaplan, D. L. (2011). Tunable self-assembly of genetically engineered silk—Elastin-like protein polymers. *Biomacromolecules*, 12(11): 3844–3850. https://doi.org/10.1021/bm201165h.

Xu, Y., and Du, Y. (2003). Effect of molecular structure of chitosan on protein delivery properties of chitosan nanoparticles. *International Journal of Pharmaceutics*, 250(1): 215–226. https://doi.org/10.1016/S0378-5173(02)00548-3.

Yadav, A., Yadav, K., and Abd-Elsalam, K. (2023). Nanofertilizers: Types, delivery and advantages in agricultural sustainability. *Agrochemicals*, 2(2): 296–336. https://doi.org/10.3390/agrochemicals2020019.

Yan, H.-B., Zhang, Y.-Q., Ma, Y.-L., and Zhou, L.-X. (2009). Biosynthesis of insulin-silk fibroin nanoparticles conjugates and *in vitro* evaluation of a drug delivery system. *Journal of Nanoparticle Research*, 11(8): 1937–1946. https://doi.org/10.1007/s11051-008-9549-y.

Yap, Y. H., Azmi, A. A., Mohd, N. K., Yong, F. S. J., Kan, S.-Y., Thirmizir, M. Z. A. et al. (2020). Green synthesis of silver nanoparticle using water extract of onion peel and application in the acetylation reaction. *Arabian Journal for Science and Engineering*, 45(6): 4797–4807. https://doi.org/10.1007/s13369-020-04595-3.

Ying, S., Guan, Z., Ofoegbu, P. C., Clubb, P., Rico, C., He, F. et al. (2022). Green synthesis of Nanoparticles: Current developments and limitations. *Environmental Technology & Innovation*, 26: 102336. https://doi.org/10.1016/j.eti.2022.102336.

You, J.-O., and Peng, C.-A. (2005). Calcium-alginate nanoparticles formed by reverse microemulsion as gene carriers. *Macromolecular Symposia*, 219(1): 147–153. https://doi.org/10.1002/masy.200550113.

Zaim, N. S. H. B. H., Tan, H. L., Rahman, S. M. A., Abu Bakar, N. F., Osman, M. S., Thakur, V. K. et al. (2023). Recent advances in seed coating treatment using nanoparticles and nanofibers for enhanced seed germination and protection. *Journal of Plant Growth Regulation*, 42(12): 7374–7402. https://doi.org/10.1007/s00344-023-11038-4.

Zhou, Y., Lin, W., Huang, J., Wang, W., Gao, Y., Lin, L. et al. (2010). Biosynthesis of gold nanoparticles by foliar broths: roles of biocompounds and other attributes of the extracts. *Nanoscale Research Letters*, 5(8): 1351–1359. https://doi.org/10.1007/s11671-010-9652-8.

Zoubi, O. M. A., Singh, A., Rajput, V. D., Varshney, A., Agrawal, S., Ghazaryan, K. et al. (n.d.). Green nanofertilizers: the need for modern agriculture, intelligent, and environmentally ly-friendly approaches. Retrieved December 16, 2024, from http://www.ecoeet.com/Green-NFs-the-Need-for-Modern-Agriculture-Intelligent-and-Environmentally,172946,0,2.html.

Zulfiqar, F., Navarro, M., Ashraf, M., Akram, N. A., and Munné-Bosch, S. (2019). Nanofertilizer use for sustainable agriculture: Advantages and limitations. *Plant Science*, 289: 110270. https://doi.org/10.1016/j.plantsci.2019.110270.

CHAPTER-8

Nanobiosensors for Smart and Sustainable Agriculture

Soumya Ranjan Mahapatra, Anish Bebarta, Akash Kumar Jena, Amiya Ranjan Tripathy and *Jyoti Prakash Sahoo**

1. Introduction

There are significant advantages for agriculture from the real-world uses of nanotechnology. As an example, the development of nanoformulations—more effective and less polluting agrochemicals—and the invention of tools like nanobiosensors, which identify biotic and abiotic stresses in plants before they affect production. Making progress in these areas will aid in tackling some of the major issues that agriculture will face in the years to come. Adapting to climate change (including changes in precipitation patterns, temperature swings, and the appearance of new plant diseases), decreasing the adverse environmental effects of agricultural operations, and overcoming crop output stagnation are some of these issues (Shang et al., 2020). The number of nanotechnological applications being tested in agriculture is constantly growing and includes, among other things, plant production, pathogen detection, and soil and water management. The unique qualities of nanoparticles (NPs) at the nanoscale, including as improved optical and mechanical capabilities and high chemical customisation, make them an ideal material for nanobiosensors. By identifying and tracking a variety of characteristics linked to plant health and environmental conditions, these cutting-edge biosensors are essential to agriculture (Shawon et al., 2020; Kashyap et al., 2019). In nanoformulations (NFs), agrochemicals including fertilizers and pesticides are made in nanoscale forms. Through the reduction of particle size, these formulations provide a number of benefits. Because they increase the effectiveness of active compounds, smaller dosages are still possible (Wang et al., 2022).

This decrease in particle size aids in lowering environmental pollution and mitigating possible adverse effects on beneficial organisms. Moreover, nanoformulations enhance the precise delivery of agrochemicals, guaranteeing they effectively reach the desired plant tissues or pests (Mittal et al., 2020) and lessen harm to non-targeted plant areas, alleviating the negative impact of chemicals on the ecosystem. Nanotechnology provides groundbreaking solutions to address agricultural challenges, enhance productivity, and minimize environmental impact. Continued research and developments in this field have the potential to revolutionize farming techniques and foster sustainable, efficient

Faculty of Agriculture and Allied Sciences, C.V. Raman Global University, Bhubaneswar - 752054, India.
* Corresponding author: jyotiprakashsahoo2010@gmail.com, jyotiprakash.sahoo@cgu-odisha.ac.in

DOI: 10.1201/9781003567684-8

agricultural practices. However, successfully integrating nanotechnology into agriculture requires educating farmers to ensure its effective use, which remains a challenge for governments around the globe (Dutta et al., 2022). The agriculture and food industries play a crucial role in the global economy, with projections indicating that food production will need to double by 2050. A considerable amount of crop losses is attributed to both biotic and abiotic stresses. Research shows that biotic factors, including pest infestations, pathogen attacks, herbicide application, and insect damage, are responsible for approximately 20 to 40% of these losses (Munir et al., 2023). A major challenge is the management of plant diseases caused by biotic factors. Traditional methods have effectively reduced pathogens, including bacteria, viruses, and fungi, by utilizing techniques like spectroscopy, chromatography, enzymatic methods, proteomics, and metabolomics (Schieber, 2018). Recently, nanotechnology has been used in many different technical processes. This includes creating and controlling new materials as well as describing and creating nanoparticles. These developments have wide-ranging uses in industries, medicine, agriculture, and pharmaceuticals (Ansari, 2023). In agriculture, the extensive application of nanocomposites can take the place of synthetic chemicals, fungicides, and pesticides in the management of plant pests (Ijaz et al., 2023).

2. Nanobiosensors and Nano Formulations in Agriculture

In modern agriculture, maintaining crop health and addressing nutritional deficiencies or disease outbreaks early is critical for economic and environmental sustainability. Precision farming, powered by advanced technologies such as drones and nanotechnology, enables targeted interventions. Nanobiosensors have become essential tools for real-time monitoring of plant health. They are capable of detecting stress markers, diseases, and nutrient deficiencies at the molecular level within individual plants. Typically, nanobiosensors are composed of three key components: a bioreceptor, a transducer, and a detector. The bioreceptor, often a biological material like antibodies or enzymes, captures chemical signals from plants. The transducer then converts these signals into measurable electrical outputs, which are processed by the detector to provide precise data. This information is critical in monitoring various plant health indicators, including reactive oxygen species, ions, and hormones, which often signal underlying issues such as drought or nutrient imbalance. Additionally, nanobiosensors are invaluable in plant breeding, pesticide monitoring, and detecting residues of chemicals, thus improving the efficiency of crop production (Siddiquee et al., 2018).

Notably, sensors utilizing carbon nanotubes, Quantum Dots (QDs), and metallic nanoparticles like gold (Au) and silver (Ag) have demonstrated remarkable potential in detecting plant pathogens, allowing for more rapid and accurate responses compared to traditional methods (Khaledian et al., 2019). Nanotechnology also finds application in agriculture through the use of nanoformulation, which enhances the efficiency of agrochemical use, particularly fertilizers and pesticides. Conventional agrochemicals often result in significant product losses due to leaching, volatilization, or degradation, which increases costs and environmental risks. Nano formulations, designed to encapsulate or control the release of active ingredients, offer a solution by providing targeted and sustained delivery of agrochemicals, reducing waste and environmental impact. For example, encapsulated nanofertilizers can release nutrients in response to specific stimuli, ensuring that plants receive them when most needed, thereby enhancing uptake efficiency. Nanoparticle-based fertilizers, with their larger surface area, provide nutrients more effectively than conventional options, significantly boosting crop yields (Khaledian et al., 2019).

Nanoparticles also help defend plants against abiotic stresses such as salinity, drought, and metal toxicity. Gold nanoparticles, for instance, act as nano-enzymes that reduce oxidative stress in plants, while cerium oxide nanoparticles (CeO NPs) enhance plant tolerance to salinity by regulating essential physiological pathways. Moreover, silica nanoparticles loaded with abscisic acid have demonstrated effectiveness in improving drought resistance in plants by prolonging the release of the hormone (Siddiqi and Husen, 2019). These technologies enable more precise and efficient use of resources, improving both productivity and sustainability. As research progresses, the integration of nanotechnology in agriculture promises to further optimize crop management and protection strategies, as well as overall crop improvement (Ray et al., 2021; Wu et al., 2020).

3. Potential of Nanobiosensors in Sustainable Agriculture

Nanobiosensors hold significant promise for promoting sustainable agriculture by enabling precise monitoring of plant health. These sensors can identify early molecular signs of stress, nutrient deficiencies, or disease, enabling timely interventions. Their accuracy reduces the need for extensive chemical applications, thus lessening environmental impact and conserving resources. Additionally, nanobiosensors can track soil conditions, improving the efficiency of fertilizer and water usage (Ficke et al., 2018). The real-time data allows farmers to make informed decisions, which enhances crop productivity while promoting sustainable and eco-friendly farming practices (Ficke et al., 2018). However, the potential of nanobiosensors in sustainable agriculture is indexed in Fig. 1.

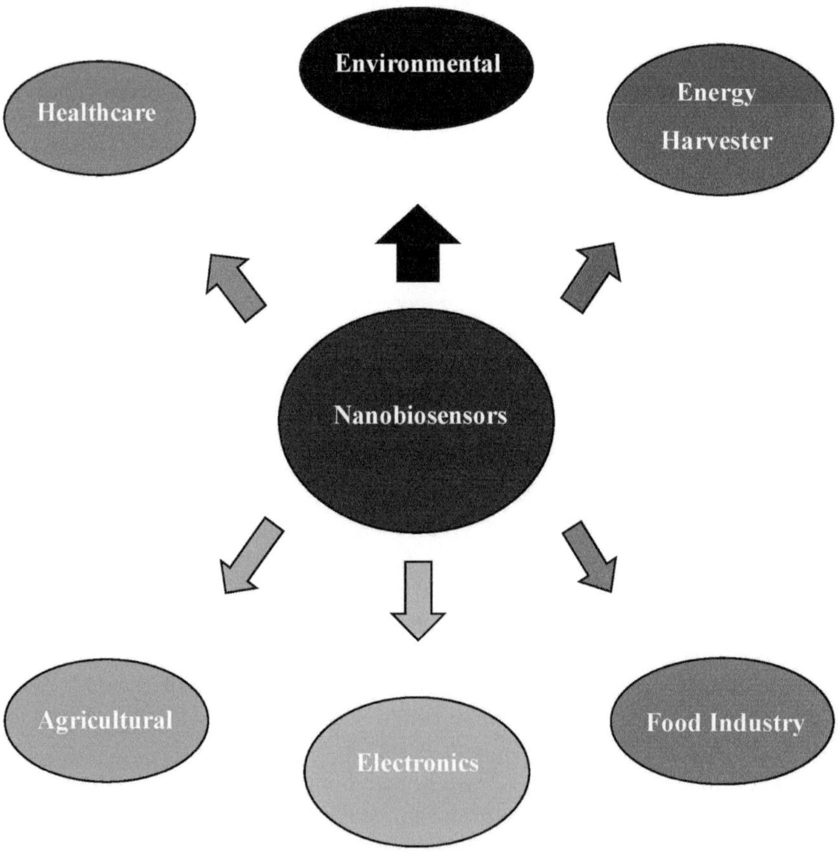

Figure 1. Potential of Nanobiosensors in Sustainable Agriculture.

3.1 Role of Nano Biosensors in Managing Plant Stress

Plant pathogens are major contributors to reduced crop yields, potentially leading to food shortages for both humans and animals. Projections indicate that food production will need to rise by 35–50% from 2012 to 2050 to sustain the projected global population of approximately 10 billion by 2050 (FAO, 2019). It is estimated that plant pathogens cause about 16% of global crop losses (Ficke et al., 2018). An example of such an impact in history is the Irish Potato Famine of 1845–1849, which occurred due to the late blight disease of potatoes caused by the pathogen *Phytophthora infestans* (Goss et al., 2014). Other significant pathogens include *Acidovorax avenae*, *Pseudomonas syringae*, *Ralstonia solanacearum*, *Pantoea stewartii*, *Botrytis cinerea*, *Pseudocercospora fijiensis*, and *Leptosphaeria maculans* (Dyussembayev et al., 2021). Traditional methods involve applying pesticides, insecticides, and fungicides, but these can lead to ecological issues and pathogen resistance over time (Ficke et al., 2018).

Nanomaterial-based biosensors, leveraging techniques like electrochemical sensing, immunosensing, and aptasensing, are becoming indispensable tools for detecting pathogens. For instance, scientists (Cebula et al., 2019) created a label-free electrochemical immunosensor for detecting Pseudomonas syringae pv. lachrymans with a Limit Of Detection (LOD) of 337 CFU/mL using a gold electrode modified with anti-Psl antibodies. Another electrochemical sensor by scientists (Lau et al., 2017) detects *Pseudomonas syringae* DC3000 with an LOD of 214 pM through DNA amplification. For detecting *Ralstonia solanacearum*, a ?coloruimetric method without prior DNA amplification was also developed (Khaledian et al., 2017). However, scientists (Lau et al., 2016) designed a versatile surface-enhanced Raman scattering (SERS)-RPA-based detection method capable of identifying multiple pathogens, including *Pseudomonas syringae* and *Botrytis cinerea*. *Pseudomonas aeruginosa*, known for causing various diseases, can be detected using an aptamer-gold nanoparticle-based lateral flow assay (Soundy and Day 2017). Additionally, gold nanoparticle-based assays have been developed for detecting pathogens such as Phytophthora infestans and Potato leafroll virus (Zhan et al., 2018; Panferov et al., 2018).

3.2 Managing Abiotic Stress

3.2.1 Monitoring Abiotic Stress with Nanobiosensors

Plants face various abiotic stresses and heavy metal contamination. These stresses can negatively impact plant physiology and biochemistry, affecting processes such as membrane composition, phytohormone responses, photosynthesis, and the production of antioxidants, ultimately leading to reduced agricultural yields (Husen 2021a, b; Ansari et al., 2022). Abiotic stress is responsible for significant crop losses, estimated at 51–82% annually (Li et al., 2021a, Li et al., 2021b; Oshunsanya et al., 2019). Effective stress management requires early detection and response. Plants have various adaptive mechanisms to counteract abiotic stress, including restoring cellular balance and repairing stress-induced damage (Kocabay et al., 2013). Under stress conditions, the levels of phytohormones such as abscisic acid (ABA), indole-3-acetic acid (IAA), Salicylic Acid (SA), ethylene (ET), and jasmonates (JAs) often increase (Mishra et al., 2016; Siddiqi and Husen, 2019).

Nanobiosensors can monitor these stress-related phytohormones. For example (Hu et al., 2020) developed an electrochemical biosensor for detecting IAA and SA in soybean seedlings under saline stress, with detection limits of 1.99 μM and 3.30 μM, respectively. This sensor used multiwall carbon nanotubes and carbon black to enhance sensitivity. Another electrochemical biosensor detected vitronectin-like proteins (VN) associated with heavy metal stress, with detection limits of 18.5 nmol/L for cadmium and 25.6 nmol/L for lead. Additional electrochemical sensors have been developed to detect phytohormones under various stress conditions (He et al., 2020; Cao et al., 2019). Creating a surface plasmon resonance-based biosensor for ABA detection with a limit of 0.33 μM, utilizing gold nanoparticle-aptamer conjugates (Husen et al., 2018, 2019).

3.2.2 Innovative Uses of Nanomaterials in Agriculture for Abiotic Stress Management

3.2.2.1 Seed Nanopriming

Seed health is crucial for agriculture, as environmental stresses can greatly impact germination and early growth stages. Stress-resistant seeds can significantly improve crop yields (Shelar et al., 2021). Seed priming enhances stress tolerance, and germination rates, and reduces the need for fertilizers and pesticides (Shelar et al., 2022; Mitra et al., 2021; Acharya et al., 2020). Nanopriming involves soaking seeds in nanoparticle solutions to boost germination and growth. This process can create nanopores for better water uptake, generate radicals to loosen cell walls, and regulate antioxidant systems. For example, Kaffir lime leaf extract-derived silver nanoparticles improved the germination of aged rice seeds (Mahakham et al., 2017). Carbon nanoparticle-based priming increased seed vigor and growth of lettuce under high salinity stress (Baz et al., 2020). Zinc-chitosan nanoparticles have antifungal properties, enhancing seed germination and fungal resistance in maize crops (Choudhary et al., 2019).

3.2.2.2 Foliar Application of Nanomaterials

Foliar application of nanomaterials involves spraying nanoparticles directly onto plant leaves to improve nutrient uptake, enhance growth, and protect against diseases, providing a targeted and efficient delivery method for agricultural inputs (Hafeez et al., 2018). Nanomaterials are also utilized in foliar applications to enhance nutrient delivery and plant growth. Nanofertilizers are becoming popular due to their efficient nutrient release and targeted delivery, which improves the uptake of nutrients and growth of the plant (Sharma et al., 2017). Various nanomaterials such as silica, iron oxide, zinc oxide, and silver nanoparticles are used in foliar applications. For instance, zinc oxide nanoparticles enhance photosynthetic efficiency and chlorophyll content, promoting plant growth (Hafeez et al., 2018). Silver nanoparticles have been found to enhance growth and yield by improving nutrient uptake and increasing stress resilience (Khan et al., 2019). Nanoparticle applications also enhance plant resilience to drought, salinity, and temperature (Khan et al., 2021).

4. Development and Applications of Nanobiosensors for Sustainable Agriculture

A transducer converts these biological signals into digital signals; a data recording unit controls the data's storage and transmission; and a biological probe, which may include antibodies for antigen binding, enzyme-substrate interactions, nucleic acid interactions, or cell-based interactions (Verma and Rani, 2021). DNA, enzymes, antibodies, biomimetic materials, and aptamers are examples of biological components that can be found in nanobiosensors. Synthetic receptors that more successfully duplicate their activities are frequently used in place of natural biological components to increase detection speed and accuracy (Neethirajan et al., 2018). Combining biological probes with different nanoparticles, including metallic, magnetic, quantum dots, graphene oxide, and carbon nanotubes, allows for the detection of analytes. Transducers can use optical signals like colorimetric detection, surface plasmon resonance, metallic fluorescence, and optical fibers, as well as electrochemical signals like amperometric, voltammetric, and potentiometric approaches (Shawon et al., 2020). Nanotechnology, which involves working with extremely small particles, addresses agricultural issues that traditional methods cannot solve. Innovations in nanomaterials and nano-devices have led to new applications in agriculture. A key development is the advanced biosensors known as nanobiosensors, which offer greater efficiency and organization than conventional biosensors. Nanobiosensors can precisely measure soil pH, moisture, various pathogens, plant hormones, metabolites, pesticides, herbicides, fertilizers, and metal ions. Their application supports sustainable agriculture by improving crop yields, optimizing input use, minimizing pollution, and reducing cultivation costs (Verma, 2017). Consequently, nanobiosensors provide accurate detection, enable continuous monitoring, and minimize both time and cost.

4.1 Types of Nanobiosensors

A biosensor is a comprehensive device that includes a biological receptor designed to recognize particular biomolecules, known as analytes, to deliver precise quantitative data. It combines physical or chemical transducers with recognition components to detect and quantify biological substances (Bhalla et al., 2016). The biological layers engage with the analytes, and the transducers then transform this interaction into quantifiable measurements (Huang et al., 2021). Biosensors can be categorized based on their transducers and recognition elements. They include types such as electrochemical (e.g., amperometric, impedimetric, potentiometric), calorimetric, piezoelectric (e.g., quartz crystal microbalance, acoustic wave), or optical (e.g., reflection, refraction, absorption,

Table 1. Types of Nanobiosensors.

Category	Examples	Advantages
Mechanical	Nanocantilevers, acoustic sensors	Extremely sensitive to mass changes; higher sensitivity with smaller sizes
Optical	Surface Plasmon Resonance (SPR), Localized SPR (LSPR), Förster Resonance Energy Transfer (FRET), Surface-Enhanced Raman Scattering (SERS), evanescent field sensors, photonic crystal sensors	High sensitivity; requires small sample volumes; excellent optical properties
Nanowire	Boron-doped silicon nanowires (SiNWs)	Analyte-independent; highly sensitive; enables real-time, quantitative detection; suitable for biological and chemical analyses
Electronic	Lab-on-a-chip systems	Integrates multiple sensors on a single chip, allowing detection of diverse target DNA sequences
PEBBLE	Poly(decyl methacrylate)-based fluorescent PEBBLE sensors	Real-time monitoring of intra- and intercellular ions and molecules; highly reversible and stable against leaching and photobleaching
Nanoshell	Gold nanoshells	Rapid assay capability; enhances chemical sensing performance
Enzyme-based	4-Hydroxyphenylpyruvate Dioxygenase (HPPD) enzyme	Precise and cost-effective methodologies

transmission, fluorescence, surface plasmon resonance, waveguide) (Malhotra and Ali, 2018). Based on their recognition elements, biosensors can be divided into several types, including whole-cell, organelle-based, aptamer-based, enzymatic, nucleic acid-based, oligonucleotide-based, immunosensors (using biomarker approaches or antigen-antibody interactions), and immunosensors (Kuswandi and Enafi, 2019). The types of nanobiosensors are indexed in Table 1.

4.1.1 Vitamin Based

Hydrogen peroxide sensors, vitamin-based sensors, and catechol sensors are examples of recent developments in nanosensor technology. Vitamin-based sensors are cutting-edge instruments that use vitamin molecules to identify and track particular compounds at the nanoscale. Utilizing the distinct properties of vitamins, these sensors provide accurate detection with potential applications in environmental monitoring, healthcare, and diagnostics. Through electrochemical oxidation and reduction reactions in electrolytes, they serve as electrochemical sensors that can identify vitamins that are soluble in water and fat (Huang et al., 2021a).

4.1.2 Catechol Based

Plant health can be evaluated with a catechol-based sensor that measures environmental characteristics like pH, conductivity, or certain ion concentrations. It finds metabolic alterations linked to plant stress and offers useful information for efficient plant care. By interacting with metal ions, catechol molecules enable these sensors to function. These interactions can be altered by physiological changes brought on by stress in plants, which can cause discernible differences in the sensor's findings. This allows for the real-time monitoring of stressors affecting plant health and facilitates prompt maintenance and intervention (Wang et al., 2023).

4.1.3 Hydrogen Peroxide Based

Aphids, spider mites, whiteflies, and fungus gnats are just a few of the pests that hydrogen peroxide ($H2O_2$) effectively controls. It can be sprayed on leaves or soaked in soil. Additionally, it aids in the control of plant diseases like black spots, powdery mildew, and downy mildew. Furthermore, via improving root development and nutrient absorption, H_2O_2 stimulates plant growth. By destroying dangerous bacteria and fungi and promoting helpful microbes, it enhances the health of the soil. Under both ideal and stressful circumstances, H_2O_2 functions as a signaling molecule in plants, supporting a variety of physiological and biochemical functions (Yan et al., 2022).

4.2 Application in the Food and Agriculture Industry

The overuse of agrochemicals like pesticides, herbicides, and food preservatives, as well as dangerous microorganisms like pathogenic bacteria, viruses, or parasites, can lead to food contamination. Foodborne infections are mostly caused by pathogenic bacteria (Inbaraj et al., 2015). Due to the inadequacy of conventional techniques, early contamination identification is challenging because there is currently little field detection of toxins and foodborne pathogens. However, the sensitivity and portability of detection have been significantly improved by the use of biosensors with nanostructures such as thin films, nanorods, nanoparticles, and nanofibers. Nanosensors and nanobiosensors are used in food microbiology to measure food ingredients, identify pathogens in food processing settings, and advise distributors and customers about food safety (Meashram et al., 2018). The nutritional value of food can be affected by a range of factors, including biological, microbiological, chemical, biochemical, and mechanical influences. Biological and microbiological issues involve contaminants such as bacteria, fungi, mites, and insects. Chemical and biochemical reactions, like the Maillard reaction and lipid oxidation, as well as mechanical damage during post-harvest processing, can also impact food quality (Thakur et al., 2018).

Additionally, mechanical factors such as germination, sprouting, and senescence can lead to spoilage (Ghaffar et al., 2020). To address these challenges, enzyme-based biosensors using glutamate dehydrogenase and glutamate oxidase have been developed to measure toxins like monosodium glutamate, helping to prevent neurotoxic effects on the human kidney, liver, and brain (Argulla and Simonian, 2016). By using precise measurements produced by nanobionics and advanced smart sensing, nanobiosensors are able to assess crop health over large fields. Sensors in nano bionics use mathematical models to identify changes in phytohormones and signaling chemicals (Butnariu and Butu, 2019). Precision farming utilizes detailed data on field and soil conditions to make informed decisions that enhance crop yield while minimizing the use of chemical fertilizers and pesticides. Smart plant systems further this by facilitating rapid analysis and effective management of costly agrochemicals, environmental stress factors, and phytopathogens, which contribute significantly to crop losses (Ghaffar et al., 2020).

Regular plant monitoring, often a laborious task, is streamlined and made more accurate through the use of smart nanobiosensors (Mittal et al., 2020). Integrating nanobiosensors with information technology can greatly benefit farmers in remote regions. Connecting nanobiosensors to Global Positioning Systems (GPS) allows for advanced monitoring and management of fertilizers, pesticides, insecticides, water levels, and physical and chemical stresses. This technology also improves pathogen detection, enabling early intervention to prevent disease in crops directly in the fields (Shang et al., 2019).

5. Role of Nanosensors in Agriculture

Advanced equipment is crucial for managing the nanostructural formulations of nanobiosensors. Techniques such as molecular assembly are controlled using sophisticated tools like Atomic Force Microscopy (AFM) and Brewster Angle Microscopy (BAM). AFM excels in providing detailed nanoscale structural information. AFM offers insights into molecular structures, surface roughness, and the composition of food materials, addressing complex issues in food science and aiding in the optimization of food structures for enhanced functionality. Additionally, force spectroscopy may be employed to identify specific intermolecular interactions that contribute to the bioactivity of food polysaccharides (Yang et al., 2007; Rodriguez et al., 2007; Otles and Yalcin, 2013; Pereira et al., 2008; Morris et al., 2011).

5.1 Working Principle of Nanosensors

To grasp the workings of nanobiosensors, it is crucial to understand nanomaterials, which are categorized into several types. These include carbon-based materials like nanotubes and nanoparticles, nanocomposites, and metals and alloys such as particles, catalysts, antimicrobial silver, and coatings with advanced tribological properties (Zhang et al., 2008). Additionally, biological nanomaterials encompass proteins, peptides, lipid nanocontainers, and DNA-templated circuits. Nano-polymers, which include linear, branched, and dendritic structures, and nano-glasses, such as amorphous materials like SiO2 and indium tin oxide, along with optical/photonics materials like LEDs, also play significant roles (Otles and Yalcin, 2013).

Nano-ceramics, composed of crystalline materials, are another key category. Nanomaterials are created through two main approaches: the bottom-up method, which builds materials from individual atoms and molecules through self-assembly—akin to the natural formation of proteins and enzymes from amino acids—and the top-down method, which breaks down larger materials into nanoscale components, including processes that convert insulating materials into conductive ones. Nanoparticles, ranging in size from 1 to 100 nm, possess distinct physico-chemical characteristics, a high surface area to mass ratio, and significant interactions within biological environments (De Corcura, 2007; Nickols-Richardson, 2007).

6. Polymer Nanocomposites

Nanocomposites are materials consisting of polymers that embed particles or fibers, with dimensions between 1 and 100 nanometers, which can be either equiaxed or elongated (Sen et al., 2020). Both inorganic (like ceramics, metals, or semiconductors) and organic (like polymers) materials can be included in this debate; metal-matrix and ceramic-matrix nanocomposites are not. In contemporary plastics manufacture, reinforcing polymers using inorganic or organic fillers is a standard procedure. The use of polymeric nanocomposites (PNCs) differs greatly from that of conventional filled polymers or polymer blends. Unlike conventional systems that utilize reinforcements in the micron range, PNCs incorporate discrete particles or fibers with dimensions in the nanometer range (less than 100 nm) (Pielichowski and Pielichowski, 2019). PNCs (polymer nanocomposites) can be grouped according to their shape. By generating a wide interfacial area in the composites, the usage of nano-sized fillers influences the interaction between the filler and the polymer, ultimately affecting the material's properties. As with traditional composites, the interfacial region starts where the fiber's characteristics diverge from those of the bulk filler and extend into the matrix until the fiber's characteristics match those of the bulk matrix (Akpan et al., 2019).

Comparisons were conducted between modified polymer-based nanocomposites and the standard polymer matrix. Although introducing different fillers can enhance the electrical, magnetic, and optical properties of conventional composites, it frequently leads to a considerable reduction in their mechanical properties (Harito et al., 2019). These improvements include greater mechanical strength, better thermal and electrical conductivity, improved fire resistance, and additional benefits. Nanocomposites can also achieve progress in applications related to optical and magnetic properties. Polymer-based nanocomposites offer significant potential in the defense sector, delivering advantages like cost efficiency and reduced weight. While the initial costs of adopting new materials can be high due to investment in equipment and processing, these costs may be lowered if the new materials can be manufactured with existing technologies. Utilizing polymer nanocomposites could enhance current technologies and adapt them for advanced material applications (Harussani et al., 2022; Ebrahimi et al., 2021; Arefi et al., 2021; Kumar and Verma, 2022).

7. Amelioration in Nanobiosensors

Nanobiosensors use various nanomaterials and are classified by the type of material they analyze and the signal transmission method. The demand is increasing for sensors that provide enhanced sensitivity, cost efficiency, rapid response, durability, effective recovery, compact size, *in situ* analysis capabilities, and user-friendly operation (Aghaei et al., 2018). In order to create nanosensor devices, nanostructured materials are essential. These consist of nano-sized thin films, polymers, and biomaterials; carbon nanotubes (CNTs) for their high surface area and superior electron conductivity; and nanoscale wires for increased detection sensitivity (Kokate et al., 2013).

Biosensors that feature predictable color changes are capable of detecting various substances, such as alpha-1-fetoprotein and hepatitis B virus antibodies in human serum, breast cancer biomarkers, *Mycobacterium tuberculosis* complex, human immunodeficiency virus type 1, DNA, and toxic metal pollutants (Priyadarshini et al., 2017; Bezzon et al., 2019). Silver nanoparticles have been incorporated into polymers for use in sensor applications (Kariuki et al., 2016). To create sensors for nitrobenzene detection, silver nanoparticles have been incorporated into a Poly-Amic Acid (PAA), polymer matrix. With low interference from related nitroaromatic chemicals, this PAA-Ag nanoparticle sensor has a sensitivity of 7.88 μA (M−1), a linear range of 10–600 mM, and a detection limit of 1.68 mm (Ayesh, 2016; Karim, 2020). The incorporation of silver nanoparticles enables the detection of various analytes by tagging them with these particles (Sepunaru et al., 2016).

8. Nanosensors for Smart Farming Technology and Sustainable Agriculture

Nanosensors, which are extremely small, can effectively monitor a variety of agricultural parameters, including soil conditions, irrigation needs, water and soil pH, nutrient levels, and the presence of diseases and pests (Becker and Makkar, 2008). By spreading these sensors across a field, farmers can gather detailed information, enabling precise actions that boost crop yields and reduce the need for excess resources such as fertilizers and pesticides. This technology supports the objectives of precision or smart farming. Made from materials like carbon nanotubes, nanosensors serve as wireless nanoantennas, collecting data from numerous points due to their tiny size (Martinez et al., 2016). External devices can compile data from nanosensors to create detailed reports and respond to critical environmental changes automatically. Nanosensors that monitor soil or plant conditions can send automatic alerts based on their readings, leading to more precise applications of fertilizers, herbicides, pesticides, and insecticides (Becker et al., 1998).

Modern nanobiosensors are now available as integrated units with a power source, sensing unit, detector, and display all in a single chip, enabling efficient detection of plant stress indicators (Kuswandi et al., 2011). In the future, nanoscale devices are expected to identify and address crop diseases, nutrient deficiencies, and other issues before visible symptoms appear. An array of nano-sensors will be deployed across farms and fields, serving as the agricultural equivalent of eyes, ears, and noses. Additionally, smart dust and gas sensors will assess pollution levels in the environment (Sassolas et al., 2012). The most effective method for this will involve real-time monitoring through autonomous sensors integrated with global positioning systems (Gruere, 2012; Hajirostamlo et al., 2015; Mukhopadhyay, 2014; Prasad et al., 2014). Nanobiosensors have emerged from the fusion of nanotechnology and biosensor technology. The integration of nanomaterials with electrical components leads to the development of microelectromechanical systems (MEMS), which serve as transducers in these sensors (Prasad et al., 2007).

9. Nanobiosensors Development for Next-Generation Diagnostic Tools

9.1 Components and Development

In agriculture, nanobiosensors amalgamate specific recognition elements such as antibodies, enzymes, or nucleic acid sequences with nanomaterials like graphene, nanowires, and nanoparticles. By employing optical, electrochemical, or thermal transducers, these components detect target analytes (such as pathogens or agrochemicals) and convert the recognition event into a measurable signal (Ramachandran et al., 2022).

9.2 Applications and Advantages

Nanobiosensors are used in agriculture for a variety of purposes, including pesticide residue identification, crop quality assessment, phytopathogen detection, and soil vitality monitoring. More effective and sustainable agronomic techniques are encouraged by their increased sensitivity, selectivity, quick reaction, and potential for downsizing, which enables real-time, on-site diagnostics (Rojas and Vincent, 2023).

9.3 Soil Health Monitoring

Precision metrics are provided to farmers to optimize irrigation and fertilization strategies via nanobiosensors , which can continually measure soil nutrient concentrations, pH levels, and moisture content. This real-time surveillance helps decrease the negative ecological effects of overuse of water and fertilizer, maintain soil health, and increase agricultural yield (Ingle et al., 2024).

9.4 Pest and Disease Detection

Nanobiosensors can detect the presence of pests and diseases at an early stage, allowing for timely intervention. By identifying specific pathogens or pest infestations, farmers can apply targeted treatments, reducing the reliance on broad-spectrum pesticides and minimizing crop damage (Dar et al., 2020).

10. Challenges

Numerous challenges remain before nanobiosensors can become viable tools for ensuring the viability and sustainability of the food and agriculture sectors. The use of nanomaterials is linked to certain problems, particularly those pertaining to toxicity and ecotoxicity, because of their special characteristics, including size, structure, high surface-to-volume ratio, and composition (Moulick et al., 2020). Their environmental impact must be rigorously assessed in terms of particle size, dosage, accumulation characteristics, immunological response, and retention time. Another challenge is the development of miniaturized versions to enhance portability.

Successful integration of nanobiosensors with information technology could provide remote food and agriculture sectors with cost reductions, increased productivity, early detection of disease outbreaks, and optimized use of natural resources like soil, water, and climate (Thakur et al., 2022). At the same time, it is critical to lower the cost of manufacturing nanobiosensors. Thus, overall stability and reusability can be improved by adding more affordable biological components, like cells, and novel matrices for stabilizing nanomaterials, such as chitosan. It can be difficult to move nanobiosensors from prototype to commercial product. Raising end-user awareness is vital, and field-scale investigations are necessary to assess and validate the practical efficacy of nanobiosensors. Many modern food producers are reluctant to reveal their usage of nanobiosensors for fear of unfavorable customer opinions that would hurt their ability to compete (Mondal et al., 2022).

11. Future Perspective

The integration of nanobiosensors in contemporary agri-food industries is expanding to enhance productivity through more efficient utilization of natural resources, thereby bolstering the sustainability of the food and agricultural sectors. Throughout the agri-food continuum—from soil analysis, resource evaluation, moisture assessment, soil pH measurement, disease management, and pathogen detection, to the identification of chemicals and adulterants harmful to humans—

nanobiosensors play a critical role up to the point of commercialization. Given their extensive range of applications, only a limited number of companies, such as IBM, Nippon, and Roche, are involved in the manufacture of nanobiosensors (Ghaffar et al., 2020). While commercialized nanobiosensors are well-documented in medical applications and diagnostics, reports on their use in agri-food industries are sparse. One of the primary challenges impeding the industry is the high cost of fabrication, automation trials, results evaluation, and field trial validation, necessitating the miniaturization of prototypes for industrial-scale production (Johnson et al., 2021). Additionally, there is a lack of a supportive market to offset these costs, leading to limited availability of commercial nanobiosensors. The development of diverse nanomaterials for biosensing various compounds through bioassays could pave the way for the commercialization of portable nanobiosensors.

In the future, these devices could be integrated with GPS systems to advance smart agriculture and precision farming. This would empower farmers to make data-driven decisions regarding irrigation, fertilization, pest control, and harvesting, ultimately reducing dependence on natural resources (Srivastava et al., 2018). The discussion could conclude with the notion that customized nanobiosensors, tailored to meet high specificity and sensitivity requirements, are likely to become a reality soon. The growing trend of using nanobiosensors in the agri-food industry is rapidly gaining momentum. These real-time nanobiosensors can monitor various physicochemical properties of soil, such as temperature, pH, moisture content, microclimate, and nutrient status (Beegum and Das, 2022).

However, creating affordable alternatives and developing broad-spectrum nanosensors could improve their commercialization and address issues related to sensitivity and specificity (Kaushik, 2022). New analytical techniques are necessary to identify, evaluate, and assess the impact of nanomaterials on ecosystems. It is important to consider factors such as size, dose, exposure duration, surface chemistry, structures, immunological response, accumulation, retention time, and other characteristics to estimate their environmental impact. Further research is needed to fully harness the potential of these specialized, rapid identification systems for adaptable, stable, and dynamic use in agroecosystems (Shawon et al., 2020).

12. Conclusion

Nanobiosensors hold transformative potential for smart and sustainable agriculture, providing precise, real-time monitoring of soil health, crop status, and environmental conditions. These sensors leverage nanotechnology to detect minute biochemical changes and pathogens, enabling early intervention and reducing the reliance on chemical inputs. By integrating with data systems, nanobiosensors support precision agriculture, enhancing crop yields while minimizing environmental impact. Furthermore, their adaptability and sensitivity offer unique advantages in addressing challenges such as resource scarcity and climate variability. As this technology matures, it promises to play a crucial role in creating more resilient, resource-efficient agricultural practices.

References

Abdel-Karim, R., Reda, Y., and Abdel-Fattah, A. (2020). Review—nanostructured materials-based nanosensors. *J. Electrochem. Soc.*, 167(3): 037554. doi:10.1149/1945-7111/ab67aa.

Acharya, P., Raturi, A., and Mahawar, H. (2020). Nanotechnology applications in seed priming for crop improvement. *Journal of Nanomaterials*, 11(6): 120–131.

Aghaei, A., Khorasanizadeh, H., and Sheikhzadeh, G. A. (2018). Measurement of the dynamic viscosity of hybrid engine oil-Cuo-MWCNT nanofluid, development of a practical viscosity correlation and utilizing the artificial neural network. *Heat Mass Transf.*, 54: 151–161. doi:10.1007/s00231-017-2112-6.

Akpan, E. I., Shen, X., Wetzel, B., and Friedrich, K. (2019). *In*: Pielichowski, K., and Majka, T. M. (Eds.). Polymer composites with functionalized nanoparticles. (Elsevier) p. 47.

Ansari, M. A. (2023). Nanotechnology in food and plant science: challenges and future prospects. *Plants,* 12(13): 2565. doi:10.3390/plants12132565.

Ansari, S. A., Husen, A., Siddiqi, K. S., and Kim, H. (2022). Nanomaterial-induced changes in plant biochemistry, physiology, and growth. *Journal of Plant Physiology*, 270: 153645.

Arefi, M., Tabatabaeian, A., and Mohammadi, M. (2021) Def. Technol. 17 64.

Arugula, M. A., and Simonian, A. L. (2016). Biosensors for detection of genetically modified organisms in food and feed. pp. 97–110. *In*: Genetically Modified Organisms in Food. Elsevier, Amsterdam, the Netherlands.

Ayesh, A. I. (2016). Metal/metal-oxide nanoclusters for gas sensor applications. *J. Nanomater.*, 1–17. doi:10.1155/2016/2359019.

Baz, H., Shawky, R., and Khairy, M. (2020). Carbon nanoparticles-based seed priming enhances salinity stress tolerance in lettuce (*Lactuca sativa* L.). *Journal of Plant Growth Regulation*, 39(4): 1483–1494.

Becker, H., Dietz, W. (1998). Microfluidic devices for μ -TAS applications fabricated by polymer hot embossing. Microfluid. *Devices Syst.*, 3515: 177–182.

Becker, K., and Makkar, H. P. S. (2008). Jatropha curcas: a potential source for tomorrow's oil and biodiesel. *Lipid Technology*, 20(5): 104–107.

Beegum, S., and Das, S. (2022). Nanosensors in agriculture. In *Agricultural Nanobiotechnology* (pp. 465–478). Woodhead Publishing.

Bezzon, V. D., Montanheiro, T. L., de Menezes, B. R., Ribas, R. G., Righetti, V. A., Rodrigues, K. F., et al. (2019). Carbon nanostructure-based sensors: a brief review on recent advances. *Adv. Mater. Sci. Eng.*, 1–21. doi:10.1155/2019/4293073.

Bhagat, R., Ingle, A. P., and Chen, H. (2023). Nanosensors and nanobiosensors for sustainable agriculture. In Nanotechnology in Agriculture and Agroecosystems (pp. 93–112). Elsevier.

Bhalla, N., Jolly, P., Formisano, N., and Estrela, P. (2016). Introduction to biosensors. *Essays Biochem.*, 60: 1–8.

Butnariu, M., and Butu, A. (2019). Plant nanobionics: Application of nanobiosensors in plant biology. *In*: Prasad, R. (Ed.). Plant Nanobionics. Nanotechnology in the Life Sciences. Springer, Cham, http://dx.doi.org/10.1007/978-3-030-16379-2_12.

Cao, Y., Yang, T., Li, H., Xu, Y., and Song, H. (2019). Electrochemical biosensor for abscisic acid detection based on aptamer-modified gold electrodes. *Sensors and Actuators B: Chemical*, 290: 348–354.

Cebula, Z., Lakatos, M., and Tóth, K. (2019). Label-free electrochemical immunosensor for detecting Pseudomonas syringae. Biosensors and Bioelectronics, 127: 276–283.

Choudhary, R., Kumar, R., Pandey, S., and Singh, D. (2019). Zinc-chitosan nanoparticles for enhancing maize seed germination and fungal resistance. *Environmental Nanotechnology, Monitoring & Management*, 11: 100218.

Dar, F. A., Qazi, G., and Pirzadah, T. B. (2020). Nano-biosensors: NextGen diagnostic tools in agriculture. *Nanobiotechnology in Agriculture: An Approach Towards Sustainability*, 129–144.

De Corcuera, J. R. (2007). Applications of Nanotechnology to Food. Insights from theNano4Food 2006 Conference, Food Safety and Quality Program Seminar April 26, University of Florida.

Dutta, P., Sharma, V., Bhardwaj, H., Agrawal, V. V., Rajesh, and Sumana, G. (2022). Fabrication of electrochemical biosensor using zinc oxide nanoflowers for the detection of uric acid. *MAPAN*, 37(3): 585–595. doi:10.1007/s12647-022-00598-7.

Dyussembayev, K., Talgat, B., Alimzhanova, M., and Bhattacharyya, D. (2021). Pathogenicity and virulence of plant pathogens. *International Journal of Molecular Sciences*, 22(8): 4152.

Ebrahimi, F., Hafezi, P., and Dabbagh, A. (2021). *Def. Technol.*, 17: 226.

FAO. (2019). The state of food and agriculture 2019: Moving forward on food loss and waste reduction. Food and Agriculture Organization of the United Nations.

Ficke, A., Cowger, C., Bergstrom, G., and Brodal, G. (2018). Understanding yield loss and pathogen biology to improve disease management: Septoria nodorum blotch – A case study in wheat. *Plant Disease*, 102(4): 696–707.

Garcia-Martinez, J. The Internet of Things Goes Nano, http://www. scientificamerican.com/article/the-internet-of-things-goes-nano/.2016.

Ghaffar, N., Farrukh, M. A., and Naz, S. (2020). Applications of nanobiosensors in agriculture. Nanoagronomy, 179–196.

Ghaffar, N., Farrukh, M. A., and Naz, S. (2020). Applications of nanobiosensors in agriculture. *In*: Javad S. (eds.). *Nanoagronomy*. Springer, Cham., 10: 179–196.

Goss, E. M., Tabima, J. F., Cooke, D. E. L., Restrepo, S., Fry, W. E., Forbes, G. A. et al. (2014). The Irish potato famine pathogen Phytophthora infestans originated in central Mexico. *Proceedings of the National Academy of Sciences*, 111(24): 8791–8796.

Gruere G. P. (2012). Implications of nanotechnology growth in food and agriculture in OECD countries. *Food Policy.*, 37: 191–198.

Hafeez, A., Razzaq, A., Khan, M. I., Iqbal, S., Hafeez, Y., and Jahanzaib, M. (2018). Zinc oxide nanoparticles as foliar application on maize: Effects on growth parameters, chlorophyll contents and antioxidant system. *Journal of Environmental Chemical Engineering*, 6(6): 6602–6613.

Hajirostamlo, B., Mirsaeedghazi, N., Arefnia, M., Shariati, M. A., and Fard, E. A. (2015). The role of research and development in agriculture and its dependent concepts in agriculture [Short Review]. *Asian J. Appl. Sci. Eng.*, 4: 1010–1016.

Harito, C., Bavykin, D. V., Yuliarto, B., Dipojono, H. K., and Walsh, F. C. (2019). *Nanoscale*, 11: 4653.

Harussani, M., Sapuan, S. M., Nadeem, G., Rafin, T., and Kirubaanand, W. (2022). *Def. Technol*. 18: 1281.

He, F., Zhang, W., Gu, X., Shen, J., Xu, F., and Wang, X. (2020). Electrochemical biosensor for phytohormone detection in plants. *Biosensors and Bioelectronics*, 150: 111940.

Hu, C., Zhang, Y., and Wu, H. (2020). Electrochemical detection of indole-3-acetic acid and salicylic acid in soybean seedlings under saline stress using carbon nanotubes. *Analytical Biochemistry*, 602: 113746.

Huang, H., Wang, D., Zhou, Y., Wu, D., Liao, X., Xiong, W., et al. (2021a). Multiwalled carbon nanotubes modified two dimensional MXene with high antifouling property for sensitive detection of ochratoxin A. *Nanotechnology*, 32(45): 455501. doi:10.1088/1361-6528/ac1a42.

Huang, X., Zhu, Y., and Kianfar, E. (2021). Nano biosensors: properties, applications and electrochemical techniques. *J. Mater. Res. Technol*. 12: 1649–1672.

Husen, A. (2021a). Plant nanobionics: Nanoparticles and nanotubes in plant systems. *Plant Physiology and Biochemistry*, 161: 269–276.

Husen, A. (2021b). Nanomaterials for enhancing plant growth and stress tolerance. *Journal of Plant Science Research*, 37(4): 459–471.

Husen, A., Siddiqi, K. S., and Kim, H. (2018). Nanomaterial effects on plant metabolism and physiology. *Journal of Nanobiotechnology*, 16(1): 80.

Husen, A., Siddiqi, K. S., and Kim, H. (2019). Nanomaterials for plant growth and development. *Journal of Plant Biology*, 62(4): 229–240.

Ijaz, U., Ahmed, T., Rizwan, M., Noman, M., Shah, A. A., Azeem, F. et al. (2023). Rice straw based silicon nanoparticles improve morphological and nutrient profile of rice plants under salinity stress by triggering physiological and genetic repair mechanisms. *Plant Physiology Biochem*. 201: 107788. doi:10.1016/j.plaphy.2023.107788.

Inbaraj, B. S., and Chen, B. (2015). Nanomaterial-based sensors for detection of foodborne bacterial pathogens and toxins as well as pork adulteration in meat products. *J. Food Drug Anal*., 24: 15–28.

Ingle, A. P., Saxena, S., Savalkar, A., Moharil, M., and Rai, M. (2024). Nano-enabled sensors and probes for plant pathogen detection and disease control in sustainable agriculture. *In Nanotechnology for Environmental Management* (pp. 244–260). CRC Press.

Johnson, M. S., Sajeev, S., and Nair, R. S. (2021, March). Role of Nanosensors in agriculture. In 2021 International Conference on Computational Intelligence and Knowledge Economy (ICCIKE) (pp. 58–63). IEEE.

Karim, M. R. (2020). Ubiquitous use of agricultural pesticides in six agro-based districts of Bangladesh and its impact on public health and environment. *Journal of Agriculture, Food and Environment*, 1(3): 47–52.

Kariuki, V. M., Fasih-Ahmad, S. A., Osonga, F. J., and Sadik, O. A. (2016). An electrochemical sensor for nitrobenzene using π-conjugated polymer-embedded nanosilver. *Analyst*, 141(7): 2259–2269. doi:10.1039/c6an00029k.

Kashyap, P. L., Kumar, S., Jasrotia, P., Singh, D. P., and Singh, G. P. (2019) Nanosensors for plant disease diagnosis: current understanding and future perspectives. *In*: Pudake, R., Chauhan, N., and Kole, C. (eds). Nanoscience for Sustainable Agriculture. Springer, Cham (Switzerland) https://doi.org/10. 1007/978-3-319-97852-9_9.

Kaushik, S. (2022). Nanosensor Technology for Smart Intelligent Agriculture. In Agricultural Biotechnology (pp. 267–299). *CRC Press*.

Khaledian, M., Baharlouei, A., and Karami, K. (2019). Gold nanoparticles functionalized with oligonucleotides for the detection of Ralstonia solanacearum in plants. *Journal of Nanotechnology in Plant Pathology*, 45(3): 215–224. https://doi.org/10.1016/j.jnpp.2019.03.005.

Khaledian, S., Ranjbar, B., and Rasaee, M. J. (2017). Colorimetric detection of Ralstonia solanacearum using gold nanoparticles without DNA amplification. *Analytical Methods*, 9(6): 917–924.

Khan, M. A., Sarfraz, W., and Ditta, A. (2024). Applications of nano-based fertilizers, pesticides, and biosensors in sustainable agriculture and food security. In *Molecular Impacts of Nanoparticles on Plants and Algae* (pp. 277–303). Academic Press.

Khan, M. I., Nazir, A., Ashraf, M., Javed, A., and Khan, M. I. (2019). Silver nanoparticles and their potential in improving plant growth under abiotic stress conditions. *Journal of Nanotechnology*, 100120.

Khan, M. R., Rizvi, T. F., Khan, A., and Raziq, S. (2021). Nanotechnology in sustainable agriculture: Recent developments and applications in crop management. *Journal of Nanobiotechnology*, 19(1): 98.

Kocabay, M., Atalay, D., and Guray, T. (2013). Abiotic stress in plants: Responses and adaptations. *Journal of Plant Research*, 126(2): 201–212.

Kokate, M., Garadkar, K., and Gole, A. (2013). One pot synthesis of magnetite–silica nanocomposites: applications as tags, entrapment matrix and in water purification. *J. Mater. Chem. A.*, 1(6): 2022–2029. doi:10.1039/c2ta00951j

Kumar, J., and Verma, R. K. (2022). *Def. Technol.*, 17: 1671.

Kuswandi, B., and Ensafi, A. A. (2019). Perspective-Paper-based biosensors: trending topic in clinical diagnostics developments and commercialization. *J. Electrochem. Soc.*, 167: 037509.

Kuswandi, B., Wicaksono, Y., Jayus, A., Abdullah, A., Heng, L. Y., and Ahmad, M. (2011). Smart packaging: sensors for monitoring of food quality and safety. Sens. Instrum. *Food Qual. Saf.*, 5(3-4): 137–146.

Lau, H. Y., Botella, J. R., and Jarman, R. H. (2016). Surface-enhanced Raman scattering for multiplex pathogen detection. *Journal of Agricultural and Food Chemistry*, 64(16): 3202–3210.

Lau, H. Y., Wu, H., and Botella, J. R. (2017). Electrochemical biosensor for pathogen detection in plants. *Biosensors and Bioelectronics*, 91: 679–683.

Li, B., Zhang, W., Guo, Z., Chen, L., and Zhang, H. (2021a). Abiotic stress management in agriculture: Advances and prospects. *Plant Science*, 306: 110854.

Li, Z., He, C., and Zhang, Q. (2021b). Nanomaterials for abiotic stress management in agriculture. *Journal of Nanoscience and Nanotechnology*, 21(5): 2564–2574.

Mahakham, W., Theerakulpisut, P., Maensiri, S., Phumying, S., and Sarmah, A. K. (2017). Kaffir lime leaf extract-derived silver nanoparticles for seed priming: Enhancing germination and seedling growth in aged rice seeds. *Journal of Plant Physiology*, 214: 45–55.

Malhotra, B. D., and Ali, M. A. (2018). Nanomaterials in biosensors: fundamentals and applications. In: Nanomaterials for Biosensors. *Elsevier*, pp. 1–74.

Meshram, B.D., Agrawal, A.K., Adil, S., Ranvir, S., and Sande, K. K. (2018). Biosensor and its application in food and dairy ındustry: A review. *Int. J. Curr. Microbiol. App. Sci.*, 7: 3305–3324.

Miguel-Rojas, C., and Pérez-de-Luque, A. (2023). Nanobiosensors and nanoformulations in agriculture: new advances and challenges for sustainable agriculture. *Emerging Topics in Life Sciences*, 7(2): 229–238.

Mishra, P., Panigrahi, J., and Behera, D. K. (2016). Plant responses to abiotic stresses and its amelioration by phytohormones. *Journal of Applied Biology and Biotechnology*, 4(6): 20–27.

Mitra, A., Rathi, B., Kumar, A., and Kaushik, N. (2021). Seed priming techniques for enhancing abiotic stress tolerance in crops. *Journal of Environmental Biology*, 42(4): 973–980.

Mittal, D., Kaur, G., Singh, P., Yadav, K., and Ali, S. A. (2020) Nanoparticle-based sustainable agriculture and food science: recent advances and future outlook. *Front. Nanotechnol.*, 2: 579954 https://doi.org/10.3389/fnano.2020.579954.

Mittal, D., Kaur, G., Singh, P., Yadav, K., and Ali, S. A. (2020). Nanoparticle-based sustainable agriculture and food science: recent advances and future outlook. *Front. Nanotechnol.*, 2: 579954.

Mondal, R., Dam, P., Chakraborty, J., Paret, M. L., Katı, A., Altuntas, S. et al. (2022). Potential of nanobiosensor in sustainable agriculture: the state-of-art. *Heliyon*, 8(12).

Morris, V. J., Woodward, N. C., and Gunning, A. P. (2011). Atomic force microscopy as a nanoscience tool in rational food design. *J. Sci. Food Agric.*, 91: 2117–2125.

Moulick, R. G., Das, S., Debnath, N., and Bandyopadhyay, K. (2020). Potential use of nanotechnology in sustainable and 'smart'agriculture: advancements made in the last decade. Plant Biotechnology Reports, 14(5): 505–513.

Martinez, M., Luz, A., and Elena M. Bennett. (2016). Trade in the US and Mexico helps reduce environmental costs of agriculture. *Environmental Research Letters*, 11.5 (2016): 055004.

Mukhopadhyay, S. S. (2014). Nanotechnology in agriculture: prospects and constraints. *Nanotechnol. Sci. Appl.*, 7: 63–71.

Munir, N., Gulzar, W., Abideen, Z., Hancook, J. T., El-Keblawy, A., and Radicetti, E. (2023). Nanotechnology improves disease resistance in plants for food security: applications and challenges. *Biocatal. Agric. Biotechnol.*, 51 102781. doi:10.1016/j.bcab.2023.102781.

Neethirajan, S., Ragavan, V., Weng, X., and Chand, R. (2018). Biosensors for sustainable food engineering: challenges and perspectives. *Biosensors*, 8: 23–31.

Negrete, J. C. (2020). Nanotechnology an option in mexican agriculture. *J. Biotechnol. Bioinfo. Res.*, 2(2): 1–3.

Nickols-Richardson, S. M. (2007). Nanotechnology: implications for food and nutrition professionals. *J. Am. Diet. Assoc.*, 107: 1494–1497.

Oshunsanya, S. O., Amodu, M. F., and Akomolafe, O. B. (2019). Abiotic stress effects on plants and methods of amelioration: A review. *African Journal of Agricultural Research*, 14(6): 310–318.

Otles, S., and Yalcin, B. (2013). Food Chemistry and Nanoscience. *Journal of Nanomaterials & Molecular Nanotechnology*, 2: 4.

Panferov, V. G., Biketov, S. F., Zherdev, A. V., and Dzantiev, B. B. (2018). Gold nanoparticle-based lateral flow immunoassay for detection of Phytophthora infestans in potato leaves. *Analytica Chimica Acta*, 1030: 142–149.

Pereira, D., Losada, P. P., Angulo, I., Greaves, W., and Cruz, J. M. (2008). Development of a Polyamide Nanocomposite for Food Industry: Morphological Structure, Processing, and Properties. *Polymer Composites*, 1–9.

Pielichowski, K., and Pielichowska, K. (2019). Thermochim. Acta 675 140.

Prakash, R., and Kaler, K. V. I. S. (2007). An integrated genetic analysis microfluidic platform with valves and a PCR chip reusability method to avoid contamination. *Microfluid. Nanofluid.*, 3: 177–187.

Prasad, A. K., Singh, R. P., Tare, V., and Kafatos, M. (2007). Use of vegetation index and meteorological parameters for the prediction of crop yield in India. *International Journal of Remote Sensing*, 28(23): 5207–5235.

Prasad, R., Kumar, V., and Prasad, K. S. (2014). Nanotechnology in sustainable agriculture: present concerns and future aspects. *Afr. J. Biotechnol.*, 13: 705–713.

Priyadarshini, P., Tiwari, K., Das, A., Kumar, D., Mishra, M. N., Desikan, P. et al. (2017). Evaluation of highly conserved hsp65-specific nested PCR primers for diagnosing Mycobacterium tuberculosis. *Int. J. Tuberc. Lung Dis.*, 21(2): 214–217. doi:10.5588/ijtld.16.034.

Ramachandran, R. P., Vellaichamy, C., and Erkinbaev, C. (2022). Smart nano-biosensors in sustainable agriculture and environmental applications. In *Food, Medical, and Environmental Applications of Nanomaterials* (pp. 527–542). Elsevier.

Ray, S., Rathod, P., and Gupta, P. (2021). Key characteristics and applications of nanobiosensors in agriculture. *Biosensor Advances*, 12(2): 158–170. https://doi.org/10.1080/bsadv.2021.12.002.

Rodriguez Patino, J. M., Caro, A. L., Rodríguez Niño, M. R., Mackie, A. R., and Gunning, A. P. (2007). Some implications of nanoscience in food dispersion formulations containing phospholipids as emulsifiers. *Food Chem.*, 102: 532–541.

Rojas, T., and Vincent, S. (2023). What's in a book? An exploration of multicultural extractions within secondary agriculture textbooks. *Journal of Agricultural Education*, 64(3): 95–108.

Sassolas, A., Prieto-Simon, B., and Marty, J. L. (2012). Biosensors for pesticide detection: New trends. *American Journal of Analytical Chemistry*, 3: 210–232.

Schieber, A. (2018). Introduction to food authentication. In *Modern Techniques for Food Authentication* (Academic Press), 1–21.

Sen, M. (2020). Nanotechnology and the environment (London: IntechOpen) pp, 1–12.

Sepunaru, L., Plowman, B. J., Sokolov, S. V., Young, N. P., and Compton, R. G. (2016). Rapid electrochemical detection of single influenza viruses tagged with silver nanoparticles. *Chem. Sci.*, 7(6): 3892–3899. doi:10.1039/c6sc00412a.

Shang, Y., Hasan, M. K., Ahammed, G. J., Li, M., Yin, H., and Zhou, J. (2019). Applications of nanotechnology in plant growth and crop protection: a review. *Molecules*, 24: 2558–2564.

Shang, J., JI, X., and Chen, X. (2020). Study on the impact of China's urbanization on agricultural ecological efficiency: Based on panel data of 13 major grain-producing regions in China from 2009 to 2018. *Chinese Journal of Eco-Agriculture*, 28(8): 1265–1276.

Sharma, N., Singhvi, R., and Joshi, A. (2017). Nano-fertilizers for sustainable agriculture: Current status and future prospects. *Journal of Environmental Management*, 203: 84–97.

Shawon, Z. B., Hoque, M. E., and Chowdhury, S.R. (2020) Nanosensors and nanobiosensors: Agricultural and food technology aspects. pp. 135–161. *In*: Pal, K., and Gomes, F., (Eds.). Nanofabrication for Smart Nanosensor Applications, Elsevier, Amsterdam.

Shawon, Z. B. Z., Hoque, M. E., and Chowdhury, S. R. (2020). Nanosensors and nanobiosensors: Agricultural and food technology aspects. pp. 135–161. *In*: Kaushik, P., and Gomes, F. (Eds.). Micro and Nano Tech. Elsevier.

Shelar, A., Sable, N., and Deshmukh, P. (2021). Seed health management under abiotic stress. *Journal of Agricultural Research and Development*, 13(2): 141–149.

Shelar, A., Sable, N., and Patil, B. (2022). Nanopriming for sustainable agriculture: Methods and future prospects. *Journal of Crop Improvement*, 36(4): 497–515.

Siddiqi, K. S., and Husen, A. (2019). *Phytohormones and abiotic stress in plants.*

Siddiquee, S., Shaarani, S. M., and Rovina, K. (2018). Zinc oxide nanocomposites for the detection of soil-borne fungal pathogens. *Plant Pathology Journal*, 52(4): 321–330. https://doi.org/10.1080/plantpath.2018.52.004.

Siddiqui, M. H., Al-Whaibi, M. H., and Ali, H. M. (2015). Role of nanotechnology in agriculture with special emphasis on plant nutrition and stress management. *Environmental and Experimental Botany*, 111: 63–73. https://doi.org/10.1016/j.envexpbot.2014.11.012.

Soundy, J., and Day, D. (2017). Selection of DNA aptamers specific for live *Pseudomonas aeruginosa. PLoS One*, 12(9): e0185385.

Srivastava, A. K., Dev, A., and Karmakar, S. (2018). Nanosensors and nanobiosensors in food and agriculture. *Environmental Chemistry Letters*, 16: 161–182.

Thakur, M., Wang, B., and Verma, M. L. (2022). Development and applications of nanobiosensors for sustainable agricultural and food industries: Recent developments, challenges and perspectives. *Environmental Technology & Innovation*, 26: 102371.

Thakur, M. S., and Ragavan, K. V. (2013). Biosensors in food processing. *J. Food Sci. Technol.*, 50: 625–641.

Thakur, S., Thakur, S., and Kumar, R. (2018). Bio-nanotechnology and its role in agriculture and food industry. *J. Mol. Genet. Med.*, 12(324): 1747–0862.

Verma, M. L. (2017). Enzymatic nanobiosensors in the agricultural and food industry. *In*: Ranjan, S., Dasgupta, N., and Lichfouse, E. (Eds.). *Nanoscience in Food and Agriculture* 4. Springer Cham, pp. 229–245, Sustain. Agri Rev., 24.

Verma, M. L., and Rani, V. (2021). Biosensors for toxic metals, polychlorinated biphenyls, biological oxygen demand, endocrine disruptors, hormones, dioxin, phenolic and organophosphorus compounds: A review. *Environ. Chem. Lett.*, 19: 1657–1666.

Wang, D., Saleh, N. B., Byro, A., Zepp, R., Sahle-Demessie, E., Luxton, T. P. et al. (2022) Nano-enabled pesticides for sustainable agriculture and global food security. Nat. Nanotechnol., 17: 347–360 https://doi.org/10.1038/s41565-022-01082-8.

Wang, H., Gu, M., Yan, M., Wu, X., Dong, Y., and Wang, G. L. (2023). Catechol exerts an in situ surface oxygen vacancy effect on BiOI: an innovative signal transduction mode for cathodic photoelectrochemistry. *J. Mater. Chem. B.*, 11(23): 5123–5130. doi:10. 1039/d3tb00566f.

Wu, Y., Ma, J., and Chen, G. (2020). Cerium oxide nanoparticles improve Arabidopsis tolerance to salinity stress by regulating ion balance. Plant Science, 294: 110449. https://doi.org/10.1016/j.plantsci.2020.110449.

Yan, Y., Chen, X., Huang, J., Huan, C., and Li, C. (2022). H2O2-induced oxidative s impairs meat quality by inducing apoptosis and autophagy via ROS/NF-κB signaling pathway in broiler thigh muscle. *Poult. Sci.*, 101(4): 101759. doi:10.1016/j.psj.2022.101759.

Yang, H., Wang, Y., Lai, S., An, H., and Li, Y. (2007). Application of atomic force microscopy as a nanotechnology tool in food science. *J. Food Sci.*, 72: R65–75.

Zhan, S., Zhang, H., and He, D. (2018). China's flexible overseas food strategy: food trade and agricultural investment between Southeast Asia and China in 1990–2015. *Globalizations*, 15(5): 702–721.

Zhang, L., GU, F. X., Chan, J. M., Wang, A. Z., and Langer, R. S. (2008). Nanoparticles in medicine; therapeutic applications and developments. *Clin. Pharmacol. Ther.*, 83: 761–769.

CHAPTER-9

Green Nanotechnology for Plant Disease Management

An Overview

Mohammed Radi,[1,2] *Mohammed Taoussi,*[1,2] *Khadija Benamar*[3] and *Rachid Lahlali*[1,*]

1. Introduction

Crop and plant diseases have considerable impacts on crop losses and food security. The Food and Agriculture Organization of the United Nations (FAO) estimates that global yearly economic losses from plant diseases amount to around US$220 billion. Pests alone accounted for around 20–40% of losses in crop output (Alexandratos, 1999). This indicates that efficient disease management is essential to satisfy the increasing food demand resulting from anticipated population expansion by 2050 (Alexandratos and Bruinsma, 2012).

The present management predominantly depends on the utilization of pesticides, including insecticides, fungicides, and herbicides. Despite numerous advantages, such as high availability, rapid efficacy, and reliability, pesticides exhibit detrimental effects on non-target organisms, contribute to insect population comeback, and facilitate resistance development (Stephenson, 2003). Moreover, it is estimated that 90% of utilized pesticides are lost during or subsequent to application (Ghormade et al., 2011; Stephenson, 2003). Consequently, there is heightened motivation to create cost-effective, high-performance insecticides that are less detrimental to the environment.

The swiftly advancing field of nanotechnology has unveiled new possibilities in agricultural sciences (Wasule et al., 2024). Nanobiotechnology holds a pivotal role among several disease management strategies for early disease detection, potential fungicides (nanofungicides), and effective delivery systems for fungicides to plants (Mishra and Singh, 2015; Rai and Ingle, 2012; Satalkar et al., 2016).

[1] Department of Plant Protection, Phytopathology Unit, Ecole Nationale d'Agriculture de Meknès, Km10, Rte Haj Kaddour, BP S/40, Meknès 50001, Morocco.

[2] Environment and Valorization of Microbial and Plant Resources Unit, Faculty of Sciences, Moulay Ismail University, Meknes, Morocco.

[3] Department of Biology, Laboratory of Microbial Biotechnology and Bioactive Molecules, Sciences and Technologies Faculty, Sidi Mohamed Ben Abdellah University, P. O. Box 2202, Fez, Morocco.

* Corresponding author: rlahlali@enameknes.ac.ma

DOI: 10.1201/9781003567684-9

In this context, nanotechnology provides novel solutions that are more efficient, sustainable, and eco-friendly (Wasule et al., 2024). Numerous strategies can be employed to efficiently generate nanoparticles, including chemical methods (Mubraiz et al., 2021), electrochemistry (Kaviya et al., 2011), radiation (Elmer and White, 2018), photochemical methods (Noor et al., 2022), Langmuir-Blodgett techniques (Sattar et al., 2022; Worrall et al., 2018), and biological approaches (Sun and Xia, 2002).

Functionalized green nanomaterials have attracted considerable interest owing to their distinctive characteristics and potential to improve plant health and production (Sundararajan et al., 2024). Furthermore, green synthesis eliminates the need for costly, hazardous, and toxic ingredients. Biological approaches have been intensively employed in recent years to produce metallic nanoparticles with diverse forms, sizes, compositions, and physicochemical properties. Green synthesis can be conducted with biological organisms, including fungi, bacteria, actinobacteria, yeasts, molds, algae, and plants along with their derivatives. In nanoparticle manufacturing, molecules found in plants or microorganisms, including proteins, enzymes, phenolic compounds, amines, alkaloids, and pigments, serve as reducing agents (Nadaroglu et al., 2017).

The present chapter emphasizes green nanotechnology's role in managing plant diseases. We will discuss mechanistic aspects of different green synthesis of nanomaterials, their mode of action against phytopathogens, and their practical application in agriculture. Further, we will outline the environmental impacts and safety concerns of nanomaterials, regulatory challenges, and future prospects regarding large-scale adoption of green nanotechnology in agriculture.

2. Synthesis and Characterization of Green Nanomaterials

Traditional methods for synthesizing nanoparticles often involve hazardous chemicals that can harm both the environment and human health. To mitigate these risks, eco-friendly approaches are being adopted, utilizing renewable natural resources with minimal ecological impact. The tenets of green chemistry promote safety, efficiency, and economic feasibility, facilitating the creation of sustainable technologies that are viable for long-term application (Alqarni et al., 2022), and due to their distinctive properties, nanoparticles are increasingly utilized across various fields, leading to heightened interest in eco-friendly synthesis methods (Alqarni et al., 2022; Khan, 2020). The green synthesis of nanoparticles primarily involves using natural agents such as herbs, microorganisms like bacteria and fungi, as well as other organic materials, including agricultural and food waste (Dikshit et al., 2021) also biological materials contain compounds such as phenols, amines, enzymes, and proteins that facilitate the reduction process for nanoparticle synthesis. This method is particularly important in areas like plant disease management, where sustainable practices are essential (Nadaroglu et al., 2017).

2.1 Methods of Green Synthesis

As recently cited, numerous physical and chemical synthesis methods necessitate high radiation levels, highly toxic reducing agents, and stabilizers, which can have harmful effects on both human health and marine ecosystems. In contrast, the green synthesis of metallic nanoparticles employs a single-step, eco-friendly bio-reduction process that requires relatively low energy to initiate. This method is also cost-effective (Dahoumane et al., 2016; El-Rafie et al., 2013; Husen and Siddiqi, 2014; Khan et al., 2015; Patel et al., 2015; Siddiqi and Husen, 2016; Wadhwani et al., 2016).

One of the most prevalent green synthesis techniques for nanoparticles is plant-mediated synthesis, which utilizes different parts of plants such as leaves, stems, and roots—as reducing agents. Plant extracts are rich in phytochemicals like flavonoids, terpenoids, and alkaloids, which serve as both reducing and capping agents in nanoparticle formation. The benefits of plant-mediated synthesis include affordability, ease of use, and the capability to produce nanoparticles in various shapes and sizes (Singh et al., 2016). Another promising eco-friendly method for synthesizing nanoparticles involves using microorganisms such as bacteria and fungi. These organisms can generate nanoparticles through their cellular metabolic pathways. The advantages of microbial synthesis include high production yields, low costs, and the ability to create nanoparticles with consistent size and shape. Enzymatic synthesis is yet another green approach that uses enzymes as reducing agents. Enzymes are biocompatible, non-toxic, and function effectively under mild conditions, making them ideal for nanoparticle production (Virkutyte and Varma, 2011). The benefits of enzymatic synthesis include high specificity, affordability, and the capacity to produce nanoparticles with uniform size and shape.

2.1.1 Biological components for synthesis

2.1.1.1 Plants

The primary method for synthesizing green routes involves using plant derivatives, which is certainly the most reliable approach for producing non-toxic and eco-friendly nanoparticles. This method offers numerous benefits, as plants are readily accessible, cost-effective, and relatively simple to implement (Katata-Seru et al., 2018). Plants possess the ability to reduce, detoxify, and aggregate metals, and they are also effective at removing metal-related pollutants. Compared to microorganisms, plant extracts are more advantageous as biological agents for nanoparticle synthesis. They are plentiful, stable, environmentally friendly, economical, and safe to use, exhibiting a wide range of metabolic functions (Hussain et al., 2016; Katata-Seru et al., 2018; Singh et al., 2016). Importantly, the creation of metallic nanoparticles is influenced by the antioxidant properties found in plant extracts (Souri and Shakeri, 2018; Vidya et al., 2017). Plants are abundant in bioactive compounds such as sugars, vitamins, and polyphenols, which aid in converting ions into atoms (Rahmat et al., 2021; Sari et al., 2023). Indeed, plant phytochemicals, including anthocyanins, polyphenols, and alkaloids, serve as both reducing and capping agents for the synthesis of metallic nanoparticles, which can occur both extracellularly and intracellularly (Abuzeid et al., 2018). Notably, the main phytochemicals involved in nanoparticle formation include terpenoids, flavones, ketones, amides, aldehydes, and carboxylic acids (Mathew et al., 2010). Polyphenolic compounds consist of an aromatic ring attached to hydroxyl groups and carbon atoms. Notably, the antioxidant activity tends to enhance with an increasing number of hydroxyl groups (Drummer et al., 2021; Fatimah, 2016; Makarov et al., 2014; Porrawatkul et al., 2017).

A plant is made up of multiple parts, similar to the human body, and several sections of these plants can be utilized for synthesis purposes. For instance, the leaves, flowers, roots, and even stems can be included (Abuzeid et al., 2018). Studies show that different parts of plants, such as leaves, can be used for various applications regarding nanoparticles (Netala et al., 2018; Vijayakumar et al., 2017), flowers (Sharma et al., 2016), fruits (Manikandan et al., 2017), seeds (Khatami et al., 2015), latex (Kumar et al., 2017), bark (Mattos et al., 2017), tubers (Kumar et al., 2017), and even cultured tissues are potential sources for nanoparticles production (Hegazy et al., 2015), When herbal derivatives are added to liquid formulations containing the desired metals, they facilitate the reduction of metal ions into corresponding nanoparticles (Singh et al., 2016).

In the last thirty years, the use of plant extracts for environmentally friendly synthesis of nanoparticles has attracted considerable interest. Many plant species have been investigated for nanoparticle production, each providing a distinct composition of organic reducing agents (Patel et al., 2022). Especially, silver and gold which were the first nanoparticles synthesized using plants, however, several widely recognized and easily accessible plants, such as *Aloe vera*, neem, lemongrass, sunflower, and alfalfa, can be employed to create nanoparticles of different metals, including silver, copper, cobalt, and zinc. Alfalfa roots have the capacity to absorb Ag (0) from an agar medium and transfer it to the plant. The leaf extract of *Pelargonium graveolens* can effectively reduce silver ions. Additionally, the leaf broth of *Azadirachta indica* can be utilized to produce pure metallic Ag/Au or bimetallic Au/Ag nanoparticles extracellularly, with sizes ranging from 50 to 100 nm (Kumar et al., 2014).

Using aqueous plant extracts for nanoparticle synthesis is a simple approach that primarily requires the plant extract as a reducing agent along with a metal-ion solution. Generally, the green synthesis process for nanoscale metals involves the following steps: obtaining the plant extract with a specific solvent, mixing it with a metal salt solution, and then performing an additional step to isolate the desired nanoscale metal. The way in which the plant leaf extract aids in the green synthesis of nanoparticles is also part of this mechanism (Makarov et al., 2014). The metal ions (I) attach to the reducing metabolites (phytochemicals) and stabilizing agents, leading to their reduction into metal atoms (II). The resulting complex of the metal ion and metabolite then interacts with other similar complexes to form a small metal nanoparticle (III).

The majority of this type of research has focused on the *ex-vivo* synthesis of nanoparticles; however, metallic nanoparticles can also be generated in living plants (*in vivo*) by reducing metal salt ions that are absorbed as soluble salts. *In vivo* synthesis of nanoparticles, including zinc, nickel, cobalt, and copper, has been observed in plants such as mustard (*Brassica juncea*), alfalfa (*Medicago sativa*), and sunflower (*Helianthus annuus*) (Marchiol, 2012). Additionally, ZnO nanoparticles have been produced using a wide range of plant leaf extracts, including coriander (*Coriandrum sativum*). (Anastas and Warner, 1998), copper leaf (*Acalypha indica*) (Gnanasangeetha and Thambavani, 2013), crown flower (*Calotropis gigantean*) (Vidya et al., 2013), Green Tea (*Camellia sinensis*) (Maensiri et al., 2008), China rose (*Hibiscus rosa-sinensis*) (Devi and Singh, 2014), and aloe leaf broth extract (*Aloe barbadensis* Miller) (Gunalan et al., 2012).

2.1.1.2 Microorganisms

Another way for producing nanoparticles with varied physical and chemical characteristics with the aid of microorganisms such as fungus, bacteria and yeast (Dao et al., 2021) and magneto tactic bacteria (Xie et al., 2009).

2.1.1.2.1 Bacteria

Bacterial species have been extensively employed in various commercial biotechnological applications, including bioremediation, genetic engineering, and bioleaching (Gericke and Pinches, 2006). Importantly, bacteria effectively convert metal ions into their nano-sized forms using various proteins and enzymes, both inside and outside of their cells. Gram-positive bacteria play a key role in producing secondary organic metabolites, such as antibiotics. These filamentous organisms require oxygen, are motile, and do not remain stationary. Their resistance to large toxic metals makes them excellent detoxifying agents, allowing them to eliminate metal ions through either intracellular or extracellular reduction processes (Li et al., 2011). Numerous microorganisms have been shown to be effective in the synthesis of numerous significant metals and metal oxide nanoparticles such as *Bacillus subtilis*, *Lactobacillus casei*, *Pseudo monas proteolytica* and *Rhodopseudomonas capsulate*, are used to produce nanoparticles of inert metals like silver and gold respectively (Al-asbahi et al., 2024; Alsamhary, 2020; Lyudmila et al., 2018). *Pseudomonas stuteri* AG259 produced silver nanoparticles in periplasmic spaces (Klaus et al., 1999). Some advantages of these precursors include their ease of manipulation based on requirements, rapid advancements, and low maintenance costs. However, drawbacks include the need for careful control and monitoring of factors such as temperature and the functional species of cell walls when using bacterial species as precursors (Iravani, 2011). Bacteria have the capability to reduce metal ions, making them significant candidates for nanoparticle production (Iravani, 2014).

The bacterial synthesis of nanoparticles has gained popularity due to the relative ease of manipulating these microorganisms (Thakkar et al., 2010). Notable bacterial strains that have been widely used for producing bioreduced silver nanoparticles with various size and shape morphologies include *Escherichia coli*, *Pseudomonas proteolytica*, *Lactobacillus casei*, *Bacillus cereus*, *Aeromonas sp. SH10*, *Phaeocystis antarctica*, *Desulfovibrio desulfuricans*, *Bacillus amyloliquefaciens*, *Bacillus indicus*, *Corynebacterium* sp. SH09, *Bacillus cecembensis*, *Enterobacter cloacae*, *Geobacter* spp., *Arthrobacter gangotriensis*, *Rhodopseudomonas capsulate* and *Shewanella oneidensis*. Similarly, several bacterial species such as *Bacillus megaterium* D01, *E. coli* DH5a, *Bacillus subtilis* 168, *Shewanella alga*, and *Plectonema boryanum* UTEX 485 are used for the preparation of gold nanoparticles (Al-asbahi et al., 2024; Esmail et al., 2022; Nagoth et al., 2024; Solís-Sandí et al., 2023; Sunkar and Nachiyar, 2012; Tariq et al., 2020).

2.1.1.2.2 Fungi

The biosynthesis of metal and metal oxide nanoparticles facilitated by fungi is an effective method for producing uniform nanoparticles with distinct shapes. This process is enhanced by the presence of various intracellular enzymes, making fungi superior biological agents for creating these nanoparticles (Chen et al., 2009). Competent fungus can generate bigger amounts of nanoparticles than bacteria (Mohanpuria et al., 2008). Furthermore, fungi have various advantages over other organisms due to the presence of enzymes, proteins, and reducing components on their cell surfaces (Narayanan and Sakthivel, 2011). The likely mechanism behind the creation of metallic nanoparticles involves enzymatic reduction (reductase) occurring either in the cell wall or within the fungal cell. Different fungal species are utilized to produce metal and metal oxide nanoparticles, including silver, gold, titanium dioxide, and zinc oxide (Narayanan and Sakthivel, 2011).

The eco-friendly production of functional and precisely designed nanoparticles using fungi is regarded as a valuable method in the field. In this regard, *Verticillium* species can perform the bioreduction of aqueous $AuCl_4^-$ ions, resulting in the formation of AuNPs that exhibit well-defined sizes and excellent uniformity (Boroumand Moghaddam et al., 2015).

The benefits of synthesizing nanoparticles using fungal species include, first, the presence of a wide variety of uniquely shaped endoenzymes, and second, their capability to produce a greater diversity of nanoparticles compared to bacteria. Additionally, these fungi possess significant reducing agents both on their cell surface and within their interior. Metals and oxides such as titanium dioxide, silver, zinc, and gold are among the various materials that can be synthesized using fungi (Narayanan and Sakthivel, 2011). Furthermore, fungus may release far more proteins, resulting in increased nanoparticle output.

2.1.1.2.3 Yeast

Yeasts are single-celled microorganisms that inhabit eukaryotic cells. A total of 1500 yeast species were detected (Yurkov et al., 2011). Numerous research teams have documented the successful creation of nanoparticles and nanomaterials using yeast. The green synthesis of various inert metal nanoparticles has proven effective with different yeast types, particularly *Saccharomyces cerevisiae* (Liu et al., 2021). Importantly, the biosynthesis of silver and gold nanoparticles has been observed using a silver-tolerant yeast strain and broth from Saccharomyces cerevisiae. A wide range of species is utilized for producing countless metallic nanoparticles. Additionally, nanoparticles can be integrated through the use of extracted yeasts and molds from eukaryotic cells, employing both endo-enzymatic and exo-enzymatic methods (Kim et al., 2024).

The choice of active metal solutions and the incubation conditions significantly affect the size of the nanoparticles produced. Due to reports of certain molds being toxic, their application in nanoparticle synthesis must be carefully regulated (Boroumand Moghaddam et al., 2015). *Candida glabrata* and *Schizosaccharomyces* species were utilized for the biosynthesis of cadmium sulfide (CdS) nanocrystals. This process leverages the ability of these yeasts to reduce cadmium ions present in their surroundings, resulting in the formation of CdS nanoparticles. Both *C. glabrata* and *S. pombe* can thrive in environments with cadmium levels reaching up to 100 mg/L, enabling them to effectively engage with cadmium ions and promote the necessary reduction for nanoparticle synthesis (Krumov et al., 2007). The biosynthesis of CdS involves converting cadmium ions into CdS through various biochemical pathways, which may include both enzymatic and non-enzymatic processes. These yeasts are capable of producing CdS nanocrystals in both *in vitro* and *in vivo* settings, demonstrating their resilience in heavy metal-rich environments (Cuéllar-Cruz et al., 2017; Moreno et al., 2019).

2.1.2 Solvents in Green Synthesis

Green synthesis of nanoparticles involves two main approaches under solvent systems, either by utilizing extracts from nature as primary components or employing water, which is abundant, non-toxic, and serves as a reducing and stabilizing agent (Shanker et al., 2016). Importantly, water enhances the reactivity of nanoparticles due to the presence of O_2 molecules, which contribute to their limited oxidation. Key examples include using gallic acid for synthesizing gold and silver nanoparticles in a water medium, and the photoablation process for gold nanoparticles also conducted in water (Shanker et al., 2016). On the other hand, alcohols like ethanol and methanol can serve as reducing and stabilizing agents in nanoparticle synthesis. Additionally, vegetable oils (e.g., olive, soybean, coconut) are non-toxic solvents that can also function as reducing agents (Hansen et al., 2023).

Ionic liquids, often derived from natural sources, are emerging as versatile solvents in green synthesis (Gharby, 2022). They can be recycled and reused, exhibiting mixed functional properties that make them effective for nanoparticle synthesis. With melting points below 100°C and stability up to 300–450°C, ionic liquids can be tailored for various applications. They act both as reducing agents and safeguards during the synthesis process (Zeng et al., 2024). They were widely employed to produce nanoparticles such as platinum, silver, aluminum, and tellurium (Van Rantwijk and Sheldon, 2007). The hydrophilic and hydrophobic properties of a liquid are determined by the nature of its positive and negative ions. According to research done by Yoosaf et al., 2007, the production of Mn_3O_4 NPs in imidazolium ionic liquids and oleyl-amine as a solvent concluded that much narrower nanoparticles were generated, which exhibited higher diffusion in ionic liquids than in the other solvent utilized (Yoosaf et al., 2007). Likewise, Sylvestre et al. developed a one-phase method for synthesizing gold and platinum nanoparticles using ionic liquids that contain –SH functional groups for the first time. These ionic liquids acted as stabilizing agents to produce small-sized crystalline structures. However, due to their resistance to decomposition, they are not deemed suitable for the green synthesis of nanoparticles (Sylvestre et al., 2004).

Regarding mechanisms, supercritical liquefaction occurs when standard dispersion mediums are subjected to heat and pressure beyond their critical points, enhancing their viscous, heat-conducting, and dense properties. Water and carbon dioxide are two common, safe solvents used in this process. As the dissolution of metal oxides decreases near the critical point, extreme saturation can lead to nanoparticle formation (Budisa and Schulze-Makuch, 2014). For instance, Kim et al., (2016) successfully prepared tungsten oxide nanoparticles using subcritical and supercritical methanol and water (Kim et al., 2016).

Nanoparticle production is simplified since the ionic liquid may act as both a reductant and a protecting agent, several metal nanoparticles, including as Au, Ag, Al, Te, Ru, Ir, and Pt, have been produced in ionic liquids (Er et al., 2017; Srivastava, 2014; Vollmer et al., 2010; Zhang and Cui, 2009). Ionic Liquids (ILs) have emerged as versatile and promising solvents for the synthesis of nanoparticles, offering unique properties that can be tailored to specific applications. ILs can be either hydrophilic or hydrophobic, depending on the nature of their constituent cations and anions. This adaptability allows researchers to fine-tune the properties of ILs for various synthesis processes, making them particularly useful in the field of nanotechnology (Dupont et al., 2002; Van Rantwijk and Sheldon, 2007; Vollmer et al., 2010; Zhang, 2006). ILs have been successfully employed in the synthesis of various metallic nanoparticles, including silver, gold, platinum, and iridium (Li et al., 2008). Researchers have developed innovative techniques such as electrochemical methods and one-phase preparation using Thiol-Functionalized Ionic Liquids (TFILs) to produce nanoparticles with specific sizes and shapes (Kim et al., 2004). For example, Dupont et al. synthesized Ir(0) nanoparticles with an average size of approximately 2 nm using 1-n-butyl-3-methylimidazolium hexafluorophosphate, a room temperature ionic liquid (Dupont et al., 2002; Van Rantwijk and Sheldon, 2007; Vollmer et al., 2010; Zhang, 2006).

One of the key advantages of using ILs in nanoparticle synthesis is their ability to produce smaller, more uniformly dispersed particles compared to conventional solvents (Van Rantwijk and Sheldon, 2007). For instance, Bussamara et al. demonstrated that manganese oxide (Mn_3O_4) nanoparticles synthesized in imidazolium ionic liquids were significantly smaller (9.9 ± 1.8 nm) and better dispersed than those produced in oleylamine (12.1 ± 3.0 nm). This improved control over particle size and dispersity is crucial for many applications in nanotechnology, where precise particle characteristics are often required (Bussamara et al., 2013).

The benefits of using ILs as solvents for nanoparticle synthesis are numerous. They provide excellent thermal stability, allowing for a broader synthesis temperature range than water (Dupont et al., 2002). ILs can dissolve various metal catalysts, polar organic compounds, and gases, making them versatile reaction media (Mendiga-Liyeva and Bostan, 2020). Additionally, their solubility properties can be modulated by altering the associated cations and anions. However, concerns about the biodegradability of some ILs have led to the development of more environmentally friendly alternatives with improved biodegradation efficiency (Welton, 2004).

In addition to ionic liquids, supercritical fluids have also shown promise in nanoparticle synthesis. Ordinary solvents can be converted into supercritical fluids at temperatures and pressures above their critical points, significantly altering properties such as density, thermal conductivity, and viscosity (Fürstner et al., 2001; Wittmann et al., 2001). Carbon dioxide is particularly attractive as a supercritical fluid due to its non-hazardous and inert nature (Ohde et al., 2001). Supercritical water, with its critical temperature of 646 K and pressure of 22.1 MPa, has also proven useful for various reactions (Pollet et al., 2011). Overall, the solvent used in green nanoparticle synthesis is determined by the synthesis process used as well as the qualities of the nanoparticles being produced. Green nanoparticle production requires careful consideration of solvent availability, toxicity, and environmental effect (Hano and Abbasi, 2022).

2.2 Nanomaterials Used in Plant Disease Management

Nanoparticles have demonstrated significant suppressive activity against plant pathogens, encompassing a wide range of materials (Fig. 1) including non-metals, metalloids, oxide metals,

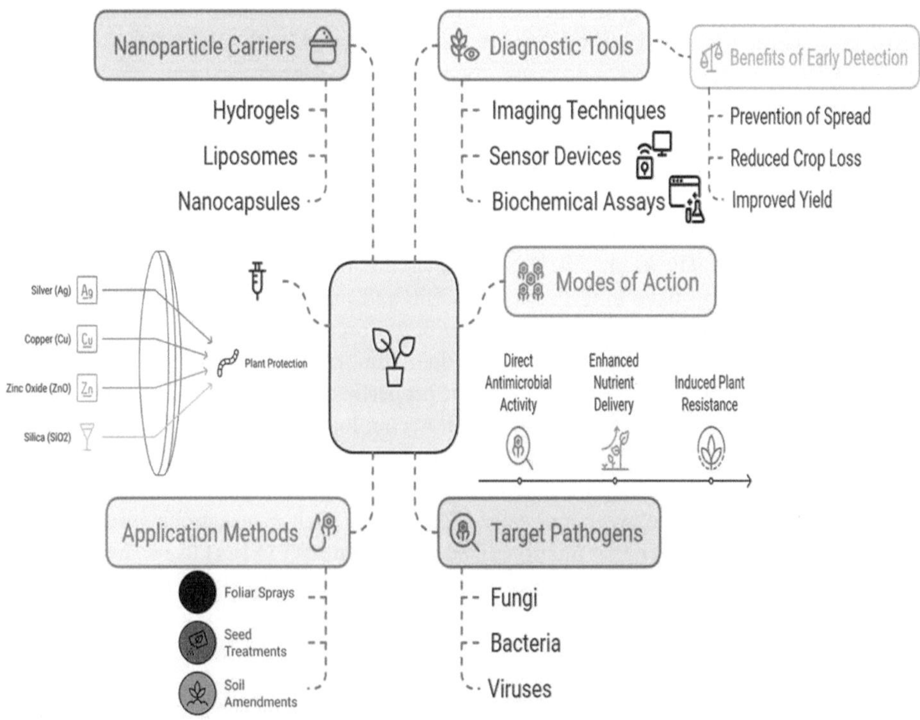

Figure 1. Multifaceted role of role of nanoparticles in alleviating plant disease management.

and nano-carbon materials (Elmer et al., 2018). Among these, silver, titanium oxide, copper, gold, zinc oxide, iron, and particularly silver have been the most extensively studied and widely utilized nanoparticles (Iravani et al., 2014; Kandi and Kandi, 2015). The mechanisms by which nanomaterials act against organisms are diverse and multifaceted. These include cell membrane perturbation, aggregation, suppression of H+-ATPase activity, inhibition of messenger RNA and protein synthesis, suppression of toxin production, blockage of microbial growth, and inhibition of nutritional flux (Dakal et al., 2016; Malerba and Cerana, 2016).

Nanoparticles have shown efficacy against numerous pathogenic microorganisms. These include fungal pathogens such as *Sclerotium rolfsii*, *Sclerotinia sclerotiorum*, *Rhizoctonia solani*, Fusarium *oxysporum* (Abdelrhim et al., 2021; Kaman and Dutta, 2019), *Colletotrichum* sp. (Goswami et al., 2020). Bacterial pathogens like *Xanthomonas oryzae* pv. Oryzae (Abdallah et al., 2019; Ogunyemi et al., 2020) have also been targeted. Moreover, nanoparticles have shown promise against viruses such as Bean Yellow Mosaic Virus (BYMV) (Elbeshehy et al., 2015), and Cucumber Mosaic Virus (CMV) (El-Sawy et al., 2017). The ability of bacteria to actively attenuate metal ions has been a major factor in their use to produce antagonistic nanoparticles for controlling insect pests. Over the past two decades, various bacterial species including *Bacillus* sp., *Corynebacterium* sp., *Pseudomonas* sp., and *Shewanella* sp. have been employed to synthesize a range of nanoparticles with metal inorganics such as Ag, Au, Al, ZnO, MnO, and TiO_2. The utilization of these bacterially generated nanoparticles has enhanced agricultural productivity while minimizing losses during crop processing, transport, and storage (Kumari et al., 2019).

2.2.1 Nanoparticle-based Solutions in Plant Nematode Disease Management

Green Silver Nanoparticles (GSN) from algal extracts have shown effective nematicidal properties against root-knot nematodes in eggplants (Bhau et al., 2016; Elmer et al., 2018). Low concentrations of GSN (4.25 mg/100 mL) were as effective as chemical nematicides in reducing nematode populations and improving plant growth, without causing phytotoxicity (Elmer et al., 2018). Nanoparticles can be designed to target parasite-infected cells, improving nematicide delivery and efficacy (Bhau et al., 2016). They can be made from biodegradable materials to reduce toxicity compared to traditional treatments. Nanoparticle-based formulations also offer advantages for developing vaccines against parasitic nematode diseases by optimizing antigen and adjuvant delivery, enhancing stability, enabling targeted delivery, and boosting immunogenicity (Elmer et al., 2018).

Bt nanoscale chitinases have demonstrated the ability to deliver nematicidal preparations to manage Caenorhabditis elegans, potentially serving as a biopesticide dispensing system for agricultural applications (Qin et al., 2020). Entomopathogenic (Bt) bacteria can be reduced in size to create Bt nanoparticles for use as biocontrol agents (Swamy and Asokan, 2013). Alternatively, Bt nanocomposites with Bt bioactives in various nanoformulations such as nanocapsules, nanoemulsion, and nanosuspension have been employed as nanopesticides (Damalas and Koutroubas, 2018). The Cry protein obtained from Bt is the main component of a large group of proteins with diverse toxic activities. These proteins are primarily toxic to *Diptera*, *Coleoptera*, and *Lepidoptera*, with some also affecting nematodes and Hymenoptera. Loading Cry proteins into nanomaterials to produce nanocomposites has proven feasible, maintaining the protein's bioactivity while enhancing stability (Devi et al., 2019).

2.2.2 Nanoparticle-based Solutions in Plant Fungal Disease Management

AgNPs have already shown promise in controlling fungal diseases in plants, potentially serving as alternatives to conventional fungicides (Menichetti et al., 2023). The effectiveness of AgNPs depends on factors such as size, concentration, and formulation. Their mechanism of action involves disrupting fungal cell membranes, inhibiting spore germination, and interfering with fungal growth and reproduction (Mansoor et al., 2021). Silver nanoparticles (AgNPs) exhibit strong antifungal activity against various plant pathogens, including *Alternaria alternata*, *Penicillium digitatum*, *Alternaria citri*, and *Raffaelea* sp. (Mansoor et al., 2021; Menichetti et al., 2023; Żarowska et al., 2019).

Nanoformulations of copper particles offer benefits such as improved efficacy, reduced environmental impact, and targeted delivery, making them promising candidates for controlling fungal plant diseases, Copper nanoparticles have been studied for their potential in managing red root-rot disease in plants, primarily caused by the fungal agent *Phoma terrestris*. Copper's fungicidal properties contribute to enhancing plant disease resistance (Ponmurugan et al., 2016). On the other hand, specific examples of bacterial nanoparticle synthesis include the use of *Bacillus* species, strain GP23, isolated from marine soil. This strain was employed to produce silver nanoparticles (AgNP) using an extracellular process, which successfully suppressed *Fusarium oxysporum*, a pathogen causing wilt disease in various crops including bananas, legumes, tobacco, sweet potatoes, cucurbits, and tomatoes (Gopinath and Velusamy, 2013). Nanocrystallization of cyclic lipopeptides from *B. subtilis* enhances their stability during storage by preventing oxidation and preserving antifungal properties through controlled release. These nanocrystals, known as solid lipid nanoparticles, effectively inhibit the growth of *Aspergillus carbonarius* and *A. niger* spores and hyphae (Lin et al., 2020).

2.2.3 Nanoparticle-based Solutions in Plant Viral Disease Management

Nanoparticles can be engineered to target and penetrate virus-infected plant cells, allowing for more efficient delivery of antiviral agents and improved therapeutic efficacy against plant viruses (Farooq et al., 2021). These nanoparticle-based formulations provide advantages over traditional treatments, including enhanced stability, targeted delivery, and increased immunogenicity (Farooq et al., 2021). Engineered nanoparticles can serve as carriers for RNA-based pesticides that target and silence key viral genes, providing protection against plant viruses (Li et al., 2023). For instance, Layered Double Hydroxide (LDH) nanoparticles loaded with RNA targeting viral genes have been shown to protect cowpea plants from common mosaic virus infection for at least 20 days (Li et al., 2023). Nanoparticles made from materials such as chitosan and silica have also demonstrated antiviral properties and the ability to induce plant defense responses against viral pathogens (Ahmed et al., 2023). These nanoparticles can be applied as foliar sprays or seed treatments to provide sustained protection against plant viruses (Mittal et al., 2020).

Silver nanoparticles (AgNPs) have shown promising antiviral activity against Tobacco Mosaic Virus (TMV) in plants. The small size, high surface area, and strong reactivity of AgNPs enable them to interact with and deactivate virus particles (Yadav and Yadav, 2018). Studies have found that AgNPs can physically inhibit TMV infection and replication in tobacco plants (Dutta et al., 2022). Nanoparticles can also serve as carriers for delivering antiviral compounds to plants, enhancing the solubility, stability, and targeted delivery of these agents while reducing their toxicity. Various nanoparticles like chitosan, silica, and metal oxides have been explored as nanocarriers loaded with antiviral compounds (Dutta et al., 2022). For example, curcumin-loaded milk protein nanoparticles have demonstrated antiviral activity against potato virus Y (Rajani et al., 2022). Additionally, chitosan nanoparticles have been shown to inactivate alfalfa mosaic virus replication and boost innate immunity in plants, while carbon-based nanomaterials can suppress tobacco mosaic virus infection and induce resistance in plants. Activating plant defense pathways using nanoparticles is a promising strategy for enhancing plant tolerance to viral diseases (Rajani et al., 2022).

2.2.4 Nanoparticle-based Solutions in Bacterial Disease Management

Biologically synthesized silver nanoparticles (AgNPs) have demonstrated potent antibacterial activity against plant pathogenic bacteria such as *Erwinia carotovora*, which causes soft rot in vegetables, and *Ralstonia solanacearum*, the causative agent of bacterial wilt in plants (Menichetti et al., 2023). The antibacterial mechanism of AgNPs involves disrupting bacterial cell membranes, inhibiting essential enzymes for bacterial growth, and interfering with DNA replication. Studies have shown that AgNPs can effectively control bacterial diseases in plants, with efficacy comparable to chemical bactericides. The size, shape, and concentration of AgNPs are crucial factors determining their antibacterial efficacy (Menichetti et al., 2023).

In a study on citrus canker control, a formulated dispersion of zinc oxide nanoparticles (Zn NPs) was applied as a foliar spray to grapefruit trees to suppress the disease caused by *Xanthomonas citri* subsp. Citri (Graham et al., 2016). The nanoformulation effectively controlled citrus canker, demonstrating the ability of engineered nanomaterials like Zn NPs to manage plant fungal diseases (Young et al., 2017). The targeted delivery and mode of action of Zn NPs contribute to their efficacy in disease suppression.

Copper nanoparticles have been evaluated for their ability to manage bacterial wilt disease caused by *Ralstonia solanacearum* in plants. Copper is an essential micronutrient that plays a role in enhancing plant defense mechanisms against bacterial pathogens. Nanoformulations of copper can potentially improve its efficacy and targeted delivery to the site of infection, leading to better control of bacterial diseases in plants. The small size and high surface area of copper nanoparticles contribute to their enhanced antibacterial activity compared to bulk copper (Wang et al., 2013).

Magnesium oxide nanoparticles (MgO NPs) have been found to induce systemic resistance in tomato plants against bacterial wilt disease caused by *Ralstonia solanacearum*. The MgO NPs can activate various plant defense pathways, including the salicylic acid and jasmonic acid signaling cascades, making the plants more tolerant to bacterial infections (Imada et al., 2016). This approach harnesses the plant's innate immunity to manage bacterial diseases in an eco-friendly manner, without directly targeting the pathogen. The size and surface properties of MgO NPs play a crucial role in their ability to induce systemic resistance in plants (Imada et al., 2016).

2.3 Characterization Techniques

Characterization of green-synthesized nanomaterials is essential for understanding their properties and potential applications. Various techniques are employed to analyze these materials, each providing unique insights into their structure and behavior.

2.3.1 Scanning Electron Microscopy (SEM)

Scanning Electron Microscopy (SEM) is widely used to provide detailed images of the surface morphology and size of nanoparticles. It utilizes reflected or scattered electrons to create high-resolution images, allowing researchers to observe the surface features and composition of the nanomaterials. SEM typically achieves resolutions of about 0.5 nm, making it a powerful tool for visualizing nanoscale structures (Titus et al., 2019). The technique can reveal important information about particle size, shape, and distribution. For green-synthesized nanoparticles, SEM analysis can confirm the formation of nanostructures and provide insights into how the biological components influence morphology (Alessio et al., 2017). SEM analysis requires careful sample preparation to obtain clear images. Samples are usually coated with a thin conductive layer, often gold, to prevent charging and improve image quality. For non-conducting samples like many biologically-synthesized nanoparticles, environmental SEM or low-voltage SEM modes may be used (Titus et al., 2019). The high magnification and depth of field offered by SEM make it invaluable for characterizing the three-dimensional aspects of nanoparticles, complementing other analytical techniques. By providing visual confirmation of nanoparticle formation and details on their external structure, SEM plays a crucial role in validating green synthesis methods (Asif et al., 2022).

2.3.2 Transmission Electron Microscopy (TEM)

Transmission Electron Microscopy (TEM) complements SEM by providing insights into the internal structure and shape of nanoparticles. It uses transmitted electrons that pass through the sample, enabling precise measurements of size and internal morphology. TEM can achieve even higher resolutions, often below 50 picometers, which is crucial for detailed structural analysis

(Titus et al., 2019). This technique is particularly valuable for examining the core-shell structures often found in green-synthesized nanoparticles, where biological components may form a coating around the metallic core (Gao, 2005). Sample preparation for TEM is critical and often more complex than for SEM. Samples must be extremely thin to allow electron transmission, typically requiring specialized preparation techniques. For green-synthesized nanoparticles, this may involve diluting the sample and depositing it on support grids (Titus et al., 2019). TEM can provide information on the crystal structure, particle size distribution, and the presence of any defects or impurities within the nanoparticles. This level of detail is essential for understanding how the green synthesis process affects the internal structure of the nanoparticles, which in turn influences their properties and potential applications (Alessio et al., 2017). It uses transmitted electrons that pass through the sample, enabling precise measurements of size and internal morphology. TEM can achieve even higher resolutions, often below 50 picometers, which is crucial for detailed structural analysis (Alessio et al., 2017; Titus et al., 2019).

2.3.3 X-ray Diffraction (XRD)

Used to determine the crystalline structure and phase of the synthesized nanoparticles, which is essential for understanding their stability and reactivity. XRD is employed to determine the crystalline structure and phase of synthesized nanoparticles (Chhantyal, 2022). This technique provides information about the material's crystalline properties, particle size, and atomic arrangement. The diffraction patterns obtained from XRD are essential for understanding the stability and reactivity of the nanomaterials, as they reveal structural deviations and defects (Alessio et al., 2017; Titus et al., 2018, 2019). The diffraction patterns obtained from XRD are essential for understanding the stability and reactivity of nanomaterials, as they reveal structural deviations and defects. These characteristics can significantly influence the nanoparticles' behavior in various applications, such as catalysis, drug delivery, or electronic devices. By analyzing XRD data, researchers can optimize synthesis methods to produce nanoparticles with desired properties and predict their performance in specific applications (Sharma et al., 2012).

2.3.4 Fourier Transform Infrared Spectroscopy (FTIR)

Fourier Transform Infrared (FTIR) spectroscopy is a powerful analytical technique for characterizing nanoparticles, particularly those synthesized through green methods. This method measures the vibration and rotation of molecules influenced by the infrared radiation at specific wavelengths, providing a molecular fingerprint that reveals details about the chemical structure and interactions of the nanoparticles (Faghihzadeh et al., 2016). It helps identify functional groups present on nanoparticles, which can influence their reactivity and compatibility with biological systems. FTIR is instrumental in identifying functional groups present on nanoparticles, which can significantly influence their reactivity and compatibility with biological systems. By measuring the absorption of infrared radiation, FTIR detects changes in molecular vibrations and rotations, providing a molecular fingerprint that helps in characterizing the chemical composition of the nanomaterials (Eid, 2022). FTIR is especially useful for identifying structural changes in molecular binding between nanoparticles and other entities, which can offer insights into the nature of their interactions and potential applications (Faghihzadeh et al., 2016).

The ability of FTIR to detect functional groups present on nanoparticles is crucial, as these groups significantly influence the reactivity and compatibility of nanomaterials with biological systems. For green-synthesized nanoparticles, FTIR can reveal the presence of biomolecules on their surface, which often play a role in their synthesis, stabilization, and biological activities (Sarwer et al., 2022). This information is valuable for understanding the mechanisms of nanoparticle formation, predicting their behavior in various applications, and assessing their potential toxicity. FTIR analysis can be performed using different methods, including transmittance FTIR, attenuated total reflectance (ATR-FTIR), and micro-spectroscopy FTIR, each providing specific advantages depending on the sample type (Eid, 2022).

3. Mechanisms of Action

3.1 Antimicrobial Properties of Nanomaterials

Nanomaterials, particularly nanoparticles, have gained significant attention for their antimicrobial properties. They exhibit effectiveness against a wide range of pathogens, including bacteria, fungi, and viruses. Metal NPs have been used as potential inhibitors of fungal pathogenic microorganisms, with more focus on NPs obtained by physical and chemical methods than on their biosynthesized counterparts (Sousa et al., 2020; Tran and Le, 2013). Their mechanisms of activity are stated as consistent with antibacterial properties, i.e., oxidative stress via ROS production, cell wall and membrane damage due to direct contact with NPs, cellular uptake of NPs and subsequent damaging interactions with cell interior and metal ion release (Aljabali et al., 2018; Sousa et al., 2020). Gupta et al., 2014 reported antifungal activity of biosynthesized AgNPs against *Candida albicans*, *Trichophyton mentagrophytes*, and *Microsporum canis*, both alone and in synergism with antibiotics (Gupta et al., 2014). AgClNPs proved antifungal activity against *C. albicans*, *Aspergillus flavus*, and *Aspergillus niger* (Paulkumar et al., 2013). Moreover, studies have also shown the antifungal activity of green AuNPs synthesized via both plant and microbial synthesis. This inhibition of fungi by biosynthesized AuNPs has already been reported against *Puccinia graminis tritci*, *Microphyton gypseum*, *A. flavus*, *A. niger*, *Tricho phyton rubrum* and *C. albicans* (Gopal et al., 2013). The antimicrobial action of these nanomaterials is attributed to several mechanisms, primarily involving their unique physicochemical properties.

3.2 Interaction of Nanomaterials with Plant Pathogens

Modern agriculture benefits from the usage of modern technology, such as nanoparticles. Their promise is mostly in pest control and focused nutrition delivery (Salem and Fouda, 2021). Sustainable agriculture aims not only to control pests but also to use resources efficiently and combat the decline of water and soil fertility. Nanoparticles (NPs) show potential in this area, particularly in developing strategies for the gradual release of fertilizers through techniques like nanocoating (Duhan et al., 2017).

The antibacterial mechanism of action of NPs is often characterized as following one of three models: Oxidative stress induction (Gurunathan et al., 2012), non-oxidative mechanisms (/Leung et al., 2014), or metal ion release (Nagy et al., 2011). These three mechanisms can occur concurrently. Similarly, current research confirms that the primary mechanisms behind nanoparticles' antibacterial effects are: damaging the bacterial cell membrane, generating ROS, penetrating the cell membrane, and triggering intracellular antibacterial effects, such as interacting with DNA and proteins (Wang et al., 2017).

3.2.1 Adhesion and Penetration

Bacterial cell walls and membranes serve as crucial protective barriers against external threats, the cell wall is particularly important for maintaining a bacterium's shape. The composition of cell membranes creates distinct adsorption pathways for nanoparticles (NPs) in Gram-positive and Gram-negative bacteria (Lesniak et al., 2013). Extensive research has shown that nanoparticles exhibit antibacterial effects on both Gram-positive and Gram-negative bacteria, starting from gram-negative bacteria, they have a thin peptidoglycan layer (about 7–8 nm thick) in their cell wall, with a negatively charged surface (Halder et al., 2015; Sirelkhatim et al., 2015). These characteristics are directly related to the antibacterial activity of NPs. The thin cell wall allows for better penetration into bacterial cells, while the negatively charged surface promotes strong electrostatic interactions between cells and NPs. This leads to the formation of Reactive Oxygen Species (ROS) and oxidative stress, resulting in bacterial cell destruction and inhibition (Mai-Prochnow et al., 2016; Pramanik et al., 2012). In contrast to Gram-negative bacteria, Gram-positive bacteria only have an outer peptidoglycan layer without the polysaccharide membrane containing structural lipopolysaccharides. This structure allows for better permeability to various substances (Cabeen and Jacobs-Wagner, 2005). For Gram-positive bacteria, the lower negative charge on their surface enhances NP penetration. This allows negatively charged superoxide radical anions and peroxide ions to enter the cell more easily, leading to cell destruction at relatively low concentrations (Padmavathy and Vijayaraghavan, 2008). Regarding fungi membrane, various metal oxide nanoparticles have shown potential effects against different fungi. Studies have demonstrated that zinc oxide NPs (12–32 nm) can disrupt membrane structure and alter permeability in fungi like *Aspergillus niger* and *Botrytis cinerea* at concentrations between 16–128 μg/ml (Jamdagni et al., 2018). Overall, NPs can enter cells through various mechanisms, including cell wall depolarization, irregular pit formation leading to cell wall degradation, receptor-mediated endocytosis, and diffusion through membrane pores (Slavin et al., 2017).

Wehling et al. studied the antibacterial activity of nanodiamonds with various surface structures and reactive groups. They found that these nanomaterials can form covalent bonds with proteins and molecules on cell walls. This interaction with intracellular components can inhibit key enzymes and proteins, disrupting bacterial metabolism and ultimately leading to cell death (Wehling et al., 2014). Research suggests that silver nanoparticles (Ag NPs) can neutralize the electrical charge on bacterial membrane surfaces, altering their permeability and ultimately causing bacterial cell death (Jung et al., 2008). The production of reactive oxygen species (ROS) suppresses the cell's antioxidant defenses and causes physical damage to the cell membrane.

Foster et al. demonstrated that titanium dioxide nanoparticles (TiO_2 NPs) can attach to bacterial cell surfaces, generating reactive oxygen species (ROS) that damage the cell membrane's structure and composition. This interference with membrane function leads to the leakage of cellular contents, ultimately causing bacterial death (Foster et al., 2011).These findings were confirmed by Scanning Electron Microscopy (SEM) in studies of Joost et al. (2015) that revealed that exposure to TiO_2 NPs causes bacterial cells to increase in volume, creates honeycomb-like changes in the cell membrane, and results in cytoplasmic leakage (Joost et al., 2015). Also, after treatment with zinc oxide nanoparticles (n-ZnO), bacterial cell surfaces develop numerous pits, and cell debris becomes visible. This leads to decreased enzymatic activity and eventually results in bacterial death (Padmavathy and Vijayaraghavan, 2011).

In the diffusion process, nanoparticles introduce reactive oxygen species (ROS) into microorganisms. For instance, graphene oxide-iron oxide NPs generate hydroxyl radicals that diffuse into bacterial cells, effectively inactivating MRSA (Pan et al., 2016). The diffusion process is efficient due to a diffusion coefficient similar to oxygen (~ 105 cm^2/s) and ROS lifetime (105–106 s) exceeding the diffusion time (~ 107 s) into bacterial cells (Pan et al., 2016). Hydrogen peroxide and hydroxyl radicals penetrate cell membranes more easily than superoxide radicals. Additionally, silver NPs smaller than 10 nm can pass through cell membrane pores, enhancing their antimicrobial activity (Mukha et al., 2013).

The adsorption mechanism involves metal ions from NPs binding to cell membrane components and metal ions bind to negatively charged functional groups like carboxyl and phosphate on the cell membrane, with different metal ions having specific binding affinities. For example, zinc ions bind to protein -SH groups. The adsorption process disrupts cell membrane function, leading to bacterial death (Padmavathy and Vijayaraghavan, 2011). Also, silver ion NPs (Ag+NPs) adsorb firmly to cell membranes due to Coulomb forces, causing protein coagulation. The surface charge of NPs, such as gold NPs, significantly affects their adsorption to membranes, with the electrical properties of the bilayer playing a crucial role (Lin et al., 2010).

3.2.2 Reactive Oxygen Species (ROS) Generation

Nanoparticles (NPs) employ a crucial antibacterial strategy through ROS-induced oxidative stress, ROS, an umbrella term for highly reactive molecules with strong oxidizing capabilities, are produced in various forms by different NPs through oxygen reduction. These forms include superoxide radical, hydroxyl radical, hydrogen peroxide, and singlet oxygen, each with distinct reactivity and dynamics. Different NPs generate specific ROS types. For instance, calcium and magnesium oxide NPs produce superoxide radicals, while zinc oxide NPs create hydrogen peroxide and hydroxyl radicals (Wang et al., 2017). Copper oxide NPs are unique in their ability to generate all four ROS types. Research shows that superoxide radicals and hydrogen peroxide induce milder stress responses and can be counteracted by the bacteria's natural antioxidants (Metryka et al., 2022). In contrast, hydroxyl radicals and singlet oxygen can rapidly kill microbes (Sharmin et al., 2021). The primary sources of ROS generation in NPs are crystal restructuring, defects, and oxygen vacancies. Typically, bacterial cells maintain a balance between ROS production and elimination. However, when ROS production becomes excessive, it shifts the cell's redox balance towards oxidation. This imbalance results in oxidative stress, leading to damage of various bacterial cell components (Li et al., 2012; Wang et al., 2017). Oxidative stress has been established as a crucial contributor to increasing the permeability of the cell membrane, which can lead to bacterial cell membrane damage (Cheloni et al., 2016). Ansari's study demonstrated that Al2O3 nanoparticles penetrated the cell membrane, becoming internalized. This interaction between the nanoparticles and the cell membrane ultimately compromised membrane integrity, likely due to oxidative stress occurring within the cell (Ansari et al., 2015). Additionally, reactive oxygen species (ROS) play a beneficial role in enhancing the expression of oxidative stress-related genes, which is a crucial mechanism in triggering bacterial cell death (Wu et al., 2011).

Nanoparticles composed of iron oxide generate ROS, which can cause oxidative stress-induced damage to proteins and DNA within fungal cells (Parveen et al., 2018). According to Haghighi et al., (2013), titanium oxide (TiO_2) at a concentration of 5.14 μg/mL inhibits *Candida albicans* biofilm formation. The antifungal effect may be attributed to oxidative stress caused by ROS generation, potentially leading to cell destruction (Haghighi et al., 2013).

3.2.3 Metal Ion Release

Metal ions released from metal oxide NPs can penetrate cell membranes and interact with functional groups of proteins and nucleic acids, such as mercapto, amino, and carboxyl groups. This interaction can damage enzyme activity, alter cell structure, disrupt normal physiological processes, and ultimately inhibit microbial growth (Wang et al., 2017). Research has shown that superparamagnetic iron oxide can directly penetrate cell membranes and interfere with transmembrane electron transfer. Additionally, heavy metal ions can indirectly act as carriers for antimicrobial substances (Hussein-Al-Ali et al., 2014). However, the effect of metal ions on intracellular pH in lipid vesicles is minimal during the antimicrobial action of metal oxide suspensions, suggesting that dissolved metal ions are not the primary mechanism of metal oxide NPs' antimicrobial activity (Yu et al., 2014).

While the interactions between metal NPs and microorganisms vary depending on the specific metal and organism, most metal NPs employ multiple mechanisms of action. These typically include oxidative stress through ROS production, metal ion release, and various non-oxidative mechanisms such as direct interactions with cell structures, intracellular penetration, and disruption of biofilm formation (Gurunathan et al., 2014), these mechanisms often rely on direct contact between NPs and bacterial cells, facilitated by electrostatic interactions, van der Waals forces, and hydrophobic interactions, thus countering an important bacterial resistance mechanism (El-Seedi et al., 2019; Slavin et al., 2017). The ability of NPs to penetrate cell walls and membranes, as well as their interactions with intracellular components like proteins, nucleic acids, and organelles, is largely dependent on their size (Mukha et al., 2013). These interactions can result in oxidative stress, enzyme inactivation, altered gene expression, changes in membrane permeability, and electrolyte imbalance (Wang et al., 2017). ROS-induced oxidative stress is one of the most studied antibacterial mechanisms of NPs. Different metal NPs produce various ROS compounds, resulting in a wide range of specific antibacterial activities (Wang et al., 2017). Excessive ROS production leads to oxidative stress in bacterial cells, causing loss of membrane integrity, DNA and enzyme damage, and ultimately triggering cell death (Cheloni et al., 2016).

3.3 Mode of Action at the Molecular Level

The antimicrobial properties on NPs can be harnessed to combat diseases caused by bacteria and fungi in plants. For instance, recent research has shed light on the impact of nanoparticles (NPs) on bacterial cells, particularly in the areas of protein synthesis, gene expression, and metabolism. Various types of NPs, including copper oxide, silver, gold, and titanium dioxide, have been shown to interfere with bacterial protein synthesis and alter gene expression (Casciaro et al., 2017; Dadi et al., 2019). For instance, CuO NPs affect proteins involved in nitrogen metabolism and electron transfer, while Ag NPs influence the expression of ribosomal subunit proteins and other enzymes (Dakal et al., 2016). Overall, the effects of NPs on bacterial DNA and gene regulation are equally significant, another examples of Au NPs that have been found to prevent the combination of ribosomal subunits with tRNA and disrupt membrane potential, leading to reduced ATP levels, this interference likely occurs due to the nanoparticles' ability to bind to and interact with cellular proteins and also nucleic acids . The disruption of ribosomal assembly can significantly impair the cell's ability to synthesize new proteins, which is crucial for various cellular functions (Enea et al., 2020). On the other hand, TiO2 NPs cause DNA compression, degeneration, and fragmentation, thereby reducing gene activity (Jayaram et al., 2019). Whole-genome analysis has confirmed that NPs can induce mutations in various areas of the bacterial genome, affecting gene expression, molecular structure activity, and RNA modification (Ewunkem et al., 2021). Ag NPs, in particular, have been shown to upregulate antioxidant genes and genes related to metal transport and reduction (Jadczak et al., 2020). NPs also play a crucial role in regulating bacterial metabolism. For example, CuO NPs regulate proteins involved in nitrogen metabolism and inhibit the activity of nitrate and nitrite reductases (Dizaj et al., 2014). Other molecular mechanism involve TiO_2 to influence bacterial adhesion rates and biofilm formation, which may in turn affect biofilm metabolism (Zhou et al., 2024). Furthermore, NPs can generate reactive oxygen species (ROS), interfering with electron transport and causing damage to macromolecules such as DNA, lipids, and proteins. Overall, the impact of NPs on bacterial cell membranes and their ability to induce oxidative stress are additional mechanisms by which they exert antibacterial effects (Casciaro et al., 2017). Iron-based NPs generate ROS through the Fenton reaction, leading to extensive damage to cellular components (Li et al., 2011). By understanding and harnessing these mechanisms, these diverse effects of NPs on bacterial cells provide potential applications in controlling the growth pathogenic microorganisms and pathogenicity (Wang et al., 2017).

4. Applications in Plant Disease Management

4.1 Nanopesticides

From an environmental protection perspective, the use of low-toxicity, eco-friendly compounds such as biopesticides is highly desirable for crop protection. Indeed, biopesticides are regarded as more beneficial than synthetic organic pesticides in minimizing harmful effects on biodiversity, living organisms, and human health (Popova et al., 2017). They can be classified into two main categories: (i) plant-sourced biopesticides and (ii) biocontrol organisms (Mnif and Ghribi, 2015; Timmusk et al., 2017). Plant-sourced nanoformulations have demonstrated the greatest impact on insects, while showing less effectiveness against fungi (Sparks et al., 2017). However, microbial-based formulations have become increasingly popular for addressing fungal and insect issues (Shi and Bode, 2018; Trivedi et al., 2017).

The U.S. Environmental Protection Agency (EPA) has registered various biopesticides, including herbicides like thaxtomin A (derived from Streptomyces spp.), fungicides such as glycerol monocarpate (sourced from animal or vegetable fat), and insecticides like citronellol (from plant essential oil) and 4-allylanisol (from basil oil) (Seiber et al., 2014). Research by Huang et al. has highlighted the effectiveness of rhizosphere bacteria against *Phytophthora nicotianae*, *Meloidogyne incognita*, and root knot black shank in tobacco crops. These biopesticides exhibit desirable characteristics such as target specificity, low toxicity to non-target organisms, and minimal environmental persistence (Huang et al., 2015). For example, the long-term acaricidal effect of essential oils from *Achillea millefolium* L., when encapsulated with chitosan nanocapsules, has been reported against adult *Tetranychus urticae* Koch due to the slow and sustained release of the essential oils used (Ahmadi et al., 2018).

Also, PONNEEM®-encapsulated tripolyphosphate cross-linked chitosan nanocarriers have shown antifeedant activity (~ 88%) and larvicidal activity (~ 90%) against *Helicoverpa armigera* (Paulraj et al., 2017). Additionally, Hedaoo et al. reported the use of renewable sources like cardanol formaldehyde to create polymeric nanocapsules for encapsulating karanja oil. The use of renewable polymeric shells provides good thermal stability to ensure the long-term preservation of core biopesticides (Hedaoo and Gite, 2014). Moreover, essential oils have been extensively studied for their herbicidal activity against weed species (Ghnaya et al., 2016; Rolli et al., 2014; Synowiec et al., 2017). These natural compounds offer several advantages over chemical herbicides, including effective herbicidal action, eco-friendliness, attractiveness to pollinators, and repellent properties for other insects (Fagodia et al., 2017; Raguso, 2016). Hazrati et al. reported a formulated nanoemulsion of *Satureja hortensis* essential oil created using a low-energy method. Carvacrol, a potent inhibitory component of *Satureja hortensis*, facilitated natural integrative control over weed species such as *Chenopodium album* and *Amaranthus retroflexus* (Hazrati et al., 2017).

The nanoemulsion of essential oils disrupts weed species in terms of germination, physiological processes, and growth. However, soil characteristics can influence the phytotoxicity and bioherbicidal activity of essential oils. For example, lemon essential oil did not demonstrate any herbicidal activity against seed germination of *Portulaca oleracea* in both soil-less and soil cultures; conversely, *Peumus boldus* essential oil (at a lower concentration of 0.125 μL/mL) exhibited significant herbicidal activity against *P. oleracea* in both sand and clay textures (Blázquez and Carbó, 2015).

Various studies have also investigated the potential of essential oils for pest control (Campolo et al., 2017; González et al., 2014; Sousa et al., 2015). Ribeiro et al. examined the effects of essential oils on the feeding behavior and oviposition of *Anticarsia gemmatalis* Hubner. Essential oils from garlic and cinnamon mint thyme provided antifeeding effects with deterrence rates of 80 and 50%, respectively, when tested on *A. gemmatalis* (Ribeiro et al., 2015). The nanocapsulation of essential oils like carvacrol and linalool has proven more effective regarding their acaricidal and oviposition activities against pests,compared to the repellent effects of unencapsulated compounds (Campos et al., 2018). However, combining biocontrol agents for managing complex diseases can be somewhat complicated due to production costs and application convenience (Liu et al., 2014). Environmental factors, including soil properties and climate conditions, can influence the stability of biopesticides, potentially leading to premature degradation. Consequently, significant research efforts have been directed toward developing different nanocarriers for the controlled targeted delivery of biopesticides (de Oliveira et al., 2014; Mattos, Tardy, et al., 2017; Song et al., 2014). But despite their high efficacy as pesticides, the commercialization of essential oils remains limited. Challenges include variations in field conditions (such as pH, UV radiation, and humidity) that need to be addressed alongside inherent physicochemical characteristics related to rainfastness and permeability. Other factors that hinder commercialization include stringent regulations, inadequate quality and quantity of materials at affordable prices, and low persistence of pesticidal effects (Pavela and Benelli, 2016). The persistence of toxins and spores derived from biopesticides in the environment remains a significant concern since spores may germinate in soil under favorable conditions (Kumar et al., 2019).

4.1.1 Formulation and Effectiveness

In agriculture, nanotechnology applications include reducing biotic stress through nanomaterial-based pesticides and encapsulated agrochemicals, enhancing plant growth and crop yield via slow-release fertilizers or water delivery systems, and developing biosensors (Hassanisaadi et al., 2022).

Nanoformulations of pesticides must possess a diverse set of characteristics, including enhanced potency and longevity, improved wettability and dispersibility, biodegradability with minimal active ingredients, and non-toxicity while maintaining effective pest control (Mariello et al., 2022).

Nanoparticles are synthesized utilizing a variety of biological, chemical, and physical approaches (Gopinath and Velusamy, 2013) but the most known Np created through traditional chemical methods have been largely overlooked in clinical and food applications due to their toxicity stemming from the chemicals used in production. In contrast, biogenically produced nanoformulations have gained prominence as they offer cost-effective and environmentally friendly alternatives that eliminate chemical toxicity (Wei et al., 2012). The most popular methods for producing nanoparticles are likely to deliver greater qualities to the enclosed substance, be cost-effective, and have little danger of toxicity (Ghiuță et al., 2018). Encapsulation is a key nanotechnology technique. The advantages of nanostructures have led to increased use of nanoencapsulation for delivering bioactive compounds in recent years (Rehman et al., 2019). As a definition, the nano/microencapsulation of biocontrol agents, both biological and chemical, is basically a process that involves encapsulating biological agents at nano or micro scales, significantly improving their effectiveness and stability. Encapsulation is essentially the process of embedding an active core through physical means and then stabilizing it using chemical methods (Mariello et al., 2022). Another crucial feature of this process is the controlled release of the formulated microbes or biomolecule (Riseh et al., 2021). Several biopolymers have been utilized to encapsulate BCAs, attracting significant scientific attention as viable and developing technologies (Baranwal et al., 2022). In this regard, different biopolymers are now exploited for encapsulating BCAs, notably chitosan, xanthan gum, alginate, gellan gum, gelatin (Klein and Poverenov, 2020).

The viability of encapsulated cells is influenced by factors such as concentration, coating type, capsule size, and initial cell count. These capsules may contain single or multiple substances, specifically designed to control the release of nuclear material at the optimal moment (Riseh et al., 2021).

Various technologies are employed for material encapsulation, including emulsion, freeze-drying, extrusion coating, spray cooling, fluidized bed coating, coacervation, spray-drying, and heat gelling. The selection of an appropriate encapsulation method depends on several factors, such as the physicochemical properties of the encapsulant and core materials, as well as their intended application. Emulsion, spray-drying, and extrusion processes have proven particularly effective for microencapsulating plant growth-promoting bacteria (Riseh et al., 2021).

Nanoscale chitinases derived from Bt have demonstrated the ability to deliver nematicidal preparations for managing *Caenorhabditis elegans*, potentially serving as a biopesticide delivery system for agricultural use (Qin et al., 2020). Nanotechnology techniques can also be used to reduce the size of entomopathogenic (Bt) bacteria, creating Bt nanoparticles for biocontrol applications (Swamy and Asokan, 2013). Additionally, Bt nanocomposites with Bt bioactives in various nanoformulations such as nanocapsules, nanoemulsions, and nanosuspensions have been employed as nanopesticides (Damalas and Koutroubas, 2018).

Another research study focused on nano-crystallization of cyclic lipopeptides extracted from *B. subtilis*, which showed an enhancement of storage stability by inhibiting oxidation and preserving anti-fungal properties through controlled release. These cyclic lipopeptide nanocrystals (solid lipid nanoparticles) effectively inhibit the growth of *Aspergillus carbonarius* and *A. niger* spores and hyphae (Lin et al., 2020). Researchers are actively pursuing the development of less toxic, highly target-specific nanopesticides that maintain high efficacy, given the rapid advancements in nanopesticide technology. Consequently, target-specific nanopesticides are expected to reduce harm to non-target crops and minimize pesticide release into the environment (Kumar et al., 2019).

4.1.2 Targeted Delivery Systems

Targeted delivery systems in nanopesticides utilize nanocarriers to ensure that the active ingredients are delivered precisely to the intended site of action, reducing exposure to non-target organisms

(Deka et al., 2021). This approach not only enhances the effectiveness of pest control but also mitigates the risks associated with pesticide use, including environmental contamination and harm to beneficial species (Athanassiou et al., 2013).

The concept of "Pesticide Delivery System" (PDS) draws inspiration from medical drug delivery methods, utilizing nanoparticles (NPs) to transport Active Ingredients (AI) to specific targets (Tsuji, 2001). PDS aims to optimize pesticide efficacy while minimizing harm to non-target organisms by controlling the concentration and duration of AI exposure (Ghormade et al., 2011). In this context, controlled distribution is critical for optimizing the release of appropriate and essential amounts of insecticides within a given time frame (Tsuji, 2001).

NPs are ideal nanocarriers due to their high loading capacity, large surface area, rapid transfer to insect bodies, and versatility in attaching various pesticide molecules. Moreover, these systems can be designed to release their payload in response to specific environmental triggers, such as pH or temperature changes, allowing for a more controlled and efficient application of pesticides (Athanassiou et al., 2018).

Various techniques, such as encapsulation, adsorption, entrapment, and covalent attachment, are used to load pesticides onto NPs (Athanassiou et al., 2018). The release characteristics depend on the AI-material bonding, nanocarrier degradation properties, and environmental conditions. Common nanocarriers include synthetic silica, polymers, titania, silver, alumina, and copper (Deka et al., 2021).

Importantly, encapsulated pesticides release more gradually, reducing application frequency and maintaining effectiveness despite degradation (Deka et al., 2021). This nanoencapsulation approach promotes safer pesticide use and reduced environmental exposure (Athanassiou et al., 2018).

Among available NPs, solid and mesoporous silica NPs show the most promise for agrochemical delivery due to their structural flexibility and ability to form pores for biomolecule loading (Athanassiou et al., 2013; Campbell et al., 2011; Jang et al., 2013; Slomberg and Schoenfisch, 2012; Wanyika et al., 2012). These characteristics make them particularly suitable for targeted delivery systems in nano-pesticides.

4.2 Nanofertilizers

4.2.1 Enhanced Nutrient Uptake

A significant portion of inorganic fertilizers applied to soil is not utilized by plants due to loss. For example, 40–70% of Nitrogen (N), 80–90% of Phosphorous (P), and 50–90% of potassium (K) fertilizers are lost or fixed in soils, leading to financial losses and potential soil nutrient imbalances as additional fertilizers are applied to compensate (El-Saadony et al., 2021). In this regard, slow-release nanofertilizers provide a promising solution to this problem, as their gradual dispersion throughout crop production can minimize nutrient loss, reduce soil degradation, and enhance soil fertility and crop productivity in the long term (Garg et al., 2023). This allows plants to absorb most of the nutrients they need without experiencing significant losses (Guo et al., 2018).

Engineered nanoparticle-based nanofertilizers, both biological and chemical, provide substantial benefits in agriculture by improving crop nutrition management, enhancing abiotic stress tolerance, and increasing agricultural productivity. These nanofertilizers contain essential nutrients like N, P, K, Fe, and Mn in a nanoparticle form, which can be delivered efficiently to plants (Nongbet et al., 2022). The nanomaterials on the surface of fertilizer particles also contribute to better control of nutrient release due to their higher surface tension compared to conventional fertilizers (El-Saadony et al., 2021). The advantages of nanofertilizers over traditional chemical fertilizers stem from their large surface area to volume ratio and superior penetration capabilities (Garg et al., 2023). They can stimulate various plant processes, including nitrogen metabolism, seed germination, photosynthesis, protein synthesis, and stress tolerance (Verma et al., 2023).

Common materials used in nanofertilizer production include zeolites, various metals and non-metals, and even plant-derived elements such as banana peels (Hussain et al., 2022). Natural zeolite, composed of over 50 minerals, has gained popularity in nanofertilizer production due to its low cost and wide availability (Eroglu et al., 2017). Many micronutrients have been produced at the nanoscale for plant development regulation (El-Saadony et al., 2021). Research by Heidari et al. (2018) demonstrated that topical application of nano and *Azospirillum brasilens* enhanced maize plant growth and yield (Heidari et al., 2018). Another study by Mokarram-Kashtiban et al. (2019) found that combining Plant Growth-Promoting Rhizobacteria (PGPR) or arbuscular mycorrhizal with small amounts of Fe-NPs significantly improved heavy metal phytoremediation and plant growth. Additionally, bioagents and nano Zn-Fe oxide were shown to enhance seed growth, photosynthesis, and osmolyte content in salt-stressed wheat plant (Mokarram-Kashtiban et al., 2019).

4.2.2 Disease Resistance

Nanofertilizers provide dual benefits by enhancing nutrient uptake and potentially boosting disease resistance in plants, they may activate plant defense mechanisms or improve overall plant health, making crops more resilient to pathogens and environmental stresses (Dutta et al., 2022). In general, it offers a feasible option for environmentally responsible farming practices, while chemical pesticides have proven highly effective and convenient in managing numerous plant diseases in recent times, their increased usage has led to ecological disruptions. These non-targeted effects pose significant threats to beneficial microorganisms and human well-being (Deveci et al., 2021; Tang et al., 2021).

Over the past 20 years, various bacterial species, including *Bacillus*, *Pseudomonas*, *Corynebacterium*, and *Shewanella*, have been used to create nanoparticles from inorganic metals such as silver, gold, aluminum, zinc oxide, manganese oxide, and titanium dioxide. The application of these bacterially synthesized nanoparticles has led to improved agricultural yields and reduced losses during crop handling, transportation, and storage (Kumari et al., 2019). For instance, a *Bacillus* species isolated from marine soil was used to produce silver nanoparticles that suppressed *Fusarium oxysporum*, a pathogen causing wilt disease in various crops. Other *Bacillus* strains have been used to create nanoparticles with antimicrobial and catalytic properties (Gopinath and Velusamy, 2013). Additionally, AgNPs generated extracellularly by *Bacillus* strain SBT8 display antibacterial action and operate as a biocatalyst (Yurtluk et al., 2018). NPs produced using silver nitrate-containing compounds utilizing *B. amyloliquefaciens* and *B. subtilis* displayed antibacterial action on gram-positive and gram-negative bacteria, as well as antifungal activity on *Candida albicans* (Ghiuță et al., 2018).

Overall, various nanoparticles, including non-metals, metalloids, oxide metals, and nanocarbon materials, have demonstrated suppressive activity against plant pathogens (Elmer et al., 2018), silver, titanium oxide, copper, gold, zinc oxide, and iron nanoparticles have been extensively tested (Iravani et al., 2014; Kandi and Kandi, 2015). They have been tested against various harmful microorganism, particularly *Sclerotinia sclerotiorum*, *Rhizoctonia solani*, *Sclerotium rolfsii*, *Fusarium oxysporum* (Abdelrhim et al., 2021; Kaman and Dutta, 2019), *Xanthomonas oryzae pv. Oryzae* (Abdallah et al., 2019; Ogunyemi et al., 2020), *Colletotrichum* sp. (Goswami et al., 2020), *Sarocladium oryzae* (Das and Dutta, 2021) and viruses such as Bean Yellow Mosaic Virus (BYMV) (Elbeshehy et al., 2015), Cucumber Mosaic Virus (CMV) (El-Sawy et al., 2017), potato virus Y (PVY) (Noha et al., 2018). In a technical manner, nanomaterials interact with living organisms through various mechanisms. They can disrupt cell membranes, form clusters, and inhibit the function of H+-ATPase enzymes. These materials also interfere with the production of messenger RNA and proteins, impede the synthesis of toxins, and prevent microbial growth. Additionally, nanomaterials can obstruct the flow of nutrients essential for cellular processes (Dakal et al., 2016; Malerba and Cerana, 2016).

5. Environmental Impact and Safety Concerns

5.1-Ecotoxicology of Nanomaterials

Despite the various advantages resulting from the application of nanoparticles in agriculture, they can lead to some ecotoxicological effects on plants, soil health and its microbiota, aquatic environment, as well as to impacts on human health.

Regarding green nanoparticles, it is important to note that few studies are existing in this context. The published works are concerned mainly the effects of green nanoparticles on plants and aquatic environments. However, effects on human health and soil beneficial microorganisms can also be expected. Regarding plants, they adsorb nanoparticles either through roots from where they are then transferred to the other organells (Tripathi et al., 2017), or through leaves where stomatal pores facilitate their penetration (Larue et al., 2014). Highlighting examples of plants toxicity caused by green nanoparticles, it was found that green ZnO nanoparticles including the *Vernonia cinerea* leaves extract have exhibited at a concentration higher than 50 mg/L, a phytotoxicity effect revealed by growth obstructions and development (Azim et al., 2022). Moreover, green Ag nanoparticles containing *Camelina sativa* L. extracts were demonstrated to cause an increase of Ag accumulation in shoots and roots of *Camelina*, resulting in a decrease of the plant photosynthetic pigments and growth (Mirmoeini et al., 2021). Besides, the use of these nanoparticles at a concentration higher than 2 g/L has led to accumulation of hydrogen peroxide, methyl glyoxal, proline, and an up-regulation of antioxidant enzymes, including superoxide dismutase, gluthatione reductase and ascorbate peroxidase, reflecting the elevation of oxidative stress (Mirmoeini et al., 2021). Furthermore, Ag nanoparticles synthesized introducing *Rosmarinus officinalis* extracts had exhibited a remarkable stress impact on wheat and tomato, through inducing production of antioxidant enzymes, and reducing chlorophyll and dry weigh (Farghaly and Nafady, 2015). In another study, green Ag nanoparticles based on *Annona squamosa* extracts displayed a toxic effect on *Phaseolus vulgaris* L when applied at high concentrations of 480 mg/L although lower concentrations had a positive effect on root elongation (Verma et al., 2020). Thus, the present works show that green nanoparticles may have a toxicity to plants, especially when applied at high concentrations.

Concerning the side effects of green nanoparticles in aquatic environments, where they can be present through runoff from agricultural fields, leaching into groundwater, or during rainfall events that wash pesticides away. They may also be carried by wind or irrigation practices. Among studies conducted in this regard, Velidandi et al. (2021) have shown that green AgCU and AgZn nanoparticles fabricated with *Annona muricata* extract have induced toxicity to *Artemia nauplii* depending on the contact period, with higher toxicity caused by AgZn nanoparticles. Indeed, after 24 h and 36 h, AgZn and AgCU induced 50% and 30%, 100% and 60% of mortality, respectively. Moreover, Saif et al., (2016) found that green CuO nanoparticles synthesized with *Pterospermum acerifolium* leaves extract, had a toxic effect to water flea *Daphnia magna*, with a median effective concentration (EC50) of 0.69 ± 0.226 mg/L. However, their toxicity was less than engineered CuO nanoparticles (EC50 = 0.102 ± 0.019 mg/L). Besides, TiO_2 nanoparticles including g *Echinophora cinerea* extract have shown a toxicity to Zebra fish at a lethal concentration (LC50) of 300 mg/L (Jafari et al., 2022). Similarly, TiO2 containing *Sesbania grandiflora* had exhibited a lethal effect to zebrafish (Srinivasan et al., 2019). ZnO nanoparticles based on papaya extract have also showed a high rate of mortality in zebrafish, especially at a high concentration of 100 mg/l for nanoparticles including papaya aqueous extract, and 20 mg/l for nanoparticles based on this plant species methanolic extract (Zavitri et al., 2023). Furthermore, *Cissus quadrangularis* Ag *nanoparticles* have induced high mortality even at low concentrations to *Ceriodaphnia cornuta* Crustaceans and *Poecilia reticulata* Fishes. Indeed, in both species, the studied nanoparticles have caused DNA damages and lesions in their structures and then their death. These studies show that green nanoparticles may have a negative effect on the environment, thus a risk assessement of green nanoparticles is crucial before their use in order to diminish their ecotoxicity.

5.2 Risk Assessment and Management

Several actions can be considered for nanoparticles risk assessement. Among the crucial elements we could pay attention to the chemical composition of the whole nanoformulation, and not just the active ingredients, since nanoparticles can be formulated by an emulsion of biological agents associated with nano-sized carrier. Moreover, it is also important to determine if the active ingredient release is slow and persistent in the environment. In the case of persistent nanoparticles, it is also essential to assess their accumulation in the food chain and in the environment (Grillo et al., 2021). Moreover, thermal and chemical stability of biogenic nanomaterials during a long time period is a crucial parameter to be considered before their commercialization. In addition, nanomaterials aggregation has to be evaluated, since it constitutes a lot of harm to cells through interactions affecting their surface properties. Besides, the investigation of the impact of these nanomaterials on plants is required, by analyzing their uptake and accumulation in plants as well as their long-term viability and impact. For this purpose, metabolomic studies, including genomics, metabolomics, transcriptomics, and proteomics, constitute a suitable tool (Omran and Baek, 2022). Further aspects are also to be taken in to account for nanomaterials risk assessement; including the surface area of nanoparticles, concentration, morphology, bioactivity, possible toxicity, and their interactions with other chemicals existing in the ecosystem which may lead to additionnal adverse effects on the surronding environments, and thereafter on human health (Mitra et al., 2023). In this respect, nanoparticles risk assessement on human health has to rely firstly on the determination of exposure routes, which can take place via drinking water including traces of nanoparticles or during nanomaterials application. Indeed, during the latter, nanoparticles may be introduced through inhalation, whereas after their application, they can be consumed orally through ingestion of contaminated water or crops. Thus, an assessement of nanoparticles persistence in contaminated areas should be carried out (Kah et al., 2021).

Regarding risk assessement management principles, it was claimed that risk prevention, risk reduction, and risk communication constitute the crucial elements. In fact, risk prevention and risk mitigation need primarily the identification of the suspected risks, then the next step of avoiding risk exposure through the use of less toxic and biodegradable nanomaterials. Also, the communication of risks through warning people of nanomaterials exposure routes, and prevention measures; will also help to a better and efficient management of nanoparticles probable effects (Okeke et al., 2022).

6. Regulatory Aspects

The regulations and guidelines for green nanoparticles in plant disease management are still evolving. However, different points are being addressed by several regulatory bodies, among which we find the European Union regulations. The European Union has strict directions concerning nanoparticles, by providing the possible best frameworks for nanoparticles risk management under the four components of REACH (Fig. 2).

Registration refers to the process of registration file submission of industries to the European agency, these files contain details on nanoparticles properties, uses and potential risks of the included substances. Evaluation, representing the process through which the European member states assess the information submitted in registration files. Authorization, reflecting the process designed to ensure that the substances are properly controlled. Restriction of chemicals, revealing the process of prohibiting the manufacture and sale of products with specific harm to humans or the environment (Kuraj, 2019). Moreover, in the US, the environmental protection agency regulates pesticides, including those utilizing nanotechnology. Indeed, all products intended for plants disease

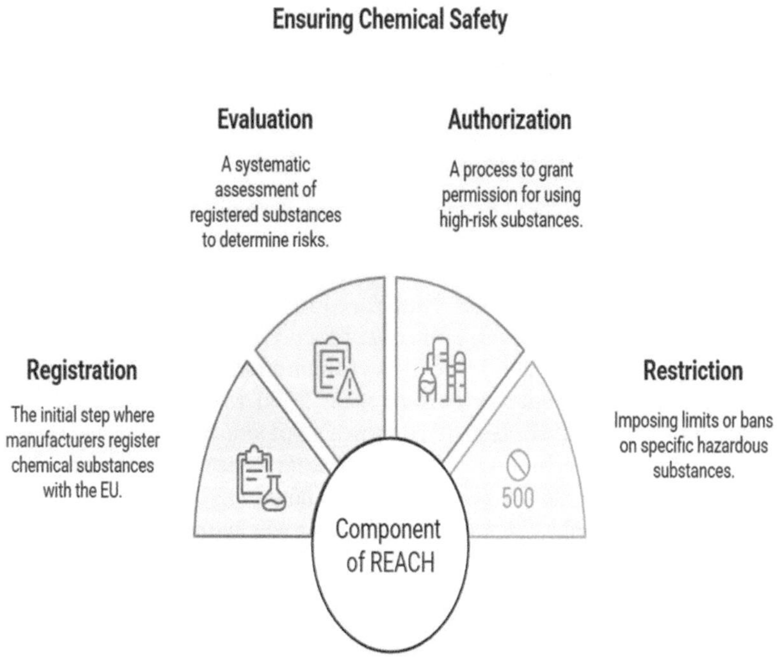

Figure 2. Components of REACH according to the European Union regulations.

management should be registered and assessed for their safety and efficiency (Lasat et al., 2018). Moreover, food and agriculture organization also provide guidelines on the use of nanotechnology on food safety which can be threatened by nanoparticles application in agricultural practices (Kumari et al., 2023). Although these current regulations and guidelines are established and governs nanotechnology application in agriculture. Future directions in policy and regulations are intended, for a better management of nanoparticles use in plants protection, and for ensuring their compliance with environmental and human health safety issues.

6.1 Current Regulations and Guidelines

6.2 Future Directions in Policy and Regulation

Future directions in policy and regulations regarding green nanomaterials use in plants disease management are likely to focus on different key areas, such as the development of standardized protocols for assessing environmental impacts of nanoparticles on different organisms including algae, terrestrial plants, and nematodes, not only on their toxicity, but also on their morphological, genetic and behavioral characteristics (Boros and Ostafe, 2020). Long term studies are also important and should be considered, since nanoparticles may not present short-term effects. However, their long-term consequences on the ecosystem can be revealed (Grün et al., 2018). Moreover, harmonized regulations constitute a good way for facilitating the global trade and enabling safety evaluations consistency (Isibor, 2024). Besides, inter-agency cooperation and involvement of expert groups, allow fostering agri-food sustainability. Likewise, public engagement is essential, and can be achieved through increasing their acceptance of green nanoparticles. This responsibility concerns scientific communities and industries for sharing knowledge, and enhancing transparency, to contribute to a strong confidence, ensuring the public that their interests and safety are the priority (Bartolucci et al., 2020). Furthermore, certification programs for products introducing green nanomaterials are also intended for promoting transparency and sustainability (Aithal and Aithal, 2021). In addition, life cycle assessments are mandatory for the evaluation of the environmental impacts of nanoparticles, from production to disposal (Pucciarelli, 2023). Finally, adaptive regulations are primordial for creating flexible regulatory frameworks able to evolve with the advancement of nanoparticles (Feitshans and Sabatier, 2022).

7. Case Studies and Practical Applications

7.1 Successful Implementations in Various Crops

The use of green nanoparticles synthesized through eco-friendly methods and involving natural materials, such as microorganisms or plant extracts, has been revealed to be an efficient tool against plants diseases caused by different pathogenic agents. In this context, different examples can be drawn. For instance, copper nanoparticles synthesized in aqueous media have exhibited a strong antifungal effect against *Fusarium oxysporum*, *Neofusicoccum* sp., and *Fusarium oxysporum*, by causing cell membranes damages, and inducing mycelium morphological changes associated with the generation of reactive oxygen species (Pariona et al., 2019). In another study, silver nanoparticles synthesized using the bacteria *Serratia* sp. have shown a potential antifungal activity against wheat pathogenic fungus *Bipolaris sorokiniana.* Indeed, at these nanoparticles' concentrations of 10, 4, and 2 µg /ml, a total inhibition of conidial germination was noted *in vitro* and *in vivo* on wheat leaves. Moreover, wheat disease caused by *Bipolaris sorokiniana* was significantly reduced in green house setting (Mishra et al., 2014). Similarly, Mishra et al. (2017) used microbial agent '*Stenotrophomonas* sp.' for the biosynthesis of silver nanoparticles. The antifungal efficacy of these nanoparticles has been demonstrated against soil born and foliar phytopathogens. In fact, a failure of *Sclerotium rolfsii* germination was observed either in PDA medium or in soil. Besides, in controlled environment, chickpea rot induced by *Sclerotium rolfsii* was successfully reduced. In addition, the mechanism of the antifungal activity of the studied nanoparticles was revealed to be related with the production of H_2O_2, the induction of phenolic acids, and the alteration of lignification.

Furthermore, Ponmurugan et al., (2016) found that green copper nanoparticles produced using *Streptomyces griseus*, displayed an effectiveness against *Panaeolus hypolateritius* infecting tea plants. Moreover, these nanoparticles resulted in the increase of leaves yield and in the improvement of soil macronutrients.

Regarding green nanoparticles synthesized using plants extracts, cerium oxide nanoparticles containing quinoa leaves extract have exerted an important activity against the fungus *Ustilago tritici* affecting wheat crops. After the use of these nanoparticles at a concentartion of 100 mg/L, the disease severity decreased to a rate of 37 and 35% in the two tested varieties of wheat (Alotaibi et al., 2023).

Moreover, silver and copper nanoparticles including *Eucalyptus globulus* leaves extract were found to be efficient against *Xanthomonas citri pv citri* responsible of Citrus Canker. Thus; in the *in vitro* test, a high inhibition zone (26.06 mm) was obtained when combining silver and copper nanoparticles, then when using silver and copper nanoparticles separately (18.26 and 15.27 mm, respectively). As in field trials, the combination of silver and copper nanoparticles has resulted in a low disease severity of 23.23%, followed by silver, and copper nanoparticles, with diseases severities of 28.31 % and 35.23%, respectively (Atiq et al., 2022). Besides, zinc oxide nanoparticles biosynthesized using Chamomile flower, olive leaves, and red tomato fruits extract, have shown a considerable activity against *Xanthomonas oryzae pv oryzae*, by affecting its motility, cell membrane structure, growth and its ability to form biofilm (Ogunyemi et al., 2019).

Overall, the use of green nanoparticles is highlighted to have successful applications aginst plant diseases caused by different pathogenic agents.

7.2 Comparative Analysis with Conventionnel Methods

In order to compare the use of green nanoparticles for plants diseases management, with the conventionnel methods including cultural practices, pesticides application, resistant varieties planting and integrated pest management, the advantages and disadvantages of each method need to be clarified.

Regarding conventionnel methods, cultural practices were widely used for the management of plants diseases, these practices comprise of crop rotation and sanitation. The first method is based on the growth of different crops in the same area in sequential years or seasons, and it presents several advantages, among which we find the elimination of pests, breaks in plants diseases cycles, an increase of soil organic matter, and an enhancement of crops yield (Selim, 2019). This method is associated with some limitations, like the restricted efficiency against soil borne pathogens which are characterized by the production of long living surviving structures mainly; oospores, sclerotia and chlamidospores, or which possess a wide range of hosts (Panth et al., 2020). Besides, crop rotation requires knowledge by farmers and technological support for successful results (Shah et al., 2021), and it is a labor intensive and time consuming (Gao et al., 2022). As for sanitation, it helps preventing the spread of different pathogens through the removal of plants infected debris (Panth et al., 2020). However, this technique could have a limited effect, thus, it should be supported by other measures, like quarantine methods (Blomme et al., 2017).

Concerning pesticides, their application has become very popular and constitutes an important process in crop production, since they allow the reduction of losses caused by pests, and lead therefore to the productivity increase (Fu et al., 2022). Nevertheless, they are related with several disadvantages. Indeed, they can also impact beneficial organisms, and threaten aquatic ecosystems, where they are able to enter through runoff or drainage. Besides, pesticides are also encountered with the problem of targeted organisms' resistance, since the latter have the ability to develop non-sensitive behavior to pesticides after their repeated application (McLaughlin et al., 2023).

Furthermore, planting resistant varieties reduces chemical use and contributes to sustainability (Ram et al., 2016). Additionally, it is an economical technique with reduced input costs and an increased yield (Li et al., 2020). On the other hand, it is important to note that conventional plant breeding's methods for the selection of resistant varieties, is time consuming (Lenaerts et al., 2019), and can be limited by mutations occurrence and an increase of the susceptibility to the environmental stresses (Ahmar et al., 2020).

As regard integrated pest management, it involves the combination of multiple strategies for plant diseases control (Richard et al., 2022). Although this technique promotes healthy crops growth with low effects on the environment, it has some disadvantages, including possible persistence of harmful species, and the gaps between the concepts of this technique and field practices which may be characterized by the use of high amounts of pesticides (Deguine et al., 2021).

With respect to green nanoparticles, they are more beneficial than chemical methods, since they include green components (Bahrulolum et al., 2021). However, their risk of toxicity needs to be well studied (de Souza et al., 2019). Thus, their use can be limited by regulatory hurdles (Lu and Ozcan, 2015). Besides, public perception of nanoparticles used in agriculture is also an important issue to be taken in to consideration (Mukherjee et al., 2019).

Overall, green nanoparticles represent a good alternative to conventional methods, but more work is needed to ensure their long-term safety and to give people trust for their use.

8. Challenges and Future Prospects

Green nanotechnology provides novel solutions for sustainable plant disease management; yet, it experiences numerous technological, economic, and regulatory obstacles that must be addressed prior to its widespread implementation. From a technical perspective, the large-scale synthesis of green nanomaterials is of paramount concern. Contemporary synthesis techniques, despite their potential, frequently exhibit poor reproducibility, and precise control over particle size is inadequately attained; this renders industrial production expensive and costly (Kah et al., 2013). Furthermore, nanoparticles exhibit restricted stability under diverse environmental conditions, hence diminishing their actual antibacterial efficacy in practical applications (Servin and White, 2016).

A more comprehensive understanding of the methods by which nanomaterials operate against infections must be established. Although their antibacterial efficacy has been established, the molecular-level interactions remain inadequately understood (Duncan, 2011). Ecotoxicological assessments of nanomaterials must also be conducted. The potential for bioaccumulation, biodegradability, and toxicity to non-target organisms necessitates long-term studies for the advancement of environmentally safe nanomaterials (Gogos et al., 2012).

Moreover, full-scale adoption relies on sensitizing farmers about the benefits accruable from green nanotechnology. This can be better achieved through the provision of education and field demonstrations to enable them to shift to more trust in the technologies concerned (Parisi et al., 2015). On the contrary, unclear regulatory guidelines appear to be one of the main obstacles. There is a need to set standards and policies concerning specific subjects that will stimulate innovation with a guarantee of taking care of consumers' and the environment's safety (Chen and Yada, 2011). In general, green nanotechnology has great potential to revolutionize plant disease management, but these prevailing challenges are only surmountable through concerted research efforts and technological development, additionally accompanied by policymaking. With such developments, this technology has the potential to play a central role in transitioning toward sustainable and resilient agriculture.

9. Conclusion

Green nanotechnology has emerged as a promising approach to the sustainable management of plant diseases through the integration of advanced nanomaterials with environmentally friendly practices. It offers new avenues for improving agricultural productivity and enhancing crop resilience by reducing injudicious consumption of harmful chemical inputs through nanopesticides, nanofertilizers, and nanosensors. Targeted delivery, early disease detection, and enhanced nutrient uptake create an important tool in a transition of agriculture toward precision. Despite these advantages, the application of green nanotechnology also presents several disadvantages in terms of safety, environmental impact, and scalability. The broad application of nanomaterials in agriculture will necessitate a proper risk assessment and established regulatory frameworks to ensure safety and efficiency in the exploitation of this technology. Other than that, economic and technological barriers are the most crucial challenges that must be overcome for the wide dissemination of these technologies, at least in developing regions. In the future, more research and development will be needed to extend these technologies to all farmers worldwide. Green nanotechnology can bridge existing knowledge and practice gaps and, therefore, drive sustainable agriculture, enabling substantial contributions to achieving food security with environmental protection worldwide.

References

Abdallah, Y., Ogunyemi, S. O., Abdelazez, A., Zhang, M., Hong, X., Ibrahim, E. et al. (2019). The green synthesis of MgO nano-flowers using Rosmarinus officinalis L.(Rosemary) and the antibacterial activities against Xanthomonas oryzae pv. oryzae. *BioMed Research International*, 2019.

Abdelrhim, A. S., Mazrou, Y. S. A., Nehela, Y., Atallah, O. O., El-Ashmony, R. M., and Dawood, M. F. A. (2021). Silicon dioxide nanoparticles induce innate immune responses and activate antioxidant machinery in wheat against Rhizoctonia solani. *Plants*, 10(12): 2758.

Abuzeid, H. M., Hashem, A. M., Kaus, M., Knapp, M., Indris, S., Ehrenberg, H. et al. (2018). Electrochemical performance of nanosized MnO2 synthesized by redox route using biological reducing agents. *Journal of Alloys and Compounds*, 746: 227–237.

Ahmadi, Z., Saber, M., Bagheri, M., and Mahdavinia, G. R. (2018). Achillea millefolium essential oil and chitosan nanocapsules with enhanced activity against Tetranychus urticae. *Journal of Pest Science*, 91: 837–848.

Ahmar, S., Gill, R. A., Jung, K.-H., Faheem, A., Qasim, M. U., Mubeen, M. et al. (2020). Conventional and molecular techniques from simple breeding to speed breeding in crop plants: recent advances and future outlook. *International Journal of Molecular Sciences*, 21(7): 2590.

Ahmed, T., Luo, J., Noman, M., Ijaz, M., Wang, X., Masood, H. A. et al. (2023). Microbe-mediated nanoparticle intervention for the management of plant diseases. *Crop Health*, 1(1): 1–11. https://doi.org/10.1007/s44297-023-00006-9.

Aithal, S., and Aithal, P. S. (2021). Green and eco-friendly Nanotechnology–concepts and industrial prospects. *International Journal of Management, Technology, and Social Sciences (IJMTS)*, 6(1): 1–31.

Al-asbahi, M. G. S. S., Al-Ofiry, B. A., Saad, F. A. A., Alnehia, A., and Al-Gunaid, M. Q. A. (2024). Silver nanoparticles biosynthesis using mixture of Lactobacillus sp. and Bacillus sp. growth and their antibacterial activity. *Scientific Reports*, 14(1): 1–12. https://doi.org/10.1038/s41598-024-59936-1.

Alessio, P., Aoki, P. H. B., Furini, L. N., Aliaga, A. E., and Constantino, C. J. L. (2017). Spectroscopic techniques for characterization of nanomaterials. In *Nanocharacterization techniques* (pp. 65–98). Elsevier.

Alexandratos, N. (1999). World food and agriculture: outlook for the medium and longer term. *Proceedings of the National Academy of Sciences*, 96(11): 5908–5914.

Alexandratos, N., and Bruinsma, J. (2012). *World agriculture towards 2030/2050: the 2012 revision.*

Aljabali, A. A. A., Akkam, Y., Al Zoubi, M. S., Al-Batayneh, K. M., Al-Trad, B., Abo Alrob, O. et al. (2018). Synthesis of gold nanoparticles using leaf extract of Ziziphus zizyphus and their antimicrobial activity. *Nanomaterials*, 8(3): 174.

Alotaibi, M. O., Alotaibi, N. M., Ghoneim, A. M., ul Ain, N., Irshad, M. A., Nawaz, R. et al. (2023). Effect of green synthesized cerium oxide nanoparticles on fungal disease of wheat plants: a field study. *Chemosphere*, 339: 139731.

Alqarni, L. S., Alghamdi, M. D., Alshahrani, A. A., and Nassar, A. M. (2022). Green nanotechnology: Recent research on bioresource-based nanoparticle synthesis and applications. *Journal of Chemistry*, 2022(1): 4030999.

Alsamhary, K. I. (2020). Eco-friendly synthesis of silver nanoparticles by Bacillus subtilis and their antibacterial activity. *Saudi Journal of Biological Sciences*, 27(8): 2185–2191.

Anastas, P. T., and Warner, J. C. (1998). Principles of green chemistry. *Green Chemistry: Theory and Practice*, 29: 14821–14842.

Ansari, M. A., Khan, H. M., Alzohairy, M. A., Jalal, M., Ali, S. G., Pal, R. et al. (2015). Green synthesis of Al 2 O 3 nanoparticles and their bactericidal potential against clinical isolates of multi-drug resistant Pseudomonas aeruginosa. *World Journal of Microbiology and Biotechnology*, 31: 153–164.

Asif, M., Yasmin, R., Asif, R., Ambreen, A., Mustafa, M., and Umbreen, S. (2022). Green synthesis of silver nanoparticles (AgNPs), structural characterization, and their antibacterial potential. *Dose-Response*, 20(1): 1–11. https://doi.org/10.1177/15593258221088709.

Athanassiou, C. G., Kavallieratos, N. G., Benelli, G., Losic, D., Usha Rani, P., and Desneux, Njj. (2018). Nanoparticles for pest control: current status and future perspectives. *Journal of Pest Science*, 91: 1–15.

Athanassiou, C. G., Kavallieratos, N. G., Evergetis, E., Katsoula, A.-M., and Haroutounian, S. A. (2013). Insecticidal efficacy of silica gel with Juniperus oxycedrus ssp. oxycedrus (Pinales: Cupressaceae) essential oil against Sitophilus oryzae (Coleoptera: Curculionidae) and Tribolium confusum (Coleoptera: Tenebrionidae). *Journal of Economic Entomology*, 106(4): 1902–1910.

Atiq, M., Mazhar, H. M. R., Rajput, N. A., Ahmad, U., Hameed, A., and Lodhi, A. (2022). Green synthesis of silver and copper nanoparticles from leaves of eucalyptus globulus and assessment of its antibacterial potential towards Xanthomonas citri pv. citri causing citrus canker. *Appl. Ecol. Environ. Res.*, 20(3), 2205–2213.

Azim, Z., Singh, N. B., Khare, S., Singh, A., Amist, N., and Yadav, R. K. (2022). Green synthesis of zinc oxide nanoparticles using Vernonia cinerea leaf extract and evaluation as nano-nutrient on the growth and development of tomato seedling. *Plant Nano Biology*, 2: 100011.

Bahrulolum, H., Nooraei, S., Javanshir, N., Tarrahimofrad, H., Mirbagheri, V. S., Easton, A. J. et al. (2021). Green synthesis of metal nanoparticles using microorganisms and their application in the agrifood sector. *Journal of Nanobiotechnology*, 19: 1–26.

Baranwal, J., Barse, B., Fais, A., Delogu, G. L., and Kumar, A. (2022). Biopolymer: A sustainable material for food and medical applications. *Polymers*, 14(5): 983.

Bartolucci, C., Antonacci, A., Arduini, F., Moscone, D., Fraceto, L., Campos, E. et al. (2020). Green nanomaterials fostering agrifood sustainability. *TrAC Trends in Analytical Chemistry*, 125: 115840.

Bhau, B. S., Phukon, P., Ahmed, R., Gogoi, B., Borah, B., Baruah, J. et al. (2016). A novel tool of nanotechnology: nanoparticle mediated control of nematode infection in plants. *Microbial Inoculants in Sustainable Agricultural Productivity: Vol. 2: Functional Applications*, 253–269.

Blázquez, M. A., and Carbó, E. (2015). Control of Portulaca oleracea by boldo and lemon essential oils in different soils. *Industrial Crops and Products*, 76: 515–521.

Blomme, G., Dita, M., Jacobsen, K. S., Pérez Vicente, L., Molina, A., Ocimati, W. et al. (2017). Bacterial diseases of bananas and enset: current state of knowledge and integrated approaches toward sustainable management. *Frontiers in Plant Science*, 8: 1290.

Boros, B.-V., and Ostafe, V. (2020). Evaluation of ecotoxicology assessment methods of nanomaterials and their effects. *Nanomaterials*, 10(4): 610.

Boroumand Moghaddam, A., Namvar, F., Moniri, M., Md. Tahir, P., Azizi, S., and Mohamad, R. (2015). Nanoparticles biosynthesized by fungi and yeast: a review of their preparation, properties, and medical applications. *Molecules*, 20(9): 16540–16565.

Budisa, N., and Schulze-Makuch, D. (2014). Supercritical carbon dioxide and its potential as a life-sustaining solvent in a planetary environment. *Life*, 4(3): 331–340. https://doi.org/10.3390/life4030331.

Bussamara, R., Melo, W. W. M., Scholten, J. D., Migowski, P., Marin, G., Zapata, M. J. M. et al. (2013). Controlled synthesis of Mn 3 O 4 nanoparticles in ionic liquids. *Dalton Transactions*, 42(40): 14473–14479.

Cabeen, M. T., and Jacobs-Wagner, C. (2005). Bacterial cell shape. *Nature Reviews Microbiology*, 3(8): 601–610.

Campbell, J. L., Arora, J., Cowell, S. F., Garg, A., Eu, P., Bhargava, S. K. et al. (2011). Quasi-cubic magnetite/silica core-shell nanoparticles as enhanced MRI contrast agents for cancer imaging. *PloS One*, 6(7): e21857.

Campolo, O., Cherif, A., Ricupero, M., Siscaro, G., Grissa-Lebdi, K., Russo, A. et al. (2017). Citrus peel essential oil nanoformulations to control the tomato borer, Tuta absoluta: chemical properties and biological activity. *Scientific Reports*, 7(1): 13036.

Campos, E. V. R., Proença, P. L. F., Oliveira, J. L., Melville, C. C., Della Vechia, J. F., De Andrade, D. J. et al. (2018). Chitosan nanoparticles functionalized with β-cyclodextrin: a promising carrier for botanical pesticides. *Scientific Reports*, 8(1): 2067.

Casciaro, B., Moros, M., Rivera-Fernandez, S., Bellelli, A., de la Fuente, J. M., and Mangoni, M. L. (2017). Gold-nanoparticles coated with the antimicrobial peptide esculentin-1a (1-21) NH2 as a reliable strategy for antipseudomonal drugs. *Acta Biomaterialia*, 47: 170–181.

Cheloni, G., Marti, E., and Slaveykova, V. I. (2016). Interactive effects of copper oxide nanoparticles and light to green alga Chlamydomonas reinhardtii. *Aquatic Toxicology*, 170: 120–128.

Chen, H., and Yada, R. (2011). Nanotechnologies in agriculture: new tools for sustainable development. *Trends in Food Science & Technology*, 22(11): 585–594.

Chen, Y., Tuan, H., Tien, C., Lo, W., Liang, H., and Hu, Y. (2009). Augmented biosynthesis of cadmium sulfide nanoparticles by genetically engineered Escherichia coli. *Biotechnology Progress*, 25(5): 1260–1266.

Chhantyal, P. (2022). The use of X-Ray diffraction for nanoparticle characterization. *AZoOptics. Https://Www. Azooptics. Com/Article. Aspx*.

Cuéllar-Cruz, M., Lucio-Hernández, D., Martínez-Ángeles, I., Demitri, N., Polentarutti, M., Rosales-Hoz, M. J. et al. (2017). Biosynthesis of micro- and nanocrystals of Pb (II), Hg (II) and Cd (II) sulfides in four Candida species: a comparative study of *in vivo* and *in vitro* approaches. *Microbial Biotechnology*, 10(2): 405–424. https://doi.org/10.1111/1751-7915.12485.

Dadi, R., Azouani, R., Traore, M., Mielcarek, C., and Kanaev, A. (2019). Antibacterial activity of ZnO and CuO nanoparticles against gram positive and gram negative strains. *Materials Science and Engineering: C*, 104: 109968.

Dahoumane, S. A., Yéprémian, C., Djédiat, C., Couté, A., Fiévet, F., Coradin, T. et al. (2016). Improvement of kinetics, yield, and colloidal stability of biogenic gold nanoparticles using living cells of Euglena gracilis microalga. *Journal of Nanoparticle Research*, 18: 1–12.

Dakal, T. C., Kumar, A., Majumdar, R. S., and Yadav, V. (2016). Mechanistic basis of antimicrobial actions of silver nanoparticles. *Frontiers in Microbiology*, 7: 1831.

Damalas, C. A., and Koutroubas, S. D. (2018). Current status and recent developments in biopesticide use. *Agriculture*, 8(1): 13.

Dao, T. N. T., Seshadri Reddy, A., Zhao, F., Liu, H., Koo, B., Moniruzzaman, M. et al. (2021). Green synthesis-based magnetic diatoms for biological applications. *ACS Sustainable Chemistry & Engineering*, 9(9): 3439–3451.

Das, A., and Dutta, P. (2021). Antifungal activity of biogenically synthesized silver and gold nanoparticles against sheath blight of rice. *Journal of Nanoscience and Nanotechnology*, 21(6): 3547–3555.

de Oliveira, J. L., Campos, E. V. R., Bakshi, M., Abhilash, P. C., and Fraceto, L. F. (2014). Application of nanotechnology for the encapsulation of botanical insecticides for sustainable agriculture: prospects and promises. *Biotechnology Advances*, 32(8): 1550–1561.

de Souza, T. A. J., Souza, L. R. R., and Franchi, L. P. (2019). Silver nanoparticles: An integrated view of green synthesis methods, transformation in the environment, and toxicity. *Ecotoxicology and Environmental Safety*, 171: 691–700.

Deguine, J.-P., Aubertot, J.-N., Flor, R. J., Lescourret, F., Wyckhuys, K. A. G., and Ratnadass, A. (2021). Integrated pest management: good intentions, hard realities. A review. *Agronomy for Sustainable Development*, 41(3): 38.

Deka, B., Babu, A., Baruah, C., and Barthakur, M. (2021). Nanopesticides: A Systematic Review of Their Prospects With Special Reference to Tea Pest Management. *Frontiers in Nutrition*, 8(August). https://doi.org/10.3389/fnut.2021.686131.

Deveci, H. A., Nur, G., Deveci, A., Kaya, I., Kaya, M. M., Küküirt, A. et al. (2021). *An overview of the biochemical and histopathological effects of insecticides*. IntechOpen.

Devi, H. S., and Singh, T. D. (2014). Synthesis of copper oxide nanoparticles by a novel method and its application in the degradation of methyl orange. *Adv Electron Electr Eng*, 4(1): 83–88.

Devi, P. S. V., Duraimurugan, P., and Chandrika, K. (2019). Bacillus thuringiensis-based nanopesticides for crop protection. In *Nano-biopesticides today and future perspectives* (pp. 249–260). Elsevier.

Dikshit, P. K., Kumar, J., Das, A. K., Sadhu, S., Sharma, S., Singh, S. et al. (2021). Green synthesis of metallic nanoparticles: Applications and limitations. *Catalysts*, 11(8): 902.

Dizaj, S. M., Lotfipour, F., Barzegar-Jalali, M., Zarrintan, M. H., and Adibkia, K. (2014). Antimicrobial activity of the metals and metal oxide nanoparticles. *Materials Science and Engineering: C*, 44: 278–284.

Drummer, S., Madzimbamuto, T., and Chowdhury, M. (2021). Green synthesis of transition-metal nanoparticles and their oxides: a review. *Materials*, 14(11): 2700.

Duhan, J. S., Kumar, R., Kumar, N., Kaur, P., Nehra, K., and Duhan, S. (2017). Nanotechnology: The new perspective in precision agriculture. *Biotechnology Reports*, 15: 11–23.

Duncan, T. V. (2011). The communication challenges presented by nanofoods. *Nature Nanotechnology*, 6(11): 683–688.

Dupont, J., de Souza, R. F., & Suarez, P. A. Z. (2002). Ionic liquid (molten salt) phase organometallic catalysis. *Chemical Reviews*, 102(10): 3667–3692.

Dupont, J., Fonseca, G. S., Umpierre, A. P., Fichtner, P. F. P., and Teixeira, S. R. (2002). Transition-metal nanoparticles in imidazolium ionic liquids: recycable catalysts for biphasic hydrogenation reactions. *Journal of the American Chemical Society*, 124(16): 4228–4229.

Dutta, P., Kumari, A., Mahanta, M., Biswas, K. K., Dudkiewicz, A., Thakuria, D. et al. (2022). Advances in nanotechnology as a potential alternative for plant viral disease management. *Frontiers in Microbiology*, 13: 935193.

Eid, M. M. (2022). Characterization of Nanoparticles by FTIR and FTIR-Microscopy. In *Handbook of consumer nanoproducts* (pp. 1–30). Springer.

Elbeshehy, E. K. F., Elazzazy, A. M., and Aggelis, G. (2015). *Silver nanoparticles synthesis mediated by new isolates of Bacillus spp ., nanoparticle characterization and their activity against Bean Yellow Mosaic Virus and human pathogens*. 6(May), 1–13. https://doi.org/10.3389/fmicb.2015.00453.

Elmer, W., Ma, C., and White, J. (2018). Nanoparticles for plant disease management. *Current Opinion in Environmental Science & Health*, 6: 66–70.

Elmer, W., and White, J. C. (2018). The future of nanotechnology in plant pathology. *Annual Review of Phytopathology*, 56(1): 111–133.

El-Rafie, H. M., El-Rafie, Mh., and Zahran, M. K. (2013). Green synthesis of silver nanoparticles using polysaccharides extracted from marine macro algae. *Carbohydrate Polymers*, 96(2): 403–410.

El-Saadony, M. T., ALmoshadak, A. S., Shafi, M. E., Albaqami, N. M., Saad, A. M., El-Tahan, A. M. et al. (2021). Vital roles of sustainable nano-fertilizers in improving plant quality and quantity-an updated review. *Saudi Journal of Biological Sciences*, 28(12): 7349–7359. https://doi.org/10.1016/j.sjbs.2021.08.032.

El-Sawy, M. M., Elsharkawy, M. M., Abass, J. M., and Kasem, M. H. (2017). Antiviral activity of 2–nitromethyl phenol, zinc nanoparticles and seaweed extract against cucumber mosaic virus (CMV) in Eggplant. *J. Virol. Antivir. Res*, 6(2).

El-Seedi, H. R., El-Shabasy, R. M., Khalifa, S. A. M., Saeed, A., Shah, A., Shah, R. et al. (2019). Metal nanoparticles fabricated by green chemistry using natural extracts: Biosynthesis, mechanisms, and applications. *RSC Advances*, 9(42): 24539–24559.

Enea, M., Pereira, E., de Almeida, M. P., Araújo, A. M., Bastos, M. de L., and Carmo, H. (2020). Gold nanoparticles induce oxidative stress and apoptosis in human kidney cells. *Nanomaterials*, 10(5): 1–17. https://doi.org/10.3390/nano10050995.

Er, H., Yasuda, H., Harada, M., Taguchi, E., and Iida, M. (2017). Formation of silver nanoparticles from ionic liquids comprising N-alkylethylenediamine: effects of dissolution modes of the silver (I) ions in the ionic liquids. *Colloids and Surfaces A: Physicochemical and Engineering Aspects*, 522: 503–513.

Eroglu, N., Emekci, M., and Athanassiou, C. G. (2017). Applications of natural zeolites on agriculture and food production. *Journal of the Science of Food and Agriculture*, 97(11): 3487–3499.

Esmail, R., Afshar, A., Morteza, M., Abolfazl, A., and Akhondi, E. (2022). Synthesis of silver nanoparticles with high efficiency and stability by culture supernatant of Bacillus ROM6 isolated from Zarshouran gold mine and evaluating its antibacterial effects. *BMC Microbiology*, 22(1): 1–10. https://doi.org/10.1186/s12866-022-02490-5.

Ewunkem, A. J., Rodgers, L., Campbell, D., Staley, C., Subedi, K., Boyd, S., and Graves, Jr., J. L. (2021). Experimental evolution of magnetite nanoparticle resistance in Escherichia coli. *Nanomaterials*, 11(3): 790.

Faghihzadeh, F., Anaya, N. M., Schifman, L. A., and Oyanedel-Craver, V. (2016). Fourier transform infrared spectroscopy to assess molecular-level changes in microorganisms exposed to nanoparticles. *Nanotechnology for Environmental Engineering*, 1(1): 1–16. https://doi.org/10.1007/s41204-016-0001-8.

Fagodia, S. K., Singh, H. P., Batish, D. R., and Kohli, R. K. (2017). Phytotoxicity and cytotoxicity of Citrus aurantiifolia essential oil and its major constituents: Limonene and citral. *Industrial Crops and Products*, 108: 708–715.

Farghaly, F. A., and Nafady, N. A. (2015). Green synthesis of silver nanoparticles using leaf extract of Rosmarinus officinalis and its effect on tomato and wheat plants. *Journal of Agricultural Science*, 7(11): 277.

Farooq, T., Adeel, M., He, Z., Umar, M., Shakoor, N., da Silva, W. et al. (2021). Nanotechnology and plant viruses: An emerging disease management approach for resistant pathogens. *ACS Nano*, 15(4): 6030–6037.

Fatimah, I. (2016). Green synthesis of silver nanoparticles using extract of Parkia speciosa Hassk pods assisted by microwave irradiation. *Journal of Advanced Research*, 7(6): 961–969.

Feitshans, I. L., and Sabatier, P. (2022). Global health impacts of nanotechnology law: Advances in safernano regulation. *Materials Today: Proceedings*, 67: 985–994.

Foster, H. A., Ditta, I. B., Varghese, S., and Steele, A. (2011). Photocatalytic disinfection using titanium dioxide: spectrum and mechanism of antimicrobial activity. *Applied Microbiology and Biotechnology*, 90: 1847–1868.

Fu, H., Tan, P., Wang, R., Li, S., Liu, H., Yang, Y., & Wu, Z. (2022). Advances in organophosphorus pesticides pollution: Current status and challenges in ecotoxicological, sustainable agriculture, and degradation strategies. *Journal of Hazardous Materials*, 424: 127494.

Fürstner, A., Ackermann, L., Beck, K., Hori, H., Koch, D., Langemann, K. et al. (2001). Olefin metathesis in supercritical carbon dioxide. *Journal of the American Chemical Society*, 123(37): 9000–9006.

Gao, F., Luan, X., Yin, Y., Sun, S., Li, Y., Mo, F., and Wang, J. (2022). Exploring long-term impacts of different crop rotation systems on sustainable use of groundwater resources using DSSAT model. *Journal of Cleaner Production*, 336: 130377.

Gao, X. (2005). *Characterization of Nanomaterials by Transmission Electron Microscopy and Related Techniques*. 42–50.

Garg, D., Sridhar, K., Stephen Inbaraj, B., Chawla, P., Tripathi, M., and Sharma, M. (2023). Nano-biofertilizer formulations for agriculture: a systematic review on recent advances and prospective applications. *Bioengineering*, 10(9): 1–33.

Gericke, M., and Pinches, A. (2006). Microbial production of gold nanoparticles. *Gold Bulletin*, 39(1): 22–28.

Gharby, S. (2022). Refining vegetable oils: chemical and physical refining. *Scientific World Journal*, 2022(Table 1). https://doi.org/10.1155/2022/6627013.

Ghiuță, I., Cristea, D., Croitoru, C., Kost, J., Wenkert, R., Vyrides, I. et al. (2018). Characterization and antimicrobial activity of silver nanoparticles, biosynthesized using Bacillus species. *Applied Surface Science*, 438: 66–73.

Ghnaya, A. Ben, Amri, I., Hanana, M., Gargouri, S., Jamoussi, B., Romane, A. et al. (2016). Tetraclinis articulata (Vahl.) Masters essential oil from Tunisia: Chemical characterization and herbicidal and antifungal activities assessment. *Industrial Crops and Products*, 83: 113–117.

Ghormade, V., Deshpande, M. V., and Paknikar, K. M. (2011). Perspectives for nano-biotechnology enabled protection and nutrition of plants. *Biotechnology Advances*, 29(6): 792–803.

Gnanasangeetha, D., and Thambavani, D. S. (2013). Biogenic production of zinc oxide nanoparticles using Acalypha indica. *Journal of Chemical, Biological and Physical Sciences (JCBPS)*, 4(1): 238.

Gogos, A., Knauer, K., and Bucheli, T. D. (2012). Nanomaterials in plant protection and fertilization: current state, foreseen applications, and research priorities. *Journal of Agricultural and Food Chemistry*, 60(39): 9781–9792.

González, J. O. W., Gutiérrez, M. M., Ferrero, A. A., and Band, B. F. (2014). Essential oils nanoformulations for stored-product pest control–Characterization and biological properties. *Chemosphere*, 100: 130–138.

Gopal, J. V., Thenmozhi, M., Kannabiran, K., Rajakumar, G., Velayutham, K., and Rahuman, A. A. (2013). Actinobacteria mediated synthesis of gold nanoparticles using *Streptomyces* sp. VITDDK3 and its antifungal activity. *Materials Letters*, 93: 360–362.

Gopinath, V., and Velusamy, P. (2013). Extracellular biosynthesis of silver nanoparticles using Bacillus sp. GP-23 and evaluation of their antifungal activity towards Fusarium oxysporum. *Spectrochimica Acta Part A: Molecular and Biomolecular Spectroscopy*, 106: 170–174.

Goswami, R., Bhattacharyya, A., and Dutta, P. (2020). *Nanotechnological Approach for Management of Anthracnose and Crown Rot Diseases of Banana.*

Graham, J. H., Johnson, E. G., Myers, M. E., Young, M., Rajasekaran, P., Das, S., and Santra, S. (2016). Potential of nano-formulated zinc oxide for control of citrus canker on grapefruit trees. *Plant Disease*, 100(12): 2442–2447.

Grillo, R., Fraceto, L. F., Amorim, M. J. B., Scott-Fordsmand, J. J., Schoonjans, R., and Chaudhry, Q. (2021). Ecotoxicological and regulatory aspects of environmental sustainability of nanopesticides. *Journal of Hazardous Materials*, 404: 124148.

Grün, A.-L., Straskraba, S., Schulz, S., Schloter, M., and Emmerling, C. (2018). Long-term effects of environmentally relevant concentrations of silver nanoparticles on microbial biomass, enzyme activity, and functional genes involved in the nitrogen cycle of loamy soil. *Journal of Environmental Sciences*, 69: 12–22.

Gunalan, S., Sivaraj, R., and Rajendran, V. (2012). Green synthesized ZnO nanoparticles against bacterial and fungal pathogens. *Progress in Natural Science: Materials International*, 22(6): 693–700.

Guo, H., White, J. C., Wang, Z., and Xing, B. (2018). Nano-enabled fertilizers to control the release and use efficiency of nutrients. *Current Opinion in Environmental Science & Health*, 6: 77–83.

Gupta, A., Bonde, S. R., Gaikwad, S., Ingle, A., Gade, A. K., and Rai, M. (2014). Lawsonia inermis-mediated synthesis of silver nanoparticles: activity against human pathogenic fungi and bacteria with special reference to formulation of an antimicrobial nanogel. *IET Nanobiotechnology*, 8(3): 172–178.

Gurunathan, S., Han, J. W., Dayem, A. A., Eppakayala, V., and Kim, J.-H. (2012). Oxidative stress-mediated antibacterial activity of graphene oxide and reduced graphene oxide in Pseudomonas aeruginosa. *International Journal of Nanomedicine*, 5901–5914.

Gurunathan, S., Han, J. W., Kwon, D.-N., and Kim, J.-H. (2014). Enhanced antibacterial and anti-biofilm activities of silver nanoparticles against Gram-negative and Gram-positive bacteria. *Nanoscale Research Letters*, 9: 1–17.

Haghighi, F., Roudbar Mohammadi, S., Mohammadi, P., Hosseinkhani, S., and Shipour, R. (2013). Antifungal activity of TiO2 nanoparticles and EDTA on Candida albicans biofilms. *Infection, Epidemiology and Microbiology*, 1(1): 33–38.

Halder, S., Yadav, K. K., Sarkar, R., Mukherjee, S., Saha, P., Haldar, S. et al. (2015). Alteration of Zeta potential and membrane permeability in bacteria: a study with cationic agents. *SpringerPlus*, 4: 1–14.

Hano, C., and Abbasi, B. H. (2022). Plant-based green synthesis of nanoparticles: Production, characterization and applications. *Biomolecules*, 12(1): 1–9. https://doi.org/10.3390/biom12010031.

Hansen, H. E., Seland, F., Sunde, S., Burheim, O. S., and Pollet, B. G. (2023). Optimum scavenger concentrations for sonochemical nanoparticle synthesis. *Scientific Reports*, 13(1): 1–9. https://doi.org/10.1038/s41598-023-33243-7.

Hassanisaadi, M., Barani, M., Rahdar, A., Heidary, M., Thysiadou, A., and Kyzas, G. Z. (2022). Role of agrochemical-based nanomaterials in plants: Biotic and abiotic stress with germination improvement of seeds. *Plant Growth Regulation*, 97(2): 375–418.

Hazrati, H., Saharkhiz, M. J., Niakousari, M., and Moein, M. (2017). Natural herbicide activity of Satureja hortensis L. essential oil nanoemulsion on the seed germination and morphophysiological features of two important weed species. *Ecotoxicology and Environmental Safety*, 142: 423–430.

Hedaoo, R. K., and Gite, V. V. (2014). Renewable resource-based polymeric microencapsulation of natural pesticide and its release study: an alternative green approach. *RSC Advances*, 4(36): 18637–18644.

Hegazy, H. S., Shabaan, L. D., Rabie, G. H., and Raie, D. S. (2015). Biosynthesis of silver nanoparticles using cell free callus exudates of *Medicago sativa* L. *Pak. J. Bot.*, 47(5): 1825–1829.

Heidari, M., Salmanpour, I., Ghorbani, H., and Asghari, H. R. (2018). Iron Chelate and Rhizobactria Changed Growth, Grain Yield, and Physiological Characteristics in Maize. *Scientia Agriculturae Bohemica*, 49(4): 245–254. https://doi.org/10.2478/sab-2018-0031.

Huang, Y., Ma, L., Fang, D. H., Xi, J. Q., Zhu, M. L., Mo, M. H. et al. (2015). Isolation and characterisation of rhizosphere bacteria active against Meloidogyne incognita, Phytophthora nicotianae and the root knot–black shank complex in tobacco. *Pest Management Science*, 71(3): 415–422.

Husen, A., and Siddiqi, K. S. (2014). Plants and microbes assisted selenium nanoparticles: characterization and application. *Journal of Nanobiotechnology*, 12: 1–10.

Hussain, I., Singh, N. B., Singh, A., Singh, H., and Singh, S. C. (2016). Green synthesis of nanoparticles and its potential application. *Biotechnology Letters*, 38: 545–560.

Hussain, N., Bilal, M., and Iqbal, H. M. N. (2022). Carbon-based nanomaterials with multipurpose attributes for water treatment: Greening the 21st-century nanostructure materials deployment. *Biomaterials and Polymers Horizon*, 1(1): 48–58.

Hussein-Al-Ali, S. H., El Zowalaty, M. E., Hussein, M. Z., Geilich, B. M., and Webster, T. J. (2014). Synthesis, characterization, and antimicrobial activity of an ampicillin-conjugated magnetic nanoantibiotic for medical applications. *International Journal of Nanomedicine*, 3801–3814.

Imada, K., Sakai, S., Kajihara, H., Tanaka, S., and Ito, S. (2016). Magnesium oxide nanoparticles induce systemic resistance in tomato against bacterial wilt disease. *Plant Pathology*, 65(4): 551–560.

Iravani, S. (2011). Green synthesis of metal nanoparticles using plants. *Green Chemistry*, 13(10): 2638–2650.

Iravani, S. (2014). Bacteria in nanoparticle synthesis: current status and future prospects. *International Scholarly Research Notices*, 2014(1): 359316.

Iravani, S., Korbekandi, H., Mirmohammadi, S. V., and Zolfaghari, B. (2014). Synthesis of silver nanoparticles: chemical, physical and biological methods. *Research in Pharmaceutical Sciences*, 9(6): 385–406.

Isibor, P. O. (2024). Regulations and policy considerations for nanoparticle safety. In *Environmental Nanotoxicology: Combatting the Minute Contaminants* (pp. 295–316). Springer.

Jadczak, P., Kulpa, D., Drozd, R., Przewodowski, W., and Przewodowska, A. (2020). Effect of AuNPs and AgNPs on the antioxidant system and antioxidant activity of lavender (*Lavandula angustifolia* Mill.) from *in vitro* cultures. *Molecules*, 25(23): 5511.

Jafari, A., Rashidipour, M., Kamarehi, B., Alipour, S., and Ghaderpoori, M. (2022). Toxicity of green synthesized TiO2 nanoparticles (TiO2 NPs) on zebra fish. *Environmental Research*, 212: 113542.

Jamdagni, P., Khatri, P., and Rana, Jjj.-S. (2018). Green synthesis of zinc oxide nanoparticles using flower extract of Nyctanthes arbor-tristis and their antifungal activity. *Journal of King Saud University-Science*, 30(2): 168–175.

Jang, H. R., Oh, H.-J., Kim, J.-H., and Jung, K. Y. (2013). Synthesis of mesoporous spherical silica via spray pyrolysis: Pore size control and evaluation of performance in paclitaxel pre-purification. *Microporous and Mesoporous Materials*, 165, 219–227.

Jayaram, D. T., Kumar, A., Kippner, L. E., Ho, P.-Y., Kemp, M. L., Fan, Y. et al. (2019). TiO 2 nanoparticles generate superoxide and alter gene expression in human lung cells. *RSC Advances*, 9(43): 25039–25047.

Joost, U., Juganson, K., Visnapuu, M., Mortimer, M., Kahru, A., Nõmmiste, E. et al. (2015). Photocatalytic antibacterial activity of nano-TiO2 (anatase)-based thin films: effects on *Escherichia coli* cells and fatty acids. *Journal of Photochemistry and Photobiology B: Biology*, 142: 178–185.

Jung, W. K., Koo, H. C., Kim, K. W., Shin, S., Kim, S. H., and Park, Y. H. (2008). Antibacterial activity and mechanism of action of the silver ion in Staphylococcus aureus and Escherichia coli. *Applied and Environmental Microbiology*, 74(7): 2171–2178.

Kah, M., Beulke, S., Tiede, K., and Hofmann, T. (2013). Nanopesticides: state of knowledge, environmental fate, and exposure modeling. *Critical Reviews in Environmental Science and Technology*, 43(16): 1823–1867.

Kah, M., Johnston, L. J., Kookana, R. S., Bruce, W., Haase, A., Ritz, V., et al. (2021). Comprehensive framework for human health risk assessment of nanopesticides. *Nature Nanotechnology*, 16(9): 955–964.

Kaman, P. K., and Dutta, P. (2019). Synthesis, characterization and antifungal activity of biosynthesized silver nanoparticle. *Indian Phytopathology*, 72: 79–88.

Kandi, V., and Kandi, S. (2015). Antimicrobial properties of nanomolecules: potential candidates as antibiotics in the era of multi-drug resistance. *Epidemiology and Health*, 37.

Katata-Seru, L., Moremedi, T., Aremu, O. S., and Bahadur, I. (2018). Green synthesis of iron nanoparticles using Moringa oleifera extracts and their applications: Removal of nitrate from water and antibacterial activity against *Escherichia coli*. *Journal of Molecular Liquids*, 256: 296–304.

Kaviya, S., Santhanalakshmi, J., Viswanathan, B., Muthumary, J., and Srinivasan, K. (2011). Biosynthesis of silver nanoparticles using Citrus sinensis peel extract and its antibacterial activity. *Spectrochimica Acta Part A: Molecular and Biomolecular Spectroscopy*, 79(3): 594–598.

Khan, F. A. (2020). Synthesis of nanomaterials: methods & technology. *Applications of Nanomaterials in Human Health*, 15–21.

Khan, M., Al-Marri, A. H., Khan, M., Shaik, M. R., Mohri, N., Adil, S. F. et al. (2015). Green approach for the effective reduction of graphene oxide using *Salvadora persica* L. root (Miswak) extract. *Nanoscale Research Letters*, 10: 1–9.

Khatami, M., Pourseyedi, S., Khatami, M., Hamidi, H., Zaeifi, M., and Soltani, L. (2015). Synthesis of silver nanoparticles using seed exudates of Sinapis arvensis as a novel bioresource, and evaluation of their antifungal activity. *Bioresources and Bioprocessing*, 2: 1–7.

Kim, D. Y., Kim, M., Sung, J. S., Koduru, J. R., Nile, S. H., Syed, A. et al. (2024). Extracellular synthesis of silver nanoparticle using yeast extracts: antibacterial and seed priming applicationss. *Applied Microbiology and Biotechnology*, 108(1). https://doi.org/10.1007/s00253-023-12920-7.

Kim, K.-S., Demberelnyamba, D., and Lee, H. (2004). Size-selective synthesis of gold and platinum nanoparticles using novel thiol-functionalized ionic liquids. *Langmuir*, 20(3): 556–560.

Kim, M., Lee, B. Y., Ham, H. C., Han, J., Nam, S. W., Lee, H. et al. (2016). Facile one-pot synthesis of tungsten oxide (WO3− x) nanoparticles using sub and supercritical fluids. *The Journal of Supercritical Fluids*, 111: 8–13.

Klaus, T., Joerger, R., Olsson, E., and Granqvist, C. G. (1999). Silver-based crystalline nanoparticles, microbially fabricated. *Proceedings of the National Academy of Sciences of the United States of America*, 96(24): 13611–13614. https://doi.org/10.1073/pnas.96.24.13611.

Klein, M., and Poverenov, E. (2020). Natural biopolymer-based hydrogels for use in food and agriculture. *Journal of the Science of Food and Agriculture*, 100(6): 2337–2347.

Krumov, N., Oder, S., Perner-Nochta, I., Angelov, A., and Posten, C. (2007). Accumulation of CdS nanoparticles by yeasts in a fed-batch bioprocess. *Journal of Biotechnology*, 132(4): 481–486. https://doi.org/10.1016/j.jbiotec.2007.08.016.

Kumar, P. N., Sakthivel, K., and Balasubramanian, V. (2017). Microwave assisted biosynthesis of rice shaped ZnO nanoparticles using Amorphophallus konjac tuber extract and its application in dye sensitized solar cells. *Mater. Sci. Pol.*, 35(1), 111–119.

Kumar, P. S. M., Francis, A. P., and Devasena, T. (2014). Biosynthesized and chemically synthesized titania nanoparticles: comparative analysis of antibacterial activity. *J. Environ. Nanotechnol.*, 3(3): 73–81.

Kumar, S., Bhattacharya, W., Singh, M., Halder, D., and Mitra, A. (2017). Plant latex capped colloidal silver nanoparticles: A potent anti-biofilm and fungicidal formulation. *Journal of Molecular Liquids*, 230: 705–713.

Kumar, S., Nehra, M., Dilbaghi, N., Marrazza, G., Hassan, A. A., & Kim, K. H. (2019). Nano-based smart pesticide formulations: Emerging opportunities for agriculture. *Journal of Controlled Release*, 294(November 2018), 131–153. https://doi.org/10.1016/j.jconrel.2018.12.012.

Kumari, B., Mallick, M. A., Solanki, M. K., Hora, A., and Mani, M. (2019). Applying nanotechnology to bacteria: an emerging technology for sustainable agriculture. In *Role of plant growth promoting microorganisms in sustainable agriculture and nanotechnology* (pp. 121–143). Elsevier.

Kumari, R., Suman, K., Karmakar, S., Mishra, V., Lakra, S. G., Saurav, G. K. et al. (2023). Regulation and safety measures for nanotechnology-based agri-products. *Frontiers in Genome Editing*, 5: 1200987.

Kuraj, N. (2019). *REACH and the environmental regulation of nanotechnology: preventing and reducing the environmental impacts of nanomaterials*. Routledge.

Larue, C., Castillo-Michel, H., Sobanska, S., Cécillon, L., Bureau, S., Barthès, V. et al. (2014). Foliar exposure of the crop Lactuca sativa to silver nanoparticles: evidence for internalization and changes in Ag speciation. *Journal of Hazardous Materials*, 264: 98–106.

Lasat, M. M., Chung, K. F., Lead, J., McGrath, S., Owen, R. J., Rocks, S. et al. (2018). Advancing the understanding of environmental transformations, bioavailability and effects of nanomaterials, an international US Environmental Protection Agency—UK Environmental Nanoscience Initiative Joint Program. *Journal of Environmental Protection*, 9(4): 385.

Lenaerts, B., Collard, B. C. Y., and Demont, M. (2019). Improving global food security through accelerated plant breeding. *Plant Science*, 287: 110207.

Lesniak, A., Salvati, A., Santos-Martinez, M. J., Radomski, M. W., Dawson, K. A., and Åberg, C. (2013). Nanoparticle adhesion to the cell membrane and its effect on nanoparticle uptake efficiency. *Journal of the American Chemical Society*, 135(4): 1438–1444.

Leung, Y. H., Ng, A. M. C., Xu, X., Shen, Z., Gethings, L. A., Wong, M. T. et al. (2014). Mechanisms of antibacterial activity of MgO: non-ROS mediated toxicity of MgO nanoparticles towards *Escherichia coli*. *Small*, 10(6): 1171–1183.

Li, M., Sun, X., Yin, M., Shen, J., and Yan, S. (2023). Recent advances in nanoparticle-mediated co-delivery system: a promising strategy in medical and agricultural field. *International Journal of Molecular Sciences*, 24(6). https://doi.org/10.3390/ijms24065121.

Li, M., Zhu, L., and Lin, D. (2011). Toxicity of ZnO nanoparticles to Escherichia coli: mechanism and the influence of medium components. *Environmental Science & Technology*, 45(5): 1977–1983.

Li, N., Bai, X., Zhang, S., Gao, Y., Zheng, L., Zhang, J., and Ma, H. (2008). Synthesis of silver nanoparticles in ionic liquid by a simple effective electrochemical method. *Journal of Dispersion Science and Technology*, 29(8): 1059–1061.

Li, W., Deng, Y., Ning, Y., He, Z., and Wang, G.-L. (2020). Exploiting broad-spectrum disease resistance in crops: from molecular dissection to breeding. *Annual Review of Plant Biology*, 71(1): 575–603.

Li, X., Xu, H., Chen, Z.-S., and Chen, G. (2011). Biosynthesis of nanoparticles by microorganisms and their applications. *Journal of Nanomaterials*, 2011(1): 270974.

Li, Y., Zhang, W., Niu, J., and Chen, Y. (2012). Mechanism of photogenerated reactive oxygen species and correlation with the antibacterial properties of engineered metal-oxide nanoparticles. *ACS Nano*, 6(6): 5164–5173.

Lin, J., Zhang, H., Chen, Z., and Zheng, Y. (2010). Penetration of lipid membranes by gold nanoparticles: insights into cellular uptake, cytotoxicity, and their relationship. *ACS Nano*, 4(9): 5421–5429.

Lin, N., Wang, C., Ding, J., Su, L., Xu, L., Zhang, B. et al. (2020). Efficacy of nanoparticle encapsulation on suppressing oxidation and enhancing antifungal activity of cyclic lipopeptides produced by Bacillus subtilis. *Colloids and Surfaces B: Biointerfaces*, 193: 111143.

Liu, H.-X., Li, S.-M., Luo, Y.-M., Luo, L.-X., Li, J.-Q., and Guo, J.-H. (2014). Biological control of Ralstonia wilt, Phytophthora blight, Meloidogyne root-knot on bell pepper by the combination of *Bacillus subtilis* AR12, Bacillus subtilis SM21 and Chryseobacterium sp. R89. *European Journal of Plant Pathology*, 139: 107–116.

Liu, X., Chen, J. Le, Yang, W. Y., Qian, Y. C., Pan, J. Y., Zhu, C. N. et al(2021). Biosynthesis of silver nanoparticles with antimicrobial and anticancer properties using two novel yeasts. *Scientific Reports*, 11(1): 1–12. https://doi.org/10.1038/s41598-021-95262-6.

Lu, Y., and Ozcan, S. (2015). Green nanomaterials: On track for a sustainable future. *Nano Today*, 10(4): 417–420.

Lyudmila, Z., Noira, V., Sayora, R., Svetlana, K., Rokhilya, J., and Javlon, T. (2018). *BioTechnology : An Indian Journal Microbial Synthesis of Silver Nanoparticles by Pseudomonas* sp. 14(4).

Maensiri, S., Laokul, P., Klinkaewnarong, J., Phokha, S., Promarak, V., and Seraphin, S. (2008). Indium oxide (In2O3) nanoparticles using Aloe vera plant extract: Synthesis and optical properties. *J. Optoelectron. Adv. Mater.*, 10(3): 161–165.

Mai-Prochnow, A., Clauson, M., Hong, J., and Murphy, A. B. (2016). Gram positive and Gram negative bacteria differ in their sensitivity to cold plasma. *Scientific Reports*, 6(1): 38610.

Makarov, V. V, Love, A. J., Sinitsyna, O. V, Makarova, S. S., Yaminsky, I. V, Taliansky, M. E. et al. (2014). "Green" nanotechnologies: synthesis of metal nanoparticles using plants. *Acta Naturae (Англоязычная Версия)*, 6(1(20)): 35–44.

Malerba, M., and Cerana, R. (2016). Chitosan effects on plant systems. *International Journal of Molecular Sciences*, 17(7): 996.

Manikandan, R., Beulaja, M., Thiagarajan, R., Palanisamy, S., Goutham, G., Koodalingam, A. et al. (2017). Biosynthesis of silver nanoparticles using aqueous extract of *Phyllanthus acidus* L. fruits and characterization of its anti-inflammatory effect against H2O2 exposed rat peritoneal macrophages. *Process Biochemistry*, 55: 172–181.

Mansoor, S., Zahoor, I., Baba, T. R., Padder, S. A., Bhat, Z. A., Koul, A. M. et al. (2021). Fabrication of silver nanoparticles against fungal pathogens. *Frontiers in Nanotechnology*, 3: 679358.

Marchiol, L. (2012). Synthesis of metal nanoparticles in living plants. *Italian Journal of Agronomy*, 7(3): e37–e37.

Mariello, M., Kim, K., Wu, K., Lacour, S. P., and Leterrier, Y. (2022). Recent advances in encapsulation of flexible bioelectronic implants: Materials, technologies, and characterization methods. *Advanced Materials*, 34(34): 2201129.

Mathew, L., Chandrasekaran, N., and Mukherjee, A. (2010). Biomimetic synthesis of nanoparticles: science, technology and applicability. *Biomimetics Learning from Nature*.

Mattos, B. D., Rojas, O. J., and Magalhães, W. L. E. (2017). Biogenic silica nanoparticles loaded with neem bark extract as green, slow-release biocide. *Journal of Cleaner Production*, 142: 4206–4213.

Mattos, B. D., Tardy, B. L., Magalhaes, W. L. E., and Rojas, O. J. (2017). Controlled release for crop and wood protection: Recent progress toward sustainable and safe nanostructured biocidal systems. *Journal of Controlled Release*, 262: 139–150.

McLaughlin, M. S., Roy, M., Abbasi, P. A., Carisse, O., Yurgel, S. N., and Ali, S. (2023). Why do we need alternative methods for fungal disease management in plants? *Plants*, 12(22): 3822.

Mendigaliyeva, S., and Bostan, D. S. (2020). Ionic liquids and deep eutectic solvents and their use for dissolving animal hair. *Advances in Chemical Engineering and Science*, 10: 40–51.

Menichetti, A., Mavridi-Printezi, A., Mordini, D., and Montalti, M. (2023). Effect of size, shape and surface functionalization on the antibacterial activity of silver nanoparticles. *Journal of Functional Biomaterials*, 14(5): 244.

Metryka, O., Wasilkowski, D., and Mrozik, A. (2022). Evaluation of the effects of Ag, Cu, ZnO and TiO2 nanoparticles on the expression level of oxidative stress-related genes and the activity of antioxidant enzymes in *Escherichia coli, Bacillus cereus* and *Staphylococcus epidermidis*. *International Journal of Molecular Sciences*, 23(9): 4966.

Mirmoeini, T., Pishkar, L., Kahrizi, D., Barzin, G., and Karimi, N. (2021). Phytotoxicity of green synthesized silver nanoparticles on Camelina sativa L. *Physiology and Molecular Biology of Plants*, 27(2): 417–427.

Mishra, S., Singh, B. R., Naqvi, A. H., and Singh, H. B. (2017). Potential of biosynthesized silver nanoparticles using *Stenotrophomonas* sp. BHU-S7 (MTCC 5978) for management of soil-borne and foliar phytopathogens. *Scientific Reports*, 7(1): 45154.

Mishra, S., Singh, B. R., Singh, A., Keswani, C., Naqvi, A. H., and Singh, H. B. (2014). Biofabricated silver nanoparticles act as a strong fungicide against Bipolaris sorokiniana causing spot blotch disease in wheat. *Plos One*, 9(5): e97881.

Mishra, S., and Singh, H. B. (2015). Biosynthesized silver nanoparticles as a nanoweapon against phytopathogens: exploring their scope and potential in agriculture. *Applied Microbiology and Biotechnology*, 99: 1097–1107.

Mitra, D., Adhikari, P., Djebaili, R., Thathola, P., Joshi, K., Pellegrini, M. et al. (2023). Biosynthesis and characterization of nanoparticles, its advantages, various aspects and risk assessment to maintain the sustainable agriculture: Emerging technology in modern era science. *Plant Physiology and Biochemistry*, 196: 103–120.

Mittal, D., Kaur, G., Singh, P., Yadav, K., and Ali, S. A. (2020). Nanoparticle-based sustainable agriculture and food science: Recent advances and future outlook. *Frontiers in Nanotechnology*, 2: 579954.

Mnif, I., and Ghribi, D. (2015). Potential of bacterial derived biopesticides in pest management. *Crop Protection*, 77: 52–64.

Mohanpuria, P., Rana, N. K., and Yadav, S. K. (2008). Biosynthesis of nanoparticles: technological concepts and future applications. *Journal of Nanoparticle Research*, 10: 507–517.

Mokarram-Kashtiban, S., Hosseini, S. M., Tabari Kouchaksaraei, M., and Younesi, H. (2019). The impact of nanoparticles zero-valent iron (nZVI) and rhizosphere microorganisms on the phytoremediation ability of white willow and its response. *Environmental Science and Pollution Research*, 26:, 10776–10789.

Moreno, A., Lucio-Hernández, D., and Cuéllar-Cruz, M. (2019). Biosynthesis of chemical compounds by Candida albicans and Candida glabrata. *Revista Iberoamericana de Micología*, 36(3): 120–128.

Mubraiz, N., Bano, A., Mahmood, T., and Khan, N. (2021). Microbial and plant assisted synthesis of cobalt oxide nanoparticles and their antimicrobial activities. *Agronomy*, 11(8): 1607.

Mukha, I. P., Eremenko, A. M., Smirnova, N. P., Mikhienkova, A. I., Korchak, G. I., Gorchev, V. F. et al. (2013). Antimicrobial activity of stable silver nanoparticles of a certain size. *Applied Biochemistry and Microbiology*, 49: 199–206.

Mukherjee, A., Maity, A., Pramanik, P., Shubha, K., Joshi, D. C., and Wani, S. H. (2019). Public perception about use of nanotechnology in agriculture. In *Advances in phytonanotechnology* (pp. 405–418). Elsevier.

Nadaroglu, H., Güngör, A. A., and Ince, S. (2017). Synthesis of nanoparticles by green synthesis method. *International Journal of Innovative Research and Reviews*, 1(1): 6–9.

Nadaroglu, H., Onem, H., and Alayli Gungor, A. (2017). Green synthesis of Ce2 O3 NPs and determination of its antioxidant activity. *IET Nanobiotechnology*, 11(4): 411–419.

Nagoth, J. A., John, M. S., Ramasamy, K. P., Mancini, A., Zannotti, M., Piras, S. et al. (2024). Synthesis of Bioactive Nickel Nanoparticles Using Bacterial Strains from an Antarctic Consortium. *Marine Drugs*, 22(2): 1–13. https://doi.org/10.3390/md22020089.

Nagy, A., Harrison, A., Sabbani, S., Munson Jr., R. S., Dutta, P. K., and Waldman, W. J. (2011). Silver nanoparticles embedded in zeolite membranes: release of silver ions and mechanism of antibacterial action. *International Journal of Nanomedicine*, 1833–1852.

Narayanan, K. B., and Sakthivel, N. (2011). Synthesis and characterization of nano-gold composite using Cylindrocladium floridanum and its heterogeneous catalysis in the degradation of 4-nitrophenol. *Journal of Hazardous Materials*, 189(1-2): 519–525.

Netala, V. R., Bukke, S., Domdi, L., Soneya, S., G. Reddy, S., Bethu, M. S. et al. (2018). Biogenesis of silver nanoparticles using leaf extract of *Indigofera hirsuta* L. and their potential biomedical applications (3-in-1 system). *Artificial Cells, Nanomedicine, and Biotechnology*, 46(sup1), 1138–1148.

Noha, K., Bondok, A. M., and El-Dougdoug, K. A. (2018). Evaluation of silver nanoparticles as antiviral agent against ToMV and PVY in tomato plants. *Sciences*, 8(01): 100–111.

Nongbet, A., Mishra, A. K., Mohanta, Y. K., Mahanta, S., Ray, M. K., Khan, M. et al. (2022). Nanofertilizers: A Smart and Sustainable Attribute to Modern Agriculture. *Plants*, 11(19): 1–20. https://doi.org/10.3390/plants11192587.

Noor, R., Yasmin, H., Ilyas, N., Nosheen, A., Hassan, M. N., Mumtaz, S. et al. (2022). Comparative analysis of iron oxide nanoparticles synthesized from ginger (Zingiber officinale) and cumin seeds (*Cuminum cyminum*) to induce resistance in wheat against drought stress. *Chemosphere*, 292: 133201.

Ogunyemi, S. O., Abdallah, Y., Zhang, M., Fouad, H., Hong, X., Ibrahim, E. et al. (2019). Green synthesis of zinc oxide nanoparticles using different plant extracts and their antibacterial activity against *Xanthomonas oryzae* pv. oryzae. *Artificial Cells, Nanomedicine, and Biotechnology*, 47(1): 341–352.

Ogunyemi, S. O., Zhang, M., Abdallah, Y., Ahmed, T., Qiu, W., Ali, M. A. et al. (2020). The bio-synthesis of three metal oxide nanoparticles (ZnO, MnO2, and MgO) and their antibacterial activity against the bacterial leaf blight pathogen. *Frontiers in Microbiology*, 11: 588326.

Ohde, H., Hunt, F., and Wai, C. M. (2001). Synthesis of silver and copper nanoparticles in a water-in-supercritical-carbon dioxide microemulsion. *Chemistry of Materials*, 13(11): 4130–4135.

Okeke, E. S., Ezeorba, T. P. C., Mao, G., Chen, Y., Feng, W., and Wu, X. (2022). Nano-enabled agrochemicals/materials: Potential human health impact, risk assessment, management strategies and future prospects. *Environmental Pollution*, 295: 118722.

Padmavathy, N., and Vijayaraghavan, R. (2008). Enhanced bioactivity of ZnO nanoparticles—an antimicrobial study. *Science and Technology of Advanced Materials*.

Padmavathy, N., and Vijayaraghavan, R. (2011). Interaction of ZnO nanoparticles with microbes—a physio and biochemical assay. *Journal of Biomedical Nanotechnology*, 7(6): 813–822.

Pan, W.-Y., Huang, C.-C., Lin, T.-T., Hu, H.-Y., Lin, W.-C., Li, M.-J. et al. (2016). Synergistic antibacterial effects of localized heat and oxidative stress caused by hydroxyl radicals mediated by graphene/iron oxide-based nanocomposites. *Nanomedicine: Nanotechnology, Biology and Medicine*, 12(2): 431–438.

Panth, M., Hassler, S. C., and Baysal-Gurel, F. (2020). Methods for management of soilborne diseases in crop production. *Agriculture*, 10(1): 16.

Pariona, N., Mtz-Enriquez, A. I., Sánchez-Rangel, D., Carrión, G., Paraguay-Delgado, F., and Rosas-Saito, G. (2019). Green-synthesized copper nanoparticles as a potential antifungal against plant pathogens. *RSC Advances*, 9(33): 18835–18843.

Parisi, C., Vigani, M., and Rodríguez-Cerezo, E. (2015). Agricultural nanotechnologies: what are the current possibilities? *Nano Today*, 10(2): 124–127.

Parveen, S., Wani, A. H., Shah, M. A., Devi, H. S., Bhat, M. Y., and Koka, J. A. (2018). Preparation, characterization and antifungal activity of iron oxide nanoparticles. *Microbial Pathogenesis*, 115: 287–292.

Patel, H., Chander, P. S., Soundarya, J., Vuppala, S., Balaji, R., and Chandrasekar, N. (2022). *A Review on Green Synthesis of Nanoparticles and Their Diverse Biomedical and Environmental Applications. Catalysts 2022, 12, 459.* s Note: MDPI stays neutral with regard to jurisdictional claims in published ....

Patel, V., Berthold, D., Puranik, P., and Gantar, M. (2015). Screening of cyanobacteria and microalgae for their ability to synthesize silver nanoparticles with antibacterial activity. *Biotechnology Reports*, 5: 112–119.

Paulkumar, K., Rajeshkumar, S., Gnanajobitha, G., Vanaja, M., Malarkodi, C., and Annadurai, G. (2013). Biosynthesis of silver chloride nanoparticles using Bacillus subtilis MTCC 3053 and assessment of its antifungal activity. *International Scholarly Research Notices*, 2013(1): 317963.

Paulraj, M. G., Ignacimuthu, S., Gandhi, M. R., Shajahan, A., Ganesan, P., Packiam, S. M. et al. (2017). Comparative studies of tripolyphosphate and glutaraldehyde cross-linked chitosan-botanical pesticide nanoparticles and their agricultural applications. *International Journal of Biological Macromolecules*, 104: 1813–1819.

Pavela, R., and Benelli, G. (2016). Essential oils as ecofriendly biopesticides? Challenges and constraints. *Trends in Plant Science*, 21(12): 1000–1007.

Pollet, P., Eckert, C. A., and Liotta, C. L. (2011). Solvents for sustainable chemical processes. *WIT Trans Ecol Environ*, 154: 21–31.

Ponmurugan, P., Manjukarunambika, K., Elango, V., and Gnanamangai, B. M. (2016). Antifungal activity of biosynthesised copper nanoparticles evaluated against red root-rot disease in tea plants. *Journal of Experimental Nanoscience*, 11(13): 1019–1031.

Popova, I. E., Dubie, J. S., and Morra, M. J. (2017). Optimization of hydrolysis conditions for release of biopesticides from glucosinolates in *Brassica juncea* and *Sinapis alba* seed meal extracts. *Industrial Crops and Products*, 97: 354–359.

Porrawatkul, P., Nuthong, W., Pimsen, R., and Thongsom, M. (2017). Green synthesis of silver nanoparticles using *Barringtonia acutangula* (L.) Gaertn leaf extract as reducing agent and their antibacterial and antioxidant activity. *Journal of Applied Science and Emerging Technology*, 16(Special): 75–81.

Pramanik, A., Laha, D., Bhattacharya, D., Pramanik, P., and Karmakar, P. (2012). A novel study of antibacterial activity of copper iodide nanoparticle mediated by DNA and membrane damage. *Colloids and Surfaces B: Biointerfaces*, 96: 50–55.

Pucciarelli, M. (2023). *Life Cycle Thinking for the sustainability assessment of nanoparticles' manufacturing and applications.*

Qin, J., Tong, Z., Zhan, Y., Buisson, C., Song, F., He, K. et al. (2020). A Bacillus thuringiensis chitin-binding protein is involved in insect peritrophic matrix adhesion and takes part in the infection process. *Toxins*, 12(4): 252.

Omran, B. A., and Baek, K.-H. (2022). Control of phytopathogens using sustainable biogenic nanomaterials: recent perspectives, ecological safety, and challenging gaps. *Journal of Cleaner Production*, 372: 133729.

Raguso, R. A. (2016). More lessons from linalool: insights gained from a ubiquitous floral volatile. *Current Opinion in Plant Biology*, 32: 31–36.

Rahmat, M., Bhatti, H. N., Rehman, A., Chaudhry, H., Yameen, M., Iqbal, M. et al. (2021). Bionanocomposite of Au decorated MnO2 via in situ green synthesis route and antimicrobial activity evaluation. *Arabian Journal of Chemistry*, 14(12): 103415.

Rai, M., and Ingle, A. (2012). Role of nanotechnology in agriculture with special reference to management of insect pests. *Applied Microbiology and Biotechnology*, 94: 287–293.

Rajani, Mishra, P., Kumari, S., Saini, P., and Meena, R. K. (2022). Role of nanotechnology in management of plant viral diseases. *Materials Today: Proceedings*, 69: 1–10. https://doi.org/10.1016/j.matpr.2022.06.355.

Ram, L. J., Silke, D.-S., Per, K., and Antoine, M. (2016). *Toward a Reduced Reliance on Conventional Pesticides in European Agriculture.*

Rehman, A., Ahmad, T., Aadil, R. M., Spotti, M. J., Bakry, A. M., Khan, I. M. et al. (2019). Pectin polymers as wall materials for the nano-encapsulation of bioactive compounds. *Trends in Food Science and Technology*, 90: 35–46.

Ribeiro, R. C., Zanuncio, T. V., de Sousa Ramalho, F., da Silva, C. A. D., Serrão, J. E., and Zanuncio, J. C. (2015). Feeding and oviposition of Anticarsia gemmatalis (Lepidoptera: Noctuidae) with sublethal concentrations of ten condiments essential oils. *Industrial Crops and Products*, 74: 139–143.

Richard, B., Qi, A., and Fitt, B. D. L. (2022). Control of crop diseases through Integrated Crop Management to deliver climate-smart farming systems for low-and high-input crop production. *Plant Pathology*, 71(1): 187–206.

Riseh, R. S., Skorik, Y. A., Thakur, V. K., Pour, M. M., Tamanadar, E., and Noghabi, S. S. (2021). Encapsulation of plant biocontrol bacteria with alginate as a main polymer material. *International Journal of Molecular Sciences*, *22*(20). https://doi.org/10.3390/ijms222011165.
International Journal of Molecular Sciences, *22*(20). https://doi.org/10.3390/ijms222011165
Rolli, E., Marieschi, M., Maietti, S., Sacchetti, G., and Bruni, R. (2014). Comparative phytotoxicity of 25 essential oils on pre-and post-emergence development of *Solanum lycopersicum* L.: A multivariate approach. *Industrial Crops and Products*, 60: 280–290.
Saif, S., Tahir, A., Asim, T., and Chen, Y. (2016). Plant mediated green synthesis of CuO nanoparticles: comparison of toxicity of engineered and plant mediated CuO nanoparticles towards Daphnia magna. *Nanomaterials*, 6(11): 205.
Salem, S. S., and Fouda, A. (2021). Green synthesis of metallic nanoparticles and their prospective biotechnological applications: an overview. *Biological Trace Element Research*, 199(1): 344–370.
Sari, K. R. P., Ikawati, Z., Danarti, R., and Hertiani, T. (2023). Micro-titer plate assay for measurement of total phenolic and total flavonoid contents in medicinal plant extracts. *Arabian Journal of Chemistry*, 16(9): 105003.
Sarwer, Q., Amjad, M. S., Mehmood, A., Binish, Z., Mustafa, G., Farooq, A. et al. (2022). Green synthesis and characterization of silver nanoparticles using myrsine africana leaf extract for their antibacterial, antioxidant and phytotoxic activities. *Molecules*, 27(21): 1–14. https://doi.org/10.3390/molecules27217612.
Satalkar, P., Elger, B. S., and Shaw, D. M. (2016). Defining nano, nanotechnology and nanomedicine: why should it matter? *Science and Engineering Ethics*, 22: 1255–1276.
Sattar, S., Siddiqui, S., Shahzad, A., Bano, A., Naeem, M., Hussain, R. et al. (2022). Comparative analysis of microbial consortiums and nanoparticles for rehabilitating petroleum waste contaminated soils. *Molecules*, 27(6): 1945.
Seiber, J. N., Coats, J., Duke, S. O., and Gross, A. D. (2014). Biopesticides: state of the art and future opportunities. *Journal of Agricultural and Food Chemistry*, 62(48): 11613–11619.
Selim, M. (2019). A review of advantages, disadvantages and challenges of crop rotations. *Egyptian Journal of Agronomy*, 41(1): 1–10.
Servin, A. D., and White, J. C. (2016). Nanotechnology in agriculture: next steps for understanding engineered nanoparticle exposure and risk. *NanoImpact*, 1: 9–12.
Shah, K. K., Modi, B., Pandey, H. P., Subedi, A., Aryal, G., Pandey, M. et al. (2021). Diversified crop rotation: an approach for sustainable agriculture production. *Advances in Agriculture*, 2021(1): 8924087.
Shanker, U., Jassal, V., Rani, M., and Kaith, B. S. (2016). Towards green synthesis of nanoparticles: from bio-assisted sources to benign solvents. A review. *International Journal of Environmental Analytical Chemistry*, 96(9): 801–835.
Sharma, D., Sabela, M. I., Kanchi, S., Mdluli, P. S., Singh, G., Stenström, T. A. et al. (2016). Biosynthesis of ZnO nanoparticles using Jacaranda mimosifolia flowers extract: synergistic antibacterial activity and molecular simulated facet specific adsorption studies. *Journal of Photochemistry and Photobiology B: Biology*, 162: 199–207.
Sharma, R., Bisen, D. P., Shukla, U., and Sharma, B. G. (2012). X-ray diffraction: a powerful method of characterizing nanomaterials. *Recent Research in Science and Technology*, 4(8), 77–79.
Sharmin, S., Rahaman, M. M., Sarkar, C., Atolani, O., Islam, M. T., and Adeyemi, O. S. (2021). Nanoparticles as antimicrobial and antiviral agents: A literature-based perspective study. *Heliyon*, 7(3): e06456. https://doi.org/10.1016/j.heliyon.2021.e06456.
Shi, Y.-M., and Bode, H. B. (2018). Chemical language and warfare of bacterial natural products in bacteria–nematode–insect interactions. *Natural Product Reports*, 35(4): 309–335.
Siddiqi, K. S., and Husen, A. (2016). Fabrication of metal nanoparticles from fungi and metal salts: scope and application. *Nanoscale Research Letters*, 11: 1–15.
Singh, P., Kim, Y.-J., Zhang, D., and Yang, D.-C. (2016). Biological synthesis of nanoparticles from plants and microorganisms. *Trends in Biotechnology*, 34(7): 588–599.
Sirelkhatim, A., Mahmud, S., Seeni, A., Kaus, N. H. M., Ann, L. C., Bakhori, S. K. M. et al. (2015). Review on zinc oxide nanoparticles: antibacterial activity and toxicity mechanism. *Nano-Micro Letters*, 7: 219–242.
Slavin, Y. N., Asnis, J., Hńfeli, U. O., and Bach, H. (2017). Metal nanoparticles: understanding the mechanisms behind antibacterial activity. *Journal of Nanobiotechnology*, 15: 1–20.
Slomberg, D. L., and Schoenfisch, M. H. (2012). Silica nanoparticle phytotoxicity to Arabidopsis thaliana. *Environmental Science and Technology*, 46(18): 10247–10254.
Solís-Sandí, I., Cordero-Fuentes, S., Pereira-Reyes, R., Vega-Baudrit, J. R., Batista-Menezes, D., and Montes de Oca-Vásquez, G. (2023). Optimization of the biosynthesis of silver nanoparticles using bacterial extracts and their antimicrobial potential. *Biotechnology Reports*, 40(October), 0–10. https://doi.org/10.1016/j.btre.2023.e00816.
Song, M., Yun, H. Y., and Kim, Y. H. (2014). Antagonistic Bacillus species as a biological control of ginseng root rot caused by Fusarium cf. incarnatum. *Journal of Ginseng Research*, 38(2): 136–145.

Souri, M., and Shakeri, A. (2018). Comparison of microwave and ultrasound assisted extraction methods on total phenol and tannin content and biological activity of *Dittrichia graveolens* (L.) Greuter and its optimization by response surface methodology. *Curr. Bioact. Compd*, 14: 23.

Sousa, F., Ferreira, D., Reis, S., and Costa, P. (2020). Current insights on antifungal therapy: Novel nanotechnology approaches for drug delivery systems and new drugs from natural sources. *Pharmaceuticals*, 13(9): 248.

Sousa, R. M. O. F., Rosa, J. S., Oliveira, L., Cunha, A., and Fernandes-Ferreira, M. (2015). Activities of Apiaceae essential oils and volatile compounds on hatchability, development, reproduction and nutrition of *Pseudaletia unipuncta* (Lepidoptera: Noctuidae). *Industrial Crops and Products*, 63: 226–237.

Sparks, T. C., Hahn, D. R., and Garizi, N. V. (2017). Natural products, their derivatives, mimics and synthetic equivalents: role in agrochemical discovery. *Pest Management Science*, 73(4): 700–715.

Srinivasan, M., Venkatesan, M., Arumugam, V., Natesan, G., Saravanan, N., Murugesan, S. et al. (2019). Green synthesis and characterization of titanium dioxide nanoparticles (TiO2 NPs) using Sesbania grandiflora and evaluation of toxicity in zebrafish embryos. *Process Biochemistry*, 80: 197–202.

Srivastava, V. (2014). *In situ* generation of Ru nanoparticles to catalyze CO2 hydrogenation to formic acid. *Catalysis Letters*, 144: 1745–1750.

Stephenson, G. R. (2003). 15 Pesticide Use and World Food Production: Risks and Benefits. *ACS Symposium Series*, 853: 261–270.

Sun, Y., and Xia, Y. (2002). Large-scale synthesis of uniform silver nanowires through a soft, self-seeding, polyol process. *Advanced Materials*, 14(11): 833–837.

Sundararajan, N., Habeebsheriff, H. S., Dhanabalan, K., Cong, V. H., Wong, L. S., Rajamani, R. et al. (2024). Mitigating global challenges: harnessing green synthesized nanomaterials for sustainable crop production systems. *Global Challenges*, 8(1): 2300187.

Sunkar, S., and Nachiyar, C. V. (2012). Biogenesis of antibacterial silver nanoparticles using the endophytic bacterium Bacillus cereus isolated from Garcinia xanthochymus. *Asian Pacific Journal of Tropical Biomedicine*, 2(12): 953–959. https://doi.org/10.1016/S2221-1691(13)60006-4.

Swamy, H. M., and Asokan, R. (2013). Bacillus thuringiensis as 'nanoparticles'-a perspective for crop protection. *Nanoscience and Nanotechnology-Asia*, 3(1): 102–105.

Sylvestre, J.-P., Poulin, S., Kabashin, A. V, Sacher, E., Meunier, M., and Luong, J. H. T. (2004). Surface chemistry of gold nanoparticles produced by laser ablation in aqueous media. *The Journal of Physical Chemistry B*, 108(43): 16864–16869.

Synowiec, A., Kalemba, D., Drozdek, E., and Bocianowski, J. (2017). Phytotoxic potential of essential oils from temperate climate plants against the germination of selected weeds and crops. *Journal of Pest Science*, 90: 407–419.

Tang, G., Tian, Y., Niu, J., Tang, J., Yang, J., Gao, Y. et al. (2021). Development of carrier-free self-assembled nanoparticles based on fenhexamid and polyhexamethylene biguanide for sustainable plant disease management. *Green Chemistry*, 23(6): 2531–2540.

Tariq, F., Ahmed, N., Afzal, M., Khan, M. A. U., and Zeshan, B. (2020). Synthesis, characterization and antimicrobial activity of bacillus subtilis-derived silver nanoparticles against multidrug-resistant bacteria. *Jundishapur Journal of Microbiology*, 13(5): 1–7. https://doi.org/10.5812/jjm.91934.

Thakkar, K. N., Mhatre, S. S., and Parikh, R. Y. (2010). Biological synthesis of metallic nanoparticles. *Nanomedicine: Nanotechnology, Biology and Medicine*, 6(2): 257–262.

Timmusk, S., Behers, L., Muthoni, J., Muraya, A., and Aronsson, A.-C. (2017). Perspectives and challenges of microbial application for crop improvement. *Frontiers in Plant Science*, 8: 49.

Titus, D., James Jebaseelan Samuel, E., and Roopan, S. M. (2018). Nanoparticle characterization techniques. In *Green Synthesis, Characterization and Applications of Nanoparticles*. Elsevier Inc. https://doi.org/10.1016/B978-0-08-102579-6.00012-5.

Titus, D., Samuel, E. J. J., and Roopan, S. M. (2019). Nanoparticle characterization techniques. In *Green synthesis, characterization and applications of nanoparticles* (pp. 303–319). Elsevier.

Tran, Q. H., and Le, A.-T. (2013). Silver nanoparticles: synthesis, properties, toxicology, applications and perspectives. *Advances in Natural Sciences: Nanoscience and Nanotechnology*, 4(3): 33001.

Tripathi, D. K., Tripathi, A., Shweta, Singh, S., Singh, Y., Vishwakarma, K., Yadav, G. et al. (2017). Uptake, accumulation and toxicity of silver nanoparticle in autotrophic plants, and heterotrophic microbes: a concentric review. *Frontiers in Microbiology*, 8: 07.

Trivedi, P., Schenk, P. M., Wallenstein, M. D., and Singh, B. K. (2017). Tiny microbes, big yields: enhancing food crop production with biological solutions. *Microbial Biotechnology*, 10(5): 999–1003.

Tsuji, K. (2001). Microencapsulation of pesticides and their improved handling safety. *Journal of Microencapsulation*, 18(2): 137–147.

Van Rantwijk, F., and Sheldon, R. A. (2007). Biocatalysis in ionic liquids. *Chemical Reviews*, 107(6): 2757–2785.

Velidandi, A., Pabbathi, N. P. P., Dahariya, S., and Baadhe, R. R. (2021). Green synthesis of novel Ag–Cu and Ag–Znbimetallic nanoparticles and their *in vitro* biological, eco-toxicity and catalytic studies. *Nano-Structures and Nano-Objects*, 26: 100687.

Verma, D. K., Patel, S., and Kushwah, K. S. (2020). Green biosynthesis of silver nanoparticles and impact on growth, chlorophyll, yield and phytotoxicity of *Phaseolus vulgaris* L. *Vegetos*, 33: 648–657.

Verma, K. K., Song, X. P., Degu, H. D., Guo, D. J., Joshi, A., Huang, H. R. et al. (2023). Recent advances in nitrogen and nano-nitrogen fertilizers for sustainable crop production: a mini-review. *Chemical and Biological Technologies in Agriculture*, 10(1): 1–14. https://doi.org/10.1186/s40538-023-00488-3.

Vidya, C., Hiremath, S., Chandraprabha, M. N., Antonyraj, M. L., Gopal, I. V., Jain, A. et al. (2013). Green synthesis of ZnO nanoparticles by Calotropis gigantea. *Int. J. Curr. Eng. Technol.*, 1(1): 118–120.

Vidya, C., Manjunatha, C., Chandraprabha, M. N., Rajshekar, M., and MAL, A. R. (2017). Hazard free green synthesis of ZnO nano-photo-catalyst using Artocarpus Heterophyllus leaf extract for the degradation of Congo red dye in water treatment applications. *Journal of Environmental Chemical Engineering*, 5(4): 3172–3180.

Vijayakumar, S., Vaseeharan, B., Malaikozhundan, B., Gopi, N., Ekambaram, P., Pachaiappan, R. et al. (2017). Therapeutic effects of gold nanoparticles synthesized using Musa paradisiaca peel extract against multiple antibiotic resistant *Enterococcus faecalis* biofilms and human lung cancer cells (A549). *Microbial Pathogenesis*, 102: 173–183.

Virkutyte, J., and Varma, R. S. (2011). Green synthesis of metal nanoparticles: biodegradable polymers and enzymes in stabilization and surface functionalization. *Chemical Science*, 2(5): 837–846.

Vollmer, C., Redel, E., Abu-Shandi, K., Thomann, R., Manyar, H., Hardacre, C. et al. (2010). Microwave irradiation for the facile synthesis of Transition-Metal nanoparticles (NPs) in Ionic Liquids (ILs) from Metal–Carbonyl Precursors and Ru-, Rh-, and Ir-NP/IL Dispersions as Biphasic Liquid–Liquid Hydrogenation Nanocatalysts for Cyclohexene. *Chemistry–A European Journal*, 16(12): 3849–3858.

Wadhwani, S. A., Shedbalkar, U. U., Singh, R., and Chopade, B. A. (2016). Biogenic selenium nanoparticles: current status and future prospects. *Applied Microbiology and Biotechnology*, 100: 2555–2566.

Wang, L., Hu, C., and Shao, L. (2017). The antimicrobial activity of nanoparticles: present situation and prospects for the future. *International Journal of Nanomedicine*, 12: 1227–1249.

Wang, X., Liu, X., and Han, H. (2013). Evaluation of antibacterial effects of carbon nanomaterials against copper-resistant Ralstonia solanacearum. *Colloids and Surfaces B: Biointerfaces*, 103: 136–142.

Wanyika, H., Gatebe, E., Kioni, P., Tang, Z., and Gao, Y. (2012). Mesoporous silica nanoparticles carrier for urea: potential applications in agrochemical delivery systems. *Journal of Nanoscience and Nanotechnology*, 12(3): 2221–2228.

Wasule, D. L., Shingote, P. R., and Saxena, S. (2024). Exploitation of functionalized green nanomaterials for plant disease management. *Discover Nano*, 19(1): 118.

Wehling, J., Dringen, R., Zare, R. N., Maas, M., and Rezwan, K. (2014). Bactericidal activity of partially oxidized nanodiamonds. *ACS Nano*, 8(6): 6475–6483.

Wei, X., Luo, M., Li, W., Yang, L., Liang, X., Xu, L. et al. (2012). Synthesis of silver nanoparticles by solar irradiation of cell-free Bacillus amyloliquefaciens extracts and AgNO3. *Bioresource Technology*, 103(1): 273–278.

Welton, T. (2004). Ionic liquids in catalysis. *Coordination Chemistry Reviews*, 248(21–24): 2459–2477.

Wittmann, K., Wisniewski, W., Mynott, R., Leitner, W., Kranemann, C. L., Rische, T. et al. (2001). Supercritical carbon dioxide as solvent and temporary protecting group for rhodium-catalyzed hydroaminomethylation. *Chemistry–A European Journal*, 7(21): 4584–4589.

Worrall, E. A., Hamid, A., Mody, K. T., Mitter, N., and Pappu, H. R. (2018). Nanotechnology for plant disease management. *Agronomy*, 8(12): 285.

Wu, B., Zhuang, W.-Q., Sahu, M., Biswas, P., and Tang, Y. J. (2011). Cu-doped TiO2 nanoparticles enhance survival of Shewanella oneidensis MR-1 under Ultraviolet Light (UV) exposure. *Science of the Total Environment*, 409(21): 4635–4639.

Xie, J., Chen, K., and Chen, X. (2009). Production, modification and bio-applications of magnetic nanoparticles gestated by magnetotactic bacteria. *Nano Research*, 2: 261–278.

Yadav, A., and Yadav, K. (2018). Nanoparticle-based plant disease management: tools for sustainable agriculture. In *Nanotechnology in the Life Sciences* (Issue December). https://doi.org/10.1007/978-3-319-91161-8_2.

Yoosaf, K., Ipe, B. I., Suresh, C. H., and Thomas, K. G. (2007). In situ synthesis of metal nanoparticles and selective naked-eye detection of lead ions from aqueous media. *The Journal of Physical Chemistry C*, 111(34): 12839–12847.

Young, M., Ozcan, A., Myers, M. E., Johnson, E. G., Graham, J. H., and Santra, S. (2017). Multimodal generally recognized as safe ZnO/nanocopper composite: A novel antimicrobial material for the management of citrus phytopathogens. *Journal of Agricultural and Food Chemistry*, 66(26): 6604–6608.

Yu, J., Zhang, W., Li, Y., Wang, G., Yang, L., Jin, J. et al. (2014). Synthesis, characterization, antimicrobial activity and mechanism of a novel hydroxyapatite whisker/nano zinc oxide biomaterial. *Biomedical Materials*, 10(1): 15001.

Yurkov, A. M., Kemler, M., and Begerow, D. (2011). Species accumulation curves and incidence-based species richness estimators to appraise the diversity of cultivable yeasts from beech forest soils. *PLoS One*, 6(8): e23671.

Yurtluk, T., Akçay, F. A., and Avcı, A. (2018). Biosynthesis of silver nanoparticles using novel Bacillus sp. SBT8. *Preparative Biochemistry and Biotechnology*, 48(2): 151–159.

Żarowska, B., Koźlecki, T., Piegza, M., Jaros-Koźlecka, K., and Robak, M. (2019). New look on antifungal activity of silver nanoparticles (AgNPs). *Polish Journal of Microbiology*, 68(4): 515–525.

Zavitri, N. G., Syahbaniati, A. P., Primastuti, R. K., Putri, R. M., Damayanti, S., and Wibowo, I. (2023). Toxicity evaluation of zinc oxide nanoparticles green synthesized using papaya extract in zebrafish. *Biomedical Reports*, 19(6): 1–11.

Zeng, Y., Shang, Z., Zheng, Z., Shi, N., Yang, B., Han, S. et al. (2024). A review of chemical modification of vegetable oils and their applications. *Lubricants*, 12(5): 180.

Zhang, H., and Cui, H. (2009). Synthesis and characterization of functionalized ionic liquid-stabilized metal (gold and platinum) nanoparticles and metal nanoparticle/carbon nanotube hybrids. *Langmuir*, 25(5): 2604–2612.

Zhang, Z. C. (2006). Catalysis in ionic liquids. *Advances in Catalysis*, 49: 153–237.

Zhou, M., Zhang, X., Quan, Y., Tian, Y., Chen, J., and Li, L. (2024). Visible light-induced photocatalytic and antibacterial adhesion properties of superhydrophilic TiO2 nanoparticles. *Scientific Reports*, 14(1): 1–14. https://doi.org/10.1038/s41598-024-58660-0.

CHAPTER-10

Plant Viral Disease Management Based on Green Nanotechnology

Esha Kumari, Sneha Kumari, Sai Santosh Das and *Jyoti Prakash Sahoo**

1. Introduction

Viruses are necessary internal parasites that proliferate by depleting their host's resources, which finally results in the host's disease. Most living things, including people, animals, and plants, are susceptible to widespread viral infections. Plant viruses are the source of many serious plant diseases that are either incurable or, in some cases, for which there is no recognized treatment. Plant diseases might be caused by about 900 distinct kinds of pathogenic viruses. Plant viruses occasionally cause pandemics and epidemics that wipe out enormous harvests, endangering the world's food supply. Viral infections are the root cause of many of the most prevalent diseases. Common plant diseases include *tobacco leaf roll*, *papaya leaf curl*, *yellow leaf curl*, and *dwarf mosaic diseases of tobacco*, *okra*, *cauliflower*, *and maize*. To meet the increasing demand for plant-based products throughout the world, plant diseases must be controlled. A few biotic vectors (fungi, insects, nematodes, etc.) and vegetative propagation are the means by which plant virus infections are transmitted from infected to healthy plants. Effective strategies for managing these biotic agents considerably lower the incidence of viral infections in plants. Chemotherapy is ineffective against viral particles that have already multiplied, but it can stop new ones from forming. Other approaches to treat viral infections include *post-transcriptional gene silencing*, or *RNA interference (RNAi)*, and the development of transgenic plants, also referred to as virus-resistant plants. Viral infections are classified and diagnosed using *Next-Generation Sequencing (NGS)* techniques. Since viral genotypes are more likely to undergo mutation, these methods are not t always enough.

Humanity has developed a variety of defence systems ever since it was realized that viruses may be dangerous. In an attempt to effectively prevent the entry and spread of various disease-causing insects and other vectors, several nations have imposed trade embargoes on agricultural items and commodities that are vital to their economy, such as grains. Even with these defences, crops are still vulnerable to a variety of insect pest infestations as well as bacterial, fungal, and viral diseases. Twenty to forty percent of the global agricultural output loss is attributed to insect attacks, while plant diseases account for 16% of yield loss (Fite and Partners, 2018). Projections assert that diseases, pests, and weeds might cause 15 to 20% of India's yearly food production to be lost. Since plant viruses are the tiniest organisms, they pose a serious threat to agricultural productivity. Viral infections account for 47% of plant diseases that reduce global food production. Phyto viruses

Faculty of Agriculture and Allied Sciences, C.V. Raman Global University, Bhubaneswar-752054, India.
* Corresponding author: jyotiprakashsahoo2010@gmail.com

DOI: 10.1201/9781003567684-10

in the host plant can cause a range of symptoms, such as twisted, mottled, and stunted fruits and foliage, due to physiological, molecular, and biochemical changes. Reduced crop growth and production result in losses, and these symptoms render the affected plants unsellable. Furthermore, fruiting, and flowering conditions may suffer from viral pathogen infection. The most common and harmful virus affecting sunflower harvests in India and other countries where sunflowers are grown is *TobaccoStripe Virus (TSV)*, which causes sunflower necrosis. Economically significant crops including cotton, legumes, tomatoes, okra, chillies, and others suffer significant losses due to gemini viruses. Using nanotechnology in this situation could be beneficial. The timely discovery of the potential of nanoparticles (NPs) has led to a rapid expansion of the scientific field of nanotechnology in recent years. Agricultural manufacturing processes are generally low-cost because nanoparticles may be used in high-sensitivity biomolecular detection, disease diagnostics, and the production of antibacterial and therapeutic compounds. Consequently, they ought to increase agricultural output.

NPs have demonstrated promise in the treatment of diseases in plants because of their large surface area, robust reactivity, and small size (less than 100 nm). The use of nano-encapsulated pesticides may provide longer crop protection with less chemical needs than traditional pesticides. NPs can transport a higher concentration of active compounds because they have a bigger active surface area than traditional carrier molecules. Moreover, nanoparticles reduce the damage they do to non-target species by increasing target selectivity. Additionally, NPs can improve a plant's general health and fortify its resilience to disease. The ethical use of nanoscale technology should enable the detection and treatment of diseases in the near future. Research on nano phytovirology is still in its infancy. Good results from continuing research suggest that NP might be utilized as an antiviral medication in addition to plant defence activators. The goal of this comprehensive analysis is to gather as much information as possible about the use of nanotechnology in the treatment of viral diseases.

2. Nanotechnology, and Nano phytovirology

"The understanding and control of matter at dimensions between approximately 1 and 100 nm, where unique phenomena enable novel applications" is the National Nanotechnology Initiative's broad definition of nanotechnology (Drexler, 1986). The surface area of nanoparticles is significantly higher than that of their bulk counterparts. Volume ratio in addition to unique physical-chemical characteristics. Enhanced biological activity, thermal conductivity, chemical stability, and catalytic reactivity are a few of these qualities, according to Taniguchi (1974) and Edelstein and Cammaratra (1996). The two most innovative and captivating scientific fields of the modern era are biology and nanotechnology (Prasad, 2009).

The establishment of environmentally benign and biosynthetic processes for the synthesis of materials based on nanoparticles is the aim of bio nanotechnology, or the fusion of biotechnology with nanotechnology (Sobha and Surendranath, 2010). Pesticides' active ingredients can conjugate, encapsulate, or function as an adsorbent with nanoparticles to protect plants. They can also function as a precise and focused distribution plan. Because nanomaterials have demonstrated antifungal, antibacterial, and antiviral effects, they can be topically administered to treated seeds and roots. A few ways that nanomaterials influence living organisms include agglutination, rupture of cell membranes, reduced H^+-ATPase function, suppression of messenger RNA and protein synthesis, reduction in microbial growth, inhibition of toxin development, and blockage of nutrition flow (Malerba and Cerana, 2016; Dakal et al., 2016). Another crucial component of plant health products is nano fertilizers, which can increase a plant's resistance to disease. However, according to Marwal and Gaur (2019), nano phytovirology is the application of nanotechnology to the early detection, diagnosis, and control of viruses and their vectors, particularly arthropods, with the goal of protecting plants against economically significant viral diseases.

Treatment for viral plant infections is challenging since the disease spreads swiftly due to vectors, mostly arthropods. The agriculture industry is at risk due to the spreading of a viral epidemic. Therefore, there is an urgent need for improved viral management, which has been achieved using a range of methods, most of which depend on the ecology of viruses. While various techniques have been developed, only a few numbers have been shown to be effective in minimizing virus-related losses in agriculture. Understanding the interactions that take place between plants, viruses, and their vector is essential to combating viral diseases. Because of their affordability and ease of use, nanotechnology-based diagnostic and treatment techniques are becoming more and more popular as a means of reducing the threat posed by plant viral infections (Marwal and Gaur, 2019). The benefits of using products with nanotechnology comprise decreased toxicity, extended shelf life, and enhanced pesticide solubility—all of which are advantageous to the environment (Worrall et al., 2018). In order to treat plant diseases, two types of nanoparticles (NPs) are used: (i) NPs that function as independent defenders, and (ii) NPs that function as nano transporters for pesticides and other chemicals that alter RNAs. Moreover, nanoparticles can be employed as biostimulants or, more specifically, as nanobiosensors to identify viruses that infect plants.

3. Nanotechnology in Plant Disease Management

Plant diseases caused by bacteria, fungi, viruses, nematodes, and other pests drastically reduce crop output. Many diseases caused by worms, bacteria, fungi, and viruses can be treated with nanotechnology. Because of their higher size-to-surface ratio, smaller nanoparticles are more adept at adhering to and piercing the surfaces of microbial cells. First, using nanosized silica-silver particles and a range of microorganisms, including *Pythium* sp., *Colletotrichum* sp., *Pseudomonas syringae*, *Xanthomonas compestris*, etc. However, Park et al. (2006) evaluated the anti microbiobial effectiveness of AgNPs. Numerous studies have looked at the nematocidal and antifungal properties of silver nanoparticles. Chitosan is another well known nanoparticle because of its low toxicity, biodegradability, and biocompatibility. It also possesses antibacterial qualities. Chickpea plants infected with *Fusarium oxysporum* f. sp. ciceri exhibited enhanced resistance to infection when treated with copper oxide (Cur-Cu) nanoparticles. The antibacterial experiment's findings show that magnesium oxide nanoparticles (MgO NPs) effectively reduce the *Ralstonia solanacearum* bacterium, which causes bacterial wilt disease in tomato plants.

MgO NPs are more effective than macroscale MgO at reducing soil-borne fungal phytopathogens, according to research comparing the two. Currently, scientists are investigating the potential application of nanoparticles as antimicrobial agents. The mechanism by which NPs suppress bacteria is not well studied. Moreover, Hajipour et al. (2012) addressed the problems of nuclear and cytoplasmic material leakage, cell wall rupture, protein and membrane degradation, and the generation of ROS. These mechanisms combat a range of plant diseases. Plant biomolecules may be impacted by ROS production caused by nanoparticles. By promoting the transcription of defence-related genes, plants can produce phenylalanine, salicylic acid, ethylene, jasmonate, and ammonia-lyase. However, the application of nanoparticles in plant disease management is indexed in Fig. 1.

4. Plant viral Disease Management Using Nano Particles

The application of silver nanoparticles (AgNPs) to bean plants led to complete inhibition of the Sun-Hemp Rosette Virus (SHRV). During a tomato experiment, compelling evidence was presented demonstrating the antiviral properties of AgNPs against Potato Virus Y (PVY) and TMV.AgNPs at a concentration of 50 ppm induced Systemic Acquired Resistance (SAR) to the viral disease, resulting in a decrease in both the rate of infection and the severity of the sickness. The activity of the antioxidant enzymes polyphenol oxidase (PPO), peroxidase (POD), and Total Soluble Proteins (TSP) can be enhanced to combat TMV and PVY in tomatoes. When tomato plants were infected with Potato Virus Y (PVY) and Tobacco Mosaic Virus (TMV), the levels of total soluble phenols and free proline in the plants increased significantly, while the amount of photosynthetic pigments reduced drastically (Noha et al., 2018). Studies have shown that chitosan nanoparticles can confer viral resistance to host plants, including those affected by snuff mosaic, potato, cucumber, peanut, and alfalfa viruses.

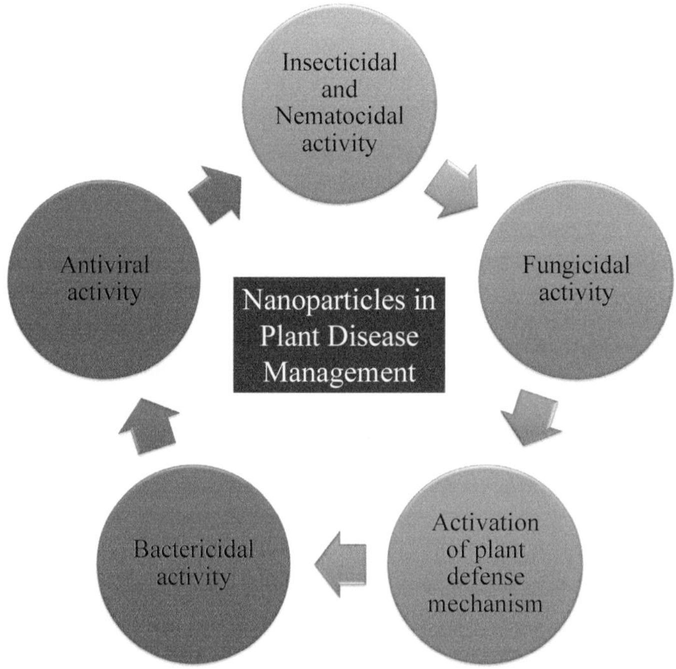

Figure 1. Nanoparticles in plant disease management.

Furthermore, the treated plants displayed a noticeable increase in plant height, weight of both roots and shoots, as well as the number of flowers and fruits. Additionally, when compared to the control group, two combinations of PGPR strains along with preparations and a chitosan molecule exhibited a significant decrease in disease severity. Tomato plants that were treated with biopreparations exhibited increased resistance to CMV infection and improved plant growth. The application of graphene-based silver nanocomposites on the surface of lettuce plants has shown antiviral effects against the Tomato Bushy Stunt Virus (TBSV), resulting in a decrease in virus concentration and severity of the disease. Upon the addition of SiO2NPs to the soil as a treatment for Cucumis sativus infected with the Papaya ringspot virus (PRSV), a notable reduction in the occurrence of symptoms was noted. In addition, the use of TiO2NPs through spraying and soil drenching led to a decrease in the infection of Broad Bean Strain Virus (BBSV) four hours after inoculation. Nanoparticles can impede activities by releasing metal ions in response to both oxidative and non-oxidative stimuli. The key components involved in the denaturation and inactivation process include RNA or DNA, diverse protein molecules, and capsid protein. The NPs may also inhibit the attachment, dissemination, replication, and fusion of the virion. Proteins undergo oxidation and generate both antiviral and non-functional properties when they come into contact with Reactive Oxygen Species (ROS).

Additionally, it is possible that the nanoparticles (NPs) will prevent the virus from locating and infecting its intended host plant. The NPs inhibit the host cells from recognizing the virus on its surface by attaching to glycoprotein receptors. The *in vitro* anti-TMV properties of ZnO nanoparticles (NPs) and SiO2 NPs were investigated by Cai et al. (2019). The data indicate that NPs cause direct harm to the TMV shell protein upon interaction with the glycoprotein membrane. The damage caused by MeNPs also impacts the process of aggregation and can potentially result in a fracture due to the direct deactivation of TMV. As stated by Cai et al. (2020), TMV particles treated with Fe3O4NPs show a similar pattern of aggregation and breaking. Silver nanoparticles (AgNPs) have the capability to attach to the capsid proteins of Tomato Mosaic Virus (ToMV) and Potato virus Y (PVY). The Transmission Electron Microscopy (TEM) analysis revealed that the barley yellow dwarf virus (PAV) was generating swollen, virus-like structures covered with gold nanoparticles,

Table 1. Various metallic, non-metallic, and bio-compound nanoparticle mechanisms of action for treating plant virus infections.

Nanoparticles (NPs)	Virus	Host plant	Mechanism of action
ZnO	TMV	Wild tobacco	Plant systemic resistance-related gene activation
ZnO	TMV	Tomato	Activation of *PAL, PR-1, Chalcone synthase*, and *POD gene*
TiO_2	BSV	Broad bean	Salicylic acid-inducible defence-related gene activation
Carboxymethyl chitosan-Ti	BYMV	Broad bean	*Enzymatic, non-enzymatic antioxidants*, and total soluble proteins activation
AgNPs	TBSV	Potato	Virus concentration and disease percentage decrease
GO-Ag	TBSV	Lettuce	*POD* and *polyphenol oxidase* activity
Au	BYDV	Barley	Viral capsid protein synthesis and viral coat protein
Fe	TMV	Wild tobacco	Plant oxidants, *SA-responsive PR-related* gene activation
SiO_2	TYLCV	Tomato	*Peroxidase* and *polyphenol oxidase* increase
NiO	CMV	Cucumber	Expression of *SA, JA*, and *ethylene related gene*
Curcumin-milk	PVY	Potato	Increase in *Polyphenol oxidase* and *peroxidase* levels

TMV: Tobacco mosaic virus; BSV: Broad bean stain virus; BYMV: Bean yellow mosaic virus; TBSV: Tomato bushy stunt virus; BYDV: Barley yellow dwarf virus; TYLCV: Tomato yellow leaf curl virus; CMV: Cucumber mosaic virus; PVY: Potato virus Y.

along with deteriorated and disappearing viral particles. Nanoliposomes, a form of nanomaterial, can be employed as a carrier for antiviral drugs. The inhibitory effect of nanoliposomes was increased by 33.6% at the protein level and by 42.0% at the gene level. In addition, Table 1 provides an index of the different methods of action of metallic, non-metallic, and bio-compound nanoparticles for the treatment of plant viral infections.

5. Nanoparticles as Biostimulants in virus-Infected Plants

Du Jardin (2015) described plant biostimulants as substances and agents that can modify a plant's physiological and biochemical processes, improving defensive mechanisms, photosynthesis, and overall plant development, in addition to insecticides and plant nutrients. According to Cai et al., (2019), studies on healthy tobacco plants have shown that ZnO, SiO_2, and Fe_2O_3 NPs stimulate plant development. Neutralizing compounds (NPs) can promote eustress in healthy and stressed plants (such as virus-infected plants) because they reduce adverse consequences. Derbalah and Elsharkawy, (2019) found that foliar spraying sick cucumber plants with NiO NPs and covering the soil improved the plants' fresh and dry weights as well as leaf yield. When treated with 50 mg/L of FeO_3 NPs and TiO_2, the fresh and dry weight of the TuMV-infected tobacco plants increased once again. Nevertheless, plants in the control group did not differ appreciably from those that did not receive treatment with nanoparticles following treatment at a 200 mg/L level.

El Shazly et al. (2017) reported that potato tubers from PVY-infected plants treated with AgNP had been determined to be of a higher grade compared to those from untreated plants. The use of NPs may have inhibited the virus's ability to spread or increased resistance. Additionally, it can be used to prevent viral contamination of seedlings. There are several mechanisms by which viral infections might propagate through seeds, as reported by Cobos et al. (2019). The seed embryo is invaded before fertilization, the seed coat is infected, the mother plant directly transfers to the embryonic tissues after fertilization, and the plant gametes are indirectly transferred. Because of their small size and antibacterial characteristics, NPs can be employed as seed antivirals by reducing the pathogen within the seed. Even though the most effective method is to completely eliminate the virus through *in vitro* cultivation, NPs can be used to reduce viral infection in seedlings. According to additional studies, NPs work better as an antiviral agent in cultures before the virus infects them (Cai et al., 2019).

6. Nanoparticles as Inhibitors of Phytoviruses

According to Galdiero et al. (2011), nanoparticles have been shown effective as viricidal agents against a variety of viruses by lowering viral infectivity. AgNPs have been shown to be a successful antiviral agent in India by Jain (2014), who showed that the Sunhemp Rosette Virus (SHRV) disease on cluster beans was fully controlled by spraying an aqueous solution of AgNPs at 50 ppm. AgNPs were biosynthesized, and Elbeshehy et al. (2015) confirmed their antiviral activity against BYMV infection *in vitro* using strains of *B. pumilus*, *B. persicus*, and *B. licheniformis* that were newly obtained. Again, they discovered that 24 hours following infection, AgNP treatment completely removed all negative BYMV signals. The percentage of infected plants, the viral load, and the severity of the disease all decreased in the treated plants.

However, when given during the pre-viral infection stage, NPs had no effect. The results showed that whereas AgNP increases the treated plant's resistance to BYMV systemically, its effects are more pronounced in the first phases of viral replication. When NPs were given three to seven days before the virus became active, the amount of virus and the percentage of infected plants decreased, while plant growth and tuber yield significantly increased. El-Shazly et al. (2017) reported that AgNPs at 0.1µg/µL were demonstrated to decrease PVY-infected potato plants both on their own and in combination with 200 µmol/L of Salicylic Acid (SA). When AgNPs were sprayed on bean plants, the Sun-Hemp Rosette Virus (SHRV) was totally eradicated (Jain, 2014). AgNPs, according to El-Dougdoug and El-Dougdoug (2018), have an antiviral effect on Potato Virus Y (PVY) and tomato TMV.

Total Soluble Proteins (TSP), polyphenol oxidase (PPO), and the antioxidant enzyme peroxidase (POD) were all elevated by AgNPs at 50 ppm. Consequently, there was a decline in the severity of the viral infection as well as in the percentage of tomatoes displaying systemic acquired resistance (SAR) to TMV and PVY. Total soluble phenols and free proline rapidly rose in tomato plants infected with PVY and TMV, but the photosynthetic pigments markedly declined (El-Dougdoug and El-Dougdoug, 2018). By delaying the onset of viral symptoms, they saw a reduction in the severity of the disease caused by the Tomato Yellow Leaf Curl Virus (TYLCV) in treated tomato plants as compared to untreated control plants. According to El-Sawy et al., (2018), tomato plants treated with SiNPs showed higher POD and PPO activity, which may indicate improved TYLCV systemic resistance. In greenhouse settings, ZnONPs have also been seen to increase the CMV resistance of brinjal plants (El-Sawy et al., 2017). However, the antiviral properties of potential nanoparticles are indexed in Table 2.

7. Adverse Effects of Nanoparticles on Plants and Prospects

Long-term exposure to nanoparticles (NPs) can impact ecosystems in both beneficial and detrimental ways. Research has shown that NPs can influence plant growth and development in various ways, depending on factors such as NP size, concentration, physical and chemical properties, and the plant species involved. NPs interact with plants through direct contact, dissolution, and co-transport with other molecules, leading to diverse effects. For instance, NPs can negatively affect photosynthesis, root growth, plant viability, and seed germination. Certain metallic nanoparticles, such as Fe_2O_3 and TiO_2, can accumulate in chloroplasts, contributing to pollution entering the food chain and potentially endangering human and animal health. When applied topically, carbon-based nanoparticles can penetrate plant leaves through stomatal openings and accumulate in the stroma and thylakoids, which can enhance photosynthetic activity. However, high concentrations of iron nanoparticles have been found to reduce chlorophyll content and photosynthesis rates in maize plants. In broccoli, ZnO-embedded NPs led to increased bioaccumulation of fungal NPs. Exposure to silver nanoparticles harmed the cell walls and vacuoles of *Oryza sativa* root cells. Treated maize plants initially showed improved photosynthetic rates and hydrogen peroxide scavenging abilities, but subsequent generations exhibited reduced biomass, chlorophyll, and photosynthetic efficiency. Accumulation of NPs can also alter soil microbiomes, affecting seed germination, plant growth,

Table 2. Antiviral properties of potential nanoparticles.

Nanoparticles (NPs)	Target virus	Host plant	References
Ag	TSWV	Potato	Shafie et al. (2018)
Ag	PVY	Potato	El-Shazly et al. (2017)
Schiff base nanosilver	TMV	Tobacco	Wang et al. (2016)
ZnO	TMV	Wild tobacco	Cai et al. (2019)
Fe_3O_4	TMV	Wild tobacco	Cai et al. (2019); El-Dougdoug and ElDougdoug (2018)
Graphene oxide-silver	TBSV	Lettuce	Elazzazy et al. (2017)
Au	BaYMV	Barley	Aref et al. (2012)
SiO_2	TYLCV	Tomato	El-Shazly et al. (2017)
Carbon Nanotubes	TMV	Wild tobacco	Cai et al. (2019); Adeel et al. (2021)
TiO_2	BBSV	Broad bean	Elsharkawy and Derbalah (2019)
NiO	CMV	Cucumber	Alkubaisi and Aref (2017); Derbalah and Elsharkawy (2019)
CeO_2	TMV	Jimsonweed and Tobacco	Eugene and Zholobak (2016); Elazzazy et al. (2017)

Tomato Spotted Wilt Virus: TSWV; Potato Virus Y: PVY; Tobacco Mosaic Virus: TMV; Tomato Bushy Stunt Virus: TBSV; Barley Yellow Mosaic Virus: BaYMV; Tomato Yellow Leaf Curl Virus: TYLCV; Broad Bean Stain Virus: BBSV; Cucumber Mosaic Virus: CMV

and reproduction. For example, while ZnO nanoparticles had no impact on soil bacteria, higher concentrations of silver nanoparticles significantly altered the abundance of certain bacterial phyla from 25 to 45%.

Nanophytovirology offers a potential strategy for long-term crop protection against plant viruses, though *in vitro* research suggests that NPs might help in virus treatment. Comprehensive *in vivo* studies are necessary to confirm these findings. DNA viruses continually threaten agricultural output, yet RNA viruses have garnered more scientific attention. The complexity of plant viral diseases is driven by interactions among the plant, virus, vector, and host environment. The precise interactions between NPs and this system remain unclear, including how NPs affect virus particles and vector behaviour. For instance, research by Patil et al. (2016) indicated that AuNPs inhibit insect development and reproduction by affecting trypsin, while studies by Jiang et al. (2017) and Benelli et al. (2018) found that specific metal and metal oxide NPs reduce membrane permeability by binding with phosphorus and sulphur in proteins and nucleic acids. NPs have shown potential in combating viruses when applied proactively to plants and, in some cases, when used in conjunction with viral inoculation. Nevertheless, more research is needed to understand the virus-host dynamics fully. While most research focuses on single NPs, nanocomposites might provide significant benefits by targeting multiple plant viruses simultaneously. Further investigation is also required to explore how NPs affect other infectious RNA molecules.

8. Conclusion

Plant pathogenic viruses are a particularly dangerous family of plant diseases that drastically lower agricultural productivity because of their quick proliferation, genetic variety, and lack of effective therapies. Nanotechnology presents one possible management approach to reduce the danger of plant viral infections. The unique physicochemical properties of NPs allow for more beneficial interactions with viruses, vectors, and host plants. When NPs come into touch with virus particles, they can physically impede and eradicate them. Because of the indirect link, the host plants may get bio-stimulated and activate their antiviral defence systems. NPs may be combined with a range of probes to generate nano biosensors that are capable of identifying and diagnosing ?phytoviruses. Additionally, by delivering double-strand breaks (dsRNA) to specific locations within the virus, materials based on nanotechnology may be produced as carrier molecules to mute the genetic material of phytoviruses. Further investigation is required to fully understand the mechanism behind the interaction between plant-virus vectors and NPs. Much more research is required in this field because nanophytovirology is still in its infancy. As supplemental management options, consideration should be given to ?bio stimulants, insect control methods, and composites of NPs with other nutrients. The possibility of NP-mediated plant virus and vector management must be taken into account as part of integrated disease control. To progress in the area of nano phyto virology, researchers from diverse domains from plant pathologists, agricultural engineers, and biologists must work together. Establishing a plant viral infection management method that is long-lasting, safe, effective, sustainable, and environmentally beneficial should thus be the primary goal of efforts.

References

Adeel, M., Shakoor, N., Ahmad, M. A., White, J. C., Jilani, G., and Rui, Y. (2021). Bioavailability and toxicity of nanoscale/bulk rare earth oxides in soil: physiological and ultrastructural alterations in *Eisenia fetida. Environmental Science: Nano*, 8(6): 1654–1666.

Afzal, O., Altamimi, A. S., Nadeem, M. S., Alzarea, S. I., Almalki, W. H., Tariq, A. et al. (2022). Nanoparticles in drug delivery: From history to therapeutic applications. *Nanomaterials*, 12(24): 4494.

Ahmad, S., Pathak, A. K., Minj, R. P., Chaudhary, S., Kumar, A., Chalotra, T. et al. (2023). A comprehensive review of nanoparticles induced stress and toxicity in plants. *Universal Journal of Green Chemistry*, 18–43.

Alkubaisi, N. A., and Aref, N. M. (2017). Dispersed gold nanoparticles potentially ruin gold barley yellow dwarf virus and eliminate virus infectivity hazards. *Applied Nanoscience*, 7(1): 31–40.

Álvarez-Martínez, A., Mc-Caughey-Espinoza, D., Magaña-Barajas, E., Buitimea-Cantúa, N., Morales-Romero, D., and Cota-Arriola, O. (2023). Viabilidad de la semilla de Ipomoea arborescens (Humb. and Bonpl. ex Willd.) G. Don. *Abanico Agroforestal*, 5: e2023–6.

Baram-Pinto, D., Shukla, S., Gedanken, A., and Sarid, R. (2010). Inhibition of HSV-1 attachment, entry, and cell-to-cell spread by functionalized multivalent gold nanoparticles. *Small*, 6(9): 1044–1050.

Baram-Pinto, D., Shukla, S., Perkas, N., Gedanken, A., and Sarid, R. (2009). Inhibition of herpes simplex virus type 1 infection by silver nanoparticles capped with mercaptoethane sulfonate. *Bioconjugate Chemistry*, 20(8): 1497–1502.

Benelli, A., Cevoli, C., and Fabbri, A. (2020). In-field hyperspectral imaging: An overview on the ground-based applications in agriculture. *Journal of Agricultural Engineering*, 51(3): 129–139.

Bohdal, O., Balles, L., Wistuba, M., Ermis, B., Archambeau, C., and Zappella, G. (2022). PASHA: Efficient HPO and NAS with progressive resource allocation. *arXiv preprint arXiv:2207.06940.*

Bringmann, L. F., Elmer, T., and Eronen, M. I. (2022). Back to basics: The importance of conceptual clarification in psychological science. *Current Directions in Psychological Science*, 31(4): 340–346.

Cai, L., Liu, C., Fan, G., Liu, C., and Sun, X. (2019). Preventing viral disease by ZnONPs through directly deactivating TMV and activating plant immunity in Nicotiana benthamiana. *Environmental Science: Nano*, 6(12): 3653–3669.

Cao, Y., Liu, X., Xiong, L., and Cai, K. (2020). Imaging and clinical features of patients with 2019 novel coronavirus SARS-CoV-2: a systematic review and meta-analysis. *Journal of Medical Virology*, 92(9): 1449–1459.

Chirkov, S., Sheveleva, A., Ivanov, P., and Zakubanskiy, A. (2018). Analysis of genetic diversity of Russian sour cherry Plum pox virus isolates provides evidence of a new strain. *Plant Disease*, 102(3): 569–575.

Cobos, A., Montes, N., López-Herranz, M., Gil-Valle, M., and Pagán, I. (2019). Within-host multiplication and speed of colonization as infection traits associated with plant virus vertical transmission. *Journal of Virology*, 93(23): 10–1128.

Cota-Arriola, O., Onofre Cortez-Rocha, M., Burgos-Hernández, A., Marina Ezquerra-Brauer, J., and Plascencia-Jatomea, M. (2013). Controlled release matrices and micro/nanoparticles of chitosan with antimicrobial potential: development of new strategies for microbial control in agriculture. *Journal of the Science of Food and Agriculture*, 93(7): 1525–1536.

Dakal, T. C., George, N., Xu, C., Suravajhala, P., and Kumar, A. (2024). Predictive and Prognostic Relevance of Tumor-Infiltrating Immune Cells: Tailoring Personalized Treatments against Different Cancer Types. *Cancers*, 16(9): 1626.

Delorey, T. M., Ziegler, C. G., Heimberg, G., Normand, R., Yang, Y., Segerstolpe, Å. et al. (2021). COVID-19 tissue atlases reveal SARS-CoV-2 pathology and cellular targets. *Nature*, 595(7865): 107–113.

Derbalah, A., Elsharkawy, M. M., Hamza, A., and El-Shaer, A. (2019). Resistance induction in cucumber and direct antifungal activity of zirconium oxide nanoparticles against *Rhizoctonia solani. Pesticide Biochemistry and Physiology*, 157: 230–236.

Derbalah, A., and Sakugawa, H. (2023). Trends in the usage of fenobucarb, the commonly used carbamate insecticide in Japan, in relation to river water concentration and risk assessment. *International Journal of Environmental Research*, 17(4): 52.

Ding, X., Liow, C. H., Zhang, M., Huang, R., Li, C., Shen, H. et al. (2014). Surface plasmon resonance enhanced light absorption and photothermal therapy in the second near-infrared window. *Journal of the American Chemical Society*, 136(44): 15684–15693.

Du Jardin, P. (2015). Plant biostimulants: Definition, concept, main categories and regulation. *Scientia horticulturae*, 196: 3–14.

Elazzazy, A. M., Elbeshehy, E. K., and Betiha, M. A. (2017). *In vitro* assessment of activity of graphene silver composite sheets against multidrug-resistant bacteria and Tomato Bushy Stunt Virus. *Tropical Journal of Pharmaceutical Research*, 16(11): 2705–2711.

El-Bendary, M., Abd-Elsalam, S., Elbaz, T., El-Akel, W., Cordie, A., Elhadidy, T. et al. (2022). Retracted Article: Efficacy of combined Sofosbuvir and Daclatasvir in the treatment of COVID-19 patients with pneumonia: A multicenter Egyptian study. *Expert Review of Anti-infective Therapy*, 20(2): 291–295.

Elbeshehy, E. K., Elazzazy, A. M., and Aggelis, G. (2015). Silver nanoparticles synthesis mediated by new isolates of Bacillus spp., nanoparticle characterization and their activity against Bean Yellow Mosaic Virus and human pathogens. *Frontiers in Microbiology*, 6: 453.

Elbeshehy, E. K., Elazzazy, A. M., and Aggelis, G. (2015). Silver nanoparticles synthesis mediated by new isolates of Bacillus spp., nanoparticle characterization and their activity against Bean Yellow Mosaic Virus and human pathogens. *Frontiers in Microbiology*, 6: 453.

Elmer, G. (2023). From the First to the Zero Person Perspective: Neutering the Mediated Life of Affinity. *Computational Culture*.

Elsharkawy, M. M., and Derbalah, A. (2019). Antiviral activity of titanium dioxide nanostructures as a control strategy for broad bean strain virus in faba bean. *Pest Management Science*, 75(3): 828–834.

Elsharkawy, M. M. (2024). Antiviral Activity of Nanoparticles in Plants. In *Nanotechnology in Plant Health* (pp. 285–299). CRC Press.

El-Shazly, M. A., Attia, Y. A., Kabil, F. F., Anis, E., and Hazman, M. (2017). Inhibitory effects of salicylic acid and silver nanoparticles on potato virus Y-infected potato plants in Egypt. *Middle East J. Agric. Res.*, 6(3): 835–848.

Fabiyi, O. A., Adebisi, O. O., Falore, S. O., Bello, T. T., and Olatunji, G. A. (2023). Assessment of Actinomyces and Pseudomonas species on Meloidogyne incognita population and growth of carrot plants in disparate soils. *Indian Phytopathology*, 76(2): 593–604.

Farooq, T., Hameed, A., and Hameed, A. (2023). Engineered nanomaterials for plant disease diagnosis and management. In *Engineered Nanomaterials for Sustainable Agricultural Production, Soil Improvement and Stress Management* (pp. 337–355). Academic Press.

Galdiero, S., Falanga, A., Vitiello, M., Cantisani, M., Marra, V., and Galdiero, M. (2011). Silver nanoparticles as potential antiviral agents. *Molecules*, 16(10): 8894–8918.

Jain, D., Daima, H. K., Kachhwaha, S., and Kothari, S. L. (2009). Synthesis of plant-mediated silver nanoparticles using papaya fruit extract and evaluation of their anti-microbial activities. Digest journal of nanomaterials and biostructures, 4(3): 557–563.

Jiang, S., Wang, J., Zhao, Y., Shang, Y., Gao, X., Li, H. et al. (2017). Sustainability of water resources for agriculture considering grain production, trade and consumption in China from 2004 to 2013. *Journal of Cleaner Production*, 149: 1210–1218.

Kah, M., Beulke, S., Tiede, K., and Hofmann, T. (2013). Nanopesticides: state of knowledge, environmental fate, and exposure modeling. *Critical Reviews in Environmental Science and Technology*, 43(16): 1823–1867.

Karniadakis, G. E., Kevrekidis, I. G., Lu, L., Perdikaris, P., Wang, S., and Yang, L. (2021). Physics-informed machine learning. *Nature Reviews Physics*, 3(6): 422–440.

Kawamoto, A., Miyata, T., Makino, F., Kinoshita, M., Minamino, T., Imada, K. et al. (2021). Native flagellar MS ring is formed by 34 subunits with 23-fold and 11-fold subsymmetries. *Nature Communications*, 12(1): 4223.

Khalid, M., and El-Sawy, H. S. (2017). Polymeric nanoparticles: Promising platform for drug delivery. International journal of pharmaceutics, 528(1-2): 675–691.

Khan, K. U., Minhas, M. U., Badshah, S. F., Suhail, M., Ahmad, A., and Ijaz, S. (2022). Overview of nanoparticulate strategies for solubility enhancement of poorly soluble drugs. *Life Sciences*, 291: 120301.

Kochkina, N. E., and Butikova, O. A. (2024). Preparation of starch/PVA nanoparticles and evaluation of their ability to stabilize Pickering emulsions. *International Journal of Biological Macromolecules*, 274: 133406.

Koo, B. K., Kang, J., Park, K. W., Rhee, T. M., Yang, H. M., Won, K. B. et al. (2021). Aspirin versus clopidogrel for chronic maintenance monotherapy after percutaneous coronary intervention (HOST-EXAM): an investigator-initiated, prospective, randomised, open-label, multicentre trial. *The Lancet*, 397(10293): 2487–2496.

Kráľová, K., and Jampílek, J. (2023). Effects of nanoparticles/nanotubes on plant growth. In *The Impact of Nanoparticles on Agriculture and Soil* (pp. 183–237). Academic Press.

Krishnaraj, C., Ramachandran, R., Mohan, K., and Kalaichelvan, P. T. (2012). Optimization for rapid synthesis of silver nanoparticles and its effect on phytopathogenic fungi. *Spectrochimica Acta Part A: Molecular and Biomolecular Spectroscopy*, 93: 95–99.

Lara, H. H., Garza-Treviño, E. N., Ixtepan-Turrent, L., and Singh, D. K. (2011). Silver nanoparticles are broad-spectrum bactericidal and virucidal compounds. *Journal of Nanobiotechnology*, 9: 1–8.

Malerba, M., and Cerana, R. (2018). Recent advances of chitosan applications in plants. *Polymers*, 10(2): 118.

Mao, R., Qiu, Y., He, J. S., Tan, J. Y., Li, X. H., Liang, J. et al. (2020). Manifestations and prognosis of gastrointestinal and liver involvement in patients with COVID-19: a systematic review and meta-analysis. The Lancet Gastroenterology and Hepatology, 5(7): 667–678.

Marques, C., Hajipour, M. J., Marets, C., Oudot, A., Safavi-Sohi, R., Guillemin, M. et al. (2023). Identification of the proteins determining the blood circulation time of nanoparticles. *ACS nano*, 17(13): 12458–12470.

Mohamed, H. I., Abd-Elsalam, K. A., Tmam, A. M., and Sofy, M. R. (2021). Silver-based nanomaterials for plant diseases management: Today and future perspectives. In *Silver Nanomaterials for Agri-Food Applications* (pp. 495–526). Elsevier.

Murphy, E., and Steenbergen, C. (2008). Mechanisms underlying acute protection from cardiac ischemia-reperfusion injury. *Physiological Reviews*, 88(2): 581–609.

Noha, K., Bondok, A. M., and El-Dougdoug, K. A. (2018). Evaluation of silver nanoparticles as antiviral agent against ToMV and PVY in tomato plants. *Sciences*, 8(01): 100–111.

Ogunyemi, O. M., Gyebi, G. A., Saheed, A., Paul, J., Nwaneri-Chidozie, V., Olorundare, O. et al. (2022). Inhibition mechanism of alpha-amylase, a diabetes target, by a steroidal pregnane and pregnane glycosides derived from Gongronema latifolium Benth. *Frontiers in Molecular Biosciences*, 9: 866719.

Park, W., Chawla, A., and O'Reilly, E. M. (2021). Pancreatic cancer: a review. *Jama*, 326(9): 851–862.

Patil, P., Peng, R. D., and Leek, J. T. (2016). What should researchers expect when they replicate studies? A statistical view of replicability in psychological science. *Perspectives on Psychological Science*, 11(4): 539–544.

Rafique, M., Sadaf, I., Rafique, M. S., and Tahir, M. B. (2017). A review on green synthesis of silver nanoparticles and their applications. *Artificial Cells, Nanomedicine, and Biotechnology*, 45(7): 1272–1291.

Rajpal, V. R., Dhingra, Y., Khungar, L., Mehta, S., Minkina, T., Rajput, V. D. et al. (2024). Exploring Metal and metal-oxide nanoparticles for nanosensing and biotic stress management in plant systems. *Current Research in Biotechnology*, 100219.

Rajput, V., Chaplygin, V., Gorovtsov, A., Fedorenko, A., Azarov, A., Chernikova, N. et al. (2021). Assessing the toxicity and accumulation of bulk-and nano-CuO in *Hordeum sativum* L. *Environmental Geochemistry and Health*, 43: 2443–2454.

Rajput, V., Minkina, T., Mazarji, M., Shende, S., Sushkova, S., Mandzhieva, S. et al. (2020). Accumulation of nanoparticles in the soil-plant systems and their effects on human health. *Annals of Agricultural Sciences*, 65(2): 137–143.

Sadeghi, B., and Gholamhoseinpoor, F. (2015). A study on the stability and green synthesis of silver nanoparticles using Ziziphora tenuior (Zt) extract at room temperature. *Spectrochimica Acta Part A: Molecular and Biomolecular Spectroscopy*, 134: 310–315.

Sathiyabama, M., Boomija, R. V., Muthukumar, S., Gandhi, M., Salma, S., Prinsha, T. K. et al. (2024). Green synthesis of chitosan nanoparticles using tea extract and its antimicrobial activity against economically important phytopathogens of rice. *Scientific Reports*, 14(1): 7381.

Shafie, A. A., Tan, Y. P., and Ng, C. H. (2018). Systematic review of economic burden of heart failure. *Heart Failure Reviews*, 23(1): 131–145.

Sofy, A. R., Sofy, M. R., Hmed, A. A., Dawoud, R. A., Refaey, E. E., Mohamed, H. I. et al. (2021). Molecular characterization of the Alfalfa mosaic virus infecting Solanum melongena in Egypt and the control of its deleterious effects with melatonin and salicylic acid. *Plants*, 10(3): 459.

Uninet, E. P. (2012). 13th Congress of the International Society for Ethnopharmacology.

Verma, A., Lamsal, K., and Verma, P. (2022). An investigation of skill requirements in artificial intelligence and machine learning job advertisements. *Industry and Higher Education*, 36(1): 63–73.

Wang, X., Yamauchi, F., Otsuka, K., and Huang, J. (2016). Wage growth, landholding, and mechanization in Chinese agriculture. *World Development*, 86: 30–45.

Worrall, L. J., Majewski, D. D., and Strynadka, N. C. (2023). Structural insights into type III secretion systems of the bacterial flagellum and injectisome. *Annual Review of Microbiology*, 77(1): 669–698.

Xin, X., Judy, J. D., Sumerlin, B. B., and He, Z. (2020). Nano-enabled agriculture: from nanoparticles to smart nanodelivery systems. *Environmental Chemistry*, 17(6): 413–425.

Zholobak, G., Dugin, S., Sybirtseva, O., Kazantsev, T., and Romanciuc, I. (2020). Determination of nitrogen and chlorophyll content in two varieties of winter wheat plants means of ground and airborne spectrometry. *Ukrainian Journal of Remote Sensing*, (26): 4–13.

CHAPTER-11

Green Nanotechnology for Protecting Plants against Bacterial Diseases

Aldo Córdoba Guerrero,[1] *Adrian Salcedo Gómez,*[2] *Daniel Mendoza Jiménez,*[2] *Fidel Landeros Jaime,*[2] *José Antonio Cervantes Chávez*[2] and *Karen Esquivel Escalante*[1,*]

1. Introduction

Green nanotechnology represents an innovative approach to addressing critical challenges in modern agriculture (Dutta and Das, 2021), particularly the management of bacterial plant diseases (Joshi et al., 2020). Traditional methods for controlling such diseases have relied heavily on chemical pesticides and antibiotics, often leading to significant environmental degradation, the development of resistant pathogens, and adverse effects on non-target organisms (De Wit, 2007). In response, green nanotechnology offers a promising alternative by integrating principles of green chemistry and nanotechnology to create eco-friendly, biocompatible nanomaterials (Dutta and Das, 2021).

Nanotechnology in agriculture has already shown its potential, especially through the development of nanofertilizers and nanopesticides, which enhance the efficiency of nutrient delivery and pest control. However, its application in combating bacterial plant diseases has gained significant attention in recent years (Acharya and Pal, 2020). The unique physicochemical properties of nanomaterials, including their high surface area, tunable particle size, and ability to interact at the molecular level, enable them to act as potent antibacterial agents (Asha and Narain, 2020). Moreover, nanoparticles can be synthesized using biological materials, such as plant extracts, bacteria, and fungi, reducing the need for toxic chemicals and lowering the environmental footprint (Maity et al., 2022).

Green synthesized nanoparticles, particularly metal-based (e.g., silver, copper) and metal oxide (e.g., zinc oxide, titanium dioxide) nanoparticles, have demonstrated strong antimicrobial activity against a variety of plant pathogens (Negrescu et al., 2022, Huston et al., 2021). These nanomaterials offer multiple mechanisms of action, including the generation of Reactive Oxygen Species (ROS),

[1] Graduate and Research Division, Engineering Faculty, Universidad Autónoma de Querétaro, Cerro de las campanas, C.P. 76010, Santiago de Querétaro, Qro, México.

[2] Natural Sciences Faculty, Universidad Autónoma de Querétaro, Carr. Chichimequillas-Anillo Vial Fray Junípero Serra, Km 8, C. P. 76000 Santiago de Querétaro, Qro, México.

* Corresponding author: karen_esq_2001@yahoo.com; https://orcid.org/0000-0003-1838-7058

DOI: 10.1201/9781003567684-11

disruption of bacterial cell membranes, and interference with vital cellular processes such as DNA replication and enzyme function. Furthermore, green-synthesized nanoparticles provide added benefits by being biocompatible and biodegradable, reducing the ecological risks often associated with conventional nanoparticles.

Despite these advances, the widespread adoption of nanotechnology in plant disease management faces several challenges. High production costs, regulatory barriers, and concerns over nanotoxicity remain significant hurdles. Moreover, long-term studies assessing the environmental fate, accumulation, and potential risks of nanoparticles in agricultural ecosystems are still needed (Huguet-Casquero et al., 2021; Muzammil et al., 2023). While green nanotechnology holds great promise, its integration into sustainable agricultural practices must be carefully managed to balance efficacy with environmental safety.

The focus of this work is to critically assess the potential of green nanotechnology as a viable strategy for managing bacterial plant diseases, exploring both the benefits and limitations of nanoparticle applications. By examining the mechanisms of action, the types of green synthesized nanoparticles, and their practical applications in disease control, this study aims to contribute to the development of sustainable and effective plant protection solutions.

2. Definition and Principles of Green Nanotechnology

Green nanotechnology is an emerging field that integrates the principles of green chemistry with nanotechnology to create environmentally friendly nanomaterials. The primary goal is to minimize the environmental and health impacts associated with the synthesis and application of nanomaterials. Green synthesis of nanoparticles using biomass extracts as reductive agents, provides a safe and environmentally friendly alternative to toxic chemicals, with potential applications in environmental, agriculture, and biomedical fields (Jadoun et al., 2021; Miu and Dinischiotu, 2022). The focus of green nanotechnology is on energy-efficient processes and the reduction of toxic chemicals, with sustainability metrics being applied to optimize and quantify the environmental impact such as atom economy, e-factor, process mass intensity, energy intensity, and life cycle analysis (García-Quintero and Palencia, 2021). This approach not only aims to protect human health and the environment but also provides a sustainable pathway for the advancement of nanotechnology (Madani et al., 2022; Huston et al., 2021; Dutta and Das, 2021; Huguet-Casquero et al., 2021).

2.1 Nanotechnology in Sustainable Agriculture and Food Security

Nanotechnology has emerged as a transformative tool in sustainable agriculture and food security. By leveraging nanoscale materials and processes, it offers innovative solutions to enhance crop productivity, improve food quality, and ensure environmental sustainability. This technology has proved to be able to significantly enhance crop productivity through the development of nanofertilizers and nanopesticides (Tang et al., 2024), which offer more efficient nutrient delivery and pest management compared to traditional methods (Ashraf et al., 2021). Another approach is the nano-priming of seeds which can improve germination rates, plant growth, and stress resistance, leading to increased crop yields and better nutritional quality of food (Gilbertson et al., 2020; Do Espirito Santo Pereira et al., 2021). Nanomaterials play a crucial role in the remediation of contaminated soils by affecting the fate, mobility, and toxicity of soil pollutants. They also contribute to the improvement of soil quality and plant growth by enhancing nutrient release and interacting with soil biota (Hofmann et al., 2020).

Also, it is possible through nanotechnology, to improve food safety and quality by advanced packaging materials that extend product shelf-life and reduce contamination (Ashraf et al., 2021). More specific studies have revealed that nano-additives can enhance the nutritional value, texture, and taste of food products (Zhou et al., 2023). Despite the potential benefits, there are significant barriers to the widespread adoption of nanotechnology in agriculture, including high initial costs, regulatory hurdles, concerns about nanotoxicity, environmental impact (Acharya and Pal, 2020, Hofmann et al., 2020), and the generation of more resistant pathogens.

2.2 Plant Diseases and Nanotechnology

As mentioned, nanotechnology has emerged as a promising approach for different applications and for managing plant diseases and has shown interesting results, offering advantages such as enhanced efficacy, reduced input requirements, and lower environmental toxicity over conventional methods (Fu et al., 2019). For example, metal-based nanoparticles (NPs) like silver nanoparticles (Ag NPs), exhibit strong antimicrobial properties that can inhibit the growth of plant pathogenic bacteria and fungi, thereby reducing plant diseases (Li et al., 2023). These nanoparticles can directly interact with pathogens or induce plant resilience to disease. Nanomaterials (NMs) can act as bioststimulants, inducing plant innate immunity and enhancing the plant's defense mechanisms against pathogens (Dong et al., 2023). In the same way, NMs can serve as carriers for active ingredients such as pesticides, micronutrients, and elicitors, ensuring targeted delivery and increased efficacy of these substances in disease management (Dutta et al., 2023). The general view points out that nanotechnology is particularly promising for managing bacterial, fungal, and viral diseases, as well as soil-borne pathogens (Alghuthaymi et al., 2021; Dutta et al., 2023; Dutta et al., 2022).

As can be seen, the NMs synthesized by conventional methods offer sustainable alternatives to chemical pesticides, reducing ecological imbalances and toxic effects associated with conventional agrochemicals. Nevertheless, there are still some concerns regarding the long-term issues of the application of these materials for sustainable agriculture. However, green synthesized nanomaterials seem to be a more adequate solution for plant disease management and food security (Maity et al., 2022) keeping all the advantages that NMs have shown and adding the benefit of having a biocompatible material by reducing chemical inputs and minimizing environmental impact (Meena et al., 2021).

3. Major Bacterial Diseases Affecting Plants

Plant bacterial diseases can be grouped into four broad categories based on the extent of damage to plant tissue and the symptoms produced in the plant, which may include vascular wilt, necrosis, soft rot, canker, and leaf spot (Zhang et al., 2020).

Vascular wilt results from the bacterial colonization of the plant's vascular system. This is due to bacteria multiplication, making a plug in the vascular system, avoiding in this way movement (translocation) of water and nutrients through the xylem of the host plant. Drooping, wilting, or death of the aerial plant structure may occur after the infection spreads and the plant is fully colonized, some examples include bacterial wilt of sweet corn, alfalfa, tobacco, tomato, and cucurbits and black rot of crucifers (Redkar et al., 2021).

Related with the specific way to get nutrients from the plant, some pathogens can cause necrosis on plant tissue, then they are called necrotrophic, this activity is carried out by secretion of different kinds of toxins and lytic enzymes, as a result the main visible damage taking on the formation of leaf spots, stem blights, or cankers. Some pathogens bring on soft rot diseases because of secretion of enzymes capable of decomposing cell wall structures, keeping down the main structural component of the plant tissue. The plant rots turns into macerated (soft and watery). These kinds of diseases frequently occur on vegetables such as potato, carrot, eggplant, squash, and tomato. On the other hand, tumor diseases are caused by bacteria that stimulate uncontrolled multiplication of plant cells, resulting in the formation of abnormally large structures due to the synthesis of cytokines having as a result a drastic change in fruit traits (Liu et al., 2024).

In the same way there are several disease transmission mechanisms related with the physiology of the pathogens. In this case, we can mention the soil borne or seed borne pathogens the first one begins with the colonization of the root system impairing the translocation of nutrients to the entire plant, that is why the first symptoms of this type of diseases are confused with nutrient deficiencies, this type of pathogens are found in the soil in a state of latency, at the moment of sensing the chemicals compounds secreted by their hosts, the infection process starts (Puławska et al., 2022; Wang and Zou, 2024). Over the years, many pathogens have been associated with economic losses, some of them are certainly seedborne in nature making for their proliferation in the plant seeds once they germinate making infections in new fields and locations. The bacterial infection brought on seeds cause a great damaging effect on the seedlings and plant development. For that reason, carrying out a way to store the contaminated seeds from one region to another easily provides an effective way of establishing bacterial pathogens into crops at an early stage (Dey and Raghuwanshi, 2024; Chalam et al., 2020). Usually, illnesses caused by bacteria bring out one main symptom, but in some cases present a range or combination of the above-mentioned symptoms. Generally, it is not difficult to recognize when a plant is infected by a bacterial pathogen.

3.1 Mechanism of Infection of Plant Pathogenic Bacteria

The pathogen mechanisms to infect plants is trough the crop tissue. The infection is carried out through the damaged tissue caused by agricultural tools, insects or simply by unfavorable weather conditions (dust, wind, heavy rain, hail, frost). At the same time bacteria can also infect plants through natural orifices or glands (stomata, hydathodes or lenticels). After penetration, the bacteria colonize the cortex and then, the xylem vessels and spread throughout the plant. The bacterial masses cut out the flow of water from the roots to the leaves, causing the wilting of the plant. Disease severity depends on susceptible hosts and strain virulence as biotic factors, but is also related with abiotic factors such as soil type, temperature, and humidity, altogether directly impacting in microorganism growth. High temperatures (30–35°C) and humidity are the main factors associated with the high incidence and severity of disease (Bellenot et al., 2022; Fujikawa and Inoue, 2020; Sakata et al., 2023).

Plant diseases not only affect the number of products harvested but also the quality, and are responsible for causing 9–15% losses. To understand the pathogenesis, population performance of the pathogens and mechanisms of the disease can help in formulating strategies to control it, communication between microorganisms is called Quorum sensing (Qs), which is the mechanism of intercellular interaction between bacteria, it involves synthesis and recognition of extracellular signaling molecules called autoinducers (AIs) (Katoch et al., 2020). While the size of a bacterial population gets in, the concentration of AIs increases at the same time, AIs connect to specific receptors that activate the stimulation or repression of target genes, the bacteria use this method to regulate processes that involve the cooperation of several bacterial cells to be effective for the accomplishment of infection. The individuals in the group, get earned from the activity of the entire assembly, this process feeds a character of multicellularity (Sharma et al., 2024; Shi et al., 2023).

3.2 Symptoms of Bacterial Diseases

Plant disease is the damage to a plant's normal health status that affects or changes its basic physiology (Joshi et al., 2020). The main signs of bacterial diseases are vascular wilt, necrosis, soft rot, mosaics, pustules on leaves and tumors. Although these types of crop diseases signs are clearly related with a specific plant pathogen, the proper identification of the microorganism related with the disease requires laboratory techniques, otherwise these kinds of diseases can cause serious and economically losses, due to a wrong diagnosis (Sakata et al., 2023). However, it is worth mentioning that diseases caused by bacteria can be mistaken for fungal diseases or nutritional, so it is until the final stage of the disease where the main symptoms are observed in old leaves (Wang and Zou, 2024).

3.2.1 Some of the most dangerous illnesses

- Black rot (*Xanthomonas campestris pv. campestris*)

Cruciferous crops are t one of the most devastating diseases resulting in large damage of yield and quality. It appears in all brassica species. Xanthomonas campestris is an important member of the Xanthomonas group of phytopathogens that causes diseases in cruciferous, it is a Gram-negative bacteria, and produces black rot, many strains have been organized into nine races based on the symptoms caused by a set of the host and nonhost plants such as: broccoli, brussels sprouts, Chinese cabbage, Romanesco, cauliflower, kohlrabi, turnip greens, pak choi, etc. (Bellenot et al., 2022; Fujikawa and Inoue, 2020). The principal symptoms usually are observed at any growth stage of plant development, mainly appear as yellow V-shaped lacerations that extend towards the base of the leaf causing wilt and necrosis. Another symptom is the blackening of veins within the lacerations and bordering areas is a diagnostic feature of this pathogen. Nevertheless, the symptoms of black rot can be related to drought stress, overwatering, or excessive fertilizers.

X. campestris is recognized as a bacterium that takes control of the water-conducting tissue of the plant with xanthan (mucilaginous sugar). It is usually spread by seeds as little as 0.03% infection can cause an epidemy. Infested seeds and transplants set up the pathogen the capacity to travel long distances. Unfortunately, symptoms may not appear in the seedbed, resulting in infected plants to be transplanted into the field (Li et al., 2024). The pathogen is then spread within a field by splashing water, wind, equipment, people, insects and so on. Once in a field, disease development is favored by warm temperature. The pathogen brings on the disease by moving from the leaves into the petiole and spreading up to the stem or into the roots and become a systemic disease. By the time the disease progresses, the veins of infected tissues become black, and the normal flow of water and nutrients is weakened. Meanwhile the symptoms on roots may not be visible on foliage, but blackened veins appear in the roots. Sometimes black rot is often followed by invasion of soft-rotting organisms, making worse the plant's general status (Li et al., 2024; Rodrigues et al., 2023).

- Bacterial canker (*Clavibacter michiganensis pv. michiganensis*)

In greenhouse settings, the main symptom is an adjustable wilting of leaves during the hot weather, finally becoming irreversible. Bacterial canker disease is one of the most destroying diseases of tomato crop (*Solanum lycopersicum*) caused by the seed-borne Gram-positive bacteria *Clavibacter michiganensis* ssp. *michiganensis*. This vascular pathogen generally invades and proliferates in the xylem through natural openings or wounds, causing wilt and canker symptoms (Nandi et al., 2018). It affects tomato seedlings and propagation stages, also in adult plants in the greenhouse. Although bacterial canker is very difficult to control once the symptoms have been observed, it is important to monitor for early symptoms of the disease, verifying it by correct and early diagnosis, and implement management strategies immediately. Although tomato is the primary host of *C. michiganensis* it has been reported to infect sweet pepper, eggplant, and many weed species of *Solanaceae* family (Wang et al., 2022).

Seedlings are more likely to become stunted and wither rapidly. Numerous small, whitish, or tan pustules may appear on leaf veins, petioles, and peduncles. In the same way, other symptoms may be seen as interveinal, pale-green water-soaked areas on leaves which turn into yellowish brown to brown necrotic areas, resembling a sunburn. At the end the disease takes control of the plant infected begins to wilt, often just the lower mature leaves or leaves just above the area of the stem that is infected. Wilted leaves demonstrate marginal necrosis. By the time the disease moves on, more wilting and leaf necrosis develop (Marwal and Gaur, 2020). Once the stems are infected, light yellow to brown streaks/cankers may appear on the stems, the affected stems are split open lengthwise; a yellow to reddish-brown staining of the vascular tissue may be detected and the pith of infected stem turns brown and may appear dry and mealy. These internal symptoms are evidence of systemic infection and spread of the pathogen (Wang at al., 2022).

- Bacterial wilt (*Ralstonia solanacearum*)

Ralstonia solanacearum is a soil-borne bacteria that causes the widespread disease known as bacterial wilt, in addition it is the causal agent of the Moko disease of banana crop and brown rot of potatoes (Eisfeld et al., 2022), it produces the devastating bacterial wilt disease in over 200 plant species, including economically important crops such as potato, tomato, peanut, eggplant, and banana. *R. solanacearum* can survive in soil and waterways from where it infects plants through the roots and colonizes the xylem tissue, blocking water flow and causing plant wilting (Eisfeld et al., 2022; Sebastià et al., 2021). The appearance of wilting is followed by the yellowing of the leaf then death of the leaf between the veins and stunted growth, the interior of the stem turns discolored and dark streaks may be visible as the final signs of disease. A cross section of an infected potato tuber may show a ring of brown discoloration and light pressure may produce an ooze (Sebastià et al., 2021).

- Bacterial leaf spot/Bacterial spot/Bacterial blight (*Pseudomonas syringae*)

Pseudomonas syringae is a seed-borne bacteria, once it is introduced into greenhouses, it can rapidly spread via splashing water. Disease development is favored by warm, moist conditions. Infected transplants carry the pathogen into production fields, then, widespread or severe symptoms usually do not develop unless the crop is sprinkler irrigated or subjected to a light frost during the production cycle. Bacterial leaf spot caused by *P. syringae* can easily be confused with *Xanthomonas blight*, except that infected leaves do not wilt and remain firmly attached to the plant. This pathogen has an extremely wide host range including woody species, vegetables, grasses, and herbaceous perennials. The disease growth is favored by high relative humidity, prolonged periods of leaf wetness, and high nitrogen levels (Newberry et al., 2017; Takikawa and Takahashi, 2014).

Initial signs of bacterial leaf spot are small leaves, water-soaked spots that are visible from both sides of the leaf. The lesions usually are delimited by leaf veins and thus have an angular, square, or rectangular appearance. However, it is frequently found only on older leaves that are protected by the plant canopy, except where sprinkler irrigation is managed. Nevertheless, fruit lesions are initiated only on green fruit, most likely because the infection occurs through fruit hairs, which are present only on immature fruit, the main symptoms are small, dark brown-to-black, raised spots. The lesions also may have a white halo, like the bird's-eye spotting seen with bacterial canker (*C. michiganensis*) (Joshi et al., 2020; Newberry et al., 2017).

- Bacterial pith necrosis (*Pseudomonas corrugata*)

Pith necrosis is caused by multiple soil-borne species of bacteria belonging to the Pseudomonas genus such as *Pseudomonas corrugata*, *P. mediterranea*, *P. cichorii*, *P. viridiflava*, and *P. fluorescens*.

P. corrugata is a ubiquitous bacterium isolated from a variety of sources and is referred to as a mild and opportunistic pathogen. However, in several cases, the infection has led to a significant loss of production with serious economic consequences, especially when specific environmental conditions are established, Generally, the percentage of infected plants by *P. corrugata* in field crops ranges from 5 to 10% even though there are cases where it is more than 50% (Joshi et al., 2020). The bacterial necrosis disease affects tomato plants as well as other important crops species in any stage of development and in different cultivation systems. The main symptoms of infections are clearer when the first clusters ripen on the tomato plants. Firstly, only slight wilting in the hottest part of the day and at times chlorosis of the apex can be observed. The principal symptom of the disease, as shown by its name, is stem pith necrosis and the pith may seem as: necrotic, dry, and slightly disaggregated in the core; hydropic, white, or dark but presenting a hardcore and necrotic in the areas near the xylem; disaggregated due to the formation of ladder-like cavities. In a progressed stage of the disease the xylem becomes necrotic, it starts from the base of the stem and works up to the leaf stems and bunches (Joshi et al., 2020; Hnamte et al., 2024; Shi et al., 2023). Table 1 presents the main symptoms and diseases caused by bacteria.

Table 1. Main symptoms and diseases caused by bacteria.

Type of Disease	Genus of bacteria	Crops disease	Reference
Stains and smut	*Pseudomonas* sp., *Xanthomonas* sp.	Tobacco fire blight Angular leaf spot of cucumber Bean halo blight Oat halo blight Apple fruit spot Walnut blight	(Dey and Raghuwanshi, 2024)
Vascular wilts	*Clavibacter* sp., *Erwinia* sp., *Pseudomonas* sp., *Xanthomonas* sp.	Bacterial wilt of alfalfa Potato ring rot Bacterial wilt and canker of tomato Bacterial wilt of cucurbits Banana Moko	(Chen et al., 2024; Dey and Raghuwanshi, 2024; Hwang et al., 2024; Keshavarz-Tohid and Ebrahimi, 2023; Mercado-Meza et al., 2023; Sakata et al., 2023)
Soft rot	*Erwinia* sp., *Pseudomonas* sp.,	Black leg of the potato Slippery surface of onion	(Dagher et al., 2020; Sakata et al., 2023; Sharifazizi et al., 2017)
Tumors or galls	*Agrobacterium* sp. *Pseudomonas* sp.	Stem galls of raspberries and blackberries Hairy root of the apple tree Olive knot	(Puławska et al., 2022; Sakata et al., 2023; Warabieda et al., 2023)
Cancros	*Pseudomonas* sp., *Xanthomonas* sp.	Citrus bacterial canker	(Lin et al., 2023; Sakata et al., 2023)

4. Types of Green Nanoparticles for Plant Bacterial Disease Control

Plant diseases inflict significant damage on crops worldwide, leading to substantial and increasing losses and economic burdens (Ristaino et al., 2021). Traditional remediation methods for combating the abovementioned diseases often involve chemical bactericides, biological control, genome edition, etc. (Sundin et al., 2016; Buttimer et al., 2017). However, the extensive use of these techniques raises concerns about environmental contamination (Kafle et al., 2022), antibiotic resistance of crops (Giacometti et al., 2021), the emergence of resistant bacterial strains (Gerhardson, 2002), and potential health risks of soil chemical accumulation (Woejko and Jaboska, 2020). Consequently, the development of new and less harassing alternatives for plant disease control is crucial.

The diverse methods for creating nanoparticles (NPs) can be categorized into two primary approaches: *top-down* (etching, photo-reduction, thermal evaporation, milling, etc.) and *bottom-up* (sol-gel, chemical vapor deposition, laser ablation, etc.), both being extensively investigated in the last few decades (Bansal et al., 2011; Abid et al., 2022). These methods have been gradually modified by green synthesis techniques to reduce energy consumption, the release of harmful chemicals such as heavy metals, and the use of complex equipment (Ying et al., 2022). This novel approach involves the use of natural biomass, including microorganisms (fungi, algae, bacteria, and certain viruses) (Bansal et al., 2011), biomolecules (lipids, pigments, proteins, polysaccharides, etc.) (Yosri et al., 2021), and plants (Rani et al., 2023), as biological reducing and stabilizing agents to create nanoparticles with different chemistry compositions (Ying et al., 2022).

Diverse NPs have shown remarkable potential for controlling plant bacterial diseases. It is well known that the NP's properties rely mainly on its size, shape, and chemical composition. NPs can be categorized into three main classes based on their composition: inorganic, organic, and hybrid, and three main classes based on the synthesis route: physical, chemical, and biosynthesis. Nevertheless, biosynthesis or green synthesis is considered a more beneficial method for the environment and agriculture, which is why, various types of green synthesized NPs have a particular focus on bacterial pathogens in different plants and animal diseases (Ansari et al., 2023; İpe et al., 2024).

4.1 Inorganic Nanoparticles

Inorganic nanoparticles are the most extensively studied class of green nanomaterials, providing significant potential for controlling plant bacterial diseases and representing an expanding area of research (Alhalili, 2023; Khan et al., 2022). These nanoparticles are primarily categorized into metal and metal oxides, the most studied nanoparticles being : silver (Ag) (Asif et al., 2022), copper (Cu) (Amjad et al., 2021), gold (Au) (Thipe et al., 2022) and selenium (Se) (Shahbaz et al., 2023), zinc oxide (ZnO) (Gadewar and Prashanth, 2024), iron oxide (Fe_3O_4) (Eldeeb et al., 2023), copper oxide (CuO) (Murugan et al., 2023), and titanium dioxide (TiO_2) (Rathi and Jeice, 2023).

Among the several inorganic nanoparticles currently being researched for plant disease treatment, the use of Ag, Se, and ZnO structures stand out. These nanoparticles, produced using plant extracts and other natural resources, exhibit strong antimicrobial and antifungal properties, making them effective in managing various plant pathogens.

Ag NPs exhibit strong antibacterial and antifungal activity, particularly against pathogens like *Staphylococcus aureus* and *Escherichia coli* (Gibaa et al., 2021). Additionally, these particles have effectively inhibited the activity of acetylcholinesterase (AChE) and butyrylcholinesterase (BChE) enzymes, suggesting their potential to protect plants from biotic stressors (Ceylan et al., 2021). Also the biocompatibility and reduced toxicity of green synthesized Ag NPs further enhances their appeal as a sustainable solution for plant disease management (Asif et al., 2022).

Green synthesized Se NPs have also been explored for their role in enhancing plant resistance to diseases. Research on *Mangifera indica* infected with mango malformation disease revealed that Se NPs at a concentration significantly improved physiological parameters like chlorophyll content and membrane stability, as well as biochemical markers such as proline and soluble sugar levels (Shahbaz et al., 2023). Moreover, Se NPs were found to increase the plant's antioxidant defense system, reducing biotic stress by enhancing both enzymatic and non-enzymatic activities (Morales-Espinoza et al., 2019).

Finally, ZnO NPs have demonstrated to inhibit the growth of fungal pathogens, effectively controlling diseases such as leaf spot, blight, and mildew (Cheema et al., 2022). The green synthesis approach has not only demonstrated to enhance the efficacy of ZnO NPs but also reduce their toxicity, making them a viable option for sustainable agriculture (Gadewar and Prashanth, 2024). Furthermore, ZnO NPs have shown potential in improving plant growth and resistance to stress, contributing to overall plant health and yield improvement (Seleiman et al., 2023).

In addition to the nanoparticles described above, Table 2 shows the current studies on other green synthesized inorganic particles investigated for the treatment of plant diseases.

4.2 Organic Nanoparticles

Organic nanoparticles synthesized through green methods are essentially composed of chemical species derived from plant extracts (Hano and Abbasi, 2021), natural polymers (Jujjavarapu and Poluri, 2020), cellulose (Carvalho et al., 2021), alginate (Bennacef et al., 2021), or other bio-based materials like lipids (Abosabaa et al., 2021). This natural origin confers them, in most cases, inherent biocompatibility, making them particularly attractive candidates for sustainable plant disease management strategies. These systems offer several advantages for plant bacterial disease control due to their inherent biodegradability, which often minimizes long-term environmental risks (Remya et al., 2022), while their potential functionalization has been demonstrated to be efficient for targeted delivery to bacterial pathogens within plant tissues (Spirescu et al., 2021).

Lipid-based nanoparticles have emerged as a promising tool for plant disease control, providing biocompatibility and the ability to encapsulate bioactive compounds like paclitaxel, curcumin, rhodomyrtone, quercetin, resveratrol, etc. (Patel et al., 2024). Plant lipids, which are essential constituents of cells, play crucial roles in structural, storage, signaling, and defensive functions. During plant-pathogen interactions, lipids are involved in both preexisting passive defense

Table 2. Inorganic green synthesized NPs applied for plant disease control.

Inorganic NPs	Source	Application	Results	References
Ag	*Avena fatua*	Antifungal activity (*Fusarium oxysporum (Fol)*)	AgNPs showed greater antifungal activity against *Fol than A. fatua* extracts, with 79.95% inhibition at 40 ppm after 96 hours.	(Qureshi et al., 2023)
Cu	Eucalyptus and Mint leaves	Fruit rot disease of chili (*Colletotrichum capsica*)	CuNPs outperformed conventional fungicides and showed better antifungal results when applied before pathogen inoculation	(Iliger et al., 2021)
Au	*Allium cepa* red peel extract	Antibacterial and antioxidant activity	AuNPs showed antimicrobial activity, against *Staphylococcus aureus* and *Candida albicans*, and effectively inhibited AChE and BChE enzymes	(İpe et al., 2024)
Se	*Melia azedarach*	Growth promoter and antifungal agent	SeNPs improved chlorophyll, membrane stability, and biochemical parameters while enhancing the antioxidant defense, reducing biotic stress from mango malformation disease	(Shahbaz et al., 2022)
ZnO	*Cinnamomum camphora* (L.) Presl	Antifungal activity	ZnONPs showed inhibitory effects on the spore germination and mycelial growth of *A. alternata* (lipid peroxidation)	(Zhu et al., 2021)
Fe_3O_4	Azadirachta indica	Antibacterial activity	NPs showed bactericidal action against *Xanthomonas campestris Xcc –C7* and inhibited the fungal growth	(Sharma et al., 2023)
CuO	*Spirodela polyrrhiza*	Antibacterial and antifungal activity	CuONPs reduce fungal mycelium by 45–50% and inhibit egg hatching	(Khan et al., 2023)
TiO_2	*Punica granatum*	Anthracnose disease in sorghum	Application of TiO_2NPs significantly increased the content of defense-related biochemicals (phenols and defense-related enzymes)	(Nabi et al., 2024)

mechanisms and pathogen-induced immune responses at the local and systemic levels (Kuniak and Gajewska, 2024).

These particles have been demonstrated to effectively deliver antimicrobial agents to the target sites within plants, enhancing their efficacy against pathogens (Pinilla et al., 2021). Lipid nanoparticles loaded with essential oils have shown significant antifungal activity against *Botrytis cinerea*, *Penicillium expansum* (Fincheira et al., 2024) and *Zataria multiflora* (Nasirzadeh Fard et al., 2023) common plant pathogens that cause gray mold in various crops.

Most protein-based nanoparticles are biodegradable and obtained from natural sources like plants. This material has been investigated to deliver agrochemicals or antimicrobial agents directly to the plant (Maximiano et al., 2022). Studies have shown that soy protein nanoparticles, when loaded with fungicides, can effectively control diseases like powdery mildew in plant crops (Huang et al., 2022). The slow release of the fungicide from the nanoparticles ensures prolonged protection against the pathogen, reducing the need for frequent applications.

Among the diverse green synthesized polymers, chitosan has been extensively studied for its antimicrobial properties and its ability to enhance plant immunity (Selvaraj et al., 2022). Chitosan nanoparticles have demonstrated significant antifungal activity against a variety of plant pathogens, including *Fusarium oxysporum* (Sravani et al., 2023) and *Phytophthora infestans* (Huang et al.,

2021). These natural polymers have also induced systemic resistance in plants like tobacco, enhancing their ability to fight off infections (Kongala and Mamidala, 2023). For instance, chitosan NPs have been shown to reduce the severity of tomato blight caused by *Alternaria solani* when applied as a foliar spray, highlighting their potential as an eco-friendly alternative to conventional fungicides (Sathiyabama et al., 2014).

Other natural polymers like cellulose, derived from plant fibers, have gained attention for their ability to control the delivery of pesticides and other agrochemicals (Zhang et al., 2022). Cellulose NPs are highly stable and have been functionalized to improve their interaction with plant surfaces. Cellulose nanocrystals, when loaded with thiamethoxam exhibited a significantly superior insecticidal activity against *Phenacoccus solenopsis* pest in comparison to technical and commercial formulations (Elabasy et al., 2020). The application of these nanoparticles significantly reduced the incidence of disease, highlighting their potential as a sustainable disease management strategy.

Finally, alginate (biopolymer derived from brown seaweed) as a nano hydrogel has been used to synthesize nanoparticles for the controlled release of agrochemical herbicides and plant nutrients (Artusio and Cas, 2021). In addition to their drug delivery capacity, composite alginate nanoformulations have been used for the encapsulation of *Bacillus velezensis* to control Pistachio Gummosis disease on pistachio trees (Moradi Pour et al., 2022). The use of alginate as a carrier not only helps in the sustained release of the active agent but also minimizes the environmental impact by reducing the amount of chemical input required to plant pest control (Martnez-Cano et al., 2022).

Other natural proteins like amplified DNA have accurately detected *Phytophthora infectants*, the pathogen causing late blight in potatoes and tomatoes, using a polymer-based microneedle patch method (Paul et al., 2019).

4.3 Hybrid Nanoparticles

Both types of green-synthesized nanoparticles discussed in the previous sections (organic and inorganic) have proved to be efficient tools for crop disease control, effectively reducing bacterial growth and enhancing crop growth rates. As mentioned earlier, inorganic particles stand out for their high antimicrobial and antifungal activity (Alhalili, 2023; Khan et al., 2022), while organic systems allow for the encapsulation and controlled release of pesticides for pest control (Hano and Abbasi, 2021). Due to these advantages, integrating both types of particles have become a topic of large research interest in recent years.

By combining organic and inorganic nanoparticles, hybrid nanoparticle systems are developed, which aim to obtain the high degradation efficiency of inorganic particles with the superior biocompatibility and stability of organic systems. This phenomenon has been reported to be obtained through synergistic effects once both materials interact (Salem et al., 2023). Table 3 presents recent research studies in which green synthesized hybrid NPs have been designed for the control of plant diseases and crop protection.

The use of hybrid nanoparticle systems has enabled the development of multifunctional conjugated systems that go beyond antimicrobial activity, promoting controlled release of nutrients, plant growth promoters, and more. These systems have tunable properties, achieved by modifying the composition and structure of both inorganic and organic particles, allowing precise customization for specific plant diseases and targeted pathogens (Salem et al., 2023). Building upon the unique properties of both inorganic and organic materials, hybrid nanoparticles represent a promising frontier in green nanotechnology for plant bacterial disease control, offering enhanced functionality and targeting capabilities compared to their counterparts.

Table 3. Green synthesized hybrid NPs systems applied for plant disease control.

Hybrid NPs	Source/Method	Application	Results	References
Ag NPs-chitosan (core-shell)	Wheat leaf waste	Antifungal and nanoprime	Improved antifungal activity against pathogenic fungi in cereals (*Fusarium oxysporum, Aspergillus niger*). Increased chlorophyll synthesis in treated seeds.	(Mondjar-Lpez et al., 2023)
Ag NPs-polysaccharides	Natural organic matter	Reduce phytotoxicity	NOM mitigated the negative effects of AgNPs and $AgNO_3$ on rice with a lower germination inhibition rate, less chlorophyll reduction, and more relative biomass.	(Huang et al., 2020)
Chitosan-ZnO NPs	Green Tomato Extract	Prevention of leaf blight (rice)	NPs effectively inhibited *Xanthomonas oryzae pv. oryzae* growth by disrupting biofilm formation, damaging cell membranes, and triggering bacterial cell death	(Abdallah et al., 2020)
Chitosan-Cu NPs	Chemical precipitation	damping off prevention	Antifungal activity against damping-off disease on chilli, cowpea, and tomato plants caused by *Rhizoctonia solani* and *Pythium aphanidermatum*	(Vanti et al., 2020)
phytochemical-capped FeO	*Cassia occidentalis L.* flower	Seed germination boost (rice)	Seeds-primed NPs showed a 50% increase in growth and higher antioxidant enzyme activity	(Afzal et al., 2021)
Chitosan-Mg NPs	*Bacillus sp.* strain RNT3	Antimicrobial activity	Nanocomposites showed remarkable antimicrobial activity against A. *oryzae* and *R. solani* and significantly inhibited the growth.	(Ahmed et al., 2021)
Ag-doped ZnO NPs	*Moringa oleifera* extract	Antimicrobial activity	composite inhibited the growth of *Rosellinia necatrix*, *Fusarium* spp., and *Sclerotinia sclerotiorum*	(Swati et al., 2020)
Alginate-Ag NPs	*Laminaria ochroleuca* alginate	Antibacterial Activity	NPs exhibited strong antibacterial activity against Gram-positive and Gram-negative bacteria	(Kaidi et al., 2022)

5. Mechanisms of Action of Green Nanoparticles Against Bacteria

5.1 Direct Antibacterial Mechanisms

Green synthesized nanoparticles have demonstrated their capacity to combat several phytopathogenic bacteria such as *Ralstonia solanacearum* and *Pseudomonas psyringae* (Khan et al., 2019, Guo et al., 2022). However, despite the intensive study on the bactericidal properties of several types of nanoparticles, their direct Mechanisms of Action (MoA) remain mostly undetermined.

Green synthesized nanoparticles made of materials such as silver (AgNPs), titanium oxide (TiO_2 NPs), zinc oxide (ZnO NPs), and magnesium oxide (MgO NPs) have demonstrated their capacity to inhibit bacterial growth *in vitro* (Thakur et al., 2019; Garibo et al., 2020; Khan et al., 2020; Pillai et al., 2020) but their MoA is only partially understood.

The interaction with the cell membrane seems to play a central role in a nanomaterial's bactericidal activity. When these nanomaterials come into contact with the surface of bacterial cells, a series of structural changes start to occur in the cell membrane, ultimately leading to a disruption of its permeability and, consequently, to an impaired respiratory process due to membrane depolarization, resulting in cell death (Rajeshkumar et al., 2019). For example, it has been observed that even brief exposure to Ag NPs causes the formation of gaps in the cell surface of bacteria, which can be observed using advanced imaging techniques like Transmission Electron Microscopy (TEM) and Scanning Electron Microscopy (SEM) (Roy et al., 2019).

Linklater et al. (2020) proposed that this effect could be determined in the large part, not by chemical interactions between the cell components and the particles, but by mechanical processes (Linklater et al., 2020). Nanoparticles of smaller sizes would create these gaps in the membrane during their translocation through the lipid bilayer by leaving an irreversible pore in its structure that leads to a compromise in its stability, while larger nanoparticles, which are unable to be internalized easily, disrupt membrane stability by exerting mechanical stress when they adhere to it. This would lead to crumpling the cytoplasmic membrane and ultimately to its rupture.

In addition to mechanical effects, nanoparticles could disrupt the function of bacterial cells by other means. In a study evaluating the bactericidal activity of ZnO NPs on *Staphylococcus aureus*, *Escherichia coli*, *Bacillus subtillis*, and *Pseudomonas aeruginosa*. Mendes et al. (2022) found that exposure to the material for a period as short as 15 minutes led to damage in the membranes of 70% of the treated cells (Mendes et al., 2022). They attributed this effect to the capacity of ZnO to cause the formation of reactive oxygen species (ROS) formed by hydroxyl radicals (^{-}OH), hydroperoxide radicals (HO_2^-), and superoxide radical anions (O_2^-), a property shared by other metallic nanoparticles. The application of metallic nanoparticles also results in elevated levels of Reactive Nitrogen Intermediates (RNI) which, in conjunction with ROS, interact with numerous targets inside the cell, including thiols, metal centers, tyrosine residues in proteins, nucleotide bases, and lipids. Some of these reactive species directly affect the activity of enzymes that inhibit bacterial respiration, arrest the growth of the cell, suppress the synthesis of DNA, and change gene expression patterns (Quinteros et al., 2016).

A proteomic study conducted to determine the influence of Ag NP on *Pseudomonas aeruginosa* (Yan et al., 2018), found evidence of several membrane proteins whose expression seemed to be up- and down-regulated by the action of the nanomaterial. These protein's primary functions were related to ion binding, transport, flagellum assembly, pore formation, antibiotic resistance, and membrane stabilization, which serve to confirm that the primary target of NPs antibacterial activity is indeed the cellular membrane. It also suggests that they act not only by interacting directly with it but by affecting the expression of genes related to its function. In addition to these proteins, Yan et al. (2018) also found a change in the expression of several enzymes classified as superoxide dismutase and oxide reductase, which are related to intracellular oxidative damage caused by metals. This, combined with the elevated ROS levels detected in the bacteria, confirms the occurrence of oxidative stress induced by the treatment (Yan et al., 2018).

This ability to regulate the expression of certain proteins plays a central role in another MoA attributed to metal nanoparticles: their capacity to interfere with the cell's energy production. In the final step of oxidative phosphorylation, enzymes called F-type ATP synthases catalyze the production of the ATP that supplies the cell's energetic needs. These enzymes serve a dual function as ATPases that create the transmembrane proton electrochemical gradient fundamental for the translocation of substances through the cell membrane (Kühlbrandt, 2019). Au NPs are known to interfere with this process by downregulating these enzymes and causing reduced levels of ATP production that lead to an arrest in the bacterium's metabolic activity, in addition to compromising the transportation of substances in and out of the cell by eliminating the proton gradient (Nisar et al., 2019).

This effect on the cell's energetic metabolism has been observed in other steps of bacterial metabolism. In their study, Mendes et al. (2022) proposed that part of the effect of the nanoparticles could come from the release of Zn^{2+} ions from the dissolution of ZnO in an aqueous solution, which inhibit the activity of glycolytic enzymes by interacting with their thiol groups (Mendes et al., 2022). A study on the effect of CuO NP on the metabolism of sulfate-producing bacteria found that strains of Desulfovibrio vulgaris suffered effects on their catabolic and anabolic processes, with a marked decrease in their capacity to reduce sulfate and produce sulfide directly associated with increasing doses of the nanoparticles, halting all catabolic activity at a concentration of 500 mg/L of CuO NP (Chen et al., 2019).

5.2 Indirect Mechanisms

In addition to the effect of nanoparticles on the bacteria themselves, when studying them in the context of agriculture, another element to consider for explaining their effectiveness against pathogens is the interaction of the nanomaterials with the plant itself. Nanoparticles act as Plant Growth-Promoting Factors (PGPFs). These are stress factors of physical, chemical, or biological origin that, in correct doses, promote the development and yield of crops while inducing the activation of the plant's immune response mechanisms against all kinds of pathogens, not only bacterial ones (Tortella et al., 2023).

When pathogen infection occurs, the accumulation of ROS in the plant tissues is one of the most significant effects that can be observed and therefore is responsible for a great deal of the stress suffered by the plant. Maintaining low levels of ROS in the system is crucial to prevent these damages, so plants respond to these conditions by unleashing cascades of stress signals that lead to the expression of antioxidant enzymes such as superoxide dismutase (SOD), catalase (CAT), and peroxidases (POX) (Abdul Malik et al., 2020; Kaur et al., 2022).

Chandra et al. (2015) conducted an experiment in which they infected plants of *Camellia sinensis* and treated them with chitosan nanoparticles (CNP) to assess the effect of the nanomaterial on the plant's response (Chandra et al., 2015). They found that plants subjected to the CNP treatment had an up-regulation in the expression of the genes that encode for SOD and CAT, as well as an increased activity of these enzymes that resulted in enhanced protection against oxidation. Similar results were observed by Abdelrhim et al. (2021) in wheat, where they reported that the application of SiO_2 NP to seedlings of this plant stimulated not only the enzymatic antioxidant machinery but also the non-enzymatic antioxidant system (Abdelrhim et al., 2021). This non-enzymatic antioxidant response consists of the production of substances such as Nitric Oxide (NO), ascorbic acid, and glutathione that work in conjunction with other systems present in the plant like the Salicylic Acid (SA) pathway, reducing the concentration of ^{-}OH, HO_2^- and O_2^- radicals.

The name phytoalexins is given to a family of compounds that are synthesized and accumulated by plants in response to pathogens. These chemicals have shown high toxicity against phytopathogenic bacteria, fungi, and nematodes. The group is characterized by its extreme diversity in terms of chemical nature, which can change significantly from one plant family to another; therefore, their classification is based on their plants of origin rather than their chemical structure (Bizuneh, 2021).

Among *phytoalexins*, those of a phenolic nature stand out for constituting most of the family. When faced with an infectious process they start accumulating in the tissues and participate in the reduction of oxidative stress, as well as the inhibition of pathogen growth, due to their antimicrobial activity (Babenko et al., 2019). Phenolic compounds like flavonoids play important roles in response to infection by disrupting bacterial membranes, impairing the formation of biofilms through interference with their quorum sensing mechanisms, and reducing their pili and fimbriae formation. Additionally, these substances can inhibit the action of bacteria-exclusive topoisomerases such as DNA gyrase that take part in the bacteria's DNA synthesis (Górniak et al., 2019).

In a study conducted using seedlings of *Arabidopsis thaliana* treated with Ag NPs, Kruszka et al. (2020) found that the treatment altered the synthesis rates of 47 different metabolites including several phytoalexins characteristic of the *Brassicaceae* family, most notably inducing the accumulation of camalexin and other related compounds such as hydroxy camalexin O-hexoside and hydroxy camalexin malonyl-hexoside, which play an important role in the defense against infections (Kruszka et al., 2020). Abdelrhim et al. (2021) found in their study that SiO_2 NPs significantly induced the accumulation of phenolic compounds as well as increased the activity of the enzyme phenylalanine ammonia-lyase (PAL) (Abdelrhim et al., 2021). PAL is the main enzyme responsible for the synthesis of phenolic compounds like flavonoids, lignin, and coumarin and is intimately related to the plant's defense system given that it is the main enzyme involved in one of the known metabolic pathways responsible for producing SA, where it catalyzes the conversion of phenylalanine into trans-cinnamic acid (Lefevere et al., 2020; Kaur et al., 2022). Given that SA is one of the most important phytohormones related to defense against abiotic stress and pathogens, the capacity of these materials to enhance the activity of enzymes related to this system increases the relevance of nanoparticles in the fight against bacteria.

The cell wall constitutes the plant's first barrier against pathogens of all kinds. Therefore, a common response against infection involves its strengthening through a process of lignification (Mishra and Baek, 2021). Considering that PAL catalyzes one of the initial steps in the phenylpropanoid pathway, which is responsible for the synthesis of lignin (Barros et al., 2016), the findings of Abdelrhim et al. (2021) could indicate that nanomaterials not only induce enzymatic defense mechanisms in treated plants but also their structural defenses (Abdelrhim et al., 2021, Mishra et al., 2017).

5.3 Role of Nanoparticle Size, Shape, and Surface Properties in Antibacterial Activity

Having established the effect of nanoparticles on bacteria and the plant´s defense system, it is important to address the characteristics of these materials that grant them their properties. ,It has been reported that there are significant differences between the capacity of a nanoparticle to act as an antibacterial agent when compared with the bulk material from where the particles were synthesized (Park et al., 2012). This could be explained because NP has a higher ratio of surface atoms over interior atoms, which means that they have a greater surface area per volume, a ratio that increases as the particle size decreases, making them more reactive than their macroscopic counterparts and increasing their capacity to interact with biological systems (Asha and Narain, 2020).

This higher reactivity is not the only consequence of the smaller size. Nanoparticles smaller than 10 nm are capable of easily diffusing through the bacterial cell wall and are generally considered to be more toxic to microorganisms than bigger particles (Linklater et al., 2020). However, experiments using particles of considerably larger sizes such as 80 nm have proven that these NP are also capable of causing damage to *Escherichia coli*, *Staphylococcus aureus*, *Pseudomonas aeruginosa*, and *Bacillus subtilis* (Mendes et al., 2022), although this may be associated with mechanical stress exerted by the particles on the cell wall, as previously discussed.

In a study conducted using colloidal suspensions of ZnO NPs, Lallo da Silva et al. (2019) found that particles of smaller sizes presented a smaller Minimum Inhibitory Concentration (MIC) against strains of *Staphylococcus aureus* and *Escherichia coli* (Lallo Da Silva et al., 2019). They also reported that the NP were less effective in inhibiting the growth of *E. coli* compared to *S. aureus*, a difference they attribute to the structural differences of Gram-positive and Gram-negative bacteria's cell walls. Gram-positive bacteria's wall is characterized by the presence of a thick layer composed of interlinking chains of peptidoglycan, while Gram-negative bacteria possess a much thinner layer of peptidoglycan in their walls but have an extra lipidic membrane and an exterior layer composed of lipopolysaccharides (LPS) and other substances (Fan et al., 2021).

Lallo da Silva et al. (2019) theorized that the reduced antimicrobial activity of their NPs was caused by the greater complexity of *E. coli's* cell wall (allo Da Silva et al., 2019). However, this is a point of discrepancy between researchers, since some posit that the peptidoglycan layer present in Gram-positive bacteria represents a greater obstacle for the internalization of NPs (Roy et al., 2019).

An alternative explanation for these discrepancies could be the characteristics of their surfaces, for example, their charge. Surface charge is an important property of nanomaterials that has considerable influence over their stability, their propensity to form aggregates, and their reactivity with the medium around them. The charge of the particle is determined by the external environment that surrounds it, which is evidenced by it changing when suspended in differently charged solutions to a charge that opposes that of the medium (Asha and Narain, 2020).

The charge of a nanoparticle is expressed in terms of its zeta potential, which is a measure of the charge gradient between the fluid that surrounds a particle in suspension and the layer of fluid that contains the oppositely charged ions that surround the particle's surface. A nanoparticle with a negative zeta potential will be attracted to objects with a positive charge and vice versa (Selvamani, 2019). Positively charged Ag NPs, for example, have been found to have a certain affinity towards the negatively charged LPS layer of Gram-negative bacteria, which could explain why some researchers have reported different results to those found by Lallo da Silva et al. (2019). The association with the membrane leads to the already discussed affectations on the wall: its depolarization and inhibition of all the fundamental processes that depend on it, ultimately leading to cell death (Rajeshkumar et al., 2019). In addition to its affinity for the cell wall, the charge of the nanoparticles could play a role in another proposed MoA. The DNA molecule is negatively charged, so positively charged nanoparticles can more easily interact with it, increasing the genotoxicity of the material (Daima and Bansal, 2015).

The hydrophobicity of the NP's surface is also an important factor in their interaction with bacterial membranes. Given the propensity of hydrophobic nanoparticles to agglomerate when suspended in aqueous solution, these tend to form larger clumps of material that face the same challenges that bigger nanoparticles undergo when interacting with the cell surfaces. On the other hand, hydrophilic nanoparticles do not face these challenges and therefore adhere more easily to membranes (Linklater et al., 2020).

Another factor of nanoparticles that determines their capacity to cause damage to bacteria is their shape. Earlier studies tried to ascertain the effect of shape on antimicrobial activity by comparing the effects of triangular, spherical, and rod-shaped Ag NPs on *E. coli* (Alshareef et al., 2017). It was reported that their truncated triangle-shaped NPs were more efficient in inhibiting the growth of the bacterium when compared to the spherical ones, and these were in turn more effective than the rod-shaped particles. After studying the structure of their nanomaterials, they attributed these results to a difference in surface area, the truncated triangles had more available surface for interacting with the bacterium's wall and in consequence, inhibiting its growth (Pal et al., 2007).

Cheon et al. (2019) used spherical, disk , and triangularly-shaped Ag NPs to evaluate their antibacterial effect (Cheon et al., 2019). They found that all three of the differently shaped NPs were capable of inhibiting the growth of *E. coli* but their results differ from the ones reported in the 2007 study since they encountered that the spherical particles were the most effective ones, followed by the disks and in the last place the triangles. They attributed this difference in results to the fact that the calculated surface areas of their nanomaterials were different from those reported by Pal et al. (2007), given that their spheres and disks had a greater surface area than their triangular particles (Pal et al., 2007).

Au NPs shaped like stars, flowers, and spheres were tested on *S. aureus*. They were unable to observe any inhibition of bacterial growth by the spherical particles but found a great increase in the lag phase of the bacterium's growth, as well as a significant reduction in its growth rate when treated with the star and flower-shaped Au NPs, with the latter being the most efficient in inhibiting S. aureus's growth (Penders et al., 2017). Given that these flower-shaped particles had a greater surface compared with the rest, this seems to confirm that the influence of shape over antimicrobial activity comes mainly from its effect on the number of exposed reactive ions present on the surface of the particle, even when considering the use of different materials.

Capping agents are trace substances added during the synthesis of nanoparticles that adsorb to their surface and modify their characteristics, providing stability while at the same time preventing the aggregation and overgrowth of the particles as well as providing new properties and enhancing some of their preexisting ones (Javed et al., 2020). There is a wide range of capping agents that go from surfactants and polymers like PVC to more naturally occurring substances like lipids, carbohydrates, and proteins depending on the intended use and method for the synthesis of the particles (Sidhu et al., 2022). Since these substances modify the properties of a particle's surface including Zeta potential, hydrophobicity, and reactive groups, as well as other characteristics such as shape and size, it is reasonable to assume that they could influence the NP's antibacterial properties.

In a study using citrate, dextran (Dex), diethyl aminoethyl-dextran (DexDEAE), carboxymethyl-dextran (DexCM) and PVP as capping agents for AgNPs, Ferreira et al., (2023) tested the antibacterial properties of the capped nanoparticles against *E. coli*, Methicillin-Resistant *S. Aureus* (MRSA), and *P. aeruginosa* and compared it with the effect of bare Ag NPs (Ferreira et al., 2023). They reported that the bare Ag NPs performed the worst among all the tested nanomaterials and observed a wide range of MIC and Minimum Biocidal Concentrations (MBC) among the coated NPs, with the Ag NP-Dex compounds being the most effective overall. They attributed the small effect of the uncoated particles to the larger particle size and the clusters formed by agglomeration and found a similar phenomenon when they examined the five coated materials that justified their respective performances.

6. Delivery and Formulation of Green Nanoparticles for Plant Protection

It has been described as a diverse landscape of green nanoparticles (NPs) with immense potential for combating plant bacterial diseases. Nevertheless, realizing the full potential of these nanomaterials is only one part of the overall picture. To ensure the proper functioning of these green-synthesized systems, it is crucial to guarantee efficient delivery (Zobir et al., 2021).

This section delves into the crucial aspects of delivery methods and controlled release strategies, drawing on recent scientific advancements to optimize the efficacy and sustainability of green NPs for plant disease control. A comprehensive exploration of various delivery techniques, encompassing foliar application, seed treatment, and soil incorporation, will be undertaken. Additionally, the advantages of encapsulation and controlled release will be explained.

6.1 Foliar Application

The foliar application of nanoparticles obtained through green synthesis principally consists of spraying a suspension of the NPs directly onto plant leaves or other above-ground plant parts. This method allows for targeted delivery, facilitating the application of these nano systems and ensuring direct interaction with the plant. It represents a simple and cost-effective technique, being easily adapted to existing spraying equipment used for conventional pesticides, eliminating the need for new equipment or modifications to crops (Hong et al., 2021). For this reason, this method is the most widely used. Figure 1 presents a diagram of the general mechanism for the foliar application of nanoparticles on plants.

Foliar applications of green synthesized nanoparticles have shown promising results in plant disease control. Silver nanoparticles were synthesized and successfully foliar applied to manage fungal diseases of *F. oxysporum* in tomato crops (Pavlicevic et al., 2023) as well as CaO NPs on *Medicago sativa* seedlings (Albayrak Turgut and Bezirganoğlu, 2022). Organic NPs like chitosan have been foliar sprayed on tomato plants colonized with arbuscular mycorrhizal fungi *Rhizophagus irregularis*, obtaining increased shoot biomass and flower number in crops, while the root application did not show quantitative effects (El Amerany et al., 2020). TiO_2 nanoparticles applied through this method have reduced anthracnose disease in sorghum by over 60% at 150 and 200 ppm, showing strong dose-dependent antifungal activity (Nabi et al., 2024). In addition to reducing the fungal activity of plant leaves and stems, foliar application has proven effective in eliminating surface contaminants. For example, foliar application of ZnO NPs has been shown to reduce Cd content in the shoots of lettuce plants (Timilsina et al., 2023).

Despite these advantages, environmental conditions such as rain, wind, and UV light can significantly reduce the effectiveness of foliar-applied NPs. UV radiation can trigger unwanted photodegradation processes, while rain may wash them off the plant surface (Hong et al., 2021). Additionally, foliar application may fail to target bacterial pathogens residing within plant tissues

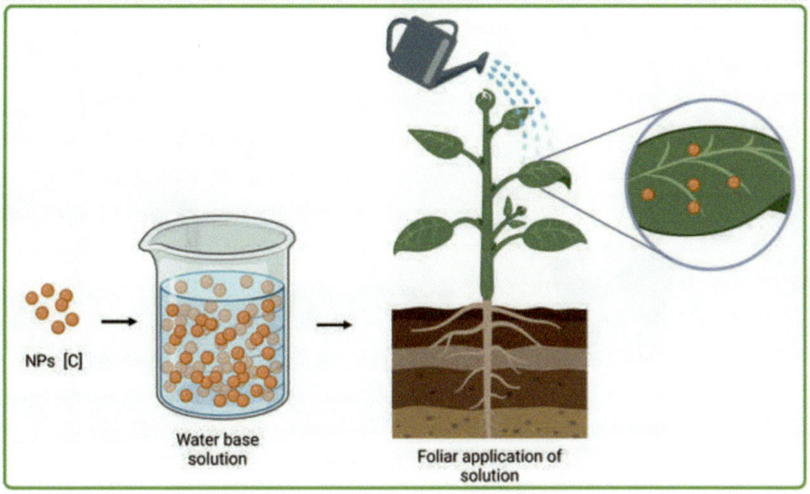

Figure 1. Foliar application of NPs to plants diagram.

or roots and could interfere with surface-level biological processes such as photosynthesis (Lu et al., 2020). These limitations highlight the need to optimize NP formulations and develop UV-resistant coatings (Sabir et al., 2022) or incorporate sticking agents (Grillo et al., 2021) to enhance the persistence of nanostructures on plant surfaces. Along with the issue of nanoparticle retention, the concentration of nanostructures has proven to be a crucial factor due to their direct contact with plant tissues. For instance, foliar-applied Ag NPs have been found to inhibit root development in *Brassica campestris*, *Oryza sativa*, and *Vigna radiata* at concentrations exceeding 4500 μg/mL (Khan and Zahoor, 2023). Additionally, hybrid CuO-citric acid particles applied to soybeans have resulted in reduced protein production in plants when solutions exceed 300 mg/L of nanoparticles. Consequently, alongside studies on antimicrobial and antifungal activity, phytotoxicity analyses must be conducted to determine the optimal formulation needed to minimize adverse effects that may impact plant growth.

Due to the direct delivery of nanoparticle solutions to the aboveground plant parts, interaction with foliar pathogens is maximized while minimizing potential harm to beneficial soil microbes compared to other application methods (Hong et al., 2021). However, it has been encountered that this application method also makes plants vulnerable to alterations through direct contact with NPs. Therefore, determining the correct formulation of these applied solutions has shown to be a crucial variable.

6.2 Seed Treatment

Seed treatment of crops with nanoparticles (NPs) is an emerging strategy currently focused on controlling plant diseases and enhancing crop productivity. These techniques involve a pre-sowing process where plant seeds are immersed in a solution containing dispersed NPs to improve seed performance and protect plants during the initial stages of growth (germination and seedling). This technique seeks to make use of NPs' antibacterial and antifungal activity, as well as the beneficial cellular stress tolerance provided during early stages (Zhao et al., 2024). Figure 2 presents a diagram of the general mechanism for the foliar application of nanoparticles on plants.

These techniques involve the use of a water-based solution with NPs to promote the initial stages of germination, without allowing the seeds to fully sprout. These controlled hydration processes have been demonstrated to enhance seed resistance to biotic and abiotic stresses, accelerate germination, and improve nutrient uptake and overall crop yield (Nile et al., 2022), leading to more controlled and faster seedling obtention.

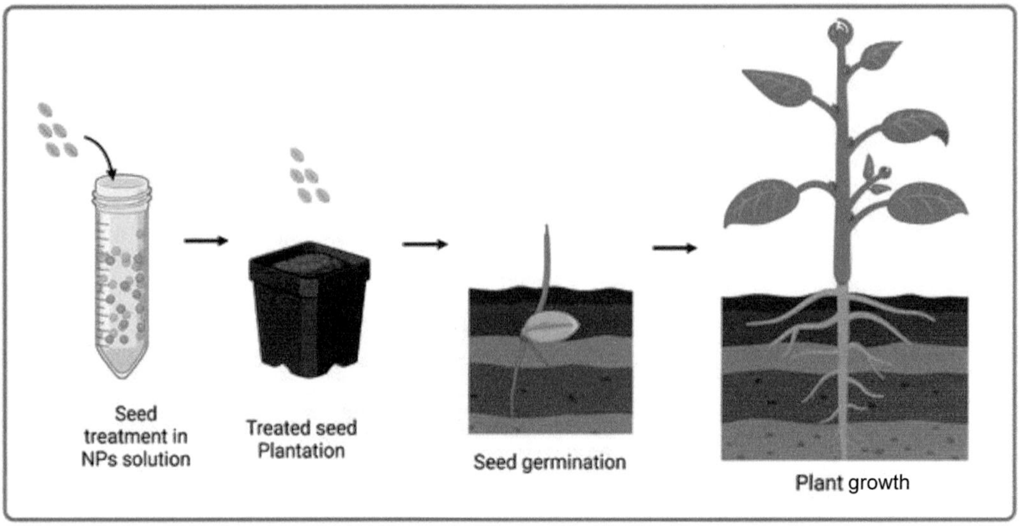

Figure 2. Seed treatment of NPs to plants diagram.

One of the main nanostructures studied using this method has been Ag NPs, which have shown improved germination and seedling vigor of aged rice seeds at concentrations of 5 and 10 ppm. This seed priming method enhanced α-amylase activity increased soluble sugar content, and elevated ROS production compared to untreated seeds (Mahakham et al., 2017). Similarly, recent studies have demonstrated the effectiveness of using FeO nanoparticles (NPs) synthesized with *Salvinia molesta* extract. These NPs applied through seed priming significantly improved tomato seed germination and seedling growth, especially at 100 ppm (max concentration at 1000 ppm). This concentration notably enhanced the seed vigor index compared, promoted seed germination, and influenced metabolic processes, including ROS generation and enzyme activity (Madusanka et al., 2024).

Despite these advantages, contact with nanoparticle solutions during the early growth stages of plants needs precise control over the concentrations and formulations of used NPs, like foliar application methods. The goal is to avoid potential over-priming (Pagano et al., 2023), which has been shown to cause metabolic damage to seeds and potentially increase their susceptibility to pathogens like in pepper (Tu et al., 2022) and *Medicago truncatula* seeds (Pagano et al., 2022).

6.3 Soil Application

Soil application of nanosized systems has emerged as a promising method for plant disease control that offers a direct route for NPs to interact with soil-borne pathogens, seeking to enhance plant health and resistance to bacterial and fungal diseases. This method consists of introducing NPs either directly or in a water solution into the planting soil (Perea Vélez et al., 2021). Figure 3 presents a diagram of the general mechanism for foliar application of nanoparticles on plants.

The direct contact of nanoparticles with the cultivation soil allows for direct interaction with microorganisms present in the soil (bacterial or fungal), without reaching direct contact with the plant's leaves and stems, contrary to what occurs in foliar application (Hong et al., 2021). Through this indirect process, it is possible to target microorganisms in the soil, while also serving as a potential source of nutrients to roots through microorganisms' degradation (Abdelaziz et al., 2022).

An example of this application method is found in several investigations, such as green synthesized Ag from neem leaf extract. These NPs significantly improved tomato plant growth and disease resistance to early blight disease caused by *Alternaria solani*. Plant soil treated with a 5 and 10 ppm solution showed increased height, fresh and dry weight, and a reduction in early

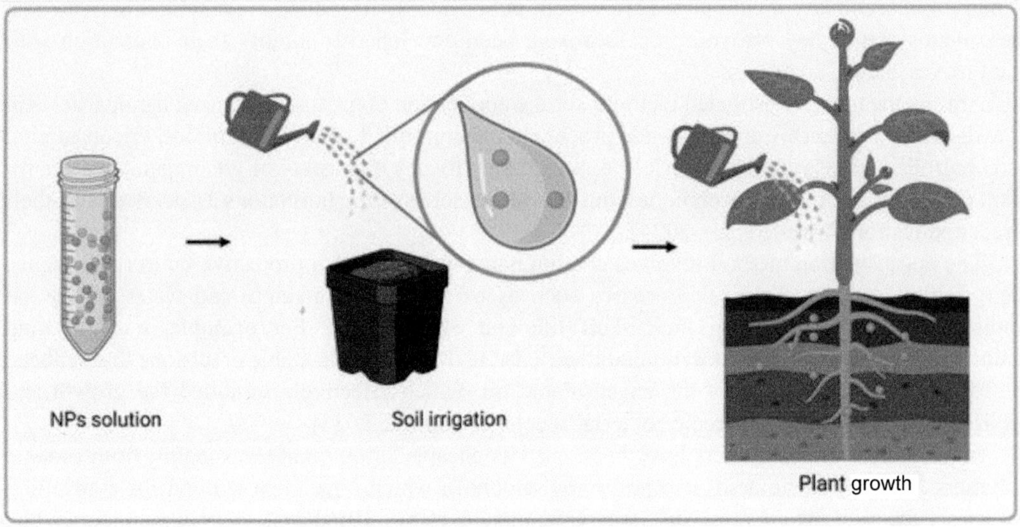

Figure 3. Soil application of NPs to plants diagram.

blight disease severity by over 70%. NPs also enhanced photosynthetic pigments and secondary metabolite accumulation (Ansari et al., 2023).

Another notable study involved the use of ZnO NPs to control Fusarium wilt disease in eggplant (Solanum melongena). Through soil application on the NPs, the severity of the crop disease was reduced by 75%, improving plant recovery after the disease and enhancing growth and metabolic indicators, including plant height, root length, biomass, chlorophyll content, and antioxidant activity (Abdelaziz et al., 2022).

The indirect contact of nanoparticles with plants through this method has enabled the use of nanostructures previously shown to be toxic to plants. For example, graphene oxide nanoparticles, when formulated at concentrations up to 1200 mg/L and applied to soil, enhanced the growth of mungbean (*Vigna radiata* L.). The addition of these particles to cultivation soil improved root and shoot length, leaf number, root nodules, pods, and seeds per pod, demonstrating their potential as growth regulators (Mirza et al., 2022).

Despite the advantages of this method compared to those previously mentioned, the primary concern with applying nanoparticles to cultivated soil is their potential for accumulation. This accumulation process can lead to long-term environmental impacts, including soil toxicity, disruption of beneficial microbial communities, and potential effects on plants and human health. Although these effects may not be immediate, over time, accumulation could alter soil chemistry, and affect nutrient availability (Rajput et al., 2020). Moreover, controlling the formulations (concentration and dispersion) of nanoparticles requires careful management to optimize this application method and mitigate adverse effects on plant and microorganism development (Javed et al., 2019).

Despite the various techniques for applying nanoparticles to plants (foliar, seed, and soil applications), they all share a factor of essential importance: the nanoparticles' concentration. The concentration of the solutions that yield the best results in terms of antimicrobial activity, growth rates, metabolite production, and other factors has been shown to depend on the specific plant organism and the formulation of the nanostructures used (including their nature, synthesis method, concentration, and application method) (Wasule et al., 2024).

6.4 Encapsulation and Controlled Release Strategies

Alongside the methods for applying nanoparticles to plants during their growth, toxicity is a critical concern in the formulation of the nanoparticles used. In these studies, it is crucial to ensure that the nanoparticles do not adversely affect plant growth, nutrient absorption, or the production of primary and secondary metabolites (Muzammil et al., 2023). To reduce the potential toxicity that these nanosystems may present, processes have been developed to modify their interaction with plant leaves, stems, and roots.

Among the most investigated methods for the application of green-synthesized nanoparticles in crop disease treatment are encapsulation processes and controlled release of particles. Encapsulation and controlled release strategies seek to optimize the efficacy and safety of green nanoparticles for plant disease control. These techniques aim to address some of the limitations associated with their direct application (Wasule et al., 2024).

The encapsulation method involves coating nanoparticles with a protective matrix to enhance their stability against chemical processes such as oxidation, degradation, and water adsorption (moisture), thereby extending their shelf life and effectiveness. For example, encapsulating lemongrass essential oil in chitosan-gelatin with ZnO NPs resulted in stable emulsions that reduced the release and degradation of the encapsulated oil, which effectively inhibited the growth and proliferation of various pathogenic bacterial species (Jovanovi, 2024).

A wide variety of polymers have been used as encapsulation matrices, ranging from organic polymers like poly(lactic acid) and poly(vinyl alcohol), which have been applied for controlling diseases such as *Phytophthora infestans* (Fukamachi et al., 2019) and degrading bacteria like *Staphylococcus aureus* (Subaşı-Zarbaliyev et al., 2023), to inorganic matrices such as polysiloxanes and clay-based materials, which have demonstrated antifungal activity against *Rhizoctonia solani* (Lin et al., 2014) and increased the efficacy of commonly used pesticides like dicamba (Granetto et al., 2022). Biopolymers such as chitosan (Mosa and El-Abeid, 2023), alginates (Moradi Pour et al., 2022), and lipid-based matrices (Pinilla et al., 2021) have also been investigated as matrix polymers, successfully encapsulating nanoparticles while simultaneously enhancing biocompatibility by using natural byproducts (from plants and animals) as matrices. An example of this is the encapsulation of Ag NPs in chitosan, which significantly showed increased activity against four fungal plant pathogens (Giri et al., 2022).

Alongside these particle encapsulation processes, the gradual degradation of polymeric matrices allows the controlled release of nanostructures into the plant environment (tissues and soil). This phenomenon enables the progressive addition of nanoparticles that come into direct or indirect contact with plants, while also reducing the frequency of application. This effect can be particularly advantageous in managing plant diseases over extended periods (L. Dos Santos et al., 2024).

This controlled release of nanoparticles depend on the correct selection of materials and encapsulation methods. Through this effect, nanoparticles can be targeted to specific plant tissues or pathogens, minimizing unintended interactions and enhancing their efficacy (Sembada and Lenggoro, 2024). Encapsulated systems like salicylic acid-chitosan nanoparticles have been studied as biostimulants for maize crop growth. These particles increased antioxidant enzyme activity, balanced ROS levels, and promoted lignin deposition in maize plants. In field trials, the controlled release of nanoparticles resulted in 59.4% control of post-flowering stalk rot disease and a 57.8% increase in maize yield compared to standard treatments (Kumaraswamy et al., 2019).

Similarly, a multi nutrient nanofertilizer was synthesized by doping calcium phosphate with potassium (K) and Nitrogen (N) from nitrate and urea. This nanofertilizer enables the slow, controlled release of essential macronutrients. The use of these nanoparticles allowed for a 40% reduction in nitrogen application without compromising the yield, highlighting its potential to enhance fertilization efficiency (Ramrez-Rodrguez, 2020).

7. Economic and Environmental Impact of Plant Bacterial Diseases

By the year 2030 more than 600 million people worldwide are predicted to face hunger, emphasizing the immense challenge of achieving the goal of zero hunger, it will, end hunger and safeguard access for all people, specifically the poor and people in susceptible situation, comprising infants, to healthy, nutritious and sufficient food all year round, safeguard the sustainability of food production systems and implement robust agricultural practices that intensification productivity and production, contribute to the preservation of ecosystems, reinforce resilience to climate change. Looking for a solution is projected consisting of food security, understanding the guarantee for residents to have enough food, with easy access and in a stable manner to guarantee their basic needs, which means that the accessibility of food must be greater than the demand in terms of energy requirements, in agreement with an established standard. The world has come back to levels of hunger not seen since 2005, and food prices keep on increasing in more countries than in 2015–2019. Along with debate, the climate crisis and the rising cost of living, civil insecurity and declining food production have influenced food deficiencies and high food prices. Investment in the agricultural sector is critical to reducing hunger and poverty, improving food security, creating jobs, and increasing the resilience to disasters and crises (Sharma et al., 2022; Peng and Berry, 2019).

Bacteria plant pathogens and pests can destroy crops, leading to reduced yields and increased production costs for farmers, resulting in higher food prices for consumers and potentially creating food shortages, influencing the global economy. Plant health is largely correlated to international trade. At the same time invasive pests and diseases can be introduced due to the association of plants and plant products across frontiers, disturbing trade and causing cost-effective losses. Protecting plant health through strict import guidelines and accreditations facilitates safeguarding local agriculture and prevents the spread of unsafe pests.

Ignoring plant health today can, in the future, result in loss of biodiversity, ecosystem degradation, and environmental contrasts. Plant disease outbreaks cause significant risks to global food security and environmental sustainability worldwide, and result in the loss of primary productivity and biodiversity that negatively impact the environmental and socio-economic conditions of the affected regions (Singh et al., 2023).

The consequences of the plant microbiome to climate change can indirectly influence the impact disease incidence. Plant microbiomes are liable to be impacted by climate change both via changes in the starting inoculant from bulk soil or the rhizosphere and by host responses, which embrace changes in host physiology, morphology, exudation models and immune responses. This is important because the plant-associated microorganisms apply large influences on host physiology, and contribute to the regulation of its metabolism, immune function and fitness in novel environments, and play a key role in stopping the colonization and growth of pathogens (Joshi et al., 2020; Katoch et al., 2020).

Climate change variables can also increase root exudation and modify its composition in crops and trees to attract beneficial microorganisms, which intensely strengthen plant growth. Pathogens can overcome the first barrier of plant defense directly, by competition with native microbial communities, or indirectly, by inducing alterations in plant biology and physiology. As climate change can instantly impact in the plant immune responses, there is a possibility that plant microbial dysbiosis can assist pathogen assault and proliferation.

8. Conclusions and Future Work

Green nanotechnology presents an innovative and sustainable approach to managing bacterial plant diseases, offering a promising alternative to conventional chemical pesticides. The unique properties of nanoparticles, such as their high surface area, tunable particle size, and ability to disrupt bacterial cells, position them as effective tools in the fight against plant pathogens. Nanomaterials synthesized through green methods, utilizing biological resources like plant extracts and microorganisms, provide an eco-friendly alternative, reducing the environmental impact typically associated with traditional nanoparticle synthesis. In some metal-based nanoparticles, such as silver, copper, and zinc oxide, have demonstrated large antibacterial activity, providing new strategies for disease management in agriculture.

Despite the promising potential of green nanotechnology, several challenges need to be addressed before it can be widely adopted. One significant hurdle is the cost of large-scale nanoparticle production, which must be optimized to compete with traditional agrochemicals. Additionally, concerns regarding nanoparticle toxicity, long-term environmental impacts, and regulatory barriers remain critical issues that must be resolved. A comprehensive understanding of the interactions between nanoparticles, plant systems, and ecosystems is crucial to ensure their safe and effective use in agricultural practices. Moreover, long-term field studies are needed to assess the environmental fate of nanoparticles and their potential impact on non-target species and soil health.

Looking forward, green nanotechnology is poised to play a transformative role in sustainable agriculture. Advancements in the field are likely to focus on optimizing the cost-effectiveness of nanoparticle production and improving synthesis techniques to make green-synthesized nanomaterials more viable commercially. Further research should also prioritize the development of hybrid nanomaterials, which combine the strengths of different types of nanoparticles to enhance their efficacy and minimize potential drawbacks. Additionally, advances in nanomaterial engineering may lead to the creation of "smart" nanoparticles that can be triggered by environmental cues or pathogen presence, ensuring more targeted and efficient delivery of antibacterial agents.

The use of nanoparticles in combination with other sustainable farming practices, such as integrated pest management, can provide a comprehensive approach to disease management. This holistic approach will not only protect crops from bacterial pathogens but also contribute to the overall health of agricultural ecosystems by reducing reliance on harmful chemicals.

Furthermore, as regulatory frameworks evolve to accommodate new technologies, it will be critical for policymakers, scientists, and industry stakeholders to collaborate in establishing guidelines that ensure the safe and sustainable use of nanotechnology in agriculture. With ongoing research and development, green nanotechnology holds the potential to revolutionize plant disease management, contributing to a more resilient, efficient, and sustainable agricultural system that meets the growing demands of global food security.

References

2022. Clavibacter michiganensis (bacterial canker of tomato). *PlantwisePlus Knowledge Bank*, 10.1079/pwkb.species.15338.

Abdallah, Y., Liu, M., Ogunyemi, S. O., Ahmed, T., Fouad, H., Abdelazez, A. et al. (2020). Bioinspired green synthesis of chitosan and zinc oxide nanoparticles with strong antibacterial activity against rice pathogen *Xanthomonas oryzae* pv. oryzae. *Molecules*, 25: 4795.

Abdelaziz, A. M., Salem, S. S., Khalil, A. M. A., El-Wakil, D. A., Fouda, H. M., and Hashem, A. H. (2022). Potential of biosynthesized zinc oxide nanoparticles to control Fusarium wilt disease in eggplant (*Solanum melongena*) and promote plant growth. *BioMetals*, 35: 601–616.

Abdelrhim, A. S., Mazrou, Y. S. A., Nehela, Y., Atallah, O. O., El-Ashmony, R. M., and Dawood, M. F. A. (2021). Silicon dioxide nanoparticles induce innate immune responses and activate antioxidant machinery in wheat against Rhizoctonia solani. *Plants*, 10: 2758.

Abdul Malik, N. A., Kumar, I. S., and Nadarajah, K. (2020). Elicitor and receptor molecules: orchestrators of plant defense and immunity. *International Journal of Molecular Sciences*, 21: 963.

Abid, N., Khan, A. M., Shujait, S., Chaudhary, K., Ikram, M., Imran, M., Haider, J. et al. (2022). Synthesis of nanomaterials using various top-down and bottom-up approaches, influencing factors, advantages, and disadvantages: A review. *Advances in Colloid and Interface Science*, 300: 102597.

Abosabaa, S. A., Arafa, M. G., and Elmeshad, A. N. (2021). Hybrid chitosan-lipid nanoparticles of green tea extract as natural anti-cellulite agent with superior *in vivo* potency: full synthesis and analysis. *Drug Delivery*, 28: 2160–2176.

Acharya, A., and Pal, P. K. (2020). Agriculture nanotechnology: Translating research outcome to field applications by influencing environmental sustainability. *NanoImpact*, 19: 100232.

Afzal, S., Sharma, D., and Singh, N. K. (2021). Eco-friendly synthesis of phytochemical-capped iron oxide nanoparticles as nano-priming agent for boosting seed germination in rice (*Oryza sativa* L.). *Environmental Science and Pollution Research*, 28: 40275–40287.

Ahmed, T., Noman, M., Luo, J., Muhammad, S., Shahid, M., Ali, M. A. et al. (2021). Bioengineered chitosan-magnesium nanocomposite: A novel agricultural antimicrobial agent against Acidovorax oryzae and Rhizoctonia solani for sustainable rice production. *International Journal of Biological Macromolecules*, 168: 834–845.

Albayrak Turgut, B., and Bezirganoğlu, I. (2022). Foliar application of CaO nanoparticles and salicylic acid on Medicago sativa seedlings enhances tolerance against Fusarium oxysporum. *Physiological and Molecular Plant Pathology*, 122: 101926.

Alghuthaymi, M. A., Rajkuberan, C., Rajiv, P., Kalia, A., Bhardwaj, K. et al. (2021). Nanohybrid antifungals for control of plant diseases: current status and future perspectives. *Journal of Fungi*, 7: 48.

Alhalili, Z. (2023). Metal oxides nanoparticles: general structural description, chemical, physical, and biological synthesis methods, role in pesticides and heavy metal removal through wastewater treatment. *Molecules*, 28: 3086.

Alshareef, A., Laird, K., and Cross, R. B. M. (2017. Shape-dependent antibacterial activity of silver nanoparticles on Escherichia coli and *Enterococcus faecium* bacterium. *Applied Surface Science*, 424: 310–315.

Amjad, R., Mubeen, B., Ali, S. S., Imam, S. S., Alshehri, S., Ghoneim, M. M., Alzarea, S. I. et al. (2021). Green Synthesis and Characterization of Copper Nanoparticles Using Fortunella margarita Leaves. *Polymers*, 13: 4364.

Ansari, M., Ahmed, S., Abbasi, A., Hamad, N. A., Ali, H. M., Khan, M. T. et al. (2023). Green synthesized silver nanoparticles: a novel approach for the enhanced growth and yield of tomato against early blight disease. *Microorganisms*, 11: 886.

Artusio, F., and Cas. (2021). Alginate Nanohydrogels as a Biocompatible Platform for the Controlled Release of a Hydrophilic Herbicide. *Processes*, 9: 1641.

Asha, A. B., and Narain, R. (2020). Nanomaterials properties. *Polymer Science and Nanotechnology.* Elsevier.

Ashraf, S. A., Siddiqui, A. J., Elkhalifa, A. E. O., Khan, M. I., Patel, M., Alreshidi, M. et al. (2021). Innovations in nanoscience for the sustainable development of food and agriculture with implications on health and environment. *Science of The Total Environment*, 768: 144990.

Asif, M., Yasmin, R., Asif, R., Ambreen, A., Mustafa, M., and Umbreen, S. (2022). Green synthesis of silver nanoparticles (AgNPs), Structural characterization, and their antibacterial potential. *Dose-Response*, 20: 155932582210887.

Babenko, L. M., Smirnov, O. E., Romanenko, K. O., Trunova, O. K., and Kosakivska, I. V. (2019). Phenolic compounds in plants: biogenesis and functions. *The Ukrainian Biochemical Journal*, 91: 5–18.

Bansal, V., Ramanathan, R., and Bhargava, S. K. (2011). Fungus-mediated biological approaches towards 'Green' synthesis of oxide nanomaterials. *Australian Journal of Chemistry*, 64: 279.

Barros, J., Serrani-Yarce, J. C., Chen, F., Baxter, D., Venables, B. J., and Dixon, R. A. (2016). Role of bifunctional ammonia-lyase in grass cell wall biosynthesis. *Nature Plants*, 2: 16050–16050.

Bellenot, C., Carrère, S., Gris, C., Noël, L. D., and Arlat, M. (2022). Genome sequences of 17 strains from eight races of *Xanthomonas campestris* pv. campestris. *Microbiology Resource Announcements*, 11: e00279–22.

Bennacef, C., Desobry-Banon, S., Probst, L., and Desobry, S. (2021). Advances on alginate use for spherification to encapsulate biomolecules. *Food Hydrocolloids*, 118: 106782.

Bizuneh, G. K. (2021). The chemical diversity and biological activities of phytoalexins. *Advances in Traditional Medicine*, 21: 31–43.

Buttimer, C., Mcauliffe, O., Ross, R. P., Hill, C., O'mahony, J., and Coffey, A. (2017). Bacteriophages and Bacterial Plant Diseases. *Frontiers in Microbiology*, 8.

Carvalho, J. P. F., Silva, A. C. Q., Silvestre, A. J. D., Freire, C. S. R., and Vilela, C. (2021). Spherical cellulose micro and nanoparticles: a review of recent developments and applications. *Nanomaterials*, 11: 2744.

Ceylan, R., Demirbas, A., Ocsoy, I., and Aktumsek, A. (2021). Green synthesis of silver nanoparticles using aqueous extracts of three Sideritis species from Turkey and evaluations bioactivity potentials. *Sustainable Chemistry and Pharmacy*, 21: 100426.

Chalam, V. C., Sharma, R., Sharma, V. D., and Maurya, A. K. (2020). Role of quarantine in management of transboundary seed-borne diseases. *In:* Tiwari, A. K. (ed.). *Advances in Seed Production and Management.* Singapore: Springer Singapore.

Chandra, S., Chakraborty, N., Dasgupta, A., Sarkar, J., Panda, K., and Acharya, K. (2015). Chitosan nanoparticles: A positive modulator of innate immune responses in plants. *Scientific Reports*, 5: 15195–15195.

Cheema, A. I., Ahmed, T., Abbas, A., Noman, M., Zubair, M., and Shahid, M. (2022). Antimicrobial activity of the biologically synthesized zinc oxide nanoparticles against important rice pathogens. *Physiology and Molecular Biology of Plants*, 28: 1955–1967.

Chen, X., Pang, C., Liu, X., Sun, J., Jin, L., Sun, Y. et al. (2024). Investigation of the antibacterial activity of benziothiazolinone against *Xanthomonas oryzae* pv. oryzae. *Pesticide Biochemistry and Physiology*, 199: 105768.

Chen, Z., Gao, S.-H., Jin, M., Sun, S., Lu, J., Yang, P. et al. (2019). Physiological and transcriptomic analyses reveal CuO nanoparticle inhibition of anabolic and catabolic activities of sulfate-reducing bacterium. *Environment International*, 125: 65–74.

Cheon, J. Y., Kim, S. J., Rhee, Y. H., Kwon, O. H., and Park, W. H. (2019). Shape-dependent antimicrobial activities of silver nanoparticles. *Int. J. Nanomedicine*, 14: 2773–2780.

Dagher, F., Olishevska, S., Philion, V., Zheng, J., and Déziel, E. (2020). Development of a novel biological control agent targeting the phytopathogen Erwinia amylovora. *Heliyon*, 6: e05222.

Daima, H. K., and Bansal, V. (2015). Chapter 10—Influence of physicochemical properties of nanomaterials on their antibacterial applications. *In:* Rai, M., and Kon, K. (eds.). *Nanotechnology in Diagnosis, Treatment and Prophylaxis of Infectious Diseases.* Boston: Academic Press.

De Wit, P. J. G. M. (2007). How plants recognize pathogens and defend themselves. *Cellular and Molecular Life Sciences*, 64: 2726–2732.

Dey, R., and Raghuwanshi, R. (2024). An insight into pathogenicity and virulence gene content of Xanthomonas spp. and its biocontrol strategies. *Heliyon*, 10: e34275.

Do Espirito Santo Pereira, A., Caixeta Oliveira, H., Fernandes Fraceto, L., and Santaella, C. (2021). Nanotechnology potential in seed priming for sustainable agriculture. *Nanomaterials*, 11: 267.

Dong, B.-R., Jiang, R., Chen, J.-F., Xiao, Y., Lv, Z.-Y., and Chen, W.-S. (2023). Strategic nanoparticle-mediated plant disease resistance. *Critical Reviews in Biotechnology*, 43: 22–37.

Dutta, D., and Das, B. M. (2021). Scope of green nanotechnology towards amalgamation of green chemistry for cleaner environment: A review on synthesis and applications of green nanoparticles. *Environmental Nanotechnology, Monitoring*, and *Management*, 15: 100418.

Dutta, P., Kumari, A., Mahanta, M., Biswas, K. K., Dudkiewicz, A., Thakuria, D. et al. (2022). Advances in nanotechnology as a potential alternative for plant viral disease management. *Frontiers in Microbiology*, 13.

Dutta, P., Kumari, A., Mahanta, M., Upamanya, Gunadhya k., Heisnam, P., Borua, S. et al. (2023). Nanotechnological approaches for management of soil-borne plant pathogens. *Frontiers in Plant Science*, 14.

Eisfeld, C., Schijven, J. F., Kastelein, P., Van Breukelen, B. M., Medema, G., Velstra, J. et al. (2022). Dose-response relationship of *Ralstonia solanacearum* and potato in greenhouse and *in vitro* experiments. *Frontiers in Plant Science*, 13.

El Amerany, F., Meddich, A., Wahbi, S., Porzel, A., Taourirte, M., Rhazi, M. et al. (2020). Foliar application of chitosan increases tomato growth and influences mycorrhization and expression of endochitinase-encoding genes. *International Journal of Molecular Sciences*, 21: 535.

Elabasy, A., Shoaib, A., Waqas, M., Shi, Z., and Jiang, M. (2020). Cellulose nanocrystals loaded with thiamethoxam: fabrication, characterization, and evaluation of insecticidal activity against *Phenacoccus solenopsis* Tinsley (Hemiptera: Pseudococcidae). *Nanomaterials*, 10: 788.

Eldeeb, B. A., El-Raheem, W. M. A., and Elbeltagi, S. (2023). Green synthesis of biocompatible Fe3O4 magnetic nanoparticles using Citrus Sinensis peels extract for their biological activities and magnetic-hyperthermia applications. *Scientific Reports*, 13: 19000.

Fan, X., Yahia, L. H., and Sacher, E. (2021). Antimicrobial properties of the Ag, Cu nanoparticle system. *Biology*, 10: 137–137.

Ferreira, A. M., Vikulina, A., Loughlin, M., and Volodkin, D. (2023). How similar is the antibacterial activity of silver nanoparticles coated with different capping agents? *RSC Advances*, 13: 10542–10555.

Fincheira, P., Espinoza, J., Levo-Raimn, M., Vera, J., Tortella, G., Brito, A. M. M. et al. (2024). Formulation of essential oils-loaded solid lipid nanoparticles-based chitosan/PVA hydrogels to control the growth of Botrytis cinerea and Penicillium expansum. *International Journal of Biological Macromolecules*, 270: 132218.

Fu, L., Wang, Z., Dhankher, O. P., and Xing, B. (2019). Nanotechnology as a new sustainable approach for controlling crop diseases and increasing agricultural production. *Journal of Experimental Botany*, 71: 507–519.

Fujikawa, T., and Inoue, Y. (2020). Genome sequences of two pathogens of cruciferous crops, *Xanthomonas campestris* pv. raphani MAFF 106181 and *X. campestris* pv. campestris MAFF 301176. *Microbiol Resour Announc*, 9.

Fukamachi, K., Konishi, Y., and Nomura, T. (2019). Disease control of Phytophthora infestans using cyazofamid encapsulated in poly lactic-co-glycolic acid (PLGA) nanoparticles. *Colloids and Surfaces A: Physicochemical and Engineering Aspects*, 577: 315–322.

Gadewar, M., and Prashanth, G. K. A. (2024). Unlocking nature's potential: Green synthesis of ZnO nanoparticles and their multifaceted applications – A concise overview. *Journal of Saudi Chemical Society*, 28: 101774.

García-Quintero, A., and Palencia, M. (2021). A critical analysis of environmental sustainability metrics applied to green synthesis of nanomaterials and the assessment of environmental risks associated with the nanotechnology. *Science of The Total Environment*, 793: 148524.

Garibo, D., Borbón-Nuñez, H. A., De León, J. N. D., García Mendoza, E., Estrada, I., Toledano-Magaña, Y. et al. (2020). Green synthesis of silver nanoparticles using Lysiloma acapulcensis exhibit high-antimicrobial activity. *Scientific Reports*, 10: 12805–12805.

Gerhardson, B. (2002). Biological substitutes for pesticides. *Trends in Biotechnology*, 20: 338–343.

Giacometti, F., Shirzad-Aski, H., and Ferreira, S. (2021). Antimicrobials and food-related stresses as selective factors for antibiotic resistance along the farm to fork continuum. *Antibiotics*, 10: 671.

Gibaa, A., Eliszewska, P., Gosiewski, T., Krawczyk, A., Duraczyska, D., Szaleniec, J. et al. (2021). Antibacterial and antifungal properties of silver nanoparticles—effect of a surface-stabilizing agent. *Biomolecules*, 11: 1481.

Gilbertson, L. M., Pourzahedi, L., Laughton, S., Gao, X., Zimmerman, J. B., Theis, T. L. et al. (2020). Guiding the design space for nanotechnology to advance sustainable crop production. *Nature Nanotechnology*, 15: 801–810.

Giri, V. P., Pandey, S., Kumari, M., Tripathi, A., Katiyar, R., White, J. C. et al. (2022). Hybridization of chitosan and biosynthesized silver nanoparticles to enhance antimicrobial activity against phytopathogens in tomato (*Solanum lycopersicum*). *ACS Agricultural Science*, and *Technology*, 2: 719–733.

Górniak, I., Bartoszewski, R., and Króliczewski, J. (2019). Comprehensive review of antimicrobial activities of plant flavonoids. *Phytochemistry Reviews*, 18: 241–272.

Granetto, M., Serpella, L., Fogliatto, S., Re, L., Bianco, C., Vidotto, F. et al. (2022). Natural clay and biopolymer-based nanopesticides to control the environmental spread of a soluble herbicide. *Science of The Total Environment*, 806: 151199.

Grillo, R., Mattos, B. D., Antunes, D. R., Forini, M. M. L., Monikh, F. A., and Rojas, O. J. (2021). Foliage adhesion and interactions with particulate delivery systems for plant nanobionics and intelligent agriculture. *Nano Today*, 37: 101078.

Guo, Y., Khan, R. a. A., Xiong, Y., and Fan, Z. (2022). Enhanced suppression of soil-borne phytopathogenic bacteria *Ralstonia solanacearum* in soil and promotion of tomato plant growth by synergetic effect of green synthesized nanoparticles and plant extract. *Journal of Applied Microbiology*, 132: 3694–3704.

Hano, C., and Abbasi, B. H. (2021). Plant-based green synthesis of nanoparticles: production, characterization and applications. *Biomolecules*, 12: 31.

Hnamte, L., Vanlallawmzuali, Kumar, A., Yadav, M. K., Zothanpuia, and Singh, P. K. (2024). An updated view of bacterial endophytes as antimicrobial agents against plant and human pathogens. *Current Research in Microbial Sciences*, 7: 100241.

Hofmann, T., Lowry, G. V., Ghoshal, S., Tufenkji, N., Brambilla, D., Dutcher, J. R., Gilbertson, L. M., Giraldo, J. P. et al. (2020). Technology readiness and overcoming barriers to sustainably implement nanotechnology-enabled plant agriculture. *Nature Food*, 1: 416–425.

Hong, J., Wang, C., Wagner, D. C., Gardea-Torresdey, J. L., He, F., and Rico, C. M. (2021). Foliar application of nanoparticles: mechanisms of absorption, transfer, and multiple impacts. *Environmental Science: Nano*, 8: 1196–1210.

Huang, X., Chen, Y., Li, J., Li, J., Gao, Q., and Mao, A. (2022). Development of a strong soy protein-based adhesive with excellent antibacterial and antimildew properties via biomineralized silver nanoparticles. *Industrial Crops and Products*, 188: 115567.

Huang, X., Li, Y., Chen, K., Chen, H., Wang, F., Han, X. et al. (2020). NOM mitigates the phytotoxicity of AgNPs by regulating rice physiology, root cell wall components and root morphology. *Environmental Pollution*, 260: 113942.

Huang, X., You, Z., Luo, Y., Yang, C., Ren, J., Liu, Y. et al. (2021). Antifungal activity of chitosan against Phytophthora infestans, the pathogen of potato late blight. *International Journal of Biological Macromolecules*, 166: 1365–1376.

Huguet-Casquero, A., Gainza, E., and Pedraz, J. L. (2021). Towards Green Nanoscience: From extraction to nanoformulation. *Biotechnology Advances*, 46: 107657.

Huston, M., Debella, M., Dibella, M., and Gupta, A. (2021). Green synthesis of nanomaterials. *Nanomaterials*, 11: 2130.

Hwang, I. S., Oh, E.-J., and Oh, C.-S. (2024). A novel virulence gene, cviA1 of Clavibacter michiganensis for necrosis development in the Nicotiana benthamiana plant. *Microbiological Research*, 285: 127743.

Iliger, K. S., Sofi, T. A., Bhat, N. A., Ahanger, F. A., Sekhar, J. C., Elhendi, A. Z. et al. (2021). Copper nanoparticles: Green synthesis and managing fruit rot disease of chilli caused by Colletotrichum capsici. *Saudi Journal of Biological Sciences*, 28: 1477—1486.

İpe, P., Baran, M. F., Baran, A., Hatipoğlu, A., Keskin, C., Yildiztekin, M. et al. (2024). Green synthesis and evaluation of antipathogenic, antioxidant, and anticholinesterase activities of gold nanoparticles (Au NPs) from *Allium cepa* L. peel aqueous extract. *Biomass Conversion and Biorefinery*, 14: 10661–10670.

Jadoun, S., Arif, R., Jangid, N. K., and Meena, R. K. (2021). Green synthesis of nanoparticles using plant extracts: a review. *Environmental Chemistry Letters*, 19: 355–374.

Javed, R., Zia, M., Naz, S., Aisida, S. O., Ain, N. U., and Ao, Q. (2020). Role of capping agents in the application of nanoparticles in biomedicine and environmental remediation: recent trends and future prospects. *Journal of Nanobiotechnology*, 18: 172–172.

Javed, Z., Dashora, K., and Mishra, M. A. (2019). Effect of accumulation of nanoparticles in soil health- a concern on future. *Frontiers in Nanoscience and Nanotechnology*, 5.

Joshi, T., Sharma, P., Joshi, T., Pandey, S. C., Pande, V., Pandey, A. et al. (2020). Chapter 3 - A spotlight on the recent advances in bacterial plant diseases and their footprint on crop production. *In:* De Mandal, S., and Bhatt, P. (eds.). *Recent Advancements in Microbial Diversity.* Academic Press.

Jovanovi 2024). Enhanced stability of encapsulated lemongrass essential oil in chitosan-gelatin and pectin-gelatin biopolymer matrices containing ZnO nanoparticles. *International Journal of Biological Macromolecules*, 275: 133335.

Jujjavarapu, S. E., and Poluri, K. M. (2020). *Green Polymeric Nanocomposites*, CRC Press.

Kafle, A., Timilsina, A., Gautam, A., Adhikari, K., Bhattarai, A., and Aryal, N. (2022). Phytoremediation: Mechanisms, plant selection and enhancement by natural and synthetic agents. *Environmental Advances*, 8: 100203.

Kaidi, S., Belattmania, Z., Bentiss, F., Jama, C., Reani, A., and Sabour, B. (2022). Synthesis and characterization of silver nanoparticles using alginate from the brown seaweed laminaria ochroleuca: structural features and antibacterial activity. *Biointerface Research in Applied Chemistry*, 12: 6046–6057.

Katoch, S., Kumari, N., Salwan, R., Sharma, V., and Sharma, P. N. (2020). Recent developments in social network disruption approaches to manage bacterial plant diseases. *Biological Control*, 150: 104376.

Kaur, S., Samota, M. K., Choudhary, M., Choudhary, M., Pandey, A. K., Sharma, A. et al. (2022). How do plants defend themselves against pathogens-Biochemical mechanisms and genetic interventions. *Physiology and Molecular Biology of Plants*, 28: 485–504.

Keshavarz-Tohid, V., and Ebrahimi, S. (2023). Phylogeny, phenotypic characteristics and pathogenicity of *Sphingomonas* sp. and Erwinia persicina as bacterial causal agents of lettuce diseases in southwest of Iran. *Physiological and Molecular Plant Pathology*, 127: 102124.

Khan, A. U., Khan, M., and Khan, M. M. (2019). Antifungal and antibacterial assay by silver nanoparticles synthesized from aqueous leaf extract of Trigonella foenum-graecum. *BioNanoScience*, 9: 597–602.

Khan, M., Khan, A. A., Parveen, A., Min, K., Yadav, V. K., Khan, A. U. et al. (2023). Mitigating the growth of plant pathogenic bacterium, fungi, and nematode by using plant-mediated synthesis of copper oxide nanoparticles (CuO NPs). *Green Chemistry Letters and Reviews*, 16.

Khan, M. E., Mohammad, A., Ali, W., and Sharwani, A. A. (2022). Recent developments in properties and applications of metal oxides. *Inorganic Anticorrosive Materials*, 10.1016/B978-0-323-90410-0.00006-4: 95–111.

Khan, M. I., Akhtar, M. N., Ashraf, N., Najeeb, J., Munir, H., Awan, T. I. et al. (2020). Green synthesis of magnesium oxide nanoparticles using Dalbergia sissoo extract for photocatalytic activity and antibacterial efficacy. *Applied Nanoscience*, 10: 2351–2364.

Khan, S., and Zahoor, M. A. (2023). The impact of silver nanoparticles on the growth of plants: The agriculture applications. *Heliyon*, 9: e16928.

Kongala, S. I., and Mamidala, P. (2023). Harpin-loaded chitosan nanoparticles induced defense responses in tobacco. *Carbohydrate Polymer Technologies and Applications*, 5: 100293.

Kruszka, D., Sawikowska, A., Kamalabai Selvakesavan, R., Krajewski, P., Kachlicki, P., and Franklin, G. (2020). Silver nanoparticles affect phenolic and phytoalexin composition of Arabidopsis thaliana. *Science of The Total Environment*, 716: 135361.

Kühlbrandt, W. (2019). Structure and Mechanisms of F-Type ATP Synthases. *Annual Review of Biochemistry*, 88: 515-549.

Kumaraswamy, R. V., Kumari, S., Choudhary, R. C., Sharma, S. S., Pal, A., Raliya, R. et al. (2019). Salicylic acid functionalized chitosan nanoparticle: A sustainable biostimulant for plant. *International Journal of Biological Macromolecules*, 123: 59–69.

Kuniak, E., and Gajewska, E. (2024). Lipids and lipid-mediated signaling in plant–pathogen interactions. *International Journal of Molecular Sciences*, 25: 7255.

L. Dos Santos, O. A., Dos Santos, M. S., Antunes Filho, S., and Backx, B. P. (2024). Nanotechnology for the control of plant pathogens and pests. *Plant Nano Biology*, 8: 100080.

Lallo Da Silva, B., Caetano, B. L., Chiari-Andréo, B. G., Pietro, R. C. L. R., and Chiavacci, L. A. (2019). Increased antibacterial activity of ZnO nanoparticles: Influence of size and surface modification. *Colloids and Surfaces B: Biointerfaces*, 177: 440–447.

Lefevere, H., Bauters, L., and Gheysen, G. (2020). Salicylic Acid Biosynthesis in Plants. *Frontiers in Plant Science*, 11.

Li, B., Shi, Q., Miao, B., and Zhang, G. (2024). Application of exopolysaccharides synthesized by Xanthomonas campestris in an efficient separation system of refractory chalcopyrite and molybdenite. *Journal of Cleaner Production*, 453: 142277.

Li, Y., Zhang, P., Li, M., Shakoor, N., Adeel, M., Zhou, P. et al. (2023). Application and mechanisms of metal-based nanoparticles in the control of bacterial and fungal crop diseases. *Pest Management Science*, 79: 21–36.

Lin, H., Liang, Y., Kaliaperumal, K., Xiong, Q., Duan, S., Jiang, Y. et al. (2023). Linoleic acid from the endophytic fungus *Diaporthe* sp. HT-79 inhibits the growth of *Xanthomonas* citri subsp. citri by destructing the cell membrane and producing reactive oxygen species (ROS). *Pesticide Biochemistry and Physiology*, 192: 105423.

Lin, Y., Liu, Q., Cheng, L., Lei, Y., and Zhang, A. (2014). Synthesis and antimicrobial activities of polysiloxane-containing quaternary ammonium salts on bacteria and phytopathogenic fungi. *Reactive and Functional Polymers*, 85: 36–44.

Linklater, D. P., Baulin, V. A., Le Guével, X., Fleury, J.-B., Hanssen, E., Nguyen, T. H. P. et al. (2020). Antibacterial action of nanoparticles by lethal stretching of bacterial cell membranes. *Advanced Materials*, 32: 2005679–2005679.

Liu, Z., Wang, J., Luo, S., Yue, Z., Li, Z., Wang, J. et al. (2024). Effects of *Xanthomonas campestris* pv. campestris on the photosynthesis of cabbage in the early stage of infection. *Scientia Horticulturae*, 324: 112620.

Lu, K., Shen, D., Liu, X., Dong, S., Jing, X., Wu, W. et al. (2020). Uptake of iron oxide nanoparticles inhibits the photosynthesis of the wheat after foliar exposure. *Chemosphere*, 259: 127445.

Madani, M., Hosny, S., Alshangiti, D. M., Nady, N., Alkhursani, S. A., Alkhaldi, H. et al. (2022). Green synthesis of nanoparticles for varied applications: Green renewable resources and energy-efficient synthetic routes. *Nanotechnology Reviews*, 11: 731–759.

Madusanka, H. K. S., Chathurika, J. a. S., Aruggoda, A. G. B., and Weerakoon, S. (2024). Effect of seed priming of green synthesized iron oxide magnetic nanoparticles using Salvinia molesta plant extract on seed germination and seedlings growth of tomato (*Solanum lycopersicum*). *International Journal Of Advance Research And Innovative Ideas In Education*, 9: 2023.

Mahakham, W., Sarmah, A. K., Maensiri, S., and Theerakulpisut, P. (2017). Nanopriming technology for enhancing germination and starch metabolism of aged rice seeds using phytosynthesized silver nanoparticles. *Scientific Reports*, 7: 8263.

Maity, D., Gupta, U., and Saha, S. (2022). Biosynthesized metal oxide nanoparticles for sustainable agriculture: next-generation nanotechnology for crop production, protection and management. *Nanoscale*, 14: 13950–13989.

Martnez-Cano, B., Mendoza-Meneses, C. J., Garca-Trejo, J. F., Macas-Bobadilla, G., Aguirre-Becerra, H., Soto-Zaraza, G. M. et al. (2022). Review and perspectives of the use of alginate as a polymer matrix for microorganisms applied in agro-industry. *Molecules*, 27: 4248.

Marwal, A., and Gaur, R. K. (2020). Disease-causing seed pathogenic microorganisms and their management practices. *In:* Tiwari, A. K. (ed.). *Advances in Seed Production and Management.* Singapore: Springer Singapore.

Maximiano, M. R., Rios, T. B., Campos, M. L., Prado, G. S., Dias, S. C., and Franco, O. L. (2022). Nanoparticles in association with antimicrobial peptides (NanoAMPs) as a promising combination for agriculture development. *Frontiers in Molecular Biosciences*, 9.

Meena, M., Zehra, A., Swapnil, P., Harish, Marwal, A., Yadav, G. et al. (2021). Endophytic Nanotechnology: An Approach to Study Scope and Potential Applications. *Frontiers in Chemistry*, 9.

Mendes, C. R., Dilarri, G., Forsan, C. F., Sapata, V. D. M. R., Lopes, P. R. M., De Moraes, P. B. et al. (2022). Antibacterial action and target mechanisms of zinc oxide nanoparticles against bacterial pathogens. *Scientific Reports*, 12: 2658–2658.

Mercado-Meza, D. Y., Guevara-González, R. G., Esquivel, K., Carbajal-Valenzuela, I., and Avila-Quezada, G. D. (2023). Green silver nanoparticles display protection against Clavibacter michiganensis subsp. michiganensis in tomato plants (*Solanum lycopersicum* L.). *Plant Stress*, 10: 100256.

Mirza, F. S., Aftab, Z.-E.-H., Ali, M. D., Aftab, A., Anjum, T., Rafiq, H. et al. (2022). Green synthesis and application of GO nanoparticles to augment growth parameters and yield in mungbean (*Vigna radiata* L.). *Frontiers in Plant Science*, 13.

Mishra, A., and Baek, K.-H. (2021). Salicylic acid biosynthesis and metabolism: a divergent pathway for plants and bacteria. *Biomolecules*, 11: 705–705.

Mishra, S., Singh, B. R., Naqvi, A. H., and Singh, H. B. (2017). Potential of biosynthesized silver nanoparticles using *Stenotrophomonas* sp. BHU-S7 (MTCC 5978) for management of soil-borne and foliar phytopathogens. *Scientific Reports*, 7: 45154–45154.

Miu, B. A., and Dinischiotu, A. (2022). New green approaches in nanoparticles synthesis: an overview. *Molecules*, 27: 6472.

Mondjar-Lpez, M., Lpez-Jimenez, A. J., and Niza, E. (2023). Chitosan coated - biogenic silver nanoparticles from wheat residues as green antifungal and nanoprimig in wheat seeds. *International Journal of Biological Macromolecules*, 225: 964–973.

Moradi Pour, M., Saberi Riseh, R., and Skorik, Y. A. (2022). Sodium alginate–gelatin nanoformulations for encapsulation of *Bacillus velezensis* and their use for biological control of pistachio gummosis. *Materials*, 15: 2114.

Morales-Espinoza, M. C., Cadenas-Pliego, G., Prez-Alvarez, M., and Hernndez-Fuentes, A. D. A. (2019). Se Nanoparticles Induce Changes in the Growth, Antioxidant Responses, and Fruit Quality of Tomato Developed under NaCl Stress. *Molecules*, 24: 3030.

Mosa, M. A., and El-Abeid, S. E. (2023). Chitosan-loaded copper oxide nanoparticles: a promising antifungal nanocomposite against *Fusarium wilt* disease of tomato plants. *Sustainability*, 15: 14295.

Murugan, B., Rahman, M. Z., and Fatimah, I. A. (2023). Green synthesis of CuO nanoparticles for biological applications. *Inorganic Chemistry Communications*, 155: 111088.

Muzammil, S., Ashraf, A., Siddique, M. H., Aslam, B., Rasul, I., Abbas, R. et al(2023). A review on toxicity of nanomaterials in agriculture: Current scenario and future prospects. *Science Progress*, 106.

Nabi, G., Anjum, T., Aftab, Z. E. H., Rizwana, H., and Akram, W. (2024). TiO2 nanoparticles: Green synthesis and their role in lessening the damage of Colletotrichum graminicola in sorghum. *Food Science*, and *Nutrition*, 10.1002/fsn3.4297.

Nandi, M., Macdonald, J., Liu, P., Weselowski, B., and Yuan, Z.-C. (2018). Clavibacter michiganensis ssp. michiganensis: bacterial canker of tomato, molecular interactions and disease management. *Molecular Plant Pathology*, 19: 2036–2050.

Nasirzadeh Fard, Y., Kelidari, H., Kazeminejad, A., Mousavi, S. J., Hedayati, M. T., Mosayebi, E. et al. (2023). Enhanced treatment in cutaneous dermatophytosis management by Zataria multiflora-loaded nanostructured lipid carrier topical gel: A randomized double-blind placebo-controlled clinical trial. *Journal of Drug Delivery Science and Technology*, 80: 104132.

Negrescu, A. M., Killian, M. S., Raghu, S. N. V., Schmuki, P., Mazare, A., and Cimpean, A. (2022). Metal oxide nanoparticles: review of synthesis, characterization and biological effects. *Journal of Functional Biomaterials*, 13: 274.

Newberry, E. A., Ritchie, L., Babu, B., Sanchez, T., Beckham, K. A., Jones, J. B. et al. (2017). Epidemiology and management of bacterial leaf spot on watermelon caused by Pseudomonas syringae. *Plant Disease*, 101: 1222–1229.

Nile, S. H., Thiruvengadam, M., Wang, Y., Samynathan, R., Shariati, M. A., Rebezov, M. et al. (2022). Nano-priming as emerging seed priming technology for sustainable agriculture—recent developments and future perspectives. *Journal of Nanobiotechnology*, 20: 254.

Nisar, P., Ali, N., Rahman, L., Ali, M., and Shinwari, Z. K. (2019). Antimicrobial activities of biologically synthesized metal nanoparticles: an insight into the mechanism of action. *JBIC Journal of Biological Inorganic Chemistry*, 24: 929–941.

Pagano, A., Folini, G., Pagano, P., Sincinelli, F., Rossetto, A., Macovei, A. et al. (2022). ROS Accumulation as a hallmark of dehydration stress in primed and overprimed medicago truncatula Seeds. *Agronomy*, 12: 268.

Pagano, A., Macovei, A., and Balestrazzi, A. (2023). Molecular dynamics of seed priming at the crossroads between basic and applied research. *Plant Cell Reports*, 42: 657–688.

Pal, S., Tak, Y. K., and Song, J. M. (2007). Does the antibacterial activity of silver nanoparticles depend on the shape of the nanoparticle? A study of the Gram-negative bacterium *Escherichia coli*. *Appl Environ Microbiol*, 73: 1712–20.

Park, S., Lee, S., Kim, B., Lee, S., Lee, J., Sim, S. et al. (2012). Toxic effects of titanium dioxide nanoparticles on microbial activity and metabolic flux. *Biotechnology and Bioprocess Engineering*, 17: 276–282.

Patel, P., Garala, K., Singh, S., Prajapati, B. G., and Chittasupho, C. (2024). Lipid-based nanoparticles in delivering bioactive compounds for improving therapeutic efficacy. *Pharmaceuticals*, 17: 329.

Paul, R., Saville, A. C., Hansel, J. C., Ye, Y., Ball, C., Williams, A. et al. (2019). Extraction of plant DNA by microneedle patch for rapid detection of plant diseases. *ACS Nano*, 13: 6540–6549.

Pavlicevic, M., Elmer, W., Zuverza-Mena, N., Abdelraheem, W., Patel, R., Dimkpa, C. et al. (2023). Nanoparticles and biochar with adsorbed plant growth-promoting rhizobacteria alleviate *Fusarium wilt* damage on tomato and watermelon. *Plant Physiology and Biochemistry*, 203: 108052.

Penders, J., Stolzoff, M., Hickey, D. J., Andersson, M., and Webster, T. J. (2017). Shape-dependent antibacterial effects of non-cytotoxic gold nanoparticles. *International Journal of Nanomedicine*, 12: 2457–2468.

Peng, W., and Berry, E. M. (2019). The concept of food security. *In:* Ferranti, P., Berry, E. M., and Anderson, J. R. (eds.) *Encyclopedia of Food Security and Sustainability.* Oxford: Elsevier.

Perea Vélez, Y. S., Carrillo-González, R., and González-Chávez, M. D. C. A. (2021). Interaction of metal nanoparticles–plants–microorganisms in agriculture and soil remediation. *Journal of Nanoparticle Research*, 23: 206.

Pillai, A. M., Sivasankarapillai, V. S., Rahdar, A., Joseph, J., Sadeghfar, F., Anuf A, R. et al. (2020). Green synthesis and characterization of zinc oxide nanoparticles with antibacterial and antifungal activity. *Journal of Molecular Structure*, 1211: 128107.

Pinilla, C. M. B., Lopes, N. A., and Brandelli, A. (2021). Lipid-based nanostructures for the delivery of natural antimicrobials. *Molecules*, 26: 3587.

Puławska, J., Kuzmanović, N., and Trzciński, P. (2022). *Agrobacterium vaccinii* sp. nov. isolated from galls on blueberry plants (*Vaccinium corymbosum*). *Systematic and Applied Microbiology*, 45: 126319.
Quinteros, M. A., Cano Aristizábal, V., Dalmasso, P. R., Paraje, M. G., and Páez, P. L. (2016). Oxidative stress generation of silver nanoparticles in three bacterial genera and its relationship with the antimicrobial activity. *Toxicology in Vitro*, 36: 216–223.
Qureshi, A. K., Farooq, U., Shakeel, Q., Ali, S., Ashiq, S., Shahzad, S. et al. (2023). The green synthesis of silver nanoparticles from avena fatua extract: antifungal activity against *Fusarium oxysporum* f.sp. lycopersici. *Pathogens*, 12: 1247.
Rajeshkumar, S., Bharath, L. V., and Geetha, R. (2019). Chapter 17 - Broad spectrum antibacterial silver nanoparticle green synthesis: Characterization, and mechanism of action. *In:* Shukla, A. K., and Iravani, S. (eds.). *Green Synthesis, Characterization and Applications of Nanoparticles.* Elsevier.
Rajput, V., Minkina, T., Mazarji, M., Shende, S., Sushkova, S., Mandzhieva, S. et al. (2020). Accumulation of nanoparticles in the soil-plant systems and their effects on human health. *Annals of Agricultural Sciences*, 65: 137–143.
Ramrez-Rodrguez, G. B. A. (2020). Engineering biomimetic calcium phosphate nanoparticles: a green synthesis of slow-release multinutrient (NPK) Nanofertilizers. *ACS Applied Bio Materials*, 3: 1344–1353.
Rani, N., Singh, P., Kumar, S., Kumar, P., Bhankar, V., and Kumar, K. (2023). Plant-mediated synthesis of nanoparticles and their applications: A review. *Materials Research Bulletin*, 163: 112233.
Rathi, V. H., and Jeice, A. R. (2023). Green fabrication of titanium dioxide nanoparticles and their applications in photocatalytic dye degradation and microbial activities. *Chemical Physics Impact*, 6: 100197.
Redkar, A., Sabale, M., and Di Pietro, A. (2021). A 'Hydrolase Switch' for vascular specialization in plant pathogenic bacteria. *Trends in Plant Science*, 26: 427–429.
Remya, R. R., Julius, A., Suman, T. Y., Mohanavel, V., Karthick, A., Pazhanimuthu, C. et al. (2022). Role of nanoparticles in biodegradation and their importance in environmental and biomedical applications. *Journal of Nanomaterials*, 2022.
Ristaino, J. B., Anderson, P. K., Bebber, D. P., Brauman, K. A., Cunniffe, N. J., Fedoroff, N. V. et al. (2021). The persistent threat of emerging plant disease pandemics to global food security. *Proceedings of the National Academy of Sciences*, 118.
Rodrigues, T. C., Gois, I. B., Fernandes, R. P. M., Blank, A. F., Sandes, R. D. D., Leite Neta, M. T. S. et al. (2023). Chemical characterization and antimicrobial activity of essential oils from Croton grewioides Baill. accessions on the phytopathogen *Xanthomonas campestris* pv. campestris. *Pesticide Biochemistry and Physiology*, 193: 105454.
Roy, A., Bulut, O., Some, S., Mandal, A. K., and Yilmaz, M. D. (2019). Green synthesis of silver nanoparticles: biomolecule-nanoparticle organizations targeting antimicrobial activity. *RSC Adv*, 9: 2673–2702.
Sabir, S., Arshad, M., Ilyas, N., Naz, F., Amjad, M. S., Malik, N. Z. et al. (2022). Protective role of foliar application of green-synthesized silver nanoparticles against wheat stripe rust disease caused by Puccinia striiformis. *Green Processing and Synthesis*, 11: 29–43.
Sakata, N., Ino, T., Hayashi, C., Ishiga, T., and Ishiga, Y. (2023). Controlling stomatal aperture, a potential strategy for managing plant bacterial disease. *Plant Science*, 327: 111534.
Salem, N., Mohammed, F., and Abdelfattah, A. (2023). Plant and microbial approaches as green methods for the synthesis of nanomaterials: synthesis, applications, and future perspectives. *Molecules*, 28.
Sathiyabama, M., Akila, G., and Charles, R. E. (2014). Chitosan-induced defence responses in tomato plants against early blight disease caused by *Alternaria solani* (Ellis and Martin) Sorauer. *Archives of Phytopathology and Plant Protection*, 47: 1963–1973.
Sebastià, P., De Pedro-Jové, R., Daubech, B., Kashyap, A., Coll, N. S., and Valls, M. (2021). The bacterial wilt reservoir host *Solanum dulcamara* shows resistance to *Ralstonia solanacearum* infection. *Frontiers in Plant Science*, 12.
Seleiman, M. F., Ahmad, A., Battaglia, M. L., Bilal, H. M., Alhammad, B. A., and Khan, N. (2023). Zinc oxide nanoparticles: A unique saline stress mitigator with the potential to increase future crop production. *South African Journal of Botany*, 159: 208–218.
Selvamani, V. (2019). Chapter 15 - stability studies on nanomaterials used in drugs. *In:* Mohapatra, S. S., Ranjan, S., Dasgupta, N., Mishra, R. K., and Thomas, S. (eds.). *Characterization and Biology of Nanomaterials for Drug Delivery.* Elsevier.
Selvaraj, M., Ramasami, N., and Jagan, E. G. (2022). The effect of chitosan nanoparticles on immune responses in plants. *Role of Chitosan and Chitosan-Based Nanomaterials in Plant Sciences.* Elsevier.
Sembada, A. A., and Lenggoro, I. W. (2024). Transport of Nanoparticles into Plants and Their Detection Methods. *Nanomaterials*, 14: 131.

Shahbaz, M., Akram, A., Raja, N. I., Mukhtar, T., Mashwani, Z. U. R., Mehak, A. et al. (2022). Green synthesis and checacterization of selenium nanoparticles and its application in plant disease management: A review. *Pakistan Journal of Phytopathology*, 34: 189–102.

Shahbaz, M., Akram, A., Raja, N. I., Mukhtar, T., Mehak, A., Fatima, N. et al. (2023). Antifungal activity of green synthesized selenium nanoparticles and their effect on physiological, biochemical, and antioxidant defense system of mango under mango malformation disease. *Plos One*, 18: e0274679.

Sharifazizi, M., Harighi, B., and Sadeghi, A. (2017). Evaluation of biological control of Erwinia amylovora, causal agent of fire blight disease of pear by antagonistic bacteria. *Biological Control*, 104: 28–34.

Sharma, A., Abrahamian, P., Carvalho, R., Choudhary, M., Paret, M. L., Vallad, G. E. et al. (2022). Future of bacterial disease management in crop production. *Annual Review of Phytopathology*, 60: 259–282.

Sharma, C., Rani, E., Kesharwani, A. K., Singh, D., Srivastava, R., and Sharma, S. N. (2023). Hydrothermally prepared phytofabricated magnetite nanoparticles for its potential antimicrobial application in disease management of agricultural crops. *Inorganic Chemistry Communications*, 158: 111599.

Sharma, S., Kumar, S., Kumar, P., and Tripathi, V. N. (2024). Quorum sensing in Gram-negative pathogens, a fresh look. *The Microbe*, 4: 100108.

Shi, H., Li, W., Zhou, Y., Wang, J., and Shen, S. (2023). Can we control potato fungal and bacterial diseases?—microbial regulation. *Heliyon*, 9: e22390.

Sidhu, A. K., Verma, N., and Kaushal, P. (2022). Role of biogenic capping agents in the synthesis of metallic nanoparticles and evaluation of their therapeutic potential. *Frontiers in Nanotechnology*, 3.

Singh, B. K., Delgado-Baquerizo, M., Egidi, E., Guirado, E., Leach, J. E., Liu, H. et al. (2023). Climate change impacts on plant pathogens, food security and paths forward. *Nature Reviews Microbiology*, 21: 640–656.

Spirescu, V. A., Chircov, C., Grumezescu, A. M., and Andronescu, E. (2021). Polymeric nanoparticles for antimicrobial therapies: An up-to-date Overview. *Polymers*, 13: 724.

Sravani, B., Dalvi, S., and Narute, T. K. (2023). Role of chitosan nanoparticles in combating *Fusarium wilt* (*Fusarium oxysporum* f. sp. ciceri) of chickpea under changing climatic conditions. *Journal of Phytopathology*, 171: 67–81.

Subaşı-Zarbaliyev, B., Kutlu, G., and Trnk, F. (2023). Polyvinyl alcohol nanoparticles loaded with propolis extract: Fabrication, characterization and antimicrobial activity. *ADMET and DMPK*, 10.5599/admet.1740.

Sundin, G. W., Castiblanco, L. F., Yuan, X., Zeng, Q., and Yang, C. H. (2016). Bacterial disease management: challenges, experience, innovation and future prospects. *Molecular Plant Pathology*, 17: 1506–1518.

Swati, Verma, R., Chauhan, A., Shandilya, M., Li, X., Kumar, R., and Kulshrestha, S. (2020). Antimicrobial potential of ag-doped ZnO nanostructure synthesized by the green method using moringa oleifera extract. *Journal of Environmental Chemical Engineering*, 8: 103730.

Takikawa, Y., and Takahashi, F. (2014). Bacterial leaf spot and blight of crucifer plants (Brassicaceae) caused by *Pseudomonas syringae* pv. maculicola and P. cannabina pv. alisalensis. *Journal of General Plant Pathology*, 80: 466–474.

Tang, Y., Zhao, W., Zhu, G., Tan, Z., Huang, L., Zhang, P. et al. (2024). Nano-pesticides and fertilizers: solutions for global food security. *Nanomaterials*, 14: 90.

Thakur, B. K., Kumar, A., and Kumar, D. (2019). Green synthesis of titanium dioxide nanoparticles using *Azadirachta indica* leaf extract and evaluation of their antibacterial activity. *South African Journal of Botany*, 124: 223–227.

Thipe, V. C., Karikachery, A. R., Akılkaya, P., Farooq, U., Genedy, H. H., Kaeokhamloed, N. et al. (2022). Green nanotechnology—An innovative pathway towards biocompatible and medically relevant gold nanoparticles. *Journal of Drug Delivery Science and Technology*, 70: 103256.

Timilsina, A., Adhikari, K., and Chen, H. (2023). Foliar application of green synthesized ZnO nanoparticles reduced Cd content in shoot of lettuce. *Chemosphere*, 338: 139589.

Tortella, G., Rubilar, O., Pieretti, J. C., Fincheira, P., De Melo Santana, B., Fernández-Baldo, M. A. et al. (2023). Nanoparticles as a promising strategy to mitigate biotic stress in agriculture. *Antibiotics (Basel)*, 12.

Tu, K., Cheng, Y., Pan, T., Wang, J., and Sun, Q. (2022). Effects of seed priming on vitality and preservation of pepper seeds. *Agriculture*, 12: 603.

Vanti, G. L., Masaphy, S., Kurjogi, M., Chakrasali, S., and Nargund, V. B. (2020). Synthesis and application of chitosan-copper nanoparticles on damping off causing plant pathogenic fungi. *International Journal of Biological Macromolecules*, 156: 1387–1395.

Wang, Y., Deng, S., Li, Z., and Yang, W. (2022). Advances in the characterization of the mechanism underlying bacterial canker development and tomato plant resistance. *Horticulturae*, 8: 209.

Wang, Y., and Zou, Q. (2024). Deep learning meta-analysis for predicting plant soil-borne fungal disease occurrence from soil microbiome data. *Applied Soil Ecology*, 202: 105532.

Warabieda, M., Kuzmanović, N., Trzciński, P., and Puławska, J. (2023). *Agrobacterium cucumeris* sp. nov. isolated from crazy roots on cucumber (*Cucumis sativus*). *Systematic and Applied Microbiology*, 46: 126402.

Wasule, D. L., Shingote, P. R., and Saxena, S. (2024). Exploitation of functionalized green nanomaterials for plant disease management. *Discover Nano*, 19: 118.

Woejko, E., and Jaboska, T. (2020). Soil biological activity as an indicator of soil pollution with pesticides – A review. *Applied Soil Ecology*, 147: 103356.

Yan, X., He, B., Liu, L., Qu, G., Shi, J., Hu, L. et al. (2018). Antibacterial mechanism of silver nanoparticles in *Pseudomonas aeruginosa*: proteomics approach†. *Metallomics*, 10: 557–564.

Ying, S., Guan, Z., Ofoegbu, P. C., Clubb, P., Rico, C., He, F. et al. (2022). Green synthesis of nanoparticles: Current developments and limitations. *Environmental Technology*, and *Innovation*, 26: 102336.

Yosri, N., Khalifa, S. a. M., Guo, Z., Xu, B., Zou, X., and El-Seedi, H. R. (2021). Marine organisms: Pioneer natural sources of polysaccharides/proteins for green synthesis of nanoparticles and their potential applications. *International Journal of Biological Macromolecules*, 193: 1767–1798.

Zhang, C., Yang, X., Yang, S., Liu, Z., and Wang, L. (2022). Eco-friendly and multifunctional lignocellulosic nanofibre additives for enhancing pesticide deposition and retention. *Chemical Engineering Journal*, 430: 133011.

Zhang, N., Yang, G., Pan, Y., Yang, X., Chen, L., and Zhao, C. (2020). A review of advanced technologies and development for hyperspectral-based plant disease detection in the past three decades. *Remote Sensing*, 12: 3188.

Zhao, L., Zhou, X., Kang, Z., Peralta-Videa, J. R., and Zhu, Y.-G. (2024). Nano-enabled seed treatment: A new and sustainable approach to engineering climate-resilient crops. *Science of The Total Environment*, 910: 168640.

Zhou, Y.-H., Mujumdar, A. S., Vidyarthi, S. K., Zielinska, M., Liu, H., Deng, L.-Z. et al. (2023). Nanotechnology for food safety and security: a comprehensive review. *Food Reviews International*, 39: 3858–3878.

Zhu, W., Hu, C., Ren, Y., Lu, Y., Song, Y., Ji, Y. et al. (2021). Green synthesis of zinc oxide nanoparticles using *Cinnamomum camphora* (L.) Presl leaf extracts and its antifungal activity. *Journal of Environmental Chemical Engineering*, 9: 106659.

Zobir, S. a. M., Ali, A., Adzmi, F., Sulaiman, M. R., and Ahmad, K. (2021). A review on nanopesticides for plant protection synthesized using the supramolecular chemistry of layered hydroxide hosts. *Biology*, 10: 1077.

CHAPTER-12

Plant Pest Control by Green Nanotechnology

Kabyashree Buragohain,[1] *Sukanya Sonowal,*[1] *Dhruba Jyoti Bordoloi,*[1] *Ankita Gogoi,*[1] *Nikha Borah,*[1] *Ratul Nath,*[1,*] *Bhaben Chowardhara,*[2] *Debojyoti Moulick*[3] and *Akbar Hossain*[4,*]

1. Introduction

Nanotechnology has emerged as a transformative force in agriculture, particularly in terms of enhancing crop protection, improving nutrient delivery, and promoting sustainable farming practices. This innovative field uses nanomaterials—substances with dimensions at the nanoscale (1–100 nanometers) to address pressing challenges such as pest management, soil degradation, and food security. The integration of nanotechnology into agricultural practices offers numerous advantages, including increased efficiency of agrochemicals, reduced environmental impact, and improved crop yields (Chowardhara et al., 2024; Pattnaik et al., 2024; Roy and Hossain, 2024).

One of the most significant applications of nanotechnology in agriculture is the development of nanopesticides. These formulations leverage nanoparticles to increase their efficacy and reduce the quantity of pesticides applied. For example, silver nanoparticles (AgNPs) have demonstrated potent insecticidal properties and are increasingly used as carriers for delivering active ingredients directly to target pests. This targeted approach minimizes the overall pesticide load on crops while maximizing pest control efficacy (Jasrotia et al., 2022). Furthermore, nanoemulsions and nanosuspensions represent advanced formulations that improve the solubility and bioavailability of active ingredients. These innovations not only increase the effectiveness of conventional pesticides but also reduce their toxicity to nontarget organisms and the environment (Aouada and De Moura, 2015). The encapsulation of agrochemicals within nanoparticles allows for controlled release, which further optimizes their application and reduces leaching into soil and water systems (Mgadi et al., 2024).

Nanotechnology also plays a crucial role in nutrient delivery systems. Traditional fertilizers often suffer from low efficiency due to rapid leaching and volatilization. In contrast, nanofilizers are designed for slow and controlled nutrient release, which enhances nutrient uptake by plants while

[1] Microbiology Laboratory, Department of Life Sciences, Dibrugarh University, Dibrugarh- 786004, Assam, India.
[2] Department of Botany, Arunachal University of Studies, Namsai-792103, Arunachal Pradesh, India.
[3] Department of Environmental Science, University of Kalyani, Nadia, West Bengal-741235, India.
[4] Bangladesh Wheat and Maize Research Institute (BWMRI), Nashipur, Dinajpur-5200, Bangladesh.
* Corresponding authors: ratulnath@dibru.ac.in; akbarhossainwrc@gmail.com

DOI: 10.1201/9781003567684-12

minimizing losses to the environment (Deb et al., 2023; Mgadi et al., 2024). This phenomenon is particularly important for essential nutrients such as nitrogen, phosphorus, and potassium. The use of nanoencapsulation techniques allows the development of fertilizers that can be tailored to release nutrients in response to specific environmental triggers, such as moisture levels or plant growth stages. This precision not only improves crop productivity but also contributes to sustainable agricultural practices by reducing the need for excessive fertilizer applications (Islam, 2020; Kashyap et al., 2022).

In addition to pest control and nutrient delivery, nanomaterials contribute significantly to soil health management. NPs can improve the soil structure, enhance microbial activity, and facilitate better water retention. For example, porous nanoparticles can encapsulate fertilizers, preventing them from being washed away by rain or irrigation while simultaneously enriching the soil with beneficial elements (Majumdar et al., 2024; Moulick et al., 2025a; Kashyap et al., 2022; Jasrotia et al., 2022). Moreover, the interaction between nanomaterials and soil microorganisms is an area of active research. While some studies suggest that nanoparticles can promote beneficial microbial communities that increase soil fertility, others indicate potential toxicity risks that require further investigation (Mgadi et al., 2024).

The integration of nanotechnology into agriculture aligns with global efforts toward sustainable farming practices. By reducing dependency on chemical inputs and enhancing the efficiency of existing agricultural methods, nanotechnology holds promise for addressing food security challenges posed by a growing global population (Kashyap et al., 2022). However, despite their potential benefits, there are significant gaps in research regarding the long-term effects of nanomaterials on ecosystems and human health. Regulatory frameworks must evolve to ensure safe application while fostering innovation in this burgeoning field (Gupta et al., 2018). Nanotechnology represents a frontier in agricultural innovation because of its ability to enhance crop protection, improve nutrient delivery systems, and promote sustainable practices. As research continues to unravel its full potential, stakeholders must prioritize safety and environmental considerations to harness these advancements effectively (Moulick et al., 2024a-c).

The use of pesticides in agriculture has been a cornerstone for enhancing crop yields and managing pest populations. However, the environmental and health concerns associated with conventional chemical pesticides have prompted a shift towards more sustainable alternatives (Roy et al., 2024; Roy et al., 2023). Environmentally friendly, biodegradable, responsive, and biocompatible pesticides—often referred to as biopesticides—offer significant advantages in promoting sustainable agricultural practices while minimizing adverse effects on ecosystems and human health. Biopesticides are typically derived from natural materials such as plants, bacteria, and fungi. This natural origin often results in lower toxicity levels than conventional pesticides, which can be harmful to nontarget organisms, including beneficial insects, birds, and mammals (US EPA, 2023). By targeting specific pests without affecting the broader ecosystem, biopesticides help maintain biodiversity and ecological balance. One of the most critical advantages of biodegradable pesticides is their ability to decompose naturally in the environment. Unlike synthetic pesticides, which can persist for long periods, leading to soil and water contamination, biodegradable options breakdown into nontoxic byproducts that can re-enter the ecosystem safely (Guerrero et al., 2023). This property significantly reduces the risk of pollution and helps protect vital natural resources such as groundwater. Compared with their chemical counterparts, biopesticides are less likely to leach into groundwater. This characteristic is crucial for protecting drinking water supplies and aquatic ecosystems from contamination (Harvesto Group, 2022).

The use of biopesticides can lead to lower health risks for both farmers and consumers. Traditional chemical pesticides are often associated with various health issues due to their toxicity. In contrast, biopesticides generally pose fewer risks during application and consumption (Shirke, 2023). This is particularly important in developing countries where farmers may have limited access to protective equipment. Biopesticides are designed to affect only specific pests rather than a broad spectrum of organisms. This specificity not only protects beneficial insects but also reduces the likelihood of developing pesticide-resistant pest populations (US EPA, 2023). Consequently, biopesticides can be integrated into pest management strategies that prioritize ecological health while maintaining effective pest control. Although the initial cost of developing biopesticides may be greater than that of traditional pesticides, their long-term economic benefits can be substantial. Biopesticides often require lower application rates because of their effectiveness at minimal doses. Additionally, their rapid degradation reduces the need for repeated applications, ultimately lowering overall costs for farmers. Biopesticides play a vital role in Integrated Pest Management (IPM) programs that combine biological control methods with minimal chemical interventions (Ghosh et al., 2020a,b; Ghosh et al., 2022). This holistic approach not only enhances pest control efficiency but also promotes sustainable farming practices that align with environmental conservation goals (Harvesto Group, 2022). As research progresses in the field of biopesticides, the development of responsive and biocompatible formulations holds great promise. Responsive pesticides can be engineered to activate under specific environmental conditions or in response to pests, further increasing their efficiency and safety (Verma et al., 2023). Additionally, advancements in biotechnology may lead to new formulations that improve the efficacy and applicability of biopesticides across diverse agricultural settings. The importance of environmentally friendly, biodegradable, responsive, and biocompatible pesticides cannot be overstated in today's agricultural landscape. These innovative solutions not only mitigate the negative impacts associated with conventional pesticides but also promote sustainable farming practices that protect human health and preserve ecological integrity. As awareness of the benefits of biopesticides grows, their adoption will likely increase, paving the way for a more sustainable future in agriculture. Morphological differences between plants treated with and without nanopesticides have been presented in Fig. 1.

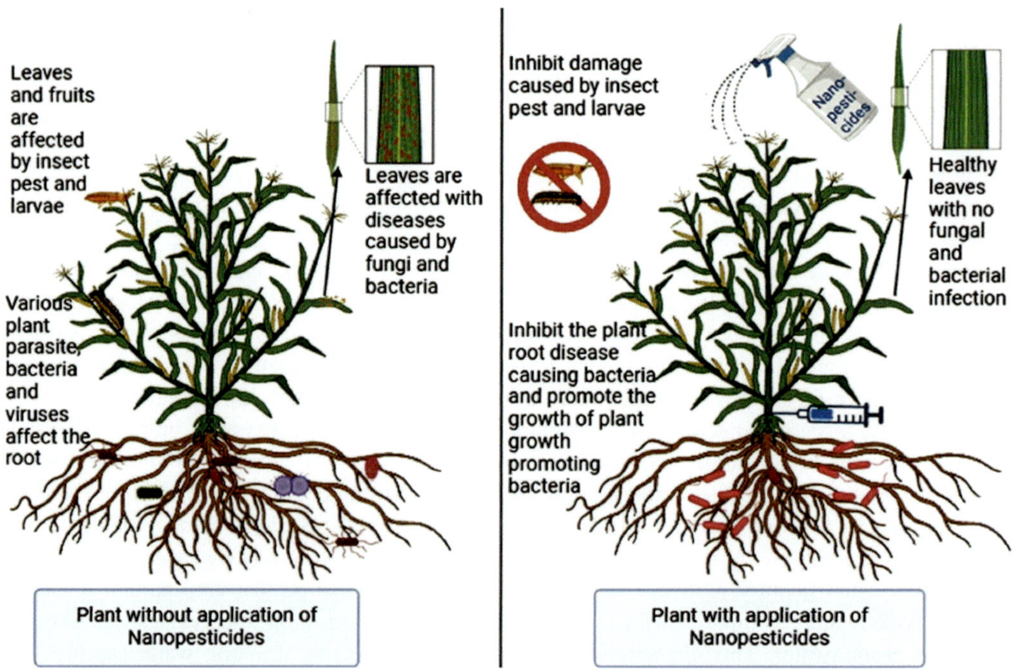

Figure 1. Morphological differences between plants treated with and without nanopesticides.

2. Mechanism of Action of Biopesticides

With the modernization of agriculture, scientists continue to make significant progress in controlling harmful insects, protecting crops, and increasing crop production. In recent decades, the widespread and often indiscriminate use of chemical pesticides has been largely adopted by both developed and developing countries, with the usage steadily increasing (Lacey and Siegel, 2000). The indiscriminate use of chemical pesticides continues to be a significant concern, as it leads to various problems post application. The residues left in the field after pesticide use can have harmful effects on living organisms, including humans, livestock, birds, and beneficial insects such as bees. Additionally, these chemical residues can also contaminate the available groundwater (Verma et al., 2021).

Pesticides derived from plants and microorganisms, along with their products, have been scientifically shown to be highly effective in combating harmful insect pest species (Verma et al., 2021). Plants and microbes have a long history of research and are considered highly promising candidates for pest management. Biopesticides are formulated pesticides derived from natural sources such as animals, plants, bacteria, and minerals. Biopesticides are sourced from bacteria, fungi, algae, nematodes, protozoa, viruses, and other compounds produced by these microbes, such as metabolites, which serve as key microbial control agents for insect pests (Mahanta et al., 2023; Tyagi et al., 2023; van Lenteren, 2012); for example, a mixture of baking soda and canola oil, which has pesticidal properties, is used in agriculture to control a range of pests, pathogens, and weeds. Biopesticides also include *Trichoderma* as biofungicides, *Phytophthora* as biohericides, and *Bacillus thuringiensis* (Bt) as bioinsecticides. Biopesticides, or biological pesticides, offer an environmentally effective solution to pest problems because they are highly specific to their target insect pests (Verma et al., 2021).

Biopesticides are a type of pesticide that uses active ingredients derived from microorganisms or natural compounds instead of synthetic chemicals. These pesticides are utilized to manage insect populations and are commonly referred to as biopesticides. They are microbial pesticides created by living microorganisms such as bacteria, fungi, and viruses. Biopesticides are categorized into three types: biochemical pesticides, microbial pesticides, and plant-incorporated protectants. The efficiency of biopesticides in controlling pests stems from their diverse modes of action, which include disrupting the gut, regulating pest growth, and altering pest metabolism. They work by denaturing proteins, causing metabolic disruptions and paralysis, activating poisoning mechanisms, inhibiting multiple sites, and secreting bioactive compounds and neuromuscular toxins (Dar et al., 2021; Fenibo et al., 2021; Sparks and Nauen, 2015). These diverse modes of action enable biopesticides to influence the development of pest resistance more effectively than chemical pesticides do. Research has shown that biopesticides are environmentally friendly, have low toxicity, are biodegradable, and have very targeted impacts, causing minimal or no harm to nontarget organisms. Unlike biopesticides, conventional pesticides largely contribute to environmental pollution, promote pest resistance, and lead to substantial postharvest contamination and bioaccumulation in food crops (Fenibo et al., 2021; Ayilara et al., 2023).

There are various types of biopesticides with different modes of action, such as sabadilla, pyrethrum, azadirachtin, and fluoroacetate, each of which has distinct effects on pests. For example, the alkaloid toxin in sabadilla disrupts nerve cell membranes. Sabadilla can kill most insects almost immediately upon application. Although pyrethrins are not toxic to humans or animals, they quickly kill insects. They work by disrupting the exchange of sodium and potassium ions in the nerve fibers of insects, which inhibits nerve impulse transmission. In contrast, azadirachtin acts as an antifeedant, reducing food consumption in pests rather than directly controlling them. It has been reported that the inhibition of the feeding behavior by azadirachtin occurs through the stimulation of deterrent receptors, which are linked to sugar receptors, leading to reduced food intake, starvation, and poor nutrition. Several studies have shown weight loss in insects such as *Spodoptera eridania*, *Periplaneta americana*, and *Helicoverpa armigera*, among others (Nidhi et al., 2022). According to the Environmental Protection Agency's classification system, which is based on the type of ingredient used, biopesticides are divided into three main categories: (1) microbial, (2) biochemical, and (3) plant-incorporated protectants (Olson, 2015).

(1) Microbial pesticides:

Microbial pesticides represent the largest category of pest specific, broad-spectrum biopesticides. These pesticides utilize microorganisms such as viruses, bacteria, fungi, and protozoans as active ingredients to manage insect pests. They are highly targeted and affect specific pest species. Additionally, microbial biopesticides are self-sustaining, host specific, and environmentally safe. One of the most commonly used microorganisms in this category is *Bacillus thuringiensis* (Bt), which is effective against a diverse range of pests, including dipterans, coleopterans, and lepidopterans (Rajamani and Negi, 2021). Table 1 shows the mode of action of different microbial pesticides. Microbial biopesticides utilize various delivery methods to effectively target crops, including live, dead organisms, and spores. Globally, microbes make up 41% of all Biological Control Agents (BCAs) used, whereas macrobiotics account for 33% of the market, with other natural products comprising the remaining 26%. These BCAs can be categorized by their origin or active components into fungal biopesticides, bacterial-based biopesticides, viral biopesticides, nematodes, and protozoan-based biopesticides (Tadesse Mawcha et al., 2024). Table 1 shows several microbe-based pesticides and their modes of action.

Bacterial pesticides:

The most commonly used microbial pesticides are bacterial biopesticides because of their versatility in combating pests. They primarily target insects but are also effective against other agricultural pests. In addition to controlling insect pests across various orders, these biopesticides can suppress the growth of bacteria and fungi that harm plants (Tadesse Mawcha et al., 2024). While approximately 100 bacteria have been identified as exo- and endopathogens of arthropods (Thacker, 2002), only a few are commercially utilized in pest management systems. Among these, *B. sphaericus, Bacillus popilliae, B. thuringiensis, Pseudomonas alcaligenes, Clostridium bifermentans, Pseudomonas aureofaciens, Serratia entomophila*, *Saccharopolyspora spinosa,* and *Streptomyces avermitilis* are considered the primary contributors. However, two bacterial species stand out as prominent pest control agents: the spore-forming soil bacterium *B. thuringiensis* (Bt) and the nonspore-forming *S. entomophila* (Chattopadhyay et al., 2017).

Mode of action of *Bacillus thuringiensis* (Bt) against pests:

Bt produces crystalline proteins (cry and cyt) that are effective against specific target insect species. When larvae ingest Cry protein, delta endotoxins are activated in the insect gut, which has an alkaline pH ranging from 9.0 to 11.0. These proteins bind to receptor sites in the gut and form pores in midgut cells (Kumar, 2012). The binding of these crystalline proteins to receptors in the insect gut determines which species are affected. The toxicity of Bti and certain toxic strains is attributed to parasporal inclusion bodies formed during sporulation. Different strains of Bt produce various insecticidal crystal proteins. When ingested by larvae, these toxins damage the gut tissues, causing paralysis. The infected larvae subsequently cease feeding and ultimately die (Fig. 2) (Pavela and Benelli, 2016). Unlike Cry proteins, Cyt toxins are directly embedded into the lipid membranes of insects. Recent studies indicate that Cyt can either enhance or mitigate resistance to mosquito-cidal Cry proteins by acting as a membrane-bound receptor for Cry toxins. Both Cry toxins and Cyt toxins function as pore-forming toxins (Bravo et al., 2007).

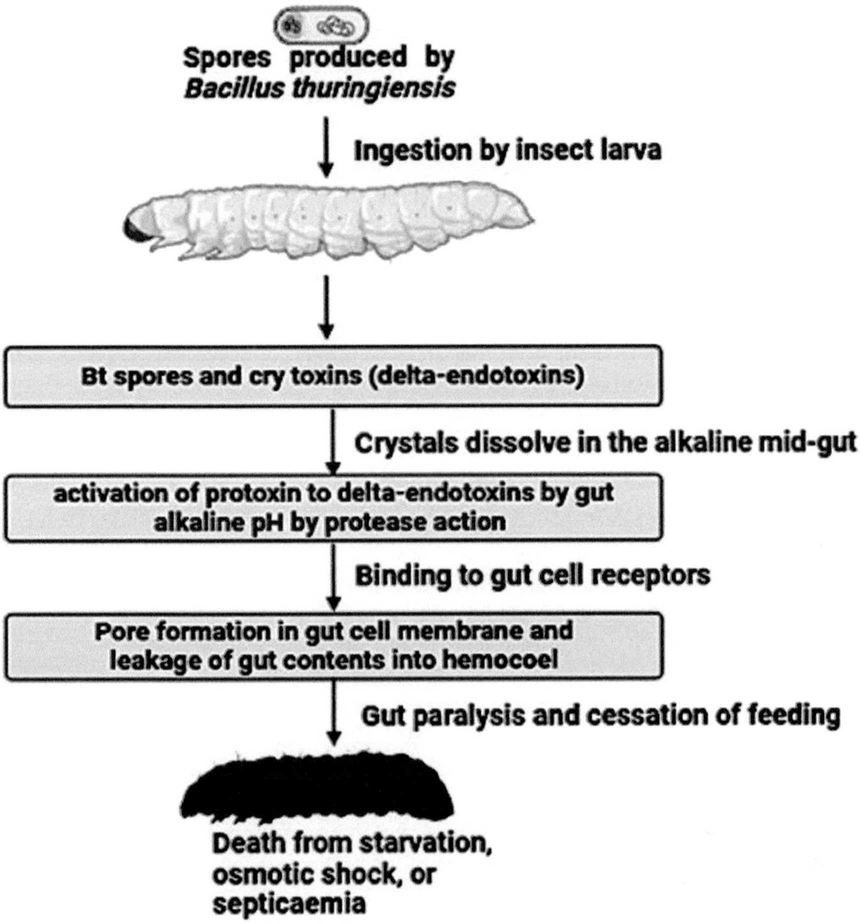

Figure 2. Mode of action of a bacterial pesticide (*B. thuringiensis*).

Fungal biopesticides:

Entomopathogenic fungi are fungi that target and infect insect hosts, and they can produce various insecticidal compounds (Butt et al., 2016). The target specificity, reproductive capacity, short lifespan, and durability of entomopathogenic fungi make them highly valuable for biological control methods (Holder and Keyhani, 2005; Sharma and Sharma, 2021; Singkaravanit et al., 2010). They primarily infect target insects through their outer coverings, but in some instances, they can also kill pests through ingestion or wounds. Fungi used as biopesticides include species such as *Trichoderma, Beauveria, Metarhizium, Paecilomyces, Fusarium, Pythium, Penicillium,* and *Verticillium* (Kachhawa, 2017). While the biochemical mechanisms of many commonly used biopesticides have been documented, the entomopathogenic fungus *Beauveria bassiana* stands out as a widely accepted model for illustrating the molecular mechanisms of biopesticide application.

The EPA has approved fungal biopesticides such as *Beauveria* and *Metarhizium* as biological control agents for managing agricultural pests. Combinations of *Beauveria bassiana* and Neem can effectively eliminate *Bemisia tabaci* (whitefly), a harmful pest affecting sweet potato and chili, with a mortality rate indicated by the LT_{50} value (Ghongade and Sangha, 2021). The entomopathogenic fungus *Metarhizium* sp. was observed to cause fungal infection (mycosis) in *Spodoptera frugiperda*, commonly known as the fall armyworm (Umaru and Simarani, 2020). Fungal biopesticides such as *Trichoderma* operate through various mechanisms, including mycoparasitism, competition, and the production of lytic enzymes and antibiotics, which can harm pests. Moreover, these biopesticides benefit plants by increasing their ability to manage stress (Lyubenova et al., 2023). Nematophagous fungi such as *Paecilomyces lilacinus* (Thom) Samson have proven effective in managing greenhouse pests such as cotton aphids, whiteflies (at their nymphal stages), and western flower thrips (at their pupal and prepupal stages). Similarly, nematode-trapping fungi such as *Arthrobotrys flagrans* (*Duddingtonia flagrans*) form traps to target the wine-infesting nematode *Xiphinema index* in response to various biotic bacterial signals. This finding highlights the synergistic interaction between fungi and bacteria in combating nematodes (Wernet and Fischer, 2023; Hezakiel et al., 2024).

Viral Biopesticides:

Virus-based biopesticides are safe for both humans and nontarget organisms (Thapa et al., 2020). Although their commercial availability is limited, nearly 50 viral insecticides have been registered. Among these, the Baculoviridae family is particularly well known and widely used as a group of viral biopesticides that target insects (Federici, 1997; Moore et al., 1987). Two types of baculoviruses are utilized for pest control: nuclear polyhedroviruses (NPVs), which have occlusion bodies shaped like polyhedra, and granuloviruses (GVs) (Copping and Menn, 2000; Dar et al., 2021). Baculoviruses, known for their high host specificity, are commonly utilized as active agents in viral biopesticides. These insect-specific viruses are effective at managing lepidopteran pests in cotton, vegetable crops, forests, and ornamental plants (Inceoglu et al., 2001; Moscardi, 1999; Hezakiel et al., 2024). The mode of action of microbe-based pesticides have been presented in Table 1.

(2) Biochemical Pesticides

Biochemical pesticides are naturally occurring substances found in the environment that manage pests through nontoxic methods. Unlike traditional pesticides, which directly target and eliminate pests, biochemical pesticides work indirectly by employing alternative mechanisms of action. For example, they interfere with the reproductive functions of their targets. Natural plant-based compounds such as terpenoids, alkaloids, phenolics, and other secondary metabolites can serve as biopesticides. Additionally, certain vegetable oils have been found to possess pesticidal properties (Ghosh et al., 2022; Ghosh et al., 2023; Abdollahdokht et al., 2022).

(3) Plant-incorporated protectants (PIPs)

Plants that contain toxin-producing genes to combat pests are referred to as Plant-Incorporated Protectants (PIPs) or genetically modified/engineered plants. These PIPs enable plants to resist harmful organisms such as bacteria, fungi, or viruses. When PIPs are specifically used to target insects, the resulting plants are called insect-resistant crops. Examples of PIPs include soybean, brinjal, cotton, corn, potato, and tomato. Transgenic plants are created by introducing the desired gene using methods such as *Agrobacterium*-mediated transfer, gene gun, or ballistic techniques (Rajamani and Negi, 2021). Pesticidal proteins derived from bacteria or fungi are incorporated into plants, enabling genetically modified plants to resist specific pests. A common example is the use of Bt protein to create plant-incorporated protectants (PIPs) through genetic engineering. Bt toxin is highly selective for its target pests, causing rapid mortality, typically within 48 hours. It is environmentally safe and does not harm vertebrates (Rajamani and Negi, 2021; Lacey and Siegel, 2000).

Understanding the molecular mechanisms underlying the action of biopesticides at each stage is crucial for developing more effective pest control strategies. Exploring how biopesticides target insect pests at the molecular level can enable synergistic combinations of biopesticides with distinct, nonoverlapping modes of action. This approach also allows for the identification and utilization of various toxic molecules within biopesticides, expanding their pesticidal capabilities.

Table 1. Mode of action of microbe-based pesticides.

Sl No.	Microbial pesticides	Mechanism of action	Effective against	References
1	*Agrobacterium radiobacter*	Toxins are produced that are harmful if ingested by certain pests.	Various insect larvae	(Verma et al., 2021)
2	*Bacillus thuringiensis*	Produces delta-endotoxins that disrupt insect midgut membranes, causing starvation and death in the larvae. It mainly focuses on lepidopteran pests, which are notorious for causing significant damage.	Various strains targeting specific insect larvae (e.g., Bt var. kurstaki for caterpillars), and Lepidopteran pests, such as stem borers in rice.	(Rajamani and Negi, 2021)
3	*Bacillus sphaericus*	Similar action to *B. thuringiensis*, primarily targeting mosquito larvae through endotoxins.	Used in mosquito control	(Tamura, 2024)
4	*Beauveria bassiana*	An entomopathogenic fungus that infects and kills insects by penetrating their cuticle and proliferating inside.	Used against aphids, whiteflies	(Ayilara et al., 2023)
5	*Metarhizium anisopliae*	Infects insects through cuticle penetration, leading to systemic infection and death.	Effective against various insect pests	(Ayilara et al., 2023)
6	Nuclear Polyhedrosis Virus (NPV)	A viral biopesticide that causes disease in insects by infecting their cells, leading to death.	Specific to certain caterpillar species	(Dhakal and Singh, 2019)
7	*Pseudomonas fluorescens*	Competes with pathogens and promotes plant growth; effective against root rot and wilts.	Effective against diseases in crops like tomato and cotton	(Dhakal and Singh, 2019)
8	Spinosad	Derived from Saccharopolyspora spinosa; acts on the nervous system of insects, causing paralysis and death.	Spinosad products like Spintor. *e.g., Saccharopolyspora spinosa*	(Tadesse Mawcha et al., 2024)
9	*Trichoderma* spp.	Mycoparasitism, antibiosis, and competition with fungal pathogens; induces plant defense mechanisms.	*Trichoderma harzianum*, *T. viride*	(Tamura, 2024)

3. Applications of Nanotechnology in Biopesticide Formulations

Nanotechnology has revolutionized various fields, including agriculture, where it plays a crucial role in the development of biopesticides. Among the innovative applications of nanotechnology in this domain are nanoinsecticides and RNA pesticides. These advances not only enhance pest control efficacy but also promote environmentally sus sustainable practises. This paper explores the applications, benefits, and challenges associated with these nanotechnology-driven solutions in biopesticide formulation.

3.1 NanoInsecticides

Nanoinsecticides are pesticide formulations that utilize nanoparticles to improve the delivery and effectiveness of Active Ingredients (AIs) against insect pests. These nanoparticles can be made from various materials, including metals (such as silver or copper), polymers, and lipids, which can encapsulate or bind to AIs, increasing their stability and solubility (Yan et al., 2020; Chen et al., 2023).

Compared with traditional pesticides, nanoinsecticides exhibit marked effectiveness due to several key mechanisms that enhance their performance. One of the primary mechanisms is their ability to achieve enhanced penetration, as the small size of nanoparticles facilitates better infiltration into insect exoskeletons, allowing direct action on target pests (Shirke, 2023). Additionally, nanoformulations can be engineered for controlled release, ensuring the gradual and sustained release of active ingredients over time, which not only prolongs their activity but also reduces the frequency of applications needed. Moreover, these advanced formulations offer remarkable target specificity by utilizing surface modifications on nanoparticles that bind specifically to pest receptors, thereby minimizing harm to nontarget organisms and ensuring precise action (Gouda et al., 2023). The advantages of nano-insecticides further underscore their potential as revolutionary pest management tools. For example, studies indicate that these formulations significantly increase the efficacy of traditional pesticides by improving their bioavailability and reducing their degradation during application and in environmental conditions (Yan et al., 2020). Their targeted nature also contributes to a reduced environmental impact, as the lower application rates required for effective pest control help decrease the overall chemical load on ecosystems (Yousef et al., 2023). Furthermore, nanoinsecticides present a promising solution to one of the major challenges in pest control: the development of resistance. Their unique modes of action and targeted delivery systems reduce the likelihood of pests developing resistance, which is a growing concern with conventional pesticides (Shirke, 2023). Collectively, these features make nano-insecticides highly effective and environmentally sustainable alternatives for modern pest management.

3.2 RNA Pesticides

RNA interference (RNAi) is a natural cellular process that regulates gene expression by degrading specific messenger RNA (mRNA) molecules. This mechanism has been harnessed in agriculture to develop RNA-based pesticides that target essential genes in pest organisms, leading to their death or reduced fitness (US EPA, 2023; Chen et al., 2023).

RNA-based pesticides represent a cutting-edge approach to pest management, with various types offering targeted and environmentally friendly pest control solutions. Among these, double-stranded RNA (dsRNA) is a prominent type that can be applied directly to plants or pests to induce gene silencing via the RNA interference (RNAi) pathway. Upon ingestion by pests, dsRNA is processed into small interfering RNAs (siRNAs), which then specifically target and degrade essential mRNAs critical for pest survival. Another type, small interfering RNA (siRNA), is similar to dsRNA but typically shorter and more specific in targeting genes within pests. However, the practical application of RNA-based pesticides is hindered by several challenges. Environmental stability is a significant concern, as dsRNA is susceptible to degradation by factors such as UV light and nucleases present on plant surfaces and in the soil (Chen et al., 2023). Additionally, the uptake efficiency of dsRNA by pests varies greatly due to barriers in their digestive systems, which can affect the effectiveness of RNA pesticides.

Nanotechnology offers innovative solutions to address these limitations and enhance the efficacy of RNA-based pesticides. Nanoparticle carriers, for example, have been developed to protect dsRNA from environmental degradation while improving its uptake by plants and pests. Polymeric nanocarriers are particularly effective, as they encapsulate dsRNA and facilitate its delivery directly into pest cells (Yan et al., 2020; Chen et al., 2023).

Another promising strategy is Spray-Induced Gene Silencing (SIGS), which involves the application of nanoparticle–dsRNA complexes directly to crops. This approach not only shields RNA molecules from degradation by environmental factors but also promotes their absorption by both plants and pests, thereby maximizing their pest control potential (Gouda et al., 2023). Together, these advancements in RNA pesticide technology and nanotechnology integration pave the way for more sustainable and precise pest management strategies.

The integration of nanotechnology into biopesticide formulations offers numerous advantages that significantly enhance modern pest management practices. One of the most critical benefits is precision targeting, as both nano-insecticides and RNA pesticides are designed to act specifically against target pests without harming beneficial organisms or disrupting ecosystems. This specificity not only protects biodiversity but also promotes ecological balance in agricultural systems. Additionally, these advanced formulations align with sustainable agricultural practices by reducing the reliance on conventional chemical pesticides and minimizing environmental contamination. Nanotechnology also enhances the efficacy of biopesticides by improving the bioavailability and stability of active ingredients, ensuring more effective and long-lasting pest control strategies.

However, despite these promising benefits, the application of nanotechnology in biopesticide formulations faces several challenges. Regulatory hurdles remain a significant barrier, as the evolving regulatory frameworks for nanotechnology-enabled products necessitate comprehensive safety assessments to address potential risks to human health and the environment. Furthermore, public perception poses another challenge, with skepticism surrounding genetically engineered products and nanotechnology applications in agriculture. To address these concerns, education and transparency are essential to build trust and foster public acceptance of these innovative solutions.

Future research in this field should prioritize optimizing nanoparticle design to develop more efficient and targeted delivery systems while ensuring safety through rigorous risk assessments. This approach will be critical for overcoming existing barriers and facilitating the widespread adoption of nanotechnology-enabled biopesticides. Ultimately, the application of nanotechnology in biopesticide formulation represents a groundbreaking advancement in sustainable agriculture, providing powerful tools to enhance pest management while safeguarding the environment. Continued research and development efforts will be vital for unlocking the full potential of these transformative technologies and addressing the challenges associated with their implementation.

4. Environmental Fate and Impact of Nanobiopesticides

Currently, the use of nanopesticides is widespread globally, particularly for agricultural applications on croplands and for controlling pests that affect both human and veterinary health (Zaheer et al., 2022). At present, nanopesticides are used extensively around the world, especially in agriculture for crop protection and in managing pests that impact human and animal health. The risk is the common belief that formulants, traditional pesticides, and active ingredients quickly separate after being applied in the field. However, this assumption needs to be verified for nanoenabled formulations, which are designed to alter the environmental behavior of active ingredients. Furthermore, there is uncertainty about whether nanopesticides can be properly assessed within the existing pesticide regulatory framework (Ale et al., 2023). Another key consideration is the intricate relationship between nanopesticides and microorganisms, which significantly influences the environmental behavior of these nanomaterials. Soil microorganisms are essential for organic matter decomposition, nutrient recycling, and plant growth promotion (Yadav et al., 2021). Specifically, the soluble compounds in a pesticide are crucial for its transport and biodegradation. As a result, it is important to examine the potential environmental behavior of nanoformulations designed to increase their solubility (Ale et al., 2023).

The eco-friendly production of stable nanoformulations for biopesticides, which offer extended insecticidal and antimicrobial activity, holds great potential for the future. The production of nanoscale materials through the use of nanoencapsulated biocidal enzymes, essential oils, plant extracts, or biological extracts is a promising and innovative alternative (Jampílek and Kráľová, 2019). Researchers should also be mindful of the environmental impact of nanobiopesticide formulations, despite their advantages over conventional biopesticides. Additionally, the potential toxicity of nanoparticles in various crops and their influence on plant metabolic activity must be carefully evaluated (Ganguli, 2019; Pan et al., 2023). The efficiency of nanobiopesticides is influenced by factors such as composition, concentration, surface charge, size, and any chemical or physical changes (Chhipa, 2019). The significant role of nanoformulations in minimizing the degradation of active ingredients, enhancing water solubility, and improving the biological availability of these ingredients is well recognized. This has been crucial in preventing widespread pest infestations, plant damage, and economic losses due to the reduced quality and yield of agricultural products and food (Ayilara et al., 2023; Chaudhary, Chaudhary, et al., 2021; Chaudhary, Khati, et al., 2021; Syafrudin et al., 2021). The use of nanobiopesticides in agriculture represents a significant advancement in pest management, promising enhanced efficacy and reduced environmental impact. However, understanding their environmental fate and potential impacts is crucial for assessing their long-term sustainability. This section explores the degradation and persistence of nanobiopesticides in the environment; their interactions with soil, water, and nontarget organisms; and ecotoxicological considerations.

Nanobiopesticides exhibit distinct degradation pathways and environmental persistence due to their nanostructured compositions, necessitating a thorough understanding of their environmental fate. Their degradation can occur via photodegradation, where sunlight exposure influences breakdown rates. Research has shown that compared with conventional formulations, nanoformulations, such as clothianidin-based products, increase the photodegradation half-life of active ingredients in water by 21%, increasing their initial efficacy but potentially prolonging their environmental presence (Djiwanti and Kaushik, 2019; Chaud et al., 2021). Additionally, microbial degradation by soil microorganisms plays a critical role in breaking down pesticides. However, nanoparticles can either inhibit or promote microbial activity, thereby altering degradation rates (Ale et al., 2023). The persistence of nanobiopesticides is shaped by interactions with environmental matrices. For example, their higher sorption rates to soil particles, up to 51% greater than those of conventional pesticides, can extend their environmental presence while reducing the risk of leaching (Chaud et al., 2021). Similarly, water stability is influenced by nanoparticle encapsulation, which can either expedite or delay degradation, raising concerns about accumulation and toxicity in aquatic ecosystems (Yousef et al., 2023).

Nanobiopesticides also interact with soil, water, and nontarget organisms in complex ways. In soils, they can influence microbial diversity and function, sometimes enhancing beneficial activity but potentially being toxic to nontarget organisms, affecting nutrient cycling and soil health (Ale et al., 2023). Certain formulations may also cause phytotoxic effects on nontarget plants, interfering with nutrient absorption through plant roots (Djiwanti and Kaushik, 2019). In aquatic environments, nanobiopesticides contribute to water quality challenges through runoff and leaching, contaminating surface and groundwater supplies (Chaud et al., 2021). Moreover, their high bioavailability can lead to bioaccumulation in aquatic organisms, with prolonged exposure potentially harming fish and other wildlife (Xu et al., 2022).

Ecotoxicological considerations are crucial for assessing the safety of nanobiopesticides. Current testing frameworks often lack the necessary specificity to address the unique properties of nanoparticles (Yousef et al., 2023). The slow-release nature of nanoformulations may result in prolonged exposure to nontarget organisms, raising concerns about chronic toxicity and long-term ecological impacts (Xu et al., 2022). Regulatory challenges further complicate this issue, as existing risk assessment protocols are not equipped to evaluate the distinct behaviors of nanoparticles. The development of tailored methodologies that consider their physicochemical properties is essential for ensuring their safe use in agriculture (Ale et al., 2023). Understanding these complex interactions and addressing regulatory gaps will be pivotal in balancing the benefits of nanobiopesticides with their environmental and ecological implications.

The environmental fate and impact of nanobiopesticides present both opportunities and challenges for sustainable agriculture. While these innovative formulations provide enhanced pest management capabilities, their degradation patterns, interactions with soil and water systems, and potential ecotoxicological effects require thorough investigation. To ensure safe integration into agricultural practices, it is essential to develop comprehensive risk assessment frameworks that adequately address the unique characteristics of nanobiopesticides. Continued research into their environmental behavior will be crucial for maximizing benefits while minimizing risks associated with these emerging technologies.

5. Challenges and Risks of Nanobiopesticides

The advent of nanotechnology in agriculture, particularly in the formulation of biopesticides, promises numerous advantages, such as enhanced efficacy, targeted delivery, and reduced environmental impact. However, the integration of nanobiopesticides into agricultural practices also presents significant challenges and potential risks to human health and the environment. This discussion explores these challenges, focusing on potential health risks associated with exposure to nanopesticides, environmental implications, and regulatory concerns.

5.1 Potential Health Risks

5.1.1 Toxicological Concerns

NPs possess unique physicochemical properties that can lead to unforeseen toxicological effects. The small size and high surface area of nanoparticles increase their reactivity and ability to penetrate biological membranes, potentially causing cellular damage (Chaud et al., 2021). Studies have indicated that exposure to certain nanopesticides can lead to acute and chronic health effects, including neurotoxicity, respiratory issues, and carcinogenic effects (Yan et al., 2020).

NPs may accumulate in living organisms, leading to increased toxicity over time. This phenomenon is referred to as bioaccumulation. This bioaccumulation can disrupt biological functions and cause long-term health issues in humans and wildlife (Djiwanti and Kaushik, 2019). The potential for nanoparticles to cross biological barriers—such as the blood–brain barrier—raises significant concerns regarding their impact on neurological health (Chaud et al., 2021).

5.1.2 Vulnerable Populations

Certain populations, including children, pregnant women, and agricultural workers, may be particularly susceptible to the adverse effects of nanopesticides. Children are more vulnerable because of their developing bodies and higher exposure levels from food and environmental sources (Yan et al., 2020). Pregnant women may face risks related to foetal development if harmful nanoparticles cross the placental barrier.

5.2 Environmental Risks

5.2.1 Ecosystem Disruption

The introduction of nanobiopesticides into agricultural ecosystems can disrupt existing ecological balances. While these products are designed to target specific pests, their unintended effects on nontarget organisms—including beneficial insects such as pollinators—can lead to decreased biodiversity (Yousef et al., 2023). The persistence of nanoparticles in the environment raises concerns about their long-term ecological impacts.

NPs can alter the soil microbial communities essential for nutrient cycling and soil health. Changes in microbial diversity can adversely affect plant growth and soil fertility (Chaud et al., 2021). Additionally, certain nanoparticles may exhibit phytotoxicity, inhibit plant growth or cause physiological stress (Djiwanti and Kaushik, 2019).

5.2.2 Water Contamination

Runoff from agricultural fields treated with nanobiopesticides poses a risk of water contamination. NPs can leach into water bodies through soil permeation or surface runoff during precipitation events (Chaud et al., 2021). This contamination can affect aquatic ecosystems by altering water quality and harming aquatic organisms through bioaccumulation.

5.3 Regulatory Challenges

5.3.1 Lack of Comprehensive Regulations

The rapid development of nanotechnology has outpaced regulatory frameworks designed to ensure safety and efficacy in pesticide use. Existing regulations often do not adequately address the unique properties and behaviors of nanoparticles (Yan et al., 2020). As a result, there is a pressing need for specific guidelines governing the testing and approval processes for nanobiopesticides.

Current risk assessment protocols may not fully account for the potential risks associated with nanoparticle exposure. There is a need for standardized methodologies that consider the unique characteristics of nanomaterials when evaluating their safety (Djiwanti and Kaushik, 2019).

5.3.2 Public Perception and Acceptance

Public perception plays a crucial role in the adoption of new agricultural technologies. Concerns regarding the safety of nanotechnology applications in agriculture can lead to resistance among consumers and stakeholders (Yousef et al., 2023). Transparent communication about the benefits and risks associated with nanobiopesticides is essential for fostering public trust.

While nanobiopesticides offer promising advancements in pest management through enhanced efficacy and reduced environmental impact, they also present significant challenges and risks that must be addressed. Potential health risks associated with exposure to nanoparticles, environmental implications related to ecosystem disruption and water contamination, and regulatory challenges highlight the need for comprehensive research and responsible management practices. To ensure the safe integration of nanobiopesticides into agricultural systems, it is crucial to develop robust regulatory frameworks that address these challenges while promoting innovation in sustainable agriculture. Ongoing research should focus on understanding the long-term effects of nanoparticle exposure on human health and ecosystems to inform best practices for their use in pest management.

6. Regulatory and Safety Assessment Challenges

The integration of nanotechnology into biopesticide formulations has the potential to revolutionize pest management in agriculture, offering enhanced efficacy, targeted delivery, and reduced environmental impact. However, this emerging technology also presents significant regulatory and safety assessment challenges that must be addressed to ensure the safe use of nanobiopesticides. This section explores these challenges, focusing on regulatory frameworks, safety assessment protocols, and the need for comprehensive risk evaluations.

6.1 Regulatory Frameworks for Nanobiopesticides

6.1.1 Current Regulatory Landscape

The rapid advancement of nanotechnology has outpaced the development of regulatory frameworks specifically tailored for nanobiopesticides. Existing pesticide regulations often do not adequately address the unique properties and behaviors of nanoparticles, leading to uncertainties regarding their safety and environmental impact (Djiwanti and Kaushik, 2019; Chaud et al., 2021). Various countries, including the United States, European Union member states, India, and Australia, have begun to develop guidelines for the regulation of nanobased agricultural products, but these efforts are still in their infancy (Yousef et al., 2023).

6.1.2 Need for Specific Regulations

There is a pressing need for regulations that specifically address the characteristics of nanoparticles used in biopesticides. Current pesticide regulations typically focus on conventional chemical formulations and may not consider the unique aspects of nanoformulations, such as their increased reactivity, bioavailability, and potential for bioaccumulation (Yan et al., 2020). Regulatory bodies must develop specific guidelines that account for the following:

(a) **Characterization of nanoparticles:** Detailed characterization is essential for understanding the physicochemical properties of nanoparticles (size, shape, surface charge) that influence their behavior in biological systems (Chaud et al., 2021).

(b) **Environmental Fate and Transport:** Assessing how nanoparticles behave in various environmental compartments—soil, water, and air—is crucial for understanding their potential ecological impacts (Djiwanti and Kaushik, 2019).

6.2 Safety Assessment Protocols

6.2.1 Risk Assessment Challenges

Risk assessment is a critical component of the regulatory process for any pesticide product. However, existing risk assessment methodologies may not be suitable for evaluating nanobiopesticides because of their unique properties (Yousef et al., 2023). Key challenges include the following:

(a) **Toxicity evaluation:** The toxicity of nanoparticles can differ significantly from that of their bulk counterparts. Traditional toxicity tests may not accurately capture the potential adverse effects of nanoparticles on human health and the environment (Yan et al., 2020). For example, nanoparticles may exhibit increased toxicity due to increased cellular uptake or interactions with biological membranes.

(b) **Ecotoxicity Testing:** Current ecotoxicity tests may require adaptation to account for the specific interactions between nanoparticles and various organisms in aquatic and terrestrial ecosystems (Chaud et al., 2021). The potential for nanoparticles to accumulate in food chains raises additional concerns about long-term ecological impacts.

(c) To ensure the safe use of nanobiopesticides, comprehensive safety assessments must be conducted that include the following:

   I. **Long-term studies:** Longitudinal studies are necessary to evaluate the chronic effects of exposure to nanobiopesticides on human health and ecosystems over time.

   II. **Multidisciplinary Approaches:** Collaboration among toxicologists, ecologists, agronomists, and regulatory scientists is essential to develop a holistic understanding of the risks associated with nanobiopesticides (Yousef et al., 2023).

6.3 Public Perception and Acceptance

6.3.1 Addressing Public Concerns

Public perception plays a crucial role in the acceptance of new agricultural technologies. Concerns regarding the safety of nanotechnology applications can lead to resistance among consumers and stakeholders (Djiwanti and Kaushik, 2019). Transparent communication about the benefits and risks associated with nanobiopesticides is essential for fostering public trust.

6.3.2 Education and Outreach

Effective education and outreach programs can help demystify nanotechnology in agriculture and address misconceptions about its safety. Engaging with farmers, consumers, and policymakers through workshops and informational campaigns can promote the understanding and acceptance of nanobiopesticides as viable pest management solutions.

The regulatory and safety assessment challenges associated with nanobiopesticides are significant but not insurmountable. The development of comprehensive regulatory frameworks tailored specifically for nanotechnology applications in agriculture is crucial to ensuring safety while promoting innovation. Additionally, adapting risk assessment protocols to account for the unique properties of nanoparticles will be essential in evaluating their potential impacts on human health and ecosystems. By addressing these challenges through collaborative efforts among researchers, regulators, industry stakeholders, and the public, it is possible to harness the benefits of nanobiopesticides while minimizing risks. Continued research into the long-term effects of nanoparticle exposure will further inform best practices for their use in sustainable agriculture.

6.4 Regulatory Frameworks for Nanobiopesticides

The introduction of nanobiopesticides into agricultural practices has the potential to enhance pest management while promoting sustainability. However, their acceptance hinges significantly on public perception, which is influenced by regulatory frameworks, safety assessments, and communication strategies. This section discusses the interplay between regulatory frameworks for nanobiopesticides and public perception, highlighting the importance of transparency, education, and stakeholder engagement.

6.4.1 Understanding Regulatory Frameworks

Current regulatory frameworks for nanobiopesticides vary widely across countries, reflecting differences in agricultural practices, environmental policies, and public health considerations. In many regions, existing pesticide regulations are being adapted to accommodate the unique properties of nanomaterials. For example:

A. India: The Department of Biotechnology (DBT) has proposed guidelines for evaluating nanoagri-inputs and products, emphasizing a case-by-case approach to assess safety and efficacy (Adholeya and Singh, 2020). These guidelines aim to ensure that nanobiopesticides meet quality standards while considering their specific characteristics compared with those of conventional pesticides.

B. United States: The Environmental Protection Agency (EPA) requires that all pesticide products containing nanomaterials undergo rigorous safety assessments before registration (Kumari et al., 2023). This includes evaluations of toxicity, environmental impact, and efficacy.

C. European Union: The EU has established comprehensive regulations under REACH (Registration, Evaluation, Authorization and Restriction of Chemicals) that address the safety of nanomaterials in various applications, including agriculture (OECD, 2013).

6.5 Challenges in Regulation

Despite these frameworks, several challenges persist:

a. Lack of consensus: There is no universally accepted regulatory standard for nanobiopesticides. Different countries have adopted varying approaches to regulation, leading to confusion among manufacturers and consumers (Yousef et al., 2023; Kumari et al., 2023).

b. Evolving Science: The rapid pace of nanotechnology development outstrips existing regulatory frameworks. Regulators must continuously adapt guidelines to account for new scientific findings related to the safety and efficacy of nanobiopesticides.

6.6 Public Perception of Nanobiopesticides

The public perception of nanobiopesticides is influenced by multiple factors, including safety concerns, environmental impacts, and trust in regulatory bodies. The potential health risks associated with nanoparticles raise significant apprehension among consumers, as studies reveal a general wariness toward technologies involving genetic modification or nanotechnology (Yan et al., 2020). Transparency in safety assessments conducted by regulatory agencies can alleviate these fears, fostering greater confidence. Additionally, growing concern about environmental sustainability highlights the importance of demonstrating that nanobiopesticides pose fewer environmental risks than conventional pesticides do (Djiwanti and Kaushik, 2019). Trust in regulatory bodies is another critical aspect, as clear and consistent communication about regulatory processes and the scientific basis for the approval of nanobiopesticides can increase public confidence.

Education and outreach programs play pivotal roles in improving public understanding and acceptance of nanobiopesticides. Stakeholder engagement initiatives, such as workshops and informational campaigns targeting farmers, consumers, and community leaders, can demystify nanotechnology in agriculture. Providing accessible and comprehensible information about how nanobiopesticides function and their advantages can encourage broader acceptance (Yousef et al., 2023; Kumari et al., 2023). Transparent communication is essential, involving open discussions about the potential risks and the safety measures undertaken during development and regulation. Highlighting the environmental and agricultural benefits of nanobiopesticides while addressing concerns directly can help foster trust among stakeholders.

The success of nanobiopesticides in sustainable agriculture hinges not only on regulatory approval but also on public perception and acceptance. Regulatory frameworks must evolve to address the unique challenges posed by nanotechnology while ensuring robust safety assessments. Public education and transparent communication strategies are crucial for building trust and acceptance of these innovative solutions. By addressing safety concerns, emphasizing environmental benefits, and fostering trust through clear communication, stakeholders can promote nanobiopesticides as a viable and sustainable option for modern agricultural practices. This underscores the interconnected nature of regulatory efforts, public perception, and acceptance in ensuring the success of nanobiopesticides.

7. Future Perspectives and Research Directions

Each year, a significant number of yields are lost by pests, but the introduction of synthetic pesticides has helped reduce these losses. The harmful effects of synthetic pesticides restrict their extensive use, emphasizing the importance of advocating for biological pesticides as safer options. Given that biopesticides have proven to be effective substitutes for chemical pesticides, it is crucial to explore their full potential for agricultural use. The lack of residual environmental effects from biopesticides can be seen as both an advantage and a disadvantage. The advantage lies in the fact that biopesticides do not persist long enough to pose risks to plants, humans, or animals, unlike synthetic pesticides, which can remain hazardous. However, this short-lived activity is also a disadvantage because biopesticides only protect crops while they are in direct contact with the pests. Pests that invade after treatment will not be impacted, possibly necessitating extra treatments, resulting in increased costs and labor for farmers. At present, the demand for and availability of biopesticides are limited, which deters both producers and users. As a result, additional research is needed to investigate biocarriers and other sustainable approaches that could increase the longevity of biopesticides in the environment. Moreover, since biopesticides have a highly specific mode of action, there is a potential risk of chemical interactions when multiple biopesticides are used on a crop affected by various pests. These findings emphasize the need for research on the compatibility of various biopesticides when used together. Moreover, most research on biopesticides has focused on crop yield rather than the nutritional quality of the harvested produce. A deeper understanding of how biopesticides impact nutritional quality could further promote their use in agriculture (Ayilara et al., 2023).

Nanotechnology has the potential to make agriculture both eco-friendly and more profitable by reducing the reliance on chemical crop protection products. The development of nanodevices and products enables the smart delivery of fertilizers, pesticides, and growth regulators, as well as the use of nanosensors for real-time monitoring of soil conditions, crop growth, and pest or disease outbreaks. The method of nanomaterial synthesis plays a crucial role, as variations in the synthesis process can affect the size, shape, and associated risks of the resulting materials. As a result, risk assessment studies are vital before these materials are used, especially since there are no established guidelines for applying such formulations to nanomaterials. It is important to analyze the toxic effects of these compounds on plants and insects. A considerable amount of research is still needed on nanoenhanced pesticides before they can be widely adopted in pest management. This involves integrating analytical techniques capable of detecting and characterizing (such as assessing size, shape, and surface properties) and quantifying the active ingredients and adjuvants present in the formulations. Future studies should prioritize the development of biodegradable and biocompatible nanoparticles for pest management. These could be produced via the use of natural polymers, plant-based extracts, or biogenic materials, ensuring minimal harm to the environment. Biodegradability is crucial to prevent the long-term buildup of substances in soil and water, reducing ecological risks. To support this, offering grants or funding to producers, entrepreneurs, researchers, and marketers can promote the production and availability of biopesticides, encouraging their wider use in agriculture. The growing number of publications and successful case studies in this area indicate robust research and strong confidence in the potential of nanopesticide-based formulations such as nanoemulsions, nanodispersions, and nanoparticles. These innovations hold great promise for developing safer and more effective chemical pesticide formulations for pest control, which could lead to significant advancements and potentially transformative changes in the field. However, owing to the potential toxicity concerns associated with nanomaterials, which remain insufficiently standardized, understood, or explored, this development is likely to undergo thorough scrutiny by both international and national safety regulators, who will likely call for more research into the environmental and human health impacts of these materials. Despite extensive research on the plant-mediated synthesis of nanoparticles for pest control, a significant gap remains between theoretical knowledge and practical, large-scale applications.

While progress in this field is gradual, the future of nanotechnology in agriculture looks promising, much like its impact in other sectors (Athanassiou et al., 2018). By combining nanotechnology with sustainable practices, effective, precise, and environmentally friendly solutions can be created. Green nanotechnology is a rapidly evolving field with significant potential in agriculture, particularly for controlling plant pests. A key advantage of green nanotechnology lies in its ability to enable targeted pest control. Researchers are working on nanoparticles capable of delivering pesticides or biological agents directly to pest-infested areas. These systems can employ stimuli-responsive triggers, such as pH changes, temperature shifts, or pest-specific enzymes, to release active compounds only where needed, improving efficiency while reducing unintended impacts on nontarget organisms.

Nanobiopesticides can be taken up, stored, and distributed throughout plants via their roots and leaves. This enables the absorption of pesticides by plants, providing effective control against pests and plant diseases. Nanobiopesticides primarily enter plants through root absorption, and they can be transported upwards to the aboveground parts via transpiration, allowing for the control of pests above the soil surface by applying the pesticides below. However, the bioavailability, effectiveness, and toxicity of nanobiopesticides are largely influenced by factors such as particle stability, number, and size distribution. Additionally, sufficient attention should be given to understanding whether the transport properties of nanobiopesticides result in residues within plants. Further research is needed on leaf-targeted bioaccumulation and the dose biotransformation mechanisms of nanodelivery to establish precise risk assessment guidelines for nanobiopesticides in the future. Moreover, the majority of nanobiopesticides currently being developed are synthesized in laboratory settings. Producing highly stable nanobiopesticides on a large scale remains challenging due to the complexity and sophistication of the preparation process, which hinders industrialization. Consequently, there are no clear benefits in terms of equipment production and investment costs for nanobiopesticides compared with traditional pesticide formulations. As a result, establishing large-scale, standardized production methods for nanobiopesticides remains a critical research challenge. Researchers must devise straightforward and sustainable techniques for fabricating nanomaterials suitable for agricultural applications.

8. Conclusion

Nanotechnology represents a significant breakthrough in the management of pathogens and pests in crops. Nanopesticides are innovative and offer new possibilities for controlling plant pests and diseases. By enhancing the properties of traditional agrochemicals—such as solubility, stability, thermal resistance, bioavailability, and reduced toxicity—nanomaterials provide notable benefits. Several nanomaterials have demonstrated potential in plant protection, with some products already reaching the market. Although nanotechnology is a relatively new development in the plant protection field, its application for insect control has already begun. Nanoencapsulated insecticides help minimize environmental loss by releasing the active ingredients only at targeted locations. By harnessing the unique properties of nanomaterials, such as their increased surface area, controlled release mechanisms, and potential for targeted delivery, nanotechnology offers the possibility of more efficient and sustainable pest management strategies. Their diverse modes of action reduce the likelihood of resistance development. However, further research is needed to address concerns about safety, handling, overuse, and disposal. The inherently multidisciplinary and collaborative nature of nanotechnology holds promise for revolutionizing plant care. Additionally, the use of biogenic methods to synthesize nanoparticles derived from plants, microbes, and other natural sources, i.e., green nanosynthesis, ensures that these technologies align with the growing demand for eco-friendly solutions in agriculture. In essence, nanobased insecticides offer a promising alternative to traditional chemical insecticides, helping reduce the burden of environmental pesticides.

The integration of green nanotechnology into plant pest control represents a promising and innovative approach for managing agricultural pests while minimizing their environmental impact. As research progresses, the full potential of green nanotechnology in plant pest control will likely be realized, offering effective alternatives to conventional pesticides, reducing chemical residues, and promoting long-term environmental sustainability. However, further investigations into the toxicity, environmental fate, and regulatory frameworks of these nanomaterials are essential to ensure their safe and responsible application in agriculture. Ultimately, green nanotechnology has the potential to revolutionize pest control, contributing to more sustainable and resilient agricultural practices in the future. Green nanotechnology offers a revolutionary method for controlling plant pests, combining precision, effectiveness, and eco-friendliness. By addressing issues related to environmental impact, scalability, and regulatory challenges, future research can help realize its full capabilities. This emerging field is poised to play a vital role in developing sustainable agricultural systems that harmonize productivity with environmental care.

References

Abdollahdokht, D., Gao, Y., Faramarz, S., Poustforoosh, A., Abbasi, M., Asadikaram, G. et al. (2022). Conventional agrochemicals towards nanobiopesticides: An overview on recent advances. *Chemical and Biological Technologies in Agriculture*, 9(1): 13. https://doi.org/10.1186/s40538-021-00281-0.

Adholeya, A., and Singh, R. (2020). *Guidelines for Evaluation of Nano-Based Agri-Input & Food Products in India.*

Ale, A., Andrade, V. S., Gutierrez, M. F., Bacchetta, C., Rossi, A. S., Orihuela, P. S. et al. (2023). Nanotechnology-based pesticides: Environmental fate and ecotoxicity. *Toxicology and Applied Pharmacology*, 471: 116560. https://doi.org/10.1016/j.taap.2023.116560.

Aouada, F. A., and Moura, M. R. d. (2015). Nanotechnology applied in agriculture: controlled release of agrochemicals. *Nanotechnologies in Food and Agriculture*, 103–118. https://doi.org/10.1007/978-3-319-14024-7_5.

Athanassiou, C. G., Kavallieratos, N. G., Benelli, G., Losic, D., Usha Rani, P., and Desneux, N. (2018). Nanoparticles for pest control: Current status and future perspectives. *Journal of Pest Science*, 91(1): 1–15. https://doi.org/10.1007/s10340-017-0898-0.

Ayilara, M. S., Adeleke, B. S., Akinola, S. A., Fayose, C. A., Adeyemi, U. T., Gbadegesin, L. A. et al. (2023). Biopesticides as a promising alternative to synthetic pesticides: A case for microbial pesticides, phytopesticides, and nanobiopesticides. *Frontiers in Microbiology*, 14. https://doi.org/10.3389/fmicb.2023.1040901.

Bravo, A., Gill, S. S., and Soberón, M. (2007). Mode of action of *Bacillus thuringiensis* Cry and Cyt toxins and their potential for insect control. *Toxicon*, 49(4): 423–435. https://doi.org/10.1016/j.toxicon.2006.11.022.

Butt, T. M., Coates, C. J., Dubovskiy, I. M., and Ratcliffe, N. A. (2016). Chapter nine - entomopathogenic fungi: new insights into host–pathogen interactions. pp. 307–364. *In*: Lovett, B., and R. J. St. Leger (Eds.). *Advances in Genetics* (Vol. 94). Academic Press. https://doi.org/10.1016/bs.adgen.2016.01.006.

Chattopadhyay, P., Banerjee, G., and Mukherjee, S. (2017). Recent trends of modern bacterial insecticides for pest control practice in integrated crop management system. *3 Biotech*, 7(1): 60. https://doi.org/10.1007/s13205-017-0717-6.

Chaud, M. V., Souto, E. B., Zielińska, A., Severino, P., Batain, F., Oliveira-Júnior, J. F. d. et al. (2021). Nanopesticides in agriculture: benefits and challenge in agricultural productivity, toxicological risks to human health and environment. *Toxics*, 9(6): 131. https://doi.org/10.3390/toxics9060131.

Chaudhary, P., Chaudhary, A., Parveen, H., Rani, A., Kumar, G., Kumar, R. et al. (2021). Impact of nanophos in agriculture to improve functional bacterial community and crop productivity. *BMC Plant Biology*, 21(1): 519. https://doi.org/10.1186/s12870-021-03298-7.

Chaudhary, P., Khati, P., Chaudhary, A., Maithani, D., Kumar, G., and Sharma, A. (2021). Cultivable and metagenomic approach to study the combined impact of nanogypsum and *Pseudomonas taiwanensis* on maize plant health and its rhizospheric microbiome. PLOS ONE, 16(4): e0250574. https://doi.org/10.1371/journal.pone.0250574.

Chen, A., Halilovic, L., Shay, J., Koch, A., Mitter, N., and Jin, H. (2023). Improving rna-based crop protection through nanotechnology and insights from cross-kingdom rna trafficking. *Current Opinion in Plant Biology*, 76: 102441. https://doi.org/10.1016/j.pbi.2023.102441.

Chhipa, H. (2019). Chapter 6—Applications of nanotechnology in agriculture. pp. 115–142. *In*: V. Gurtler, A. S. Ball, and S. Soni (Eds.). Methods in Microbiology (Vol. 46). Academic Press. https://doi.org/10.1016/bs.mim.2019.01.002.

Chowardhara, B., Saha, B., Awasthi, J. P., Deori, B. B., Nath, R., Roy, S. et al. (2024). An assessment of nanotechnology-based interventions for cleaning up toxic heavy metal/metalloid-contaminated agroecosystems: Potentials and Issues. *Chemosphere*.359: 142178. https://doi.org/10.1016/j.chemosphere.2024.142178.

Copping, L. G., and Menn, J. J. (2000). Biopesticides: A review of their action, applications and efficacy. *Pest Management Science*, 56(8): 651–676. https://doi.org/10.1002/1526-4998(200008)56:8<651::AID-PS201>3.0.CO;2-U.

Dar, S. A., Wani, S. H., Mir, S. H., Showkat, A., Dolkar, T., and Dawa, T. (2021). Biopesticides: Mode of Action, Efficacy and Scope in Pest Management. Journal of Advanced Research in Biochemistry and Pharmacology, 4(1), Article 1.

Deb, P., Mandal, A., Harendra, Choudhury, S. R., Das, A., Hazra, S. et al. (2023). Improving plant nutrient use efficiency for climate-resilient agriculture. pp. 209–243. *In*: Hasanuzzaman, M. (Eds.). *Climate-resilient agriculture, vol 2: Agro-biotechnological advancement for crop production*. Cham: Springer International Publishing. https://doi.org/10.1007/978-3-031-37428-9_10.

Dhakal, R., and Singh, D. (2019). Biopesticides: A Key to Sustainable Agriculture. *International Journal of Pure & Applied Bioscience*, 7. https://doi.org/10.18782/2320-7051.7034.

Djiwanti, S. R., and Kaushik, S. (2019). Nanopesticide: future application of nanomaterials in plant protection. *Plant Nanobionics*, 255–298. https://doi.org/10.1007/978-3-030-16379-2_10.

Federici, B. A. (1997). Baculovirus Pathogenesis. pp. 33–59. *In*: Miller, L. K. (Ed.). The Baculoviruses. Springer US. https://doi.org/10.1007/978-1-4899-1834-5_3.

Fenibo, E. O., Ijoma, G. N., and Matambo, T. (2021). Biopesticides in sustainable agriculture: a critical sustainable development driver governed by green chemistry principles. *Frontiers in Sustainable Food Systems*, 5. https://doi.org/10.3389/fsufs.2021.619058.

Ganguli, P. (2019). Chapter 16—Patenting issues in the development of nanobiopesticides. *In*: Koul, O. (Ed.). *Nano-Biopesticides Today, and Future Perspectives* (pp. 367–395). Academic Press. https://doi.org/10.1016/B978-0-12-815829-6.00017-6.

Ghongade, D. S., and Sangha, K. S. (2021). Efficacy of biopesticides against the whitefly, Bemisia tabaci (Gennadius) (Hemiptera: Aleyrodidae), on parthenocarpic cucumber grown under protected environment in India. *Egyptian Journal of Biological Pest Control*, 31(1): 19. https://doi.org/10.1186/s41938-021-00365-x.

Ghosh, D., Brahmachari, K., Brestic, M., Ondrisik, P., Hossain, A., Skalicky, M. et al. (2020a). Integrated weed and nutrient management improve yield, nutrient uptake and economics of maize in the rice–maize cropping system of Eastern India. *Agronomy*, 10(12): 1906. https://doi.org/10.3390/agronomy10121906.

Ghosh, D., Brahmachari, K., Das, A., Sarkar, S., Dinda, N. K., Moulick, D. et al. (2023). Residual effect of organic manure on weed population and nitrogen uptake in greengram under rice–maize-greengram system. Indian *Journal of Weed Science*, 55(4): 374–379. http://dx.doi.org/10.5958/0974-8164.2023.00070.9.

Ghosh, D., Brahmachari, K., Sarkar, S., Dinda, N. K., Das, A., and Moulick, D. (2022). Impact of nutrient management in rice–maize-greengram cropping system and integrated weed management treatments on summer greengram productivity. *Ind. J. Weed Sci.*, 54(1): 25–30. https://doi.org/10.5958/0974-8164.2022.00004.1.

Ghosh, D., Brahmachari, K., Skalicky, M., Hossain, A., Sarkar, S., Dinda, N. K. et al. (2020b). Nutrients supplementation through organic manures influence the growth of weeds and maize productivity. *Molecules*, 25(21): 4924. https://doi.org/10.3390/molecules25214924.

Ghosh, D., Brahmachari, K., Skalický, M., Roy, D., Das, A., Sarkar, S. et al. (2022). The combination of organic and inorganic fertilizers influence the weed growth, productivity and soil fertility of monsoon rice. *PloS one*, 17(1): p.e0262586. https://doi.org/10.1371/journal.pone.0262586.

Gouda, M. R., and Ashwini, L. M. B. (2023). Emerging Role of Entomopathogenic Fungi in Nanoparticle Synthesis and their Application in Pest Control. https://meddocsonline.org/journal-of-plant-biology-and-crop-research/emerging-role-of-entomopathogenic-fungi-in-nanoparticle-synthesis-and-their-application-in-pest-control.pdf.

Guerrero Ramírez, J. R., Ibarra Muñoz, L. A., Balagurusamy, N., Frías Ramírez, J. E., Alfaro Hernández, L., and Carrillo Campos, J. (2023). Microbiology and Biochemistry of Pesticides Biodegradation. *Int. J. Mol. Sci.* 2023 Nov 4; 24(21): 15969. doi: 10.3390/ijms242115969. PMID: 37958952; PMCID: PMC10649977.

Gupta, N., Upadhyaya, C. P., Singh, A. P., Abd-Elsalam, K. A., and Prasad, R. (2018). Applications of silver nanoparticles in plant protection. Nanotechnology in the Life Sciences, 247–265. https://doi.org/10.1007/978-3-319-91161-8_9.

Harvesto Group. (2022). Eco-Friendly Pest Control: The Benefits of Using Bio-Pesticides in Agriculture. Retrieved from https://www.harvestogroup.com/post/eco-friendly pest-control-the-benefits-of-using-biopesticides-in-agriculture.

Hezakiel, H. E., Thampi, M., Rebello, S., and Sheikhmoideen, J. M. (2024). Biopesticides: A Green Approach Towards Agricultural Pests. *Applied Biochemistry and Biotechnology*, 196(8): 5533–5562. https://doi.org/10.1007/s12010-023-04765-7.

Holder, D. J., and Keyhani, N. O. (2005). Adhesion of the Entomopathogenic Fungus Beauveria (Cordyceps) bassiana to Substrata. *Applied and Environmental Microbiology*, 71(9): 5260–5266. https://doi.org/10.1128/AEM.71.9.5260-5266.2005.

Inceoglu, A. B., Kamita, S. G., Hinton, A. C., Huang, Q., Severson, T. F., Kang, K. et al. (2001). Recombinant baculoviruses for insect control. *Pest Management Science*, 57(10): 981–987. https://doi.org/10.1002/ps.393.

Islam, M. T. (2020). Application of Nanomaterials in Plant Protection (Doctoral dissertation, Doctoral dissertation, Bangabandhu Sheikh Mujibur Rahman Agricultural University Gazipur City, Bangladesh).

Jampílek, J., and Kráľová, K. (2019). Chapter 17 - Nanobiopesticides in agriculture: State of the art and future opportunities. pp. 397–447. *In*: Koul, O. (Ed.). Nano-Biopesticides Today, and Future Perspectives. Academic Press. https://doi.org/10.1016/B978-0-12-815829-6.00018-8.

Jasrotia, P., Nagpal, M., Mishra, C. N., Sharma, A. K., Kumar, S., Kamble, U. et al. (2022). Nanomaterials for postharvest management of insect pests: current state and future perspectives. *Frontiers in Nanotechnology*, 3. https://doi.org/10.3389/fnano.2021.811056.

Kachhawa, D. (2017). Microorganisms as a biopesticides. *J. Entomol. Zool. Stud.*, 5: 468–473.

Kashyap, P. L., Kumar, S., Kaul, N., Aggarwal, S. K., Jasrotia, P., Bhardwaj, A. K. et al. (2022). Nanotechnology for wheat and barley health management: current scenario and future prospectus. *New Horizons in Wheat and Barley Research*, 337-363. https://doi.org/10.1007/978-981-16-4134-3_12.

Kumar, S. (2012). Biopesticides: A need for food and environmental safety. *Journal of Biofertilizers and Biopesticides*, 3. https://doi.org/10.4172/2155-6202.1000e107.

Kumari, R., Suman, K., Karmakar, S., Mishra, V., Lakra, S. G., Saurav, G. K. et al. (2023). Regulation and safety measures for nanotechnology-based agri-products. *Frontiers in Genome Editing*, 5. https://doi.org/10.3389/fgeed.2023.1200987.

Lacey, L. A., and Siegel, J. P. (2000). Safety and ecotoxicology of entomopathogenic bacteria. pp. 253–273. *In*: Charles, J.-F., Delécluse, A., and Roux, C. N.-L. (Eds.). *Entomopathogenic Bacteria: From Laboratory to Field Application*. Springer Netherlands. https://doi.org/10.1007/978-94-017-1429-7_14.

Lyubenova, A., Rusanova, M., Nikolova, M., and Slavov, S. B. (2023). Plant extracts and *Trichoderma* spp: Possibilities for implementation in agriculture as biopesticides. *Biotechnology & Biotechnological Equipment*, 37(1): 159–166. https://doi.org/10.1080/13102818.2023.2166869.

Mahanta, S., Shree, J., Santra, S. C., Moulick, D., and Hossain, A. (2023). Deciphering of mycogenic nanoparticles by spectroscopic methods. pp. 93–117. *In*: *Kuddus*, M. (Eds.). *Myconanotechnology and Application of Nanoparticles in Biology*. Academic Press. https://doi.org/10.1016/B978-0-443-15262-7.00003-6.

Majumdar, A., Moulick, D., and Srivastava, S. (2024). 'Save Soil' by managing soil nutrient losses, agronomic practices and crop-microbial interaction: World Soil Day 2022. Frontiers in Sustainable Food Systems. 8: 1360937.

Mgadi, K., Ndaba, B., Roopnarain, A., Rama, H., and Adeleke, R. (2024). Nanoparticle applications in agriculture: overview and response of plant-associated microorganisms. *Frontiers in Microbiology*, 15. https://doi.org/10.3389/fmicb.2024.1354440.

Moore, N. F., King, L. A., & Possee, R. D. (1987). Viruses of insects. International Journal of Tropical Insect Science, 8(3), 275–289. https://doi.org/10.1017/S1742758400005257

Moscardi, F. (1999). Assessment of the application of baculoviruses for control of lepidoptera. *Annual Review of Entomology*, 44(Volume 44, 1999), 257–289. https://doi.org/10.1146/annurev.ento.44.1.257.

Moulick, D., Hossain, A., Barek, V., Chowardhara, B., Mukherjee, A., Pattnaik, B. K. et al.. (2024c) .Heavy metal stress in the agro-environment: consequences, adaptations and remediation. *International Journal of Environmental Science and Technology*, 21: 927–9340. https://doi.org/10.1007/s13762-024-05657-x.

Moulick, D., Khan, F. R., Siddiqui, N., Pradhan, J., Khamari, P., and Mahanta, S. (2024b). Applications of nanotechnology in abiotic stress management and biofortification. pp. 50–66. *In*: Roy, S., and Hossain, A. (Eds.). *The Nanotechnology Driven Agriculture*. CRC Press.

Moulick, D., Majumdar, A., Choudhury, A., Das, A., Chowardhara, B., Pattnaik, B. K. et al. (2024a.). Emerging concern of nanopollution in agro-ecosystem: flip side of nanotechnology. *Plant Physiol Biochem.*, 211: 108704. https://doi.org/10.1016/j.plaphy.2024.108704.

Moulick, D., Mukherjee, A., Mahanta, S., Chowdhara, B., Pattnaik, B. K., Mukhopadhyay, A. K. et al. (2025a). Nanocomposites and nanobiochar: a novel solution to alleviate heavy metal stress in crops. In *Nanomaterials and Nano-Biochar in Reducing Soil Stress* (pp. 183-224). Apple Academic Press.

Nidhi, Mushtaq, S., Saji, G., and Gupta, V. K. (2022). Mode of action of biopesticides against pests and future prospects of biopesticides and nanobiopesticides. *Journal of Advanced Scientific Research*, 13(11): Article 11. https://doi.org/10.55218/JASR.2022131101.

OECD. (2013). Regulatory Frameworks for Nanotechnology in Foods and Medical Products: Summary Results of a Survey Activity. OECD Publishing.

Olson, S. (2015). An analysis of the biopesticide market now and where it is going. *Outlooks on Pest Management*, 26(5): 203–206. https://doi.org/10.1564/v26_oct_04.

Pan, X., Guo, X., Zhai, T., Zhang, D., Rao, W., Cao, F. et al. (2023). Nanobiopesticides in sustainable agriculture: Developments, challenges, and perspectives. *Environmental Science: Nano*, 10(1): 41–61. https://doi.org/10.1039/D2EN00605G.

Pattnaik, B. K., Sahu, C., and Moulick, D. (2024). Nanoparticles in soil–water–plant inference. pp. 85–105. *In*: Roy, S., and Hossain, A. (eds.). *The Nanotechnology Driven Agriculture*. CRC Press.

Pavela, R., and Benelli, G. (2016). Essential oils as ecofriendly biopesticides? Challenges and Constraints. *Trends in Plant Science*, 21(12): 1000–1007. https://doi.org/10.1016/j.tplants.2016.10.005.

Rajamani, M., and Negi, A. (2021). Biopesticides for pest management. pp. 239–266. *In*: Venkatramanan, V., Shah, S. and Prasad, R. (eds.). *Sustainable Bioeconomy: Pathways to Sustainable Development Goals*. Springer. https://doi.org/10.1007/978-981-15-7321-7_11.

Roy, D., Ghosh, D., Chander, S., Saha, B., and Moulick, D. (2024). Herbicide Contamination in Ground Water and Their Probable Remediation. *In Environmental Contaminants*. Apple Academic Press.

Roy, S., and Hossain, A. (eds.). 2024. The Nanotechnology Driven Agriculture: The Future Ahead. CRC Press.

Roy, D., Ghosh, S., Datta, D., Sreekanth, D., Pawar, D., Mukherjee, P. K. et al. (2023). Climate resilient weed management for crop production. pp. 125–156. *In*: Mirza Hasanuzzaman (eds.). *Climate-Resilient Agriculture*, Vol 2: Agro-Biotechnological Advancement for Crop Production. Cham: Springer International Publishing. https://doi.org/10.1007/978-3-031-37428-9_7.

Sharma, R., and Sharma, P. (2021). Fungal entomopathogens: A systematic review. *Egyptian Journal of Biological Pest Control*, 31(1): 57. https://doi.org/10.1186/s41938-021-00404-7.

Shirke A. (2023). Biopesticide Importance and Advantages in Indian Agriculture. Retrieved from https://www.linkedin.com/pulse/biopesticide-importance-advantages-indian-abhijeet-shirke-sm83f.

Singkaravanit, S., Kinoshita, H., Ihara, F., and Nihira, T. (2010). Cloning and functional analysis of the second geranylgeranyl diphosphate synthase gene influencing helvolic acid biosynthesis in Metarhizium anisopliae. *Applied Microbiology and Biotechnology*, 87(3): 1077–1088. https://doi.org/10.1007/s00253-010-2556-9.

Sparks, T. C., and Nauen, R. (2015). IRAC: Mode of action classification and insecticide resistance management. Pesticide Biochemistry and Physiology, 121: 122–128. https://doi.org/10.1016/j.pestbp.2014.11.014.

Syafrudin, M., Kristanti, R. A., Yuniarto, A., Hadibarata, T., Rhee, J., Al-onazi, W. A. et al. (2021). Pesticides in Drinking Water—A Review. *International Journal of Environmental Research and Public Health*, 18(2), Article 2. https://doi.org/10.3390/ijerph18020468.

Tadesse Mawcha, K., Malinga, L., Muir, D., Ge, J., and Ndolo, D. (2024). Recent Advances in Biopesticide Research and Development with a Focus on Microbials. *F1000Research*, 13: 1071. https://doi.org/10.12688/f1000research.154392.3.

Tamura, H. (2024). Bacterial Pesticides: Mechanism of Action, Possibility of Food Contamination, and Residue Analysis Using MS. *Journal of Pesticide Science*, 49(3): 135–147. https://doi.org/10.1584/jpestics.D24-006.

Thacker, J. R. M., and Jonathan R. M. (2002). An introduction to arthropod pest control. Cambridge University Press. https://cir.nii.ac.jp/crid/1130282270675101312.

Thapa, A., Tamang, J., and Rai, M. (2020). Viral Insecticides In Sustainable Agriculture: *A. TTPP*, 2020, 329.

Tyagi, S., Yadav, P., Chakraborty, A., Majumdar, A., Moulick, D., Santra, S. C. et al. (2023). Myconanotechnologies: an approach towards sustainable agriculture. pp. 139–157. *In*: Kuddus, M. (eds.). *Myconanotechnology and Application of Nanoparticles in Biology*. Academic Press. https://doi.org/10.1016/B978-0-443-15262-7.00005-X.

Umaru, F. F., and Simarani, K. (2020). Evaluation of the potential of fungal biopesticides for the biological control of the Seed Bug, Elasmolomus pallens (Dallas) (Hemiptera: Rhyparochromidae). Insects, 11(5), Article 5. https://doi.org/10.3390/insects11050277.

US EPA (2023). What are Biopesticides? Retrieved from https://www.epa.gov/ingredients-used-pesticide-products/what-are-biopesticides.

van Lenteren, J. C. (2012). The state of commercial augmentative biological control: Plenty of natural enemies, but a frustrating lack of uptake. BioControl, 57(1): 1–20. https://doi.org/10.1007/s10526-011-9395-1.

Verma, D. K., Guzmán, K. N. R., Mohapatra, B., Talukdar, D., Chávez-González, M. L., Kumar, V. et al. (2021). Recent trends in plant- and microbe-based biopesticide for sustainable crop production and environmental security. pp. 1–37. *In*: Prasad, R., Kumar, V., Singh, J., and Upadhyaya, C. P. (eds.). Recent Developments in Microbial Technologies. Springer Nature. https://doi.org/10.1007/978-981-15-4439-2_1.

Verma, Neelendra Singh, Kuldeep, Dinesh, Chouhan, Monika, Prajapati, Rajkumar and Singh, Satish. (2023). A Review on Eco-Friendly Pesticides and Their Rising Importance in Sustainable Plant Protection Practices. *International Journal of Plant & Soil Science*, 35: 200–214. 10.9734/ijpss/2023/v35i224126.

Wernet, V., and Fischer, R. (2023). Establishment of Arthrobotrys flagrans as biocontrol agent against the root pathogenic nematode Xiphinema index. Environmental Microbiology, 25(2): 283–293. https://doi.org/10.1111/1462-2920.16282.

Xu, Z., Tang, T., Qin, L., Yu, J., Zhang, C., Zhao, X. et al. (2022). Environmental risks and the potential benefits of nanopesticides: a review. *Environmental Chemistry Letters*, 20(3): 2097–2108. https://doi.org/10.1007/s10311-021-01338-0.

Yadav, J., Jasrotia, P., Kashyap, P. L., Bhardwaj, A. K., Kumar, S., Singh, M. et al. (2021). Nanopesticides: Current status and scope for their application in agriculture. *Plant Protection Science*, 58(1): 1–17. https://doi.org/10.17221/102/2020-PPS.

Yan, S., Ren, B., and Shen, J. (2020). Nanoparticle-mediated double-stranded rna delivery system: a promising approach for sustainable pest management. Insect Science, 28(1): 21–34. https://doi.org/10.1111/1744-7917.12822.

Yousef, H. A., Fahmy, H. M., Arafa, F. N., Abd Allah, M. Y., Tawfik, Y. M., El Halwany, K. K. et al. (2023). Nanotechnology in pest management: advantages, applications, and challenges. International Journal of Tropical Insect Science, 43(5): 1387–1399. https://doi.org/10.1007/s42690-023-01053-z.

Zaheer, T., Ali, M. M., Abbas, R. Z., Atta, K., Amjad, I., Suleman, A. et al. (2022). Insights into Nanopesticides for Ticks: The Superbugs of Livestock. *Oxidative Medicine and Cellular Longevity*, 2022(1): 7411481. https://doi.org/10.1155/2022/7411481.

CHAPTER-13

Multi-Omics of Nanotechnology-Mediated Abiotic Stress Resilience in Crops

Saher Fatima,[1] *Shaharyar,*[1] *Saira Zaheer,*[1] *Huma Arshad,*[1]
Sunbal Khalil Chaudhari,[2] *Murtaza Hasan*[3,*] and *Ghazala Mustafa*[1,*]

1. Introduction

Globally, cereals are considered staple crops and reliable sources of calories, especially in underdeveloped nations where they provide 60% of total caloric intake (Haider et al., 2023). However, these crops, essential for nutrition and the economy, are highly vulnerable to climate change due to limited extreme adaptation features (Mao et al., 2023). The IPCC predicts a disruption in rainfall patterns, potentially leading to an increase in worldwide temperatures of up to 2°C by 2050 and over 5°C by 2100 compared to pre-industrial levels (Reynolds et al., 2021).

Drought and flooding, two major abiotic stresses, are occurring more frequently due to climate change, causing over 70% of global crop productivity losses (Francini and Sebastiani, 2019). From 1980 to 2015, droughts reduced maize production and wheat yield by 40% and 21%, respectively (Zhu et al., 2022). Flooding leads to hypoxia or anoxia in plants, which severely hampers growth, development, and survival (Chugh et al., 2022). Cereals suffer oxidative stress and protein carbonylation due to drought and flooding, which also affect the protein function. Energy metabolism is disrupted by ethanol and lactic acid buildup under flooding, while malondialdehyde (MDA) builds up and damages membranes. The over-activation of defense pathways by transcription factors such as WRKY and NAC, acting as stress responsive factors, might result in decreased biomass and growth inhibition. Although plants have evolved morphological, physiological, and molecular pathways to manage stress, they still struggle with increasingly frequent droughts and floods (Villalobos-López et al., 2022). To ensure sustainable agriculture and achieve the SDG of "Zero Hunger," scientists are developing stress-resistant crops (Toulotte et al., 2022).

Recent innovative technologies like nanomaterials have become a viable tool for sustainable agriculture in response to abiotic challenges caused by climate change (Rizwan et al., 2017). Not only

[1] Department of Plant Sciences, Quaid-i-Azam University, Islamabad 45320, Pakistan.
[2] Institute of Molecular Biology and Biotechnology, The University of Lahore, Sargodha Campus, Sargodha 42100, Pakistan.
[3] Department of Biotechnology, Faculty of Chemical and Biological Sciences, The Islamia University of Bahawalpur, 63100, Pakistan.
* Corresponding authors: mghazala@qau.edu.pk; murtaza@iub.edu.pk

DOI: 10.1201/9781003567684-13

plant morphology but physiology is also impacted with the help of nanoparticles (NPs) in accordance with the kind of plant variety, application technique, exposure duration, and dosage. By regulating transcriptome, metabolomics, and proteomics responses, they trigger many pathways, such as the MAPK pathway, which in turn governs genes associated with plant hormones and LEA proteins. NPs raise the amounts of flavonoids and improve antioxidant enzymes (POD, SOD, APX-GSH), as well as transcription factors such as HSFs, bZIP, bHLH, and MYB. In addition, they affect the amino acid metabolism, tricarboxylic acid cycle, starch breakdown, glycolysis, and proteins involved in photosynthesis, stress response, and ATP synthesis (Chen et al., 2024). Managing crop stress using nanoparticles has been employed to lessen the negative impacts caused by temperature, salinity, flooding, and drought. Nanoparticles are helpful for plants to overcome abiotic stresses at lower concentration as compared to their bulk or ionic compartments (Khan et al., 2022). Like SiO_2-NPs are anti-stressors, by improving their antioxidant activity while removing excessive Pb from *O. basilicum* (Ullah et al., 2023). AgNP exposure at different levels causes a significant increase in plant biomass and can be used for their antioxidant and antibacterial activities as compared to control (Zulfiqar et al., 2022). Another study exhibits positive effects of SiO_2 NPs at higher concentrations and somewhat reduce the growth at minor concentrations even in normal growth of mung beans (Zaheer et al., 2023). The potential of Fe_2O_3-NPs to improve the performance of *T. aestivum and Oryza sativa* by enhancing the antioxidant enzyme activities and chlorophyll content (Zafar et al., 2024; Ullah et al., 2024). Assessing NPs responsive alterations in gene expressions and proteome profiles of crop plants, especially under stress conditions, constitutes a vigorous approach regarding the production of stress tolerant crops (Mustafa et al., 2022).

However, due to a lack of knowledge about their safe utilization, the growing use of engineered nanomaterials (ENMs) in agriculture poses potential dangers to people, animals, and plants. As a result, their use in food production has been limited. The need for more precise mechanistic data has been brought to light by recent research on how plants react to engineered nanomaterials (ENMs). Hence, due to advancements in analytical techniques, the study of ENM-biota has moved from low-throughput bioassays to high-throughput omics methods (Majumdar et al., 2021). Omics methods are essential to understand how plant stress tolerance is enhanced by engineered nano-materials in nano-toxicity studies and in precision agriculture. This chapter combines data on how abiotic stress is increased by using ENMs, in association with metabolomics, transcriptomics, and proteomics.

2. Role of Omics in Understanding NPs Influence on Plant Cellular Responses

Water is crucial for growth of plants, though its needs vary among different species. Mechanisms like drought avoidance, escape, and tolerance are common in plants to deal with dry environments. Plants may also go dormant, fasten their life cycle, store water in succulent tissues, grow enlarged root systems, or decrease the surface area on their leaves by increasing stomatal control to prevent evapotranspiration (Ngara et al., 2021). Contrarily, flooding hinders metabolic pathways and mitochondrial respiration by retarding oxygen diffusion (Zhou et al., 2020). Nanoparticles (NPs) enhance resistance to drought and floods by a number of uses, such as leaf therapy, hydroponics, seed therapy or irrigation (Fiol et al., 2021). NPs are taken up by leaves via cuticles or stomata, while roots absorb them by apoplastic or symplastic routes which depends on their size (Avellan et al., 2019).

Transcriptomics, metabolomics, and proteomics collectively provide information about how NPs affect genes, metabolites, and proteins that are sensitive to abiotic stresses. These pressures have a major impact on crop yields, particularly those of cereals (Khater et al., 2023; Li and Keller, 2023). For example, proteomics demonstrates how NPs boost ROS-detoxifying enzymes like SOD for oxidative stress protection and ABA-responsive genes encourage stomatal closure to conserve water (Table 1). Even though these molecular pathways are intricate, integrated multi-omics methods

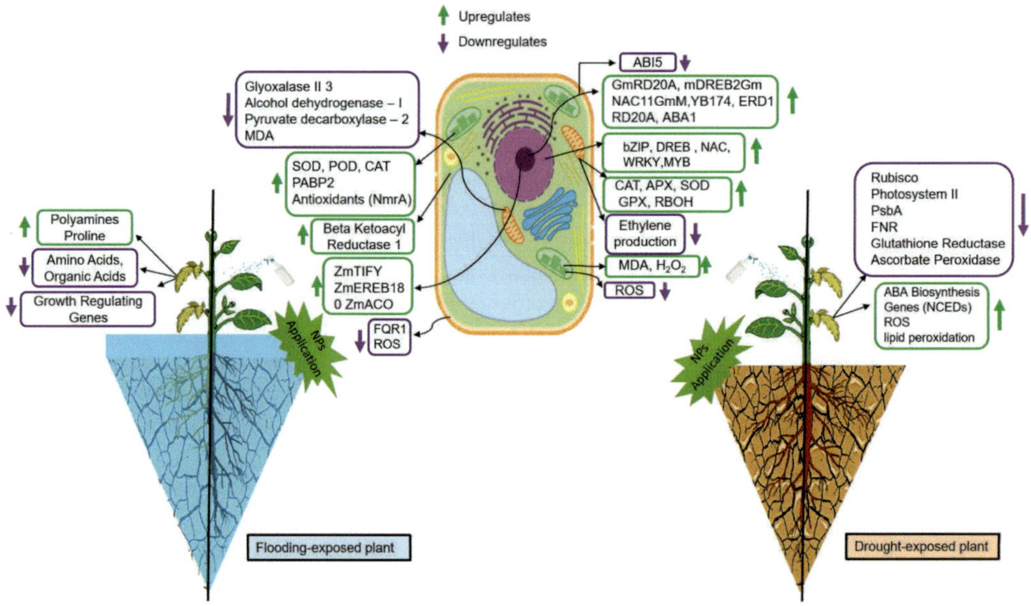

Figure 1. Diagrammatic illustration of how the application of nanoparticles (NP) affects plant responses to floods and drought stress. The downregulation and overexpression of important regulatory elements are highlighted in the diagram, which depicts the alterations at the molecular, physiological, biochemical levels. The varied responses that nanoparticles cause to improve stress tolerance in different environmental settings are illustrated by the display of specific transcription factors (e.g., bZIP, DREB, NAC, WRKY, and MYB) and cellular components involved in stress adaption.

help us understand them better. Some of the molecular pathways that are targeted by nanoparticles during drought and flooding are discussed below (Fig. 1).

Transcriptomic Response to Nanotechnology mediated Abiotic Stress

The large-scale examination of an organism's expressed RNAs at a certain point in its growth stage or in reaction to an external stimulus is known as transcriptomics. Even though mRNAs are translated into proteins, non-coding sequences (for example, miRNAs) also perform essential regulatory roles in normal plant growth, including stress responses. Transcripts can be studied by utilizing a wide range of techniques, that include RNA sequencing (RNA-seq), RNA microarray, reverse transcriptase quantitative PCR (RT-qPCR, Expressed Sequence Tags (ESTs), northern blotting and hybridization techniques, and Serial Analysis of Gene Expression (SAGE). The two methods that are currently most frequently employed in investigations of the plant transcriptome are microarrays and RNA-seq approaches, with each having advantages and disadvantages (Lowe et al., 2017). Among cereal crops, wheat (Triticum aestivum) is a significant food crop, which makes it crucial to investigate the molecular mechanisms behind wheat's abiotic, particularly drought stress response. Recent studies have utilized transcriptome analysis to understand the genetic mechanisms underlying drought resistance in wheat and explored the potential of nanoparticles (NPs) to enhance this tolerance. Truman and Deguo are the names of two wheat cultivars used in the study by (Niu et al., 2023), under drought stress, respectively. These genes were linked to antioxidant mechanisms, flavonoid biosynthesis, and plant hormone signaling controlled by transcription factors bHLH and bZIP. Their activity improves resistance to drought, providing wheat with the ability to adapt to stress.

Various studies have demonstrated that metal and metal oxide nanoparticles, such as Fe, Cu, ZnO and Co NPs, significantly affect the expression of drought-responsive genes through pathways like ABA-dependent and ABA-independent. The plants roots and shoots showed increased expression of drought tolerant marker genes including GmRD20A, GmDREB2, GmERD1, GmFDL19, GmNAC11,

GmWRKY27, GmMYB118, and GmMYB174 by NPs application (Linh et al., 2020). These recent studies by Kandhole et al. (2022), and Yang et al. (2018) have explored the potential of nanoparticles to enhance drought tolerance and demonstrated that NPs upregulate Transcription Factors (TFs) and genes such as GmDREB2 (belonging to dehydration responsive elements family), GmWRKY27, and ZIP1, that are crucial for drought response in plants, i.e., the transcription factor GmDREB2 binds to DRE/CRT sites to regulate gene expression in response to drought, salt, and cold stress, improving osmotic balance, GmWRKY27 promotes the secondary metabolites synthesis, lignin, and hormone signaling, under stress, and ZIP1 maintains homeostasis and nutritional absorption by controlling the transport of iron and zinc and also influences photosynthesis and metabolism which provides stress resistance. Raeesi Sadati et al. (2021) and Sun et al. (2020) found that NPs stimulate antioxidant enzymes, aiding oxidative stress mitigation. Therefore the transcriptome analysis has proved to be effective in understanding what genes and transcription factors are regulating the abiotic stress responses.

Research by Mustafa et al. (2015) explain that oxidative stress caused due to flooding is alleviated by silver nanoparticles (AgNPs). The bZIP transcription factors are upregulated that cause the genes CAT, APX, and SOD to become functional. These genes encode antioxidant enzymes that are necessary for detoxicate Reactive Oxygen Species (ROS) in waterlogged environments. Moreover, bZIP control stomatal closure and water balance by influencing the abscisic acid (ABA) pathways to cease loss of water during stress. To alter the hypoxia caused by water logging, nanoparticles also stimulate transcription factors like the ERF (Ethylene Response Factors), which regulates the expression of the ACO and ACS genes. These two genes are vital for the synthesis of ethylene. Ethylene regulation increases plant life and survival in submerged conditions by avoiding flood-induced cell death (Ban et al., 2023). The WRKY transcription factors are important for managing oxidative stress and pathogen protection; two processes that are made worse by waterlogging. WRKY helps maintain the cellular homeostasis which is done by controlling the stress responsive genes expression. Nanoparticles further increase the expression of the WRKY gene, which is why they help in stress mitigation. By activation of these secondary metabolites like flavonoids and phenols acting as antioxidants are produced to help mitigate oxidative damage. As a result, NPs promote greater recovery and development, and improve the plant's ability to tolerate oxidative stress during difficult circumstances (Munir et al., 2023).

1. Proteomic Response to Nanotechnology Mediated Abiotic Stress

A thorough understanding of proteomics is essential for comprehending the complex behavior of proteins in stressed plants, by identifying the important proteins that help plants to adapt in environmental challenges (Chaturvedi et al., 2016) Understanding the protein expression regulation by nanoparticles enlighten the pathways that improve plant resistance under stress conditions. Flooding and nanoparticle exposure genetically modifies some target proteins to enhance plant resistance to stress. Mustafa and Komatsu (2016), used gel-free proteomic analysis that showed flooding had a significant impact on 1541 proteins, while 1360, 1856, and 1668 proteins were changed for 5 nm, 30–60 nm, and 135 nm NPs application respectively. Venn diagram functional analysis explains that proteins involved in amino acid metabolism, cell functions (e.g., transport and organization), and protein metabolism (e.g., synthesis and degradation) showed notable changes. In particular, 519 protein metabolism proteins were impacted by 30–60 nm NPs. Furthermore, proteins involved in redox and nucleotide metabolism were affected by 5 nm NPs, signaling and photosystem proteins were improved by 30–60 nm NPs, and transport-related proteins were raised by 30% by 135 nm NPs. This demonstrates how exposure to nanoparticles and flooding stress in soybean proteomes interact in intricate, size-dependent ways providing stress management. In addition, wheat treated with AgNPs accumulated metabolism and cell defense-related proteins in the shoot and root, indicating a defense mechanism against the stress of nanoparticles (Vannini et al., 2014). This experiment explains that in plants treated with AgNP, there was a differential expression of 12 proteins in the shoots and 27 in the roots. Energy generation is supported by essential metabolic enzymes such as malate dehydrogenase, glyceraldehyde-3-phosphate dehydrogenase, and enolase. Oxidative stress is lessened by antioxidants like SOD, peroxidase, and GST. While expansins and actin-related proteins preserve the integrity of the cell wall, HSP70 and HSP90 fix misfolded proteins. Signaling proteins like calmodulins and ABC transporters control stress reactions and detoxification, whereas defense proteins like PR proteins and protease inhibitors improve protection. By regulating oxidative damage, preserving structures, and managing energy, these proteins work together to help plants adapt to stress caused by nanoparticles.

Proteomic analysis of soybeans treated with silver nanoparticles revealed that by regulating the production of stress-responsive proteins and chaperones, nanoparticles inhibit excessive protein breakdown and support metabolic stability in situations where there is a water shortage, suggesting that silver NPs may improve growth under flooding stress by controlling protein quality. When silver nanoparticles were combined with nicotinic acid and KNO_3 on soybean roots during flooding stress, 6340 proteins were identified, 351 of which had notable expression alterations. The primary findings showed that 60 of the proteins that were differentially expressed had to do with protein metabolism. Notably, there were increases in stress response proteins, including methyltransferases and protein degradation, i.e., RRM domain-containing proteins, which suggested improved adaptation mechanisms. 14 proteins increased that were related to transport, including aldo-ket-reductase domain-containing proteins. These proteins improved signaling pathways for stress and also for nutrients. These changes in proteins have a strong chemical reaction that increases soybeans' ability to withstand flooding (Hashimoto et al., 2020). Furthermore, nanoparticles such as titanium dioxide (TiO_2) and silver enhance protein synthesis, as a result stabilizing cellular structures and sustaining energy levels by ATP production pathways during stressed circumstances (Mustafa et al., 2022). BS NFs ZnO might improve soybean growth by increasing the protein folding due to HSP70 accumulation and redox metabolism through detoxification of hydrogen peroxide. While soybeans treated with CS NWs ZnO increase the oxidative stress and loss of protein folding (Mustafa et al., 2023).

These results highlight that silver nanoparticles control the quality of protein in soybeans and various metabolic pathways in wheat, eventually supporting plants to grow and develop in stressful conditions. Aluminum oxide nanoparticles (NPs) increased proteins in soybean roots recovering from floods, these proteins were involved in protein synthesis, transport, and development, and their function suggested that they help in stress recovery (Mustafa et al., 2015). In plant proteomics studies copper nanoparticles, or CuNPs are also reported to have an impact on functional proteins. Yasmeen et al. (2018), studied wheat seedlings treated with copper nanoparticles that showed CuNPs decreased proteins linked to photosynthesis and tetrapyrrole production and increased proteins involved in the tricarboxylic acid cycle and glycolysis, i.e., ribulose bisphosphate carboxylase (RuBisCO) essential for CO_2 fixation in photosynthesis, chaperonins helping in proper folding of protein to ensure accurate protein functioning, and proteasome subunits involved in protein degradation, for removing the damaged proteins. This is because plants lower the production of tetrapyrrole-linked and photosynthesis-related proteins in response to stress circumstances like drought or flooding, as photosynthesis becomes less efficient and they increase the TCA cycle and glycolysis to produce ATP, which is vital for sustaining metabolic processes and driving important operations. These mechanisms supply the building blocks needed to create osmolytes, which aid in cell water retention. Furthermore, the energy produced maintains antioxidant mechanisms that counteract hazardous reactive oxygen species (ROS), enhancing the plant's survival. A study that used Gel-free proteomics analysis in wheat grains treated with NPs was conducted that demonstrated the influence of CuNPs on metabolism. There was an increase in glycolysis and starch breakdown proteins, increasing energy production for stress adaptation, when treated with NPs (Zaheer et al., 2024). Cerium Oxide (CeO_2) nanoparticles have been shown to stabilize proteins, such as heat shock proteins (HSP70), and antioxidant enzymes, including ascorbate peroxidase and catalase. They prevent cellular damage under adverse conditions like drought by providing the plant's defense against oxidative stress (Preetha et al., 2023). Moreover, in plant roots zinc oxide (ZnO) nanoparticles are reported to down regulate proteins associated with the oxidation-reduction cascade, such as quinone reductase, galactose oxidase, SKU5-like proteins, and GDSL motif lipase 5 (Zhao et al., 2012). To summarize, NPs influence protein quality control, metabolism, stress signaling, and antioxidant defenses under abiotic stresses.

2. Metabolomic Response to nanotechnology mediated abiotic stress

Metabolomics focuses on the secondary metabolite profiles and physiological processes that are modified under stressed circumstances. It is a large-scale exploration of tiny molecules i.e., metabolites. This field is developing rapidly and three important methods being employed in plant metabolomics research that help in providing a precise knowledge of stress responses at various developmental stages are targeted analysis, metabolite profiling, and metabolic fingerprinting. The use of these techniques is driving the area of plant metabolomics ahead. Depending on the dose and type of NP used, nanoparticles (NPs) have an impact on secondary plant metabolism and they can reduce the effect of drought and flooding by either generating oxidative bursts or ROS scavenging (Iqbal et al., 2020). Reduced detrimental effects? of stressors on plant growth are dependent on the NP-induced influence of the metabolite profile (Ali et al., 2021). Many studies during abiotic stress gave evidence about nanoparticles efficiency in improving metabolites and causing resistance against these stresses. As described earlier flooding is the major abiotic stress hindering plant growth by causing an oxygen shortage and the generating toxic compounds. NPs help mitigate these consequences by enhancing plant survival during flooding and lowering stress-induced damage. According to recent research by Lu et al., 2024, rice resilience to floods was enhanced when chitosan oligosaccharides increased proline levels and reduced traumatic acid, as a result lessening physical damage. This study also revealed that metabolites which serve as anaerobic respiration energy stores, including glucose and sucrose, also proline and gamma-aminobutyric acid (GABA), which stabilize proteins and regulate pH levels, are influenced by NPs.

Likewise, during flooding period in spinach, H2O2 and MDA levels were decreased by Nitric Oxide (NO) treatment, lowering oxidative damage, and increases plant resilience to stress (Kiymaci et al., 2024). Drought stress has the same oxidative damage on plants as a result of ROS over accumulation. Several studies have demonstrated that NPs help mitigate these effects by improving key metabolite levels during drought too. For instance, in wheat and rice, Si NPs raise the proline content and also aid in the synthesis of glycine betaine, trehalose, and other osmolytes that promote water retention and stabilize cell structure. NPs also encourage the buildup of carbohydrates (mannitol and trehalose) for preserving membranes and balancing water content. Drought itself also causes an increase in secondary metabolites, such as flavonoids and phenolic compounds, which serve as signaling molecules and oxidative defense but NPs boost them for greater tolerance. Protective proteins such as Late Embryogenesis Abundant (LEA) proteins are synthesized by NPs, which in turn increase drought tolerance, by raising level of abscisic acid (ABA), and causing stomatal closure to limit the water loss (Elshyb et al., 2021). ZnO NPs are also reported to enhance proline, soluble sugars, and antioxidant enzyme activity (Sadati et al., 2024). Additionally, plant resistance to drought stress is increased by selenium (Se) and nanoceria NPs, which also boost stress-related proteins and antioxidant activities (Zeeshan et al., 2024; Djanaguiraman et al., 2018). Hence, metabolic content was greatly influenced by nanoparticle addition, which offered insights into a potential strategy for not only these certain crops but also for other commercial cereal crops that are susceptible to varying environmental conditions. Some of the metabolic responses of plants to NP-influenced stress mitigation are given in the table below (Table 1).

3. Cellular Pathways

There are various signaling pathways that are influenced by nanoparticles during abiotic stress conditions, in which ABA and ethylene signaling pathways play an important role. The study of these pathways by omics makes it easier to be acknowledged by researchers. ABA and ethylene are crucial signaling molecules that act as stress hormones. The ROS interaction with gibberellic

Table 1. Nanoparticles influence on Plants' Omics during Abiotic stress.

Plant	NPs	Stress	Response	References
Sorghum	Nanoceria (10 mg L–1)	Drought	Amount of H_2O_2 (36%), SOD (41%) decrease.	(Djanaguiraman et al., 2018)
Soybean	Selenium	Drought	Boosted the levels of osmolytes such as soluble sugar and proline.	(Zeeshan et al., 2024).
Lallemantia iberica	TiO2NPs	Severe Drought	Increase flavonoid and phenolic compounds Decrease MDA level.	(Alabdallah et al., 2021)
Wheat	Selenium	Drought	Significant increase in the accumulation of total soluble sugars (33%) and free amino acids (118%),	(Nawaz et al., 2015)
Rice	50 mg kg^{-1} SiO_2	Drought	Lignin, total phenolics, and proline in leaves increase by 11–126%, while JA content in stem and SA content in leavesincrease, respectively, by 142.5% and 37.6%.	(Wang et al., 2023)
Finger millet	Ca-U NPs	Drought	Increased proline by 35.4%	(Mishra et al., 2023)
Spinach	NO	Flooding	Proline increased, help stress tolerance and osmotic adjustment.	(Kıymacı et al., 2024)
Rice	Chitosan Oligosaccharides	Flooding	Increase proline, tryptophan. Decrease traumatic acid, IAA, JA, melatonin	(Lu et al., 2024)
Maize	O-CMC-NPs	Flooding	Decrease MDA content	(Wu et al., 2024)
Soybean	Aluminum oxide	Flooding	1360, 1856, and 1668 proteins for 5 nm, 30–60 nm, and 135 nm NPs application were changed respectively, that were related to protein metabolism, photosystem and redox homeostasis.	(Mustafa et al., 2016)
Soybean	Silver NPs with KNO_3 and nicotinic acid	Flooding	increase in stress response proteins including methyltransferases and protein degradation i.e. RRM domain-containing protein, was observed. 14 transport-related proteins including aldo-ket-reductase domain-containing protein also rose, indicating enhanced signaling pathways for nutrients and stress.	(Hashimoto et al., 2020)
Wheat	CuNPs	Abiotic stress	decreased levels of proteins linked to photosynthesis and tetrapyrrole production and higher levels of proteins involved in the tricarboxylic acid cycle and glycolysis i.e. Ribulose bisphosphate carboxylase (RuBisCO), Chaperonins, and Proteasome subunits involved in protein degradation, removing damaged proteins.	(Yasmeen et al., 2018)
Sorghum	CeO_2 Nps	Drought	elevate the plant's defense against oxidative stress, by stabilizing proteins, such as heat shock proteins (HSP70) and antioxidant enzymes, like catalase and ascorbate peroxidase, and preventing cellular damage	(Preetha et al., 2023)
Soybean	AgNPs	Flooding	upregulating bZIP transcription factors which activate genes SOD, CAT, and APX and regulate ABA pathway	(Mustafa et al., 2015)
Soybean	Fe, Co, Cu, ZnO NPs	Drought	increase expression of drought tolerance marker genes (GmRD20A, GmDREB2, GmERD1, GmFDL19, GmNAC11, GmWRKY27, GmMYB118, and GmMYB174).	(Linh et al., 2020)

acid and abscisic acid influences the seed dormancy and programmed cell death in the aleuronic cell layer through seed germination (Arshad and Mustafa, 2023). Nanoparticles influence these molecules to provide resistance against abiotic stresses. Transcriptome analysis reveals that through ethylene response factors (ERFs) and ABA-Responsive Elements (ABREs), the expression of stress-responsive genes is modified to provide resistance. ABA and ethylene, both interact with each other and provide the desired reaction against abiotic stresses. Nanoparticles improve the ethylene level which increases the synthesis of ABA, creating a positive feedback loop. This feedback mechanism supports stress tolerance and glucose metabolism. Transcription factors ERFs modify ABA-responsive genes during ethylene signaling, as a result ABA-related stress responses like ROS detoxification and stomatal regulation are produced during both stresses (drought/flooding). The rice ABA signaling genes OsABI5, Oshox22, OsNAC5, and DSM2 increase yield potential, which were discovered through ABA biosynthesis under drought stress (Basu et al., 2016). Upregulation of ABA-responsive genes, such as AREB, CRK1, and NCED3, and downregulation of genes linked to stress, such as DDF2 and ZFHD1 was indicated by the application of NPs. SnRK2 activated by NPs aid in stomatal closure and in controlling water balance. Genes indulged in ROS signaling, like APX2 and RBOH1, were improved by NPs thus providing antioxidant defenses. These responses help in the regulation of oxidative stress and enhance plant adaptability to water stress by specifically targeting ABA-dependent pathways (Al-Khayri et al., 2023). *Brassica* species also showed more significant accumulation of amino acid, inhibition of proteins and total sugar content at different concentrations, due to enhanced antioxidant enzyme and reduced metal uptake (Shehzad et al., 2023). The callose deposition and composition of the cell wall also plays a fundamental role in improving resistance in *Brassica* species (Shehzad et al., 2023). Biosynthetic enzymes, 1-aminocyclopropane-1-carboxylic acid (ACC) synthases (ACSs) and ACC oxidases (ACOs) majorly control the production of ethylene. Silver and silicon NPs are reported to reduce ethylene production by inhibiting these mentioned enzymes (Mahajan et al., 2022). All these cascades trigger a series of adaptive reactions for lowering ROS, and ethylene reduction delays senescence, promotes stress mitigation during drought.

There is evidence that plants respond to flooding stress by activating these pathways controlled by ABA and ethylene. During flooding stress, ethylene production is increased that enhances adaptive responses of plants for example, elongation of stems and formation of adventitious roots to avoid hypoxia to provide flood tolerance. However, excessive ethylene production causes premature senescence and tissue damage; thus, nanoparticles influence ERFs to fine tune ethylene and reduce its harmful effects. Iron oxide and silicon nanoparticles are reported to modulate ethylene production (Koleva et al., 2022) and reduce these detrimental effects of excessive ethylene. Thus, they promote the plant's resistance to floods by preserving water balance and eliminating hypoxic conditions. Moreover, in maize, ABA is also reported to strengthen the plant's antioxidant defenses during flooding (Zhao et al., 2021). The above studies hold promising insights that abiotic stresses can be mitigated by nanotechnology in cereal plants too.

Enzymatic and non-enzymatic processes are employed by plants to maintain cell redox equilibrium under stressful circumstances (Hasanuzzaman et al., 2020). Contrarily, as explained above, during drought, the ABA signaling pathway influences stomatal closure, among other functions and controls how plants react to environmental stressors. These enzymatic and non-enzymatic pathways work together with ABA signaling controlling oxidative stress responses to ameliorate overall stress tolerance. The utilization of potential NPs reduces the adverse effects of drought stress by enhancing these mechanisms. Decreases in biomass and RWC of soybean plants under drought stress were observed but the application of Se NP mitigated declines in carotenoids and total chlorophyll contents, thus improving photosynthetic efficiency. Chloroplast structure and stomatal aperture were also improved by NPs, countering the damage. Moreover, Electrolyte Leakage (EL) and ROS levels, such as H_2O_2 and $O_2^{\bullet -}$, were decreased by the application of Se NPs, indicating lower oxidative damage. The activities of antioxidant enzymes like catalase (CAT), peroxidase (POD), and Glutathione Reductase (GR), improved the glutathione (GSH/GSSG) ratio and proline content (Zeeshan et al., 2024). Nanoceria neutralizes ROS such as hydrogen peroxide and superoxide radicals, and reduces their levels, promotes plant growth by improving reproductive success with an increase in pollen germination and seed output, photosynthetic efficiency is also increased. They protect plant cell membranes by reducing lipid peroxidation to enhance their resistance to drought stress (Djanaguiraman et al., 2018). In eggplant seedlings, silver nanoparticles (AgNPs) alleviated proline accumulation, hydrogen peroxide (H_2O_2), and malondialdehyde (MDA), enhancing antioxidant enzyme activity. Thus, AgNPs manage stress and improve plant health by mitigating oxidative damage and supporting cellular integrity. In general, AgNPs have a promising environmentally safe solution for providing plant resilience under stressed conditions (Alabdallah et al., 2021).

Al_2O_3 NPs down-regulate the ROS-detoxifying enzyme FQR1, which lessens the oxidative damage in soybeans. Al_2O_3-NPs activate an antioxidant that detoxifies ROS, which is NmrA-Like. Protein synthesis is supported by up-regulation of Poly (A)-Binding Protein 2 (PABP2) by Al_2O_3 NPs. Moreover, AgNPs down-regulate glyoxalase II 3, which detoxifies harmful methylglyoxal, oxidative stress caused by ethanol production is reduced through these NPs by downregulating pyruvate decarboxylase 2 (PDC) and alcohol dehydrogenase 1 (ADH). AgNPs also up-regulate Beta Ketoacyl Reductase 1 (BKR1) through wax production. All these responses provide stress tolerance collectively (Almutairi, 2019).

O-CMC-NPs lessen the damage caused during flooding in maize by increased activity of antioxidant enzymes such as SOD, POD, and CAT, a drop in MDA concentration and lipid peroxidation was also observed. Stress-responsive genes such as ZmTIFY, ZmEREB180, and ZmACO, which are implicated in the tolerance to waterlogging, are controlled by O-CMC-NPs encouraging its recovery of maize from floods (Wu et al., 2024).

3. Conclusion and Future Perspectives

This chapter highlights that abiotic stress tolerance in crops could be greatly increased by integrating nanotechnology with omics techniques such as transcriptomics, proteomics, and metabolomics (Fig. 1). At the molecular, cellular, and physiological levels, nanoparticles (NPs) modulate gene expression, enhance antioxidant responses, and regulate hormone transmission to promote stress tolerance. The broad use of NPs in agriculture is, however, limited by concerns about their toxicity and long-term environmental safety. To ensure sustainable use, future research should focus on evaluating the ecotoxicity and long-term impacts of nanoparticles (Fig. 1). Furthermore, investigating synergistic combinations of different NPs may provide enhanced resilience to stress. Our knowledge of NP-plant interactions will be further improved by integrating multi-omics data, and the impact of nanoparticles will be maximized by customizing them according to crop types, size, dosage, and environmental factors. Precision agriculture will be able to increase productivity, water efficiency, and nutrient uptake due to these developments, which will ultimately support sustainable agricultural methods and global food security in the face of mounting environmental difficulties.

References

Alabdallah, Nadiyah, M., MD Hasan, Abdalrhaman M. Salih, Roushdy, S. S., Aisha S. Al-Shammari, Sumayah, I. Alsanie et al. (2021). Silver nanoparticles improve growth and protect against oxidative damage in eggplant seedlings under drought stress. *Plant, Soil & Environment*, 67(11).

Ali, E. F., Ahmed Mohamed El-Shehawi, Ibrahim, O. H. M., Abdul-Hafeez, E. Y., Moussa, M. M. and Hassan, F. A. S. (2021). A vital role of chitosan nanoparticles in improvisation the drought stress tolerance in *Catharanthus roseus* (L.) through biochemical and gene expression modulation. *Plant Physiology and Biochemistry*, 161: 166–175.

Al-Khayri Jameel Mohammed, Ramakrishnan Rashmi, Rutwick Surya Ulhas, Wudali N. Sudheer, Akshatha Banadka, Praveen Nagella, Mohammed Ibrahim Aldaej et al. (2023). The Role of Nanoparticles in Response of Plants to

Abiotic Stress at Physiological, Biochemical, and Molecular Levels. Plants (Basel). 2023 Jan 7;12(2):292. doi: 10.3390/plants12020292. PMID: 36679005; PMCID: PMC9865530.

Almutairi, Zainab M. (2019). Plant molecular defense mechanisms promoted by nanoparticles against environmental stresses.

Arshad, H., & Mustafa, G. (2023). Hormonal Response in Plants Influenced by Reactive Oxygen Species. In *Reactive Oxygen Species: Prospects in Plant Metabolism* (pp. 147–160). Singapore: Springer Nature Singapore.

Avellan, Astrid, Jie Yun, Yilin Zhang, Eleanor Spielman-Sun, Jason M. Unrine, Juergen Thieme et al. (2019). Nanoparticle size and coating chemistry control foliar uptake pathways, translocation, and leaf-to-rhizosphere transport in wheat. *ACS nano*, 13(5): 5291–5305.

Ban, Yihui, Jiayuan Tan, Yang Xiong, Xiantong Mo, Yinghe Jiang, and Zhouying Xu. (2023). Transcriptome analysis reveals the molecular mechanisms of Phragmites australis tolerance to CuO-nanoparticles and/or flood stress induced by arbuscular mycorrhizal fungi. *Journal of Hazardous Materials*, 442: 130118.

Basu, Supratim, Venkategowda Ramegowda, Anuj Kumar, and Andy Pereira. (2016). Plant adaptation to drought stress. *F1000Research*, 5.

Chaturvedi, Palak, Arindam Ghatak, and Wolfram Weckwerth. (2016). Pollen proteomics: from stress physiology to developmental priming. *Plant Reproduction*, 29: 119–132

Chen, Lingling, Lan Zhu, Xiaohui Liu, Lu Chen, Han Zhou, Huixin Ma et al. (2024). Recent omics progress in nanobiotechnology for plant abiotic stress tolerance improvement. *The Crop Journal.*

Chugh, Vishal, Dasmeet Kaur, Shalini Purwar, Prashant Kaushik, Vijay Sharma, Hitesh Kumar et al. (2022). Applications of molecular markers for developing abiotic-stress-resilient oilseed crops. *Life*, 13(1): 88.

Djanaguiraman, Maduraimuthu, Remya Nair, Juan Pablo Giraldo, and Pagadala Venkata Vara Prasad. (2018). Cerium oxide nanoparticles decrease drought-induced oxidative damage in sorghum leading to higher photosynthesis and grain yield. *ACS omega*, 3(10): 14406–14416.

Elshayb, Omnia M., Abdelwahed M. Nada, Heba M. Ibrahim, Heba E. Amin, and Ayman M. Atta. (2021). Application of silica nanoparticles for improving growth, yield, and enzymatic antioxidant for the hybrid rice ehr1 growing under water regime conditions. *Materials*, 14(5): 1150.

Fiol, Diego F., María C. Terrile, Jesica Frik, Florencia A. Mesas, Vera A. Álvarez, and Claudia A. Casalongué. (2021). Nanotechnology in plants: recent advances and challenges. *Journal of Chemical Technology & Biotechnology*, 96(8): 2095–2108.

Francini Alessandra, and Luca Sebastiani. (2019). Abiotic stress effects on performance of horticultural crops. *Horticulturae*, 5(4): 67.

Haider, Ghulam, Muhammad Ansar Farooq, Tariq Shah, Saadatullah Malghani, Masood Iqbal Awan, Muhammad Habib-ur-Rahman et al. (2023). Cereal responses to nutrients and avenues for improving nutrient use efficiency. In *Cereal Crops* (pp. 79–106). CRC Press.

Hasanuzzaman, Mirza, M. H. M. Borhannuddin Bhuyan, Khursheda Parvin, Tasnim Farha Bhuiyan, Taufika Islam Anee, Kamrun Nahar et al. (2020). Regulation of ROS metabolism in plants under environmental stress: A review of recent experimental evidence. *International Journal of Molecular Sciences*, 21(22): 8695.

Hashimoto, Takuya, Ghazala Mustafa, Takumi Nishiuchi, and Setsuko Komatsu. (2020). Comparative analysis of the effect of inorganic and organic chemicals with silver nanoparticles on soybean under flooding stress. *International Journal of Molecular Sciences*, 21(4): 1300.

Iqbal, Mohammed Shariq, Akhilesh Kumar Singh, Satarudra Prakash Singh, and Mohammad Israil Ansari. (2020). Nanoparticles and plant interaction with respect to stress response. *Nanomaterials and Environmental Biotechnology*, 1–15.

Kandhol, Nidhi, Mukesh Jain, and Durgesh Kumar Tripathi. (2022). Nanoparticles as potential hallmarks of drought stress tolerance in plants. *Physiologia Plantarum*, 174(2): e13665.

Khater, Mahmoud A., Mona G. Dawood, Mohamed E. El-Awadi, Mervat S. Sadak, and Abd El-Fattah D. Badr. (2023). Middle East Journal of Applied Sciences Volume: 13| Issue: 03| July–Sept.| 2023.

Khan, I., Hasan, M., Kausar, R., Shehzad, J., and Mustafa, G. (2022). Plant molecular responses to nanoparticle stress. In *Plant and nanoparticles* (pp. 239–264). Singapore: Springer Nature Singapore.

Kıymacı, Gülbanu, Banu Çiçek Arı, Önder Türkmen, Mustafa Paksoy, Ünal Kal, Deniz Metin, and Rufeyde Nur Özen. (2024). Morph-physiological and biochemical properties effects of exogenous nitric oxide applications on spinach under flooding stress conditions.

Koleva, Lyubka, Aisha Umar, Nasim Ahmad Yasin, Anis Ali Shah, Manzer H. Siddiqui, Saud Alamri et al. (2022). Iron oxide and silicon nanoparticles modulate mineral nutrient homeostasis and metabolism in cadmium-stressed Phaseolus vulgaris. *Frontiers in Plant Science*, 13: 806781.

Li, Weiwei, and Arturo A. Keller. (2023). Assessing the Impacts of Cu and Mo Engineered Nanomaterials on Crop Plant Growth Using a Targeted Proteomics Approach. *ACS Agricultural Science & Technology*, 4(1): 103–117.

Linh, Tran My, Nguyen Chi Mai, Pham Thi Hoe, Le Quynh Lien, Ninh Khac Ban, Le Thi Thu Hien et al. (2020). Metal-based nanoparticles enhance drought tolerance in soybean. *Journal of Nanomaterials*, 2020(1): 4056563.

Lowe, Rohan, Neil Shirley, Mark Bleackley, Stephen Dolan, and Thomas Shafee. (2017). Rohan, Neil Shirley, Mark Bleackley, Stephen Dolan, and Thomas Shafee. Transcriptomics technologies. *PLoS Computational Biology*, 13(5): e1005457.

Lu, Haoyu, Mei Wang, Shangfeng Zhou, Ke Chen, Lifeng Wang, Zhenxie Yi et al. (2024). Chitosan oligosaccharides mitigate flooding stress damage in rice by affecting antioxidants, Osmoregulation, and Hormones. *Antioxidants*, 13(5): 521.

Mahajan, Shreya, Janvi Kadam, Pranjali Dhawal, Siddhivinayak Barve, and Shruti Kakodkar. (2022). Application of silver nanoparticles in *in-vitro* plant growth and metabolite production: revisiting its scope and feasibility. *Plant Cell, Tissue and Organ Culture (PCTOC)*, 150(1): 15–39.

Majumdar, Sanghamitra, and Arturo A. Keller. (2021). Omics to address the opportunities and challenges of nanotechnology in agriculture. *Critical Reviews in Environmental Science and Technology*, 51(22): 2595–2636.

Mao, Hude, Cong Jiang, Chunlei Tang, Xiaojun Nie, Linying Du, Yuling Liu, Peng Cheng et al. (2023). Wheat adaptation to environmental stresses under climate change: Molecular basis and genetic improvement. *Molecular Plant*.

Mishra, Dhruv, Manoj Kumar Chitara, Viabhav Kumar Upadhayay, Jagat Pal Singh, and Preeti Chaturvedi. (2023). Plant growth promoting potential of urea doped calcium phosphate nanoparticles in finger millet (*Eleusine coracana* (L.) Gaertn.) under drought stress. *Frontiers in Plant Science*, 14: 1137002.

Munir, Neelma, Wafa Gulzar, Zainul Abideen, Mirza Hasanuzzaman, Ali El-Keblawy, and Fengliang Zhao. (2023). Plant–Nanoparticle Interactions: Transcriptomic and Proteomic Insights. *Agronomy*, 13(8): 2112.

Mustafa, Ghazala, and Setsuko Komatsu. (2016). Insights into the response of soybean mitochondrial proteins to various sizes of aluminum oxide nanoparticles under flooding stress. *Journal of Proteome Research*, 15(12): 4464–4475.

Mustafa, Ghazala, and Setsuko Komatsu. (2022). Proteomics of plant-nanoparticle interaction mechanism. In *Plant and Nanoparticles* (pp. 67–84). Singapore: Springer Nature Singapore.

Mustafa, Ghazala, Katsumi Sakata, and Setsuko Komatsu. (2015). Proteomic analysis of flooded soybean root exposed to aluminum oxide nanoparticles. *Journal of Proteomics*, 128: 280–297.

Mustafa, Ghazala, Katsumi Sakata, Zahed Hossain, and Setsuko Komatsu. (2015). Proteomic study on the effects of silver nanoparticles on soybean under flooding stress. *Journal of Proteomics*, 122: 100–118.

Mustafa, G., Miyagawa, S., Hasan, M., Yamaguchi, H., Hitachi, K., Tsuchida, K. et al. (2023). Bio-synthesized nanoflowers and chemically synthesized nanowires zinc-oxide induced changes in the redox and protein folding in soybean seedlings: A proteomic analysis. *Journal of Plant Growth Regulation*, 42(4): 2570–2584.

Mustafa, G., Farooq, A., Riaz, Z., Hasan, M., and Amna. (2022). Nano-proteomics of stress tolerance in crop plants. *Sustainable Agriculture Reviews 53: Nanoparticles: A New Tool to Enhance Stress Tolerance*, 373–397.

Nawaz, Fahim, R. Ahmad, M. Y. Ashraf, E. A. Waraich, and S. Z. Khan. (2015). Effect of selenium foliar spray on physiological and biochemical processes and chemical constituents of wheat under drought stress. *Ecotoxicology and Environmental Safety*, 113: 191–200.

Ngara, Rudo, Tatenda Goche, Dirk Z. H. Swanevelder, and Stephen Chivasa. (2021). Sorghum's whole-plant transcriptome and proteome responses to drought stress: A review. *Life*, 11(7): 704.

Niu, Yufei, Jingyu Li, Fanting Sun, Taiyu Song, Baojia Han, Zijie Liu et al. (2023). Comparative transcriptome analysis reveals the key genes and pathways involved in drought stress response of two wheat (*Triticum aestivum* L) varieties. *Genomics*, 115(5): 110688.

Preetha, Jaganathan Sakthi Yazhini, Duraisampath Sriram, Paramasivam Premasudha, Ramesh Namdeo Pudake, and Muthukrishnan Arun. (2023). Cerium oxide as a nanozyme for plant abiotic stress tolerance: An overview of the mechanisms. *Plant Nano Biology*, 100049.

Raeesi Sadati, Seyede Yalda, Sodabeh Jahanbakhsh Godehkahriz, Ali Ebadi, and Mohammad Sedghi. (2021). Study of expression pattern of some transcription factors in wheat under drought stress and zinc nanoparticles. *Plant Genetic Researches*, 7(2): 135–144.

Reynolds, Matthew, P., Janet, M. Lewis, Karim Ammar, Bhoja R. Basnet, Leonardo Crespo-Herrera, José Crossa, Kanwarpal S. Dhugga et al. (2021). Harnessing translational research in wheat for climate resilience. *Journal of Experimental Botany*, 72(14): 5134–5157.

Rizwan, Muhammad, Shafaqat Ali, Muhammad Farooq Qayyum, Yong Sik Ok, Muhammad Adrees, Muhammad Ibrahim et al. (2017). Effect of metal and metal oxide nanoparticles on growth and physiology of globally important food crops: A critical review. *Journal of Hazardous Materials*, 322: 2–16.

Sadati, Seyedeh Yalda Raeisi, Sodabeh Jahanbakhsh Godehkahriz, Ali Ebadi, and Mohammad Sedghi. (2022). Zinc oxide nanoparticles enhance drought tolerance in wheat via physio-biochemical changes and stress genes expression. *Iranian Journal of Biotechnology*, 20(1): e3027.

Shehzad, J., Mustafa, G., Arshad, H., Ali, A., Naveed, N. H., Riaz, Z. et al. (2023). Morpho-physiological and biochemical responses of Brassica species toward lead (Pb) stress. *Acta Physiologiae Plantarum*, 45(1): 8.

Shehzad, J., Khan, I., Zaheer, S., Farooq, A., Chaudhari, S. K., and Mustafa, G. (2023). Insights into heavy metal tolerance mechanisms of Brassica species: Physiological, biochemical, and molecular interventions. *Environmental Science and Pollution Research*, 30(50): 108448–108476.

Sun Luying, Fengbin Song, Junhong Guo, Xiancan Zhu, Shengqun Liu, Fulai Liu, and Xiangnan Li. (2020). Nano-ZnO-induced drought tolerance is associated with melatonin synthesis and metabolism in maize. *International Journal of Molecular Sciences*, 21(3): 782.

Toulotte, Justine M., Chrysoula K. Pantazopoulou, Maria Angelica Sanclemente, Laurentius ACJ Voesenek, and Rashmi Sasidharan. (2022). Water stress resilient cereal crops: Lessons from wild relatives. *Journal of Integrative Plant Biology*, 64(2): 412–430.

Ullah, H., Khan, I., Mustafa, G., Shehzad, J., Hasan, M., Saeed, U. et al. (2023). Molecular characterization and mitigative role of silicon dioxide nanoparticles in Ocimum basilicum under lead (Pb) stress. *Silicon*, 15(6): 2551–2565.

Ullah, J., Gul, A., Khan, I., Shehzad, J., Kausar, R., Ahmed, M. S. et al. (2024). Green synthesized iron oxide nanoparticles as a potential regulator of callus growth, plant physiology, antioxidative and microbial contamination in *Oryza sativa* L. *BMC Plant Biology*, 24(1): 939.

Vannini, Candida, Guido Domingo, Elisabetta Onelli, Fabrizio De Mattia, Ilaria Bruni, Milena Marsoni et al. (2014). Phytotoxic and genotoxic effects of silver nanoparticles exposure on germinating wheat seedlings. *Journal of Plant Physiology*, 171(13): 1142–1148.

Villalobos-López, Miguel Angel, Analilia Arroyo-Becerra, Anareli Quintero-Jiménez, and Gabriel Iturriaga. (2022). Biotechnological advances to improve abiotic stress tolerance in crops. *International Journal of Molecular Sciences*, 23(19): 12053.

Wang, Jing, Xuesong Cao, Chuanxi Wang, Zhenggao Xiao, Feiran Chen, Yan Feng et al. (2023). Transcriptomics and metabolomics reveal the mechanisms of enhanced constitutive resistance in rice (*Oryza Sativa* L.) by silica nanomaterials. *Environmental Science: Nano*, 10(10): 2831–2848.

Wu, Hao, Peng-rui Du, Xiao-ran Miao, Rui-quan Hou, Sheng-nan Li, Muhammad Zeeshan et al. (2024). O-Carboxymethyl chitosan nanoparticles: A novel approach to enhance water stress tolerance in maize seedlings. *International Journal of Biological Macromolecules*, 277: 134459.

Yang, Kwang-Yeol, Stephanie Doxey, Joan E. McLean, David Britt, Andre Watson, Dema Al Qassy et al. (2018). Remodeling of root morphology by CuO and ZnO nanoparticles: effects on drought tolerance for plants colonized by a beneficial pseudomonad. *Botany*, 96(3): 175–186.

Yasmeen, Farhat, Naveed Iqbal Raja, Noshin Ilyas, and Setsuko Komatsu. (2018). Quantitative proteomic analysis of shoot in stress tolerant wheat varieties on copper nanoparticle exposure. *Plant Molecular Biology Reporter*, 36: 326–340.

Zafar, S., Farooq, A., Batool, S., Tariq, T., Hasan, M., & Mustafa, G. (2024). Green synthesis of iron oxide nanoparticles for mitigation of chromium stress and anti-oxidative potential in Triticum aestivum. *Hybrid Advances*, *5*, 100156.

Zaheer, Saira, Junaid Shehzad, Sunbal Khalil Chaudhari, and Ghazala Mustafa. (2024). 2 The Nanotechnology Application of to Accelerate Plant Adaptation to Abiotic Stresses. *Advanced Nanotechnology in Plants: Methods and Applications*, 13.

Zaheer, S., Shehzad, J., Chaudhari, S. K., Hasan, M., and Mustafa, G. (2023). Morphological and biochemical responses of Vigna radiata L. seedlings towards green synthesized SiO2 NPs. *Silicon*, 15(14): 5925–5936.

Zeeshan, Muhammad, Xin Wang, Abdul Salam, Hao Wu, Shengnan Li, Shiqi Zhu et al. (2024). Selenium nanoparticles boost the drought stress response of soybean by enhancing pigment accumulation, oxidative stress management and ultrastructural integrity. *Agronomy*, 14(7): 1372.

Zhao, Lijuan, Bo Peng, Jose A. Hernandez-Viezcas, Cyren Rico, Youping Sun, Jose R. Peralta-Videa et al. (2012). Stress response and tolerance of Zea mays to CeO2 nanoparticles: cross talk among H2O2, heat shock protein, and lipid peroxidation. *ACS nano*, 6(11), 9615–9622.

Zhao, Yancui, Wenying Zhang, Salah Fatouh Abou-Elwafa, Sergey Shabala, and Le Xu. (2021). Understanding a mechanistic basis of ABA involvement in plant adaptation to soil flooding: The current standing. *Plants*, 10(10): 1982.

Zhou, Wenguan, Feng Chen, Yongjie Meng, Umashankar Chandrasekaran, Xiaofeng Luo, Wenyu Yang et al. (2020). Plant waterlogging/flooding stress responses: From seed germination to maturation. *Plant Physiology and Biochemistry*, 148: 228–236.

Zhu, Xiufang, Tingting Liu, Kun Xu, and Changxiu Chen. (2022). The impact of high temperature and drought stress on the yield of major staple crops in northern China. *Journal of Environmental Management*, 314: 115092.

Zulfiqar, H., Amjad, M. S., Mehmood, A., Mustafa, G., Binish, Z., Khan, S. et al. (2022). Antibacterial, antioxidant, and phytotoxic potential of Phytosynthesized silver nanoparticles using Elaeagnus umbellate fruit extract. *Molecules*, 27(18): 5847.

CHAPTER-14

Green-Engineered Nanoparticles for Smart and Sustainable Agriculture

Ahmed M. Galal,[2] *Thomas Mounir,*[2] *Karim Ehab Lasheen,*[2] *Hala Hisham,*[2] *Farah Esmail Mohamed*[2] and *Sozan E. El-Abeid*[1,*]

1. Introduction

Green organic and inorganic synthesis includes enzymes, exopolysaccharides, plant extracts, and microbes. The process of producing environmentally friendly nanoparticles by the utilization of microorganisms is known as the synthesis of microbially based nanoparticles, and is biocompatible and non-toxic. These preparation methods can be considered as improvements in long-term eco-friendly and low-cost methods for agriculture and pharmaceutical applications. These methods can prepare a variety of nanoparticles using different strains without extensive chemical or physical modification. Due to the above-mentioned benefits, the adoption of plant and microbially synthesized nanoparticles has been progressing significantly in all relative fields and is regarded as an eco-friendly green method without any harmful effects in smart agriculture (El-Abeid et al., 2024a). Nanoparticles are produced artificially, and organically, and can exist in aggregations or unbound conditions (Thunugunta et al., 2015). These nanoscale particles exhibit unique optical, magnetic, and electrical properties due to their quantum confinement, high surface energy, and large specific surface area (Pal et al., 2019). While other techniques exist for fabricating nanoscale materials, the two predominant ways are top-down and bottom-up synthesis (Huston et al., 2021). The top-down and bottom-up approaches are done by breaking large materials (superstructure) into small nano-sized particles (atomic level) and combining individual atoms to create bulk-large (superstructure) nanomaterials (Sivaraj et al., 2020). However, as a result of growing requirements for nanoparticles, the industry is derived to find processes safer and less using hazardous chemicals and energy by increasing awareness of safe, non-toxic creation of nanoparticles in ecological methods (Thunugunta et al., 2015).

One of these methods is called green synthesis as this method is used by many plant species, and microorganisms such as fungi, bacteria, and algae. This method is better for nanoparticle synthesis as they are cheaper, eco-friendly, and compatible bio alternative (Fatima et al., 2024). With the

[1] Nanotechnology & Advanced Nano-Materials Laboratory (NANML), Mycology and Disease Survey Research Department, Plant Pathology Research Institute, Agricultural Research Center, Giza 12619, Egypt.
[2] Biotechnology department, faculty of science, Ain Shams University, Cairo, Egypt.
* Corresponding author: sozanelabeid@yahoo.com

DOI: 10.1201/9781003567684-14

incredible increase in population, massive changes in climate and environment, and the problems that are affecting the planet, science comes in handy to offer help via its branches that when combined to reach a solution (Aslam et al., 2022; Bhandari et al., 2023; Kansotia et al., 2024). Nanotechnology is a field of science that is concerned with studying and manipulating materials at a nanometer scale. Biotechnology, a multidisciplinary science that aims to find solutions to problems, provides a valuable product and many other different aims by using biological systems and technological techniques. From here, nanobiotechnology can be explained in an example of biosynthesis of metal nanoparticles. Contributing to different fields such as medicine, environment, and agriculture (Aslam et al., 2022; Palit and Hussain, 2020; Prasad, 2019). Emphasizing agriculture, nanoparticles (NPs) can be used for many purposes as biotic and abiotic stress tolerance, nanofertilizers, nanopesticides, crop productivity enhancement, decreasing contamination, and other applications. Choosing the suitable NPs differs between plant species, according to the environmental conditions, and depends upon other factors (Balusamy et al., 2023; Gupta et al., 2023).

Since there are limitations for everything, nanoparticles could sometimes be toxic and can adversely affect the plant and environment. Introducing green nanotechnology helps in the acknowledgement of sustainable methods and more eco-friendly methods for the synthesis of NPs solving vast problems and offering privileges to agriculture (Gupta et al., 2023; Palit and Hussain, 2020; Singh et al., 2021).

This chapter will address green synthesis methods; applications of green nanoparticles in different fields such as environmental, agricultural, and industrial; innovation in green nanotechnology; and comparison between green nanoparticles and other methods.

2. Structure and Function of Nanoparticle

Among metal nanoparticles, ZnO-NPs, and silver nanoparticles have attracted considerable interest owing to their distinctive physicochemical characteristics and numerous potential uses (Fig. 1), particularly in antibacterial activities. Ag NPs exhibit remarkable antiviral, antifungal, and antibacterial activities. These activities are largely dependent on the size and shape of the particles. Smaller or triangular-shaped nanoparticles tend to be more active. Their size and shape influence their properties, for instance, surface plasmon resonance, which is crucial for their interaction with microbial cells. Similar to this, ZnO-NPs demonstrate morphology-dependent characteristics. Different forms, such as rod, wire, and flower shapes, show varying antibacterial actions. Their synthesis techniques employed play a crucial role in determining the properties and functionality of ZnONPs. Tailored synthesis techniques enable the production of nanoparticles with specific characteristics suited to their targeted applications (Zeghoud et al., 2022).

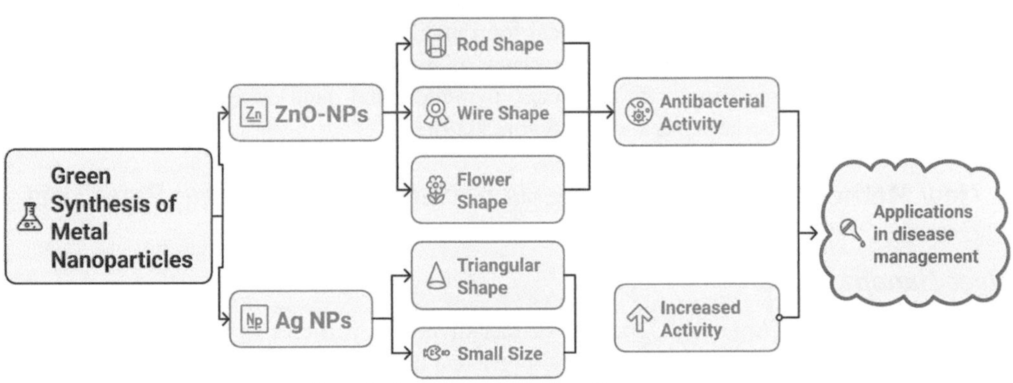

Figure 1. Structure and function of nanoparticle.

Importance of Functional Group Analysis: The introduction emphasizes that functional group analysis is a crucial aspect of organic chemistry education, particularly for identifying unknown organic compounds. This analysis helps in understanding the chemical behavior and properties of various organic substances (Othman et al., 2024).

- **Structure and function of silver nanoparticle**

Silver nanoparticles exhibit a range of applications, largely influenced by their size and shape (Syafiuddin et al., 2017). Their extensive surface area facilitates bacterial interactions, resulting in significantly elevated antibacterial activities. Silver-NPs adhere to the membranes or cell walls of target microorganisms via electrostatic interactions between positively charged silver ions and negatively charged cell membranes (Abbasi et al., 2014; Choi et al., 2010). In the medical field, silver nanoparticles with diverse shapes, including spheres, flowers, triangles, and rods, have been utilized (Wei et al., 2015; Zhang et al., 2016). The physicochemical properties of Ag NPs involve their shape, surface area, and surface charge, among other factors (Sriram et al., 2012). The relationship between surface energy and surface area indicates that the biological properties of smaller particles are likely to be enhanced (Abbas et al., 2024). Shape and size were shown to have a substantial impact on the antibacterial activities of AgNPs synthesized using a wet chemical approach for the treatment of gram-negative bacteria. The antibacterial activity of large spherical Ag NPs is inferior to that of triangular-shaped Ag NPs, while small spherical Ag NPs exhibit higher antibacterial activity (Raza et al., 2016). Spherical facets (100) of Ag NPs have lower antibacterial activity compared to the triangular facets (111) (Raza et al., 2016; Alshareef et al., 2017).

When it comes to *Escherichia coli*, *Pseudomonas aeruginosa*, and *Staphylococcus aureus*, Ag nanospheres interact more with the bacteria than the triangular nanoplates. Ag nanospheres showed complete antifungal characteristics, spherical-shaped Ag NPs at 9 nm and quasi-spherical-shaped Ag NPs at 21 nm. Antiviral activity is observed in 5–20 nm Ag NPs when produced using HEPES buffer. The capacity of wound healing is high for 22 nm Ag NPs (Gao et al., 2013; Ahmad et al., 2019; Osonga et al., 2020; Kodintcev, 2022). Due to their optical properties, Ag NPs have several industrial applications, especially in sensors (Willets and Van Duyne, 2007). These special properties, such as surface plasmon resonance, is excited by a particular wavelength of light. The peculiar property of spherical Ag NPs is that they can shift the SPR peak wavelength from 400 nm-violets to 530 nm-green by changing the particle's size and the refractive index localized near its surface (Abou El-Nour et al., 2010; Abbasi et al., 2014). Due to their sizes from 4 to 12 nm, Ag NPs produced in ceramics and glasses have different electrical conductivities. Based on the investigation of the direct electrical resistivity of the AgNP layer at 80–300 K, a linear relation of the surface resistivity was obtained as a function of temperature between 120 and 300 K. The melting range of Ag NPs is between 4 to 50 nm, and their thermal characteristics are studied at diameters between 3 to 6 nm. Ag NPs are less stable in a spherical form than in a hexagonal shape (Syafiuddin et al., 2017). TEM and SEM images reveal the presence of Ag NPs in various forms like decahedra, prisms, spheres, flowers, nano-wires, nano-bars, pyramids, and nano-cubes (Zhang et al., 2014; Pryskoka et al., 2015; Khodashenas and Ghorbani, 2019).

3. New Methods for Green Synthesis of Nanoparticles Using Plant Leaf Extract

Silver Nanoparticles

Aloe vera was used by (Anju et al., 2021) to extract silver nanoparticles by following the protocol of (Logaranjan et al., 2016). First, wash the aloe vera leaf using water then get rid of the skin which is green colored, and cut this skin into small sized pieces, finally, the grinding of the aloe vera gel was to make a solution then filtering it to use after being put in a flask containing $AgNO_3$, in the end, adding ammonia solution to dilute the extract of the plant to optimize the pH of the medium. At room temperature, the mixture of the reaction is left there, and the appearance of a yellowish-orange color means the silver nanoparticles which are spherical-shape are formed. But (Anju et al., 2021) modified this method by adding sodium borohydride to facilitate the formation of spherical-shaped AgNPs.

Melkamu and Bitew, 2021 followed the protocol of (Ananda Murthy et al., 2020) but made a few modifications to leaf extract preparation by Melkamu and Bitew (2021) left the fresh leaves of *Hagenia abyssinica* after washing multiple times with tap water to dry for 20 days to ensure the removal of moisture and took 50 gm of plant powdered leaves and put it in 450 ml of distilled water instead of deionized water and at the end they kept the mixture to warm at 65 degree Celsius using hot plate on magnetic stirrer, filtered the mixture with whatmann no.1 filter and left the filtrate for storage at 4 degree Celsius for further experiments.

The AgNPs were synthesized by (Melkamu and Bitew, 2021) by following the protocol of (Ahmed et al., 2016) but (Melkamu and Bitew, 2021) made a little modification as the silver nanoparticles were synthesized by using silver nitrate to HAPLE which is used as a reducing and capping agent with ratio 1:2 while silver nitrate is used as a precursor. They prepared an aqueous solution of $AgNO_3$ in the volumetric flask to synthesize the silver nanoparticles, the Ag ions were reduced by a volume of HAPLE which was mixed with the solution of silver nitrate and the formation of the $AgNO_3$ nanoparticles occurred while being in optimum incubation and temperature conditions (40 degree Celsius for 1.5 hr) in the incubation process which leads to a brown colored solution which appears indicating the synthesis of the silver nanoparticles is done.

Zinc nanoparticles

(*Olea europaea*) & (*Lycopersicon esculentum* M.) and (*Matricaria chamomilla* L.) extracts were prepared following the protocol of (Prakash and Paul, 2012; Rajeshkumar et al., 2018) but (Ogunyemi et al., 2019) with little modifications as these (chamomile flower, olive leave and reddish tomatoes) respectively, were placed in double-distilled, rinsed and then air-dried. Subsequently, 2 grams of each tomato were extracted in 200 ml of water at an average temperature of 65° Celsius for 4 hrs. These extracts were left to cool down at 37°C (room temperature) and filtered using Whatman filter paper number 1. The extracts were then used to synthesize ZnONPs according to Sutradhar and Saha (2016), with some modifications by Ogunyemi et al. (2019), They prepared a 1:1 molar ratio solution of zinc oxide and plant extracts in separate flasks. The mixture was heated and stirred continuously for 4 hrs at 100 rpm. The resulting nanoparticle solution was centrifuged at 10,000 g for 20 min to purify it, discarding the supernatants and collecting the pellets containing the nanoparticles. The pellets were washed using d.H2O then lyophilized and storing them at –80°C.

The following method is according to (Vijayakumar et al., 2019) but (Loganathan et al., 2022) made a little modification as they added 80 ml of 2M solution of zinc acetate dehydrate into 20 ml of *Pterolobium hexapetalum*-aquase leaf extract (Ph-ALE) then this mix was stirred for 3 hrs at 80°C. Then waiting for 1 hr to add a solution of sodium hydroxide of 2 M drop by drop to make the PH level at 12. This results in a change of color and a formation of white precipitate particles indicating the ZnO-NPs formation. These white precipitate particles were collected and dried the put in the muffle furnace to be calcinated (350°C at 3 hr).

- Green nanotechnology provides tools for transforming biological systems into sustainable methods for synthesizing nanomaterials, all the while steering clear of associated toxicity. Green techniques utilize biological sources, as the physicochemical production of these nanoparticles involves numerous hazardous chemicals and harsh conditions (Nasrollahzadeh et al., 2019). There cannot be a precise definition of what nanotechnology for green innovation, or green nanotechnology, should include as green innovation is not yet well defined. According to published research on the subject, green nanotechnology serves as the basis for safe, low-net-environmental-impact processes that employ renewable resources, are energy-efficient, reduce waste, and emit fewer greenhouse gases. In addition to being used in environmentally friendly production processes, green nanotechnology can be viewed as aiding in the development of sustainable solutions to global issues including energy scarcity, clean water scarcity, and several other environmental concerns.

- Green nanotechnology is associated with other concepts like sustainable engineering and manufacturing, green chemistry, and green manufacturing. Green nanotechnology may perform multiple kinds of functions and affect a product's entire value chain. It can also be an enabling technology, acting as a tool to help advance product or technology development. Together with other technologies and fields like biology and life sciences, materials and environmental sciences, electronics, and computing, nanotechnology is being employed more and more for green applications, resulting in products that combine several technological advancements.
- Applications of green nanotechnology, however, have the potential to lower final product carbon emissions and save energy. Green nanotechnology can help in air purification, waste and byproduct reduction, wastewater treatment, and site rehabilitation. It is crucial to understand that the great majority of synthetic methods used to create nanoparticles frequently include the use of hazardous and environmentally damaging compounds. A wide range of potentially harmful chemical-reducing agents, including sodium citrate, hydrazine, and sodium borohydride, are utilized in the production of metallic nanoparticles. The manufacture of a wide range of nanomaterials involves several demanding physical processes, such as pyrolysis and attrition, which have both immediate and long-term detrimental effects on the environment. Therefore, to mitigate or eliminate hazardous nanoparticulate pollutants that are undetectable throughout the whole production process, and to develop nanoparticle production methods that utilize minimal or no harmful ingredients (Solomon et al., 2024).
- In general, harmful chemical-based reducing agents are not used in green nanotechnology procedures to create nanoparticles. Comparing this method to other non-chemical methods, such as using bacteria, fungi, and algae to produce nanoparticles, reveals an important advantage. Innovative research has shown that phytochemicals from soy, tea, cumin, cinnamon, and grapes can be used to create precisely targeted, tumor-specific gold nanoparticles in a completely harmless manner. Green nanotechnology methods combine all the benefits of conventional green synthetic processes. They are often quick and inexpensive because they utilize agricultural byproducts and ensure environmentally benign qualities (Dhingra et al., 2010; Vágó et al., 2016; Zambre et al., 2012).

  Encapsulating tea and cinnamon phytochemicals (T-AuNPs and Cin-AuNPs) to create tumor-specific gold nanoparticles is a pioneering endeavor in which green nanomedicine nanoparticles have been used to improve optical contrast and computed tomographic X-ray imaging for cancer diagnosis. Mangiferin functionalized gold nanoparticles (MGF-AuNPs) are a pioneer in precision medicine using green nanotechnology. While targeting pro-tumor macrophages, these nanoparticles have immunomodulatory effects that cure cancer (Chanda et al., 2011; Nune et al., 2009).
- The abundance of phytochemicals with biocompatible qualities in our planet's biodiversity can be utilized to develop innovative green nanotechnologies. These can be used for nanoparticle synthesis, stability, and application of nanomaterials in various product formulations. The manufacture of nanoparticles by the use of microbes is another non-chemical reduction technique. The capacity of numerous bacterial, fungal, yeast, and algal species to produce AuNPs through biological processes has also been documented. It has also been shown that irradiating metal precursors with a number of radiation sources, such as sunshine, microwaves, and ultrasonic waves, can produce nanoparticles (Eckelman et al., 2008; Thipe et al., 2022).
- Green nanotechnology (GNT)'s potential is multifaceted, tackling global issues including water treatment, eco-friendly agriculture, and renewable energy. It is essential to the advancement of the Sustainable Development Goals (SDGs) of the UN (Fig. 2). Nanotechnology, for instance, improves the effectiveness of insecticides and fertilizers in agriculture, supporting

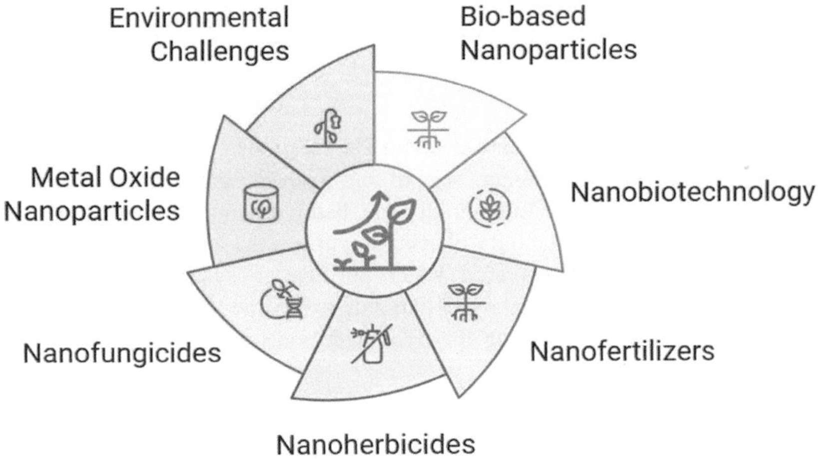

Figure 2. Advancements in green nanotechnology for Agriculture.

SDG 2 (Zero Hunger). This initiative enhances conversion and storage technologies within the energy sector, contributing to the achievement of the Sustainable Development Goal (SDG7), which focuses on affordable and clean energy. Moreover, advancements in water purification through nanotechnology contribute to the attainment of SDG6 (Clean Water and Sanitation). Furthermore, GNT advocates for clean manufacturing practices and waste reduction to achieve SDG12 (Responsible Consumption and Production). It facilitates the conservation of terrestrial ecosystems, the safeguarding of marine environments (SDG14), and the advancement of climate action (SDG13). In alignment with SDG 17, collaboration among governments, businesses, and civil society is essential for the attainment of these objectives (Kumar et al., 2024).

- The application of nanomaterials such as quantum dots, nanowires, and perovskites plays a crucial role in enhancing the efficiency of solar cells within the domain of green nanotechnologies. These materials allow for the construction of flexible, lightweight, and transparent solar cells, as well as higher power conversion efficiency and lower manufacturing costs. Ghosh and Subudhi (2022) claim that GNT facilitates the development of more energy-efficient and energy-storage-capable batteries, supercapacitors, and fuel cells by utilizing nanomaterials such as metal oxides, nanocarbons, and nanocomposites. Longer device lifespans, increased energy storage capacity, and the potential for lightweight and adaptable energy storage systems are all results of this. By employing thermal insulation coatings use nanocoatings to reduce heat transmission, nanoscale sensors, and control systems to accurately monitor and optimize energy use in transportation systems, industrial operations, and smart grids, GNT also contributes to increased energy efficiency (Thipe et al., 2022).
    - Techniques for Using Nanotechnology in Green Innovation Support for green nanotechnology has been included in broader initiatives to "green" the economic trajectory in several nations. Green nanotechnology works within a complicated environment of laws, regulations, and related initiatives for green growth as well as for research, technology, and innovation in general. The development of green nanotechnology for application in processes or products is largely influenced by framing conditions, including research, environmental, and corporate policies, as well as regulations and standards. Significant R&D efforts that are more applications-oriented and motivated by major challenges are part of the policy to support green nanotechnology. Innovative public-private partnerships are emerging as a result of policy interventions aimed at sharing and reducing the risks associated with the development and commercialization of applications employing green nanotechnology. Investments are being made to support efforts to guarantee that nanotechnology is produced responsibly, and more and more national and international projects are examining ethical

and social concerns as well as Environmental Health and Safety (EHS). Initiatives have also been launched to guarantee the economic and environmental viability of nanotechnology-based industrial methods (Grieger, et al., 2010).

- Future Prospects:
  - Advanced Technologies for Water Treatment The creation of cutting-edge nanomaterials for water purification will be the main focus of future developments in green nanotechnology. For instance, nanostructured membranes are being developed to greatly increase their lifespan and efficiency by not only filtering impurities but also self-cleaning and resisting fouling. Large-scale solar-powered water purification systems are anticipated to be made possible by nano photocatalysts like zinc oxide and titanium dioxide, providing an environmentally benign substitute for areas with poor energy availability (Rathod et al., 2024).
  - Intelligent Farming With its clever delivery mechanisms, green nanotechnology has the potential to completely transform agriculture. By enabling regulated and targeted release, fertilizer and pesticide nanoencapsulation will lower waste and contamination in the environment. By placing nanosensors in the soil or plants, farmers will be able to keep an eye on pest activity, water requirements, and nutrient levels in real-time, increasing crop productivity and sustainability (Bhandari et al., 2023; Maksimović and Omanovic-Miklicanin, 2017).
  - Renewable Energy Systems of the Future Innovations in green nanotechnology will have a substantial positive impact on the energy sector: The development of next-generation solar panels with increased efficiency and reduced production costs is being investigated using perovskite nanoparticles. Higher storage capacity and quicker charging cycles are anticipated from energy storage innovations including hydrogen fuel cells, supercapacitors, and nanostructured batteries. The performance and longevity of solar panels and wind turbine blades will continue to be improved by nanocoatings (Osman et al., 2024).

4. Comparison of Properties And Structure of Silver Nanoparticles Synthesized by Different Methods

- **Biological Synthesis**

In biological synthesis, more so plant-mediated, Ag NPs range from about 11–90 nm and are often spherical. They show a variety of beneficial properties such as antibacterial and antioxidant effects. For example, nanoparticles around 33.8 nm exhibit both antibacterial and antioxidant properties. While those ranging between 10–90 nm exhibit antibacterial capabilities. Meanwhile, some of the smaller particles, such as those ranging from 4–32nm, are effective photo-catalysts. Additionally, microbial synthesis also produces spherical Ag NPs, generally sized between 20 and 90 nm. These nanoparticles are recognized for their optical properties and various health benefits, including antibacterial and anti-proliferative effects (Abbas et al., 2024).

- **Chemical Synthesis**

Chemical methods offer a range of sizes and structures. For example, the bromide-mediated polyol process yields 20 nm nanowires with conductive properties, while the sol-gel method produces particles as small as 7 nm, known for their catalytic activity. Hydrothermal methods yield various sizes (from 5 to 70 nm) with excellent antibacterial and antifungal properties. In chemical reduction methods, the size can vary significantly, from 6.18 nm to 250 nm, with properties including antimicrobial effects and optical capabilities. Interestingly, nanoparticles formed through the polyol process can take on different shapes, such as hexagonal and icosahedral, enhancing their functionality in various applications (Abbas et al., 2024).

- **Physical Methods**

Physical synthesis techniques, like sputtering and laser ablation, produce much smaller nanoparticles, often below 10 nm. These methods can create wormlike and spherical structures, often used for catalytic and antimicrobial purposes. For instance, laser ablation yields 20 to 50 nm spherical particles that are effective against microbes (Abbas et al., 2024).

- **Photochemical synthesis**

Introduces even more diversity, with sizes ranging from 4 to 220 nm. These nanoparticles can take on various shapes, including prisms and triangular plates, showcasing their unique optical properties (Abbas et al., 2024).

In summary, the synthesis method significantly influences the size, shape, and properties of silver nanoparticles. Biological methods tend to produce spherical particles with strong antibacterial effects, while chemical and physical methods provide a broader range of sizes and structures, each with specific applications in fields like medicine and catalysis. Understanding these differences is crucial for optimizing the use of Ag NPs in various technologies Fig. (3).

Researchers have reported several methods for measuring ZnO-NPs' morphological impacts, which have a large influence on their antibacterial activity. The characteristics of ZnO-NPs are influenced by the conditions under which they are synthesized. Based on previously reported unique synthesis mechanisms, ZnO-NPs can take on a diverse array of shapes, including nanorods, nanowires, nanotubes, nanospheres, nanoneedles, drums, nano rings, spirals, polyhedrons, flowers, disks, plates, stars, and boxes. Each of these shapes demonstrates morphologically dependent physicochemical properties (Sun et al., 2016). Consequently, it is feasible to acquire essential ZnO-NP structures that optimize antibacterial activity by modifying key variables including precursor types,

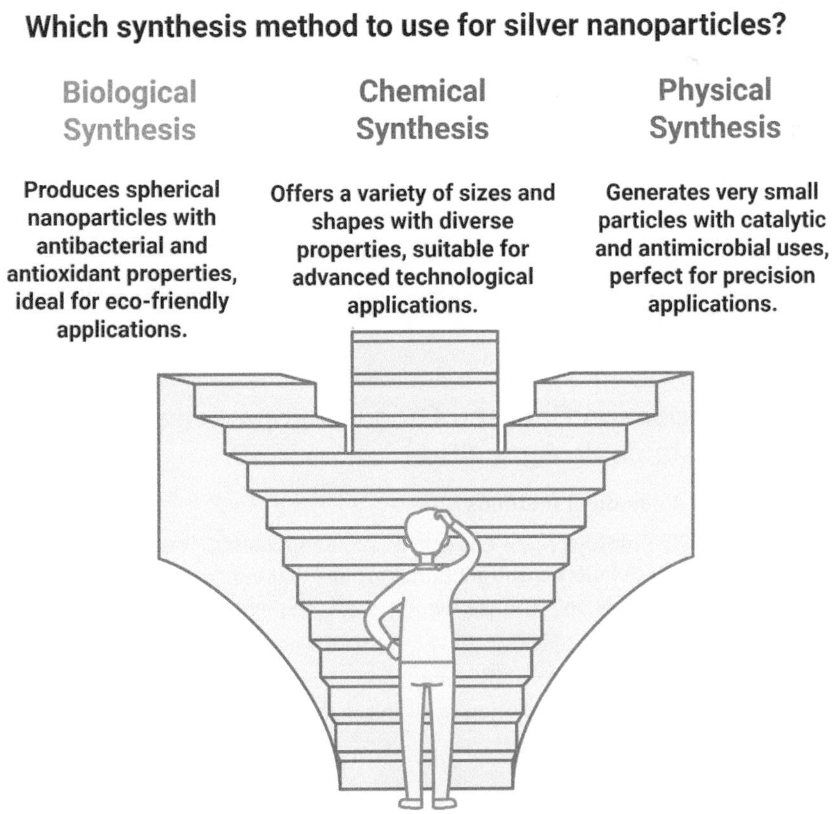

Figure 3. Comparison of properties and structure of nanoparticles synthesized by different methods.

solvents, pH, and temperature (Zhou et al., 2023; Song et al., 2019). Under controlled development conditions, like the surface, surface activity governs the shape. The shape-dependent activity was described using the proportion of active sides in the nanoparticles. Different active surfaces are produced by different synthesis and growth methods for nanoparticles. Nevertheless, studies have revealed that different morphologies with entirely different active properties have different levels of antibacterial activity (Ramani et al., 2014). The morphologies of ZnO-nanostructures will influence their internalization, as evidenced by the fact that rods and wires pass through bacterial cell walls more readily than spherical ZnO-NPs. However, compared to those in rod or spherical shape, ZnO-NPs in a floral form showed more biocidal activity against *S. aureus* and *E. coli* (Yang et al., 2009). The shapes of ZnO nanoparticles as observed by SEM include: (A) spherical ZnO, (B) bicone ZnO, (C) hexagonal sections of ZnO, and (D) plates of ZnO (Sun et al., 2016; Abbas et al., 2024; Asif et al., 2023). The global menace of pests and diseases remains a significant concern as they prevented the crops from growing; also slumped returns from agricultural production and impaired food security (Alghuthaymi et al., 2015). Farmers typically use different kinds of antidotes including fungicides, bactericides, and chemicals to arrest plant diseases resulting from various phytopathogens. Many such chemicals when used on a large scale showed ecotoxicological effects, and overuse leads to increased resistance among the pathogenic microbes towards such agrochemicals. They are sometimes absorbed in the food chain and deposited in the human body (Jayaseelan et al., 2012). Green-produced NPs provide a means of fighting postharvest decay that is new, simple, safe, and inexpensive. The risk of using a wide range of nanotechnology applications for controlling plant diseases then significantly increases. In general, there are two ways used in applying NPs in the prevention of plant diseases. In a direct application, nanoparticles are applied to the leaves of plants or seeds during their growth to prevent the onset of plant diseases. The second approach involves the development of nanoparticles to serve as carriers for pesticides, herbicides, and fertilizers, which are beneficial for nutrient applications in flooded conditions (Alghuthaymi et al., 2015). In agriculture, nanoparticles are utilized as biomarkers or rapid testing kits for the efficient detection of phytobacteria, viruses, and fungi. They also function to evaluate environmental conditions such as drought and salinity levels, the concentration of heavy metals as well as nutrients in a given soil (Alghuthaymi et al., 2015). To determine the presence of *Xanthomonas axonopodis pv. vesicatoria* which causes bacterial spot disease in plants (Yao et al., 2009). employed fluorescent silica nanoparticles together with antibodies. In the same manner (Singh et al., 2010) deployed Surface Plasmon Resonance (SPR) to test the nanogold-based immunosensors that have been developed to detect Karnal bunt disease (*Tilletia indica*) affecting wheat crops. Additionally (Wang et al., 2010), constructed an electrochemical sensor made from a gold electrode and copper nanoparticles that assesses oil seed salicylic acid to identify or detection of the harmful fungus *Sclerotinia sclerotiorum* (Saratale et al., 2018).

5. Efficiency Comparison Between Green Nanoparticles and other Methods in Sustainable Agriculture

- **Green synthesis and chemical methods**
    - **Green Synthesis:** Employs plant extracts as reducing agents, leading to the formation of metal nanoparticles (MNPs) with functional groups that enhance agricultural applications, such as carboxylic acids and amines, which promote plant growth and disease resistance (El-Abeid et al., 2024b).
    - **Chemical Methods:** Typically utilize synthetic chemicals that can be harmful to both the environment and human health, often resulting in the use of excess reagents and hazardous waste (Othman et al., 2024).

- **Traditional Methods in Sustainable Agriculture**

Traditional methods in sustainable agriculture have played a fundamental role in mitigating environmental impact while improving food security. These methods include crop rotation, intercropping, and organic fertilization, which improve soil health and biodiversity, hence increasing the resilience against pests and diseases. As an example, crop rotation interferes with the life cycles of pests, improving nutrient availability, which sustains crop yields and reduces dependence on chemical fertilizers and pesticides. Moreover, practices like composting transform organic waste into nutrient-rich amendments to the soil, which is one of the best examples of waste valorization--a very important concept in sustainability (Ghosh, 2018). These methods have been quite successful, but the introduction of green nanoparticles added a new twist to traditional methods. The use of the properties of nanomaterial could allow farmers to increase the effectiveness of nutrients and decrease the use of toxic chemicals (Bathe, et al., 2019). Therefore, a comparative analysis of the latter two methods could be a promising pathway to increasing agricultural output using more efficient and sustainable ways.

- **Efficiency and Environmental Impact Comparison of Conventional Agricultural Practices**

The efficiency and environmental impact of conventional agricultural practices have always been of great concern in the search for sustainable agriculture. Traditional methods often rely on high levels of chemical fertilizers and pesticides; while these are quite effective at boosting crop yields in the short term, they can indeed have very damaging consequences for nutrient runoff and soil degradation that affect the surrounding ecosystems. These practices lead to the pollution of water bodies, especially through nutrient leaching, which promotes eutrophication in aquatic systems; hence, the need for innovative solutions is essential (Fellet et al., 2021). Green nanoparticles are incorporated to replace these conventional nanoparticles, which promise better performance in NUE and with reduced ecological hazards. New studies also highlight that nanomaterials have the potential to offer more efficient fertilizer delivery to lower waste and reduce possible soil and water toxicity (Mura, et al., 2013). Green nanotechnology would better guarantee that agricultural output increases will correspond to more environmentally sustainable manners to preserve ecological integrity.

- **Green Nanomaterials in Agriculture**
  - Safety and Cost-Effectiveness: Green synthesized nanomaterials (NMs) are emerging as safer substitutes to chemically synthesized NMs that pose health risks. Such green methods are also rather cost-effective, making them suitable for widespread agricultural applications (Singh et al., 2023).
  - Diverse Applications: Green NMs in agriculture are used to improve crop yield, control pests, and maintain soil health, showing their multifunctionality (Singh et al., 2023).

While these advances hold great promise, challenges remain in the scaling of these methods to widespread agricultural use and the ability to ensure consistent quality and efficacy of the synthesized materials.

Tyler et al., 2023 focused on a greener one-pot synthesis method for monoterpene-functionalized lactide oligomers, which can enhance sustainability in agriculture by providing eco-friendly alternatives for producing biopolymers, potentially adding functional properties to agricultural products. O'Brien et al., 2020 discussed the use of Deep Eutectic Solvents (DESs) in green synthesis, specifically for rhodanine derivatives via Knoevenagel condensation, promoting environmentally friendly methods that could be applicable in sustainable agriculture and functional compound development. Lonarić and Molnar (2022) presented a novel method for synthesizing functional carbonaceous 3-D micro/nanostructures from agricultural by-products like rice husk and wheat straw using femtosecond laser irradiation, promoting sustainable agriculture and adding functionality for various applications in electronics, sensing, and environmental fields.

Tavangar et al., 2013 discussed various eco-friendly synthetic methods, including mechanochemistry, photochemical processes, ultrasound, microwave irradiation, and solvent-free thermal heating, which contribute to sustainable agriculture by reducing hazardous substances and enhancing the environmental and economic viability of chemical production. Kumar (2024) discussed green synthetic protocols, such as microwave irradiation, for synthesizing oxadiazole derivatives, which can enhance sustainable agriculture by developing biologically active compounds with potential therapeutic activities, thereby contributing to eco-friendly practices and reducing environmental impact. The key differences in functional groups used in green syntheses for sustainable agriculture compared to traditional methods primarily revolve around the reduction of harmful chemicals and waste. Green synthesis emphasizes eco-friendly practices, utilizing biological materials and minimizing the environmental impact, while traditional methods often rely on toxic reagents and generate significant waste.

6. Environmental Impact

- **Green Approaches** were focused on reducing solvent use and waste generation, aligning with the principles of green chemistry (Othman et al., 2024). For instance, the synthesis of photocatalysts using biocompatible materials minimizes toxic byproducts.
- **Conventional Practices**: Often lead to ecological damage due to the extensive use of agrochemicals, which can disrupt ecosystems and pose health risks.
- While green synthesis provides significant advantages in sustainability and safety, it is essential to consider that traditional methods may still provide higher yields and faster results in certain contexts, highlighting the need for a balanced approach in agricultural practices (Othman et al., 2024).

7. Application of green nanoparticles in Sustainable Agriculture

Nanofertilizers

Environmentally friendly materials known as bio/nanofertilizers could potentially supplement plants with essential nutrient substances and aid plant growth. To enhance plant growth, micro mono-bacteria, also known as biofertilizers consist of living, or inactive, microorganisms. There are three types of nanofertilizers: (i) nanomaterials including copper, manganese, molybdenum, nickel, zinc, and iron; (ii) nanoparticles such as carbon nanotubes, silica, and titanium dioxide that encourage plant growth; and (iii) phosphorus, calcium, nitrogen, and potassium. Significantly, research has suggested that phosphate nanofertilizers, in comparison to normal fertilizers, had a 20% increase in the yield of seeds and a remarkable 32% in the yield of soybeans (Liu and Lal, 2014). Also, calcium phosphate nanoparticles decorated with nitrogen and potassium may result in a 40% reduction of nitrogen usage by the plants, in comparison with other conventional fertilizers.

Bio and nanofertilizers are said to assist in plant development in several ways, including a reduction in biotic and abiotic stresses on plants, increasing the solubility of nutrients, enhancing phytohormones biosynthesis in soil, improving plant growth by nitrogen-fixing, enhancing soil fertility, among others (El-Ramady et al., 2018). In summary, green nanomaterials appear to be a developing avenue for use as bio or nanofertilizers , which can aid seed germination, increase plant growth, and improve the fresh weight of the plant. Nanofertilizers do have different biochemical effects on plants, which in turn enhances the plant's ability to endure stresses triggered by drought, diseases, and insect pests. For example, boosting the plant's immune responses to disease, and promoting auxin production in the plant. They also help improve soil fertility and structures by increasing the solubility of the nutrients in the soil chemicals which in return support healthy plants. The processes used to produce these environmentally friendly fertilizers are both simple and contribute to enhancing plant growth (Osman et al., 2024).

Pesticides

As indicated by (Wang et al., 2018), diseases in plants are detrimental issues that lead to an overall net loss of approximately US$ 220 billion annually and cuts across global agricultural practices. These ailments are caused by a plethora of organisms such as protozoa, insects, fungi, worms, parasites, bacteria, and viruses that affect crops. Encapsulating pesticides into nanopesticides is done to control the release, improve selectivity, and increase stability. Athanassiou et al. (2018) discovered that nanopesticides come with certain distinct advantages that entail: (i) dissolving active ingredients that were previously insoluble, (ii) animating the release characteristics of the pesticide, (iii) improving mobility because of the small size, (iv) increase resistance of pesticide against degradation, (v) extending the lifetime of effective action and (vi) decrease the use of harmful materials. Therefore, due to greater water solubility, controlled release systems, increased movement, and lower toxicity, nanopesticides have a lot of potential to mitigate the challenges of harmful plant diseases. Such a broad range of advantages provides a viable and encouraging solution to the contemporary agricultural challenges that the world faces (Osman et al., 2024).

8. Innovation in Green Nanotechnology in Sustainable Agriculture

Green nanoparticles, synthesized using eco-friendly materials, are an exciting new alternative to traditional methodologies involving agrochemicals. The comparative efficiency of green nanoparticles vis-à-vis other agricultural techniques, with particular emphasis on their unique properties of functional groups that make this type of nanoparticle especially effective in enhancing plant growth and plant resilience against pests.

- **Overview of Sustainable Agriculture and How Technology Can Be Used to Improve Efficiency**

Sustainable agriculture is an important strategy for food production, which is based on ecological health, economic viability, and social equity. The integration of technology into these practices is, therefore, crucial for enhancing efficiency by better use of resources and minimizing environmental impact. Advanced techniques in precision farming include the use of data analytics and GPS technology to maximize crop yields while minimizing waste from inputs, which is important in the quest for sustainable food production. Moreover, the research related to green nanotechnology is a very promising path of innovation. Green nanoparticles can improve nutrient delivery and pest control methods, hence contributing to healthier crops by using fewer pesticides. The circular bioeconomy model changes agricultural waste into useful resources, which not only reduces waste but also contributes toward sustainable production systems. Ultimately, technology enhances the resiliency and sustainability of agricultural practices against global challenges.

- **Green Nanoparticles in Sustainable Agriculture**

The incorporation of green nanoparticles in sustainable agriculture marks a milestone in both agronomy and environmental stewardship. Using biologically sourced materials, such as plant extracts or waste products, researchers can synthesize nanoparticles that enhance agricultural productivity while minimizing ecological footprints. For instance, silver nanoparticles produced from food waste, such as carrot residues, have been shown to possess remarkable antimicrobial properties that protect crops from pathogens without introducing harmful chemicals into the ecosystem (Ghosh, 2018). This innovative approach not only valorizes food waste but also aligns with the principles of circular economy in agriculture. Furthermore, the employment of such green nanotechnology reduces the energy and resource intensiveness typically associated with conventional nanoparticle synthesis methods, as highlighted in studies comparing biological synthesis to traditional manufacturing (Ghosh, 2018). Thus, green nanoparticles emerge as a viable solution to improve crop resilience while promoting sustainable agricultural practices.

9. Mechanisms of Action and Functional Group Utilization in Green Nanoparticles

The mechanism of action and utilization of functional groups in green nanoparticles are crucial for effectiveness and versatility in application within sustainable agriculture (Fig. 4). Green nanoparticles synthesized by generally environmentally benign processes employ functional groups like hydroxyl, carboxyl, and amino that allow interactions with a variety of biological substrates. For example, silver nanoparticles exhibit excellent antimicrobial activity due to their surface functional groups, which improve the binding with microbial membranes, leading to disruption and death of the cells. This mechanism indicates that the role of nanoparticles is to reduce pathogen loads on crops for healthier agricultural practices (Ghosh, 2018). Besides, encapsulation techniques in the case of corn silk extract in chitosan nanoparticles enable better bioavailability and controlled release of the active compound, enhancing its antioxidant and antibacterial properties (Mady, 2021). This will give a new dimension to using green nanoparticles as an eco-friendly technology in agriculture, where resources would be optimally utilized to reduce environmental footprints.

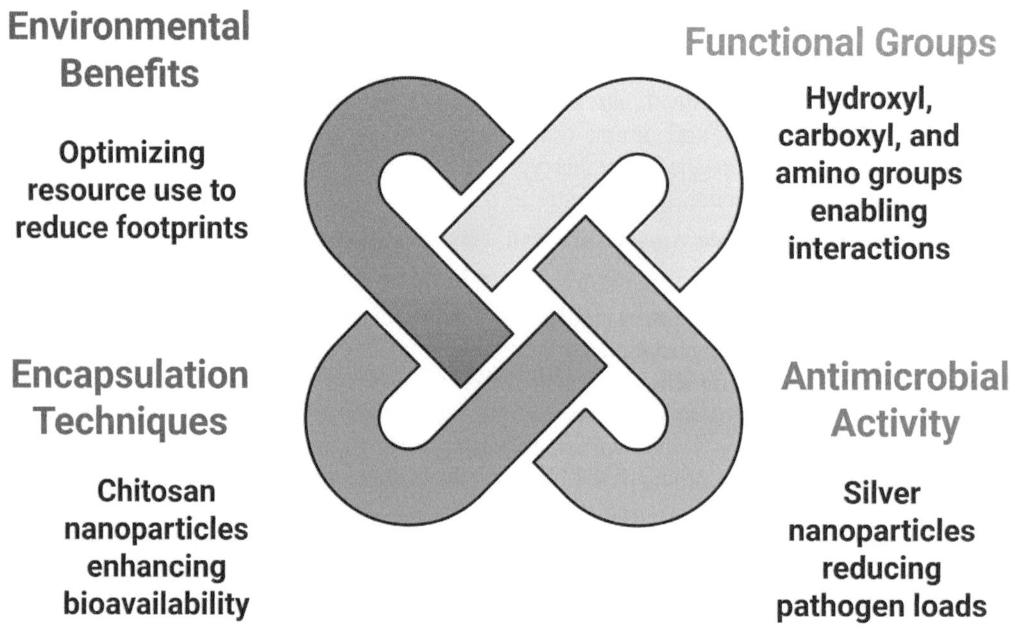

Figure 4. Role of green nanoparticles in sustainable agriculture.

10. Limitations of Traditional Methods

It highlights the drawbacks of conventional methods, which often involve the use of excessive reagents. These traditional approaches not only lead to significant waste generation but also pose challenges in waste disposal, raising environmental concerns. There should be alternative methods that align with the principles of green chemistry. This includes reducing the volume of reagents used and minimizing waste, thereby making the analysis more environmentally friendly (Othman et al., 2024).

11. Conclusion

The investigation of green nanoparticles, as a novel approach in sustainable agriculture, demonstrates a significant advantage compared to conventional methods. The use of nanotechnology increases the efficiency of pesticide and fertilizer applications and reduces environmental hazards from conventional agricultural practices. Evidence has shown that these nanomaterials can contribute to the efficient use of resources and leave non-toxic residues in ecosystems, hence facilitating the development of more viable and sustainable agricultural practices. The unique properties can also be utilized for precision in targeting agricultural inputs, hence overcoming crop challenges from biotic and abiotic stressors. Overall, the integration of green nanoparticles into agriculture provides fulfillment to the growing demands of food production but most importantly contributes constructively to environmental health; hence, green nanoparticles present a promising path for future advances in agriculture. Summarize findings and implications for future research and agricultural practices. In comparision there is a great advantage accrued to the integration of nanotechnology into green nanoparticles in relation to sustainable agriculture. The findings indicate that green nanoparticles enhance not only the efficiency of nutrient delivery but also improve pest management solutions with a more targeted approach and less environmental impact. These results are likely to preface a revolutionary change in agricultural practices, with the adoption of green nanotechnology leading to increased crop yields and reduced chemical inputs. Future research should be directed at the long-term impact of nanoparticles on soil health and biodiversity, ensuring that any risks are mitigated while optimizing the formulation of such materials. Further, cross-disciplinary studies could lead to a better understanding of functional groups within nanoparticles, opening up new uses for these materials in ways that will continue to promote sustainable agriculture. Ultimately, the integration of these findings holds the potential to revolutionize food production while adhering to ecological stewardship principles.

References

Abbas, R., Luo, J., Qi, X., Naz, A., Khan, I. A., Liu, H. et al. (2024). Silver nanoparticles: synthesis, structure, properties and applications. *Nanomaterials*, 14(17): 1425. https://doi.org/10.3390/nano14171425.

Abbasi, E., Milani, M., Fekri Aval, S., Kouhi, M., Akbarzadeh, A., Tayefi Nasrabadi, H. et al. (2014). Silver nanoparticles: Synthesis methods, bio-applications and properties. *Critical Reviews in Microbiology*, 1–8. https://doi.org/10.3109/1040841X.2014.912200.

Abou El-Nour, K. M. M., Eftaiha, A., Al-Warthan, A., and Ammar, R. A. A. (2010). Synthesis and applications of silver nanoparticles. *Arabian Journal of Chemistry*, 3(3): 135–140. https://doi.org/10.1016/j.arabjc.2010.04.008.

Ahmad, S., Munir, S., Zeb, N., Ullah, A., Khan, B., Ali, J. et al. (2019). Green nanotechnology: a review on green synthesis of silver nanoparticles — an ecofriendly approach. *International Journal of Nanomedicine*, 14: 5087–5107. https://doi.org/10.2147/IJN.S200254.

Ahmed, S., Saifullah, Ahmad, M., Swami, B. L., and Ikram, S. (2016). Green synthesis of silver nanoparticles using Azadirachta indica aqueous leaf extract. *Journal of Radiation Research and Applied Sciences*, 9(1): 1–7. https://doi.org/10.1016/j.jrras.2015.06.006.

Alghuthaymi, M. A., Almoammar, H., Rai, M., Said-Galiev, E., and Abd-Elsalam, K. A. (2015). Myconanoparticles: synthesis and their role in phytopathogens management. *Biotechnology & Biotechnological Equipment*, 29(2): 221–236. https://doi.org/10.1080/13102818.2015.1008194.

Alshareef, A., Laird, K., and Cross, R. B. M. (2017). Chemical synthesis of copper nanospheres and nanocubes and their antibacterial activity against *Escherichia coli* and *Enterococcus* sp. Acta Metallurgica Sinica (English Letters), 30(1): 29–35. https://doi.org/10.1007/s40195-016-0518-7.

Ananda Murthy, H. C., Desalegn Zeleke, T., Ravikumar, C. R., Anil Kumar, M. R., and Nagaswarupa, H. P. (2020). Electrochemical properties of biogenic silver nanoparticles synthesized using Hagenia abyssinica (Brace) JF. Gmel. medicinal plant leaf extract. Materials Research Express, 7(5). https://doi.org/10.1088/2053-1591/ab9252.

Anju, T. R., Parvathy, S., Valiya Veettil, M., Rosemary, J., Ansalna, T. H., Shahzabanu, M. M. et al. (2021). Green synthesis of silver nanoparticles from Aloe vera leaf extract and its antimicrobial activity. *Materials Today: Proceedings*, 43: 3956–3960. https://doi.org/10.1016/j.matpr.2021.02.665.

Asif, N., Amir, M., and Fatma, T. (2023). Recent advances in the synthesis, characterization and biomedical applications of zinc oxide nanoparticles. In Bioprocess and Biosystems Engineering (Vol. 46, Issue 10, pp. 1377–1398). *Springer Science and Business Media Deutschland GmbH*. https://doi.org/10.1007/s00449-023-02886-1.

Aslam, A. A., Aslam, M. S., and Aslam, A. A. (2022). An overview on green synthesis of nanoparticles and their advanced applications in sustainable agriculture. In *International Journal of Applied Chemical and Biological Sciences* (Vol. 3, Issue 2). www.visnav.in/ijacbs/.

Athanassiou, C. G., Kavallieratos, N. G., Benelli, G., Losic, D., Usha Rani, P., and Desneux, N. (2018). Nanoparticles for pest control: current status and future perspectives. *Journal of Pest Science*, 91(1): 1–15. https://doi.org/10.1007/s10340-017-0898-0.

Balusamy, S. R., Joshi, A. S., Perumalsamy, H., Mijakovic, I., and Singh, P. (2023). Advancing sustainable agriculture: a critical review of smart and eco-friendly nanomaterial applications. *Journal of Nanobiotechnology*, 21(1): 372. https://doi.org/10.1186/s12951-023-02135-3.

Bathe, M., Chrisey, L. A., Herr, D. J., Lin, Q., Rasic, D., Woolley, A. T. et al. (2019). Roadmap on biological pathways for electronic nanofabrication and materials. *Nano Futures*, 3(1): 012001.

Bhandari, G., Dhasmana, A., Chaudhary, P., Gupta, S., Gangola, S., Gupta, A. et al. (2023). A perspective review on green nanotechnology in agro-ecosystems: opportunities for sustainable agricultural practices & environmental remediation. *Agriculture*, 13(3): 668. https://doi.org/10.3390/agriculture13030668.

Chanda, N., Shukla, R., Zambre, A., Mekapothula, S., Kulkarni, R. R., Katti, K. et al. (2011). An effective strategy for the synthesis of biocompatible gold nanoparticles using cinnamon phytochemicals for phantom CT imaging and photoacoustic detection of cancerous cells. *Pharmaceutical Research*, 28(2): 279–291. https://doi.org/10.1007/s11095-010-0276-6.

Choi, O., Yu, C.-P., Esteban Fernández, G., and Hu, Z. (2010). Interactions of nanosilver with Escherichia coli cells in planktonic and biofilm cultures. Water Research, 44(20): 6095–6103. https://doi.org/10.1016/j.watres.2010.06.069.

Dhingra, R., Naidu, S., Upreti, G., and Sawhney, R. (2010). Sustainable nanotechnology: through green methods and life-cycle thinking. *Sustainability*, 2(10): 3323–3338. https://doi.org/10.3390/su2103323.

Eckelman, M. J., Zimmerman, J. B., and Anastas, P. T. (2008). Toward green nano. *Journal of Industrial Ecology*, 12(3): 316–328. https://doi.org/10.1111/j.1530-9290.2008.00043.x.

El-Abeid, S. E., Mosa, M. A., Boudaden, J., Ibrahim, D. S., Attia, E. M., Shaban, W. M. et al. (2024) a. Nanobiosensors: a powerful technology for early detection of plant parasitic nematodes. *Sensing and Imaging*, 25(1): 23.

El-Abeid, S. E., Mosa, M. A., El-Tabakh, M. A., Saleh, A. M., El-Khateeb, M. A., and Haridy, M. S. (2024b). Antifungal activity of copper oxide nanoparticles derived from Zizyphus spina leaf extract against Fusarium root rot disease in tomato plants. *Journal of Nanobiotechnology*, 22(1): 28.

El-Ramady, H., El-Ghamry, A., Mosa, A., and Alshaal, T. (2018). Nanofertilizers vs. Biofertilizers: New Insights. *Environment, Biodiversity and Soil Security*, 2(1): 40–50. https://doi.org/10.21608/jenvbs.2018.3880.1029.

Fatima, Z., Saleem, R., Khan, R. R. M., Liaqat, M., Pervaiz, M., Saeed, Z. et al. (2024). Green synthesis, properties, and biomedical potential of gold nanoparticles: A comprehensive review. In Biocatalysis and Agricultural Biotechnology (Vol. 59). Elsevier Ltd. https://doi.org/10.1016/j.bcab.2024.103271.

Fellet, G., Pilotto, L., Marchiol, L., and Braidot, E. (2021). Tools for nano-enabled agriculture: Fertilizers based on calcium phosphate, silicon, and chitosan nanostructures. *Agronomy*, 11(6): 1239.

Gao, M., Sun, L., Wang, Z., and Zhao, Y. (2013). Controlled synthesis of Ag nanoparticles with different morphologies and their antibacterial properties. *Materials Science and Engineering*: C, 33(1): 397–404. https://doi.org/10.1016/j.msec.2012.09.005.

Ghosh, P. (2018). Investigations into the levels of horticultural food wastes in Western Australia and a waste valorisation strategy using waste Daucus carota for producing high-value Silver nanoparticles with antimicrobial properties (Doctoral dissertation, Murdoch University).

Ghosh, S., and Subudhi, S. (2022). Developments in fuel cells and electrochemical batteries using nanoparticles and nanofluids. *Energy Storage*, 4(3): e288.

Grieger, K. D., Fjordbøge, A., Hartmann, N. B., Eriksson, E., Bjerg, P. L., and Baun, A. (2010). Environmental benefits and risks of zero-valent iron nanoparticles (nZVI) for in situ remediation: risk mitigation or trade-off?. *Journal of Contaminant Hydrology*, 118(3-4): 165–183.

Gupta, A., Rayeen, F., Mishra, R., Tripathi, M., and Pathak, N. (2023). Nanotechnology applications in sustainable agriculture: An emerging eco-friendly approach. *Plant Nano Biology*, 4: 100033. https://doi.org/10.1016/j.plana.2023.100033.

Huston, M., Debella, M., Dibella, M., and Gupta, A. (2021). Green synthesis of nanomaterials. In *Nanomaterials* (Vol. 11, Issue 8). MDPI AG. https://doi.org/10.3390/nano11082130.

Jayaseelan, C., Rahuman, A. A., Kirthi, A. V., Marimuthu, S., Santhoshkumar, T., Bagavan, A. et al. (2012). Novel microbial route to synthesize ZnO nanoparticles using Aeromonas hydrophila and their activity against pathogenic bacteria and fungi. Spectrochimica Acta Part A: Molecular and Biomolecular Spectroscopy, 90: 78–84. https://doi.org/10.1016/j.saa.2012.01.006.

Kansotia, K., Naresh, R., Sharma, Y., Sekhar, M., Sachan, P., Baral, K. et al. (2024). Nanotechnology-driven solutions: transforming agriculture for a sustainable and productive future. *Journal of Scientific Research and Reports*, 30(3): 32–51. https://doi.org/10.9734/jsrr/2024/v30i31856.

Khodashenas, B., and Ghorbani, H. R. (2019). Synthesis of silver nanoparticles with different shapes. *Arabian Journal of Chemistry*, 12(8): 1823–1838. https://doi.org/10.1016/j.arabjc.2014.12.014.

Kodintcev, A. N. (2022). Characterization and potential applications of silver nanoparticles: an insight on different mechanisms. *Chimica Techno Acta*, 9(4). https://doi.org/10.15826/chimtech.2022.9.4.02.

Kumar, A., Tyagi, P. K., Tyagi, S., and Ghorbanpour, M. (2024). Integrating green nanotechnology with sustainable development goals: a pathway to sustainable innovation. In *Discover Sustainability* (Vol. 5, Issue 1). Springer Nature. https://doi.org/10.1007/s43621-024-00610-x.

Kumar, V. (2024). Eco-Friendly Approaches to Chemical Synthesis. B P International, pp. 14–28. ISBN 978-81-970279-6-3.

Liu, R., and Lal, R. (2014). Synthetic apatite nanoparticles as a phosphorus fertilizer for soybean (Glycine max). Scientific Reports, 4(1): 5686. https://doi.org/10.1038/srep05686.

Loganathan, S., Manimaran, K., Mutamimurugan, K., Prakash, D. G., and Subashini, R. (2022). Green Synthesis of Zinc Oxide Nanoparticles (ZnO-NPs) by Pterolobium hexapetalum (Roth) Santapau & Wagh Aqueous Leaf Extract and Its Assessment of Biocompatible, Antibacterial, DPPH Radical Scavenging, Anticancer and Larvicidal Activities: An Effective Eco-Friendly Approach. https://doi.org/10.21203/rs.3.rs-1793777/v1.

Logaranjan, K., Raiza, A. J., Gopinath, S. C. B., Chen, Y., and Pandian, K. (2016). Shape- and size-controlled synthesis of silver nanoparticles using aloe vera plant extract and their antimicrobial activity. *Nanoscale Research Letters*, 11(1). https://doi.org/10.1186/s11671-016-1725-x.

Lonarić, M., and Molnar, M. (2022). Green Synthesis of 2-thioxothiazolidin-4-one Derivatives in Deep Eutectic Solvents via Knoevenagel Condensation. *Letters in Organic Chemistry*, 19(10): 890–901.

Mady, M. (2021). The Antioxidant and Antimicrobial Activities of Cornsilk Extract Encapsulated in Polysaccharide Nanoparticles.

Maksimović, M., and Omanović-Mikličanin, E. (2017, October). Green internet of things and green nanotechnology role in realizing smart and sustainable agriculture. In VIII international scientific agriculture symposium "AGROSYM (pp. 2290–2295).

Melkamu, W. W., and Bitew, L. T. (2021). Green synthesis of silver nanoparticles using Hagenia abyssinica (Bruce) J.F. Gmel plant leaf extract and their antibacterial and anti-oxidant activities. Heliyon, 7(11). https://doi.org/10.1016/j.heliyon.2021.e08459.

Mura, S., Seddaiu, G., Bacchini, F., Roggero, P. P., and Greppi, G. F. (2013). Advances of nanotechnology in agro-environmental studies. *Italian Journal of Agronomy*, 8(3): e18–e18.

Nasrollahzadeh, M., Sajjadi, M., Sajadi, S. M., and Issaabadi, Z. (2019). *Green Nanotechnology* (pp. 145–198). https://doi.org/10.1016/B978-0-12-813586-0.00005-5.

Nune, S. K., Chanda, N., Shukla, R., Katti, K., Kulkarni, R. R., Thilakavathy, S. et al. (2009). Green nanotechnology from tea: phytochemicals in tea as building blocks for production of biocompatible gold nanoparticles. *Journal of Materials Chemistry*, 19(19): 2912. https://doi.org/10.1039/b822015h.

O'Brien, D. M., Atkinson, R. L., Cavanagh, R., Pacheco, A. A., Larder, R., Kortsen, K. et al. (2020). A 'greener'one-pot synthesis of monoterpene-functionalised lactide oligomers. *European Polymer Journal*, 125: 109516.

Ogunyemi, S. O., Abdallah, Y., Zhang, M., Fouad, H., Hong, X., Ibrahim, E. et al. (2019). Green synthesis of zinc oxide nanoparticles using different plant extracts and their antibacterial activity against *Xanthomonas oryzae* pv. oryzae. *Artificial Cells, Nanomedicine, and Biotechnology*, 47(1): 341–352. https://doi.org/10.1080/21691401.2018.1557671.

Osman, A. I., Zhang, Y., Farghali, M., Rashwan, A. K., Eltaweil, A. S., Abd El-Monaem, E. M. et al. (2024). Synthesis of green nanoparticles for energy, biomedical, environmental, agricultural, and food applications: A review. In Environmental Chemistry Letters (Vol. 22, Issue 2, pp. 841–887). Springer Science and Business Media Deutschland GmbH. https://doi.org/10.1007/s10311-023-01682-3.

Osonga, F. J., Akgul, A., Yazgan, I., Akgul, A., Eshun, G. B., Sakhaee, L. et al. (2020). Size and shape-dependent antimicrobial activities of silver and gold nanoparticles: a model study as potential fungicides. *Molecules*, 25(11): 2682. https://doi.org/10.3390/molecules25112682.

Othman, K. A., Omer, R. A., Mohammad, A. A. A., and Safin, D. A. (2024). Green approaches for efficient functional group analysis of organic compounds. *Journal of Chemical Education*, 101(2): 567–577.

Pal, G., Rai, P., and Pandey, A. (2019). Green synthesis of nanoparticles: A greener approach for a cleaner future. Green Synthesis, Characterization and Applications of Nanoparticles, 1–26. https://doi.org/10.1016/B978-0-08-102579-6.00001-0.

Palit, S., and Hussain, C. M. (2020). Green Nanomaterials: A Sustainable Perspective (pp. 23–41). https://doi.org/10.1007/978-981-15-3560-4_2.

Prakash, M. B., and Paul, S. (2012). Green synthesis of silver nanoparticles using vinca roseus leaf extract and evaluation of their antimicrobial activities. https://api.semanticscholar.org/CorpusID:17880043.

Prasad, R. (Ed.). (2019). Plant Nanobionics: Volume 1, Advances in the Understanding of Nanomaterials Research and Applications.

Pryskoka, A. O., Rudenko, A. V., Reznichenko, L. S., Gruzina, T. G., Ulberg, Z. R., and Chekman, I. S. (2015). The antimicrobial activity of silver nanoparticles in vitro. Vìsnik Farmacìï, 0(2(82)): 54–58. https://doi.org/10.24959/nphj.15.2002.

Rajeshkumar, S., Kumar, S. V., Ramaiah, A., Agarwal, H., Lakshmi, T., and Roopan, S. M. (2018). Biosynthesis of zinc oxide nanoparticles usingMangifera indica leaves and evaluation of their antioxidant and cytotoxic properties in lung cancer (A549) cells. *Enzyme and Microbial Technology*, 117: 91–95. https://doi.org/10.1016/j.enzmictec.2018.06.009.

Ramani, M., Ponnusamy, S., Muthamizhchelvan, C., and Marsili, E. (2014). Amino acid-mediated synthesis of zinc oxide nanostructures and evaluation of their facet-dependent antimicrobial activity. Colloids and Surfaces B: Biointerfaces, 117: 233–239. https://doi.org/10.1016/j.colsurfb.2014.02.017.

Rathod, S., Preetam, S., Pandey, C., and Bera, S. P. (2024). Exploring synthesis and applications of green nanoparticles and the role of nanotechnology in wastewater treatment. *Biotechnology Reports*, 41: e00830. https://doi.org/10.1016/j.btre.2024.e00830.

Raza, M., Kanwal, Z., Rauf, A., Sabri, A., Riaz, S., and Naseem, S. (2016). Size- and Shape-dependent antibacterial studies of silver nanoparticles synthesized by wet chemical routes. *Nanomaterials*, 6(4): 74. https://doi.org/10.3390/nano6040074.

Saratale, R. G., Saratale, G. D., Shin, H. S., Jacob, J. M., Pugazhendhi, A., Bhaisare, M. et al. (2018). New insights on the green synthesis of metallic nanoparticles using plant and waste biomaterials: current knowledge, their agricultural and environmental applications. *Environmental Science and Pollution Research*, 25(11): 10164–10183. https://doi.org/10.1007/s11356-017-9912-6.

Singh, N. B., Chaudhary, R. G., Desimone, M. F., Agrawal, A., and Shukla, S. K. (2023). Green synthesized nanomaterials for safe technology in sustainable agriculture. *Current Pharmaceutical Biotechnology*, 24(1): 61–85.

Singh, R. P., Handa, R., and Manchanda, G. (2021). Nanoparticles in sustainable agriculture: An emerging opportunity. *Journal of Controlled Release*, 329: 1234–1248. https://doi.org/10.1016/j.jconrel.2020.10.051 .

Singh, S., Singh, M., Agrawal, V. V., and Kumar, A. (2010). An attempt to develop surface plasmon resonance based immunosensor for Karnal bunt (Tilletia indica) diagnosis based on the experience of nano-gold based lateral flow immuno-dipstick test. *Thin Solid Films*, 519(3): 1156–1159. https://doi.org/10.1016/j.tsf.2010.08.061.

Sivaraj, A., Kumar, V., Sunder, R., Parthasarathy, K., and Kasivelu, G. (2020). Commercial Yeast Extracts Mediated Green Synthesis of Silver Chloride Nanoparticles and their Anti-mycobacterial Activity. *Journal of Cluster Science*, 31(1): 287–291. https://doi.org/10.1007/s10876-019-01626-4.

Solomon, N. O., Simpa, P., Adenekan, O. A., and Obasi, S. C. (2024). Sustainable nanomaterials' role in green supply chains and environmental sustainability. *Engineering Science & Technology Journal*, 5(5): 1678–1694.

Song, X., Qi, Y., Zhang, M., Zhang, G., and Zhan, W. (2019). Application and optimization of drag reduction characteristics on the flow around a partial grooved cylinder by using the response surface method. *Engineering Applications of Computational Fluid Mechanics*, 13(1): 158–176. https://doi.org/10.1080/19942060.2018.1562382.

Sriram, M. I., Kalishwaralal, K., Barathmanikanth, S., and Gurunathani, S. (2012). Size-based cytotoxicity of silver nanoparticles in bovine retinal endothelial cells. *Nanoscience Methods*, 1(1): 56–77. https://doi.org/10.1080/17458080.2010.547878.

Sun, Y., Chen, L., Bao, Y., Zhang, Y., Wang, J., Fu, M. et al. (2016a). The applications of morphology controlled ZnO in Catalysis. Catalysts, 6(12): 188. https://doi.org/10.3390/catal6120188.

Sutradhar, P., and Saha, M. (2016). Green synthesis of zinc oxide nanoparticles using tomato (Lycopersicon esculentum) extract and its photovoltaic application. *Journal of Experimental Nanoscience*, 11(5): 314–327. https://doi.org/10.1080/17458080.2015.1059504.

Syafiuddin, A., Salmiati, Salim, M. R., Beng Hong Kueh, A., Hadibarata, T., and Nur, H. (2017). A review of silver nanoparticles: research trends, global consumption, synthesis, properties, and future challenges. *Journal of the Chinese Chemical Society*, 64(7): 732–756. https://doi.org/10.1002/jccs.201700067.

Tavangar, A., Tan, B., and Venkatakrishnan, K. (2013). The influence of laser-induced 3-D titania nanofibrous platforms on cell behavior. *Journal of Biomedical Nanotechnology*, 9(11): 1837–1846.

Thipe, V. C., Karikachery, A. R., Çakılkaya, P., Farooq, U., Genedy, H. H., Kaeokhamloed, N. et al. (2022). Green nanotechnology—An innovative pathway towards biocompatible and medically relevant gold nanoparticles. *Journal of Drug Delivery Science and Technology*, 70: 103256.

Thunugunta, T., Reddy, A. C., and Lakshmana Reddy, D. C. (2015). Green synthesis of nanoparticles: Current prospectus. *Nanotechnology Reviews*, 4(4): 303–323. https://doi.org/10.1515/NTREV-2015-0023.

Tyler, J. L., Katzenburg, F., and Glorius, F. (2023). A focus on sustainable method development for greener synthesis. *Chemical Science*, 14(27): 7408–7410.

Vágó, A., Szakacs, G., Sáfrán, G., Horvath, R., Pécz, B., and Lagzi, I. (2016). One-step green synthesis of gold nanoparticles by mesophilic filamentous fungi. *Chemical Physics Letters*, 645: 1–4. https://doi.org/10.1016/j.cplett.2015.12.019.

Vijayakumar, S., Vaseeharan, B., Sudhakaran, R., Jeyakandan, J., Ramasamy, P., Sonawane, A. et al. (2019). Bioinspired zinc oxide nanoparticles using lycopersicon esculentum for antimicrobial and anticancer applications. *Journal of Cluster Science*, 30(6): 1465–1479. https://doi.org/10.1007/s10876-019-01590-z.

Wang, Y., Xia, R., Hu, H., and Peng, T. (2018). Biosynthesis, characterization and cytotoxicity of gold nanoparticles and their loading with N-acetylcarnosine for cataract treatment. *Journal of Photochemistry and Photobiology B: Biology*, 187: 180–183. https://doi.org/10.1016/j.jphotobiol.2018.08.014.

Wang, Z., Wei, F., Liu, S.-Y., Xu, Q., Huang, J.-Y., Dong, X.-Y. et al. (2010). Electrocatalytic oxidation of phytohormone salicylic acid at copper nanoparticles-modified gold electrode and its detection in oilseed rape infected with fungal pathogen Sclerotinia sclerotiorum. *Talanta*, 80(3): 1277–1281. https://doi.org/10.1016/j.talanta.2009.09.023.

Wei, L., Lu, J., Xu, H., Patel, A., Chen, Z.-S., and Chen, G. (2015). Silver nanoparticles: synthesis, properties, and therapeutic applications. *Drug Discovery Today*, 20(5): 595–601. https://doi.org/10.1016/j.drudis.2014.11.014.

Willets, K. A., & Van Duyne, R. P. (2007). Localized Surface Plasmon Resonance Spectroscopy and Sensing. *Annual Review of Physical Chemistry*, 58(1): 267–297. https://doi.org/10.1146/annurev.physchem.58.032806.104607.

Yang, H., Liu, C., Yang, D., Zhang, H., and Xi, Z. (2009). Comparative study of cytotoxicity, oxidative stress and genotoxicity induced by four typical nanomaterials: the role of particle size, shape and composition. *Journal of Applied Toxicology*, 29(1): 69–78. https://doi.org/10.1002/jat.1385.

Yao, K. S., Li, S. J., Tzeng, K. C., Cheng, T. C., Chang, C. Y., Chiu, C. Y. et al. (2009). Fluorescence silica nanoprobe as a biomarker for rapid detection of plant pathogens. *Advanced Materials Research*, 79–82: 513–516. https://doi.org/10.4028/www.scientific.net/AMR.79-82.513.

Zambre, A., Upendran, A., Shukla, R., Chanda, N., Katti, K. K., Cutler, C. et al. (2012). CHAPTER 6. *Green Nanotechnology – a Sustainable Approach in the Nanorevolution* (pp. 144–156). https://doi.org/10.1039/9781849735469-00144.

Zeghoud, S., Hemmami, H., Seghir, B. B., Amor, I. B., Kouadri, I., Rebiai, A. et al. (2022). A review on biogenic green synthesis of ZnO nanoparticles by plant biomass and their applications. *Materials Today Communications*, 33, 104747.

Zhang, T., Song, Y.-J., Zhang, X.-Y., & Wu, J.-Y. (2014). Synthesis of silver nanostructures by multistep methods. *Sensors*, 14(4): 5860–5889. https://doi.org/10.3390/s140405860.

Zhang, X.-F., Liu, Z.-G., Shen, W., and Gurunathan, S. (2016). Silver nanoparticles: synthesis, characterization, properties, applications, and therapeutic approaches. *International Journal of Molecular Sciences*, 17(9): 1534. https://doi.org/10.3390/ijms17091534.

Zhou, X.-Q., Hayat, Z., Zhang, D.-D., Li, M.-Y., Hu, S., Wu, Q. et al. (2023). Zinc Oxide Nanoparticles: Synthesis, Characterization, Modification, and Applications in Food and Agriculture. *Processes*, 11(4): 1193. https://doi.org/10.3390/pr11041193.

CHAPTER-15

Green Nanotechnology for Supporting Global Nutrition and Food Security

Hanane Choubbane,[1,2] *Khadija El hazzam*[1,3] and *Fatima El Amerany*[4,*]

1. Introduction

1.1 Overview of Green Nanotechnology

Green nanotechnology as well as "Greentech" or "Cleantech" have seamlessly integrated into our daily vocabulary. As they gain popularity as marketing tools for various products and scientific initiatives, it becomes increasingly important to clarify what these terms truly signify.

While efforts to quantify the "greenness" of specific technologies are still in their infancy, understanding these elements is crucial for enhancing their impact. Key concepts such as energy, environment, pollution, sustainability, global warming, and resources frequently emerge in qualitative analyses; however, the critical numerical data needed to support these discussions is often lacking (Madaleno et al., 2022).

At the core of these issues lies the connection between the quality of human life and the thermodynamic and electromagnetic properties of the environment that sustains it. Unfortunately, this connection has been weakening over the past 250 years. Looking at the future, the quality of life for all living beings will depend on our ability to forge a new, cohesive relationship between technology and nature.

For a "green" nanoscientist or engineer, taking a step back to reflect on broader scientific questions can lead to a clearer focus on specific projects. This, in turn, can enhance productivity, reduce wasted empirical efforts, and facilitate the effective utilization of the rapidly advancing developments in optical, electrical, thermal, chemical, and mechanical nanoscience for energy and environmental applications.

[1] Laboratoire de Recherche en Développement Durable et Santé, Faculté des Sciences et Techniques, Université Cadi Ayyad, Marrakech, Morocco.

[2] Centre d'Agrobiotechnologie et Bioingénierie, Unité de Recherche Labellisée CNRST Centre AgroBiotech, URL-CNRST 05, Université Cadi Ayyad, Marrakech, Morocco.

[3] AgroBioSciences Department (AgBS), College of Agriculture and Environmental Science (CAES), Mohammed VI Polytechnic University, Benguerir 43150, Morocco.

[4] Laboratory of Natural Resources and Environment, Polydisciplinary Faculty of Taza, Department of Biology, Chemistry, and Geology, Sidi Mohamed Ben Abdellah University, PO Box 1223, 35000 Taza, Morocco.

* Corresponding author: el.amerany.fatima@gmail.com, fatima.elamerany1@usmba.ac.ma

DOI: 10.1201/9781003567684-15

The success of our future endeavors will hinge on an expanding coalition of traditional chemists, physicists, agricultural experts, and environmental scientists. Their efforts must evolve and align with the principles of the "new-green" paradigm. While some architects and engineers have already embraced this framework, a broader understanding of these issues is essential among those making design decisions.

Furthermore, as we navigate these emerging needs, it is crucial to consider the economic implications of any new technologies. The world requires innovative products and enhanced versions of the existing ones that can be produced at scale and at a low cost, utilizing the resources available to us. Reducing costs involves more than just the raw materials; it also encompasses the energy and time spent in manufacturing, as well as challenges related to supply and installation.

Humans need to play a significant role in the "green" equation, and energy is merely one component. The perception of what it means to be "green" is evolving, yet lingering memories of unattractive, cumbersome structures and uncomfortable lifestyles persist, as do many disappointments regarding past initiatives. Previously, the focus on nature often overlooked economics, human needs, and political realities; however, that narrow view is gradually fading.

1.2 Importance of Nutrition and Food Security

The advancement of agriculture stands as one of the most effective means to alleviate poverty and provide sustenance for an anticipated global population of 9.7 billion by 2050. Agricultural growth has proven to be a more effective avenue for income generation among the impoverished compared to other sectors. It plays a critical role in economic development and food security, especially in regions adversely affected by climate change.

Current food systems also significantly impact public health, as approximately 70% of agricultural water use leads to unsustainable levels of pollution and waste. Alarmingly, around three billion individuals suffer from health issues linked to inadequate nutrition and poor food quality (Chowardhara et al., 2024).

While the Green Revolution successfully boosted agricultural production around the world, achieving sustainable food practices continues to be a challenge, particularly in developing countries that often lack access to advanced technology and mechanized farming methods. Today's agricultural practices must prioritize efficiency, focusing on maximizing output while minimizing resource consumption; this includes cultivating water-efficient and drought-resistant crops (Yu et al., 2024).

Moreover, the agricultural sector has embraced various innovative technologies aimed at improvement. Among them are biotechnology, nanotechnology, and particularly molecular biology, with genome editing standing out as a cutting-edge development in the field (Vats et al., 2022).

Nanotechnology holds significant potential across various fields such as medicine, agriculture, drug delivery, and toxicology, all aimed at enhancing human well-being (Pang et al., 2025). In agriculture, it stands alongside the innovative CRISPR/Cas genome editing technology, offering promising solutions for increasing food production both effectively and economically (Alam et al., 2024).

The impact of nanotechnology extends to drug delivery, tissue culture, the food industry, and biomedical engineering (Acharya et al., 2024; Huang et al., 2024). For instance, titanium and silver nanoparticles (NPs) are utilized as antimicrobial agents in food packaging and storage, while zinc and zinc oxide NPs contribute to improving the nutritional quality of food. Moreover, platinum and gold nanowires serve as advanced biosensors in food analysis (Karnwal et al., 2024).

The widespread acceptance of nanotechnology across these diverse sectors has garnered significant investments, supported by academia, leading food corporations, and increasing government funding. In this chapter, we will explore different axes, beginning with, enhancing crop production, improving food safety and quality and concluding with sustainability and environmental impact.

2. Green Nanotechnology in Agriculture: Enhancing Crop Production

Green synthesis techniques utilize natural sources, including plants, microbes (i.e., bacteria), minerals, or specific enzymes, to produce nanofertilizers, nanosensors, and nanopesticides. These approaches are praised for being environmentally friendly and sustainable. Although these green methods provide various pathways for green NPs production, they depend on characteristics such as the size and the specific agricultural needs. Many green materials and products with improved quality have been manufactured based on the use of NPs and natural precursors. These green nanotechnologies are non-toxic and eco-friendly and have been acknowledged to boost crop development and yield.

2.1 Nanofertilizers

The application of green nanofertilizers in agricultural fields aims largely to reduce mineral losses during fertilization, supply essential minerals for plant growth, and increase yield and fruit quality. For example, Hamed et al. (2024) noticed that the biosynthesis of nanobiofertilizer capsules from biomaterials, such as chitosan and alginate, can control the release of NPK and beneficial microorganisms (*Pseudomonas Fluorescence*) and enhance plant growth. Some nanofertilizers are produced naturally by microorganisms to lower the toxicity of minerals in the soil. Kumar et al. (2023) showed that iron-based nanofertilizers can be synthesized through bioreduction of iron ions by microorganisms (e.g., fungi and bacteria). This method not only generates efficient nanofertilizers but also reduces the ecological impact typically associated with traditional chemical synthesis. There are other green NPs, such as Zn NPs and Cu NPs that are produced by pouring zinc acetate salt and copper sulfate salt, respectively, into basil extract (Abbasifar et al., 2020). The application of either 4000 ppm of Zn NPs or 2000 ppm of Cu NPs shows a promising effect on basil growth; it increases leaf number, root volume, stem length, leaf dimensions, internode length, fresh and dry weight, photosynthesis, phenolic contents, flavonoid contents, and antioxidant activity. Similar results have been observed in wheat seedlings after applying Zn NPs produced from *Withania coagulans* extract (Nazir et al., 2024). In addition, other works showed that applying sulfur NPs generated from *Moringa oleifera* leaves extract increases plant resistance to salt stress by inducing osmolyte contents and the expression levels of stress-responsive genes such as Chlorophyll a-b binding protein of LHCII type 1-like (*Lhcb1*) (Khalifa et al., 2024).

Beneficial and greener NPs can also be produced by using substances produced by microorganisms. For instance, chitosan NPs are produced by mixing chitosan, produced from shrimp waste, and extracellular proteins of *Penicillium oxalicum* (Saad et al., 2022).

The recent 7 works published recently are summarized in Table 1.

2.2 Nanosensors

Nanosensors are increasingly utilized in agriculture to monitor various factors such as soil humidity, pesticide residues, toxic pollutants, nutrient levels, and plant stress, enabling early detection to minimize yield loss. For example, a nanosensor known as a single-walled carbon nanotube (SWNT) has been injected into the leaf midveins of *Brassica rapa* plants exposed to different stressors, including light, high temperature, pathogen infection, and mechanical damage (Ang et al., 2024). This approach allows for early detection of Salicylic Acid (SA), a key stress signal. The SWNT nanosensor revealed distinct patterns and temporal variations in the production of hydrogen peroxide (H_2O_2) and SA, with each stressor generating a unique response. Specifically, early H_2O_2 waveforms were found to contain stress-specific information. These results underscore the potential of sensor multiplexing to uncover complex stress signaling pathways in plants, providing a promising avenue for cultivating crops resilient to climate change and enabling early-stage stress detection. In another application, the transformation of silver ions into silver NPs (Ag NPs) utilizing neem plant extract as an agent for reduction and stabilization has demonstrated high sensitivity,

Table 1. Some green nanofertilizers produced recently and their effect on plants.

Green NPs	Diameters	Application modes	Plant species	Effects	References
Nanooxides	5 to 9 nm	Foliar spray	Squash (cv. Eskandarani F1)	Increase plant growth and yield.	Shebl et al. (2019)
Zinc NPs	59 to 80 nm	Foliar spray	Basil (*Ocimum basilicum* L.)	Increase yield, leaf quality, and photosynthetic performance.	Abbasifar et al. (2020)
Copper NPs	71 to 93 nm	Foliar spray			
Phosphorous-Containing Hydroxyapatite NPs	229.6 nm	Foliar spray	Pomegranate (*Punica granatum* L.)	Reduce chlorophyll biosynthesis, increase membrane injury and tail DNA.	Abdelmigid et al. (2022)
Sulfur NPs	20 to 30 nm	Soaking seeds Foliar spray seedlings	Tomato (*Solanum lycopersicum*L. Cv Nadar and Naqeeb)	Increase germination percentage, plant growth, and chlorophyll biosynthesis.	Ansari et al. (2023)
	15 to 16 nm	Foliar spray	Faba bean (*Vicia faba* L. cv. Giza-716)	Increase plant growth, chlorophyll fluorescence, enzyme activities, osmolyte contents, and the expression levels of stress-responsive genes in plants subjected to salt stress.	Khalifa et al. (2024)
Chitosan NPs	ND	Foliar spray	Wheat (*Triticum aestivum* L.)	Increase the number of spikelets/spike, grain yield, as well as nutrient (i.e., nitrogen and potassium) concentrations in wheat.	Saad et al. (2022)
	118.7 nm	Foliar spray	Rice (*Oryza sativa* L.)	Suppress the impact of *Pyricularia grisea* and *Xanthomonas oryzae* on leaves.	Sathiyabama et al. (2024)

exhibiting a detection limit of 39.1 nm/mM (Varghese Alex et al., 2020). This sensitivity makes Ag NPs ideal for use in eco-friendly sensor devices, particularly in pesticide detection. Additionally, Ag NPs synthesized using *Ocimum sanctum* leaf extract have shown high sensitivity in detecting mercury ions accumulated in water (Pattnaik et al., 2023). As these innovations progress, they have the potential to transform agricultural practices, ensure food security, and promote environmental health through more precise and timely interventions.

2.3 Nanopesticides

Pesticides, while widely used for their effectiveness in controlling pests and diseases, are increasingly recognized as a significant source of environmental pollution. Over 1000 bioactive compounds are used in the manufacturing of insecticides, herbicides, and fungicides, which have substantially improved agricultural productivity and helped in the prevention of infectious diseases (Poudel et al., 2022). However, the widespread use of these chemicals also brings significant health risks, including various chronic conditions such as cardiovascular diseases and even certain types of cancer (Shekhar et al., 2024). In response to the environmental and health concerns posed by traditional pesticides, recent advancements have shifted towards greener alternatives, such as the green synthesis of nanopesticides. This innovative approach aims to enhance pesticide efficacy while simultaneously reducing or even preventing the environmental pollution caused by conventional pesticides.

Recent research has demonstrated the promising potential of nanotechnology for pest control, especially through the use of NPs synthesized from biodegradable materials and natural extracts. One study conducted by Zheng et al. (2023) successfully synthesized thermally stable Chlorfenapyr NPs (about 100 nm in size) using chitosan and carboxymethyl chitosan as crosslinkers. These NPs exhibited remarkable pest-killing efficiency, achieving a 92% control rate against *Spodoptera frugiperda*, a major maize pest, when applied via irrigation or leaf spraying. Similarly, Zavala-Zapata et al. (2024) developed silver NPs (64 ppm) using pecan shell extract and silver nitrate, which demonstrated complete mortality (100%) in laboratory conditions and partial mortality (80%) in greenhouse trials against *Diaphorina citri*, a key pest in citrus cultivation, within just three days of application. Additionally, another study documented the green synthesis of 240 nm nanocapsules using *Eucalyptus* extracts, which caused 100% mortality of *Myzus persicae* (a significant aphid pest) within two days of treatment at a concentration of 50 $mg.ml^{-1}$ (Khoshraftar et al., 2019).

These findings highlight the growing potential of green-synthesized nanopesticides to improve pest control strategies while mitigating the adverse environmental and health impacts associated with conventional pesticide use.

3. Improving Food Safety and Quality

The global food industry faces a significant challenge in ensuring the safety and quality of food products, as the demand for more efficient, sustainable, and innovative solutions continues to grow. Nanotechnology has emerged as a promising field that can address these concerns, providing a range of applications that can improve food safety and quality (Kim et al., 2018; Ramachandraiach and Hong, 2021; Zambrano-Zaragoza et al., 2018).

One of the key areas where nanomaterials have shown potential is in food packaging. Incorporating nanomaterials into polymer matrices can enhance the thermal, barrier, and mechanical properties of packaging materials, thereby extending the shelf-life and preserving the freshness of food products.

3.1 Green Nanotechnology in Food Packaging

The food packaging industry has undergone a significant transformation in recent years, driven by the advancements in nanotechnology, which involves the manufacture and use of materials at the nanoscale (up to 100 nm), and opened up new possibilities for the development of innovative food packaging materials with enhanced properties (Bradley et al., 2011; Sharma et al., 2017).

One of the most active areas of research and development in food nanotechnology is the application of nanomaterials in food packaging. Thus, they can be used to improve the barrier properties of packaging materials, enhancing their ability to protect food from oxygen, humidity, and other environmental factors that can compromise freshness and shelf-life (Mihindukulasuriya and Lim, 2014; El Amerany et al., 2022). Additionally, nanomaterials can be incorporated into packaging to provide antimicrobial properties, reducing the risk of foodborne illnesses and extending the shelf-life of perishable foods (Duncan, 2011).

The increasing focus on sustainability and environmental consciousness has also led to the development of "green" nanotechnology for food packaging applications. Green nanotechnology refers to the design, production, and use of nanomaterials and nano-enabled products in a way that minimizes harm to human health and the environment (Bradley et al., 2011; Duncan, 2011; Mihindukulasuriya and Lim, 2014). One of the key considerations in the development of green nanotechnology for food packaging is the potential for chemical migration from the packaging material into the food. To address this concern, researchers have explored the use of naturally-derived nanomaterials, such as those derived from cellulose, chitosan, or other biopolymers, which have been shown to have low migration rates and improved biodegradability (Kumar et al., 2017; Li et al., 2022; Liu et al., 2021).

Additionally, green nanotechnology has the potential to contribute to the circular bioeconomy by enabling the valorization of agro-industrial wastes and food processing by-products. Through the use of nanomaterials, these waste streams can be converted into high-value packaging materials, reducing the environmental impact and creating new economic opportunities.

3.2 Innovative Solutions and Advancement in Food Packaging

Nanotechnology-based approaches have provided new opportunities for the development of innovative food sensing and packaging solutions, addressing long-standing challenges in the food sector. The application of green nanotechnology in food packaging has led to the development of innovative solutions that address environmental concerns while enhancing the functional properties of packaging materials. Nanomaterials can be used to reinforce mechanical strength, enhance gas barrier properties, increase water repellence, and provide antimicrobial and scavenging activity to food packaging materials (Ganjeh et al., 2024). The application of nanotechnology in food packaging has mainly focused on parameters related to food quality, safety, and stability. Some of the key developments in this field include:

Enhanced Barrier and Mechanical Properties: The use of cellulose-based nanomaterials, such as cellulose nanocrystals and nanofibers, as reinforcing agents in bio-based packaging films, improve their mechanical strength and gas barrier properties along with food preservation capabilities, and have gained a welcoming response from market waste (Patel et al., 2018). These nanomaterials derived from renewable and biodegradable sources, align with the principles of green nanotechnology (Li et al., 2022; Mustafa and Andreecu, 2020).

Antimicrobial and Antioxidant Properties: The incorporation of natural antimicrobial agents, such as essential oils or metal oxide NPs, into packaging materials to extend the shelf-life of food products and reduce the risk of microbial contamination contributing to a more sustainable food supply chain (Dey et al., 2022; Mahmoudpour et al., 2020; Sharma et al., 2017; El Amerany et al., 2022). Nanotechnology enables the creation of packaging materials and antioxidant properties, which can prevent oxidation, thereby improving food safety and quality.

Sensing and Detection: The development of intelligent packaging systems is another area where green nanotechnology has made significant advancements. The incorporation of nanomaterials allows the development of packaging sensors that can detect freshness or food spoilage and the presence of harmful compounds, such as allergens, toxins, and other contaminants, providing real-time monitoring of food safety to consumers and reducing food waste (Akhilesh et al., 2024).

Environmental and Health Implications: Nanotechnology has emerged as a transformative approach in the development of packaging materials, particularly in the food and pharmaceutical sectors. This technology offers innovative solutions to enhance product quality, safety, and shelf life while addressing concerns related to chemical contaminants. By improving barrier properties, providing antimicrobial and antioxidant functions, and enabling advanced sensing capabilities, nanotechnology addresses key challenges in the packaging industry. While nanotechnology provides significant benefits, the potential risks associated with nanomaterials must be carefully managed to ensure their safe and effective use, there are concerns regarding the environmental, health, and safety implications of nanomaterials. These need to be thoroughly evaluated to ensure consumer safety and regulatory compliance.

These advancements in green nanotechnology for food packaging have the potential to significantly impact the industry, contributing to the development of more sustainable, efficient, and environmentally friendly packaging solutions that address the growing concerns about plastic pollution and food waste (Altaf et al., 2022; Wahab et al., 2021; Trajkovska Petkoska et al., 2021; Wyser et al., 2016).

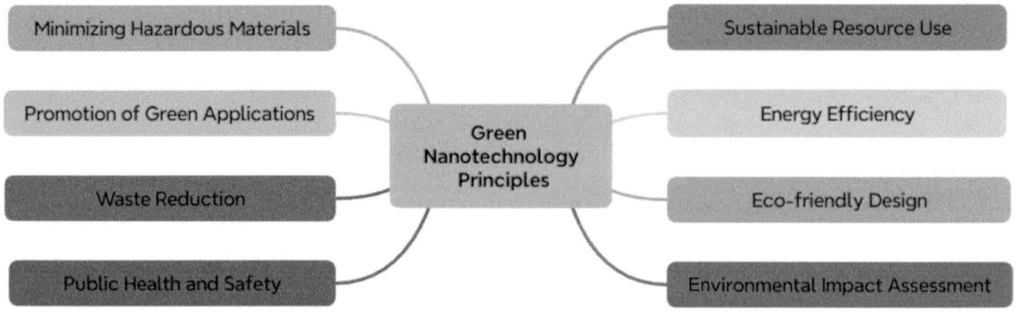

Figure 1. Green Nanotechnology Principles.

The field of green nanotechnology has emerged as a promising approach to address the environmental challenges in the food packaging industry.

4. Sustainability and Environmental Impact

Green nanotechnology's main objective is minimizing environmental damage by using fewer chemicals and producing less waste. Additionally, it might encourage more environmentally friendly farming methods. This entails maximizing the benefits for sustainability, health, and safety while limiting the usage of energy and harmful materials. Nanotechnology has enormous potential benefits for a number of industries, including food systems and agriculture (Pramanik et al., 2020).

4.1 Green Nanotechnology Principles

The use of green nanotechnology in agriculture and the food production chain has basic principles to guarantee the decrease of chemical dependence levels, the nutrient losses in fertilization, and the improvement of yield production by enhancing nutrient availability and limiting pest damages (Neme et al., 2021). These principles mainly concern raw materials, the processing, and the final product. The application of green nanotechnology presents eight different principles, i.e., sustainable resource use, minimizing hazardous materials, energy efficiency, promotion of green applications, eco-friendly design, waste reduction, environmental impact assessment, and public health and safety (Fig. 1).

4.1.1 Sustainable Resource Use

The resources that can bring together economic interests, environmental objectives, and human needs in policies and actions are known as sustainable resources. The incorporation of sustainable resources into green nanotechnology applications, specifically through the utilization of abundant, bio-based, biodegradable, active, cost-effective, biocompatible, and renewable materials, is a major step in minimizing its impact on the environment (Primožič et al., 2021).

Natural polymers are considered suitable sustainable resources that can play an interesting role in different applications. For instance, natural protein-based nanocomposite is considered an eco-friendly alternative to synthetic material, offering biodegradability and unique properties (Jafarzadeh et al., 2023). On similar lines, polysaccharides fully satisfy the properties of sustainable resources, including their widespread availability, low cost, absence of toxicity, biodegradability, and biocompatibility (Merino et al., 2017). Chitosan is another abundant renewable resource on Earth. The chemical and biological characteristics of chitosan are noteworthy. It has been extensively researched in the fields of environmental sciences, engineering, food, agriculture, and biomedical (Merino et al., 2017; El Amerany et al., 2023; Ait Hamdan et al., 2023; Elouali et al., 2024; El Amerany et al., 2025). Similarly, cellulose is a suitable option for usage as a reinforcing nanomaterial since it is fibrous, robust, and water-insoluble, as well as the most abundant renewable biopolymer produced in nature (Torres-Giner et al., 2020). Alginate is also a water-soluble, anionic linear polysaccharide known for its high availability, biodegradability, biocompatibility, and non-toxic properties (Merino et al., 2017).

The use of nature-derived polymers in nano-encapsulated fertilizers and pesticides is a sustainable strategy to lessen the environmental damage and toxicity while improving the effectiveness of these products through regulated release mechanisms, which is advantageous for the health of both humans and animals (Singh et al., 2024). Moreover, in food fortification nature-derived polymers can enhance the bioavailability of nutrients, and smaller amounts can be utilized efficiently, which reduces the demand for raw materials and minimizes environmental degradation associated with their extraction and processing (Katya et al., 2023). In food packaging, bionanomaterials generated from natural polymers are used to provide smart packaging technology that extends the shelf life and reduces microbial contamination, hence boosting food safety. Additionally, compared to traditional non-biodegradable polymers like plastics, their biodegradability and biocompatibility qualities help to reduce environmental waste (Primožič et al., 2021; Singh et al., 2024; Wypij et al., 2023).

4.1.2 Minimizing Hazardous Materials

Minimizing hazardous materials in agriculture can be effectively achieved through the adoption of nanoscale agrochemicals, which enhance the efficiency of agricultural inputs while reducing the amount of sprayed agrochemicals by smart delivery of active ingredients, minimizing nutrient losses in fertilization, increasing yields through optimized water and nutrient management, and diminishing the reliance on traditional chemical fertilizers and pesticides that contribute to environmental pollution (Hossain et al., 2020; Merino et al., 2017; Singh et al., 2021; Singh et al., 2024). Besides, nanopesticides offer notable environmental benefits compared to conventional pesticides, including a 43% reduction in toxicity to non-target species and a 22% decrease in the probability of harmful active ingredients seeping into the soil (Sridhar et al., 2023).

The integration of nanotechnology in food processing offers a promising approach to reduce reliance on chemical additives, which are linked to various health concerns. By utilizing nanoscale materials, the efficiency of nutrient absorption is enhanced, promoting improved health outcomes without introducing harmful substances. Green nanotechnology also plays a vital role in advancing food safety by allowing for the detection and elimination of contaminants at the nanoscale. This capability ensures that fortified foods remain free from dangerous materials, ultimately safeguarding consumer health (Katya et al., 2023). Similarly, adopting bio-based packaging helps decrease dependence on harmful synthetic materials, lowers environmental impact, and reduces the potential health risks linked to chemical leaching (Primožič et al., 2021).

4.1.3 Energy Efficiency

Energy efficiency is a key principle that green nanotechnology aims to address. Indeed, nanoscale agrochemicals like nanofertilizers and nanopesticides are engineered to enhance the efficiency of nutrient and active ingredient release. This precise, targeted delivery system reduces the overall quantity of chemicals necessary, which, in turn, lowers the energy required for their production, transport, and application. As a result, the use of these advanced materials contributes to a more sustainable and energy-efficient agricultural system (Singh et al., 2021). As well, the processes used in nanotechnological food fortification demand less energy than conventional methods. This improved energy efficiency helps reduce the carbon footprint of food production, supporting global initiatives aimed at mitigating climate change (Katya et al., 2023).

4.1.4 Promotion of Green Applications

Green applications rely heavily on technological advancements to drive innovation. Digital tools like sensors, drones, and data analytics enable farmers to optimize resource usage, including water and fertilizers, while minimizing waste and human intervention (Sridhar et al., 2023). In fact, the integration of digital technologies in agriculture can result in a reduction of costs by at least 23%, along with a nearly 5% decrease in medium-level waste. Additionally, it helps lower greenhouse gas emissions, minimizes environmental impacts, and contributes to both social and economic sustainability (Sridhar et al., 2023).

Advancements in the green application of nanotechnology are revolutionizing plant improvement and genomic transformation efforts. This includes the development of novel NPs sourced from biomass, which hold the potential to greatly enhance the value of agricultural products (Hossain et al., 2020).

In food systems, the use of biobased nanomaterial encapsulation technology enables highly efficient bioactive delivery systems. These systems offer enhanced control over the release, stability, and protection of active ingredients, safeguarding them from external environmental factors (Merino et al., 2017). In this regard, the incorporation of nanomaterials into food packaging facilitates the creation of smart and active packaging systems that enhance sensory attributes, such as taste and texture, as well as the nutritional value of food products. Furthermore, this integration improves the mechanical and barrier properties of packaging materials, increasing their resistance to water and oxygen, and making them more durable and efficient in preserving food (Torres-Giner et al., 2020). These innovative systems offer the ability to provide real-time insights into the freshness and safety of food, actively interacting with the product to enhance quality control (Lamri et al., 2021). For instance, these systems can include antimicrobial agents that prevent spoilage, reducing the reliance on chemical preservatives that may pose health risks (Primožič et al., 2021).

Nanotechnological processing plays a pivotal role in enhancing food quality, nutritional value, and safety by enabling the development of nano-fortified functional foods that meet contemporary health demands. Additionally, the use of nano-scale materials allows for the precise delivery of nutraceutical ingredients, offering significant benefits in the treatment of obesity and related comorbidities. This highlights the potential of nanotechnology in addressing major health challenges. Furthermore, recent advancements in nanomaterial-assisted techniques have improved the taste and appeal of healthier foods, leading to higher satisfaction and opening new opportunities for research and development in food nanoscience, and making nutritious options more attractive to consumers (Katya et al., 2023). Similarly, advancements in nanotechnology have expanded the use of nanomaterials as innovative tools for developing nano-sensing technologies, which play a crucial role in the effective detection of food contaminants. This holds significant importance from a public health standpoint (Dhiman et al., 2023).

4.1.5 Eco-Friendly Design

The application of green nanotechnology is positioned as both an eco-friendly and economically sustainable solution for agriculture (Hossain et al., 2020). The creation of biodegradable and compostable nanomaterials is a core principle of green nanotechnology applications. This concept guarantees that the materials decompose into non-toxic substances over time, significantly reducing their long-term environmental impact (Torres-Giner et al., 2020). Indeed, the degradation of these materials provides several benefits, including enhancing compost quality and preventing the build-up of toxic waste commonly associated with traditional practices. As a result, naturally occurring polymers are increasingly favored over synthetic alternatives for such applications due to their eco-friendly properties, low cost, availability, and their ability to reduce the long-term environmental impact of nanomaterials and pollution from synthetic materials (Merino et al., 2017; Torres-Giner et al., 2020).

Biodegradable polymers, sourced from natural materials, play a key role in various applications such as food packaging, nanofertilizers, nanopesticides, food fortification, and nutrient delivery systems. These polymers include polysaccharides, proteins, polyesters, and lipids are crucial for preserving food quality and preventing contamination (Yılmaz, 2024). Therefore, green chemistry principles are applied through the development and use of biodegradable nanomaterials, aligning with the design of safer chemicals and processes. Therefore resulting that nanotechnology can contribute to sustainability.

4.1.6 Waste Reduction

The incorporation of green nanotechnology in agriculture enhances the efficiency of various farming practices with minimal effort while providing greater yields and minimizing waste production (Sharma and Kar, 2019).

In the food system, the fortification of foods using green nanotechnology can improve shelf life and stability, which helps in reducing food spoilage, resulting in reduction of waste. Consequently, it not only preserves valuable resources but also mitigates the negative impact and greenhouse gas emissions generated from decomposed food waste (Katya et al., 2023). In addition, green nanotechnology can reduce the reliance on chemical preservatives and additives in fortified foods, which limits environmental contamination with harmful chemicals, benefiting ecosystems and helping to preserve soil and water quality (Katya et al., 2023; Primožič et al., 2021). For food packaging, the durability of used nanomaterials leads to reduced packaging waste as products remain intact for longer periods, minimizing the need for excessive packaging materials, and allowing for their easier recycling and composting, which reduces landfill waste and promotes a more sustainable waste management approach (Torres-Giner et al., 2020).

Overall, modern green technologies uphold sustainability and enhance food security by boosting production and minimizing food waste (Neme et al., 2021).

4.1.7 Environmental Impact Assessment

The incorporation of Life Cycle Assessment (LCA) in green nanotechnology plays a crucial role in identifying potential risks and optimizing processes to minimize adverse effects. For example, in food packaging, a deeper understanding of life cycle risk assessment, along with environmental health and safety guidelines, is essential to address the inhalation potential risks of used nanomaterials. This approach aims to guarantee that the advantages of green nanotechnology, such as improved agricultural productivity and enhanced food safety, are achieved without compromising environmental sustainability or human health (Torres-Giner et al., 2020).

Figure 2 gives an overview of various stages of nanomaterials LCA from production to disposal. The life cycle of any product begins with the production process, this stage typically

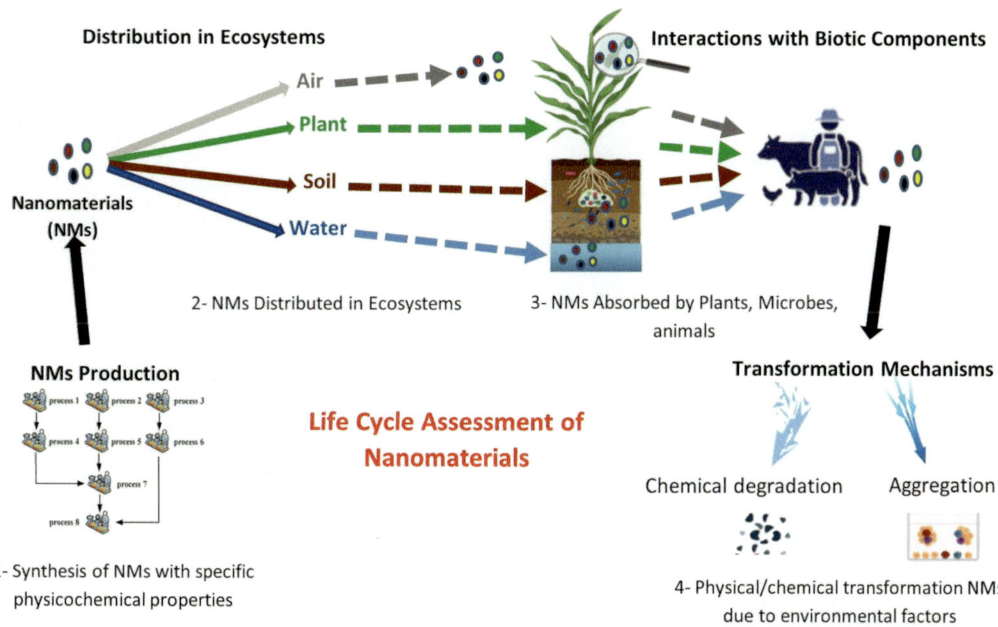

Figure 2. Life cycle and behavior of green nanotechnology products in ecosystems. NMs: Nanomaterials.

includes the raw materials, manufacturing process, energy consumption, and waste generation. The produced nanomaterials are distributed in the ecosystems (soil, water, air). The quantity and the potential exposure in the environmental space depend largely on the application methods and the nanomaterial formula. Based on their potential exposure, they interact with soil, water, plants, animals, or microbes. Their mechanisms of interaction can be chemical interactions by adsorption, absorption, and transformation, or biological uptake by plants or animals. The final stage of nanomaterials LCA concerns physical, chemical, or biological transformations of nanomaterials in the ecosystems, which affect their potential risk, mobility, and persistence. The main transformation processes are the degradation of NPs by sunlight, water, or microbes or aggregation; presented with the clumping of materials together in certain conditions. The transformation process causes changes in the properties of nanomaterials over time (e.g., size, surface charge, toxicity), which affect their implications for agriculture and food system (Ciambelli et al., 2019; Hischier and Walser, 2012; Nizam et al., 2021; Pourzahedi et al., 2018).

4.1.8 Public Health and Safety

The advancements of green nanotechnology are largely enforced considering their impact on public health and the safety of formulated products. For instance, biofortification is a considerable green nanotechnology to resolve malnutrition issues in developing countries. This strategy has the potential to improve micronutrient intake, and nutritional status, and promote better health outcomes (Raut et al., 2024). Moreover, innovations in food packaging can improve food preservation and safety (Torres-Giner et al., 2020).

Nevertheless, to effectively govern the function of green nanotechnology in agriculture and food systems, swift, dependable, and efficient communication is essential (Scott-Fordsmand et al., 2021). Consumers have the right to be fully informed about the origin and the processes used in the production of the final products (Dhiman et al., 2023). Transparency is a crucial obligation for companies employing green nanotechnology, particularly in food production. Therefore, collaboration among industry stakeholders, regulatory authorities, and consumers is essential to foster trust and support informed decision-making regarding the use of nanotechnology in food products (Kalita and Baruah, 2019). Indeed, communication with authorities, professional users, and consumers should be grounded in the most up-to-date scientific knowledge (Scott-Fordsmand et al., 2021).

4.2 Ethical Considerations

Green nanotechnology holds the potential to revolutionize nutrition and food security, while it is crucial to carefully consider its social and ethical implications. This vigilance is necessary to guarantee that the advantages of nanotechnology are achieved without affecting public health and environmental safety. By addressing these concerns, we can ensure responsible and sustainable integration into agricultural practices and food systems (Fraceto et al., 2016).

4.2.1 Safety and Toxicity

The assessment of toxicity is a critical ethical consideration for nanomaterials before they are commercialized and applied in the field. This evaluation is crucial for protecting human, animal, and environmental health, ensuring that the technology is safe for widespread adoption and use (Pramanik et al., 2020). Certain nanomaterials may persist in the environment and accumulate within ecosystems, potentially impacting wildlife and plant life (Biswas et al., 2022). It is crucial to ensure that the advantages of using nanomaterials in food do not result in environmental degradation or fail to address potential long-term impacts (Biswas et al., 2022; Dhiman et al., 2023). It should be noted that, the long-term effects of NPs on human health and the environment are not fully understood (Torres-Giner et al., 2020).

In food packaging, concerns have been raised about the migration of NPs from packaging into food. This potential migration raises important questions about toxicity and safety. Current migration tests and risk assessments remain inconclusive, emphasizing the need for comprehensive evaluation before widespread implementation (Primožič et al., 2021).

Moreover, the potential toxicity of nanomaterials to non-target organisms, such as beneficial microorganisms and insects, requires thorough evaluation. It is crucial to ensure that nanomaterials used in food and agriculture do not pose unintended risks to the environment and biodiversity (Iavicoli et al., 2017). Thus, it is crucial to adopt a balance between the benefits of green nanotechnology with the potential environmental and human health risks (Jebarani D. and Alex, 2023; Kalita and Baruah, 2019). Besides, ongoing monitoring and assessment of the effects of NPs on ecosystems are essential to mitigate any negative impacts. This requires continuous research to gain a deeper understanding of the long-term effects of nanomaterials, including their behavior in various biological systems and potential accumulation in living organisms. Additionally, the development of effective analytical tools is crucial to ensure their safety (Jebarani and Alex, 2023).

4.2.2 Regulation and Standardization

The implementation of green nanotechnology requires clear regulations and standards to guide their development and use in food and agriculture. This includes establishing labeling requirements and conducting safety assessments (Ashraf et al., 2021).

At the international level, various initiatives are underway to conduct risk analysis of nanomaterials in the food and agricultural sectors (Merino et al., 2017). The regulatory approaches for nanomaterials used in food and feed production vary internationally, reflecting different strategies and perspectives (Amenta et al., 2015; Rathee et al., 2022). To date, except for the European Union (EU), no country has established a regulatory framework for the mandatory labeling of nanomaterials (Merino et al., 2017; Rathee et al., 2022).

The EU includes specific and robust provisions in its legislation. In fact, nanomaterials related to the agri-food sector are governed by clear regulations, and additional legislation is currently being implemented to better address nanomaterials. This includes the Novel Food Regulation and the annexes to the REACH (Registration, Evaluation, Authorization, and Restriction of Chemicals) Regulation (Acharya and Pal, 2020; Merino et al., 2017).

The United States has established multiple organizations to regulate nanomaterials and safeguard against potential risks to health and the environment. The key agencies, such as the Environmental Protection Agency (EPA) and the Food and Drug Administration (FDA), manage nanomaterials within existing regulatory frameworks, including the Toxic Substances Control Act (TSCA) and the Federal Insecticide, Fungicide, and Rodenticide Act (FIFRA) (Acharya and Pal, 2020).

In Asia, the Ministry of Agriculture of Japan established regulatory approaches, which involve stringent testing protocols for nutrient release rates and environmental impacts (Ragkos et al., 2021). In this regard, China has developed a comprehensive regulatory framework for fertilizer management to promote sustainable productivity, alongside ongoing financial support for innovative products that align with environmental objectives (Acharya and Pal, 2020).

In food systems, a regulatory framework is essential for the use of NPs in food packaging and food fortification. Its implementation ensures economic viability, consumer acceptance, and adherence to regulatory considerations for food products (Katya et al., 2023; Lamri et al., 2021). Likewise, for nanosensors developed using green nanotechnology, the regulatory framework should incorporate guidelines for toxicity testing and risk assessment to ensure that any nanobiohybrids used in sensors are safe for both human health and the environment (Jebarani and Alex, 2023).

Overall, the definitions of nanomaterials vary across different countries, leading to diverse risk management approaches. To date, aside from the EU, no other country has established a regulatory framework mandating the labeling of nanomaterials (Kalita and Baruah, 2019).

4.2.3 Cost

Nanotechnology holds significant promise across various industries, including the food sector, but its development and implementation can be costly, with challenges in scaling from laboratory research to commercial applications. Despite these financial hurdles, the market impact of nanotechnology has been substantial (Primožič et al., 2021). It has evolved into a multibillion-dollar global industry, with the demand for the global nanotechnology market projected to increase from USD 68.0 billion in 2023 to USD 183.7 billion by 2028, growing at a compound annual growth rate (CAGR) of 22.0% from 2023 through 2028. The food sector alone has seen considerable growth, with the nanofood market valued at USD 15 billion in 2020 with an estimated CAGR of 12.7% (BBC Research, 2023). The nano-fertilizer market is also experiencing a rapid expansion, projected to grow from USD 401.8 million in 2022 to USD 1, 675.61 million by 2032 at a CAGR of 15.4% (Singh et al., 2024).

North America and Europe are leading the way, with significant growth also occurring in the Asia-Pacific and Latin American markets. Furthermore, this expansion has generated approximately two million jobs, underscoring the economic potential and future growth of the industry (Merino et al., 2017).

The integration of green nanotechnology in food packaging aligns with circular bioeconomy strategies by repurposing agro-industrial waste and food processing by-products. This approach not only minimizes waste but also adds value to materials that would otherwise be discarded, fostering recycling and enhancing resource efficiency (Torres-Giner et al., 2020). This economic advantage is vital for both producers and consumers, as it contributes to making products more affordable and promotes greater sustainability (Lamri et al., 2021). Likewise, as consumer awareness of health and nutrition grows, the demand for nano-fortified foods is likely to rise. This increased demand may enable manufacturers to price these products at a premium, potentially resulting in higher profit margins. However, the cost-effectiveness will ultimately depend on consumer acceptance and their willingness to pay for these enhanced products (Katya et al., 2023).

4.2.4 Public Perception

Public perception and consumer acceptance of the adoption of green nanotechnology in food and agriculture remain mixed, with some consumers expressing openness to its potential benefits while others remain skeptical, particularly about "nanofoods." A significant challenge lies in the lack of scientific knowledge among consumers, leading to concerns about the safety and health implications of consuming products containing nanomaterials (Hossain et al., 2020). This wariness is compounded by a general uncertainty surrounding nanotechnology and its potential risks (Singh et al., 2021).

While attitudes towards nanotechnology range from neutral to slightly positive, the lack of understanding often leads to hesitation, particularly when consumers are uncertain about regulatory oversight or the safety of these products (Katya et al., 2023).

Transparent communication and education regarding the benefits and risks of nanotechnology are essential to building public trust and alleviating these concerns (Katya et al., 2023; Pramanik et al., 2020). Companies that employ nanotechnology in food production must prioritize providing clear and accessible information to build consumer confidence and potentially shift public attitudes toward broader acceptance (Kalita and Baruah, 2019). For example, the EU enforces comprehensive regulations and frameworks for nanomaterials under Regulation (EU) 2019/1009, by emphasizing biosafety and effectiveness for human, animal, and environmental health (Singh et al., 2024). By applying these frameworks, responsible and sustainable use of nanotechnology is ensured (Dhiman et al., 2023; Pramanik et al., 2020).

References

Abbasifar, A., Shahrabadi, F., and ValizadehKaji, B. (2020). Effects of green synthesized zinc and copper nano-fertilizers on the morphological and biochemical attributes of basil plant. *Journal of Plant Nutrition*, 43(8): 1104–1118.

Abdelmigid, H. M., Morsi, M. M., Hussien, N. A., Alyamani, A. A., Alhuthal, N. A., and Albukhaty, S. (2022). Green synthesis of phosphorous-containing hydroxyapatite nanoparticles (nHAP) as a novel nano-fertilizer: preliminary assessment on pomegranate (*Punica granatum* L.). *Nanomaterials*, 12(9): 1527.

Acharya, A., and Pal, P. K. (2020). Agriculture nanotechnology: Translating research outcome to field applications by influencing environmental sustainability. *NanoImpact*, 19: 100232. https://doi.org/10.1016/j.impact.2020.100232.

Acharya, B., Behera, A., Behera, S., and Moharana, S. (2024). Recent advances in nanotechnology-based drug delivery systems for the diagnosis and treatment of reproductive disorders. In ACS Applied Bio Materials (Vol. 7, pp. 1336–1361). American Chemical Society. doi: 10.1021/acsabm.3c01064.

Ait Hamdan, Y., El Amerany, F., Desbrieres, J., Aghrinane, A., Oudadesse, H., and Rhazi, M. (2023). The evolution of the global COVID-19 epidemic in Morocco and understanding the different therapeutic approaches of chitosan in the control of the pandemic. *Polymer Bulletin*, 80(10): 10633–10659.

Ansari, M., Ahmed, S., Abbasi, A., Khan, M. T., Subhan, M., Bukhari, N. A. et al. (2023). Plant mediated fabrication of silver nanoparticles, process optimization, and impact on tomato plant. *Scientific Reports*, 13(1): 18048.

Ang, M. C. Y., Saju, J. M., Porter, T. K., Mohaideen, S., Sarangapani, S., Khong, D. T. et al. (2024). Decoding early stress signaling waves in living plants using nanosensor multiplexing. *Nature Communications*, 15(1): 2943.

Akhilesh, K., Shweta, K., Arpit, S., and Anuradha, S. (2024). Biosensors for Food Spoilage Detection: A Comprehensive Review of Current Advances. *Journal of Food Chemistry & Nanotechnology*, 10. doi: 10.17756/jfcn.2024-s1-010.

Alam, O., Khan, S. U., Gul, S., Gul, H., and Ullah, I. (2024). *Innovative Nanotechnology in CRISPR-Based Crop Genome Editing* (pp. 121–151). doi: 10.1007/978-3-031-76000-6_6.

Altaf, A., Usmani, Z., Dar, A. H., and Dash, K. K. (2022). A comprehensive review of polysaccharide-based bionanocomposites for food packaging applications. *Discover Food*, 2(1): 1–13. doi: 10.1007/S44187-022-00011-X/TABLES/2.

Amenta, V., Aschberger, K., Arena, M., Bouwmeester, H., Moniz, F. B., Brandhoff, P. et al. (2015). Regulatory aspects of nanotechnology in the agri/feed/food sector in EU and non-EU countries. *Regulatory Toxicology and Pharmacology*, 73(1): 463–476. https://doi.org/10.1016/j.yrtph.2015.06.016.

Ashraf, S. A., Siddiqui, A. J., Elkhalifa, A. E. O., Khan, M. I., Patel, M., Alreshidi, M. et al. (2021). Innovations in nanoscience for the sustainable development of food and agriculture with implications on health and environment. *Science of The Total Environment*, 768: 144990. https://doi.org/10.1016/j.scitotenv.2021.144990.

BBC Research. (2023, October). Global Nanotechnology Market Poised to Reach 183.7 Billion by 2028. https://www.bccresearch.com/pressroom/nan/global-nanotechnology-market-poised-to-reach-1837-billion-by-2028?srsltid=AfmBOopSa9WEnASx8FqDJ7jMNm6hZLwW9NgU43nK8dx4eYSMSZK2SX8W.

Biswas, R., Alam, M., Sarkar, A., Haque, M. I., Hasan, M. M., and Hoque, M. (2022). Application of nanotechnology in food: Processing, preservation, packaging and safety assessment. *Heliyon*, 8(11): e11795. https://doi.org/10.1016/j.heliyon.2022.e11795.

Bradley, E. L., Castle, L., and Chaudhry, Q. (2011). Applications of nanomaterials in food packaging with a consideration of opportunities for developing countries. *Trends in Food Science & Technology*, 22(11): 604–610. doi: 10.1016/J.TIFS.2011.01.002.

Chowardhara, B., Saha, B., Awasthi, J. P., Deori, B. B., Nath, R., Roy, S. et al. (2024). An assessment of nanotechnology-based interventions for cleaning up toxic heavy metal/metalloid-contaminated agroecosystems: Potentials and issues. *Chemosphere*, *359*. doi: 10.1016/j.chemosphere.2024.142178.

Ciambelli, P., La Guardia, G., and Vitale, L. (2019). *Nanotechnology for Green Materials and Processes*. 179: 97–116. https://doi.org/10.1016/b978-0-444-64337-7.00007-0.

Dey, A., Pandey, G., and Rawtani, D. (2022). Functionalized nanomaterials driven antimicrobial food packaging: A technological advancement in food science. In Food Control (Vol. 131). Elsevier Ltd. doi: 10.1016/j.foodcont.2021.108469.

Dhiman, S., Singh, P., Kumar, A., and Mukherjee, G. (2023). Application of Nanomaterials in Detection of Food Contaminants. In *Nutritional Science and Technology* (pp. 295–316). John Wiley & Sons, Ltd. https://doi.org/10.1002/9781394229116.ch13.

Duncan, T. V. (2011). Applications of nanotechnology in food packaging and food safety: Barrier materials, antimicrobials and sensors. *Journal of Colloid and Interface Science*, 363(1): 1–24. doi: 10.1016/J.JCIS.2011.07.017.

El Amerany, F., Aboudamia, F. Z., Janah, I., Taourirte, M., and Rhazi, M. (2022). Polymer-Based Nanoparticles (NPs): A Promising Approach for Crop Productivity. pp. 119–154. *In: Plant and Nanoparticles*. Singapore: Springer Nature Singapore.

El Amerany, F., Meddich, A., Wahbi, S., Taourirte, M., and Rhazi, M. (2023). Application of compost, as a peat substitute, and foliar spray of chitosan and/or roots inoculation with beneficial fungus increases tomato plants productivity and fruit size. *Gesunde Pflanzen*, 75(5): 1485–1495.

El Amerany, F., Ait Ali, O., and Rhazi, M. (2025). Chitosan reduces the toxicity of frass from black soldier flies (*Hermetia illucens*) used to cultivate Pepper (*Capsicum annuum* L.). *Euro-Mediterranean Journal for Environmental Integration*. https://doi.org/10.1007/s41207-024-00719-5.

Elouali, S., Ait Ali Ouydir, H., Ait Hamdan, Y., Eladlani, N., and Rhazi, M. (2024). Chitosan from Periplaneta americana L.: a sustainable solution for heavy metals removal. *Euro-Mediterranean Journal for Environmental Integration*, 1–12.

Fraceto, L., Grillo, R., Medeiros, G., Scognamiglio, V., Rea, G., and Bartolucci, C. (2016). Nanotechnology in Agriculture: Which Innovation Potential Does It Have? *Frontiers in Environmental Science*, 4. https://doi.org/10.3389/fenvs.2016.00020.

Ganjeh, A. M., Saraiva, J. A., Pinto, C. A., Casal, S., and Gonçalves, I. (2024). Recent advances in nano-reinforced food packaging based on biodegradable polymers using layer-by-layer assembly: A review. In Carbohydrate Polymer Technologies and Applications (Vol. 7). Elsevier Ltd. doi: 10.1016/j.carpta.2023.100395.

Hamed, R., Jodeh, S., and Alkowni, R. (2024). Nano bio fertilizer capsules for sustainable agriculture. *Scientific Reports*, 14(1): 13646.

Hischier, R., and Walser, T. (2012). Life cycle assessment of engineered nanomaterials: State of the art and strategies to overcome existing gaps. *The Science of the Total Environment*, 425: 271–282. https://doi.org/10.1016/j.scitotenv.2012.03.001.

Hossain, A., Kerry, R. G., Farooq, M., Abdullah, N., and Tofazzal Islam, M. (2020). Application of Nanotechnology for Sustainable Crop Production Systems. pp. 135–159. *In:* Thangadurai, D., Sangeetha, J., and Prasad, R. (eds.). *Nanotechnology for Food, Agriculture, and Environment*. Springer International Publishing. https://doi.org/10.1007/978-3-030-31938-0_7.

Huang, Q., Chen, Y., Zhang, W., Xia, X., Li, H., Qin, M. et al. (2024). Nanotechnology for enhanced nose-to-brain drug delivery in treating neurological diseases. In Journal of Controlled Release (Vol. 366, pp. 519–534). Elsevier B.V. doi: 10.1016/j.jconrel.2023.12.054.

Iavicoli, I., Leso, V., Beezhold, D., and Shvedova, A. (2017). Nanotechnology in agriculture: Opportunities, toxicological implications, and occupational risks. *Toxicology and Applied Pharmacology*, 329: 96–111. https://doi.org/10.1016/j.taap.2017.05.025.

Jafarzadeh, S., Zargar, M., Garavand, F., Forough, M., Amjadi, S., Kouzegaran, V. J. et al. (2023). Plant protein-based nanocomposite films: A review on the used nanomaterials, characteristics, and food packaging applications. *Critical Reviews in Food Science and Nutrition*, 63(29): 9667–9693. https://doi.org/10.1080/10408398.2022.2070721.

Jebarani D., A. H., and Alex, S. A. (2023). Nanobiohybrids: A promising approach for sensing diverse environmental water pollutants. In *Nanobiohybrids for Advanced Wastewater Treatment and Energy Recovery*. IWA Publishing. https://doi.org/10.2166/9781789063592_0147.

Kalita, D., and Baruah, S. (2019). Chapter 11—The impact of nanotechnology on food. pp. 369–379. *In:* do Nascimento, R. F., Ferreira, O. P., Paula, A. J. D., and V. de O. S. Neto (eds.). *Nanomaterials Applications for Environmental Matrices*. Elsevier. https://doi.org/10.1016/B978-0-12-814829-7.00011-2.

Karnwal, A., Kumar Sachan, R. S., Devgon, I., Devgon, J., Pant, G., Panchpuri, M. et al. (2024). Gold Nanoparticles in Nanobiotechnology: From Synthesis to Biosensing Applications. In ACS Omega (Vol. 9, pp. 29966–29982). American Chemical Society. doi: 10.1021/acsomega.3c10352.

Katya M., Aguilar-Pérez, R. P.-S., Gustavo Ruiz-Pulido, Dora I. Medina, and Iqbal, H. M. N. (2023). Insight of nanotechnological processing for nano-fortified functional foods and nutraceutical—Opportunities, challenges, and future scope in food for better health. *Critical Reviews in Food Science and Nutrition*, 63(20): 4618–4635. https://doi.org/10.1080/10408398.2021.2004994.

Khalifa, A. M., Safhi, F. A., and Elsherif, D. E. (2024). Green synthesis of a dual-functional sulfur nanofertilizer to promote growth and enhance salt stress resilience in faba bean. *BMC Plant Biology*, 24(1): 607.

Khoshraftar, Z., Safekordi, A. A., Shamel, A., and Zaefizadeh, M. (2019). Synthesis of natural nanopesticides with the origin of Eucalyptus globulus extract for pest control. *Green Chemistry Letters and Reviews*, 12(3): 286–298.

Kim, D. Y., Kadam, A., Shinde, S., Saratale, R. G., Patra, J., and Ghodake, G. (2018). Recent developments in nanotechnology transforming the agricultural sector: a transition replete with opportunities. pp. 849–864. *In:* Journal of the Science of Food and Agriculture (Vol. 98). John Wiley and Sons Ltd. doi: 10.1002/jsfa.8749.

Kumar, N., Kaur, P., and Bhatia, S. (2017). Advances in bio-nanocomposite materials for food packaging: a review. *Nutrition and Food Science*, 47(4): 591–606. doi: 10.1108/NFS-11-2016-0176/FULL/XML.

Kumar, V., Kaushik, N. K., Tiwari, S. K., Singh, D., and Singh, B. (2023). Green synthesis of iron nanoparticles: Sources and multifarious biotechnological applications. *International Journal of Biological Macromolecules*, 253: 127017.

Lamri, M., Bhattacharya, T., Boukid, F., Chentir, I., Dib, A. L., Das, D. et al. (2021). Nanotechnology as a Processing and Packaging Tool to Improve Meat Quality and Safety. *Foods*, 10(11). https://doi.org/10.3390/foods10112633.

Li, J., Zhang, F., Zhong, Y., Zhao, Y., Gao, P., Tian, F. et al. (2022). Emerging Food Packaging Applications of Cellulose Nanocomposites: A Review. In Polymers (Vol. 14). MDPI. doi: 10.3390/polym14194025.

Liu, Y., Ahmed, S., Sameen, D. E., Wang, Y., Lu, R., Dai, J. et al. (2021). A review of cellulose and its derivatives in biopolymer-based for food packaging application. In Trends in Food Science and Technology (Vol. 112, pp. 532–546). Elsevier Ltd. doi: 10.1016/j.tifs.2021.04.016.

Madaleno, M., Dogan, E., and Taskin, D. (2022). A step forward on sustainability: The nexus of environmental responsibility, green technology, clean energy and green finance. *Energy Economics*, *109*. doi: 10.1016/j.eneco.2022.105945.

Mahmoudpour, M., Torbati, M., Mousavi, M. M., de la Guardia, M., and Ezzati Nazhad Dolatabadi, J. (2020). Nanomaterial-based molecularly imprinted polymers for pesticides detection: Recent trends and future prospects. *TrAC - Trends in Analytical Chemistry*, 129: 115943. doi: 10.1016/j.trac.2020.115943.

Merino, D., Casalongué, C., and Alvarez, V. (2017). Polysaccharides as Eco-Nanomaterials for Agricultural Applications. In *Handbook of Ecomaterials*. https://doi.org/10.1007/978-3-319-68255-6_124.

Mihindukulasuriya, S. D. F., and Lim, L. T. (2014). Nanotechnology development in food packaging: A review. *Trends in Food Science & Technology*, 40(2): 149–167. doi: 10.1016/J.TIFS.2014.09.009.

Mustafa, F., and Andreescu, S. (2020). Nanotechnology-based approaches for food sensing and packaging applications. *RSC Advances*, 10(33): 19309–19336. doi: 10.1039/D0RA01084G.

Nazir, M. A., Hasan, M., Mustafa, G., Tariq, T., Ahmed, M. M., Golzari Dehno, R. et al. (2024). Zinc oxide nano-fertilizer differentially effect on morphological and physiological identity of redox-enzymes and biochemical attributes in wheat (*Triticum aestivum* L.). *Scientific Reports*, 14(1): 13091.

Neme, K., Nafady, A., Uddin, S., and Tola, Y. B. (2021). Application of nanotechnology in agriculture, postharvest loss reduction and food processing: Food security implication and challenges. *Heliyon*, 7(12): e08539. https://doi.org/10.1016/j.heliyon.2021.e08539.

Nizam, N., Hanafiah, M., and Woon, K. (2021). A Content Review of Life Cycle Assessment of Nanomaterials: Current Practices, Challenges, and Future Prospects. *Nanomaterials*, 11. https://doi.org/10.3390/nano11123324.

Pang, K. K., Yu, J., Yin, Y., Liu, X., Zou, B., and Tang, S. J. (2025). Assessing the Impact of Nanotechnology on the Well-being of Human Life. *Journal of Information Technology Management*, 17: 73–85. doi: 10.22059/jitm.2025.99924.

Patel, A., Patra, F., Shah, N., and Khedkar, C. (2018). Application of Nanotechnology in the Food Industry: Present Status and Future Prospects. *Impact of Nanoscience in the Food Industry*, 1–27. doi: 10.1016/B978-0-12-811441-4.00001-7.

Pattnaik, C., Mishra, R., Sahu, A. K., Sahoo, L. N., Sahoo, N. K., Tripathy, S. K. et al. (2023). Green synthesis of glucose-capped stable silver nanoparticles: a cost-effective sensor for the selective detection of Hg 2+ ions in aqueous solutions. *Sensors & Diagnostics*, 2(3): 647–656.

Poudel, D. K., Niraula, P., Aryal, H., Budhathoki, B., Phuyal, S., Marahatha, R. et al. (2022). Plant-Mediated Green Synthesis of Ag NPs and Their Possible Applications: A Critical Review. *Journal of Nanotechnology*, 2022(1): 2779237.

Pourzahedi, L., Pandorf, M., Ravikumar, D., Zimmerman, J., Seager, T., Theis, T. et al. (2018). Life cycle considerations of nano-enabled agrochemicals: Are today's tools up to the task? *Environmental Science. Nano*, 5: 1057–1069. https://doi.org/10.1039/C7EN01166K.

Pramanik, P., Krishnan, P., Maity, A., Mridha, N., Mukherjee, A., and Rai, V. (2020). Application of Nanotechnology in Agriculture. pp. 317–348. *In:* Dasgupta, N., Ranjan, S., and Lichtfouse, E. (eds.). *Environmental Nanotechnology Volume 4*. Springer International Publishing. https://doi.org/10.1007/978-3-030-26668-4_9.

Primožič, M., Knez, Ž., and Leitgeb, M. (2021). (Bio)Nanotechnology in Food Science—Food Packaging. *Nanomaterials*, 11(2). https://doi.org/10.3390/nano11020292.

Ragkos, A., Hayashi, K., Serra, J., Shibata, H., Michalis, E., Eguchi, S. et al. (2021). Contrasting Considerations among Agricultural Stakeholders in Japan on Sustainable Nitrogen Management. *Sustainability*, 13: 4866. https://doi.org/10.3390/SU13094866.

Ramachandraiah, K., and Hong, G. P. (2021). Polymeric nanomaterials for the development of sustainable plant food value chains. In Food Bioscience (Vol. 41). Elsevier Ltd. doi: 10.1016/j.fbio.2021.100978.

Rathee, S., Nayak, V., Singh, K. R., and Ojha, A. (2022). Nanofortification of vitamin B-complex in food matrix: Need, regulations, and prospects. *Food Chemistry: Molecular Sciences*, 4: 100100. https://doi.org/10.1016/j.fochms.2022.100100.

Raut, D., Afrayeem, S., Singh, V., Jejal, A., Sachan, P., Sahoo, S., and Pandey, S. (2024). Enhancing Nutrition, Crop Resilience, and Food Security through Biofortification. *International Journal of Environment and Climate Change*, 14: 241–253.

Saad, A. M., Alabdali, A. Y. M., Ebaid, M., Salama, E., El-Saadony, M. T., Selim, S. et al. (2022). Impact of green chitosan nanoparticles fabricated from shrimp processing waste as a source of nano nitrogen fertilizers on the yield quantity and quality of wheat (*Triticum aestivum* L.) cultivars. *Molecules*, 27(17): 5640.

Sathiyabama, M., Boomija, R. V., Muthukumar, S., Gandhi, M., Salma, S., Prinsha, T. K. et al. (2024). Green synthesis of chitosan nanoparticles using tea extract and its antimicrobial activity against economically important phytopathogens of rice. *Scientific Reports*, 14(1): 7381.

Scott-Fordsmand, J. J., Amorim, M. J. de B., de Garidel-Thoron, C., Castranova, V., Hardy, B., Linkov, I. et al. (2021). Bridging international approaches on nanoEHS. *Nature Nanotechnology*, 16(6): 608–611. https://doi.org/10.1038/s41565-021-00912-5.

Sharma, S., and Kar, D. (2019). *An Insight into Plant Nanobionics and Its Applications* (pp. 65–82). https://doi.org/10.1007/978-3-030-12496-0_3.

Sharma, C., Dhiman, R., Rokana, N., and Panwar, H. (2017). Nanotechnology: An untapped resource for food packaging. *Frontiers in Microbiology*, 8(SEP): 243298. doi: 10.3389/FMICB.2017.01735/BIBTEX.

Shebl, A., Hassan, A. A., Salama, D. M., Abd El-Aziz, M. E., and Abd Elwahed, M. S. (2019). Green synthesis of nanofertilizers and their application as a foliar for *Cucurbita pepo* L. *Journal of Nanomaterials*, 2019(1): 3476347.

Shekhar, C., Khosya, R., Thakur, K., Mahajan, D., Kumar, R., Kumar, S. et al. (2024). A systematic review of pesticide exposure, associated risks, and long-term human health impacts. *Toxicology Reports*, 101840.

Singh, H., Sharma, A., Bhardwaj, S. K., Arya, S. K., Bhardwaj, N., and Khatri, M. (2021). Recent advances in the applications of nano-agrochemicals for sustainable agricultural development. *Environmental Science: Processes & Impacts*, 23(2): 213–239. https://doi.org/10.1039/D0EM00404A.

Singh, S., Singh, R., Singh, K., Katoch, K., Zaeen, A. A., Birhan, D. A. et al. (2024). Smart fertilizer technologies: An environmental impact assessment for sustainable agriculture. *Smart Agricultural Technology*, 8: 100504. https://doi.org/10.1016/j.atech.2024.100504.

Sridhar, A., Ponnuchamy, M., Kumar, P. S., Kapoor, A., Vo, D.-V. N., and Rangasamy, G. (2023). Digitalization of the agro-food sector for achieving sustainable development goals: A review. *Sustainable Food Technology*, 1(6): 783–802. https://doi.org/10.1039/d3fb00124e.

Torres-Giner, S., Prieto, C., and Lagaron, J. M. (2020). Nanomaterials to enhance food quality, safety, and health impact. *Nanomaterials*, 10(5). https://doi.org/10.3390/nano10050941.

Trajkovska Petkoska, A., Daniloski, D., D'Cunha, N. M., Naumovski, N., and Broach, A. T. (2021). Edible packaging: Sustainable solutions and novel trends in food packaging. *Food Research International*, 140: 109981. doi: 10.1016/J.FOODRES.2020.109981.

Varghese Alex, K., Tamil Pavai, P., Rugmini, R., Shiva Prasad, M., Kamakshi, K., and Sekhar, K. C. (2020). Green synthesized Ag nanoparticles for bio-sensing and photocatalytic applications. *ACS omega*, 5(22): 13123–13129.

Vats, S., Kumawat, S., Brar, J., Kaur, S., Yadav, K., Magar, S. G. et al. (2022). Opportunity and challenges for nanotechnology application for genome editing in plants. In *Plant Nano Biology* (Vol. 1). Elsevier B.V. doi: 10.1016/j.plana.2022.100001.

Wahab, A., Rahim, A. A., Hassan, S., Egbuna, C., Manzoor, M. F., Okere, K. J. et al. (2021). Application of nanotechnology in the packaging of edible materials. *Preparation of Phytopharmaceuticals for the Management of Disorders: The Development of Nutraceuticals and Traditional Medicine*, 215–225. doi: 10.1016/B978-0-12-820284-5.00004-6.

Wypij, M., Trzcińska-Wencel, J., Golińska, P., Avila-Quezada, G. D., Ingle, A. P., and Rai, M. (2023). The strategic applications of natural polymer nanocomposites in food packaging and agriculture: Chances, challenges, and consumers' perception. *Frontiers in Chemistry*, 10. https://doi.org/10.3389/fchem.2022.1106230.

Wyser, Y., Adams, M., Avella, M., Carlander, D., Garcia, L., Pieper, G. et al. (2016). Outlook and Challenges of Nanotechnologies for Food Packaging. *Packaging Technology and Science*, 29(12): 615–648. doi: 10.1002/PTS.2221.

Yılmaz, B. B. (2024). Perspectives of Biodegradable Nanocoatings in Food Packaging: pp. 116–172. *In:* R. Garg & A. Anjum (eds.). *Advances in Chemical and Materials Engineering*. IGI Global. https://doi.org/10.4018/979-8-3693-3136-1.ch005.

Yu, K., Zhao, S., Sun, B., Jiang, H., Hu, L., Xu, C. et al. (2024). Enhancing Food Production Through Modern Agricultural Technology. In Plant, Cell and Environment. John Wiley and Sons Inc. doi: 10.1111/pce.15299.

Zambrano-Zaragoza, M. L., González-Reza, R., Mendoza-Muñoz, N., Miranda-Linares, V., Bernal-Couoh, T. F., Mendoza-Elvira, S. et al. (2018). Nanosystems in edible coatings: A novel strategy for food preservation. In International Journal of Molecular Sciences (Vol. 19). MDPI AG. doi: 10.3390/ijms19030705.

Zavala-Zapata, V., Ramírez-Barrón, S. N., Sánchez-Borja, M., Aguirre-Uribe, L. A., Delgado-Ortiz, J. C., Sánchez-Peña, S. R. et al. (2024). Insecticide efficacy of green synthesis silver nanoparticles on diaphorina citri kuwayama (Hemiptera: Liviidae). *Insects*, 15(7): 469.

Zheng, Q., Wu, J., Yan, W., Zhu, S., Miao, X., Wang, R. et al. (2023). Green synthesis of a chlorfenapyr chitosan nanopesticide for maize root application: Reducing environmental pollution and risks to nontarget organisms. *International Journal of Biological Macromolecules*, 253: 126988.

CHAPTER-16

Nanoparticle-Delivering Biostimulants for High-Yield, High-Quality and Stress-Resilient Crops

Mohsen Mohamed Elsharkawy

1. Introduction

There are several reasons for stressing on the production of food and the distribution system, such as rapid climate change, water shortages, population increase, and soil pollution (Usman et al., 2020; Ijaz et al., 2023). Food security is now the century's greatest problem due to the reduction of fertile land caused by the growth in industrialization and urbanization (Kopittke et al., 2019). A more robust and productive strategy is indispensable for supplying the world's food requirements. While food supplies have been developed considerably, technical innovations are still necessary for its sustainability (Husaini and Sohail, 2023). Additionally, sustainable agricultural production has been at risk due to the inefficient use of agrochemicals. Plant sustainability is negatively impacted by the use of synthetic fertilizers because of their low nutritional effectiveness, elevated environmental concerns, increased production costs, and decreased farmer earnings (Ahmed et al., 2023). To fulfil future global food demands responsibly, a high-tech alteration in stress management systems is required. Several scientific areas have linked with agri-nanotechnology to provide novel procedures that enhance the agricultural characteristics of crops (Fraceto et al., 2016; Ijaz et al., 2023). Using effectively controlled inputs, such as pesticides, fertilizers, and agrochemicals, facilitated by agri-nanotechnology is essential for maximizing crop yield. A range of applications of nanotechnology could improve agricultural yield and quality.

The concept of biostimulants is broad and insufficiently clear. Biostimulants are natural medications that promote the overall health, vitality, and development of plants while also protecting them against diseases. They are suitable for both agricultural and horticulture crops. Biostimulants can be added to fertilizers to improve nutrient absorption, plant growth, and resistance to different stressors. Biostimulants can include enzymes, proteins, amino acids, vitamins, nanomaterials and other substances.

In addition to coating seeds, nanomaterials may be used as nanobiosensors, nanofertilizers, nanocarriers, and nanopesticides to effectively control pests (Table 1). These new approaches based

---

Agricultural Botany Department, Faculty of Agriculture, Kafrelsheikh University, Kafr Elsheikh 33516, Egypt.
Email: mohsen.abdelrahman@agr.kfs.edu.eg, https://orcid.org/0000-0003-4731-056X

DOI: 10.1201/9781003567684-16

Table 1. Utilization of nanomaterials in stress-resilient crops.

Application	Nanomaterial	Citation
Nano-biosensors for Chili leaf curl betasatellite (ChLCB) detection	MWCNTs-Zn NPs	Tahir et al., 2017
Nanofertilizer	Carbon Nanoparticles	Xin et al., 2022
Stimulation of plant-microbe interactions	Pristine and sulfidized zinc oxide nanoparticles	Chen et al., 2023
Nanopesticides	ZnO NPs, CuO NPs, AgNPs and FeO NPs	Jiang et al., 2021; Kim et al., 2022
Nano-carriers	Polyethylene oxide nanofibers	Campaña and Arias 2020
Nano-coatings	Titanium dioxide nanoparticles	Bundele et al., (2023)
Nano-clays	Nanocomposite powder with titanium Dioxide and montmorillonite clay	James et al., (2023)
Nanobubble irrigation	Nanobubble water	He et al., (2022)
Nano material for increased storage	Silver nanoparticles	Nayab and Akhtar (2023)

on nanotechnology have the ability to transform farming methods and aid in the creation of resilient and sustainable systems for farming. These measures can lead to a higher yield by using fewer resources and minimizing environmental collateral harm.

2. Interaction Between Nanoparticles and Plants

Scientific and technological developments are helping the agriculture industry by providing novel ideas and answers to serious problems. Since the development of agri-nanotechnology, new effective and pollutant-free nanoformulations for environmentally friendly farming are consistently being developed (Fraceto et al., 2016). The innovative substances have the capacity to change plant behaviors as soon as they reach the complicated plant-soil system, which may be simply used to analyze the consequences. Substantial comprehension of beneficial or detrimental interactions between NMs (nanomaterials) with plants is imperative. Therefore, they may offer novel approaches to developing superior items based on nanomaterials. Additionally, NM doses in natural environments are less than the declared hazardous. Detailed safety evaluations are still needed (Batley et al., 2013).

Cerium-oxide nanoparticles have been shown to improve plants' photosynthetic efficiency by lowering the buildup of reactive oxygen species during stressful situations (Wu et al., 2017, 2018). Additionally, wheat seeds immersed in MWCNTs (multiwall carbon nanotubes) solution have been shown to expedite root development and enhance vegetative biomass (AboDalam et al., 2022). TiO_2 and SiO_2 NPs were applied to soybean seeds to boost nitrate reductase activity, which promotes seed germination; however, the simultaneous use of both nanomaterials was more advantageous (Behl et al., 2022). Likewise, immersing watermelon seeds in Fe_2O_3 nanoparticles enhanced germination, stimulated plant development, and finally increased fruiting (Guleria et al., 2022). Additionally, immersing seeds in an oxidized MWCNT solution increased the uniformity, germination rate, and root and shoot development of Indian mustard (Szöllősi et al., 2020). Whenever mesoporous silica nanoparticles are introduced through a nutritional medium, nanoparticles concentrate in the chloroplast, which helps the seedling's growth by enhancing photosynthesis (Sun et al., 2016). Additionally, treating seeds with TiO_2 nanoparticles (0.25%) accelerated nitrogen absorption and photosynthesis and increased spinach's dry weight (Feng et al., 2022). Likewise, carbon nanotubes sprayed on tomato leaves can enhance growth (McGehee et al., 2017). The reliability of carbon nanotubes from an environmental standpoint is still debateable and has to be confirmed for field use.

The size, makeup of chemicals, functional groups, and coating type all have a significant impact on the absorption of NMs. Plant physiology is impacted by the interaction and absorption of NMs, which results in molecular alterations (Jin et al., 2017). MWCNTs (Multiwalled-carbon nanotubes) have an intriguing impact on tobacco cell development, increasing it by 55–64%. This effect is due to a special molecular process. Cell wall expansion, cell division, and tobacco aquaporin correspond to the bulk of the active genes (Khodakovskaya et al., 2012). The key to greater development and vigor is NMs' capacity to penetrate the seeds' outer layer and permit water uptake. Furthermore, a promising tactic is the seed priming approach, which further supports the ability of high-yield crops by using nanotechnology prior to sowing (Acharya et al., 2020). The growing process of several essential crops, including maize, soybeans, and barley, is impacted by the application of MWCNTs. TEM analysis and Raman spectroscopy verified that MWCNTs congregate inside the endosperm of treated seeds (Lahiani et al., 2013). Plant development was facilitated by NPs' interaction with cellular machinery after they were delivered to various plant parts. MSN (Mesoporous silica)-based NPs can promote photosynthesis by binding to chloroplasts, which raises the amount of total protein, chlorophyll, and seed germination. The fact that none of the plants experienced stress at the greatest amount of MSNs (2.000 mg/l) suggests that using them as a smooth release mechanism is safer (Sun et al., 2016). Silica is vital for nutrition because the insufficient supply of it weakens plants and increases their vulnerability to stress (Rafi et al., 1997). Additionally, silica treatment increases biomass in plants under stress and minimizes the effect of drought stress (Ahmad et al., 2007). Maize responds well to silica NPs derived from rice husk. Maize plants benefited from increased silica absorption and extended roots, leading to improved drought resistance. Root sectioning shows that silica NPs are directly absorbed and accumulated in maize epidermal cells (Suriyaprabha et al., 2012). The influence of silica NPs on crops was widely studied (Rastogi et al. 2019). TiO_2 NP has been shown to enhance plant development in *Brassica napus*, resulting in improved seed vigor. Large radicles and enhanced plumule development were the most noticeable effects (Mahmoodzadeh et al., 2013). Nanoparticles can have both good and detrimental effects depending on their concentration and size. Plants obtain smaller TiO_2 NPs more efficiently than bigger particles (Larue et al., 2012). Furthermore, these nanoparticles can influence miRNA levels, triggering plant growth (Boykov et al., 2019). Fe NPs had a favorable influence on *Capsicum annum* development at lower doses, boosting chloroplast numbers. However, larger quantities of these NPs were harmful (Yuan et al., 2018). Metal-oxide nanoparticles, including Cu and ZnO, have been shown to impact the germination of black gram seed (Raja et al., 2019). PVP-protected platinum nanoparticles improve germination of seeds and growth in peas (Rahman et al., 2020). Chitosan NP improves seed germination and development by increasing IAA levels in wheat even at low concentrations (Li et al., 2019).

There is insufficient research on the effects of nanocoating strategies on the agrochemical sector to determine their degree of technological readiness. It is also challenging to evaluate the benefits of the technology in a complete manner because its advantages are not consistently disclosed. Recent research indicates that using electrospun cobalt NPs coating and biosynthesized AgNPs coatings has exciting potential to increase germination efficiency (Mahakham et al., 2017; Krishnamoorthy and Rajiv, 2017). The development of tobacco callus and physiological indicators were more significantly impacted by the use of zinc as ZnO than by other forms of zinc. The form of ZnO provides more zinc after application to the callus. This promoted the formation of callus and increased the amount of protein (Mazaheri-Tirani and Dayani, 2020; Wiszniewska, 2021). ZnO NPs applied topically raised the amounts of phosphorus, total soluble leaf protein, and chlorophyll in cluster beans (Choudhary et al., 2017). The utilization of dextran-coated ZnO-NPs in wheat plants demonstrated and improved Zn content and growth performance in comparison to the conventionally used zinc supplements. Additionally, it has been revealed that nanomaterials are highly effective in improving plant development through upgrading and speeding up seed germination and increasing fertilization capacity (Wu and Li, 2022).

3. Nanofertilizers

Nowadays, a wide range of issues affect agriculture worldwide, including insufficient water availability, declining soil organic-matter, crop output stagnation, nutritional deficiencies, scarcity of workers, reduced land area as a result of urbanization and land degradation, and nutritional deficiencies in the soil (Godfray et al., 2010). The use of nanotechnology and nanoscience is expanding rapidly, and new techniques are always being suggested to provide innovative and appealing materials for crop management and production (Mittal et al., 2020). It is an important concept in improving the field of precision agriculture, which allows farmers to use fertilizers and other inputs effectively. In an attempt to boost food production and safeguard crops, the growing population has led to the massive manufacturing of conventional fertilizers, which decreases soil fertility and product quality. In addition to exacerbating the problems of the vulnerable ecology, these fertilizers have harmful effects on human health. Innovative approaches, like nanotechnology, are essential for sustainable agriculture.

Nano-based fertilization methods can improve plant growth and development through direct nutrient absorption, tailored administration, and gradual nutrient release (Rehman et al., 2022). AuNPs and AgNPs are commonly used in nanofertilizers because of their ease of preparation, bioconjugation, and favorable outcomes under different experimental circumstances. Heavy metal nanoparticles are used for their distinctive characteristics and possibly advantageous effects to improve plant development and absorption of nutrients, rather than meeting nutritional needs (Kanwar et al., 2019). NPs tailored to the demands of multiple plant-related mechanisms have been developed, depending on the utilization of different metals (Ali et al., 2020; Perea Vélez et al., 2021). Nanoscale iron has been shown to dramatically improve germination, chlorophyll content, and various physiological indicators in maize and rice (Prerna et al., 2021). Spraying of TiNPs at a suitable concentration on leaves can improve photosynthesis and other physiological processes in plants (Larue et al., 2018). ZnNPs enhance the germination in several crops, including wheat, peanuts, onions, and soybeans (Faizan et al., 2020). Cu-CNFs boost plant sustenance and development by improving water absorption, protein, and chlorophyll contents (Ashfaq et al., 2017). Another approach is to encapsulate Cu-Zn-CNFs in starch and polyvinyl alcohol polymeric composites. The novel Cu-Zn-CNFs composites supported plant growth (Kumar et al., 2018). Graphene and GO-based carbon compounds lead to considerable improvements in plant growth. Plants treated with graphene oxide showed shorter seminal roots, but there were also more adventitious roots (Jiao et al., 2016). Graphene quantum dots were found to improve the growth of coriander and garlic plants by boosting the growth of shoots, leaves, roots, flowers, and fruits (Chakravarty et al., 2015).

Nanofertilizers (NF), also known as smart fertilizers or eco-friendly fertilizers, have the ability to improve application amounts and reduce nutrient loss, particularly of nitrogen and phosphorus (Dimkpa and Bindraban, 2017). It provides a regulated and progressive release of nutrients to the intended location, decreasing the risk of environmental and water pollution (Dwivedi et al., 2016). NFs supply nutrients to plants and improve fertilizer effectiveness in small amounts (Rameshaiah et al., 2015). Encapsulating fertilizers in nanoparticle form can enhance nutrient uptake, which in turn lowers nutrient loss, enhances crop quality and output, and reduces the chance of environmental deterioration. It has also shown that using NF topically might lessen plant stress (Tarafdar et al., 2012).

Based on the plant nutritional requirements, NF may be classified into three categories including macronutrient NF, micronutrient NF, and nanoparticulate NF. A mixture of components, including potassium, calcium, magnesium, nitrogen, and phosphorus, make up macronutrient nanofertilizers. Developed hydroxyapatite NPs with a size of 16 nm considerably improved soybean growth and yield compared to the control group. Studies indicate that nanoparticles can effectively fertilize agricultural crops (Liu and Lal, 2015). Compared to macronutrients, plant micronutrients—which include molybdenum, iron, copper, nickel, zinc, and manganese—are needed in far lesser quantities for crops to develop healthily. Composite formulations of micronutrients are more efficient in supplying adequate nutrients and have a lower negative impact on the environment. NPs contain micronutrients that improve fertilizer usage efficiency and nutrient bioavailability in comparison to other traditional salts and ions (Dimkpa and Bindraban, 2017). CuO, ZnO, and B_2O_3 are three micronutrient NPs that have been effectively utilized for agronomic reinforcement to reduce the stress of drought in soybean (Dimkpa et al., 2017). Nanoparticulate NF, including TiO_2, CNTs, and SiO_2, promote plant development. Combining TiO_2 and SiO_2 enhanced growth, seed germination, and nitrogen fixation in soybeans (Lu et al., 2002).

Through releasing nutrients gradually over time, nanomaterials can help reduce fertilizer loss from runoff and infiltration. Apatite nanoparticles (NPs) increased soybean yield and growth rate by decreasing the mobility of soil phosphorus and maintaining its bioavailability (Liu and Lal, 2014). Additionally, the use of surfactant-modified zeolites increases availability and existence by causing a gradual release of phosphate. It is possible to directly apply a nanoscale formulation of nutrients such as iron and copper to plants to promote plant development (Dasauni and Nailwal, 2022). The benefits of applying these NF have been thoroughly covered in literature; for instance, the use of micronutrients NF is indirectly linked to improved absorption of macronutrients (Dimkpa et al., 2018; Predoi et al., 2020). Limited field research has indicated that modest concentrations of nanoscale micronutrients could enable crops to develop more quickly in the face of abiotic stress (Dimkpa et al., 2019). Despite the limitations, field research has indicated that modest concentrations of nanoscale micronutrients improve crop development and help overcome abiotic stress conditions (Arsic et al., 2020). Three distinct types of nanohybrids—Zn-doped hydroxyapatite-urea, hydroxyapatite-urea, and Mg-doped hydroxyapatite-urea—can be synthesized in two steps for use as slow-releasing fertilizers containing nitrogen. The results clearly show that 50% nitrogen dosage nanohybrids preserved wheat crop production and nitrogen absorption efficiency comparable to 100% nitrogen dose urea fertilizer, which is a step forward to reduce ammonia in agriculture (Sharma et al., 2022). The creation of phosphorus-containing hydroxyapatite nanoparticles with coffee ground pomegranate peel and extracts suggests a new green synthesis method for phosphorus nanofertilizer manufacture (Abdelmigid et al., 2022). Studies have been conducted in laboratory settings, greenhouses, and fields to examine the effects of NF on plant growth and production. Although micronutrients are required in trace levels that are easily controlled, as opposed to macronutrients, which are required in huge quantities for field applications, the technical readiness level for this method is often greater. Nonetheless, the impact of applying nanoscale micronutrients is likewise less than that of applying macronutrients.

NM can be used as a NF since it is smaller than the pore size of plant cell walls, allowing for better nutrient absorption. NF can be applied directly to soil or leaves. There are two main applications for NF. Foliar entrance occurs first, followed by root entry. Spraying the trough cuticle, lenticels, hydathodes, stromata, and wounds results in foliar penetration, whereas the lateral root, root hairs, root tips, and rhizodermis offer root access. Foliar fertilization with NF is suitable for poor soil and weather conditions. This method of administration allows nutrients to enter the plant's leaves directly, eliminating fertilizer waste. Foliar use of NF increases NUE (Nutrient Utilization-Efficiency) and promotes fast crop development (Mahil et al., 2019). The base of trichomes, nano-pores, and stromal openings in leaves enable NP absorption and transit inside the leaf. It has been shown that 43 nm nanoparticles are firmly embedded in the leaf's inner compartment, indicating that the NF is an efficient way to improve nutrient absorption.

NF is recommended for increased NUE due to its superior capacity to transfer nutrients through plasmodesmata's nanosized pores, which typically carry ions between cells (Iqbal et al., 2019). Additionally, plant root entrance exhibits notable nutrient absorption passage, such as being far more permeable to nano fertilizer than to manure or traditional fertilizer. Fluorescently tagged monodispersed-mesoporous silica NPs explains the uptake and translocation process. NPs enter the root via apoplastic and symplastic routes, and then go to the xylem tissue to reach the plant's areal parts, including the stem and leaves (Sun et al., 2014). For example, the movement patterns of CeO_2 and ZnO NPs from the entry into the ryegrass roots to the vascular tissues in soybeans are explained (López-Moreno et al., 2010). These particles enter the cell and proceed either apoplastically or symplastically, passing through the plasmodesmata before arriving to the cytoplasm. They start to go to various cytoplasmic organelles and take part in diverse cell metabolic processes in the cytoplasm. Additionally, it has been shown that plants grown in the hydroponic medium absorb more NPs than sand-grown plants (Solanki et al., 2015). As a result, NFs have incredible ability to move and supply nutrients from cell to cell through a 50–60 nm plasmodesmata channel. Consequently, greater NUE and reduced nutrient loss during nanofertilizer delivery lead to increased field crop and plant yield and improved nutritional quality (Iqbal et al., 2019).

4. Plant Microbiome Reshaping

Nanomaterials can enhance soil sustainability and productivity by boosting nutrient absorption, water retention, and plant pathogen protection (da Silva Júnior et al., 2022). The significance of plant-microbe interactions is widely recognized in agriculture (Trivedi et al., 2022). Regulating soil and phyto-microbiome can improve tolerance to salty irrigation water and prevent diseases. Nanomaterials including Ag, CuO, and TiO_2 can alter the soil microbial population, affecting extremely important soil and plant characteristics. Carbon-based nanomaterials penetrate roots and shoots with varying size, concentration, and solubility, leading to variations in plant metabolism, increased biomass, and optimized grain and fruit yield. Additionally, these nanomaterials show promise in improving photosynthetic efficiency, plant growth, seed germination, and the synthesis of biomolecules, including chemicals with potential medical applications. Multiwalled-carbon nanotubes and graphene-based nanomaterials have a proliferative impact under stress by reducing the detrimental effects of drought and salt stress in plants. Furthermore, by encapsulating fungicides, they have been employed as antifungal agents. Carbon-based nanomaterials affect soil microorganisms, requiring careful evaluation of their impact on plant growth, and disease resistance (Wagay et al., 2019). Using AgNPs, ZnONPs, CuONPs, at a high dose (100 mg/kg), can have an impact on the soil microbiota, but small doses had no significant effect (Zhang et al., 2020). Few researches have explored the use of nanomaterials to engineer soil or phytomicrobiomes. The dynamic character and diversity of soil caused by microbes pose hurdles for practical application of this strategy (Asadishad et al., 2018). Although the technology is not yet ready for use, it plays a significant role in sustainable agriculture.

Improving soil may boost crop output and food quality, especially in places subjected to soil deterioration and environmental stress. Traditional cultivation procedures in agricultural fields do not rely substantially on nanotechnology (Al-Mamun et al., 2021; Wang et al., 2021; Das et al., 2022). However, a variety of nanomaterials can be utilized for soil conditioning (Al-Mamun et al., 2021, Das et al., 2021). The hydrothermally developed nano mineral-based soil conditioner, which is obtained from potassium-rich feldspar, has determined that it is capable of increasing soil density, enhancing soil pH, and reducing the toxicity of aluminium and cadmium (Liu et al., 2017). Utilizing biodegradable cross-linkers can increase water retention ability (Kiran et al., 2019). Applying chitosan NPs could enhance soil quality and increase wheat crop productivity (Zein El-Abdeen and Farroh, 2019). However, most investigations on nano-based conditioning products have been undertaken at laboratory conditions, with limited feasibility evaluations for field-scale usage (Das et al., 2021). The conventional approaches for improving soil properties have been utilized for centuries but replacing them with nanomaterials is challenging due to high criteria and limited options for precision agriculture (Ijaz et al., 2023). Since the technology is currently in its early stages, extensive feasibility testing is required before it can move forward.

5. Nanopesticides

Reducing crop losses from different diseases, pests, and environmental stressors is crucial for maintaining the sustained production of crops. Pesticides using nanotechnology can aid in accomplishing these goals. The worldwide pesticides market is expected to increase continuously (Singh and Mazumdar, 2022). Applying pesticides to the field may have a low chance (0.1%) of reaching their intended location. Nano-formulations, which aid in the targeted administration of pesticides, can raise this precision to a new level. The potential of zinc compounds in inhibiting wheat head scab and the synthesis of deoxynivalenol was examined (Savi et al., 2015). Pre-sowing seed administration of metal nanoparticles led to the onset of the defensive response in wheat against eyespot disease (Panyuta et al., 2016). Numerous types of silver nanoparticles have been shown to reduce plant pathogenic fungi, including *Bipolaris sorokiniana* and *Magnaporthe grisea* (Jo et al., 2009). The AgNPs treatments greatly decreased the fungal invasion. The EC50 of AgNPs needed to reduce colonization by 50% was greater for *B. sorokiniana* than for *M. grisea*. AgNPs can inhibit certain infections, according to the results of the aforementioned studies; however, the results varied and were dependent on the type and quantity of AgNPs utilized. To establish resistance in various crops, chitosan is important. Biofunctionalized MnNPs inhibit watermelon *Fusarium* wilt by interfering with Fon infection, activating host defensive mechanisms, and altering the soil microbial ecology. Watermelon roots and shoots treated with Bio-MnNPs at 100 $\mu g\ mL^{-1}$ had higher Mn contents, which boosted growth and inhibited the Fon infection leading to decreased *Fusarium* wilt. Delivering pesticides to precise sites is essential to avoid fungal invasion. To manage *M. grisea*, an enzyme-responsive β-glucanase nanodelivery is used to modify the chlorothalonil (CHT) in the plant vascular system (Kaziem et al., 2021). This results in the synthesis of CHT@MSNs-β-glucans by attaching β-glucans in yeast cell walls. In comparison to the commercial CHT product (CHT-WP), the results demonstrated that CHT@MSNs-β-glucans exhibited higher protection from alkaline and acidic environments and resistance against *M. grisea* (Kaziem et al., 2021). Chitosan's sustainable and innocuous properties have made it the preferred material for producing nanoparticles in a variety of fields (El-Naggar et al., 2022). It has been discovered that *Phomopsis asparagi, Aspergillus niger*, *Rhizopus oryzae*, *Alternaria alternata*, and *Rhizopus stolonifer* are all immune to the normal polymer form chitosan (Ing et al., 2012). Likewise, the potential of Cu-chitosan nanoparticles to stimulate plant development and improve maize resistance against *Curvularia lunata* has been investigated (Choudhary et al., 2017).

Nanopesticides are currently ready for use and offer a practical way to produce crops in a sustainable manner. To describe the new materials' complexity, including their selectivity and lack of sensitivity to environmental stimulation, more investigation is required. The infield delivery of nanomaterials is fraught with difficulties, such as the change of the structure of nano-formulations. These factors must be carefully evaluated and managed appropriately. In agriculture, nanomaterials are being investigated for enhanced pesticide effectiveness, precise application, and regulated release. However, there is a concern regarding the possible impact of nanopesticides on plant development. This calls for careful study and risk evaluation to guarantee secure and environmentally sound use while minimizing negative effects on ecosystems and health.

Nanotechnology shows significant potential in preventing plant pathogenic-bacteria and fungi by using the antimicrobial characteristics of metal-based nanoparticles (Li et al., 2023). Global food safety is supported by nanopesticides, which provide a more effective and sustainable method of pest management with decreased detrimental effects on the environment.

6. Stimulation of Plant Resistance Against Abiotic Stress

Stressors including drought, heat, frost, and salt cause significant financial losses in the agricultural sector. Climate change continues to have a significant impact on agricultural production by making stressors more frequent and intense, which increases the risk of sustainable cultivation of crops (Vema et al., 2022). Numerous nanomaterials have been employed to mitigate the effects of environmental stressors, consequently promoting plant growth and photosynthetic efficiency (Hassanisaadi et al., 2022; Shalaby et al., 2022). To maintain ionic homeostasis, *Fritillaria imperialis* successfully withstands freezing stress by triggering tolerance mechanisms such as the upregulation of Ca2+ signaling proteins, LEA, NHX1, and P5CS, as well as the induced overexpression of OsCNGC6. Antioxidant systems, such as phenols, SOD, anthocyanins, catalase, and ascorbate peroxidase, are triggered to mitigate the high accumulation of H_2O_2 induced by freezing stress, providing protection against oxidative stress, water stress, and photosynthetic deteriorate during freezing (Hajihashemi et al., 2020). Additionally, in biological applications, nanoparticles—more especially, carbonaceous nanomaterials, or CNMs—have created opportunities for tissue repair (Mamidi et al., 2022).

Plants face several environmental challenges, including drought, UV radiation, salt stress, high temperatures, and salt stress, due to their widespread presence. Plants respond to stress by altering molecular pathways, expressing stress-responsive genes, and producing antioxidative enzymes to scavenge under harsh environments (Rejeb et al., 2014). Plants produce a variety of organic osmolytes, including trehalose and polyols, as well as distinct amino acids, like glycine and proline, to protect themselves against osmotic stress. Leaf application of SiO_2 nanoparticles enhanced the growth of sugarcane under chilling stress by preserving the efficient quantum-yield of cyclic electron transport during photoprotection and photosynthesis (Elsheery et al., 2020).

Some nanoparticles have antioxidative effects on plants by scavenging ROS generated under threats or stress. Exposure to 0.01% TiO_2 NP boosted biomass and chlorophyll content by activating antioxidant enzymes. This led to lower hydrogen peroxide and malondialdehyde levels (Abdel Latef et al., 2018). Nano-TiO_2 was shown to activate the production of important non-coding RNA, which is crucial for abiotic stress tolerance (Frazier et al., 2014). Water deficiency is a major concern in plant cultivation, resulting in decreased vigor and crop damage. Nano-TiO_2 boosts the status of plant hydration by activating the NR (nitrate reductase) enzyme, leading to increased osmolyte accumulation. Elevated NR enzyme leads to synthesis of Nitric Oxide (NO) and triggers proline synthesis (Khan et al., 2020). TiO_2 nanoparticles can protect plants from stress through both enzymatic and non-enzymatic mechanisms. TiO_2 nanoparticles regulate enzymes, including glutamine synthase and glutamate hydrogenase, leading to increased nutrient accumulation and essential oil synthesis (Ahmad et al., 2018). The hypothesized defense of aromatic plants against stress is by altering their composition and profile of essential oil production (Gohari et al., 2020). The results revealed that enhanced resistance to water stress by triggering antioxidant defense mechanisms (Karamian et al., 2020).

NPs help plants develop defense mechanism in an effective manner. Toxic metals bond to the surface of NPs due to their huge surface area and tiny size, decreasing their availability. Abiotic stress factors include drought, temperature changes, salinity, alkalinity, and metal and mineral toxicity. Nanoparticles can operate as nano-enzymes to combat oxidative stress, similar to antioxidant enzymes (Sharifi et al., 2020). Photosynthesis, a crucial metabolic activity in plants, is particularly vulnerable to osmotic and oxidative stress. Therefore, reducing these stresses can help preserve proper functioning. Under stress circumstances, Rubisco, photosystem, and ATP are the main targets of the photosynthetic machinery. Silica nanoparticles were shown to boost plant transpiration, total chlorophyll, water usage efficiency, and carbonic anhydrase activity in squash plants during abiotic stress (Siddiqui et al., 2014). TiO_2 affects photoreduction activity and hinders linolenic acid in the electron-transport chain. It also lowered the rate of oxygen formation in the chloroplasts (Mingyu et al., 2008). Abiotic stress is characterized by the production of ROS by different cell organelles. Additionally, plants have the enzyme machinery necessary to combat the oxidative stress that the environment places on them. However, plants confront the effects of such factors when stress surpasses the defensive system. NMs reduce this kind of stress by supplying amino acids and free nutrients, building up osmolytes, and activating certain genes.

The function of polyamines as essential biostimulants for plant development and resilience to stress is another topic of research. Enhancing crop productivity and stress tolerance may be possible by studying the molecular signaling of PAs in plants (Tyagi et al., 2022). Furthermore, compared to post-treatment application, foliar application of zinc-oxide nanoparticles prior to heat stress reduces heat-induced damage and preserves outstanding plant growth, suggesting that this is a useful tactic for reducing heating-related damage (Kareem et al., 2022).

Nanoceria can have a wide range of effects on plants, both beneficial and detrimental, depending on the type of plant, growing circumstances, coating, exposure concentration, and surface charge (Milenković et al., 2019). The family of CeO NPs (nanoceria) has several uses in the biomedical sector as an antioxidant (Liu and Shi, 2019). Even though NPs can have certain detrimental effects on plants, when we consider their advantages, they outweigh their disadvantages and can be used to improve plant health. As we previously discussed, these NPs may interact at the nano–bio interface to increase plants' resistance to different abiotic stresses by controlling vital pathways (Saxena et al., 2016). For example, high ROS generation from abiotic stress reduces the plant's ability to photosynthesize and may result in the oxidation of biomolecules (Wakeel et al., 2020). The distinct redox potential of nano-Ce, which is based on the easy transition between Ce^{3+} and Ce^{4+} oxidation states, makes it ideal for mitigating ROS negative impact (Collin et al., 2014). Additionally, with a low Ce^{3+}/Ce^{4+} ratio, a CeO NP resembles catalase activity and creates hydrogen peroxide, whereas at a high Ce^{3+}/Ce^{4+} ratio, it mimics superoxide dismutase (Wang et al., 2012; Ma et al., 2015; Pulido-Reyes et al., 2015). The oxidative mitigating effect is developed to other stressors, such as heat, dark freezing, and excessive light. Additionally, it decreases photosystem II abundance, chlorophyll level and photochemical capacity (Gao et al., 2020). The physiology of plants is also threatened by salt stress through a process similar to oxidative stress. It was discovered that CeO NPs might help rapeseed plants recover from the oxidative stress. NaCl (100 mM) had the impact by altering physiobiochemical characteristics, including higher plant biomass, chlorophyll content, and an effective photosystem. CeO nanoparticles reduced ROS levels and lipid peroxidation, leading to enhanced carbon absorption and pollen germination (Djanaguiraman et al., 2018). CeO NPs can reduce the effects of salinity stress on seedlings. Modifying pathways such as antioxidative enzymes, ion binding, Ca^{2+} signaling, and terpene production can reduce oxidative stress and improve salinity tolerance (An et al., 2020).

Si (Silicon) is the second most plentiful mineral on Earth after oxygen, making it a valuable resource in agriculture. Si is classified as a semi-essential element for plants, meaning it is not only necessary for existence but also provides significant benefits (Luyckx et al., 2017). Nanoparticles can interact with plants, leading to physiological and morphological changes that promote stress tolerance. It promotes plant growth, improves biomass and physiology, alters tissue differentiation and aids in stress adaptation. At different concentrations, SiNPs demonstrated anti-stress effects against drought stress in *Crataegus* sp.; the responses of the seedlings differed according to the concentration administered at different phases of drought stress, i.e., mild to severe. Improved photosynthetic capacity, membrane-electrolyte leakage, increased water content, and elevated amounts of proline, carotenoids, and chlorophylls are some of these impacts (Ashkavand et al., 2015). The utilization of Nano-SiO_2 mediated proline accumulation, boosted tolerance in a saline environment, preserved the ionic balance, strengthened the antioxidant system, and raised levels of several phytopropanoids, all of which resulted in osmotic adjustments (Soleymanzadeh et al., 2020). Furthermore, it has been demonstrated that SiO_2 NPs under salt stress improve transpiration rate, stomatal conductance, water usage efficiency, and decreased chlorophyll degradation, resulting in resistance to external insults (Haghighi and Pessarakli, 2013). The main factor responsible for the notable alterations in the epicuticular wax layer is salt stress. Contrary to the salt-stressed plant, strawberry plants treated with nano-Si showed improved epicuticular wax thickness and structure (Avestan et al., 2019). A research on *Capsicum annum* plants found that nano-Si significantly reduced salt stress compared to their bulk counterparts (Tantawy et al., 2015). SiO_2 NPs had a concentration-dependent influence on potato plants under various salt stress levels, resulting in considerable alterations compared to their bulk. At lower concentrations (50 mg/L), NPs improved stress relief, whereas higher concentrations (100 mg/L) resulted in considerable alterations compared to the bulk counterpart (Gowayed et al., 2017). These results significantly contradict the notion that NMs are only harmful and offer a more comprehensive understanding of the beneficial and detrimental effects of NPs. Therefore, before using NPs in any sector, it is necessary to first examine their features.

Plant hormones are essential in times of external stress and provide improved response to various environmental changes (Verma et al., 2016; Raza et al., 2019). Gibberellin, a plant growth hormone, helps plants respond to abiotic challenges including drought, shade, floods, and cold temperatures by slowing growth and focusing on stress resistance (Colebrook et al., 2014). NMs can also be used to deliver plant hormones to the plant; for example, the *A. thaliana* plant was used to test mesoporous-silica nanoparticles that entrap abscisic acid. It clearly demonstrated the extended hormone release and enhanced the seedlings' capacity to resist drought (Sun et al., 2018).

7. Challenges

NMs are becoming the most challenging pollutant class to regulate because of their increased use and concentration in many environments, which pose serious environmental hazards (Jogaiah et al., 2021). The specificity of a nanoparticle's attachment to a biological target and its solubility determine how hazardous it is. The antimicrobial qualities of metal-based ENMs have the ability to trigger an entirely unanticipated series of alterations in the microorganism populations inside environments which could harm both humans and plants (Chauhan et al., 2022). Even though CNMs degrade easily, fungi that break down wood preferentially utilize fullerene. Following their accumulation in microbes, the fullerene NPs are passed throughout the food chain due to feeding interactions. It is possible that prolonged exposure to these NPs might have unanticipated effects on the food chain even in the absence of an acute harmful effect (Sager et al., 2022). There is no effect on human microbiota or gut health from the little quantity of carbon nanotubes. However, more research should be done to clarify the promoting mechanism of nanomaterials and their effects.

8. Conclusion

In the present situation, meeting the nutritional requirements of the world's rapidly expanding population is imperative. To satisfy the world's food needs, a resilient and sustainable agricultural production system is necessary. Extreme climatic changes and growing population demands have rendered conventional chemical-based agricultural methods impractical. According to recent research, nanobiotechnology has the potential for sustainable agricultural production because of its improved genome editing efficiency, targeted nutrient delivery, pest management effectiveness, and potential applications in smart plant sensors. This chapter provides an in-depth mechanistic understanding of the possible uses of synthetic nanomaterials to increase agricultural yield and resilience to stress. Additionally, we discussed the degree of technological preparedness on nano-based approaches to give an understandable sense of the way we currently perceive in this field. The regulatory requirements to reduce related risks and promote public acceptance are examined, along with the current obstacles and ramifications of scaling up nanobiotechnology in crop production, in order to achieve research goals that support a sustainable nano-enabled agri-tech uprising. This chapter concludes by emphasizing the significance of nano-enabled materials for enhancing crop health and illustrating how they might help address concerns about food safety worldwide.

References

Abdel Latef, A. A. H., Srivastava, A. K., El-sadek, M. S. A., Kordrostami, M., and Tran, L. S. P. (2018). Titanium dioxide nanoparticles improve growth and enhance tolerance of broad bean plants under saline soil conditions. *Land. Degrad. Develop.*, 29: 1065–1073. doi: 10.1002/ldr.2780.

Abdelmigid, H. M., Morsi, M. M., Hussien, N. A., Alyamani, A. A., Alhuthal, N. A., and Albukhaty, S. (2022). Green synthesis of phosphorous-containing hydroxyapatite nanoparticles (nHAP) as a novel nano-fertilizer: preliminary assessment on pomegranate (*Punica granatum* L.). *Nanomaterials*, 12: 1527, https://doi. org/10.3390/nano12091527.

AboDalam, H., Devra, V., Ahmed, F.K., Li, B., and Abd-Elsalam, K.A. (2022). Rice wastes for green production and sustainable nanomaterials: an overview, pp. 707–728. *In:* Agri-Waste Microbes Prod. Sustain. Nanomater. Elsevier, 2022, https://doi. org/10.1016/B978-0-12-823575-1.00009-3.

Acharya, P., Jayaprakasha, G. K., Crosby, K. M., Jifon, J. L., and Patil, B. S. (2020). Nanoparticle-mediated seed priming improves germination, growth, yield, and quality of watermelons (*Citrullus lanatus*) at multi-locations in texas. *Sci. Rep.*, 10: 5037. doi: 10.1038/s41598-020-61696-7.

Ahmad, B., Shabbir, A., Jaleel, H., Khan,M.M. A., and Sadiq, Y. (2018). Efficacy of titaniumdioxide nanoparticles in modulating photosynthesis, peltate glandular trichomes and essential oil production and quality in *Mentha piperita* L. *Curr. Plant. Biol.*, 13: 6–15. doi: 10.1016/j.cpb.2018.04.002.

Ahmad, F., Aziz, T., Maqsood, M. A., Tahir, M. A., and Kanwal, S. (2007). Effect of silicon application on wheat (*Triticum aestivum* L.) growth under water deficiency stress. *J. Food Agric.*, 7: 1–7. doi: 10.9755/ejfa.v12i1.5170.

Ahmed T., Noman, M., Gardea-Torresdey, J. L., White, J. C., and Li, B. (2023). Dynamic interplay between nano-enabled agrochemicals and the plant-associated microbiome. *Trends Plant Sci.*, 28: 1310–1325, https://doi. org/10.1016/j.tplants.2023.06.001.

Ali, S., Perveen, S., Ali, M., Jiao, T., Sharma, A.S., Hassan, H. et al. (2020). Bioinspired morphology-controlled silver nanoparticles for antimicrobial application. *Mater. Sci. Eng. C.*, 108: 110421, https://doi. org/10.1016/J. MSEC.2019.110421.

Al-Mamun, M. R., Hasan, M. R., Ahommed, M. S., Bacchu, M. S., Ali, M. R., and Khan, H. M. Z. (2021). Nanofertilizers towards sustainable agriculture and environment, *Environ. Technol. Innov.* 23, 101658, https:// doi.org/10.1016/J.ETI.2021.101658.

An, J., Hu, P., Li, F., Wu, H., Shen, Y., White, J. C. et al. (2020). Emerging investigator series: molecular mechanisms of plant salinity stress tolerance improvement by seed priming with cerium oxide nanoparticles. *Environ. Sci. Nano* 7: 2214–2228. doi: 10.1039/D0EN00387E.

Arsic, M., Le Tougaard, S., Persson, D.P., Martens, H.J., Doolette, C.L., Lombi, E. et al. (2020). Bioimaging techniques reveal foliar phosphate uptake pathways and leaf phosphorus status. *Plant Physiol* 183: 1472, https:// doi.org/10.1104/PP.20.00484.

Asadishad, B., Chahal, S., Akbari, A., Cianciarelli, V., Azodi, M., Ghoshal, S. et al. (2018). Amendment of agricultural soil with metal nanoparticles: effects on soil enzyme activity and microbial community composition. *Environ. Sci. Technol.*, 52: 1908–1918, https://doi.org/10.1021/acs.est.7b05389.

Ashfaq, M., Verma, N., and Khan, S. (2017). Carbon nanofibers as a micronutrient carrier in plants: efficient translocation and controlled release of Cu nanoparticles. *Environ. Sci.: Nano*, 4: 138–148, https://doi. org/10.1039/C6EN00385K.

Ashkavand, P., Tabari, M., Zarafshar, M., Tomášková, I., and Struve, D. (2015). Effect of SiO_2 nanoparticles on drought resistance in hawthorn seedlings. *Fore. Res. Paper*, 76: 350–359. doi: 10.1515/frp-2015-0034.

Avestan, S., Ghasemnezhad, M., Esfahani, M., and Byrt, C. S. (2019). Application of nano-silicon dioxide improves salt stress tolerance in strawberry plants. *Agronomy*, 9: 246. doi: 10.3390/agronomy9050246.

Batley, G. E., Kirby, J. K., and McLaughlin, M. J. (2013). Fate and risks of nanomaterials in aquatic and terrestrial environments. *Acc. Chem. Res.*, 46: 854–862. doi: 10.1021/ar2003368.

Behl, T., Kaur, I., Sehgal, A., Singh, S., Sharma, N., Bhatia, S. et al. (2022). The dichotomy of nanotechnology as the cutting edge of agriculture: nano-farming as an asset versus nanotoxicity, *Chemosphere*, 288: 132533,https://doi. org/10.1016/J.CHEMOSPHERE.2021.132533.

Boykov, I. N., Shuford, E., and Zhang, B. (2019). Nanoparticle titanium dioxide affects the growth and microRNA expression of switchgrass (*Panicum virgatum*). *Genomics*, 111: 450–456. doi: 10.1016/j.ygeno.2018.03.002.

Bundele, M., Rane, N., Lande, V., Dani, A., and Shende, S. (2023). Green synthesis of TiO_2 nanoparticle from *Plumeria rubra* L. leaves for anticorrosive application. *Mater. Today Proc.*, 72: 1685–1691, https://doi.org/10.1016/J. MATPR.2022.09.469.

Campaña, J. M., Arias, M. (2020). Nanofibers as a delivery system for arbuscular mycorrhizal fungi. *ACS Appl. Polym. Mater.*, 2: 5033–5038, https://doi. org/10.1021/acsapm.0c00874.

Chakravarty, D., Erande, M. B., and Late, D. J. (2015). Graphene quantum dots as enhanced plant growth regulators: effects on coriander and garlic plants. *J. Sci. Food Agric.*, 95: 2772–2778, https://doi.org/10.1002/JSFA.7106.

Chauhan, M., Kaur, G., Sharma, B., Chaudhary, G. R., Chauhan, M., Kaur, G. et al. (2022). Antimicrobial applications of engineered metalbased nanomaterials. *Biomed. Transl. Res.*, 495–521, https://doi.org/10.1007/978-981-16-9232-1_27.

Choudhary, R. C., Kumaraswamy, R. V., Kumari, S., Sharma, S. S., Pal, A., Raliya, R. et al. (2017). Cu-chitosan nanoparticle boost defense responses and plant growth in maize (*Zea mays* L.), *Sci. Rep.*, 7: 1–11, https://doi.org/10.1038/s41598-017-08571-0.

Colebrook, E. H., Thomas, S. G., Phillips, A. L., and Hedden, P. (2014). The role of gibberellin signalling in plant responses to abiotic stress. *J. Experiment. Boil.*, 217: 67–75. doi: 10.1242/jeb.089938.

Collin, B., Auffan, M., Johnson, A. C., Kaur, I., Keller, A. A., Lazareva, A. et al. (2014). Environmental release, fate and ecotoxicological effects of manufactured ceria nanomaterials. *Environ. Sci. Nano*, 1: 533–548. doi: 10.1039/C4EN00149D.

Das, D., Sahoo, S., Das, R., and Datta, S.C. (2022). Advances in clay research for sustainable agriculture. *Soil Manag. Sustain. Agric.*, 3–47, https://doi.org/10.1201/9781003184881-2.

Das, P., Gogoi, N., Sarkar, S., Patil, S. A., Hussain, N., Barman, S. et al. (2021). Nano-based soil conditioners eradicate micronutrient deficiency: soil physicochemical properties and plant molecular responses. *Environ. Sci.: Nano*, 8: 2824–2843, https://doi.org/10.1039/D1EN00551K.

Dasauni, K., and Nailwal, T. K. (2022). Nano–delivery system: in the agriculture sector. *NanoEnabled Agrochem. Agric.* 467–484, https://doi.org/10.1016/B978-0-323-91009-5.00014-8.

da Silva Júnior, A. H., Mulinari, J., de Oliveira, P. V., de Oliveira, C. R. S., and Reichert Júnior, F. W. (2022). Impacts of metallic nanoparticles application on the agricultural soils microbiota. *J. Hazard. Mater. Adv.*, 7: 100103, https:// doi.org/10.1016/J.HAZADV.2022.100103.

Dimkpa, C. O., and Bindraban, P. S. (2017). Nanofertilizers: new products for the industry? *J. Agric. Food Chem.*, 66: 6462–6473. doi: 10.1021/acs.jafc.7b02150.

Dimkpa, C. O., Bindraban, P. S., Fugice, J., Agyin-Birikorang, S., Singh, U., and Hellums, D. (2017). Composite micronutrient nanoparticles and salts decrease drought stress in soybean. *Agro. Sustain. Dev.*, 37: 5. doi: 10.1007/s13593-016-0412-8.

Dimkpa, C. O., Singh, U., Bindraban, P. S., Elmer, W. H., Gardea-Torresdey, J. L., and White, J.C. (2018). Exposure to weathered and fresh nanoparticle and ionic Zn in soil promotes grain yield and modulates nutrient acquisition in wheat (*Triticum aestivum* L.). *J. Agric. Food Chem.*, 66: 9645–9656, https://doi.org/10.1021/ACS.JAFC.8B03840.

Dimkpa, C. O., Singh, U., Bindraban, P. S., Elmer, W. H., Gardea-Torresdey, J. L., and White, J. C. (2019). Zinc oxide nanoparticles alleviate drought-induced alterations in sorghum performance, nutrient acquisition, and grain fortification. *Sci. Total Environ.*, 688: 926–934, https://doi.org/10.1016/J.SCITOTENV.2019.06.392.

Djanaguiraman, M., Nair, R., Giraldo, J. P., and Prasad, P. V. V. (2018). Cerium oxide nanoparticles decrease drought-induced oxidative damage in sorghum leading to higher photosynthesis and grain yield. *ACS Omega*, 3: 14406–14416. doi: 10.1021/acsomega.8b01894.

Dwivedi, S., Saquib, Q., Al-Khedhairy, A. A., and Musarrat, J. (2016). Understanding the role of nanomaterials in agriculture," in Microbial Inoculants in Sustainable Agricultural Productivity (New Delhi: Springer), 271–288. doi: 10.1007/978-81-322-2644-4_17.

El-Naggar, N. E. A., Saber, W. E. I. A., Zweil, A. M., and Bashir, S. I. (2022). An innovative green synthesis approach of chitosan nanoparticles and their inhibitory activity against phytopathogenic *Botrytis cinerea* on strawberry leaves. *Sci. Rep.*, 12: 1–20, https://doi.org/10.1038/s41598-022-07073-y.

Elsheery, N. I., Sunoj, V. S. J., Wen, Y., Zhu, J. J., Muralidharan, G., and Cao, K. F. (2020). Foliar application of nanoparticles mitigates the chilling effect on photosynthesis and photoprotection in sugarcane. *Plant Physiol. Biochem. PPB.*, 149: 50–60, https://doi.org/10.1016/J.PLAPHY.2020.01.035.

Faizan, M., Hayat, S., and Pichtel, J. (2020). Effects of zinc oxide nanoparticles on crop plants: a perspective analysis. *Sustain. Agric. Rev.*, 41: 83–99, https://doi.org/10.1007/978-3-030-33996-8_4.

Feng, Y., Kreslavski, V. D., Shmarev, A. N., Ivanov, A. A., Zharmukhamedov, S. K., Kosobryukhov, A. et al. (2022). Effects of iron oxide nanoparticles (Fe_3O_4) on growth, photosynthesis, antioxidant activity and distribution of mineral elements in wheat (*Triticum aestivum*) plants. *Plants*, 11: 1894, https://doi.org/10.3390/PLANTS11141894/S1.

Fraceto, L. F., Grillo, R., de Medeiros, G. A., Scognamiglio, V., Rea, G., and Bartolucci, C. (2016). Nanotechnology in agriculture: which innovation potential does it have? *Front. Environ. Sci.*, 4: 20. doi: 10.3389/fenvs.2016.00020.

Frazier, T. P., Burklew, C. E., and Zhang, B. (2014). Titaniumdioxide nanoparticles affect the growth and microRNA expression of tobacco (*Nicotiana tabacum*). *Funct. Integ. Genom.*, 14: 75–83. doi: 10.1007/s10142-013-0341-4.

Gao, G., Tester, M. A., and Julkowska, M. M. (2020). The use of high-throughput phenotyping for assessment of heat stress-induced changes in *Arabidopsis*. *Plant Phenom.*, 2020, 14. doi: 10.34133/2020/3723916.

Godfray, H. C., Beddington, J. R., Crute, I. R., Haddad, L., Lawrence, D., and Muir, J. F. (2010). Food security: the challenge of feeding 9 billion people. *Science*, 327: 812–818. doi: 10.1126/science.1185383.

Gohari, G., Mohammadi, A., Akbari, A., Panahirad, S., Dadpour, M. R., Fotopoulos, V. et al. (2020). Titanium dioxide nanoparticles (TiO_2 NPs) promote growth and ameliorate salinity stress effects on essential oil profile and biochemical attributes of *Dracocephalum moldavica*. *Sci. Rep.*, 10: 912. doi: 10.1038/s41598-020-57794-1.

Gowayed, M. H., Al-Zahrani, H. S., and Metwali, E. M. (2017). Improving the salinity tolerance in potato (*Solanum tuberosum*) by exogenous application of silicon dioxide nanoparticles. *Int. J. Agric. Biol.*, 19: 183–194. doi: 10.17957/IJAB/15.0262.

Guleria, G., Thakur, S., Shandilya, M., Sharma, S., Thakur, S., and Kalia, S. (2022). Nanotechnology for sustainable agro-food systems: the need and role of nanoparticles in protecting plants and improving crop productivity. *Plant Physiol. Biochem.*, https://doi.org/10.1016/J.PLAPHY.2022.12.004.

Haghighi, M., and Pessarakli, M. (2013). Influence of silicon and nano-silicon on salinity tolerance of cherry tomatoes (*Solanum lycopersicum* L.) at early growth stage. *Sci. Horticult.*, 161: 111–117. doi: 10.1016/j.scienta.2013.06.034.

Hajihashemi, S., Brestic, M., Landi, M., and Skalicky, M. (2020). Resistance of *Fritillaria imperialis* to freezing stress through gene expression, osmotic adjustment and antioxidants. *Sci. Rep.*, 10: 1–13, https://doi.org/10.1038/s41598-020-63006-7.

Hassanisaadi, M., Barani, M., Rahdar, A., Heidary, M., Thysiadou, A., and Kyzas, G. Z. (2022). Role of agrochemical-based nanomaterials in plants: biotic and abiotic stress with germination improvement of seeds. *Plant Growth Regul.*, 97: 375–418, https://doi.org/10.1007/S10725-021-00782-W.

He, J., Liu, Y., Wang, T., Chen, W., Liu, B., Zhou, Y. et al. (2022). Effects of nanobubble in subsurface drip irrigation on the yield, quality, irrigation water use efficiency and nitrogen partial productivity of watermelon and muskmelon. *Int. Agrophysics.*, 36: 163–171, https://doi.org/10.31545/INTAGR/150413.

Husaini, A. M., and Sohail, M. (2023). Robotics-assisted, organic agricultural-biotechnology based environment-friendly healthy food option: beyond the binary of GM versus Organic crops. *J. Biotechnol.*, 361: 41–48.

Ijaz, M., Khan, F., Ahmed, T., Noman, M., Zulfiqar, F., Rizwan, M. et al. (2023). Nanobiotechnology to advance stress resilience in plants: Current opportunities and challenges. *Materials Today Bio.*, 22: 100759, https://doi.org/10.1016/j.mtbio.2023.100759.

Ing, L. Y., Zin, N. M., Sarwar, A., and Katas, H. (2012). Antifungal activity of chitosan nanoparticles and correlation with their physical properties. *Int. J. Biomater.*, https://doi.org/10.1155/2012/632698.

Iqbal, M., Raja, N. I., Hussain, M., Ejaz, M., and Yasmeen, F. (2019). Effect of silver nanoparticles on growth of wheat under heat stress. *Ira. J. Sci. Technol. Transac. A Sci.*, 43: 387–395. doi: 10.1007/s40995-017-0417-4.

James, A. A., Rahman, M. R., Huda, D., Rahman, M. M., Uddin, J., Bin Bakri, M. K. et al. (2023). Optimization of novel nanocomposite powder for simultaneous removal of heavy metals from palm oil mill effluent (POME) by response surface methodology (RSM). *Environ. Dev. Sustain.*, 1–27, https://doi.org/10.1007/s10668-022-02849-8.

Jiang, H., Lv, L., Ahmed, T., Jin, S., Shahid, M., Noman, M. et al. (2021). Effect of the nanoparticle exposures on the tomato bacterial wilt disease control by modulating the rhizosphere bacterial community. *Int. J. Mol. Sci.*, 23: https://doi.org/10.3390/IJMS23010414.

Jiao, J., Yuan, C., Wang, J., Xia, Z., Xie, L., Chen, F. et al. (2016). The role of graphene oxide on tobacco root growth and its preliminary mechanism. *J. Nanosci. Nanotechnol.*, 16: 12449–12454, https://doi.org/10.1166/JNN.2016.12987.

Jin, Y., Fan, X., Li, X., Zhang, Z., Sun, L., Fu, Z. et al. (2017). Distinct physiological and molecular responses in *Arabidopsis thaliana* exposed to aluminum oxide nanoparticles and ionic aluminum. *Environ. Pollut.*, 228: 517–527. doi: 10.1016/j.envpol.2017. 04.073.

Jo, Y.-K., Kim, Byung H., and Jung, Geunhwa. (2009) antifungal activity of silver ions and nanoparticles on phytopathogenic fungi, Plant Dis., 93: 1037–1043, https://doi.org/10.1094/PDIS-93-10-1037.

Jogaiah, S., Paidi, M. K., Venugopal, K., Geetha, N., Mujtaba, M., Udikeri, S. S. et al. (2021). Phytotoxicological effects of engineered nanoparticles: an emerging nanotoxicology. *Sci. Total Environ.*, 801: 149809, https://doi.org/10.1016/J.SCITOTENV.2021.149809.

Kanwar, M.K., Sun, S., Chu, X., and Zhou, J. (2019). Impacts of metal and metal oxide nanoparticles on plant growth and productivity. *Nanomater Plant Potential*, 379–392, https://doi.org/10.1007/978-3-030-05569-1_15.

Karamian, R., Ghasemlou, F., and Amiri, H. (2020). Physiological evaluation of drought stress tolerance and recovery in *Verbascum sinuatum* plants treated with methyl jasmonate, salicylic acid and titanium dioxide nanoparticles. Plant Biosys. *Int. J. Deal. Aspect. Plant. Biol.*, 154: 277–287. doi: 10.1080/11263504.2019.1591535.

Kareem, H.A., Hassan, M.U., Zain, M., Irshad, A., Shakoor, N., Saleem, S. et al. (2022). Nanosized zinc oxide (n-ZnO) particles pretreatment to alfalfa seedlings alleviate heat-induced morpho-physiological and ultrastructural damages. *Environ. Pollut.*, 303: 119069, https://doi.org/10.1016/J.ENVPOL.2022.119069.

Kaziem, A.E., Yang, L., Lin, Y., Kazem, A.E., Xu, H., and Zhang, Z.X. (2021). Pathogenic invasion responsive carrier based on mesoporous silica/β-glucan nanoparticles for smart delivery of fungicides. *ACS Sustain. Chem. Eng.*, 9: 9126–9138, https://doi. org/10.1021/acssuschemeng.1c02962.

Khan, M. N., AlSolami, M. A., Basahi, R. A., Siddiqui, M. H., Al-Huqail, A. A., Abbas, Z. K. et al. (2020). Nitric oxide is involved in nano-titanium dioxide-induced activation of antioxidant defense system and accumulation of osmolytes under water-deficit stress in *Vicia faba* L. *Ecotoxicol. Environ Saf.*, 190: 110152. doi: 10.1016/j. ecoenv.2019.110152.

Kim, S. W., Jung, J. H., Lamsal, K., Kim, Y. S., Min, J. S., and Lee, Y. S. (2012). Antifungal effects of silver nanoparticles (AgNPs) against various plant pathogenic fungi. *Mycobiology*, 40: 53, https://doi.org/10.5941/ MYCO.2012.40.1.053.

Kiran, Tiwari, R., Krishnamoorthi, S., and Kumar, K. (2019). Synthesis of cross-linker devoid novel hydrogels: swelling behaviour and controlled urea release studies. *J. Environ. Chem. Eng.*, 7: 103162, https://doi. org/10.1016/J.JECE.2019.103162.

Khodakovskaya, M. V., De Silva, K., Biris, A. S., Dervishi, E., and Villagarcia, H. (2012). Carbon nanotubes induce growth enhancement of tobacco cells. *ACS Nano*, 6: 2128–2135. doi: 10.1021/nn204643g.

Kopittke, P. M., Menzies, N. W., Wang, P., McKenna, B. A., and Lombi, E. (2019). Soil and the intensification of agriculture for global food security. *Environ. Int.*, 132: 105078.

Krishnamoorthy, V., and Rajiv, S. (2017). Potential seed coatings fabricated from electrospinning hexaaminocyclotriphosphazene and cobalt nanoparticles incorporated polyvinylpyrrolidone for sustainable agriculture. *ACS Sustain. Chem. Eng.*, 5: 146–152, https://doi.org/10.1021/acssuschemeng.6b01088.

Kumar, R., Ashfaq, M., and Verma, N. (2018) Synthesis of novel PVA–starch formulation supported Cu–Zn nanoparticle carrying carbon nanofibers as a nanofertilizer: controlled release of micronutrients. *J. Mater. Sci.*, 53: 7150–7164, https://doi.org/10.1007/S10853-018-2107-9

Lahiani, M. H., Dervishi, E., Chen, J., Nima, Z., Gaume, A., Biris, A. S. et al. (2013). Impact of carbon nanotube exposure to seeds of valuable crops. *ACS Appl. Mater. Interfaces.*, 5: 7965–7973. doi: 10.1021/am402052x.

Larue, C., Laurette, J., Herlin-Boime, N., Khodja, H., Fayard, B., and Flank, A. M. (2012). Accumulation, translocation and impact of TiO_2 nanoparticles in wheat (*Triticum aestivum* spp.): influence of diameter and crystal phase. *Sci. Total Environ.* 431: 197–208. doi: 10.1016/j.scitotenv.2012.04.073

Larue, C., Baratange, C., Vantelon, D., Khodja, H., Surblé, S., Elger, A. et al. (2018) Influence of soil type on TiO_2 nanoparticle fate in an agro-ecosystem. *Sci. Total Environ.* 630: 609–617, https://doi.org/10.1016/J. SCITOTENV.2018.02.264.

Li, R., He, J., Xie, H., Wang, W., Bose, S. K., Sun, Y., et al. (2019). Effects of chitosan nanoparticles on seed germination and seedling growth of wheat (*Triticum aestivum* L.). *Int. J. Boil. Macromolec.*, 126, 91–100. doi: 10.1016/j.ijbiomac.2018.12.118.

Li, Y., Zhang, P., Li, M., Shakoor, N., Adeel, M., Zhou, P. et al. (2023) Application and mechanisms of metal-based nanoparticles in the control of bacterial and fungal crop diseases. *Pest Manag. Sci.*, 79: 21–36, https://doi. org/10.1002/PS.7218.

Liu, R., and Lal, R. (2014). Synthetic apatite nanoparticles as a phosphorus fertilizer for soybean (*Glycine max*). *Sci. Rep.*, 4: 5686, https://doi.org/10.1038/SREP05686.

Liu, S., Qi, X., Han, C., Liu, J., Sheng, X., Li, H. et al. (2017). Novel nano-submicron mineral-based soil conditioner for sustainable agricultural development. *J. Clean. Prod.*, 149: 896–903, https://doi.org/10.1016/J. JCLEPRO.2017.02.155.

Liu, Y., and Shi, J. (2019). Antioxidative nanomaterials and biomedical applications. *Nano Today*, 27:146–177. doi: 10.1016/j.nantod.2019.05.008

López-Moreno, M. L., de la Rosa, G., Hernández-Viezcas, J. Á., Castillo-Michel, H., Botez, C. E., Peralta-Videa, J. R., et al. (2010). Evidence of the differential biotransformation and genotoxicity of ZnO and CeO_2 nanoparticles on soybean (*Glycine max*) plants. *Environ. Sci. Technol.*, 44: 7315–7320. doi: 10.1021/es903891g.

Lu, C., Zhang, C., Wen, J., Wu, G., and Tao, M. (2002). Research of the effect of nanometer materials on germination and growth enhancement of *Glycine max* and its mechanism. *Soybean Sci.*, 21: 168–171.

Luyckx, M., Hausman, J. F., Lutts, S., and Guerriero, G. (2017). Silicon and plants: current knowledge and technological perspectives. *Front. Plant. Sci.*, 8: 411. doi: 10.3389/fpls.2017.00411.

Ma, C., Rui, Y., Liu, S., Li, X., Xing, B., and Liu, L. (2015). Phytotoxic mechanism of nanoparticles: destruction of chloroplasts and vascular bundles and alteration of nutrient absorption. *Sci. Rep.*, 5: 11618. doi: 10.1038/ srep11618.

Mahakham, W., Sarmah, A. K., Maensiri, S., and Theerakulpisut, P. (2017). Nanopriming technology for enhancing germination and starch metabolism of aged rice seeds using phytosynthesized silver nanoparticles. *Sci. Rep.*, 7: 1–21, https://doi.org/10.1038/s41598-017-08669-5.

Mahil, T., Baburai, N., and Aravinda, K. (2019). Foliar application of nanofertilizers in agricultural crops- A review. *J. Farm Sci.*, 32: 239–249.

Mahmoodzadeh, H., Nabavi, M., and Kashefi, H. (2013). Effect of nanoscale titanium dioxide particles on the germination and growth of canola (*Brassica napus*). *J. Ornamen. Horticult. Plants*, 3: 25–32.

Mamidi, N., Velasco Delgadillo, R. M., Barrera, E. V., Ramakrishna, S., and Annabi, N. (2022). Carbonaceous nanomaterials incorporated biomaterials: the present and future of the flourishing field, Compos. Part B Eng. 243 (110150), https://doi.org/10.1016/J.COMPOSITESB.2022.110150.

Mazaheri-Tirani, M., and Dayani, S. (2020). *In vitro* effect of zinc oxide nanoparticles on *Nicotiana tabacum* callus compared to ZnO micro particles and zinc sulfate ($ZnSO_4$), *Plant Cell, Tissue Organ Cult* 140: 279–289, https://doi.org/10.1007/S11240-019-01725-0.

McGehee, D. L., Lahiani, M. H., Irin, F., Green, M. J., and Khodakovskaya, M. V. (2017). Multiwalled carbon nanotubes dramatically affect the fruit metabolome of exposed tomato plants. *ACS Appl. Mater.* Interfaces, 9: 32430–32435, https://doi.org/10.1021/ACSAMI.7B10511.

Milenković, I., Mitrović, A., Algarra, M., Lázaro-Martínez, J. M., Rodríguez- Castellón, E., Milenković, V. et al. (2019). Interaction of carbohydrate coated cerium-oxide nanoparticles with wheat and pea: stress induction potential and effect on development. *Plant*, 8: 478. doi: 10.3390/plants8110478.

Mingyu, S., Chao, L., Chunxiang, Q., Lei, Z., Liang, C., and Hao, H. (2008). Nano-anatase relieves the inhibition of electron transport caused by linolenic acid in chloroplasts of spinach. *Biol. Trace Ele. Res.*, 122: 73–81. doi: 10.1007/s12011-007-8055-x.

Mittal, D., Kaur, G., Singh, P., Yadav, K., and Ali, S. A. (2020). Nanoparticle-Based Sustainable Agriculture and Food Science: Recent Advances and Future Outlook. *Front. Nanotechnol.*, 2: 579954. doi: 10.3389/fnano.2020.579954.

Nayab, D. E., and Akhtar, S. (2023). Green synthesized silver nanoparticles from eucalyptus leaves can enhance shelf life of banana without penetrating in pulp. *PLoS One*, 18: e0281675, https://doi.org/10.1371/JOURNAL.PONE.0281675.

Panyuta, O., Belava, V., Fomaidi, S., Kalinichenko, O., Volkogon, M., and Taran, N. (2016). The effect of pre-sowing seed treatment with metal nanoparticles on the formation of the defensive reaction of wheat seedlings infected with the eyespot causal agent. *Nanoscale Res. Lett.*, 11: 1–5, https://doi.org/10.1186/s11671-016-1305-0.

Perea Vélez, Y. S., Carrillo-González, R., González-Chávez, M., and del C. A. (2021) Interaction of metal nanoparticles–plants–microorganisms in agriculture and soil remediation. *J. Nanoparticle Res.*, 239(23): 1–48, https://doi.org/ 10.1007/S11051-021-05269-3, 2021.

Predoi, D., Ghita, R.V., Iconaru, S. L., Cimpeanu, C. L., and Raita, S. M. (2020). Application of nanotechnology solutions in plants fertilization. *Urban Hortic. - Necessity Futur*, 2020 https://doi.org/10.5772/INTECHOPEN.91240.

Prerna, D. I., Govindaraju, K., Tamilselvan, S., Kannan, M., Vasantharaja, R., Chaturvedi, S. et al. (2021). Influence of nanoscale micro-nutrient α-Fe2O3 on seed germination, seedling growth, translocation, physiological effects and yield of rice (*Oryza sativa*) and maize (*Zea mays*). *Plant Physiol. Biochem.*, 162: 564–580, https://doi.org/10.1016/J.PLAPHY.2021.03.023.

Pulido-Reyes, G., Rodea-Palomares, I., Das, S., Sakthivel, T. S., Leganes, F., Rosal, R. et al. (2015). Untangling the biological effects of ceriumoxide nanoparticles: the role of surface valence states. *Sci. Rep.*, 5: 15613. doi: 10.1038/srep15613.

Rafi, M. M., Epstein, E., and Falk, R. H. (1997). Silicon deprivation causes physical abnormalities in wheat (*Triticum aestivum* L.). *J. Plant Physiol.*, 151: 497–501.doi: 10.1016/S0176-1617(97)80017-X.

Rahman, M. S., Chakraborty, A., Mazumdar, S., Nandi, N. C., Bhuiyan, M. N., and Alauddin, S. M. (2020). Effects of poly (vinylpyrrolidone) protected platinum nanoparticles on seed germination and growth performance of *Pisum sativum*. *NanoStruc. NanoObj.*, 21: 100408. doi: 10.1016/j.nanoso.2019.100408.

Raja, K., Sowmya, R., Sudhagar, R., Moorthy, P. S., Govindaraju, K., and Subramanian, K. S. (2019). Biogenic ZnOand Cu nanoparticles to improve seed germination quality in blackgram (*Vigna mungo*). *Mater. Lett.*, 235: 164–167. doi: 10.1016/j.matlet.2018.10.038.

Rameshaiah, G. N., Pallavi, J., and Shabnam, S. (2015). Nano fertilizers and nano sensors–an attempt for developing smart agriculture. *Int. J. Engineer. Res. Gen. Sci.*, 3: 314–320.

Rastogi, A., Tripathi, D. K., Yadav, S., Chauhan, D. K., Živčák, M., Ghorbanpour, M. et al. (2019). Application of silicon nanoparticles in agriculture. *3Biotech*, 9, 90. doi: 10.1007/s13205-019-1626-7.

Raza, A., Mehmood, S. S., Tabassum, J., and Batool, R. (2019). Targeting plant hormones to develop abiotic stress resistance in wheat. *In Wheat*, 9,13, 961872.doi: 10.1007/978-981-13-6883-7_22.

Rehman, A., Qunyi, T., Usman, M., Manzoor, M.F., Zhao, L., Feng, J. et al. (2022). Nano-enabled agrochemicals for sustainable agriculture, Nano-Enabled Agrochem. *Agric*, 291–306, https://doi.org/10.1016/B978-0-323-91009-5.00020-3.

Rejeb, I. B., Pastor, V., and Mauch-Mani, B. (2014). Plant responses to simultaneous biotic and abiotic stress: molecular mechanisms. *Plants*, 3: 458–475. doi: 10.3390/plants3040458.

Sager, T. M., Umbright, C. M., Mustafa, G. M., Roberts, J. R., Orandle, M. S., Cumpston, J. L. et al. (2022). Pulmonary Toxicity and Gene Expression Changes in Response to Whole-Body Inhalation Exposure to Multi-Walled Carbon Nanotubes in Rats, vol. 34: 2022, pp. 200–218, https://doi.org/10.1080/08958378.2022.2081386.

Savi, G. D., Piacentini, K. C., de Souza, S. R., Costa, M. E. B., Santos, C. M. R., and Scussel, V. M. (2015). Efficacy of zinc compounds in controlling Fusarium head blight and deoxynivalenol formation in wheat (*Triticum aestivum* L.). *Int. J. Food Microbiol.*, 205: 98–104, https://doi.org/10.1016/J.IJFOODMICRO.2015.04.001.

Saxena, R., Tomar, R. S., and Kumar, M. (2016). Exploring nanobiotechnology to mitigate abiotic stress in crop plants. *J. Pharma. Sci. Res.*, 8: 974.

Shah, M. A., and Khan, A. A. (2014). Use of diatomaceous earth for the management of stored-product pests. *Int. J. Pest Manag.*, 60: 100–113. doi: 10.1080/09670874.2014.918674.

Shalaby, T.A., Bayoumi, Y., Eid, Y., Elbasiouny, H., Elbehiry, F., Prokisch, J. et al. (2022). Can nanofertilizers mitigate multiple environmental stresses for higher crop productivity? Sustainability, 14: 3480, https://doi.org/10.3390/SU14063480.

Sharifi, M., Faryabi, K., Talaei, A. J., Shekha, M. S., Ale-Ebrahim, M., and Salihi, A. (2020). Antioxidant properties of gold nanozyme: a review. *J. Mol. Liq.*, 297: 112004. doi: 10.1016/j.molliq.2019.112004.

Sharma, B., Shrivastava, M., Afonso, L. O. B., Soni, U., and Cahill, D. M. (2022). Zinc- and magnesium-doped hydroxyapatite nanoparticles modified with urea as smart nitrogen fertilizers. ACS Appl. *Nano Mater.*, 5: 7288–7299, https://doi.org/10.1021/acsanm.2c01192.

Siddiqui, M. H., and Al-Whaibi, M. H. (2014). Role of nano-SiO_2 in germination of tomato (*Lycopersicum esculentum* seeds Mill.). *Saudi. J. Biol. Sci.*, 21: 13–17. doi: 10.1016/j.sjbs.2013.04.005.

Singh, P., and Mazumdar, P. (2022). Microbial pesticides: trends, scope and adoption for plant and soil improvement. *Biopesticides* (2022) 37–71, https://doi.org/10.1016/B978-0-12-823355-9.00023-7.

Solanki, P., Bhargava, A., Chhipa, H., Jain, N., and Panwar, J. (2015). Nano-fertilizers and their smart delivery system. *In*: Rai, M. (ed.). *Nanotechnologies in Food and Agriculture* (Cham: Springer), 81–101.doi: 10.1007/978-3-319-14024-7_4.

Soleymanzadeh, R., Iranbakhsh, A., Habibi, G., and Ardebili, Z. O. (2020). Selenium nanoparticle protected strawberry against salt stress through modifications in salicylic acid, ion homeostasis, antioxidant machinery, and photosynthesis performance. *Acta. Biolog. Cracoviensia. Botan.*, 62: 33–42.

Sun, D., Hussain, H. I., Yi, Z., Rookes, J. E., Kong, L., and Cahill, D. M. (2016). Mesoporous silica nanoparticles enhance seedling growth and photosynthesis in wheat and lupin. *Chemosphere*, 152: 81–91.doi: 10.1016/j.chemosphere.2016.02.096.

Sun, D., Hussain, H. I., Yi, Z., Rookes, J. E., Kong, L., and Cahill, D. M. (2018). Delivery of abscisic acid to plants using glutathione responsive mesoporous silica nanoparticles. *J. Nanosci. Nanotech.*, 18: 1615–1625. doi: 10.1166/jnn.2018.14262.

Sun, D., Hussain, H., Yi, Z., Siegele, R., Cresswell, T., Kong, L. et al. (2014). Uptake and cellular distribution, in four plant species, of fluorescently labeled mesoporous silica nanoparticles. *Plant. Cell. Rep.*, 33: 1389–1402.doi: 10.1007/s00299-014-1624-5.

Suriyaprabha, R., Karunakaran, G., Yuvakkumar, R., Rajendran, V., and Kannan, N. (2012). Silica nanoparticles for increased silica availability in maize (*Zea mays* L.) seeds under hydroponic conditions. *Curr. Nanosci.*, 8: 902–908. doi: 10.2174/157341312803989033.

Szőllősi, R., Molnár, Á., Kondak, S., and Kolbert, Z. (2020). Dual effect of nanomaterials on germination and seedling growth: stimulation vs. Phytotoxicity. *Plants*, 9: 1745, https://doi.org/10.3390/PLANTS9121745.

Tahir, M. A., Hameed, S., Munawar, A., Amin, I., Mansoor, S., Khan, W. S. et al. (2017). Investigating the potential of multiwalled carbon nanotubes based zinc nanocomposite as a recognition interface towards plant pathogen detection. *J. Virol. Methods.*, 249: 130–136, https://doi.org/10.1016/J.JVIROMET.2017.09.004.

Tantawy, A. S., Salama, Y. A. M., El-Nemr, M. A., and Abdel-Mawgoud, A. M. R. (2015). Nano silicon application improves salinity tolerance of sweet pepper plants. *Int. J. Chem.Tech. Res.*, 8: 11–17.

Tarafdar, J. C., Xiong, Y., Wang, W. N., Quinl, D., and Biswas, P. (2012). Standardization of size, shape and concentration of nanoparticle for plant application. *Appl. Biol. Res.*, 14: 138–144.

Trivedi, P., Batista, B. D., Bazany, K. E., and Singh, B. K. (2022). Plant–microbiome interactions under a changing world: responses, consequences and perspectives. *New Phytol.*, 234: 1951–1959, https://doi.org/10.1111/NPH.18016.

Tyagi, A., Ali, S., Ramakrishna, G., Singh, A., Park, S., Mahmoudi, H. et al. (2022) Revisiting the role of polyamines in plant growth and abiotic stress resilience: mechanisms, crosstalk, and future perspectives. *J. Plant Growth Regul.* 1–25, https://doi.org/10.1007/S00344-022-10847-3.

Usman M., Farooq, M., Wakeel, A., Nawaz, A., Cheema, S. A., ur Rehman, H. et al. (2020) Nanotechnology in agriculture: current status, challenges and future opportunities. *Sci. Total Environ.*, 721: 137778.

Vema, V. K., Sudheer, K. P., Rohith, A. N., and Chaubey, I. (2022). Impact of water conservation structures on the agricultural productivity in the context of climate change. *Water Resour. Manag.*, 36: 1627–1644, https://doi.org/10.1007/S11269-022-03094-4.

Verma, V., Ravindran, P., and Kumar, P. P. (2016). Plant hormonemediated regulation of stress responses. *BMC Plant. Boil.*, 16: 86. doi: 10.1186/s12870-016-0771-y.

Wagay, J. A., Singh, S., Raffi, M., Rahman, Q. I., and Husen, A. (2019). Effect of carbon-based nanomaterials on rhizosphere and plant functioning. *In*: *Nanomater. Plant Potential, Springer International Publishing*, 2019, pp. 553–575, https://doi.org/10.1007/978-3-030-05569-1_22.

Wakeel, A., Xu, M., and Gan, Y. (2020). Chromium-induced reactive oxygen species accumulation by altering the enzymatic antioxidant system and associated cytotoxic, genotoxic, ultrastructural, and photosynthetic changes in plants. *Int. J. Molec. Sci.*, 21: 728. doi: 10.3390/ijms21030728.

Wang, L., Rinklebe, J., Tack, F.M.G., and Hou, D. (2021). A review of green remediation strategies for heavy metal contaminated soil. *Soil Use Manag.*, 37: 936–963, https://doi.org/10.1111/SUM.12717.

Wang, Q., Ma, X., Zhang, W., Pei, H., and Chen, Y. (2012). The impact of cerium oxide nanoparticles on tomato (*Solanum lycopersicum* L.) and its implications for food safety. *Metallomics*, 4: 1105–1112. doi: 10.1039/c2mt20149f.

Wiszniewska, A. (2021). Priming strategies for benefiting plant performance under toxic trace metal exposure. *Plants*, 10 623, https://doi.org/10.3390/PLANTS10040623.

Wu, H., and Li, Z. (2022). Recent advances in nano-enabled agriculture for improving plant performance. *Crop J.*, 10 1–12, https://doi.org/10.1016/J.CJ.2021.06.002.

Wu, H., Shabala, L., Shabala, S., and Giraldo, J. P. (2018). Hydroxyl radical scavenging by cerium oxide nanoparticles improves Arabidopsis salinity tolerance by enhancing leaf mesophyll potassium retention. *Environ. Sci.: Nano,* 5: 1567–1583, https://doi.org/10.1039/C8EN00323H.

Wu, H., Tito, N., and Giraldo, J. P. (2017). Anionic cerium oxide nanoparticles protect plant photosynthesis from abiotic stress by scavenging reactive oxygen species. *ACS Nano*, 11: 11283–11297, https://doi.org/10.1021/ACSNANO.7B05723.

Yuan, J., Chen, Y., Li, H., Lu, J., Zhao, H., and Liu, M. (2018). New insights into the cellular responses to iron nanoparticles in *Capsicum annuum*. *Sci. Rep.*, 8: 3228. doi: 10.1038/s41598-017-18055-w.

Yu, Z., Sun, X., Song, H., Wang, W., Ye, Z., Shi, L. et al. (2015). Glutathioneresponsive carboxymethyl chitosan nanoparticles for controlled release of herbicides. *Mater. Sci. Appl.*, 6: 591. doi: 10.4236/msa.2015.66062.

Zein El-Abdeen, A., and Farroh, Y. (2019). Preparation and characterization of nano organic soil conditioners and it's effected on sandy soil properties and wheat productivity. *Nat. Sci.*, 17: 115–128, https://doi.org/10.7537/marsnsj170219.13.

Zhang, P., Guo, Z., Zhang, Z., Fu, H., White, J. C., and Lynch, I. (2020). Nanomaterial transformation in the soil–plant system: implications for food safety and application in agriculture. *Small*, 16: 2000705, https://doi.org/10.1002/SMLL.202000705.

CHAPTER-17

Plant-Derived Nanoparticles for Stress-Resilient and Sustainable Agriculture

Israt Jahan,[1] *Meskul Zannat,*[1] *Arup Ratan Basak,*[2] *Arpita Rani Roy,*[3] *Srishti Chauhan*[4] and *Prodipto Bishnu Angon*[4,*]

1. Introduction

One of the biggest problems of the modern era will be feeding the world's population safely and sustainably. By 2030, there will be over 8.5 billion people on the planet and 50% more food will be needed to feed them all, according to United Nations research (Jiang et al., 2022). However, several factors impede agricultural development. Numerous stressors, both biotic (such as bacteria, fungi, nematodes, herbivores, and weeds) and abiotic (drought, high temperatures, salinity, and waterlogging), are encountered by the plant during its growth, all of which have an adverse effect on its development and productivity (He et al., 2018; Cirillo et al., 2022). Cytotoxic metabolites, such as reactive oxygen species, are produced when cells are subjected to extreme environmental circumstances. These metabolites disrupt the homeostatic balance of cells and cause damage to their membranes (Khan and Khan, 2017). Therefore, a new technique with great efficiency and little pollution is required to solve this agricultural conundrum. In the last couple of decades, enormous advancements have been seen in nanotechnology.

Plant and agrarian scientific publications have made great strides in developing effective and environmentally sustainable methods for producing plant-based nanoparticles to counteract the detrimental impacts of various stresses (Chakraborty et al., 2022). Nanoparticles' enormous surface/volume ratio, which results from their nanoscale size, drastically changes their properties and characteristics as compared to their regular counterparts (Nejatzadeh, 2021). Given the numerous issues confronting agriculture, including severe environmental stress, climate change, and food security, nanotechnology can significantly increase agricultural yield and quality while reducing the harmful effects of pesticides and fertilizers on the environment.

1 Department of Genetics and Plant Breeding, Bangladesh Agricultural University, Mymensingh 2202, Bangladesh.
2 Institute of Genetics and Biotechnology, Hungarian University of Agriculture and Life Science (MATE), Gödöllő, 2100, Hungary.
3 Faculty of Agriculture, Bangladesh Agricultural University, Mymensingh 2202, Bangladesh.
4 Institute of Agronomy, Hungarian University of Agriculture and Life Science (MATE), Gödöllő, 2100, Hungary.
* Corresponding author: angonbishnubau@gmail.com

DOI: 10.1201/9781003567684-17

Furthermore, by promoting physiological processes during plant growth and enhancing the quality of soil and agricultural products, nanotechnology can mitigate a variety of abiotic challenges (Jiang et al., 2022).

Nanoparticles may impact the plant system in both positive and negative ways. Numerous nanoparticles can cause oxidative damage to biomolecules at extremely high concentrations, which can harm or even kill plant cells. Nonetheless, when present at ideal concentrations (low nanomolar concentrations), nanoparticles act as essential modulators of plant growth and development (Tariq et al., 2021).

It is possible to create nanoparticles chemically, physically, or biologically. Although the production of nanoparticles with the required size and high purity has been accomplished with success using physical and chemical approaches, these procedures are expensive and involve hazardous chemicals. As biological methods of nanoparticle production are quick, economical, and environmentally friendly, many researchers have focused on them in the modern period (Chakraborty et al., 2022). For the biosynthesis of a variety of nanoparticles, such as iron, platinum, palladium, gold, silver, selenium, zinc, copper, and others, certain plant parts—such as seeds, flowers, fruits, stems, bark, leaves, peels, and roots—have been extensively used. It has also been documented that these parts can rewire numerous plant physio-biochemical processes under stressful circumstances (Kumari et al., 2022).

At an alarming rate, abiotic stresses have been endangering ecological stability. Abiotic factors that impede global agricultural production include drought, salinity, high temperature, and heavy metal toxicity (Khan et al., 2021). Many nanoparticles have been shown to be crucial in regulating how plants react to a range of environmental factors, including drought, salt, heat stress, and heavy metal toxicity (Iqbal et al., 2019; Wahid et al., 2020; Sreelakshmi et al., 2020; Mustafa et al., 2021; Venkatachalam et al., 2017). Applying nanoparticles improves a plant's capacity to adapt to stress by changing important metabolic pathways and strengthening defences against free radicals and oxidative stress (Jalil and Ansari, 2019). Nanoparticles (NPs) have gained attention recently because of their great performance, affordability, and environmental friendliness in sustainable agriculture. The objectives of this study are (i) to investigate how plant-derived nanoparticles can be used to control biotic stress; (ii) to assess their capacity to reduce abiotic stress; (iii) to discuss how NPs can be used in sustainable agriculture; (iv) to determine the constraints and difficulties of using NPs.

2. Derivation of NPs from Plant

NPs are made from biological sources using "green synthesis." This approach is considered environmentally safe since it uses natural resources instead of dangerous chemicals. The biological route of preparation of NPs makes use of plants, bacteria, fungi or algae. The hazardous effects of green-produced NPs are negligible this facilitates the use of the NPs in various sectors of the economy (Melesse et al., 2024). Plant and microbial-mediated synthesis techniques are less costly and more ecologically friendly than chemical and physical techniques. This is because the energy requirement of the synthesis process is very low, and expensive precursors have been removed from the synthesis process (Jeevanandam et al., 2022). This technology makes the synthesis of NPs from plants more environmentally beneficial since it is more soluble, reasonable, and ecologically harmless (Singh et al., 2023)

In severe environments, NPs can boost the amount of chlorophyll and activate certain enzymes, which enhances photosynthesis efficiency and total plant development (El-Saadony et al., 2022). To assist plants in dealing with abiotic stress, NPs boost the synthesis of anti-stress chemicals and activate defense enzymes (Al-Khayri et al., 2023). When plants are under stress, NPs like cerium oxide help scavenge Reactive Oxygen Species (ROS), protect them from oxidative damage, and increase photosynthetic efficiency (Wu et al., 2017).

In the production of palladium (Pd), platinum (Pt), gold (Au), and silver (Ag), plant extracts exhibit a variety of electrical, magnetic, optical, mechanical, and chemical characteristics (Shafey, 2020).

In their works, Vanitha et al. (2022), and Mittal et al. (2013) explore the development of copper (Cu), iron (Fe), selenium (Se), and zinc (Zn) derived extracts as well as the conversion of metal ions into their respective NPs. Barks, fruits, seeds, leaves, stems, flowers, roots, and fruit skins are among the plant components from which these NPs are extracted (Kumar and Rajeshkumar,2018). Plant extracts contain phenolic chemicals, alkaloids, terpenoids, and co-enzymes that function as reducing agents to turn metal ions into NPs (Mittal et al 2013).

2.1 Roots

Plants' ability to absorb carbon, divide and release carbon into the soil, absorb water and nutrients from the soil, and control biogeochemical cycles are all directly impacted by the structure and function of their roots (Warren et al., 2015). This is due to the fact that the numerous aspects and therapeutic qualities of roots that affect the effectiveness of their utilization in green nanotechnology. In addition, the composition of phytochemicals in plant roots affects some properties of the formed NPs. It was for this reason, the bioactive constituents in the root extract of *Strobilanthes glutinosus* were again confirmed by the generation of silver NPs with special absorbance peaks this proved successful synthesis (Javed et al., 2023). Furthermore, some phenolic compounds found in root extracts stabilize the NPs by reducing agglomeration and improving their stability in solution, in addition to helping to reduce metal ions (Kuppusamy et al., 2016).

The green production of silver NPs using dried *Zingiber officinale* root is a simple, cost-effective, and ecologically friendly method with potential antibacterial qualities (Vijaya et al., 2017). The root extract of *Plumbago zeylanica* was used to swiftly and efficiently create bimetallic Ag and Au NPs having antibacterial and antibiofilm activity against a range of bacterial species (Salunke et al., 2014). *Astragalus tribuloides Delile* root extract and *Panax ginseng* may be able to kill harmful bacteria when used to make Ag and Au NPs in a way that does not harm the environment (Shingh et al., 2015; Sharifi-Rad et al., 2020).

Superior antibacterial effectiveness against bacteria that are gram-negative was demonstrated by TiO2 NPs produced from titanium tetrabutoxide and *Kniphofia foliosa* root extract (Bekele et al., 2020). Zinc oxide (ZnO) NPs may be made using *Salvadora persica* extract from roots as a reducing agent (Verma et al., 2020). FTIR, or Fourier transform infrared, shows that the plant's bioactive chemicals adhere to the surfaces of iron oxide NPs when *Mimosa pudica* root extract is separated for NP synthesis (Niraimathee et al., 2016). Additionally, there was evidence that these NPs function as elicitors in plant systems, enhancing the medicinal qualities of plants by inducing the production of secondary metabolic products (Fatima et al., 2021).

2.2 Leaf

There has been a lot of interest in the green synthesis of metal NPs using plant extracts as an affordable and ecologically friendly alternative to conventional chemical techniques (Makarov et al., 2014). The synthesis is easily scalable because it is usually carried out at room temperature and pressure (Mittal et al., 2013). Flavonoids present in the phytochemical content which plays a crucial role for synthesizing NPs and can also enhance the bioactivity of NPs leaf extracts are very preferred for this process (Jeevanandam et al., 2016). The process entails reducing metal ions, such as Ag, adding plant extract, and then watching for a color shift that verifies the synthesis of NPs (Mubayi et al., 2012). Numerous techniques may be used to characterize the resulting NPs, which have potential use in medical and antibacterial therapy (Mittal et al., 2013; Mubayi et al., 2012). For example, higher antibacterial activity was demonstrated by AgNPs made from plant leaf extracts, such as those of bananas, neem, and black tulsi. To confirm the production of NPs, the procedure comprises decreasing metal ions, such as Ag, adding plant extract, and then observing for a color shift (Mubayi et al., 2012). The resultant NPs may be characterized using a variety of methods and may find application in antimicrobial and medicinal therapy (Mittal et al., 2013; Mubayi et al., 2012).

An aqueous extract from the *Capsicum chinense* plant leaf can generate stable, easily synthesizable Ag and Au NPs and may emanate antibacterial properties (Lomelí-Rosales et al., 2022). *Pelargonium odoratissimum* leaves may be utilized to create ZnONPs with antioxidant, antibacterial, and anti-inflammatory qualities (Abdelbaky et al., 2022).

2.3 Stem

There are many studies documenting the reduction of metal ions into NPs using plant stem extracts. For instance, stem extract from *Momordica charantia* has been used to create Ag NPs having antibacterial qualities (Akinsiku et al., 2019). AgNPs with antifungal and antibacterial properties were also produced using stem bark extract from Syzygium alternifolium and bananas (Das and Velusamy, 2013; Yugandhar et al., 2015). AgNP generation is usually indicated by any change in the reaction solution's color, which may be verified by UV-visible spectroscopy. According to Akinsiku et al. (2019), AgNPs' surface plasmon resonance range is 400–450 nm. Clinacanthus nutans leaf and stem extracts are used to create AgNPs, which are more stable, less poisonous, and comparatively less expensive for a variety of uses (Yusuf et al., 2020).

According to Uddin et al. (2021), biological techniques of producing nickel oxide NPs utilizing stem extract from *Berberis balochistanica* are inexpensive, simple to use, and non-toxic. These NPs may find use in agricultural settings. Additionally, it was shown that ZnONPs produced using a hydrothermal technique from Swertia chirayita stem extract had an extremely small particle size (17 nm), a greater photocatalytic effectiveness (88.67%), and antibacterial activity against *Staphylococcus aureus* and *Escherichia coli*.

2.4 Bud

One of the effective approaches to increasing crop protection is claiming NPs from the extracts of the buds. As can be seen , a variety of plant extracts, especially those from buds, provide a rich supply of phytochemicals for the environmentally conscious production of NPs. For example, the following noble metal NPs, AgNPs, PbNPs, CuNPs, CuONPs, and AuNPs have been synthesized and reported in works by Anand and Bhagat (2019) and Kumari et al. (2022) as a result of the existence of polyphenols and flavonoids reducing agents in *Syzygium aromaticum* buds. In the context of safe synthesis for these NPs, sesquiterpenoids initially reported in *Tussilago farfara* flower bud extract was used as an efficient reductant (Lee et al., 2019). This extract from flower buds of *Couroupita guianensis* synthesizes Ag NPs in a highly stable form with desired antibacterial effects (Kumar et al., 2016). Metal-based NPs improve a plant's resistance to disease by inhibiting plant pathogenic bacteria and fungi, but also decreasing yield, crop quality, and growth (Li et al., 2022). Synthetically synthesizing these NPs enhances an array of plant developments and their tolerance to both biotic stresses, restores the bioavailability of such vital nutrients besides capitalizing on the inherent chemicals within plant extractions (Nguyen et al., 2021). NPs have shown themselves to be both germination promoters and disease preventers, which makes them viable options for creating effective and environmentally responsible plant cultivation methods.

2.5 Flower

A sustainable method for a number of uses is the green production of metallic NPs using floral solution (Kumar et al., 2020). Antibacterial NPs are formed via an ability to stabilize and reduce metal ions that extract from saffron petals (Azizian-Shermeh et al., 2019). These floral phytochemicals can be used to reduce and stabilize metallic NPs like Ag and Au (Kumar et al., 2020). The floral extract of *Tecomella undulata* may be used to produce AgNPs with a regulated dimension of 12.5 nm (Chaudhuri and Malodia, 2017). Tin chant of AgNPs employing the floral extract of *Cassia alata* yielded spherical AgNPs that had particle size between 20 and 100 nm and exhibited bioactivity against fungal skin infections (Muruganandham et al., 2023). Additionally, R. Tetraphylla flower extracts efficiently create AgNPs with antibacterial, antifungal, and mitotic qualities, providing an economical and ecologically sound way to create NPs (P et al., 2019). For agricultural application, coccine and ZnONPs from Canna indica flower extract can be harnessed for increasing mass and germination rate, bleaching, and enhancing seed germination (Nguyen et al., 2021). To the same, CuO NPs derived from herbal tea blossoms reveal promising catalytic behavior in both aryl and heteroaryl C-heteroatom cross coupling processes potentially paving way for the novel bio-based catalysts (Veisi et al., 2021). Additionally, using techniques like FTIR spectroscopy, Transmission Electron Microscopy (TEM), and UV-visible spectroscopy, the consistency and morphology of flower-mediated NPs will be examined to determine their effectiveness in protecting crops. This will help determine the stability, size, and shape of the generated NPs (Mittal et al., 2013). According to Velsankar et al. (2020), using floral species in the production of NPs might result in the development of novel nanomaterials that improve crop protection, thereby advancing sustainable farming.

2.6 Fruit

The use of fruit extract to green manufacture metallic NPs has several advantages for the environment, including the elimination of the need for stabilization and reduction of chemicals (Kumar et al., 2020). The production of non-cytotoxic Au NPs can benefit from the use of fruit peel extracts as reducing and stabilizing ingredients (Pechyen et al., 2021). Ag NPs, which are spherical and usually range in size from between two and two hundred nm, are made from fruit peel extracts.This process is more economical and ecologically friendlier than chemical procedures (Lazulfa et al., 2023). Numerous shaped and sized gold NPs have been successfully produced using citrus fruits, such as lemon, orange, and sweet orange (Sujitha and Kannan, 2013; Yang et al., 2019). Ag NPs derived from *Crataegus douglasii* fruit extract have promising antibacterial activity against *Staphylococcus aureus* and *Escherichia coli* (Ghaffari-Moghaddam and Hadi-Dabanlou, 2014).

In addition to reducing harmful compounds and improving antibacterial activity, orange fruit peel extract efficiently synthesizes ZnO NPs (Thi et al., 2020). Furthermore, *Crataegus pinnatifida* fruit extract may be used to easily produce gold and silver chloride NPs, which may be used as novel anti-inflammatory and antibacterial medications (Kang et al., 2017). Phytofabricated selenium NPs made from *Emblica officinalis* fruit extract could be biocompatible, antioxidant, and antibacterial in a range of applications (Gunti et al., 2019). Additionally, AgNPs made from fruit extract from *Sambucus nigra* showed anti-inflammatory properties both *in vitro* and *in vivo* on psoriasis lesions (David et al., 2014). Additionally, TiO_2 and Al_2O_3 NPs with anti-inflammatory and antioxidant properties are made from *Terenna asiatica* fruit extract (Bokhary et al., 2022; Venkatappa et al., 2023). The Newcastle Disease Virus may be inhibited in its growth by CuO NPs made from *Syzygium alternifolium* fruit extract, which makes them a potentially greener and more affordable nanomedicine option (Yugandhar et al., 2018).

2.7 Seed

Since it offers a cost-effective alternative to chemical stabilizers and reducing agents, the single-step ecological production of metal NPs using extracts from seeds is popular (Bhardwaj et al., 2022). Plant phytochemicals, such as proteins, reducing sugars, and polyphenol compounds, are essential for the synthesis of Au and AgNPs (Zuhrotun et al., 2023). Temperature and reaction time had no discernible effects on the size and shape of gold NPs; however, extract concentration and pH have a major influence (Pinto et al., 2017). For instance, according to Ahmed Fragoon et al. (2012), adding more black seed extract increased the yields of AuNPs by strengthening the bonds between biomolecules and developing NPs. *Coffea arabica* extract-synthesized gold NPs exhibit stable, monodisperse, and quasi-spherical shapes, suggesting possible uses in a number of industries (Keijok et al., 2019). The environmentally benign, nontoxic, and inexpensive nature of AgNPs made from *Sinapis arvensis* and *Trifolium resupinatum* seed exudates makes them a viable new class of antifungal drugs and perfoms *against Rhizoctonia solani and Neofusicoccum parvum* (Khatami et al., 2015; Khatami et al., 2016). Moreover, the ecologically sound production of zinc oxide NPs using powdered activated carbon and *Peganum harmala* seed extract significantly improves the removal of Cr (VI) from solutions of water (Fazlzadeh et al., 2017). The photocatalytic activity of pomegranate seed extract in the production of Fe_2O_3 NPs has shown promise (Bibi et al., 2018). These researches demonstrate how plant-based NPs synthesis offers environmentally benign substitutes for conventional chemical synthesis techniques in a range of agricultural applications. The roles of the various NP types that are generated from plant components are mentioned in Table 1.

3. Plant-derived Nanoparticles for Evicting Biotic Stress

Plant-derived nanoparticles (PDNPs) have drawn interest over the past decade due to their amazing and enormous potential as medicinal agents and the creation of simpler, safer, and less costly biomedical equipment. Antioxidant, antibacterial, anticancer, antifungal, anti-inflammatory, and antiviral properties are among their many medicinal uses. This method for creating nanoparticles is effective and environmentally friendly (Khan ands Khan, 2023). Numerous biotic stresses significantly reduce agricultural productivity. Although many traditional pesticides, such as fungicides, herbicides, and bactericides, can eradicate weeds and infections, they can also pollute the environment and pose health hazards. The use of metal nanoparticles with improved characteristics in green plant extracts has raised a lot of concerns. When plant extracts with therapeutic potential are combined with variety of metals, the result is nanoparticles with increased activity (Khan and Khan, 2023; Verma et al., 2021).

3.1 Bacteria

The living environment is full of bacteria, some of which are pathogenic to plants, and some can infiltrate the body to endanger human health. Moreover, bacterial resistance is a global health concern and research on novel therapeutic medications with broad-spectrum antibacterial activity is essential (Table 2). In plant-derived nanoparticles, the bioactive compounds present in plants enable the reduction of metallic ions into finely structured metal nanoparticles characterized by distinctive sizes and notable antibacterial properties. Studies have demonstrated that in pathogenic bacteria, the controlled release of metals from MNPs produces Reactive Oxygen Species (ROS) including superoxide anion ($O_2^{\cdot -}$) and hydroxyl radical (•OH) (Behzad et al., 2021). These nanoparticles increase the stress on bacterial cells to break down their resistance by simultaneously targeting multiple biomolecules in bacteria (Nguyen et al., 2022). Sometimes, nanoparticles produced through chemical synthesis may pose a significant risk of increased harm to human health and the environment. Furthermore, the cost of producing nanoparticles is probably going to be high. In the meanwhile, one of the best and most sustainable methods is to create biogenic nanoparticles by using plant materials. Specifically, the abundance of phytochemicals (such as alkaloids, flavonoids, reducing sugars, and polyphenols) present in the plant extract can boost both the antimicrobial activity and stability of nanoparticles (Bharathi et al., 2020).

As mentioned earlier, one of the most important problems of the past few years is antibiotic resistance, which will only get worse. Because of the quick evolution of bacterial genomes, bacteria have become resistant to antimicrobial treatments. Since NPs and other compounds have been mixed with various organic and inorganic substances to improve antibacterial efficacy, they may be a viable option in the fight against such resistant pathogenesis in the pursuit of a novel (Hano and Abbasi, 2021).

It is crucial to remember that the length of the inhibition zone (measured in mm) varies depending on the type of nanoparticle, the bacterial strain, and the concentration (μg/mL) used. This variability underscores the nanoparticle efficacy's concentration-dependent nature and suggests that dosage and nanoparticle type optimization is crucial for achieving maximum antibacterial activity.

Table 1. Different types of NPs synthesized from plant parts and their functions.

Different NP Types	NPs synthesized from		Dimensions and form	Function	Source
	Plants	Parts of the plant used			
ZnONPs	*Mussaenda frondos*	Stem	Hexagonal, 5–20 nm	notable antioxidant, anti-cancer, anti-inflammatory, and antidiabetic actions	Jayappa et al., 2020
	Peganum harmala	Seeds	Non-uniform, 40 nm	Adsorption of Chromium (VI)	Fazlzadeh et al., 2017
NiONPs	*Rheum turkestanicum*	Root	Hexagonal, 12–15 nm	Photocatalytic activity, capable of eliminating about 70% tumors' cells	Saheb et al., 2019
	Ocimum sanctum	Leaf	Round to polyhedral, 13–36 nm	Adsorption of pollutant	Pandian et al., 2015
AgNPs	*Ficus carica*	Fruit	Spherical, 10–30 nm	notable enhanced antioxidant activity	Kumar et al., 2016
	Cleome viscose	Fruit	Spherical, 20–50 nm	Anticancer and antibacterial activity	36
	Polianthus tuberosa	Bud	Oval, 50±2 nm	Mosquito larvicidal activity	Rawani, 2017
	Tussilago farfara	Flower bud	Spherical, 13.57±3.26 nm	Anticancer and antibacterial activity	Lee et al.,2019
	Zingiber officinale	Root	Spherical, 15 nm	antibacterial activity against foodborne pathogens	Velmurugan et al.,2014
	Coffea arabica	Seeds	Ellipsoidal and spherical, 20–30 nm	Antibacterial activity against *S. aureus* and *E. coli* was shown to be effective.	Dhand et al., 2016
Fe_2O_3NPs	*Punica granatum*	Seeds	Semi-spherical, 25-55 nm	Photo-catalytic activity	Fazlzadeh et al.,2017
CuONPs	*Syzygium alternifolium*	Fruit	Sphere-shaped, 2–69 nm	Powerful antiviral activity against the Newcastle Disease Virus (NDV)	Yugandhar et al.,2018
MgONPs	*Rosmarinus officinalis*	Flower	Round. 8.8 nm	Antibacterial activities against *Xanthomonas oryzae*	Abdallah et al.,2019
Cr_2O_3NPs	*Callistemon viminalis*	Flower	Cubic-like platelet, 92.2 nm	Work as reducing and/or oxidizing agent	Sone et al.,2016
CeO_2NPs	*Hibiscus sabdariffa*	Flower	Spherical, 3.9 nm	effective chelating agent	Thovhogi et al.,2015
La_2O_3NPs	*Eucalyptus globulus*	Leaf	irregular particle shape, 50.37 nm	antioxidant, antidiabetic, and anti-inflammatory properties	Maheshwaran et al.,2021
TiO_2NPs	*Withania somnifera*	Root	Square and spherical aggregates, 50–90 nm	significantly inhibit and destroy preformed *C. albicans*, *P. aeruginosa*, *E. coli*, MRSA, *S. marcescens*, and *L. monocytogenes* biofilms.	Al-Shabib et al., 2020
SeNPs	*Leucas lavandulifolia*	Stem	Ciecular, 56–75 nm	Antimicrobial action	Kirupagaran et al., 2016
		Leaf	Round, 56–75 nm	Antimicrobial properties	Kirupagaran et al., 2016

Table 2. Plant-derived nanoparticles with affected bacterial species.

Plant name	Scientific name	Used Plant part	Nanoparticles	Affected Bacterial Species	References
Pineapple	*Ananas comosus*	Leaf extract	Ag (Silver)	*S. aureus, S. pneumoniae, P. mirabilis* and *E. coli*	(Emeka et al., 2014)
Mugwort	*Artemisia marschalliana*	Aerial parts	Ag	*S. aureus, P. aeruginosa, A. baumannii* and *B. cereus*	(Salehi et al., 2016)
Marigold	*Tagetes erecta*	Flower	Ag	*E. coli, P. aeruginosa, S. aureus*	(Padalia et al., 2015)
Lemon balm	*Melissa officinalis*	Leaf extract	Ag	*S. aureus* and *E. coli*	(de Jesús Ruíz-Baltazar et al., 2017)
Betel palm	*Areca catechu*	Plant extract	Au (Gold)	*E. coli, K. pneumonia, P. aeruginosa, Enterobacter* and *S. aureus*	(Rajan et al., 2015)
Blumea	*Blumea eriantha*	Whole plant	Au, Iron (Fe)	*B. cereus, B. subtilis, S. aureus, E. coli, B. cereus, B. subtilis, S. aureus, E. coli*	(Chavan et al., 2020)
	Artemisia haussknechtii	Leaf extract	Ag, Copper (Cu)	*E. coli, S. Aureus, E. coli, S. aureus, S. marcescens*	(Alavi and Karimi, 2018)
Common dandelion	*Taraxacum officinale*	Leaf Extract	Ag	*X. axonopodis, P. syringae*	(Saratale et al., 2018)
Sweet wormwood	*Artemisia annua*	Stem bark	ZnO	*E. coli, S. aureus, P. aeruginosa, S. epidermidis, B. subtilis*	(Wang et al., 2020)
Bitter leaf	*Blumea eriantha*	Leaf Extract	$Zn_{1-x}Cu_xO$	*S. aureus, E. coli, P. aeruginosa*	(Wang et al., 2020)
Globe artichoke	*Cynara scolymus*	Leaf Extract	ZnO	*S. aureus, E. coli, P. aeruginosa*	(Rajapriya et al., 2020)
Coltsfoot	*Tussilago farfara*	Flower bud	Ag	*E. coli, E. faecalis, P. aeruginosa, S. aureus*	(Lee et al., 2019)

3.2 Fungus

Different fungal pathogens impact the agricultural economy by bringing seasonal diseases and ultimately lowering crop quality. Synthesizing nanoparticles from plant extracts represents one of the most promising and environmentally sustainable approaches to enhancing antifungal applications in both medical and agricultural sectors (Table 3). As previously addressed, metallic nanoparticles derived from plants also use the same method to eliminate fungal species by releasing metal ions from their nanostructure, generating free radicals and ROS. Fungal cell death is caused by ROS, which also causes protein damage, inhibition of DNA replication, and inhibition of enzymes (Nguyen et al., 2022). The antifungal efficacy of nanoparticles is influenced by various factors, such as their size distribution, morphology, composition, degree of crystallinity, level of agglomeration, and surface chemistry (Cruz-Luna et al., 2021).

Table 3. Plant-derived nanoparticles with antifungal effect.

Plant name	Scientific name	Used Plant part	Nanoparticles	Affected Fungal Species	References
Maize	*Zea mays*	Corn Cob Xylan	Ag	*Candida albicans, C. parapsilosis* and *Cryptococcus neoformans*	(Viana et al., 2020)
Chinaberry	*Melia azedarach*	Leaf extract	Ag	*Verticillium dahliae*	(Jebril et al., 2020)
Jujube	*Ziziphus nummularia*	Leaf extract	Zno	*Candida albicans, Candida glabrata* and *Cryptococcus neoformans*	(Padalia and Chanda, 2017)
Aloe vera	*Aloe vera*	Leaf	Ag	*Aspergillus* spp. and *Rhizopus* spp.	(Medda et al., 2015)
Wild mustard	*Sinapis arvensis*	Seed exudates	Ag	*Neofusicoccum parvum*	(Khatami et al., 2015)
Nodeweed	*Synedrella nodiflora*	Leaf extract	Ag and Au	*Aspergillus spp.* and *Penicillium spp.*	(Vijayan et al., 2018)
Night-flowering jasmine	*Nyctanthes arbortristis*	Flower extract	ZnO	*Alternaria alternata, Aspergillus niger , Botrytis cinerea, Fusarium oxysporum* and *Penicillium expansum*	(Jamdagni et al., 2018)
Podlech	*Seripheidium quettense*	Whole plant	Ag	*A. niger, A. fumigatus, A. flavus* and *Mucor* spp.	(Qasim Nasar et al., 2019)
Common nasturtium	*Tropaeolum majus*	Leaf Extract	Ag	*Penicilium notatum, A. niger, C. albicans, Trichoderma viridiae* and *Mucor* spp.	(Valsalam et al., 2019)
Bonpland's croton	*Croton sparsiflorus*	Leaf extract	Ag	*Mucor* spp., *Tricoderma* spp. and *A. nigar*	(Kathiravan et al., 2015)
Globe artichoke	*Cynara scolymus*	Leaf extract	ZnO	*C. albicans* and *Candida tropicalis*	(Rajapriya et al., 2020)
Intellect plant	*Celastrus paniculatus*	Leaf extract	Cu	*F. oxysporum*	(Mali et al., 2020)
Redroot pigweed	*Amaranthus retroflexus*	Leaf extract	Ag	*Macrophomina phaseolina, F. oxysporum* and *A. alternata*	(Bahrami-Teimoori et al., 2017)

3.3 *Virus*

Recently, extensive research has focused on metal nanoparticles synthesized using plant extracts, which have demonstrated nontoxicity and biocompatibility. These nanoparticles exhibit potential antiviral properties and are promising for targeted drug delivery applications. (Naikoo et al., 2021). Nanoparticles can control plant virus infections by both directly deactivating the virions and indirectly by triggering plant defense mechanisms to prevent the spread of infection (Table 4). The use of nanotechnology to deliver double-stranded RNA (dsRNA) in environmental contexts holds significant potential for activating RNA interference (RNAi) in plants, thereby elevating their resistance to viral infections (Carvalho et al., 2022). The external layer of viruses, predominantly

Table 4. Plant-derived nanoparticles with antiviral effect.

Plant name	Scientific name	Used Plant part	Nanoparticles	Affected Plant pathogenic virus	References
Ceylon cinnamon	*Cinnamon zeylanicum*	Plant barks	Ag	*Leek yellow stripe virus* (LYSV) on Garlic	(Abdelkader et al., 2024)
Basil	*Ocimum basilicum*	Plant extract	Ag	*Cucumber Mosaic Virus* (CMV) in Squash	(Abdelkhalek et al., 2022)
Orange		Peel extract	CuO	Tobacco mosaic virus (TMV)	(Liu et al., 2024)

composed of capsid proteins, features a highly reactive surface. This surface readily interacts with metallic ions and serves as binding sites for the components dispersed in nanoscale (Wasule et al., 2024). Engineered nanoparticles, including both metal nanoparticles (NPs) and metal oxide NPs, exhibit several effects on viral pests. They reduce % of virus infections and the growth of lesions while limiting the rate of pathogenesis. Additionally, these nanoparticles inhibit viral replication by reducing polymerase activity and destabilize the viral shell proteins. They also inactivate the glycoprotein envelope of the virus, aggregate the virus-cell wall, and damage virus-like particles, making them effective in controlling viral pests (Kumar et al., 2022).

Many plants derived nanomaterials are being employed as control agents to combat different phytopathogens. However, their ability for combating viral infections is still currently limited. Table 3 highlights the antiviral effects of plant-derived nanoparticles synthesized from different plant parts. For instance, Ceylon cinnamon (*Cinnamon zeylanicum*) bark-derived silver nanoparticles (AgNPs) were effective against *Leek yellow stripe virus* (LYSV) in garlic+. Notably, foliar application of this biogenic AgNPs 48 hours post-inoculation inhibited viral activity by 90%. (Abdelkader et al., 2024). Similarly, biosynthesized AgNPs from basil (*Ocimum basilicum*) plant extract significantly inhibited *Cucumber Mosaic Virus* (CMV) in squash, improving antioxidant enzyme activity, free radical scavenging, and phenolic content (Abdelkhalek et al., 2022). Additionally, orange peel extract was used to synthesize copper oxide nanoparticles (CuONPs), which controlled *Tobacco Mosaic Virus* (TMV) by damaging viral particles, enhancing plant immunity, and promoting plant growth (Liu et al., 2024). These findings underscore the potential of plant-based nanoparticles as antiviral agents to combat phytopathogens.

3.4 Insects

Insect pests damage plant crops both directly and indirectly by serving as vectors of some viral diseases (Table 5) (Arora and Sandhu, 2017). When compared to traditional physical and chemical methods of insect management, the green synthesis of nanoparticles using plant materials has various advantages because it eliminates the need for toxic chemicals and high-energy inputs (Benelli, 2018). Like earlier biotic stressors, the green synthesis process utilizes a variety of metals, mostly silver, to create nanoparticles.

While the table focuses on affected pests and nanoparticle sources, it is important to note that the duration of mortality and mortality rates (LC_{50} values) may vary across different life stages of insects. These findings underscore the potential of plant-based nanoparticles as eco-friendly alternatives for pest control, providing targeted and stage-specific insecticidal activity.

Table 5. Plant-derived nanoparticles against insect pests.

Plant name	Scientific name	Used Plant part	Nanoparticles	Affected Insect Pest	References
Herbaceous seepweed	*Suaeda maritima*	Leaf extract	Ag	*Spodoptera litura*	(Suresh et al., 2018)
Christmas candlestick	*Leonotis nepetifolia*	Leaf extract	Ag	*Spodoptera litura* and *Helicoverpa armigera*	(Manimegalai et al., n.d.)
Ferns	*Cyclosorous interruptus, Christella dentata* and *Nephrolepis cordifolia*	Whole body	Ag	*Spodoptera litura*	(Selvaraj et al., 2018)
Indian snakeroot	*Rauvolfia serpentine*	Leaf extract	NiO (Nickel Oxide)	*Tuta absoluta*	(Rahman et al., 2021)
Rice	Oriza sativa	Dried husk	Si (Silica)	Aphis gossypii, Phenacoccus solenopsis	(Pavitra et al., 2018)

4. Plant derived nanoparticles in evicting abiotic stress

The frequency of abiotic stressors, such as heat, drought, salinity, and heavy metal toxicity has increased in the world's ecosystems, putting plants at risk (Imran et al.,2021; Gechev and Petrov, 2020; Zhang et al.,2022). Besides, cold is another significant stress, as many crops cannot thrive in extremely cold climates (Bhat et al.,2022; Atayee and Noori, 2020). These abiotic pressures are caused by human activities and fast industrialization, which represent significant dangers via climate change (Shah et al.,2024; Khalid et al.,2022). Globally, drought and heat stress simultaneously limit crop development, growth, production, and agricultural sustainability (Bhardwaj et al.,2021, Sinha et al.,2021). Additionally, reduced soil water potential due to high salinity (EC more than 15 ds) causes plant roots to absorb less water (Dos Santos et al.,2022, Abdalla et al.,2022). Heavy metals also cause a hamper of plants that slows their development and makes them less effective in what they do (Bharti and Sharma, 2022; Rashid et al.,2023). Furthermore, cold stress affects plant growth, development, and yields (Hwarari et al.,2022). According to these reasons, plant derived nanoparticles are commonly used to create stress-resistant plants that can endure various abiotic stresses (Castillo-Henríquez et al.,2020; Kumari et al.,2022; Bhattacharya and Saha, 2023). The regulating functions of plant-derived nanoparticles under abiotic stress are shown in Fig. 1 and Table 6.

4.1 Drought stress

Drought stress is a mix of multiple variables that hinder plant development, growth, and output by producing an osmotic and ionic imbalance (Mukarram et al., 2021). The morpho-physiological impacts of drought stress on plants include decreased rates of photosynthesis and transpiration, stomata closure due to conductance of stomatal cells, and changes in the stress hormone response (Muhammad et al., 2023). One possible strategy to mitigate the detrimental effects of drought stress on plants is the administration of NPs generated from plants (Figure 2). For instance, *Sorghum bicolor* seeds were primed with αFe_2O_3NPs, which can be obtained from *Aspalathus linearis* (rooibos), enhancing plant structure and protecting sorghum plants from drought-induced oxidative

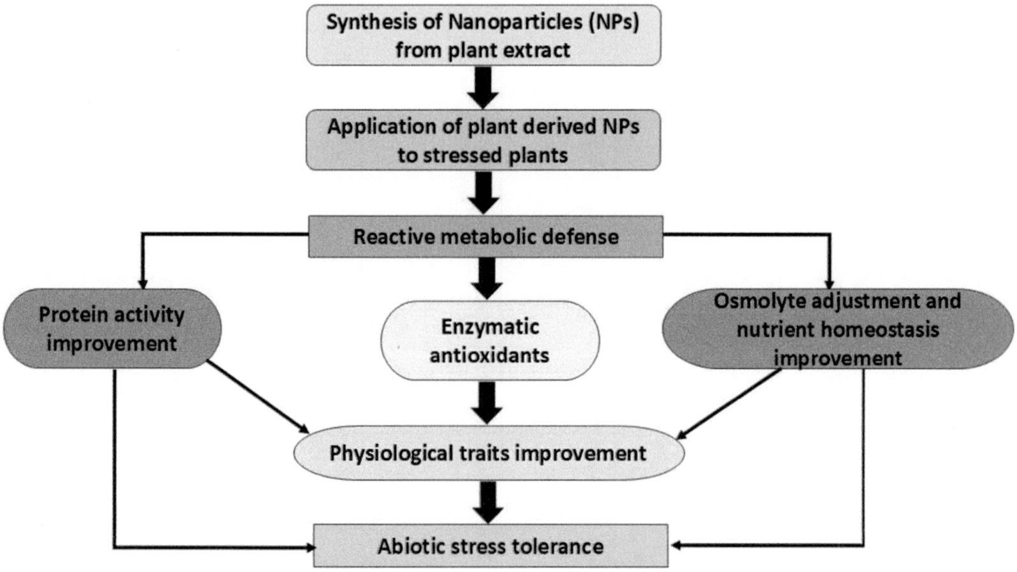

Figure 1. The regulating functions of plant-derived nanoparticles under abiotic stress.

damage. (Ndou et al., 2023). FeNPs were synthesized from *Chaetomorpha antennina* and these NPs effectively increase *Setaria italica*'s resistance to drought stress by positively modulating osmolyte concentrations and enzymes that served as antioxidants, such as CAT, SOD, and POX (Sreelakshmi et al., 2021). *Cucumis melo* seedlings' capacity to withstand drought was enhanced by the application of ZnONPs derived from *Artemisia annua* leaves, which also boosted their absorption of nutrients and vegetative development (Rehman et al., 2023). Moreover, *Triticum aestivum* was treated with SeNPs (30 mgL^{-1}) derived from *Allium sativum* bud extract, which considerably enhanced growth indices while lowering lipid peroxidation and ionic leakage, thereby decreasing the cellular damage generated by drought stress (Ikram et al., 2020). The expression of genes sensitive to drought such as- *GMYB174, GMWRKY27, GMERD1, GMFDL19, GMRD20A, GMDREB2, GMYB118, GMNAC11* are significantly increased by the administration of NPs (Kandhol et al., 2022). Plants treated with NPs showed increased expression of the *GmWRKY27* gene, suggesting that NPs are linked to the control of ABA production and stomata function under drought stress (Linh et al., 2020). Furthermore, the application of NPs increased APX, CAT, and SOD activities and decreased drought-induced oxidative damage (Sutulienė et al., 2021; Zahedi et al., 2020). The content of phenolic compounds under drought stress was significantly raised by the exogenous delivery of ZnO-NPs (25 and 50 mg/L), indicating that NPs also have a large potential to encourage the accumulation of phenolic compounds in plants (El-Zohri et al., 2021). In addition, NPs treatment increases the synthesis of gibberellins (GA) and Indole Acetic Acid (IAA), which promotes plant growth in drought-stressed environments (Li et al., 2021). According to Sun et al. (2020), NPs increased melatonin production, which led to improved drought tolerance. This suggests that NPs help crops fight drought stress by boosting endogenous hormones. TiO_2 nanoparticles enhance photosynthetic performance and efficiency in drought-stressed plants by boosting nitrate reductase activity, nitrogen absorption, and RuBisCO (ribulose-1,5-bisphosphate carboxylase/oxygenase) activity (Ali et al., 2021). Additionally, NPs improved gas exchange properties and increased the absorption of Ca, Mg, N, K, and Fe to give more photosynthesis under drought stress (Faraji and Sepehri, 2020).

Table 6. Functions of different plant derived nanoparticles under different abiotic stress.

Plants used for synthesis	Extract taken from that plant part	Name of the NPs	Plants used for study	Given Abiotic stress	Concentration of applied NPs	Studied Plant Responses	References
Carica papaya	whole plant	ZnONPs	*Carthamus tinctorius*	Salinity (250 mM)	17 mg L^{-1}	Increased productivity, such as the number of pods a plant produces, increases proline content and antioxidant enzyme activity while lowering ROS production.	Yasmin et al., 2021
Allium cepa	whole plant	AgNPs and CuNPs	*Triticum aestivum*	Drought stress (field capacity of 40%, 60%, and 80%)	CuNPs (0, 3, 5, and 7 mg/l) and AgNPs (30 mg/l) in MS medium	Produced greater output and drought tolerance	Ahmed et al., 2021
Buddleja asiatica	leaf	TiO_2NPs	*Triticum aestivum*	Salinity (100 and 150 mM)	NPs at 20, 40, 60, and 80 mg/l	Reduced salinity stress by the antioxidant defense system (SOD) being activated.	Mustafa et al., 2021
Albizia lebbeck	leaf	AgNPs	*Solanum melongena*	Drought stress ((80%, 50%, 35%, and 20% field capacity)	0, 0.1, 0.2, 0.5 μmol	Activated enzymatic defense systems to reduce drought stress.	Alabdallah et al., 2021
Chaetomorpha antennina	whole algal extract	FeNPs	*Setaria italica*	Drought (10% PEG)	5, 10, 15, 20, 50, 90 and 120 mg L^{-1}	enhanced growth traits, a variety of metabolic processes, and photosynthetic efficiency, elevated osmolyte accumulation and the activity of antioxidant enzymes such as CAT, SOD, and POX.	Sreelakshmi et al., 2021
Triticum aestivum	leaf	AgNPs	*Triticum aestivum*	Salinity (100 mM)	300 ppm	lowered lipid peroxidation and ROS accumulation while maintaining ionic equilibrium. enhanced GPX, APX, GR, and SOD activities, elevated concentrations of GSH and AsA, two non-enzymatic antioxidants, as well as increased proline metabolism and nitrogen absorption.	Wahid et al., 2020
Adiantum lunulatum	leaf	FeONPs	*Oryza sativa*	Heavy metal (Arsenic) stress (50 μM)	Fertigation with 100 mg/l FeONPs	Enhanced activity of antioxidant enzymes decreased oxidative stress	Bidi et al., 2021

Musa paradisica	leaf	TiO_2NPs	*Vigna radiata*	Heavy metal (arsenic; 10 μM)	0.1%	prevented ROS-induced issues by reducing arsenic accumulation, raising protein content, and activating antioxidant enzymes including SOD, CAT, and APX.	Katiyar et al., 2020
Capparis spinosa	whole plant	AgNPs	*Triticum aestivum*	Salinity (25 and 100 mM)	1 mg L^{-1}	reduced ABA concentration, enhanced IBA and BAP contents, and improved growth characteristics and photosynthetic responses.	Abou-Zeid and Ismail, 2018; Wahid et al., 2020
Cocos nucifera	husk	Fe_3O_4NPs	*Oryza sativa*	Heavy metal (cadmium; 0.01% *w/w*)	20 mg	improved chlorophyll content, PSII quantum efficiency, plant biomass, and agricultural output.	Maqsood et al., 2023; Zhou et al., 2023
Prosopis cineraria	leaf	FeONPs	*Triticum aestivum*	Drought stress	Priming with 12.5, 25, 50, 75, and 100 ppm	Diminished consequences of drought stress	Singh et al., 2022
Moringa oleifera	leaf	AgNPs	*Triticum aestivum*	Heat (3 hours each day; 35–40°C)	50, 75, 100, and 25 mg L^{-1}	reduced the amounts of MDA and H_2O_2 and strengthened the antioxidant defense system.	Iqbal et al., 2019; Shukla et al., 2024
Agathosma betulina	whole plant	ZnONPs	*Sorghum bicolor*	Salinity (400 mM of NaCl)	5 and 10 mg/l were used to prime	enhanced osmolytes and antioxidants, which lessened the negative consequences of salinityc	Rakgotho et al., 2022
Triticum aestivum	leaf	AuNPs	*Triticum aestivum*	Salinity (100 mM)	300 ppm	kept the ionic ratio of $K+$: $Na+$ constant in the root and shoot systems, enhanced absorption of nitrogen, and antioxidant compounds such as GSH and AsA that are both enzymatic and non-enzymatic.	Wahid et al., 2022
Cuminum Cyminum and Zingiber officinale	leaf	FeONPs	*Triticum aestivum*	Drought stress	0.9, 1.2, 0.6, and 0.3 mM	Water stress reduction via enzymatic and physiological mechanisms	Noor et al., 2022
Hordeum vulgare	leaf	SeNPs	*Hordeum vulgare*	Salinity (100 and 200 mM)	100 ppm	Flavonoids, phenolic compounds, and photosynthetic pigments were found in higher concentrations, and the buildup of MDA and H_2O_2 was inhibited.	Habibi and Aleyasin, 2020

Table 6 contd. ...

...Table 6 contd.

Plants used for synthesis	Extract taken from that plant part	Name of the NPs	Plants used for study	Given Abiotic stress	Concentration of applied NPs	Studied Plant Responses	References
Mentha arvensis	leaf	Au-Ag alloy NPs	*Mentha piperita*	Salinity (0, 50, 100, and 150 mg/l)	applied 100 parts per million of AuAg alloy.	Salt stress reduction via increased osmolytes and enzymatic antioxidants	Aliakbarpour et al., 2020
Hevea barsiliensis	bark	Fe_3O_4NPs	*Oryza sativa*	Heavy metal (cadmium; 15.0 mg kg^{-1})	0.5 g	decreased oxidative damage brought on by cadmium, preserved nutritional homeostasis, and enhanced biomass of plants and photosynthetic pigments.	Sebastian et al., 2018; Ahmed et al., 2021
Aloe vera	whole plant	AgNPs	*Spinacia oleracea*	Salinity (0, 50, 100, and 150 mM)	100, 80, 40, 20, and 10 ppm	reduced stress from salt and enhanced germination	Ali Naghi Zadeh et al., 2020
Ochradenus Arabicus	whole plant	AgNPs	*Maerua oblongifolia*	Salinity (NaCl) at 0, 50, 100, and 200 mM	0, 10, 20, and 30 mg/l	Oxidative damage was decreased by both enzymatic and non-enzymatic antioxidant mechanisms.	Shaikhaldein et al., 2022
Hibiscus sabdarifa	leaf	AgNPs	Roselle plants	Salinity stress	0, 15, 30, 45, 60, 75, and 90 mg/l of AgNPs	Salt stress was decreased by both enzymatic and non-enzymatic antioxidant mechanisms.	Sadat-Hosseini et al., 2022

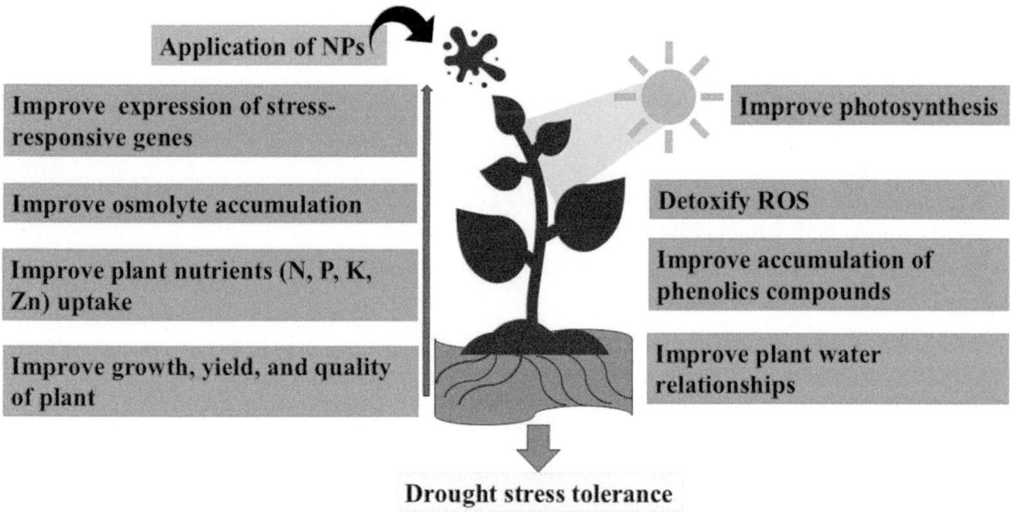

Figure 2. Beneficiary roles of NPs during drought stress.

4.2 Salinity Stress

A major concern to agricultural output is the negative effects of salt stress on growth, development, germination, and production of plants (Trejo-Téllez, 2023). Plants acquire toxic concentrations of chloride (Cl^-) and sodium (Na^+) ions when soil salinity levels are excessive, which adversely impacts their growth and yield (Shahid, 2020). For optimal survival, plants balance their development and stress tolerance by excluding excess salt, which helps to maintain ionic equilibrium and stimulates pathways that signal phytohormones (Fu and Yang, 2023). Plant extract-derived nanoparticles may alter the phytohormone content in response to abiotic stress and promote the growth of plant abiotic stress resistance (Kumari et al., 2022). The function of plant derived NPs in mitigating plant stresses caused by salt has been the subject a lot of investigation . In *Carthamus tinctorius* under salt stress, ZnONPs (17 mgL^{-1}) from *Carica papaya* extract improved antioxidant metabolism (Yasmin et al., 2021). Additionally, nitric oxide generation, nitrogen metabolism, ion balance, and defense mechanisms were retained by gold nanoparticles (AuNPs) biosynthesized from wheat leaves (*Triticum aestivum* cv. Pusa Kiran), increasing wheat's resilience to salt stress (Wahid et al., 2022). Zinc oxide nanoparticles (ZnO NPs) derived from *Ochradenus arabicus* shoot extract increased *Salvia officinalis* shoot growth and decreased salt toxicity (Alenezi et al., 2022).

Research by Yazıcılar et al., 2021 has shown that calcium oxide nanoparticles (CaONPs) derived from *Punica granatum* fruit extract considerably minimize the adversities produced by salt stress on *Triticale* callus . Remarkably, certain metabolic activities linked to plant eco-physiological output are stimulated by nanoparticles, which may help plants withstand salt and regulate ion homeostasis (Abideen et al., 2022). For example, when applied topically, biosynthesized AuNPs raised non-enzymatic antioxidant contents (AsA and GSH), improved nitrogen (N) metabolic activity, and enhanced enzymatic antioxidant activities (SOD, APX, GPX, and GR). Additionally, they dramatically reduced lipid peroxidation and ROS production under salt stress (Wahid et al., 2022). *Ocimum basilicum* leaf-derived silver nanoparticles (SNPs) at concentrations of 50, 100, and 200 μM enhanced the ionic ratio of K^+: Na^+, the nutrient content, including K, P, and N, and promoted the uptake of total soluble proteins, free amino acids, and cysteine in *Triticum aestivum* under salinity stress (Ragab and Saad-Allah, 2020).

Several further studies have shown that using phytogenic NPs enhances growth and physiological processes, hence reducing the negative consequences of salt stress (Sheikhalipour et al., 2021; El-Badri et al., 2021). Plant growth parameters, biomass production, and the activities of ROS-detoxifying enzymes (SOD and CAT) were all improved in *Vigna unguiculata* when ZnONPs generated from *Phoenix dactylifera* leaflet extract were applied topically under salt stress (Alabdallah and Alzahrani, 2020a). ZnONPs made from *Phoenix dactylifera* leaflet extract applied to *Abelmoschus esculentus* had a similar outcome (Alabdallah and Alzahrani, 2020b). Furthermore, exogenously applied zinc nanoparticles (ZnNPs) derived from *Sorghum bicolor* leaf extract improved *Abelmoschus esculentus* growth and photosynthetic traits under salt stress by upregulating the antioxidant defense system at a dose of 0.3% (Zafar et al., 2021). Likewise, TiO2NPs made from *Buddleja asiatica* leaf extract reduced the effects of salt stress on *Triticum aestivum* by improving many morpho-physiological and biochemical traits (Mustafa et al., 2021).

4.3 Heavy Metal Stress

Higher levels of pollutants, such as heavy metals, have been found in the environment due to accelerated globalization, posing a threat to natural systems and human health (Wei et al., 2023). When these substances are incorporated into soil, they reduce soil fertility (Nazie et al., 2020) and negatively affect plant development, photosynthesis, and nutrient availability (Bharti and Sharma, 2022). Nonetheless, under heavy metal stress, phytogenic NPs may regulate the overall growth and physiological functions of plants (Adeyemi et al., 2022). Besides, NPs also boost plant development when subjected to heavy metal stress, enhancing plant resilience to heavy metal stress (Faizan et al., 2023). For example, TiO_2NPs synthesized from *Trianthema portulacastrum* and *Chenopodium quinoa* leaf extract effectively prevented Cd^{2+} ion assimilation, improving plant height, grain output, spike length, and chlorophyll content (Irshad et al., 2021). Besides, TiO_2NPs, produced from *Musa paradisica* leaf extract, dramatically decreased As^{3+} ion buildup in arsenic-stressed *Vigna radiata*, increased protein content, and boosted antioxidant activity, facilitating ROS detoxification (Katiyar et al., 2020). Cd-stressed *Oryza sativa* treated with Fe_3O_4NPs (0.5 mgg^{-1}) made from *Cocos nucifera* husk extract increased chlorophyll content, plant biomass, and photosystem II (PSII) quantum efficiency, as well as higher agricultural yields (Maqsood et al., 2023). Under heavy metal stress (lead and cadmium), ZnONPs derived from *Ulva lactuca* markedly enhanced photosynthetic pigments and plant development in *Leucaena leucocephala* (Venkatachalam et al., 2017). In *Helianthus annus*, SNPs (100 μM) from *Ocimum basilicum* leaf extract increased resistance to stress caused by Mn (100 mM), increased the amount of crude protein, total amino acids, and cysteine, and decreased ROS-mediated lipid peroxidation (Saad-Allah and Ragab, 2020).

4.4 Heat Stress

Climate change and global warming have increased the frequency of extreme heat stress events, which has a detrimental effect on plant production (Zahra et al., 2023). Heat stress impairs nutrient transport, digestion, and absorption, negatively impacting plant growth, development, and reproduction (Mishra et al., 2023; Sampath et al., 2023). Conversely, green synthesized nanoparticles could help lessen the harm that heat stress causes to plants (Azmat et al., 2022; Dikshit et al., 2021). For instance, by increasing antioxidant defense and lowering MDA and H_2O_2 levels, AgNPs made from *Moringa oleifera* leaf extract significantly reduced the negative effects of heat stress in *Triticum aestivum* (Iqbal et al., 2019). The potential of ginger-derived nanoparticles (NPs) to effectively transport therapeutic chemicals has been studied; this might be used to boost plants' tolerance to heat stress (Zhang et al., 2019). In heat-stressed conditions, they could also serve as carriers of essential nutrients and growth regulators, improving the efficacy of foliar applications (Fan et al., 2023).

4.5 Cold Stress

The geographic range of many plant species and agricultural productivity are significantly impacted by abiotic stresses such as cold stress (Aslam et al., 2022). Furthermore, cold stress disrupts meiosis, tapetal programmed cell death, pollen viability, and fertilization, rendering plants infertile (Huang et al., 2022). However, the use of plant-derived nanoparticles (NPs) in cold stress reduction has garnered a lot of attention lately due to their potential to boost plant resistance against abiotic stresses. For example, Cerium oxide nanoparticles (SeNPs), synthesized from plant biomass like *Salvia miltiorrhiza*, exhibit strong ROS scavenging abilities (Wu et al., 2017). When applied to cucumber, SeNPs improved photosynthetic efficiency and overall growth under cold stress conditions, demonstrating their potential as abiotic stress protection (Liu et al., 2021).

5. Limitations

Plant extract-based nanoparticle production is ecologically friendly and non-toxic, making it suited for many applications (Melese et al., 2024). Furthermore, using plant extracts prevents the requirement for toxic chemicals often employed in traditional nanoparticle production, keeping to the principles of green chemistry (Mazzini et al., 2024). Notwithstanding the many advantages that green synthesis techniques for NPs from plant extract provide, they also come with some drawbacks and difficulties. The main disadvantage of green nanotechnology is its low manufacturing yield (Khan, 2020). Besides, the production of eco-friendly nanomaterials in considerable numbers for industries has proven sluggish (Ying et al., 2022). Additionally, green synthesis faces limitations due to endemic or seasonal plants, energy usage, long reaction times, and lower conversion rates of nanoparticles compared to chemical methods (Kumari et al., 2022). It is more challenging to replicate and standardize synthesis processes when various plant species or even separate plant parts produce nanoparticles with diverse sizes, properties, and shapes (Kumari et al., 2024; Hazel and Gobi, 2022). Moreover, Kumari et al., 2024 highlighted that, although plants may operate as efficient reducing agents, the phytochemical composition might vary greatly depending on environmental variables and plant maturity, altering the synthesis result. On the other hand, genetic variability in plant extracts used in NP synthesis leads to a lack of understanding of NPs assessment, necessitating high-throughput instrumentation for purification and characterization (Kumari et al., 2022).

Green synthesis is simple and does not need high pressure or temperature conditions, making it suitable for laboratory-scale synthesis (Salem and Fouda, 2021; Huston et al., 2021). Despite the ease of laboratory-scale synthesis, producing plant extracts for industrial applications may be difficult owing to large biomass needs and resource-intensive extraction processes (Muhib et al., 2024). This challenge is worsened by the necessity for constant quality in the raw materials, which may be difficult to attain when sourcing from varied ecological regions (Hazel and Gobi, 2022; Barhoum et al., 2020). Another major worry is the existence of toxic or undesirable chemicals in plant extracts, which might be accidentally absorbed into the nanoparticles. For instance, whereas green-synthesized nanoparticles are often hailed as non-toxic, the existence of lingering phytochemicals may provide hazards that need careful consideration (Makki and Chandrasekharaiah,2023; Mustapha et al.,2022). Additionally, the organic substances that were employed to create these nanoparticles may deteriorate with time, changing their characteristics and efficacy and affecting their long-term stability (Muhib et al., 2024).

Last but not the least, it is important to consider the environmental effects of collecting plant resources to produce nanoparticles. In cases where some plants are sourced in big amounts, ecological imbalances may result, especially if the plants are not grown properly (Álvarez-Chimal and Arenas-Alatorre, 2023). This situation raises doubts regarding the long-term sustainability of depending entirely on plant extracts for nanoparticle production, particularly in the context of rising demand for nanomaterials across numerous sectors (CHIHAI et al., 2022). Moreover, there is a lack of defined methods for synthesizing and using plant derived NPs, which hampers their regulatory clearance and broad adoption (Tinnirello et al., 2023).

References

Abdalla, M., Ahmed, M. A., Cai, G., Zarebanadkauki, M., and Carminati, A. (2022). Coupled effects of soil drying and salinity on soil–plant hydraulics. *Plant Physiology*, 190(2): 1228–1241.

Abdallah, Y., Ogunyemi, S. O., Abdelazez, A., Zhang, M., Hong, X., Ibrahim, E. et al. (2019). The green synthesis of MgO nano-flowers using *Rosmarinus officinalis* L. (Rosemary) and the antibacterial activities against *Xanthomonas oryzae* pv. oryzae. *BioMed. Research International*, 2019(1): 5620989.

Abdelbaky, A., El-Mageed, T., Babalghith, A., Selim, S., and Mohamed, A. (2022). Green synthesis and characterization of ZnO Nanoparticles Using *Pelargonium odoratissimum* (L.) Aqueous Leaf Extract and Their Antioxidant, Antibacterial and Anti-inflammatory Activities. *Antioxidants*, 11.

Abdelkader, H. S., Kheder, A. A., Amin, H. A., and Shafie, R. M. (2024). A comparative study of the antiviral effects of biogenic silver nanoparticles and nanosilica (nSiO2) against Leek yellow stripe virus on *Allium sativum* L. *European Journal of Plant Pathology*. https://doi.org/10.1007/s10658-024-02965-3.

Abdelkhalek, A., El-Gendi, H., Alotibi, F. O., Al-Askar, A. A., Elbeaino, T., Behiry, S. I. et al. (2022). Ocimum basilicum-Mediated Synthesis of Silver Nanoparticles Induces Innate Immune Responses against Cucumber Mosaic Virus in Squash. *Plants*, 11(20). https://doi.org/10.3390/plants11202707.

Abideen, Z., Hanif, M., Munir, N., and Nielsen, B. L. (2022). Impact of nanomaterials on the regulation of gene expression and metabolomics of plants under salt stress. *Plants*, 11(5): 691.

Abou-Zeid, H., and Ismail, G. (2018). The role of priming with biosynthesized silver nanoparticles in the response of Triticum aestivum L to salt stress. *Egyptian Journal of Botany*, 58(1): 73–85.

Adeyemi, J. O., Oriola, A. O., Onwudiwe, D. C., and Oyedeji, A. O. (2022). Plant extracts mediated metal-based nanoparticles: synthesis and biological applications. *Biomolecules*, 12(5): 627.

Ahmed, F., Javed, B., Razzaq, A., and Mashwani, Z. U. R. (2021). Applications of copper and silver nanoparticles on wheat plants to induce drought tolerance and increase yield. *IET Nanobiotechnology*, 15(1): 68–78.

Ahmed, T., Noman, M., Manzoor, N., Shahid, M., Abdullah, M., Ali, L. et al. (2021). Nanoparticle-based amelioration of drought stress and cadmium toxicity in rice via triggering the stress responsive genetic mechanisms and nutrient acquisition. *Ecotoxicology and Environmental Safety*, 209: 111829.

Akinsiku, A.A., Ajanaku, K.O., Adebisi, A. A., Edobor-Osoh, A., Aladesuyi, O., Samson, T. O. et al. (2019, April). Momordica charantia stem extract mediated biogenic synthesis of silver nanoparticles: optical and antimicrobial efficacy. In *IOP Conference Series: Materials Science and Engineering* (Vol. 509, No. 1, p. 012018). IOP Publishing.

Alabdallah, N. M., and Alzahrani, H. S. (2020a). The potential mitigation effect of ZnO nanoparticles on [*Abelmoschus esculentus* L. Moench] metabolism under salt stress conditions. *Saudi Journal of Biological Sciences*, 27(11): 3132–3137.

Alabdallah, N. M., and Alzahrani, H. S. (2020b). Impact of ZnO nanoparticles on growth of cowpea and okra plants under salt stress conditions. *Biosciences Biotechnology Research Asia*, 17(2): 329–340.

Alabdallah, N. M., Hasan, M., Salih, A. M., SS, R., Al-Shammari, A. S., Alsanie, S. I. et al. (2021). Silver nanoparticles improve growth and protect against oxidative damage in eggplant seedlings under drought stress. *Plant, Soil & Environment*, 67(11).

Alavi, M., and Karimi, N. (2018). Characterization, antibacterial, total antioxidant, scavenging, reducing power and ion chelating activities of green synthesized silver, copper and titanium dioxide nanoparticles using Artemisia haussknechtii leaf extract. *Artificial Cells, Nanomedicine, and Biotechnology*, 46(8): 2066–2081. https://doi.org/10.1080/21691401.2017.1408121.

Alenezi, N. A., Al-Qurainy, F., Tarroum, M., Nadeem, M., Khan, S., Salih, A. M. et al. (2022). Zinc oxide nanoparticles (ZnO NPs), biosynthesis, characterization and evaluation of their impact to improve shoot growth and to reduce salt toxicity on salvia officinalis in vitro cultivated. *Processes*, 10(7): 1273.

Ali Naghi Zadeh, A., Mahmoudi Zarandi, M., Khoshroo, S. M. R., and Hasan Zadeh, F. (2020). The effect of green silver nanoparticles on spinach (Spinacia oleracea) seed germination under salt stress. *Iranian Journal of Biological Sciences*, 15(3): 1–10.

Ali, E. F., El-Shehawi, A. M., Ibrahim, O. H. M., Abdul-Hafeez, E. Y., Moussa, M. M., and Hassan, F. A. S. (2021). A vital role of chitosan nanoparticles in improvisation the drought stress tolerance in *Catharanthus roseus* (L.) through biochemical and gene expression modulation. *Plant Physiology and Biochemistry*, 161: 166–175.

Aliakbarpour, F., Vaziri, A., and Mohseni, M. (2020). Green synthesis of Au-Ag nanoparticles using Mentha piperita and effects of Au-Ag alloy nanoparticles on the growth of Mentha piperita under salinity stress. *J. Plant Process Funct*, 9: 37.

Al-Khayri, J., Rashmi, R., Ulhas, R., Sudheer, W., Banadka, A., Nagella, P. et al. (2023). The role of nanoparticles in response of plants to abiotic stress at physiological, biochemical, and molecular levels. *Plants*, 12.

Al-Shabib, N. A., Husain, F. M., Qais, F. A., Ahmad, N., Khan, A., Alyousef, A. A. et al. (2020). Phyto-mediated synthesis of porous titanium dioxide nanoparticles from Withania somnifera root extract: broad-spectrum attenuation of biofilm and cytotoxic properties against HepG2 cell lines. *Frontiers in Microbiology*, 11: 1680.

Álvarez-Chimal, R., and Arenas-Alatorre, J. Á. (2023). Green synthesis of nanoparticles. A biological approach.

Anand, R., and Bhagat, M. (2019). Silver nanoparticles (agnps): as nanopesticides and nanofertilizers. *MOJ Biology and Medicine*, 4(1): 19–20.

Arora, R., and Sandhu, S. (2017). Breeding insect resistant crops for sustainable agriculture. In *Breeding Insect Resistant Crops for Sustainable Agriculture*. Springer Singapore. https://doi.org/10.1007/978-981-10-6056-4.

Aslam, M., Fakher, B., Ashraf, M. A., Cheng, Y., Wang, B., and Qin, Y. (2022). Plant low-temperature stress: Signaling and response. *Agronomy*, 12(3): 702.

Atayee, A. R., and Noori, M. S. (2020). Alleviation of cold stress in vegetable crops. *J. Sci. Agric.*, 4: 38–44.

Azizian-Shermeh, O., Valizadeh, M., Taherizadeh, M., and Beigomi, M. (2019). Phytochemical investigation and phytosynthesis of eco-friendly stable bioactive gold and silver nanoparticles using petal extract of saffron (*Crocus sativus* L.) and study of their antimicrobial activities. *Applied Nanoscience*, 1–14.

Azmat, A., Tanveer, Y., Yasmin, H., Hassan, M. N., Shahzad, A., Reddy, M. et al. (2022). Coactive role of zinc oxide nanoparticles and plant growth promoting rhizobacteria for mitigation of synchronized effects of heat and drought stress in wheat plants. *Chemosphere*, 297: 133982.

Bahrami-Teimoori, B., Nikparast, Y., Hojatianfar, M., Akhlaghi, M., Ghorbani, R., and Pourianfar, H. R. (2017). Characterisation and antifungal activity of silver nanoparticles biologically synthesised by *Amaranthus retroflexus* leaf extract. *Journal of Experimental Nanoscience*, 12(1): 129–139. https://doi.org/10.1080/17458080.2017.1279355.

Banerjee, P., Satapathy, M., Mukhopahayay, A., and Das, P. (2014). Leaf extract mediated green synthesis of silver nanoparticles from widely available Indian plants: synthesis, characterization, antimicrobial property and toxicity analysis. *Bioresources and Bioprocessing*, 1: 1–10.

Barhoum, A., Jeevanandam, J., Rastogi, A., Samyn, P., Boluk, Y., Dufresne, A. et al. (2020). Plant celluloses, hemicelluloses, lignins, and volatile oils for the synthesis of nanoparticles and nanostructured materials. *Nanoscale*, 12(45): 22845–22890.

Behzad, F., Naghib, S. M., kouhbanani, M. A. J., Tabatabaei, S. N., Zare, Y., and Rhee, K. Y. (2021). An overview of the plant-mediated green synthesis of noble metal nanoparticles for antibacterial applications. *Journal of Industrial and Engineering Chemistry*, 94: 92–104. https://doi.org/10.1016/J.JIEC.2020.12.005.

Bekele, E., Gonfa, B., Zelekew, O., Belay, H., and Sabir, F. (2020). Synthesis of titanium oxide nanoparticles using root extract of kniphofia foliosa as a template, characterization, and its application on drug resistance bacteria. *Journal of Nanomaterials*, 2020: 1–10.

Benelli, G. (2018). Mode of action of nanoparticles against insects. In *Environmental Science and Pollution Research* (Vol. 25, Issue 13, pp. 12329–12341). Springer Verlag. https://doi.org/10.1007/s11356-018-1850-4.

Bharathi, D., Preethi, S., Abarna, K., Nithyasri, M., Kishore, P., and Deepika, K. (2020). Bio-inspired synthesis of flower shaped iron oxide nanoparticles (FeONPs) using phytochemicals of *Solanum lycopersicum* leaf extract for biomedical applications. *Biocatalysis and Agricultural Biotechnology*, 27: 101698. https://doi.org/10.1016/J.BCAB.2020.101698.

Bhardwaj, A., Devi, P., Chaudhary, S., Rani, A., Jha, U. C., Kumar, S. et al. (2021). 'Omics' approaches in developing combined drought and heat tolerance in food crops. *Plant Cell Reports*, 1–41.

Bhardwaj, K., Chopra, C., Bhardwaj, P., Dhanjal, D., Singh, R., Najda, A. et al. (2022). Biogenic Metallic Nanoparticles from Seed Extracts: Characteristics, Properties, and Applications. *Journal of Nanomaterials*.

Bharti, R., and Sharma, R. (2022). Effect of heavy metals: An overview. *Materials Today: Proceedings*, 51: 880–885.

Bhat, K. A., Mahajan, R., Pakhtoon, M. M., Urwat, U., Bashir, Z., Shah, A. A. et al. (2022). Low temperature stress tolerance: An insight into the omics approaches for legume crops. *Frontiers in Plant Science*, 13: 888710.

Bhattacharya, S., and Saha, J. (2023). Green synthesis and application of biogenic nanomaterials as a blueprint in mitigation of abiotic stress in crop plants: a conceptual review. *Advancement of GI-Science and Sustainable Agriculture: A Multi-dimensional Approach*, 155–182.

Bibi, I., Nazar, N., Ata, S., Sultan, M., Ali, A., Abbas, A. et al. (2019). Green synthesis of iron oxide nanoparticles using pomegranate seeds extract and photocatalytic activity evaluation for the degradation of textile dye. *Journal of Materials Research and Technology*.

Bidi, H., Fallah, H., Niknejad, Y., and Tari, D. B. (2021). Iron oxide nanoparticles alleviate arsenic phytotoxicity in rice by improving iron uptake, oxidative stress tolerance and diminishing arsenic accumulation. *Plant Physiology and Biochemistry*, 163: 348–357.

Bokhary, K., Maqsood, F., Amina, M., Aldarwesh, A., Mofty, H., and Al-Yousef, H. (2022). Grapefruit extract-mediated fabrication of photosensitive aluminum oxide nanoparticle and their antioxidant and anti-inflammatory potential. *Nanomaterials*, 12.

Carvalho, R., Jones, J. B., and Paret, M. L. (2022). Utility of nanoparticles in management of plant viruses. In *Nanotechnology-Based Sustainable Alternatives for the Management of Plant Diseases* (pp. 233–241). Elsevier. https://doi.org/10.1016/B978-0-12-823394-8.00001-9.

Castillo-Henríquez, L., Alfaro-Aguilar, K., Ugalde-Álvarez, J., Vega-Fernández, L., Montes de Oca-Vásquez, G., and Vega-Baudrit, J. R. (2020). Green synthesis of gold and silver nanoparticles from plant extracts and their possible applications as antimicrobial agents in the agricultural area. *Nanomaterials*, 10(9): 1763.

Chakraborty, S., Singh, A., and Roychoudhury, A. (2022). Biogenic nanoparticles and generation of abiotic stress-resilient plants: A new approach for sustainable agriculture. *Plant Stress*, 6: 100117.

Chaudhuri, S., and Malodia, L. (2017). Phytosynthesis and characterisation of silver nanoparticles synthesised from flower extract of roheda (*Tecomella undulata* G. Don). *life science journal*, 2: 65–73.

Chavan, R. R., Bhinge, S. D., Bhutkar, M. A., Randive, D. S., Wadkar, G. H., Todkar, S. S. et al. (2020). Characterization, antioxidant, antimicrobial and cytotoxic activities of green synthesized silver and iron nanoparticles using alcoholic Blumea eriantha DC plant extract. *Materials Today Communications*, 24: 101320. https://doi.org/10.1016/J.MTCOMM.2020.101320.

Chihai, R., Săracu, A. F., and Ungureanu, C. V. (2022). Green synthesis of metal nanoparticles using microalgae: a review. *The Annals of "Dunarea de Jos" University of Galati. Fascicle IX, Metallurgy and Materials Science*, *45*(3): 16–19.

Cirillo, V., Molisso, D., Aprile, A. M., Maggio, A., and Rao, R. (2022). Systemin peptide application improves tomato salt stress tolerance and reveals common adaptation mechanisms to biotic and abiotic stress in plants. *Environmental and Experimental Botany*, 199: 104865.

Cruz-Luna, A. R., Cruz-Martínez, H., Vásquez-López, A., and Medina, D. I. (2021). Metal nanoparticles as novel antifungal agents for sustainable agriculture: current advances and future directions. *Journal of Fungi*, *7*(12): 1033. https://doi.org/10.3390/JOF7121033.

Das, J., and Velusamy, P. (2013). Antibacterial effects of biosynthesized silver nanoparticles using aqueous leaf extract of Rosmarinus officinalis L. *Materials Research Bulletin*, 48(11): 4531–4537.

de Jesús Ruíz-Baltazar, Á., Reyes-López, S. Y., Larrañaga, D., Estévez, M., and Pérez, R. (2017). Green synthesis of silver nanoparticles using a Melissa officinalis leaf extract with antibacterial properties. *Results in Physics*, 7: 2639–2643. https://doi.org/10.1016/J.RINP.2017.07.044.

Dhand, V., Soumya, L., Bharadwaj, S., Chakra, S., Bhatt, D., and Sreedhar, B. J. M. S. (2016). Green synthesis of silver nanoparticles using Coffea arabica seed extract and its antibacterial activity. *Materials Science and Engineering: C*, 58: 36–43.

Dikshit, P. K., Kumar, J., Das, A. K., Sadhu, S., Sharma, S., Singh, S. et al. (2021). Green synthesis of metallic nanoparticles: Applications and limitations. *Catalysts*, 11(8): 902.

Dos Santos, T. B., Ribas, A. F., de Souza, S. G. H., Budzinski, I. G. F., and Domingues, D. S. (2022). Physiological responses to drought, salinity, and heat stress in plants: a review. *Stresses*, 2(1): 113–135.

El-Badri, A. M., Batool, M., Wang, C., Hashem, A. M., Tabl, K. M., Nishawy, E. et al. (2021). Selenium and zinc oxide nanoparticles modulate the molecular and morpho-physiological processes during seed germination of Brassica napus under salt stress. *Ecotoxicology and Environmental Safety*, 225: 112695.

El-Saadony, M., Saad, A., Soliman, S., Salem, H., Desoky, E., Babalghith, A. et al. (2022). Role of nanoparticles in enhancing crop tolerance to abiotic stress: a comprehensive review. *Frontiers in Plant Science*, 13.

El-Zohri, M., Al-Wadaani, N. A., and Bafeel, S. O. (2021). Foliar sprayed green zinc oxide nanoparticles mitigate drought-induced oxidative stress in tomato. *Plants*, 10(11): 2400.

Emeka, E. E., Ojiefoh, O. C., Aleruchi, C., Hassan, L. A., Christiana, O. M., Rebecca, M. et al. (2014). Evaluation of antibacterial activities of silver nanoparticles green-synthesized using pineapple leaf (Ananas comosus). *Micron*, 57(C): 1–5. https://doi.org/10.1016/J.MICRON.2013.09.003.

Faizan, M., Alam, P., Rajput, V. D., Faraz, A., Afzal, S., Ahmed, S. M. et al. (2023). Nanoparticle mediated plant tolerance to heavy metal stress: what we know?. *Sustainability*, 15(2): 1446.

Fan, J., Yu, H., Lu, X., Xue, R., Guan, J., Xu, Y. et al. (2023). Overlooked spherical nanoparticles exist in plant extracts: from mechanism to therapeutic applications. *ACS Applied Materials & Interfaces*, 15(7): 8854–8871.

Faraji, J., and Sepehri, A. (2020). Exogenous nitric oxide improves the protective effects of TiO_2 nanoparticles on growth, antioxidant system, and photosynthetic performance of wheat seedlings under drought stress. *Journal of Soil Science and Plant Nutrition*, 20: 703–714.

Fatima, K., Abbas, S. R., Zia, M., Sabir, S., Khan, R. T., Khan, A. A. et al. (2021). Induction of secondary metabolites on nanoparticles stress in callus culture of artemisia annua l. *Brazilian Journal of Biology*, 81(2): 474–483.

Fazlzadeh, M., Khosravi, R., and Zarei, A. (2017). Green synthesis of zinc oxide nanoparticles using Peganum harmala seed extract, and loaded on Peganum harmala seed powdered activated carbon as new adsorbent for removal of Cr(VI) from aqueous solution. *Ecological Engineering*, 103: 180–190.

Fazlzadeh, M., Khosravi, R., and Zarei, A. (2017). Green synthesis of zinc oxide nanoparticles using Peganum harmala seed extract, and loaded on Peganum harmala seed powdered activated carbon as new adsorbent for removal of Cr (VI) from aqueous solution. *Ecological Engineering*, 103: 180–190.

Fragoon, A., Li, J., Zhu, J., and Zhao, J. (2012). Biosynthesis of controllable size and shape gold nanoparticles by black seed (*Nigella sativa*) extract. *Journal of Nanoscience and Nanotechnology*, 12(3): 2337–2345.

Fu, H., and Yang, Y. (2023). How plants tolerate salt stress. *Current Issues in Molecular Biology*, 45(7): 5914–5934.

Gechev, T., and Petrov, V. (2020). Reactive oxygen species and abiotic stress in plants. *International Journal of Molecular Sciences*, 21(20): 7433.

Geremew, A., Carson, L., Woldesenbet, S., Wang, H., Reeves, S., Brooks, N. et al. (2023). Effect of zinc oxide nanoparticles synthesized from Carya illinoinensis leaf extract on growth and antioxidant properties of mustard (*Brassica juncea*). *Frontiers in Plant Science*, 14.

Ghaffari-Moghaddam, M., and Hadi-Dabanlou, R. (2014). Plant mediated green synthesis and antibacterial activity of silver nanoparticles using Crataegus douglasii fruit extract. *Journal of Industrial and Engineering Chemistry*, 20: 739–744.

Gunti, L., Dass, R., and Kalagatur, N. (2019). Phytofabrication of selenium nanoparticles from emblica officinalis fruit extract and exploring its biopotential applications: antioxidant, antimicrobial, and biocompatibility. *Frontiers in Microbiology*, 10.

Habibi, G., and Aleyasin, Y. (2020). Green synthesis of Se nanoparticles and its effect on salt tolerance of barley plants. *Int. J. Nano Dimens*, 11(2): 145–157.

Hano, C., and Abbasi, B. H. (2021). Plant-Based Green Synthesis of Nanoparticles: Production, Characterization and Applications. *Biomolecules 2022, 12*(1), 31. https://doi.org/10.3390/BIOM12010031.

Hazel, D., and Gobi, N. (2022). One-Pot Facile Green Synthesis of Iron Oxide Nanoparticles Using Aqueous Stem Extract of Amaranthus Campestris and Comparison of its Characteristics with Chemically Synthesized Iron Oxide Nanoparticles.

He, M., He, C.Q., and Ding, N.Z. (2018). Abiotic stresses: general defenses of land plants and chances for engineering multistress tolerance. *Frontiers in Plant Science*, 9: 1771.

Hernández-Díaz, J., Garza-García, J., Zamudio-Ojeda, A., León-Morales, J., López-Velázquez, J., and García-Morales, S. (2020). Plant-mediated synthesis of nanoparticles and their antimicrobial activity against phytopathogens.*Journal of The Science of Food and Agriculture*.

Huang, B., Fan, Y., Cui, L., Li, C., and Guo, C. (2022). Cold stress response mechanisms in anther development. *International Journal of Molecular Sciences*, 24(1): 30.

Huston, M., DeBella, M., DiBella, M., and Gupta, A. (2021). Green synthesis of nanomaterials. *Nanomaterials*, 11(8): 2130.

Hwarari, D., Guan, Y., Ahmad, B., Movahedi, A., Min, T., Hao, Z. et al. (2022). ICE-CBF-COR signaling cascade and its regulation in plants responding to cold stress. *International Journal of Molecular Sciences*, 23(3): 1549.

Ikram, M., Raja, N.I., Javed, B., Mashwani, Z. U. R., Hussain, M., Hussain, M. et al. (2020). Foliar applications of bio-fabricated selenium nanoparticles to improve the growth of wheat plants under drought stress. *Green Processing and Synthesis*, 9(1): 706–714.

Iliger, K., Sofi, T., Bhat, N., Ahanger, F., Sekhar, J., Elhendi, A. et al. (2021). Copper nanoparticles: green synthesis and managing fruit rot disease of chilli caused by colletotrichum capsici. *Saudi Journal of Biological Sciences,* 28(2): 1477–1486.

Imran, Q. M., Falak, N., Hussain, A., Mun, B. G., and Yun, B. W. (2021). Abiotic stress in plants; stress perception to molecular response and role of biotechnological tools in stress resistance. *Agronomy*, 11(8): 1579.

Iqbal, M., Raja, N. I., Mashwani, Z. U. R., Hussain, M., Ejaz, M., and Yasmeen, F. (2019). Effect of silver nanoparticles on growth of wheat under heat stress. *Iranian Journal of Science and Technology, Transactions A: Science*, 43: 387–395.

Irshad, M. A., ur Rehman, M. Z., Anwar-ul-Haq, M., Rizwan, M., Nawaz, R., Shakoor, M. B. et al. (2021). Effect of green and chemically synthesized titanium dioxide nanoparticles on cadmium accumulation in wheat grains and potential dietary health risk: A field investigation. *Journal of Hazardous Materials*, 415: 125585.

Jalil, S. U., and Ansari, M. I. (2019). Nanoparticles and abiotic stress tolerance in plants: synthesis, action, and signaling mechanisms. *Plant signaling Molecules*, 549–561.

Jamdagni, P., Khatri, P., and Rana, J. S. (2018). Green synthesis of zinc oxide nanoparticles using flower extract of Nyctanthes arbor-tristis and their antifungal activity. *Journal of King Saud University - Science*, 30(2): 168–175. https://doi.org/10.1016/J.JKSUS.2016.10.002.

Javed, R., Ijaz, S., Hameed, H., Nazish, M., Sharif, M. S., Afreen, A. et al. (2023). Phytochemical-mediated biosynthesis of silver nanoparticles from strobilanthes glutinosus: exploring biological applications. *Micromachines*, 14(7): 1372.

Jayappa, M.D., Ramaiah, C.K., Kumar, M.A.P., Suresh, D., Prabhu, A., Devasya, R. P. et al. (2020). Green synthesis of zinc oxide nanoparticles from the leaf, stem and in vitro grown callus of *Mussaenda frondosa* L.: characterization and their applications. *Applied nanoscience*, 10: 3057–3074.

Jebril, S., Khanfir Ben Jenana, R., and Dridi, C. (2020). Green synthesis of silver nanoparticles using Melia azedarach leaf extract and their antifungal activities: *In vitro* and *in vivo*. *Materials Chemistry and Physics*, 248: 122898. https://doi.org/10.1016/J.MATCHEMPHYS.2020.122898.

Jeevanandam, J., Chan, Y. S., and Danquah, M. K. (2016). Biosynthesis of metal and metal oxide nanoparticles. *ChemBioEng Reviews*, 3(2): 55–67.

Jeevanandam, J., Kiew, S. F., Boakye-Ansah, S., Lau, S. Y., Barhoum, A., Danquah, M. K. et al. (2022). Green approaches for the synthesis of metal and metal oxide nanoparticles using microbial and plant extracts. *Nanoscale*, 14(7): 2534–2571.

Jiang, Y., Zhou, P., Zhang, P., Adeel, M., Shakoor, N., Li, Y. et al. (2022). Green synthesis of metal-based nanoparticles for sustainable agriculture. *Environmental Pollution*, 309: 119755.

Kandhol, N., Jain, M., and Tripathi, D. K. (2022). Nanoparticles as potential hallmarks of drought stress tolerance in plants. *Physiologia Plantarum*, 174(2): e13665.

Kang, J., Kim, Y., Singh, P., Huo, Y., Soshnikova, V., Markus, J. et al. (2017). Biosynthesis of gold and silver chloride nanoparticles mediated by Crataegus pinnatifida fruit extract: *in vitro* study of anti-inflammatory activities. *Artificial Cells, Nanomedicine, and Biotechnology*, 46: 1530–1540.

Kathiravan, V., Ravi, S., Ashokkumar, S., Velmurugan, S., Elumalai, K., and Khatiwada, C. P. (2015). Green synthesis of silver nanoparticles using Croton sparsiflorus morong leaf extract and their antibacterial and antifungal activities. *Spectrochimica Acta Part A: Molecular and Biomolecular Spectroscopy*, 139: 200–205. https://doi.org/10.1016/J.SAA.2014.12.022.

Katiyar, P., Yadu, B., Korram, J., Satnami, M. L., Kumar, M., and Keshavkant, S. (2020). Titanium nanoparticles attenuates arsenic toxicity by up-regulating expressions of defensive genes in *Vigna radiata* L. *Journal of Environmental Sciences*, 92: 18–27.

Keijok, W., Pereira, R., Alvarez, L., Prado, A., Da Silva, A., Ribeiro, J. et al. (2019). Controlled biosynthesis of gold nanoparticles with Coffea arabica using factorial design. *Scientific Reports*, 9.

Khalid, M. F., Iqbal Khan, R., Jawaid, M. Z., Shafqat, W., Hussain, S., Ahmed, T. et al. (2022). Nanoparticles: the plant saviour under abiotic stresses. *Nanomaterials*, 12(21): 3915.

Khan, M. F., and Khan, M. A. (2023). Plant-Derived Metal Nanoparticles (PDMNPs): Synthesis, Characterization, and Oxidative Stress-Mediated Therapeutic Actions. *Future Pharmacology*, 3(1): 252–295. https://doi.org/10.3390/futurepharmacol3010018.

Khan, M. I. R., and Khan, N. A. (Eds.). (2017). *Reactive oxygen species and antioxidant systems in plants: role and regulation under abiotic stress*. Singapore: Springer Singapore.

Khan, M. I. R., Palakolanu, S. R., Chopra, P., Rajurkar, A. B., Gupta, R., Iqbal, N. et al. (2021). Improving drought tolerance in rice: Ensuring food security through multi-dimensional approaches. *Physiologia Plantarum*, 172(2): 645–668.

Khan, S. H. (2020). Green nanotechnology for the environment and sustainable development. *Green Materials for Wastewater Treatment*, 13–46.

Khatami, M., Nejad, M., Salari, S., and Almani, P. (2016). Plant-mediated green synthesis of silver nanoparticles using Trifolium resupinatum seed exudate and their antifungal efficacy on Neofusicoccum parvum and Rhizoctonia solani.. *IET nanobiotechnology*, 104: 237–43.

Khatami, M., Pourseyedi, S., Khatami, M., Hamidi, H., Zaeifi, M., and Soltani, L. (2015). Synthesis of silver nanoparticles using seed exudates of Sinapis arvensis as a novel bioresource, and evaluation of their antifungal activity. *Bioresources and Bioprocessing*, 2: 1–7.

Khatami, M., Pourseyedi, S., Khatami, M., Hamidi, H., Zaeifi, M., and Soltani, L. (2015). Synthesis of silver nanoparticles using seed exudates of sinapis arvensis as a novel bioresource, and evaluation of their antifungal activity. *Bioresources and Bioprocessing*, 2(1): 1–7. https://doi.org/10.1186/S40643-015-0043-Y/FIGURES/6.

Kirupagaran, R., Saritha, A., and Bhuvaneswari, S. (2016). Green synthesis of selenium nanoparticles from leaf and stem extract of Leucas lavandulifolia Sm. and their application. *Journal of Nanoscience and Technology*, 224–226.

Kumar, A., Choudhary, A., Kaur, H., Guha, S., Mehta, S., and Husen, A. (2022). Potential applications of engineered nanoparticles in plant disease management: a critical update. *Chemosphere*, 295. https://doi.org/10.1016/j.chemosphere.2022.133798.

Kumar, B., Smita, K., Cumbal, L., and Debut, A. (2016). Ficus carica (Fig) fruit mediated green synthesis of silver nanoparticles and its antioxidant activity: a comparison of thermal and ultrasonication approach. *BioNanoScience*, *6:* 15–21.

Kumar, H., Bhardwaj, K., Kuča, K., Kalia, A., Nepovimova, E., Verma, R. et al. (2020). Flower-based green synthesis of metallic nanoparticles: applications beyond fragrance. *Nanomaterials*, 10.

Kumar, R., Murthy, J., Rao, N., and Bhargava, Y. (2016). Evaluation of silver nanoparticles synthetic potential of Couroupita guianensis Aubl., flower buds extract and their synergistic antibacterial activity. *3 Biotech*, 6.

Kumar, S., and Rajeshkumar, S. (2018). Plant-Based Synthesis of Nanoparticles and Their Impact. 33–57.

Kumari, M., Sadhu, P., Talele, C., and Shah, N. (2024). Green nanotechnology: how plants can help synthesize nanoparticles for biomedical and environmental purposes. *Journal of Natural Remedies*, 1021–1034.

Kumari, S., Khanna, R. R., Nazir, F., Albaqami, M., Chhillar, H., Wahid, I. et al. (2022). Bio-synthesized nanoparticles in developing plant abiotic stress resilience: a new boon for sustainable approach. *International Journal of Molecular Sciences*, 23(8): 4452.

Kuppusamy, P., Yusoff, M. M., Maniam, G. P., and Nagarajan, G. (2016). Biosynthesis of metallic nanoparticles using plant derivatives and their new avenues in pharmacological applications – an updated report. *Saudi Pharmaceutical Journal*, 24(4): 473–484.

Lazulfa, I., and Wikantyasning, E. (2023). Systematic Literatur Review: Synthesis of Silver Nanoparticles (Agnps) Using Secondary Metabolites of Fruit Peel Extract. *Medical Sains : Jurnal Ilmiah Kefarmasian.*

Lee, Y. J., Song, K., Cha, S. H., Cho, S., Kim, Y. S., and Park, Y. (2019). Sesquiterpenoids from Tussilago farfara flower bud extract for the eco-friendly synthesis of silver and gold nanoparticles possessing antibacterial and anticancer activities. *Nanomaterials*, 9(6): 819.

Li, M., Zhang, P., Adeel, M., Guo, Z., Chetwynd, A.J., Ma, C. et al. (2021). Physiological impacts of zero valent iron, $Fe3O4$ and $Fe2O3$ nanoparticles in rice plants and their potential as Fe fertilizers. *Environmental Pollution*, 269: 116134.

Li, Y., Zhang, P., Li, M., Shakoor, N., Adeel, M., Zhou, P. et al. (2022). Application and mechanisms of metal-based nanoparticles in the control of bacterial and fungal crop diseases. *Pest Management Science.*

Linh, T. M., Mai, N. C., Hoe, P. T., Lien, L. Q., Ban, N. K., Hien, L. T. T. et al. (2020). Metal-based nanoparticles enhance drought tolerance in soybean. *Journal of Nanomaterials*, 2020(1): 4056563.

Liu, J., Fu, C., Li, G., Khan, M. N., and Wu, H. (2021). ROS homeostasis and plant salt tolerance: plant nanobiotechnology updates. *Sustainability*, 13(6): 3552.

Liu, S., Tian, W., Liu, Z., Wei, X., Yuan, K., Du, W. et al. (2024). Biosynthesis of cupric oxide nanoparticles: its antiviral activities against TMV by directly destroying virion and inducing plant resistance. *Phytopathology Research*, 6(1). https://doi.org/10.1186/s42483-024-00250-z.

Lomelí-Rosales, D., Zamudio-Ojeda, A., Reyes-Maldonado, O., López-Reyes, M., Basulto-Padilla, G., López-Naranjo, E. et al. (2022). Green synthesis of gold and silver nanoparticles using leaf extract of capsicum chinense plant. *Molecules*, 27.

Maheshwaran, G., Muneeswari, R. S., Bharathi, A. N., Kumar, M. K., and Sudhahar, S. (2021). Eco-friendly synthesis of lanthanum oxide nanoparticles by Eucalyptus globulus leaf extracts for effective biomedical applications. *Materials Letters*, 283: 128799.

Makarov, V. V., Love, A. J., Sinitsyna, O. V., Makarova, S. S., Yaminsky, I. V., Taliansky, M. E. et al. (2014). "Green" nanotechnologies: synthesis of metal nanoparticles using plants. *Acta Naturae (англоязычная версия)*, 6(1(20)): 35–44.

Makki, H. J., and Chandrasekharaiah, K. S. (2023). Synthesis and characterization of citrus limonum copper nanoparticles and its in vitro evaluation of cytotoxicity against jurkat cells. *Pharmacognosy Magazine*, 19(3): 499–508.

Mali, S.C., Dhaka, A., Githala, C.K., and Trivedi, R. (2020). Green synthesis of copper nanoparticles using *Celastrus paniculatus* Willd. leaf extract and their photocatalytic and antifungal properties. *Biotechnology Reports*, 27: e00518. https://doi.org/10.1016/J.BTRE.2020.E00518.

Manimegalai, T., Raguvaran, K., Kalpana, M., and Maheswaran, R. (n.d.). *Green synthesis of silver nanoparticle using Leonotis nepetifolia and their toxicity against vector mosquitoes of Aedes aegypti and Culex quinquefasciatus and agricultural pests of Spodoptera litura and Helicoverpa armigera*. https://doi.org/10.1007/s11356-020-10127-1/ Published.

Maqsood, M. F., Shahbaz, M., Khalid, F., Rasheed, Y., Asif, K., Naz, N. et al. (2023). Plant-based nanoparticles in ROS mitigation and abiotic stress tolerance: a review. *Plant Stress*, 100281.

Mazzini, L. F., Bousada, G. M., Puiatti, G. A., and Moreira, R. P. (2024). Green Synthesis of Iron Nanozymes Using Eucalyptus grandis Extract and Their Application as Peroxidase-Like Catalysts for H2O2 Quantification. *Journal of the Brazilian Chemical Society*, 35(5): e-20230171.

Medda, S., Hajra, A., Dey, U., Bose, P., and Mondal, N. K. (2015). Biosynthesis of silver nanoparticles from Aloe vera leaf extract and antifungal activity against *Rhizopus* sp. and *Aspergillus* sp. *Applied Nanoscience (Switzerland)*, *5*(7): 875–880. https://doi.org/10.1007/S13204-014-0387-1/FIGURES/7.

Melese, A., Wubet, W., Hussen, A., Mulate, K., and Hailekiros, A. (2024). A review on biogenic synthesized zinc oxide nanoparticles: synthesis, characterization, and its applications. *Reviews in Inorganic Chemistry*, (0).

Mishra, S., Spaccarotella, K., Gido, J., Samanta, I., and Chowdhary, G. (2023). Effects of heat stress on plant-nutrient relations: An update on nutrient uptake, transport, and assimilation. *International Journal of Molecular Sciences*, 24(21): 15670.

Mittal, A., Chisti, Y., and Banerjee, U. (2013). Synthesis of metallic nanoparticles using plant extracts. *Biotechnology advances*, 31 2: 346–56 .

Mubayi, A., Chatterji, S., Rai, P. M., and Watal, G. (2012). Evidence based green synthesis of nanoparticles. *Adv. Mat. Lett.*, 3(6): 519–525.

Muhammad, M., Waheed, A., Wahab, A., Majeed, M., Nazim, M., Liu, Y. H. et al. (2023). Soil salinity and drought tolerance: An evaluation of plant growth, productivity, microbial diversity, and amelioration strategies. *Plant Stress*, 100319.

Muhib, M. M. A., Mubassira, M., Khanam, S., Abeed-Ul-Haque, S., and Hasan, M. F. (2024). A systematic review on green synthesis of silver nanoparticles, characterization and applications. *World Journal of Advanced Research and Reviews*, 21(3): 1221–1238.

Mukarram, M., Choudhary, S., Kurjak, D., Petek, A., and Khan, M. M. A. (2021). Drought: Sensing, signalling, effects and tolerance in higher plants. *Physiologia plantarum*, 172(2): 1291–1300.

Muruganandham, M., Sivasubramanian, K., Velmurugan, P., Arumugam, N., Almansour, A., Kumar, R. et al. (2023). Phytocrystallization of silver nanoparticles using Cassia alata flower extract for effective control of fungal skin pathogens. *Green Processing and Synthesis*, 12.

Mustafa, H., Ilyas, N., Akhtar, N., Raja, N. I., Zainab, T., Shah, T. et al. (2021). Biosynthesis and characterization of titanium dioxide nanoparticles and its effects along with calcium phosphate on physicochemical attributes of wheat under drought stress. *Ecotoxicology and Environmental Safety*, 223: 112519.

Mustafa, N., Raja, N. I., Ilyas, N., Ikram, M., Mashwani, Z. U. R., and Ehsan, M. (2021). Foliar applications of plant-based titanium dioxide nanoparticles to improve agronomic and physiological attributes of wheat (*Triticum aestivum* L.) plants under salinity stress. *Green Processing and Synthesis*, 10(1): 246–257.

Mustapha, T., Misni, N., Ithnin, N. R., Daskum, A. M., and Unyah, N. Z. (2022). A review on plants and microorganisms mediated synthesis of silver nanoparticles, role of plants metabolites and applications. *International Journal of Environmental Research and Public Health*, 19(2): 674.

Naikoo, G. A., Mustaqeem, M., Hassan, I. U., Awan, T., Arshad, F., Salim, H. et al. (2021). Bioinspired and green synthesis of nanoparticles from plant extracts with antiviral and antimicrobial properties: A critical review. *Journal of Saudi Chemical Society*, 25(9): 101304. https://doi.org/10.1016/J.JSCS.2021.101304.

Nazir, F., Fariduddin, Q., and Khan, T. A. (2020). Hydrogen peroxide as a signaling molecule in plants and its crosstalk with other plant growth regulators under heavy metal stress. *Chemosphere*, 252: 126486.

Ndou, N., Rakgotho, T., Nkuna, M., Doumbia, I.Z., Mulaudzi, T., and Ajayi, R. F. (2023). Green synthesis of iron oxide (Hematite) nanoparticles and their influence on Sorghum bicolor growth under drought stress. *Plants*, 12(7): 1425.

Nejatzadeh, F. (2021). Effect of silver nanoparticles on salt tolerance of Satureja hortensis l. during *in vitro* and *in vivo* germination tests. *Heliyon*, 7(2).

Nguyen, D., Le, H., Nguyen, T., Nguyen, T., Bach, L., Nguyen, T. et al. (2021). Multifunctional ZnO nanoparticles bio-fabricated from Canna indica L. flowers for seed germination, adsorption, and photocatalytic degradation of organic dyes. *Journal of Hazardous Materials*, 420: 126586.

Nguyen, D., Nguyen, H., Le, N., Nguyen, K., Nguyen, H., Le, H. et al. (2021). Copper nanoparticle application enhances plant growth and grain yield in maize under drought stress conditions. *Journal of Plant Growth Regulation,* 41(1): 364–375.

Nguyen, N.T.T., Nguyen, L.M., Nguyen, T.T.T., Nguyen, T.T., Nguyen, D.T.C., and Tran, T.Van. (2022). Formation, antimicrobial activity, and biomedical performance of plant-based nanoparticles: a review. *Environmental Chemistry Letters 2022, 20*(4): 2531–2571. https://doi.org/10.1007/S10311-022-01425-W.

Niraimathee, V., Subha, V., Ravindran, R., and Renganathan, S. (2016). Green synthesis of iron oxide nanoparticles from Mimosa pudica root extract. *International Journal of Environment and Sustainable Development*, 15: 227.

Noor, R., Yasmin, H., Ilyas, N., Nosheen, A., Hassan, M.N., Mumtaz, S. et al. (2022). Comparative analysis of iron oxide nanoparticles synthesized from ginger (Zingiber officinale) and cumin seeds (*Cuminum cyminum*) to induce resistance in wheat against drought stress. *Chemosphere*, 292: 133201.

P,V., Udayabhanu, U., G., N., Hs, L., and N, C. (2019). Plant-mediated green synthesis of Ag nanoparticles using *Rauvolfia tetraphylla* (L.) flower extracts: Characterization, Biological activities and screening of their catalytic activity in formylation reaction. *Scientia Iranica.*

Padalia, H., and Chanda, S. (2017). Characterization, antifungal and cytotoxic evaluation of green synthesized zinc oxide nanoparticles using Ziziphus nummularia leaf extract. *Artificial Cells, Nanomedicine, and Biotechnology*, 45(8): 1751–1761. https://doi.org/10.1080/21691401.2017.1282868.

Padalia, H., Moteriya, P., and Chanda, S. (2015). Green synthesis of silver nanoparticles from marigold flower and its synergistic antimicrobial potential. *Arabian Journal of Chemistry*, 8(5): 732–741. https://doi.org/10.1016/J.ARABJC.2014.11.015.

Pandian, C. J., Palanivel, R., and Dhananasekaran, S. (2015). Green synthesis of nickel nanoparticles using Ocimum sanctum and their application in dye and pollutant adsorption. *Chinese journal of Chemical Engineering*, 23(8): 1307–1315.

Pavitra, G., Sushila, N., Sreenivas, A.G., Ashok, J., and Sharanagouda, H. (2018). Biosynthesis of Green Silica nanoparticles and its effect on cotton aphid, aphis gossypii glover and mealybug, Phenacoccus solenopsis Tinsley. *International Journal of Current Microbiology and Applied Sciences*, 7(10): 1450–1460. https://doi.org/10.20546/IJCMAS.2018.710.162.

Pechyen, C., Ponsanti, K., Tangnorawich, B., and Ngernyuang, N. (2021). Waste fruit peel – Mediated green synthesis of biocompatible gold nanoparticles. *Journal of Materials Research and Technology*, 14: 2982–2991.

Pinto, R., Lucas, J., Morais, M., Santos, S., Silvestre, A., Marques, P. et al. (2017). Demystifying the morphology and size control on the biosynthesis of gold nanoparticles using Eucalyptus globulus bark extract. *Industrial Crops and Products*, 105: 83–92.

Qasim Nasar, M., Zohra, T., Khalil, A.T., Saqib, S., Ayaz, M., Ahmad, A. et al. (2019). Seripheidium quettense mediated green synthesis of biogenic silver nanoparticles and their theranostic applications. *Green Chemistry Letters and Reviews*, 12(3): 310–322. https://doi.org/10.1080/17518253.2019.1643929.

Rafique, Z., Sadaf, I., Tahir, M., Rafique, Z., Nabi, G., Iqbal, T. et al. (2019). Novel and facile synthesis of silver nanoparticles using Albizia procera leaf extract for dye degradation and antibacterial applications. *Materials science & engineering. C, Materials for Biological Applications*, 99: 1313–1324.

Ragab, G. A., and Saad-Allah, K. M. (2020). Green synthesis of sulfur nanoparticles using Ocimum basilicum leaves and its prospective effect on manganese-stressed *Helianthus annuus* (L.) seedlings. *Ecotoxicology and Environmental Safety*, 191: 110242.

Rahman, Md. A., Parvin, A., Khan, Md. S.H., War, A.R., Lingaraju, K., Prasad, R. et al. (2021). Efficacy of the green synthesized nickel-oxide nanoparticles against pulse beetle, *Callosobruchus maculatus* (F.) in black gram (*Vigna mungo* L.). *International Journal of Pest Management*, 67(4): 306–314. https://doi.org/10.1080/09670874.2020.1773572.

Rajan, A., Vilas, V., and Philip, D. (2015). Studies on catalytic, antioxidant, antibacterial and anticancer activities of biogenic gold nanoparticles. *Journal of Molecular Liquids*, 212: 331–339. https://doi.org/10.1016/J.MOLLIQ.2015.09.013.

Rajapriya, M., Sharmili, S. A., Baskar, R., Balaji, R., Alharbi, N. S., Kadaikunnan, S. et al. (2020). Synthesis and Characterization of Zinc Oxide Nanoparticles Using Cynara scolymus Leaves: Enhanced Hemolytic, Antimicrobial, Antiproliferative, and Photocatalytic Activity. *Journal of Cluster Science*, 31(4): 791–801. https://doi.org/10.1007/s10876-019-01686-6.

Rakgotho, T., Ndou, N., Mulaudzi, T., Iwuoha, E., Mayedwa, N., and Ajayi, R. F. (2022). Green-synthesized zinc oxide nanoparticles mitigate salt stress in Sorghum bicolor. *Agriculture*, 12(5): 597.

Rashid, A., Schutte, B. J., Ulery, A., Deyholos, M. K., Sanogo, S., Lehnhoff, E. A. et al. (2023). Heavy metal contamination in agricultural soil: environmental pollutants affecting crop health. *Agronomy*, 13(6): 1521.

Rawani, A. (2017). Mosquito larvicidal activity of green silver nanoparticle synthesized from extract of bud of *Polianthus tuberosa* L. *Int. J. Nanotechnol. Appl.*, 11(1): 17–28.

Rehman, A., Weng, J., Li, P., Shah, I. H., ur Rahman, S., Khalid, M. et al. (2023). Green synthesized zinc oxide nanoparticles confer drought tolerance in melon (*Cucumis melo* L.). *Environmental and Experimental Botany*, 212: 105384.

Saad-Allah, K. M., and Ragab, G. A. (2020). Sulfur nanoparticles mediated improvement of salt tolerance in wheat relates to decreasing oxidative stress and regulating metabolic activity. *Physiology and Molecular Biology of Plants*, 26(11): 2209-2223.

Sadat-Hosseini, M., Naeimi, A., Boroomand, N., Aalifar, M., and Farajpour, M. (2022). Alleviating the adverse effects of salinity on Roselle plants by green synthesized nanoparticles. *Scientific Reports*, 12(1): 18165.

Saha, R., Subramani, K., Sikdar, S., Fatma, K., and Rangaraj, S. (2021). Effects of processing parameters on green synthesised ZnO nanoparticles using stem extract of *Swertia chirayita*. *Biocatalysis and Agricultural Biotechnology*, 33: 101968.

Saheb, M., Hosseini, H.A., Hashemzadeh, A., Elahi, B., Hasanzadeh, L., Oskuee, R.K. et al. (2019). Photocatalytic and biological attributes of green synthesized nickel oxide nanoparticles by Rheum turkestanicum (RT) root extract. *ChemistrySelect*, 4(8): 2416–2420.

Salehi, S., Sadat Shandiz, S. A., Ghanbar, F., Darvish, M. R., Ardestani, M. S., Mirzaie, A. et al. (2016). Phytosynthesis of silver nanoparticles using <em>Artemisia marschalliana</em> Sprengel aerial part extract and assessment of their antioxidant, anticancer, and antibacterial properties. *International Journal of Nanomedicine*, 11: 1835–1846. https://doi.org/10.2147/IJN.S99882.

Salem, S.S., and Fouda, A. (2021). Green synthesis of metallic nanoparticles and their prospective biotechnological applications: an overview. *Biological Trace Element Research*, 199(1): 344–370.

Salunke, G., Ghosh, S., Kumar, R., Khade, S., Vashisth, P., Kale, T. et al. (2014). Rapid efficient synthesis and characterization of silver, gold, and bimetallic nanoparticles from the medicinal plant Plumbago zeylanica and their application in biofilm control. *International Journal of Nanomedicine*, 9: 2635–2653.

Sampath, V., Shalakhti, O., Veidis, E., Efobi, J. A. I., Shamji, M. H., Agache, I. et al. (2023). Acute and chronic impacts of heat stress on planetary health. *Allergy*, 78(8): 2109–2120.

Saratale, R.G., Benelli, G., Kumar, G., Kim, D.S., and Saratale, G.D. (2018). Bio-fabrication of silver nanoparticles using the leaf extract of an ancient herbal medicine, dandelion (*Taraxacum officinale*), evaluation of their antioxidant, anticancer potential, and antimicrobial activity against phytopathogens. *Environmental Science and Pollution Research*, 25(11): 10392–10406. https://doi.org/10.1007/S11356-017-9581-5/FIGURES/7.

Sebastian, A., Nangia, A., and Prasad, M. N. V. (2018). A green synthetic route to phenolics fabricated magnetite nanoparticles from coconut husk extract: Implications to treat metal contaminated water and heavy metal stress in Oryza sativa L. *Journal of Cleaner Production*, 174: 355–366.

Sebastian, A., Nangia, A., and Prasad, M. N. V. (2019). Cadmium and sodium adsorption properties of magnetite nanoparticles synthesized from *Hevea brasiliensis* Muell. Arg. bark: Relevance in amelioration of metal stress in rice. *Journal of hazardous materials*, *371*: 261–272.

Selvaraj, P., Jayaseeli, P.J.R., and Ansilin, M. (2018). *Pest control potential of Cyclosorous interruptus, Christella dentata and Nephrolepis cordifolia on the biology of Spodoptera litura (Fab.)*.

Shafey, A. (2020). Green synthesis of metal and metal oxide nanoparticles from plant leaf extracts and their applications: A review. *Green Processing and Synthesis*, 9: 304–339.

Shah, I. H., Manzoor, M. A., Jinhui, W., Li, X., Hameed, M. K., Rehaman, A. et al. (2024). Comprehensive review: Effects of climate change and greenhouse gases emission relevance to environmental stress on horticultural crops and management. *Journal of Environmental Management*, 351: 119978.

Shahid, S. (2020). A DNA methylation reader with an affinity for salt stress.

Shaikhaldein, H.O., Al-Qurainy, F., Nadeem, M., Khan, S., Tarroum, M., Salih, A. M. et al. (2022). Assessment of the impacts of green synthesized silver nanoparticles on *Maerua oblongifolia* shoots under *in vitro* salt stress. *Materials*, 15(14): 4784.

Sharifi-Rad, M., Pohl, P., Epifano, F., and Álvarez-Suarez, J. (2020). Green synthesis of silver nanoparticles using astragalus tribuloides delile. Root Extract: Characterization, Antioxidant, Antibacterial, and Anti-Inflammatory Activities. *Nanomaterials*, 10.

Sheikhalipour, M., Esmaielpour, B., Gohari, G., Haghighi, M., Jafari, H., Farhadi, H. et al. (2021). Salt stress mitigation via the foliar application of chitosan-functionalized selenium and anatase titanium dioxide nanoparticles in stevia (*Stevia rebaudiana Bertoni*). *Molecules*, 26(13): 4090.

Shukla, G., Singh, A., Chaudhary, N., Singh, S., Basnal, N., and Gaurav, S. S. (2024). Metal nanoparticles to improve the heat resilience in wheat (*Triticum aestivum* L.). *Nanotechnology*, 35(20): 205101.

Singh, A., Basnal, N., Shukla, G., Chaudhary, N., Singh, S., and Gaurav, S. S. (2022). Evaluation of efficacy of Phyto-synthesized iron oxide nanoparticles in contributing drought resilience in wheat (*Triticum aestivum* L.). *Nanotechnology*, 33(48): 485101.

Singh, H., Desimone, M., Pandya, S., Jasani, S., George, N., Adnan, M. et al. (2023). Revisiting the green synthesis of nanoparticles: uncovering influences of plant extracts as reducing agents for enhanced synthesis efficiency and its biomedical applications. *International Journal of Nanomedicine*, 18: 4727–4750.

Singh, P., Kim, Y., Wang, C., Mathiyalagan, R., and Yang, D. (2015). The development of a green approach for the biosynthesis of silver and gold nanoparticles by using Panax ginseng root extract, and their biological applications. *Artificial Cells, Nanomedicine, and Biotechnology*, 44: 1150–1157.

Sinha, R., Fritschi, F. B., Zandalinas, S. I., and Mittler, R. (2021). The impact of stress combination on reproductive processes in crops. *Plant Science*, 311: 111007.

Sone, B. T., Manikandan, E., Gurib-Fakim, A., and Maaza, M. (2016). Single-phase α-Cr2O3 nanoparticles' green synthesis using Callistemon viminalis' red flower extract. *Green Chemistry Letters and Reviews*, 9(2): 85–90.

Sreelakshmi, B., Induja, S., Adarsh, P. P., Rahul, H. L., Arya, S. M., Aswana, S. et al. (2021). Drought stress amelioration in plants using green synthesized iron oxide nanoparticles. *Materials Today: Proceedings*, 41: 723–727.

Sujitha, M.V., and Kannan, S. (2013). Green synthesis of gold nanoparticles using Citrus fruits (Citrus limon, Citrus reticulata and Citrus sinensis) aqueous extract and its characterization. *Spectrochimica Acta Part A: Molecular and Biomolecular Spectroscopy*, 102: 15–23.

Sun, L., Song, F., Guo, J., Zhu, X., Liu, S., Liu, F., and Li, X. (2020). Nano-ZnO-induced drought tolerance is associated with melatonin synthesis and metabolism in maize. *International Journal of Molecular Sciences*, 21(3): 782.

Suresh, U., Murugan, K., Panneerselvam, C., Rajaganesh, R., Roni, M., Aziz, A. T. et al. (2018). Suaeda maritima -based herbal coils and green nanoparticles as potential biopesticides against the dengue vector Aedes aegypti and the tobacco cutworm Spodoptera litura. *Physiological and Molecular Plant Pathology*, 101: 225–235. https://doi.org/10.1016/j.pmpp.2017.01.002.

Sutulienė, R., Ragelienė, L., Samuolienė, G., Brazaitytė, A., Urbutis, M., and Miliauskienė, J. (2021). The response of antioxidant system of drought-stressed green pea (*Pisum sativum* L.) affected by watering and foliar spray with silica nanoparticles. *Horticulturae*, 8(1): 35.

Tariq, M., Choudhary, S., Singh, H., Siddiqui, M. A., Kumar, H., Amir, A. et al. (2021). Role of nanoparticles in abiotic stress. *Technology in Agriculture*, 323.

Thi, T., Nguyen, T., Thi, D., Thi, K., Phan, B., and Pham, K. (2020). Green synthesis of ZnO nanoparticles using orange fruit peel extract for antibacterial activities. *RSC Advances*, 10: 23899–23907.

Thovhogi, N., Diallo, A., Gurib-Fakim, A., and Maaza, M. (2015). Nanoparticles green synthesis by Hibiscus sabdariffa flower extract: main physical properties. *Journal of Alloys and Compounds*, 647: 392–396.

Tinnirello, V., Rabienezhad Ganji, N., De Marcos Lousa, C., Alessandro, R., and Raimondo, S. (2023). Exploiting the opportunity to use plant-derived nanoparticles as delivery vehicles. *Plants*, 12(6): 1207.

Trejo-Téllez, L.I. (2023). Salinity Stress Tolerance in Plants. *Plants*, 12(20): 3520.

Uddin, S., Safdar, L., Anwar, S., Iqbal, J., Laila, S., Abbasi, B. et al. (2021). Green Synthesis of Nickel Oxide Nanoparticles from Berberis balochistanica Stem for Investigating Bioactivities. *Molecules*, 26.

Valsalam, S., Agastian, P., Arasu, M.V., Al-Dhabi, N.A., Ghilan, A.K.M., Kaviyarasu, K. et al. (2019). Rapid biosynthesis and characterization of silver nanoparticles from the leaf extract of *Tropaeolum majus* L. and its enhanced in-vitro antibacterial, antifungal, antioxidant and anticancer properties. *Journal of Photochemistry and Photobiology B: Biology*, 191: 65–74. https://doi.org/10.1016/J.JPHOTOBIOL.2018.12.010.

Vanitha, G., Manikandan, R., Sathiyamoorthi, K., and Dhinakaran, B. (2022). Review on green synthesis of nanoparticles using various strong electrolytic metal solutions mediated by various plant parts. *Journal of Nanoscience and Technology*.

Veisi, H., Karmakar, B., Tamoradi, T., Hemmati, S., Hekmati, M., and Hamelian, M. (2021). Biosynthesis of CuO nanoparticles using aqueous extract of herbal tea (*Stachys Lavandulifolia*) flowers and evaluation of its catalytic activity. *Scientific Reports*, 11.

Velmurugan, P., Anbalagan, K., Manosathyadevan, M., Lee, K.J., Cho, M., Lee, S.M. et al. (2014). Green synthesis of silver and gold nanoparticles using Zingiber officinale root extract and antibacterial activity of silver nanoparticles against food pathogens. *Bioprocess and Biosystems Engineering*, 37: 1935–1943.

Velsankar, K., Preethi, R., Ram, P., Ramesh, M., and Sudhahar, S. (2020). Evaluations of biosynthesized ag nanoparticles via allium sativum flower extract in biological applications. *Applied Nanoscience*, 10(9): 3675–3691.

Venkatachalam, P., Jayaraj, M., Manikandan, R., Geetha, N., Rene, E. R., Sharma, N. C. et al. (2017). Zinc oxide nanoparticles (ZnONPs) alleviate heavy metal-induced toxicity in Leucaena leucocephala seedlings: a physiochemical analysis. *Plant Physiology and Biochemistry*, 110: 59–69.

Venkatachalam, P., Jayaraj, M., Manikandan, R., Geetha, N., Rene, E.R., Sharma, N.C. et al. (2017). Zinc oxide nanoparticles (ZnONPs) alleviate heavy metal-induced toxicity in Leucaena leucocephala seedlings: a physiochemical analysis. *Plant Physiology and Biochemistry*, 110: 59–69.

Venkatappa, M., Udagani, C., Gowda, S., Venkataramaiah, S., Casini, R., Moussa, I. et al. (2023). Green Synthesised TiO2 Nanoparticles-Mediated Terenna asiatica: Evaluation of Their Role in Reducing Oxidative Stress, Inflammation and Human Breast Cancer Proliferation. *Molecules*, 28.

Verma, M.L., Dhanya, B. S., Thakur, M., Jeslin, J., and Jana, A. K. (2021). Plant derived nanoparticles and their biotechnological applications. In *Comprehensive Analytical Chemistry* (Vol. 94, pp. 331–362). Elsevier B.V. https://doi.org/10.1016/bs.coac.2021.01.011.

Verma, P., Khan, F., and Banerjee, S. (2020). Salvadora persica root extract-mediated fabrication of ZnO nanoparticles and characterization. *Inorganic and Nano-Metal Chemistry*, 51: 427–433.

Viana, R.L.S., Fidelis, G.P., Medeiros, M.J.C., Morgano, M.A., Alves, M.G.C.F., Passero, L.F.D. et al. (2020). Green Synthesis of Antileishmanial and Antifungal Silver Nanoparticles Using Corn Cob Xylan as a Reducing and Stabilizing Agent. *Biomolecules,* 10(9): 1235. https://doi.org/10.3390/BIOM10091235.

Vijaya, J., Jayaprakash, N., Kombaiah, K., Kaviyarasu, K., Kennedy, J., Ramalingam, J. et al. (2017). Bioreduction potentials of dried root of Zingiber officinale for a simple green synthesis of silver nanoparticles: Antibacterial studies.*Journal of photochemistry and photobiology. B, Biology*, 177: 62–68.

Vijayan, R., Joseph, S., and Mathew, B. (2018). Eco-friendly synthesis of silver and gold nanoparticles with enhanced antimicrobial, antioxidant, and catalytic activities. *IET Nanobiotechnology*, 12(6): 850–856. https://doi.org/10.1049/IET-NBT.2017.0311.

Wahid, I., Kumari, S., Ahmad, R., Hussain, S.J., Alamri, S., Siddiqui, M.H. et al. (2020). Silver nanoparticle regulates salt tolerance in wheat through changes in ABA concentration, ion homeostasis, and defense systems. *Biomolecules*, 10(11): 1506.

Wahid, I., Rani, P., Kumari, S., Ahmad, R., Hussain, S.J., Alamri, S. et al. (2022). Biosynthesized gold nanoparticles maintained nitrogen metabolism, nitric oxide synthesis, ions balance, and stabilizes the defense systems to improve salt stress tolerance in wheat. *Chemosphere*, 287: 132142.

Wang, D., Cui, L., Chang, X., and Guan, D. (2020). Biosynthesis and characterization of zinc oxide nanoparticles from Artemisia annua and investigate their effect on proliferation, osteogenic differentiation and mineralization in human osteoblast-like MG-63 Cells. *Journal of Photochemistry and Photobiology B: Biology*, 202: 111652. https://doi.org/10.1016/J.JPHOTOBIOL.2019.111652.

Warren, J., Hanson, P., Iversen, C., Kumar, J., Walker, A., and Wullschleger, S. (2015). Root structural and functional dynamics in terrestrial biosphere models--evaluation and recommendations. *The New phytologist*, 205 1: 59–78.

Wasule, D.L., Shingote, P.R., and Saxena, S. (2024). Exploitation of functionalized green nanomaterials for plant disease management. In *Discover Nano* (Vol. 19, Issue 1). Springer. https://doi.org/10.1186/s11671-024-04063-z.

Wei, Y., Wu, Y., and Wang, X. (2023). Novel adsorbents and adsorption methods for pollutant removal. *Toxics*, 11(12): 954.

Worrall, E., Hamid, A., Mody, K., Mitter, N., and Pappu, H. (2018). Nanotechnology for plant disease management. *Agronomy*, 8(12): 285.

Wu, H., Tito, N., and Giraldo, J. (2017). Anionic cerium oxide nanoparticles protect plant photosynthesis from abiotic stress by scavenging reactive oxygen species. *ACS nano*, 11 11, 11283–11297.

Yang, Y., He, J., Li, Q., Gao, L., Hu, J., Zeng, R. et al. (2019). Self-healing of electrical damage in polymers using superparamagnetic nanoparticles. *Nature nanotechnology*, 14(2): 151–155.

Yasmin, H., Mazher, J., Azmat, A., Nosheen, A., Naz, R., Hassan, M.N. et al. (2021). Combined application of zinc oxide nanoparticles and biofertilizer to induce salt resistance in safflower by regulating ion homeostasis and antioxidant defence responses. *Ecotoxicology and Environmental Safety*, 218: 112262.

Yasmin, H., Mazher, J., Azmat, A., Nosheen, A., Naz, R., Hassan, M.N. et al. (2021). Combined application of zinc oxide nanoparticles and biofertilizer to induce salt resistance in safflower by regulating ion homeostasis and antioxidant defence responses. *Ecotoxicology and Environmental Safety*, 218: 112262.

Yazıcılar, B., Böke, F., Alaylı, A., Nadaroglu, H., Gedikli, S., and Bezirganoglu, I. (2021). In vitro effects of CaO nanoparticles on Triticale callus exposed to short and long-term salt stress. *Plant Cell Reports*, 40: 29–42.

Ying, S., Guan, Z., Ofoegbu, P.C., Clubb, P., Rico, C., He, F. et al. (2022). Green synthesis of nanoparticles: Current developments and limitations. *Environmental Technology & Innovation*, 26: 102336.

Yugandhar, P., Haribabu, R., and Savithramma, N. (2015). Synthesis, characterization and antimicrobial properties of green-synthesised silver nanoparticles from stem bark extract of *Syzygium alternifolium* (Wt.) Walp. *3 Biotech*, 5: 1031–1039.

Yugandhar, P., Vasavi, T., Jayavardhana Rao, Y., Uma Maheswari Devi, P., Narasimha, G., and Savithramma, N. (2018). Cost effective, green synthesis of copper oxide nanoparticles using fruit extract of *Syzygium alternifolium* (Wt.) Walp., characterization and evaluation of antiviral activity. *Journal of Cluster Science*, 29: 743–755.

Yusuf, S., Mood, C., Ahmad, N., Sandai, D., Lee, C., and Lim, V. (2020). Optimization of biogenic synthesis of silver nanoparticles from flavonoid-rich *Clinacanthus nutans* leaf and stem aqueous extracts. *Royal Society Open Science*, 7.

Zafar, S., Hasnain, Z., Aslam, N., Mumtaz, S., Jaafar, H. Z., Wahab, P. E. M. et al. (2021). Impact of Zn nanoparticles synthesized via green and chemical approach on okra (*Abelmoschus esculentus* L.) growth under salt stress. *Sustainability*, 13(7): 3694.

Zahedi, S.M., Moharrami, F., Sarikhani, S., and Padervand, M. (2020). Selenium and silica nanostructure-based recovery of strawberry plants subjected to drought stress. *Scientific reports*, 10(1): 17672.

Zahra, N., Hafeez, M.B., Ghaffar, A., Kausar, A., Al Zeidi, M., Siddique, K.H. et al. (2023). Plant photosynthesis under heat stress: Effects and management. *Environmental and Experimental Botany*, 206: 105178.

Zhang, H., Zhu, J., Gong, Z., and Zhu, J. K. (2022). Abiotic stress responses in plants. *Nature Reviews Genetics*, 23(2): 104–119.

Zhang, M., Yang, C., Yan, X., Sung, J., Garg, P., and Merlin, D. (2019). Highly biocompatible functionalized layer-by-layer ginger lipid nano vectors targeting p-selectin for delivery of doxorubicin to treat colon cancer. *Advanced Therapeutics*, 2(12): 1900129.

Zhou, P., Zhang, P., He, M., Cao, Y., Adeel, M., Shakoor, N. et al. (2023). Iron-based nanomaterials reduce cadmium toxicity in rice (*Oryza sativa* L.) by modulating phytohormones, phytochelatin, cadmium transport genes and iron plaque formation. *Environmental Pollution*, 320: 121063.

Zuhrotun, A., Oktaviani, D., and Hasanah, A. (2023). Biosynthesis of Gold and silver nanoparticles using phytochemical compounds. *Molecules*, 28.

CHAPTER-18

Microbial-Derived Nanoparticles for Stress-Resilient and Sustainable Agriculture

Swapna Battana, Srinivasan Kameswaran* and *Mandala Ramakrishna*

1. Introduction

Crop productivity has declined over the past few decades despite rising food demands brought on by a growing human population (Guleria et al., 2023; Singh et al., 2022). Reports state that 653 million people will still be undernourished in 2030, eight hundred million populaces with hungriness, and two billion people have a deficiency of micronutrients (FAO et al., 2024). According to a 2011 estimate from the Food and Agriculture Organization, 1.3 billion tons of food are lost annually worldwide. Given that the global human population is anticipated to surpass ten billion people by 2050, food security will therefore continue to be a significant concern (Rosegrant et al., 2024). Plant growth, development, agricultural output, and productivity are all negatively impacted by climate change, which is seen as a significant limiting factor. Climate change-related abiotic stressors such as drought, salt, soil erosion, flooding, and nutrient depletion are among the main causes of stunted plant growth, which lowers crop plant output and quality (Khalid et al., 2019).

Stress is defined as any unfavorable circumstance or material that restricts or interferes with a plant's growth, development, or metabolism. Plant height, leaf area, and crop development all decline under drought stress. Cold stress lowers crop output by slowing plant growth and efficiency. Salinity stress can negatively affect cytosolic metabolism, cell growth, membrane function, and the formation of radicals generating water stress in plants. Global precipitation may improve due to greater CO_2 concentrations, leading to more rainfall, which could hurt agricultural growth. Crops that are subjected to significant water stress have higher crude protein content but a lower amylose percentage. Higher osmotic potential and ion toxicity work together to prevent seed germination and impair growth, which in turn impacts the quality and quantity of agricultural production. Physiological adaptation and integrated cellular or molecular responses are among the escape-avoidance and tolerance mechanisms that plants evolve in response to abiotic stress.

Department of Botany, Vikrama Simhapuri University College, Kavali-524201, SPSR Nellore District, Andhra Pradesh, India.

* Corresponding author: swapnaivsr@gmail.com

DOI: 10.1201/9781003567684-18

Farming fields frequently experience a combination of pressures. Numerous abiotic and biotic stressors that plants encounter in the natural environment can significantly lower their production and total harvest (Da Costa et al., 2021; Ramegowda et al., 2024). There may be more coupled stress interactions in reality than the 120 interactions that Priya et al. (2023) thoroughly surveyed and documented.

Abiotic stressors, especially drought and high temperatures, frequently co-occur in field conditions, significantly lowering crop yields and endangering the world's food supply. Plant membrane damage and reduced photosynthesis have resulted from reduced water intake, elevated canopy temperature, and the buildup of reactive oxygen species brought on by drought and high temperatures. Additionally, there has been evidence of negative effects on reproductive processes (Kamatchi et al., 2024).

Macrophomina phaseolina and abiotic stressors such as heat, salinity, and drought—all of which are prevalent in the changing climate were studied by Ranjan et al. in 2023. It has been proposed that these stressors intensify the pathogen's effects on plants, resulting in notable reductions in crop output.

One potential and workable way to address this issue is to create plants that can withstand stress. Nevertheless, inter-generic or inter-specific hybridization has not improved crop plant stress tolerance as intended using traditional breeding techniques. To boost food production while reducing the usage of fertilizer and pesticides to lessen abiotic stress, new and innovative agricultural techniques that have the potential to transform contemporary agriculture are therefore needed. More people are also realizing that to solve environmental issues and maintain long-term ecological balance, current agricultural techniques must change to more effective and sustainable ones (Ahmed et al., 2023). Thus, a technical revolution in creating crop management systems that are resilient to stress is required to meet future global food demands sustainably.

The broad-spectrum uses of nanotechnology in many industries, especially plant biotechnology, and agriculture, have recently attracted a lot of attention to this new and quickly growing field of study. In particular, nanotechnology can aid in the development of secure and sustainable agricultural practices by offering practical answers to issues facing the sector (Lowry et al., 2019; Shang et al., 2019). In addition to reducing the use of hazardous chemicals, the use of NPs in agriculture has garnered interest since it has improved the nutritional content and storage time frame of food products (Ansari, 2023). For improving plant development and yield under stressful situations, they have been widely used as nano sensors, nano pesticides, and nano fertilizers (Mandal and Sarkar, 2024; Vyas et al., 2024). Unfortunately, the use of hazardous chemicals and the procedures used to manufacture them provide challenges for NPs made chemically. The green process known as "microbial synthesis," which uses bacteria, fungi, algae, and viruses to create clean, biocompatible, and environmentally friendly nanoparticles (NPs) to create stress-resistant crops, has been embraced by manufacturers recently (Noman et al., 2020a, 2020b).

Nanoparticles are a tool used in nanotechnology to increase crop productivity and food quality while reducing abiotic stressors.

2. Microorganisms for Nanoparticle Synthesis

Prolonged studies have been conducted to use microorganisms as bio-reactors to manufacture a variety of NPs. Enzymes generated during cellular activity are used by microorganisms to transform the target ions into metal nanoparticles. While marine yeast (*R. diobovatum*) biosynthesized PbS NPs were reported by Sintubin et al., 2009; Khosravi-Darani et al., 2019 reported employing Lactobacillus used for the production of silver nanoparticles. Similar findings were made by other researchers who used other fungi to produce Fe_3O_4, SiO_2, and TiO_2 (Bharde et al., 2006, Bansal et al., 2005). Various NPs were produced using biomolecules such as viruses, membranes, diatoms, and nucleic acids as templates. Using microwave heating and DNA as templates, Kundu et al. revealed synthesizing zinc oxide nanoparticles with a variety of morphologies (Kundu and Nithiyanantham, 2014).

Microbe-based NP synthesis is a cost-effective and environmentally benign process (Kim et al., 2018; Karunakaran et al., 2024). "*Pseudomonas deceptionensis, Bacillus licheniformis, Pyrobaculum islandicum, Pseudomonas aeruginosa, Pyrococcus furiosus*", and other bacteria have been used to produce gold, silver, zinc, iron, nickel, and other important nanomaterials (Sanaeimehr et al., 2018). Fungal, bacterial, and algal organisms are examples of microorganisms that grow quickly, are simple to control, and are inherently suited to thrive in a variety of environments. Distinctive nanomaterials like exopolysaccharides, nanocellulose, and nanowires can be produced by bacteria and microalgae (Li et al., 2017).

Microorganisms and a variety of methods, including extracellular synthesis, bioreduction, and green synthesis, are used in the synthesis of microbe-mediated NPs to create NPs with desired characteristics (Reddy et al., 2022). These NPs have demonstrated encouraging outcomes in managing a variety of plant diseases brought on by different pathogens, such as "*Xanthomonas oryzae* pv. *Oryzae*", *Alternaria solani,* and *Fusarium oxysporum* (Abdel-Aziz et al., 2020; Joshi et al., 2019, Namburi et al., 2021).

Microbial enzymes catalyze the reduction of metal ions to nanoparticles after they are initially attached to the surface or become trapped within the microbial system.

2.1 Synthesis of Nanoparticles by Bacteria

The synthesis of NPs using platinum, silver, gold, palladium, magnetite, and titanium dioxide is thought to be possible in bacteria. Three distinct methods are frequently used for nanoparticle production extracellularly through the bacteria: (1) the use of bacterial biomass, (2) using the supernatant of bacterial cultures, and (3) "using the cell-free extract". Extracellular enzymes that rely on "nicotinamide adenine dinucleotide (NADH) and its reduced counterpart (NADPH)" operate as reducing agents, with NADH and NADPH serving as electron donors. In the instance of nanoparticle synthesis intracellularly, enzymes, and positively charged metal ions interact to assist bacteria absorb metallic ions from the surrounding environment and keep them in the cell. The reduction of metal ions to NPs is also significantly influenced by sugar molecules (Ovais et al., 2018).

Mycelial growth suppression of approximately 80% in PDA and PDB growth media, at 20 µg/ml NiONPs has been demonstrated by silver nanoparticles (AgNPs) produced by *Pseudomonas poae* strain CO an endophytic bacteria derived from *Allium sativum* plants (Ibrahim et al., 2020).

The seed coat is broken and penetrated by smaller nanoparticles, which then release ions to encourage development. According to Sable et al. (2018), bajra (*Pennisetum glaucum*) seeds treated with 0.2 ml of AgNPs made from *Bacillus subtilis* spizizenni bacterial supernatant germination was increased by 48% above the control.

According to Abdelkhalek et al. (2020), viral infections considerably decreased the amount and quality of agricultural yields and were a major contributor to the plant's growth restrictions. The most deadly plant virus is Bean Yellow Mosaic Virus (BYMV), which drastically reduces crop output and yield quality. AgNPs biosynthesized by *Rhizobium leguminosarum* filtrate can inhibit the bean yellow mosaic virus (BYMV), as shown by Abdelkhalek et al. in 2022. Table 1 summarizes the various bacterial nanoparticles for crop improvement.

2.2 Synthesis of Nanoparticles by Actinomycetes

A type of microorganism known as actinobacteria shares characteristics with both bacteria and fungi, and are recognized to have a variety of uses in agricultural production systems. Making a number of growth-promoting substances and metabolites, such as antibiotics, which enable the host plants to endure biotic and abiotic stressors, is one of the primary roles.

Table 1. Various bacterial nanoparticles for crop improvement.

Bacteria	Nanoparticle synthesized	Production	Application	Reference
Escherichia coli	AgNP	Extracellular	Improved plant growth, antimicrobial agent	Kalishwaralal et al., 2010
Lactobacillus casei ssp. CCM 7088	AgNP	Extracellular	Improved plant growth, antifungal activity	Singh et al., 2014; Khosravi-Darani et al., 2019
Serratia sp. BHU-S4	AgNP	Extracellular	fungicide	Mishra et al., 2014
Bacillus licheniformis Dahb1	AgNP	Extracellular	Antifungal activity	Shanthi et al., 2016
Pseudomonas stutzeri	AgNP	Intracellular	Improved plant growth, antimicrobial agent	Gudkov et al., 2020
Lactobacillus acidophilus DSMZ 20079T	CdSNP	Extracellular	Nano biosensors	Riaz et al., 2020
Pseudomonas poae strain CO	AgNPs	Intracellular	Antifungal activity	Ibrahim et al., 2020

Additionally, actinobacteria aid in the creation of nanoparticles. Actinomycetes are known to release much more enzymes, proteins, small molecules with reducing characteristics, and secondary metabolites than bacteria and fungi. This, in turn, greatly increases the production of metal nanoparticles (Subramanian et al., 2016).

Silver nanoparticles made with terrestrial soil Streptomyces sp.-SBU3 have been shown by Krishnakumar and Bai (2015) to be effective in plant protection against bacterial and fungal pathogens.

Streptomyces diastaticus actinobacterial metabolites were utilized in creating biogenic silver nanoparticles (Ag-NPs), which were found to be an effective pesticide by Kalpana et al., 2024. With its low toxicity to non-target organisms, this biogenic insecticide provides an economical and sustainable solution for future pest management plans.

2.3 Synthesis of Nanoparticles by Fungi and Yeast

"Mycosynthesis" is the term used to describe the fungus that produces NPs. Because of their high metal tolerance, high biomass production, easy handling, and secretion of huge amounts of nanoparticle stabilizing agents, fungi are recommended for the biological synthesis of nanoparticles (Chauhan et al., 2023). NPs can also be produced by fungal species either extracellularly or intracellularly. Bio reduction is the mechanism by which fungal cells react to metal or metal oxide stress by modifying the ions into not many harmful NPs. Mucilaginous substance, which can bind dangerous metals, helps the metal ions bind to the cell surface during intracellular production. They are then converted to NPs after being transported inside the cell by certain transporters. The concentration changes where the nanoparticle created is well-adjusted by substrate entrance causing the NPs to disperse outside the cell after synthesis. NPs aggregate Sand cause damage if they stay inside the cell (Anjum et al., 2023). Metal ions are detoxified into NPs by secretory metabolites, which are membrane-associated tiny peptides and proteins produced during extracellular synthesis. In reaction to the metal stressors, the fungus *Aspergillus* and *Penicillium* create the benzoquinones and toluquinones derivatives. According to Jha and Prasad (2016), these metabolites set off the redox reactions that result in the creation of NPs.

Spagnoletti et al., (2019) used a low-cost, environmentally friendly synthesis method to produce stable silver nanoparticles from the exudation of "the soil fungus *Macrophomina phaseolina*". To determine if silver nanoparticles could be used for seed priming, the impact of different concentrations on the germination of soybean was investigated. Promising results were obtained

Table 2. Different Fungal nanoparticles in agriculture.

Fungi and Yeast	Nanoparticle synthesized	Application	References
Penicillium chrysogenum	Se NP	Growth stimulant and bactericidal	Amin et al., 2021; Abdelsalam et al., 2023
Aspergillus flavus and Candida albicans	Se NP	fungicide	Bafghi et al., 2021
Fusarium oxysporum	Se NP	antifungal	Islam et al., 2022
Aspergillus terreus	Se NP	antimicrobial	Saied et al., 2023
Daedalea Mushroom	ZnONP	antifungal agent	Kamal et al., 2023

when silver nanoparticles made from the marine endophytic fungus *Fusarium equiseti* were used as a plant growth stimulator in Vigna radiata (Haji Basheerudeen et al., 2021).

Sorghum and Zea mays germination rate and seedling establishment was enhanced by seed priming with silver nanoparticles synthesised using yeast extract (Kim et al., 2024). A summary of the several fungal-derived nanoparticles for agricultural improvement is provided in Table 2.

2.4 Synthesis of Nanoparticles by Algae

Both macro and microalgae are at the forefront of creating nanoparticles that can effectively serve a variety of multipurpose purposes. Because they are autotrophic, algae use sunlight to turn carbon dioxide into biomass. They can be used as nano bio factories producing metallic nanoparticles and are found in a variety of habitats. They also create a wide range of metabolites, such as lipids, proteins, carbohydrates, vitamins, and various pigments (Kokabi and Yousefzadi, 2022). Because of their unique physiological and biochemical characteristics, algae are considered effective biotic systems for producing nanoparticles. They include important active constituents such as polysaccharides, proteins, pigments, phenols, and antioxidants, which make them the primary source of useful natural substances (El-Naggar et al., 2020). Using algae species to biologically synthesize NPs has become popular recently. The main reasons for this enormous and growing interest are (i) its ease of handling, (ii) its ability to uptake and accumulate metal ions, (iii) low cost synthesis; and (iv) its eco-safe, quick, and healthy synthesizing approaches (Khan et al., 2022).

Through extraction, mixing with a metal solution, and constant stirring in regulated settings, nanoparticles could be biologically produced from macro and microalgae. In terms of agricultural sustainability, NPs have been shown to boost output while lowering the quantity of chemicals sprayed for the protection of crops against diseases, pests, and managing environmental constraints (Singh et al., 2021; Hayat et al., 2023; Thabet and Alqudah, 2024). Moreover, metal reduction occurs during synthesis when metal ions interact with cellular constituents such as carbohydrates, minerals, lipids, oils, polyunsaturated fatty acids, antioxidants, and pigments (phycobilins or chlorophylls) to change them into non-valent states.

The use of algae in agriculture is significant. By using seaweed (macroalgae) as fertilizers, less phosphorus and nitrogen runoff is produced than when cattle dung is used. The thalloid red alga *Galaxaura elongata* was used to create very stable gold nanoparticles that demonstrated antibacterial action (Abdel-Raouf et al., 2017). According to Vijayakumar et al. (2022), green-produced ZnO nanoparticles from *Laminaria* (brown macroalgae) also had a substantial impact on mung bean (*Vigna radiata*) fruit germination, root lengths, and shoot lengths. Mustafa et al. (2024) also showed a noteworthy effect on plant growth and development while administering copper nanoparticles (NPs) as nanofertilizers to cowpeas using foliar and soil treatment methods.

Abdel Latef et al. (2018) found that broad bean plants under salinity stress treated with lower concentrations (0.01%) of titanium dioxide nanoparticles were more effective in boosting several metabolites, including soluble sugars, antioxidative enzymes and amino acids.

Table 3. Different Algal nanoparticles in agriculture.

Algae	Type of Nanoparticle	Activity	References
Chlorella vulgaris	Ag NPs	Acts as an antifungal agent in plants	Thakkar et al., 2010
Haematococcus candid and *Chlorella* sp.	Ag NPs	Repressed fungal growth	El-Moslamy et al., 2016
Sargassum ilicifolium	Ag NPs	Improved seed germination	Roy and Anantharaman, 2018
Turbinaria conoides and *Sargassum tenerrimum*	Au NPs	Helps in remediation	Terra et al., 2019
Cyanobacteria	ZnO NPs	Nano biofertilizer with antioxidative properties	Terra et al., 2019
Cyanobacteria	Ag NPs	Antibacterial activity	Annamalai et al., 2016
Ulva intestinalis	INPs	Nano biofertilizer	Mondal et al., 2024

Rice disease-causing bacteria have been suppressed and plant growth has been enhanced by AgNPs made from red seaweed extracts from *Gracilaria crassa* and *Grateloupia lithophila* (Roseline et al., 2023).

According to Roselin et al. (2019), green synthesized AgNPs with extracts of "*Gracilaria corticata, Gracillaria edulis, Hypnea musciformis*, and *Spyridia hypnoides*" were found to have antifungal action against *Ustilaginoidea virens* and to be effective against "*Xanthomonas axonopodis* pv. *citri* and *X. oryzae* pv. *Oryzae*". Iron nanoparticles (INPs) containing *Ulva intestinalis* biomass promoted plant growth, which enhanced rice crop yields and quality (Mondal et al., 2024). A summary of the several algal nanoparticles for agricultural improvement is provided in Table 3.

To comprehend the effects of algal nanoparticles at various concentrations on different plant growth and metabolic processes, physiological and molecular research must be conducted, as the aforementioned contradictions and gaps are abundantly evident.

2.5 Synthesis of Nanoparticles by Virus

Viral coat proteins are used to create Virus-Like Particles (VLPs), a biodegradable and biocompatible nanomaterial. It has been demonstrated that numerous insect and plant virus coat proteins can come together to form VLPs. VLPs are viral protein complexes that self-assemble. Unlike entomopathogen viruses, they do not contain genetic information, making them safer alternatives to viruses. Targeting the release of pesticides and other beneficial compounds into the soil can significantly increase their effectiveness and efficiency, reducing crop losses brought on by pests and parasitic organisms. For treating parasitic nematodes in plants, Tobacco Mild Green Mosaic Virus (TMGMV) and Red Clover Necrotic Mosaic Virus (RCNMV) have demonstrated good delivery of nematicide and movement in soil (Caparco et al., 2023). Knowledge of the characteristics of both soil and the nanoparticles is necessary for utilizing viral nanoparticles in their effective soil delivery". The mobility of tobacco mosaic virus with a genetically encoded lysine on its surface (TMV-Lys), potato virus X, TMGMV, and virus-like particles of physalis mottle virus (PhMV)" was evaluated by Venkateswaran et al. in 2023. TMGMV can be utilized as a nanocarrier for nematicides and pesticides due to its increased soil mobility.

3. Applications of Nanoparticles in Agriculture

Nanoparticles generated from microbes are frequently utilized to create goods like nano pesticides, nano fungicides, nano biosensors, and nano fertilizers. These nano-based products can shield crops from environmental stresses, lower chemical contamination, and improve the quality and output of agricultural goods.

In agriculture, nanoparticles (NPs) are being used more and more to enhance plant development and growth. One such method that makes use of NPs to boost crop resistance, seed germination, and seedling vigor is nano-priming.

Zinc oxide nanoparticles (ZnO NPs) were synthesized using a straightforward process using a bacterial biosurfactant that was obtained from a recently identified strain of *Klebsiella* sp. RGUDBI03. These ZnO NPs improved the germination of *Oryza sativa* and *Cicer arietinum* seeds when used for seed priming. Additionally, in earthworms (*Eudrilus eugeniae*) and red blood cells, the ZnO NP priming solution showed no cytotoxicity, suggesting that it is non-cytotoxic and environmentally benign for use in the next field tests (Das et al., 2024).

The use of magnesium oxide nanoparticles made from *Turbinaria ornata* as a nano priming agent in green gram (*V. radiata*) increased the proportion of seeds that germinated and the vigor of the seedlings in comparison to hydropriming (Anand et al., 2020). Utilizing *Trichoderma citrinoviride* extract, myco-synthesis of silver and titanium oxide nanoparticles was employed as a seed priming agent in *Solanum lycopersicum*. Tomato seed germination and development were improved by this study (Sonawane et al., 2021).

Tomato plants' antioxidant enzyme activity was elevated by zinc oxide nanoparticles application (Wang et al., 2018). The creation of ZnO nanoparticles mediated by the seaweed *Turbinaria ornata* showed enhanced effects on germination and growth metrics according to Itroutwar et al., 2020. At low concentrations of ZnO nanoparticles, the study also helped to improve the yield in rice crops and characteristics, including the weight of grains, the thickness of seed, seed length, and breadth. It suggests that a promising strategy for advancing agriculture with necessary micronutrients in the future will be the use of ZnO nanoparticles as nutrients in crops.

To boost crop productivity even in the face of environmental stress and to lessen nutrient stress, Nano Fertilization has developed as an innovative and hopeful substitute for traditional fertilizers (Saraiva et al., 2023; Azam et al., 2022; Salama et al., 2022; El-Ramady et al., 2022; El-Bialy et al., 2023). Plants can also receive nutrients from Nanofertilizers (NFs), which are chemicals with a considerably smaller particle size (nanoscale < 100 nm). The tiny size of nano fertilizers makes it easier for them to get through biological barriers and go through the vascular system of plants. Higher absorption rates as well as enhanced nutrient availability and utilization efficiency are characteristics of nutrients in the form of nanoparticles (NPs). According to Shalaby et al. (2022), nano fertilizers can efficiently supply the nutritional needs of plants to support crop development and yield at doses up to 100 times lower than those of conventional mineral fertilizers.

Although NFs are a promising approach for their beneficial effects on agriculture, they need to be cautiously made and assessed for sustainability, environmental, and safety standards because certain nanoparticles have shown toxic impacts on microbes in soil and plants (El-Ramady et al., 2022). A solution to the problems with nano fertilizers is the development of Bio-Nano Fertilizers (B-NFs), which are made up of nanoparticles applied in combination with Plant Growth-Promoting Rhizobacteria (PGPR) or organic compounds and bio stimulants derived from microbes, plants. These biostimulants serve as capping agents, improving the "biocompatibility, stability, efficacy, and safety" of NFs (Manivannan et al., 2021).

3.1 Role of Microbial-derived NPs in Stress Resilience

Abiotic stressors, such as drought, salt, and extremes in temperature, have a growing impact on the global ecological system. Over the past few years, experts from a wide range of study fields have become more interested in nanotechnology. These nanomaterials have produced several unique characteristics that set them apart from their macro-sized counterparts because of their nanoscale size. Nanomaterials were more reactive, and soluble, and had a higher surface area than bigger compounds. Therefore, these materials can play a positive role in recovering from the negative impacts of these pressures and supporting sustainable agriculture globally. The ability of microbe-derived nanoparticles to withstand stress is drawing attention to their usage in agriculture.

The potential role of several microbe-derived nanoparticles in the defense system against abiotic stress has been investigated. These nanomaterials improve nutrient absorption, enzyme reactions, and the relationship between plant roots and microorganisms that promote growth during abiotic stress, all of which help crops become more resilient to stress. The results of these preliminary studies are encouraging and herald a new era of applying nanomaterials to increase agricultural yields in hazardous settings.

Using nanoparticles to increase crop growth and yields is a significant step in mitigating the effects of abiotic stressors.

3.1.1 Delivery of Microbe-derived Nanoparticles to Crop Plants

There are several methods for applying nanoparticles to improve crop growth, such as seed priming, foliar spray, and root application. Whether nanoparticles translocate to various parts of the plant and how best to apply them are both very intriguing concerns. The impacts of nanoparticles depend on several variables, including their size, composition, application technique, environmental parameters, rhizosphere and phyllosphere surroundings, and certain crop species. Among other crucial qualities, nanoparticles must have specific chemical, physical, and dose properties to pass through cell membranes and barriers.

Exchange of ions, surface precipitation, and adsorption are a few of the often-seen surface chemical reactions for nanoparticle absorption (Singh et al., 2021). The first interaction between nanoparticles and plants, for example, occurs at the plant cell wall. Particle mobility is controlled and the plant's filtration capabilities are enhanced by the many pores in its cell wall, which range in diameter from 5 to 20 nm. Nanoparticles with a diameter in this range can easily enter plant cells. Larger nanoparticles, on the other hand, interact with cell wall constituents to create new holes and entrance points (Alnaddaf et al., 2023; Farooq et al., 2024). It has been observed that tiny nanoparticles reach roots due to a variety of factors, including capillary pressures, osmotic pressure, and transit by the epidermis.

The walls of root cells contain semipermeable epidermal cells with tiny pores that are inaccessible to large nanoparticles. To enter epidermal cells, certain nanoparticles have been found to cause the cells to develop new holes. According to Arumugam et al. (2023), and Rehman et al. (2024) chemical precipitation, ion exchange, chelation, and endocytosis are the processes by which plants absorb metal nanoparticles (Hong et al., 2021; Zhang et al., 2024). Ion transporters, including ion channels, carrier proteins, and aquaporins, have also been noted as substitutes for plants' absorption of metal nanoparticles (Yadav et al., 2014; Ahmad et al., 2022; Omar et al., 2023). Application of a mathematical framework called "Lipid Exchange Envelope Penetration", helps in better understanding of how nanoparticles penetrate cells by learning more about the exchange of lipids that takes place between the cell envelope and the nanoparticles (Wong et al., 2016; Lew et al., 2018; Arya et al., 2021).

Nanoparticles are absorbed by plants through their cellular structure's nucleus, cytoplasm, and epidermis. If roots do not participate in the nanoparticle absorption, the absorption of nutrients may be altered. Nanoparticles can enter the coat through intercellular gaps between parenchymatous cells, and later diffuse within the cotyledon before directly entering the seed coat (Tondey et al., 2021; Savassa et al., 2021). The nanoparticles use both apoplast and symplast paths to reach the region of interest.

Research on the transport of nanoparticles (Rani et al., 2023; Sembada et al., 2024; Talebi et al., 2024) has found that the appropriate membrane carrier proteins usually facilitate the nanoparticles' passage through xylem channels. Furthermore, for nanoparticles to reach the interior system of the leaf, they must surpass the stomata and cuticle. Nanoparticles less than 5 nm in size travel through cuticles and those beyond 5 nm in size use the other pathway. Like roots, leaves have an internal transportation system. Ruttkay-Nedecky et al. (2017) state that nanoparticles penetrate leaves through stomata, hydathodes, cuticles, and injured tissues. Phloem tubes then carry the particles to the intended organ or locations by using apoplastic and symplastic pathways, which include fruit, roots, and shoots. Endocytosis, the development of pores, and the involvement of certain carriers accompany the penetration of nanoparticles (Zhang et al., 2024). According to Sidhu et al. (2024), nanoparticles interact with soil and soil microbes before being absorbed by plants.

3.1.2 Microbe-derived NPs in Mitigation of Salinity Stress

One of the abiotic factors that might restrict crop yield, particularly in dry or semi-arid settings, is salinity in the soil or water. An estimate from 2011 states that high salinity affects more than 800 million hectares of land worldwide. According to the same study, by 2050, salinity will have an impact on almost 17 million hectares of agricultural land. Low soil osmotic potential, nutritional imbalance, the specific ion effect, or combinations of these variables are linked to the detrimental impact of salinity on crops.

According to Mohamed et al. (2017), seed priming with silver nanoparticles made from "*Aspergillus terreus* MALEX fungi" showed increased salt stress tolerance in wheat. Adding 2–5 mM Ag NPs to plants under salt stress enhanced their biomass, boosted the antioxidative enzyme activities, and reduced the reactive oxygen species turn out.

AgNP seed priming has been shown by Khan et al. (2020) to improve crop plants' physiological and biochemical responses to salinity. This could increase the output of crops in salt-degraded areas worldwide. In maize, nano-priming with TiO_2 reduces salinity damage and may be a useful alternative method to lessen the negative effects of salinity stress (Shah et al., 2021). Silver nanoparticles (AgNPs) made from the aqueous extract of the marine algae *Iyengaria stellata* have been shown by Patel et al. (2024) to minimize the salt stress effects on tomato and enhance their general growth.

3.1.3 Microbe-derived NPs for Drought Stress Tolerance

One of the main environmental stressors that significantly lowers crop productivity is drought. Water deficit circumstances may worsen due to climate change's influence on rainfall variability, rising water demand, and declining water availability. Since drought stress is influenced by several variables, including annual precipitation and the soil's water-withholding ability, determining the extent of the damage is difficult (Fahad et al., 2017; Lobell et al., 2011). According to a study citing global data analysis published between 1980 and 2015, drought may have decreased yield in wheat in addition to maize by 21 and 40%, respectively (Daryanto et al., 2016). Among the main consequences of drought are low germination (Queiroz et al., 2019), decreased growth of seedlings (Sarshad et al., 2021), decreased availability of nutrients, decreased photosynthesis (Sanjari et al., 2021), the quantity and size of individual leaves (Hammad et al., 2017), the fresh and dry weight of plants (Goodarzian Ghahfarokhi et al., 2015), etc. Likewise, dehydration can cut coffee yield by up to 80% (Jin et al., 2018). Several physiological changes, such as stomatal closure and decreased transpiration, are among the ways that plants react to drought (Prakash et al., 2017). Furthermore, plants alter the expression of certain genes to increase their resistance to drought stress.

Using an extract from the algae *Arthrospira platensis*, zinc oxide nanoparticles (ZnO-NPs) were created. The underlying mechanisms by which ZnO-NPs modulated oxidative stress during drought in rice cultivars (cv. Swarna Sub1) were investigated. The consequences of the drought were lessened due to a decrease in lipid peroxidation and protein carbonylation (Das et al., 2024).

3.1.4 Microbe-derived NPs for Flood Resistance

Both quantity and quality are lost when rice is submerged, especially when it is mature and seeding. Shoot weight, yield of grains, harvest index and other components of yield are all negatively impacted by submersion, which also considerably postpones flowering and maturity (Ferrero et al., 2020). Grain weight, fullness, and number per panicle decreases are the primary reasons for lower grain yields due to submergence (Rascio and Navari-Izzo, 2011). Immersion in the vegetative phase, critically reduced yield quality metrics.

3.1.5 Microbe-derived NPs Intervention for Heavy Metal Tolerance

Abiotic stress brought on by soil pollution with heavy metals can change plant metabolism and degrade agricultural productivity both qualitatively and quantitatively, raising questions about food security (Jamla et al., 2021). Among these heavy metals, hexavalent chromium (Cr (VI)) has harmful effects on human health. CuNPs produced by the *Klebsiella pneumoniae* strain were found to have a positive effect on soil contaminated with Cr by reducing chromium accumulation (Noman et al., 2020a). Chatterjee et al. (2020) investigated the sequestration of Cr (VI) in water using "superparamagnetic iron oxide" nanoparticles (Fe_3O_4) that were produced utilizing the fungus *Aspergillus niger* BSC-1.

There have been attempts to sequester and limit the accumulation of cadmium (Cd) in plants because it is a carcinogenic heavy metal with a long half-life. In their investigation of biogenic CuNPs on wheat plants, Noman et al. (2020b) found that copper nanoparticles made from the "copper-resistant *Shigella flexneri* SNT22" bacteria had a notable impact on cadmium absorption and transport.

Iron oxide nanoparticles (FeO-NPs) were demonstrated in another study to mitigate the effects of salt and cadmium stress on wheat (Manzoor et al., 2021). The bacterial strain *Pantoea ananatis* RNT4 was used to create FeO-NPs. Furthermore, FeO-NPs have been demonstrated to lower the salinity levels of wheat grains along with promoting plant growth. When compared to CuNPs, FeO-NP application dramatically decreased the amount of Cd that plants absorbed in wheat.

The effects of Cd and drought stress on rice were also mitigated by FeO-NPs made with Bacillus strain RNT1 and ranging in size from 18 to 94 nm (Ahmed et al., 2021). The presence of NPs was found to down-regulate the "Cd transporter genes OsHMA2, OsHMA3, and OsLCT1" expression, according to gene expression experiments they conducted. The production of SiNPs by *Trichoderma pseudoharzianum* was utilized to combat the simultaneous stresses including cadmium toxicity in addition to *Ralstonia solanacearum*-induced biotic stress, which demonstrated notable stress reduction following soil application in tomato crops. With a reduction in Cd bioavailability, which lowers absorption and translocation into plants, plants displayed an increased stress tolerance index. Similarly, pathogen cell disruption brought about by therapy resulted in a better disease tolerance index (Khan et al., 2023).

3.1.6 Microbe-derived NPs for Temperature Stress Tolerance

One of the main abiotic stressors that restrict plant development and yield is temperature. Plants that are exposed to extremely high or low temperatures experience serious negative consequences and occasionally even perish. Because of this, plants have developed complex defenses against temperature fluctuations. For example, respiration, transpiration, membrane thermostability, and osmotic control are among the physiological systems that are negatively impacted by heat stress (Sharma et al., 2022; Posch et al., 2019). Plants under high-temperature stress typically have shorter life cycles and lower productivity due to a reduction in photosynthetic efficiency (Lal et al., 2021; Zahra et al., 2023). The impact of low-temperature occurrences, which are made worse by climate change, cannot be disregarded, even though heat stress has received a lot of attention. By altering water, nutrient intake and translocation, the structure of root, photosynthetic activity, and hormonal signaling, cold stress also has a major impact on plant development (Ahmad et al., 2024).

Biosynthesized selenium nanoparticles, produced by "Lactobacillus acidophilus ML14", improved the resistance of "*Fusarium culmorum* and *Fusarium graminearum*", to heat stress as well as drought, thereby promoting wheat growth and productivity (El-Saadony et al., 2021).

The application of zinc oxide nanoparticles on leaves externally upstreamed the release of antioxidative enzymes and osmolytic substances to reduce the heat stress effects in mung beans, as demonstrated by Kareem et al. 2022a, b. In addition to increasing the gas exchange characteristics and chlorophyll levels, foliar-applied nanoZnO during high-temperature stress also increased the number of fruits per individual plant and seeds per fruit leading to higher productivity.

The usage of NPs also keeps the levels of photosynthetic pigments including carotenoid, chlorophyll a, and chlorophyll b stable during cold stress. They help plants that have been subjected to cold stress recover quickly and fortify their defense mechanisms. By lowering the number of free radicals produced by plastids, accelerating light absorption and electron transport in the chloroplast, and enhancing rubisco carboxylation, NPs lessen the depressing effects of cold stress on crops. In photosynthetic processes, nanoparticles boost the activity of enzymes. In cold-stressed plants, carbon nanotubes enhance photosynthetic pigments. They decrease the generation of free radicals and increase the pace of electron transport in the chloroplast. Even when the water supply is restricted, they increase the expression of the genes for water channels, which make the water available for photosynthesis.

3.1.7 Microbial Nanoparticles For Biotic Stress Tolerance

Global food security and sustainable agricultural productivity are seriously threatened by crop biotic stresses. One possible strategy for sustainable crop production is nanotechnology. Studies on the resistance of crop plants to diseases mediated by nanoparticles was initiated recently. Nanotechnology can transfer hormones and biomolecules, induce disease-resistance genes, control the amount of secondary metabolites, and produce transgenics that have increased crop protection (Dong et al., 2023).

Relative to commercial pesticides, ions and bulk particles, NPs are nearly 1.6 times more effective at controlling crop diseases, according to studies (Cao and Wang, 2022).

Microbially synthesized NPs have a strong potential for reducing the effects of various environmental stresses on plants (Fig. 1), according to these investigations. Nevertheless, more investigation may show that microbially produced NP is preferable to chemical synthesis and other environmentally friendly methods, such as using plants.

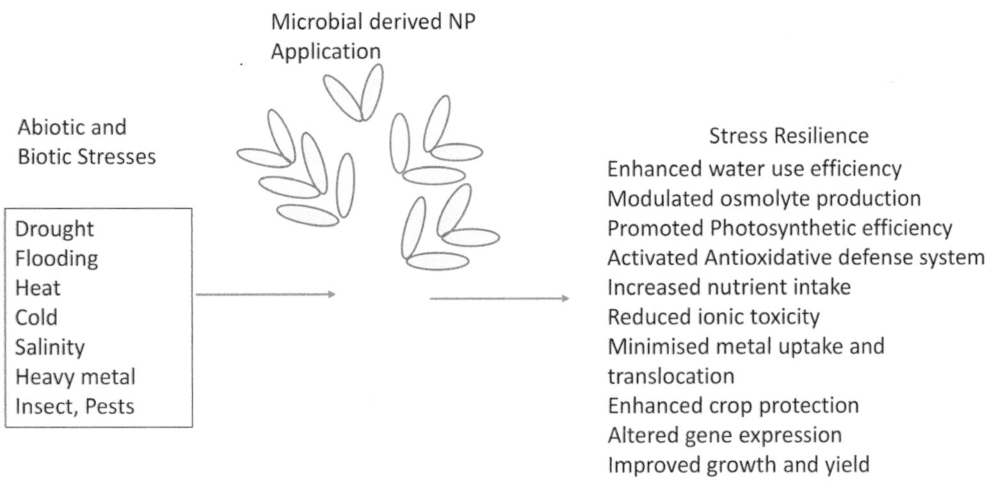

Figure 1. Microbial-derived nanoparticles in plant stress resilience.

3.2 Role of Microbial-derived NPs in Sustainable Agriculture

"Sustainable agriculture" is the primary priority for all researchers to boost agricultural output and feed all living things on the globe. To improve soil fertility and agricultural productivity, different types of microbes are currently being used to create nanoparticles with a range of sizes, shapes, components, and other characteristics. In contemporary and sustainable agriculture, the combination of microbial nanotechnology with farming methods opens new avenues for development and output.

Under the green method, the microbial population can function as a crop system's nano fertilizer in addition to increasing production (Bhardwaj et al., 2023). By using nanocomposites to create

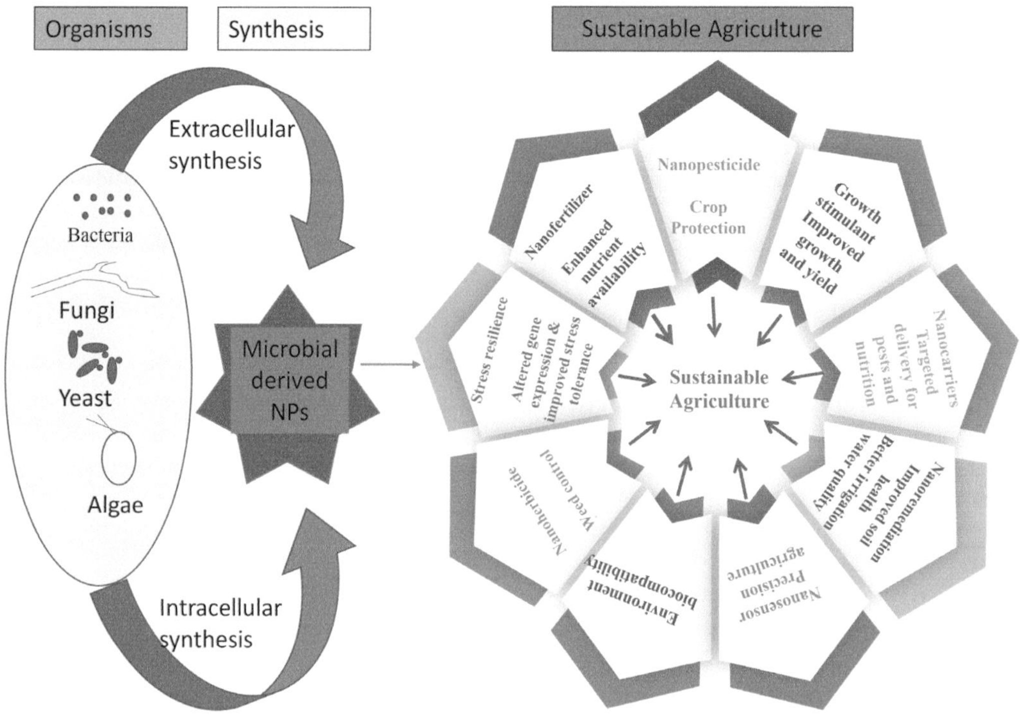

Figure 2. Role of microbe-derived nanoparticles in sustainable agriculture.

smart fertilizers, it may be possible to decrease the likelihood of pollution in soil, water, and air while simultaneously providing a targeted, consistent, and coordinated delivery of plant nutrients.

Sustainable patterns of "production and consumption (SDG 12)" as well as the "sustainable use of water resources (SDG 14)" are two of the Sustainable Development Goals (SDGs) of the United Nations. Accordingly, nanocomposite-based fertilizers (NCPFs) could be a means of achieving both agricultural and environmental sustainability as well as SDGs 12 and 14 (Chakraborty et al., 2023).

Microbial-derived NPs have been shown in numerous studies to be less persistent in the environment and less harmful to non-target creatures than chemical pesticides (Koul et al., 2021; Raj et al., 2021). These advantages have led scientists to look into the possibility of using microbe-mediated NPs as an environmentally responsible and sustainable substitute for treating plant diseases (Fig. 2).

4. Future Perspectives and Conclusions

Microbial synthesis approaches offer a non-toxic, economical, and environmentally sustainable way to produce metal nanoparticles than other conventional methods including physical and chemical ones. One of the main obstacles to the manufacture of nanoparticles using microorganisms is the need for laborious purification procedures to recover internal nanoparticles. In the agricultural industry, nanoparticles are used extensively for crop protection, crop improvement, and other purposes. However, the toxicity of nanoparticles might be affected by their interaction with other compounds and nutrients that are essential for plants. Therefore, before nanoparticles are commercialized, thorough research on several topics is required, including nanoparticle shape, concentration, exposure duration, environmental conditions, etc., to ensure the safety utilization of microbe-derived nanoparticles in farming.

References

Abdel Latef, A. A. H., Srivastava, A. K., El-sadek, M.S.A., Kordrostami, M., and Tran, L. S. P. (2018). Titanium dioxide nanoparticles improve growth and enhance tolerance of broad bean plants under saline soil conditions. *Land Degradation & Development*, 29(4): 1065–1073.

Abdel-Aziz, M. M., Emam, T. M., and Elsherbiny, E. A. (2020). Bioactivity of magnesium oxide nanoparticles synthesized from cell filtrate of endobacterium *Burkholderia rinojensis* against Fusarium oxysporum. *Materials Science and Engineering: C*, 109: 110617.

Abdelkhalek, A., Yassin, Y., Abdel-Megeed, A., Abd-Elsalam, K. A., Moawad, H., and Behiry, S. I. (2022). Rhizobium leguminosarum bv. viciae-mediated silver nanoparticles for controlling Bean Yellow Mosaic Virus (BYMV) infection in faba bean plants. *Plants*, 12(1): 45.

Abdelkhalek, A., and Hafez, E. (2020). Plant viral diseases in Egypt and their control. *Cottage industry of biocontrol agents and their applications: practical aspects to deal biologically with pests and stresses facing strategic crops*, pp. 403–421.

Abdel-Raouf, N., Al-Enazi, N. M., and Ibraheem, I. B. (2017). Green biosynthesis of gold nanoparticles using *Galaxaura elongata* and characterization of their antibacterial activity. *Arabian Journal of Chemistry*, 10: S3029–S3039.

Abdelsalam, A., El-Sayed, H., Hamama, H. M., Morad, M. Y., Aloufi, A. S., and Abd El-Hameed, R. M. (2023). Biogenic selenium nanoparticles: anticancer, antimicrobial, insecticidal properties and their impact on soybean (*Glycine max* L.) Seed Germination and Seedling Growth. *Biology*, 12(11): 1361.

Ahmad, W., Coffman, L., Weerasooriya, A. D., Crawford, K., and Khan, A. L. (2024). The silicon regulates microbiome diversity and plant defenses during cold stress in *Glycine max* L. *Frontiers in Plant Science*, 14: 1280251.

Ahmed, T., Noman, M., Rizwan, M., Ali, S., Ijaz, U., Nazir, M. M., ALHaithloul, H. A. S. et al. (2022). Green molybdenum nanoparticles-mediated bio-stimulation of *Bacillus* sp. strain ZH16 improved the wheat growth by managing in planta nutrients supply, ionic homeostasis and arsenic accumulation. *Journal of Hazardous Materials*, 423: 127024.

Ahmed, T., Masood, H. A., Noman, M., Al-Huqail, A. A., Alghanem, S. M., Khan, M. M. et al. (2023). Biogenic silicon nanoparticles mitigate cadmium (Cd) toxicity in rapeseed (*Brassica napus* L.) by modulating the cellular oxidative stress metabolism and reducing Cd translocation. *Journal of Hazardous Materials*, 459: p.132070.

Ahmed, T., Noman, M., Manzoor, N., Shahid, M., Abdullah, M., Ali, L. et al. (2021). Nanoparticle-based amelioration of drought stress and cadmium toxicity in rice via triggering the stress responsive genetic mechanisms and nutrient acquisition. *Ecotoxicol. Environ. Saf.*, 209: 111829.

Al-Khayri, J. M., Rashmi, R., Surya Ulhas, R., Sudheer, W. N., Banadka, A., Nagella, P. et al. (2023). The role of nanoparticles in response of plants to abiotic stress at physiological, biochemical, and molecular levels. *Plants*, 12(2): 292.

Alnaddaf, L. M., Al-Khayri, J. M., and Jain, S. M. (2023). Introduction: impact of nanotechnology on plant cell biology. pp. 1–15. *In: Nanomaterial Interactions with Plant Cellular Mechanisms and Macromolecules and Agricultural Implications*. Cham: Springer International Publishing.

Amin, M. A., Ismail, M. A., Badawy, A. A., Awad, M. A., Hamza, M. F., Awad, M.F. et al. (2021). The Potency of fungal-fabricated selenium nanoparticles to improve the growth performance of *Helianthus annuus* L. and control of cutworm Agrotis ipsilon. *Catalysts*, 11(12): 1551.

Anand, K. V., Anugraga, A. R., Kannan, M., Singaravelu, G., and Govindaraju, K. (2020). Bio-engineered magnesium oxide nanoparticles as nano-priming agent for enhancing seed germination and seedling vigour of green gram (*Vigna radiata* L.). *Materials Letters*, 271: 127792.

Anjum, S., Vyas, A., Sofi, T.A., Mirza, U., Bera, S., Chakraborty, S. et al. (2023). Fungal-Based Nanoparticles. Microbial Processes for Synthesizing Nanomaterials. *Springer Nature Singapore*. pp. 81–111.

Annamalai, J., Shanmugam, J., and Nallamuthu, T. (2016). Salt stress enhancing the production of Phytochemicals in Chlorella vulgaris and Chlamydomonas reinhardtii. *J. Algal Biomass Util*, 7(1): 37–44.

Ansari, M.A. 2023. Nanotechnology in food and plant science: challenges and future prospects. *Plants*, 12(13): 2565.

Arumugam, V., and Vasudevan, V. (2023). Impact of metal oxide nanoparticles against heavy metal stress in plants. pp. 177–202. *In: Nanometal Oxides in Horticulture and Agronomy*. Academic Press.

Arya, S.S., Tanwar, N., and Lenka, S. K. (2021). Prospects of nano-and peptide-carriers to deliver CRISPR cargos in plants to edit across and beyond central dogma. *Nanotechnology for Environmental Engineering*, 6(2): 22.

Azam, M., Bhatti, H.N., Khan, A., Zafar, L., and Iqbal, M. (2022). Zinc oxide nano-fertilizer application (foliar and soil) effect on the growth, photosynthetic pigments and antioxidant system of maize cultivar. *Biocatalysis and Agricultural Biotechnology*, 42: 102343.

Bafghi, M. H., Darroudi, M., Zargar, M., Zarrinfar, H., and Nazari, R. (2021). Biosynthesis of selenium nanoparticles by Aspergillus flavus and Candida albicans for antifungal applications. *Micro & Nano Letters*, 16(14): 656–669.

Banerjee, K., Pramanik, P., Maity, A., Joshi, D. C., Wani, S.H., and Krishnan, P. (2019). Methods of using nanomaterials to plant systems and their delivery to plants (mode of entry, uptake, translocation, accumulation, biotransformation and barriers). pp. 123–152. *In: Advances in phytonanotechnology*. Academic Press.

Bansal, V., Rautaray, D., Bharde, A., Ahire, K., Sanyal, A., Ahmad, A. et al. (2005). Fungus-mediated biosynthesis of silica and titania particles. *Journal of Materials Chemistry*, 15(26): 2583–2589.

Begum, S. R., and Jayawardana, N. U. (2023). Green synthesized metal nanoparticles as an ecofriendly measure for plant growth stimulation and disease resistance. *Plant Nano Biology*, 3: 100028.

Bharde, A., Rautaray, D., Bansal, V., Ahmad, A., Sarkar, I., Yusuf, S. M. et al. (2006). Extracellular biosynthesis of magnetite using fungi. *Small*, *2*(1): 135–141.

Bhardwaj, I., Kumar, V., Bhardwaj, N., Verma, R., Bhardwaj, Y., and Kumari, T. (2023). A glimpse into the performance and synthesis of microbial nanoparticles and its new advances in soil enrichment and plant nutrition: a review. *Nanotechnology for Environmental Engineering*, 8(4): 943–964.

Cao, X., and Wang, Z. (2022). Application of nano-agricultural technology for biotic stress management: mechanisms, optimization, and future perspectives. *Environmental Science: Nano*, 9(12): 4336–4353.

Caparco, A. A., González-Gamboa, I., Hays, S. S., Pokorski, J. K., and Steinmetz, N. F. (2023). Delivery of nematicides using TMGMV-derived spherical nanoparticles. *Nano Letters*, 23(12): 5785–5793.

Chakraborty, R., Mukhopadhyay, A., Paul, S., Sarkar, S., and Mukhopadhyay, R. (2023). Nanocomposite-based smart fertilizers: A boon to agricultural and environmental sustainability. *Science of the Total Environment*, 863: 160859.

Chatterjee, S., Mahanty, S., Das, P., Chaudhuri, P., and Das, S. (2020). Biofabrication of iron oxide nanoparticles using manglicolous fungus Aspergillus niger BSC-1 and removal of Cr (VI) from aqueous solution. *Chem. Eng. J.*, 385: 123790.

Chauhan, A., Anand, J., Parkash, V., and Rai, N., (2023). Biogenic synthesis: A sustainable approach for nanoparticle synthesis mediated by fungi. Inorg. *Nano-Met. Chem*., 53(5): 460–473.

Da Costa, M. V. J., Ramegowda, Y., Ramegowda, V., Karaba, N. N., Sreeman, S. M., and Udayakumar, M. (2021). Combined drought and heat stress in rice: responses, phenotyping and strategies to improve tolerance. *Rice Science*, 28(3): 233–242.

Daryanto, S., Wang, L., and Jacinthe, P. A. (2016). Global synthesis of drought effects on maize and wheat production. *PloS one*, 11(5): e0156362.

Das, A., Bagchi, S., Pal, S., Ganguly, A., Sil, S. K., and Adak, M. K. (2024). Utilizing Arthrospira platensis for the fabrication of zinc oxide nanoparticles: Analysis and assessment for enhancing drought tolerance in Sub1A QTL bearing rice seedlings. *Plant Nano Biology*, 10: 100101.

Das, I., and Borah, D. (2024). Microbial biosurfactant-mediated green synthesis of zinc oxide nanoparticles (ZnO NPs) and exploring their role in enhancing chickpea and rice seed germination. *Discover Nano*, 19(1): 1–21.

Dong, B. R., Jiang, R., Chen, J. F., Xiao, Y., Lv, Z. Y., and Chen, W. S. (2023). Strategic nanoparticle-mediated plant disease resistance. *Critical Reviews in Biotechnology*, 43(1): 22–37.

El-Bialy, S.M., El-Mahrouk, M.E., Elesawy, T., Omara, A.E.D., Elbehiry, F., El-Ramady, H. et al. (2023). Biological nanofertilizers to enhance growth potential of strawberry seedlings by boosting photosynthetic pigments, plant enzymatic antioxidants, and nutritional status. *Plants*, 12(2):302.

El-Moslamy, S., Kabeil, S., and Hafez, E. J. O. (2016). Bioprocess Development for Chlorella vulgaris cultivation and biosynthesis of anti-phytopathogens silver nanoparticles. *J. Nanomater. Mol. nanotechnol.*, 5: 1. *J. Nanomater. Mol. Nanotechnol*, *9*: 2.

El-Naggar, N. E. A., Hussein, M. H., Shaaban-Dessuuki, S. A., and Dalal, S. R. (2020). Production, extraction and characterization of Chlorella vulgaris soluble polysaccharides and their applications in AgNPs biosynthesis and biostimulation of plant growth. *Scientific Reports*, 10(1): 3011.

El-Ramady, H., Prokisch, J., El-Baily, S., Elasawi, T. H., Elmahrouk, M., Omara, A. E. D. et al. (2022). Biological nanofertilizer for horticultural crops: A diagrammatic mini-review. *Environment, Biodiversity and Soil Security*, 6(2022): 339–348.

El-Saadony, M. T., Saad, A. M., Najjar, A. A., Alzahrani, S. O., Alkhatib, F. M., Shafi, M. E. et al. (2021). The use of biological selenium nanoparticles to suppress Triticum aestivum L. crown and root rot diseases induced by Fusarium species and improve yield under drought and heat stress. *Saudi Journal of Biological Sciences*, 28(8): 4461–4471.

Fahad, S., Bajwa, A. A., Nazir, U., Anjum, S. A., Farooq, A., Zohaib, A. et al. (2017). Crop production under drought and heat stress: plant responses and management options. *Frontiers in Plant Science*, 8: 1147.

FAO, IFAD, UNICEF, WFP and WHO. (2024). The State of Food Security and Nutrition in the World 2024 – Financing to end hunger, food insecurity, and malnutrition in all its forms. *Rome*. https://doi.org/10.4060/cd1254en.

Farooq, A., Khan, I., Shehzad, J., Hasan, M., and Mustafa, G. (2024). Proteomic insights to decipher nanoparticle uptake, translocation, and intercellular mechanisms in plants. *Environmental Science and Pollution Research*, 31(12): 18313–18339.

Ferrero, D. M., Piattoni, C. V., Asencion Diez, M. D., Rojas, B. E., Hartman, M. D., Ballicora, M. A. et al. (2020). Phosphorylation of ADP-glucose pyrophosphorylase during wheat seeds development. *Frontiers in Plant Science*, 11: 1058.

Goodarzian Ghahfarokhi, M., Mansurifar, S., Taghizadeh-Mehrjardi, R., Saeidi, M., Jamshidi, A. M., and Ghasemi, E. (2015). Effects of drought stress and rewatering on antioxidant systems and relative water content in different growth stages of maize (Zea mays L.) hybrids. *Archives of Agronomy and Soil Science*, 61(4): 493–506.

Gudkov, S. V., Shafeev, G. A., Glinushkin, A. P., Shkirin, A. V., Barmina, E. V., Rakov, I. I. et al. (2020). Production and use of selenium nanoparticles as fertilizers. *ACS Omega*, 5(28): 17767–17774.

Guleria, G., Thakur, S., Shandilya, M., Sharma, S., Thakur, S., and Kalia, S. (2023). Nanotechnology for sustainable agro-food systems: The need and role of nanoparticles in protecting plants and improving crop productivity. *Plant Physiology and Biochemistry*, 194: 533–549.

Haji Basheerudeen, Mushtaq, S.A., Soundhararajan, R., Nachimuthu, S.K., and Srinivasan, H. (2021). Marine endophytic fungi mediated Silver nanoparticles and their application in plant growth promotion in *Vigna radiata* L. *International Journal of Nano Dimension*, 12(1).

Hamida, R. S., Ali, M. A., Redhwan, A., and Bin-Meferij, M. M. (2020). Cyanobacteria–a promising platform in green nanotechnology: a review on nanoparticles fabrication and their prospective applications. *International Journal of Nanomedicine*, pp. 6033–6066.

Hammad, H. M., Farhad, W., Abbas, F., Fahad, S., Saeed, S., Nasim, W. et al. (2017). Maize plant nitrogen uptake dynamics at limited irrigation water and nitrogen. *Environmental Science and Pollution Research*, 24: 2549–2557.

Hayat, F., Khanum, F., Li, J., Iqbal, S., Khan, U., Javed, H. U. et al. (2023). Nanoparticles and their potential role in plant adaptation to abiotic stress in horticultural crops: A review. *Scientia Horticulturae*, 321: 112285.

Hong, J., Wang, C., Wagner, D. C., Gardea-Torresdey, J. L., He, F., Rico, C. M. et al. (2021). Foliar application of nanoparticles: mechanisms of absorption, transfer, and multiple impacts. *Environ Sci Nano*, 8(5): 1196–210.

Ibrahim, E., Zhang, M., Zhang, Y., Hossain, A., Qiu, W., Chen, Y. et al. (2020). Green-synthesization of silver nanoparticles using endophytic bacteria isolated from garlic and its antifungal activity against wheat Fusarium head blight pathogen *Fusarium graminearum*. *Nanomaterials*, 10(2): 219.

Ibrahim, E., Zhang, M., Zhang, Y., Hossain, A., Qiu, W., Chen, Y. et al. (2020). Green-synthesization of silver nanoparticles using endophytic bacteria isolated from garlic and its antifungal activity against wheat Fusarium head blight pathogen Fusarium graminearum. *Nanomaterials*, 10(2): 219.

Islam, S. N., Naqvi, S. M. A., Raza, A., Jaiswal, A., Singh, A. K., Dixit, M. et al. (2022). Mycosynthesis of highly fluorescent selenium nanoparticles from Fusarium oxysporum, their antifungal activity against black fungus *Aspergillus niger*, and *in-vivo* biodistribution studies. *3 Biotech*, 12(11): 309.

Itroutwar, P. D., Govindaraju, K., Tamilselvan, S., Kannan, M., Raja, K., and Subramanian, K. S. (2020). Seaweed-based biogenic ZnO nanoparticles for improving agro-morphological characteristics of rice (*Oryza sativa* L.). *Journal of Plant Growth Regulation*, 39: 717–728.

Jamla, M., Khare, T., Joshi, S., Patil, S., Penna, S., and Kumar, V. (2021). Omics approaches for understanding heavy metal responses and tolerance in plants. *Current Plant Biology*, 27: 100213.

Jha, A. K., and Prasad, K. (2016). Understanding mechanism of fungus mediated nanosynthesis: a molecular approach. Advances and Applications Through Fungal Nanobiotechnology. *Springer*. pp. 1–23.

Jin, N., Ren, W., Tao, B., He, L., Ren, Q., Li, S., and Yu, Q. (2018). Effects of water stress on water use efficiency of irrigated and rainfed wheat in the Loess Plateau, China. *Science of the Total Environment*, 642: 1–11.

Joshi, S. M., De Britto, S., Jogaiah, S., and Ito, S. I. (2019). Mycogenic selenium nanoparticles as potential new generation broad spectrum antifungal molecules. *Biomolecules*, 9(9): 419.

Kalishwaralal, K., Deepak, V., Pandian, S. R. K., Kottaisamy, M., BarathManiKanth, S., Kartikeyan, B. et al. (2010). Biosynthesis of silver and gold nanoparticles using Brevibacterium casei. *Colloids and surfaces B: Biointerfaces*, 77(2): 257–262.

Kalpana, M., Raguvaran, K., Manimegalai, T., Kalaivani, S., Devapriya, P., and Maheswaran, R. (2024). Actinobacteria mediated synthesis of silver nanoparticles using Streptomyces diastaticus and their biological efficacy against human vector mosquitoes and agricultural insect pests. *Journal of Natural Pesticide Research*, 10: 100091.

Kamal, A., Saba, M., Kamal, A., Batool, M., Asif, M., Al-Mohaimeed, A. M. et al. (2023). Bioinspired green synthesis of bimetallic iron and zinc oxide nanoparticles using mushroom extract and use against *Aspergillus niger*; The Most Devastating Fungi of the Green World. *Catalysts*, 13(2): 400.

Kamatchi, K. M., Anitha, K., Kumar, K. A., Senthil, A., Kalarani, M. K., and Djanaguiraman, M. (2024). Impacts of combined drought and high-temperature stress on growth, physiology, and yield of crops. *Plant Physiology Reports*, 29(1):28–36.

Kareem, H. A., Hassan, M. U., Zain, M., Irshad, A., Shakoor, N., Saleem, S. et al. (2022b). Nanosized zinc oxide (n-ZnO) particles pretreatment to alfalfa seedlings alleviate heat-induced morpho-physiological and ultrastructural damages. *Environmental Pollution*, 303: 119069.

Kareem, H. A., Saleem, M. F., Saleem, S., Rather, S. A., Wani, S.H., Siddiqui, M. H. et al. (2022a). Zinc oxide nanoparticles interplay with physiological and biochemical attributes in terminal heat stress alleviation in mungbean (*Vigna radiata* L.). *Frontiers in Plant Science*, 13: 842349.

Karunakaran, A., Fathima, Y., Singh, P., Beniwal, R., Singh, J., and Ramakrishna, W. (2024). Next-generation biofertilizers: nanoparticle-coated plant growth-promoting bacteria biofertilizers for enhancing nutrient uptake and wheat growth. *Agriculture*, 14(4): 517.

Khalid, M. F., Hussain, S., Ahmad, S., Ejaz, S., Zakir, I., Ali, M. A. et al. (2019). Impacts of abiotic stresses on growth and development of plants. pp. 1–8. *In: Plant Tolerance to Environmental Stress*. CRC press.

Khan, F., Shahid, A., Zhu, H., Wang, N., Javed, M. R., Ahmad, N. et al. (2022). Prospects of algae-based green synthesis of nanoparticles for environmental applications. *Chemosphere*, 293: 133571.

Khan, I., Raza, M. A., Awan, S.A., Shah, G. A., Rizwan, M., Ali, B. et al. (2020). Amelioration of salt induced toxicity in pearl millet by seed priming with silver nanoparticles (AgNPs): The oxidative damage, antioxidant enzymes and ions uptake are major determinants of salt tolerant capacity. *Plant Physiology and Biochemistry*, 156: 221–232.

Khan, R. A. A., Alam, S. S., Najeeb, S., Ali, A., Ahmad, A., Shakoor, A. et al. (2023). Mitigating Cd and bacterial wilt stress in tomato plants through trico-synthesized silicon nanoparticles and Trichoderma metabolites. *Environ. Pollut.*, 122041.

Khosravi-Darani, K., da Cruz, A., Mozafari, M., Abdi, Z., and Ahmadi, N. (2019). Biosynthesis of metal nanoparticles by probiotic bacteria. *Letters in Applied NanoBioScience*, 8(3): 619–626.

Kim, D. Y., Kim, M., Sung, J.S., Koduru, J. R., Nile, S. H., Syed, A. et al. (2024). Extracellular synthesis of silver nanoparticle using yeast extracts: Antibacterial and seed priming applications. *Applied Microbiology and Biotechnology*, 108(1): 150.

Kim, T. Y., Kim, M. G., Lee, J. H., and Hur, H. G. (2018). Biosynthesis of nanomaterials by Shewanella species for application in lithium ion batteries. *Frontiers in Microbiology*, 9: 2817.

Kokabi, M., and Yousefzadi, M. (2022). Algal nanoparticles and their potential application in agriculture. pp. 189-198. *In: Nano-enabled Agrochemicals in Agriculture*. Academic Press.

Koul, B., Poonia, A. K., Yadav, D., and Jin, J. O. (2021). Microbe-mediated biosynthesis of nanoparticles: Applications and future prospects. *Biomolecules*, 11(6): 886.

Krishnakumar, S., and Bai, V. D. M. (2015). Extracellular biosynthesis of silver nanoparticles using terrestrial Streptomyces sp-SBU3 and its antimicrobial efficiency against plant pathogens. *Int. J. Tech. Chem. Res.*, 1: 112–118.

Kundu, S., and Nithiyanantham, U. (2014). DNA-mediated fast synthesis of shape-selective ZnO nanostructures and their potential applications in catalysis and dye-sensitized solar cells. *Industrial & Engineering Chemistry Research*, 53(35): 13667–13679.

Lal, M. K., Tiwari, R. K., Gahlaut, V., Mangal, V., Kumar, A., Singh, M.P. et al. 2021. Physiological and molecular insights on wheat responses to heat stress. *Plant Cell Reports*, pp. 1–18.

Lew, T. T. S., Wong, M. H., Kwak, S. Y., Sinclair, R., Koman, V. B., and Strano, M. S. (2018). Rational design principles for the transport and subcellular distribution of nanomaterials into plant protoplasts. *Small*, 14(44): 1802086.

Li, C., Zhou, L., Yang, H., Lv, R., Tian, P., Li, X. et al. (2017). Self-assembled exopolysaccharide nanoparticles for bioremediation and green synthesis of noble metal nanoparticles. *ACS Applied Materials & Interfaces*, 9(27): 22808–22818.

Lobell, D. B., Schlenker, W., and Costa-Roberts, J. (2011). Climate trends and global crop production since 1980. *Science*, 333(6042): 616–620.

Lowry, G.V., Avellan, A., and Gilbertson, L. M. (2019). Opportunities and challenges for nanotechnology in the agri-tech revolution. *Nature Nanotechnology*, 14(6): 517–522.

Mandal, M., and Sarkar, A. (2024). Green syntheses of nanoparticles from plant growth–promoting microorganisms and their application in the agri-food industries. pp. 185–204. *In: Nanotechnology and Nanomaterials in the Agri-Food Industries*. Elsevier.

Manivannan, N., Aswathy, S., Malaikozhundan, B., and Boopathi, T. (2021). Nano-zinc oxide synthesized using diazotrophic Azospirillum improves the growth of mung bean, Vigna radiata. *International Nano Letters*, 11: 405–415.

Manzoor, N., Ahmed, T., Noman, M., Shahid, M., Nazir, M.M., Ali, L. et al. (2021). Iron oxide nanoparticles ameliorated the cadmium and salinity stresses in wheatplants, facilitating photosynthetic pigments and restricting cadmium uptake. *Sci. Total Environ.*, 769: 145221.

Mishra, S., Singh, B. R., Singh, A., Keswani, C., Naqvi, A. H., and Singh, H. B. (2014). Biofabricated silver nanoparticles act as a strong fungicide against Bipolaris sorokiniana causing spot blotch disease in wheat. *Plos one*, 9(5): e97881.

Mohamed, A.K.S., Qayyum, M.F., Abdel-Hadi, A.M., Rehman, R.A., Ali, S., and Rizwan, M. (2017). Interactive effect of salinity and silver nanoparticles on photosynthetic and biochemical parameters of wheat. *Archives of Agronomy and Soil Science*, 63(12): 1736–1747.

Mondal, A., Paul, S., Pal, R., and Paul, S. (2024). Nano iron loaded algal biomass: For better yield, amino acid and iron content in rice–A 'nano-phycofertilizer'. *Algal Research*, p. 103573.

Mustafa, M., Azam, M., Bhatti, H. N., Khan, A., Zafar, L., and Abbasi, A. M. R. (2024). Green fabrication of copper nano-fertilizer for enhanced crop yield in cowpea cultivar: A sustainable approach. *Biocatalysis and Agricultural Biotechnology*, 56: 102994.

Namburi, K. R., Kora, A. J., Chetukuri, A., and Kota, V. S. M. K. (2021). Biogenic silver nanoparticles as an antibacterial agent against bacterial leaf blight causing rice phytopathogen *Xanthomonas oryzae* pv. oryzae. *Bioprocess and Biosystems Engineering*, 44(9): 1975–1988.

Noman, M., Ahmed, T., Hussain, S., Niazi, M. B. K., Shahid, M., Song, F. et al. (2020b). Biogenic copper nanoparticles synthesized by using a copper-resistant strain Shigella flexneri SNT22 reduced the translocation of cadmium from soil to wheat plants. *J. Hazard. Mater.*, 398: 123175.

Noman, M., Shahid, M., Ahmed, T., Tahir, M., Naqqash, T., Muhammad, S. et al. (2020a). Green copper nanoparticles from a native Klebsiella pneumoniae strain alleviated oxidative stress impairment of wheat plants by reducing the chromium bioavailability and increasing the growth. *Ecotoxicol. Environ. Saf.*, 192: 110303.

Omar, R. A., Talreja, N., Ashfaq, M., and Chauhan, D. (2023). Nanostructure-based smart fertilizers and their interaction with plants. pp. 399–430. *In: Nanofertilizers for Sustainable Agroecosystems: Recent Advances and Future Trends*. Cham: Springer Nature Switzerland.

Ovais, M., Khalil, A. T., Raza, A., Islam, N. U., Ayaz, M., Saravanan, M. et al. (2018). Multifunctional theranostic applications of biocompatible green-synthesized colloidal nanoparticles. *Applied Microbiology and Biotechnology*, 102: 4393–4408.

Patel, D., Patel, B., Patani, A., Yadav, V. K., Alharbi, S. A., Alarfaj, A. A. et al. (2024). Biogenic silver nanoparticles derived from the marine brown algae Iyengaria stellata for plant growth promotion under saline conditions. *Physiologia Plantarum*, 176(6): e14638.

Posch, B. C., Kariyawasam, B. C., Bramley, H., Coast, O., Richards, R.A., Reynolds, M. P. et al. (2019). Exploring high temperature responses of photosynthesis and respiration to improve heat tolerance in wheat. *Journal of Experimental Botany*, 70(19): 5051–5069.

Prakash, M., Sunilkumar, B., Sathiyanarayanan, G., and Gokulakrishnan, J. (2017). Screening for drought tolerance in mungbean. *Legume Research-An International Journal*, 40(3): 423–428.

Priya, P., Patil, M., Pandey, P., Singh, A., Babu, V. S., and Senthil-Kumar, M. (2023). Stress combinations and their interactions in plants database: a one-stop resource on combined stress responses in plants. *The Plant Journal*, 116(4): 1097–1117.

Priyadharshini, R. I., Prasannaraj, G., Geetha, N., and Venkatachalam, P. (2014). Microwave-mediated extracellular synthesis of metallic silver and zinc oxide nanoparticles using macro-algae (*Gracilaria edulis*) extracts and its anticancer activity against human PC3 cell lines. *Applied Biochemistry and Biotechnology*, 174: 2777–2790.

Queiroz, M.S., Oliveira, C.E., Steiner, F., Zuffo, A.M., Zoz, T., Vendruscolo, E.P. et al. (2019). Drought stresses on seed germination and early growth of maize and sorghum. *Journal of Agricultural Science*, 11(2): 310.

Raj, N. B., Swamy, M. K., Purushotham, B., and Sukrutha, S. K. (2021). Applications of microbe-based nanoparticles in agriculture: present state and future challenges. *Microbial Nanobiotechnology: Principles and Applications*, pp. 343–382.

Ramegowda, V., Senthil, A., and Senthil-Kumar, M. (2024). Stress combinations and their interactions in crop plants. *Plant Physiology Reports*, 29(1): 1–5.

Ramírez Anguiano, A. C., Velasco Ramírez, A. P., Zamudio Ojeda, A., Jiménez Torres, H. D., Velázquez Juárez, G., Velázquez López, J. M. et al. (2023). Effects of Nanomaterials/Nanocomposites on Trace Element Uptake and Phytotoxicity. pp. 127–156. *In: Nanomaterials and Nanocomposites Exposures to Plants: Response, Interaction, Phytotoxicity and Defense Mechanisms*. Singapore: Springer Nature Singapore.

Rani, S., Kumari, N., and Sharma, V. (2023). Uptake, translocation, transformation and physiological effects of nanoparticles in plants. *Archives of Agronomy and Soil Science*, 69(9): 1579–1599.

Ranjan, S., Mirchandani, R., and Senthil-Kumar, M. (2024). Abiotic stress impact on the interaction between *Macrophomina phaseolina* and crop plants. *Plant Physiology Reports*, 29(1): 18–27.

Rascio, N., and Navari-Izzo, F. (2011). Heavy metal hyperaccumulating plants: how and why do they do it? And what makes them so interesting?. *Plant science*, 180(2): 169–181.

Reddy, K. V., Sree, N. R. S., Kumar, P. S., and Ranjit, P. (2022). Microbial enzymes in the biosynthesis of metal nanoparticles. pp. 329–350. *In: Ecological Interplays in Microbial Enzymology*. Singapore: Springer Nature Singapore.

Rehman, A., Khan, S., Sun, F., Peng, Z., Feng, K., Wang, N. et al. (2024). Exploring the nano-wonders: unveiling the role of Nanoparticles in enhancing salinity and drought tolerance in plants. *Frontiers in Plant Science*, 14: 1324176.

Riaz, S., Raza, Z. A., and Majeed, M. I. (2020). Preparation of cadmium sulfide nanoparticles and mediation thereof across poly (hydroxybutyrate) nanocomposite. *Polymer Bulletin*, 77(2): 775–791.

Rosegrant, M.W., Sulser, T.B., Dunston, S., Mishra, A., Cenacchi, N., Gebretsadik, Y. et al. (2024). Food and nutrition security under changing climate and socioeconomic conditions. *Global Food Security*, 41: 100755.

Roseline, T. A., Murugan, M., Sudhakar, M. P., and Arunkumar, K. (2019). Nanopesticidal potential of silver nanocomposites synthesized from the aqueous extracts of red seaweeds. *Environmental Technology & Innovation*, 13: 82–93.

Roseline, T. A., Sudhakar, M. P., and Arunkumar, K. (2023). Aqueous extraction of red seaweed bioactive compounds and synthesis of silver nanoparticles for agriculture applications. *Journal of Agriculture and Food Research*, 14: 100769.

Roy, S., and Anantharaman, P. (2018). Agricultural and biomedical application of Silver Nanoparticles synthesized by *Halimeda gracilis* Harvey ex J. Agardh. *International Journal of Environment, Agriculture and Biotechnology*, 3(5): 1739–1747.

Roy, S., and Anantharaman, P. (2018). Biosynthesis of silver nanoparticles by *Sargassumilicifolium* (Turner) C. *Agardh* with their antimicrobial activity and potential for seed germination. *Journal of Applied Physics and Nanotechnology*, 1(1): 2.

Ruttkay-Nedecky, B., Krystofova, O., Nejdl, L., and Adam, V. (2017). Nanoparticles based on essential metals and their phytotoxicity. *Journal of Nanobiotechnology*, 15: 1–19.

Sable, S.V., Ranade, S., and Joshi, S. (2018). Role of AgNPs in the enhancement of seed germination and its effect on plumule and radicle length of *Pennisetum glaucum*. *IET Nanobiotechnol.*, 12(Issue 7): 922–926.

Saied, E., Mekky, A.E., Al-Askar, A.A., Hagag, A.F., El-bana, A.A., Ashraf, M. et al. (2023). Aspergillus terreus-mediated selenium nanoparticles and their antimicrobial and photocatalytic activities. *Crystals*, 13(3): 450.

Salama, D. M., Abd El-Aziz, M. E., Shaaban, E.A., Osman, S. A., and Abd El-Wahed, M. S. (2022). The impact of nanofertilizer on agro-morphological criteria, yield, and genomic stability of common bean (*Phaseolus vulgaris* L.). *Scientific Reports*, 12(1): 18552.

Sanaeimehr, Z., Javadi, I., and Namvar, F. (2018). Antiangiogenic and antiapoptotic effects of green-synthesized zinc oxide nanoparticles using Sargassum muticum algae extraction. *Cancer Nanotechnology*, 9: 1–16.

Sanjari, S., Shobbar, Z.S., Ghanati, F., Afshari-Behbahanizadeh, S., Farajpour, M., Jokar, M. et al. (2021). Molecular, chemical, and physiological analyses of sorghum leaf wax under post-flowering drought stress. *Plant Physiology and Biochemistry*, 159: 383–391.

Saraiva, R., Ferreira, Q., Rodrigues, G. C., and Oliveira, M. (2023). Nanofertilizer use for adaptation and mitigation of the agriculture/Climate change dichotomy effects. *Climate*, 11(6): 129.

Sarshad, A., Talei, D., Torabi, M., Rafiei, F., and Nejatkhah, P. (2021). Morphological and biochemical responses of *Sorghum bicolor* (L.) Moench under drought stress. *SN Applied Sciences*, 3(1): 81.

Savassa, S. M., Castillo-Michel, H., del Real, A. E. P., Reyes-Herrera, J., Marques, J. P. R., and de Carvalho, H. W. (2021). Ag nanoparticles enhancing Phaseolus vulgaris seedling development: understanding nanoparticle migration and chemical transformation across the seed coat. *Environmental Science: Nano*, 8(2): 493–501.

Sembada, A.A., and Lenggoro, I. W. (2024). Transport of nanoparticles into plants and their detection methods. *Nanomaterials*, 14(2): 131.

Shah, T., Latif, S., Saeed, F., Ali, I., Ullah, S., Alsahli, A. A. et al. (2021). Seed priming with titanium dioxide nanoparticles enhances seed vigor, leaf water status, and antioxidant enzyme activities in maize (*Zea mays* L.) under salinity stress. *Journal of King Saud University-Science*, 33(1): 101207.

Shalaby, T. A., El-Bialy, S. M., El-Mahrouk, M. E., Omara, A. E. D., El-Beltagi, H. S., and El-Ramady, H. (2022). Acclimatization of in vitro banana seedlings using root-applied bio-nanofertilizer of copper and selenium. *Agronomy*, 12(2): 539.

Shang, Y., Hasan, M. K., Ahammed, G. J., Li, M., Yin, H., and Zhou, J. (2019). Applications of nanotechnology in plant growth and crop protection: a review. *Molecules*, 24(14): 2558.

Shanthi, S., Jayaseelan, B.D., Velusamy, P., Vijayakumar, S., Chih, C.T., and Vaseeharan, B. 2016. Biosynthesis of silver nanoparticles using a probiotic Bacillus licheniformis Dahb1 and their antibiofilm activity and toxicity effects in Ceriodaphnia cornuta. *Microbial Pathogenesis*, *93*: pp.70–77.

Sharma, N., Thakur, M., Suryakumar, P., Mukherjee, P., Raza, A., Prakash, C. S. et al. (2022). 'Breathing Out'under heat stress—respiratory control of crop yield under high temperature. *Agronomy*, 12(4): 806.

Sidhu, A. K., Sharma, M., Bhickchand Agrawal, S., Pradip Bhavsar, P., and Samota, M. K. (2024). Nanomaterial strategies for enhancing plant resilience in the face of temperature stress. *CABI Agriculture and Bioscience*, 5(1): 60.

Singh, S., Kapoor, D., Khasnabis, S., Singh, J., and Ramamurthy, P. C. (2021). Mechanism and kinetics of adsorption and removal of heavy metals from wastewater using nanomaterials. *Environmental Chemistry Letters*, 19(3): 2351–2381.

Singh, S., Singh, B. K., Yadav, S. M., and Gupta, A. K. (2015). Applications of nanotechnology in agricultural and their role in disease management. *Res. J. Nanosci. Nanotechnol.*, 5(1): 1–5.

Singh, A., Sengar, R. S., Rajput, V. D., Minkina, T., and Singh, R. K. (2022). Zinc oxide nanoparticles improve salt tolerance in rice seedlings by improving physiological and biochemical indices. *Agriculture*, 12(7): 1014.

Singh, A., Tiwari, S., Pandey, J., Lata, C., and Singh, I. K. (2021). Role of nanoparticles in crop improvement and abiotic stress management. *Journal of Biotechnology*, 337: 57–70.

Sintubin, L., De Windt, W., Dick, J., Mast, J., Van Der Ha, D., Verstraete, W. et al. (2009). Lactic acid bacteria as reducing and capping agent for the fast and efficient production of silver nanoparticles. *Applied microbiology and biotechnology*, 84: 741–749.

Sonawane, H., Arya, S., Math, S., and Shelke, D. (2021). Myco-synthesized silver and titanium oxide nanoparticles as seed priming agents to promote seed germination and seedling growth of Solanum lycopersicum: A comparative study. *International Nano Letters*, 11(4): 371–379.

Spagnoletti, F. N., Spedalieri, C., Kronberg, F., and Giacometti, R. (2019). Extracellular biosynthesis of bactericidal Ag/AgCl nanoparticles for crop protection using the fungus Macrophomina phaseolina. *Journal of Environmental Management*, 231: 457–466.

Subramanian, K.S., Muniraj, I., and Uthandi, S. (2016). Role of actinomycete-mediated nanosystem in agriculture. *Plant Growth Promoting Actinobacteria: A New Avenue for Enhancing the Productivity and Soil Fertility of Grain Legumes*, pp. 233–247.

Talebi, S.M. 2024. Uptake of nanomaterials by plants and translocation within plants. In *Microbiome and Nano-Cross-Talk* (pp. 19-41). Academic Press.

Terra, A. L. M., Kosinski, R. D. C., Moreira, J. B., Costa, J. A. V., and Morais, M. G. D. 2019. Microalgae biosynthesis of silver nanoparticles for application in the control of agricultural pathogens. *Journal of Environmental Science and Health, Part B*, 54(8): 709–716.

Thabet, S.G., and Alqudah, A.M. 2024. Unraveling the role of nanoparticles in improving plant resilience under environmental stress condition. *Plant and Soil*, pp. 1–18.

Thakkar, K. N., Mhatre, S. S., and Parikh, R. Y. (2010). Biological synthesis of metallic nanoparticles. *Nanomedicine: nanotechnology, biology and medicine*, 6(2): 257–262.

Tondey, M., Kalia, A., Singh, A., Dheri, G. S., Taggar, M. S., Nepovimova, E. et al. (2021). Seed priming and coating by nano-scale zinc oxide particles improved vegetative growth, yield and quality of fodder maize (*Zea mays*). *Agronomy*, 11(4): 729.

Venkateswaran, U. P., Caparco, A. A., González-Gamboa, I., Caballero, R. M., Schuphan, J., and Steinmetz, N. F. (2023). Plant viral nanocarrier soil mobility as a function of soil type and nanoparticle properties. *ACS Agricultural Science & Technology*, 3(7): 583–592.

Vijayakumar, S., Chen, J., Kalaiselvi, V., Tungare, K., Bhori, M., González-Sánchez, Z. I. et al. (2022). Marine polysaccharide laminarin embedded ZnO nanoparticles and their based chitosan capped ZnO nanocomposites: Synthesis, characterization and *in vitro* and *in vivo* toxicity assessment. *Environmental Research*, 213: 113655.

Vyas, T.K., More, B., and Mehta, M.P. 2024. Improving stress resilience in plants by nanoparticles. pp. 73–96. *In: Improving Stress Resilience in Plants*. Academic Press.

Wang, X. P., Li, Q. Q., Pei, Z. M., and Wang, S. C. (2018). Effects of zinc oxide nanoparticles on the growth, photosynthetic traits, and antioxidative enzymes in tomato plants. *Biologia Plantarum*, 62: 801–808.

Wong, M. H., Misra, R. P., Giraldo, J. P., Kwak, S. Y., Son, Y., Landry, M.P. et al. (2016). Lipid exchange envelope penetration (LEEP) of nanoparticles for plant engineering: a universal localization mechanism. *Nano Letters*, 16(2): 1161–1172.

Yadav, T., Mungray, A. A., and Mungray, A. K. (2014). Fabricated nanoparticles: current status and potential phytotoxic threats. *Reviews of Environmental Contamination and Toxicology volume: With Cumulative and Comprehensive Index Subjects Covered Volumes 221–230*, pp. 83–110.

Yang, X., Wang, B., Chen, L., Li, P., and Cao, C. (2019). The different influences of drought stress at the flowering stage on rice physiological traits, grain yield, and quality. *Scientific Reports*, 9(1): 3742.

Zahra, N., Hafeez, M. B., Ghaffar, A., Kausar, A., Al Zeidi, M., Siddique, K. H. et al. (2023). Plant photosynthesis under heat stress: Effects and management. *Environmental and Experimental Botany*, 206: 105178.

Zhang, H. Y., and Su, W. H. (2024). Classification, uptake, translocation, and detection methods of nanoparticles in crop plants: a review. *Environmental Science: Nano*.,11: 1847–1870.

CHAPTER-19

Green Nanofertilizers

Types, the Mode of Action and Pollution Mitigation for Sustainable Agriculture and Agroecosystems

Kun Chen,[1,#] *Mozammel Hoque,*[1,#] *Naser Khan,*[1] *Philip Kwong,*[1] *Mohammad Boshir Ahmed*[1,2,3,*] and *Volker Hessel*[1,2,4,*]

1. Introduction

Agriculture is the main source of livelihood for many people in developing countries, with nearly two-thirds of the global population dependent on it. The high demand for agricultural products puts significant pressure on the need to develop new techniques that enhance crop yield and productivity while reducing the negative impact of chemical fertilizers on the soil (Tran et al., 2022). Fertilizers are essential for boosting agricultural yields, but excessive application can disrupt soil ecology and reduce the area available for crop production. To address the increasing demand for food, farmers are increasingly depending on chemical fertilizers, which have become a vital component of modern farming systems. These fertilizers provide various macronutrients like nitrogen, phosphorus, and potassium, along with essential micronutrients such as iron, zinc, and boron. Recently, there has been a rise in the use of chemical nutrients to achieve higher yields. However, this practice can lead to significant issues, including groundwater pollution after harvest. Additionally, overuse of chemical fertilizers can adversely affect soil structure by reducing organic matter content, leading to nutrient imbalances and soil acidification (Ojeniyi, 2000). It also negatively affects microorganism activities. Furthermore, it has not been readily available to farmers over the years, and the costs are high. Another issue is that most inorganic fertilizer in the soil is gradually depleted by crops (Ogbomo, 2011). The primary concern is to use fertilizers more efficiently by substituting them with nanofertilizers. Nanoparticles exhibit exceptional physicochemical properties due to their small size, which results in a larger surface area-to-mass ratio, increased reactivity, enhanced ionizing

[1] School of Chemical Engineering, Adelaide University, Adelaide 5005, Australia.
[2] ARC Centre of Excellence Plants for Space, Adelaide University, Adelaide, 5064, Australia.
[3] Waite Research Institute, Adelaide University, Adelaide 5005, Australia.
[4] School of Engineering, University of Warwick, Conventry, United Kingdom.
# Equally contributed
* Corresponding authors: mohammadboshir.ahmed@adelaide.edu.au; volker.hessel@adelaide.edu.au

DOI: 10.1201/9781003567684-19

capability, chemical stability, improved absorbability, pH resistance, and thermal stability. These characteristics make nanomaterials when combined with fertilizers, more effective in promoting plant growth and development. However, there are notable distinctions between nanofertilizers and green nanofertilizers, primarily concerning their synthesis methods. Nanofertilizers can be produced using various techniques, including green (biological) synthesis, physical methods, and chemical processes. In contrast, green nanofertilizers are specifically synthesized using biological entities such as plants, microorganisms, or enzymes (Gosavi et al., 2020). Green nanofertilizers are considered eco-friendly and more sustainable than traditional methods because they use fewer chemicals. This leads to less chemical contamination and a smaller environmental footprint. Sustainable agriculture aims to reduce the use of agrochemicals to protect the environment from pollution, which can cause problems such as water contamination, soil degradation, and harm to beneficial organisms. Therefore, green nanofertilizers could provide a practical solution to the challenges encountered in sustainable agriculture (Batool et al., 2024). Green nanofertilizers provide advanced nutrient delivery systems, including targeted, slow, and controlled release to plants. This enhances crop productivity while promoting environmental sustainability compared to traditional chemical fertilizers. Various types of green nanofertilizers are increasingly used today, including biogenic nanofertilizers, bio-based nanofertilizers, biopolymer-based nanofertilizers, bio-nanocomposites, metal-based nanofertilizers, as well as silica and zeolite-based nanofertilizers. Because of their high surface area-to-volume ratio and nanoscale size, nanofertilizers provide enhanced availability and absorption for plants. These fertilizers typically have a particle size of less than 1-100 nm in at least one dimension, which allows for improved uptake by plants from the soil or leaves. As a result, they contribute to the production of more photosynthates and biomass, both of which are essential for healthy crop growth. Green nanofertilizers also possess several advantages, including easier application, smaller quantity requirements, targeted delivery, controlled and slow-release action, and a reduction in chemical usage. They result in comparatively lower salt accumulation in the soil than conventional fertilizers. Consequently, they effectively meet crop nutrient requirements while enhancing soil health and nutrient bioavailability. Moreover, Phosphorus (P) released from conventional products often encounters fixation in the soil due to cations (such as Ca^{2+}, Al^{3+}, and Fe^{3+}) present. This fixation transforms P into a less accessible form and significantly decreases its release efficiency, which typically ranges from only 10 to 15% (Khan et al., 2009). Additionally, the high solubility of these fertilizers negatively impacts the environment, as the phosphorus in them is highly mobile. This mobility can lead to phosphorus entering nearby water bodies through run-off or seepage, resulting in eutrophication (Daniel et al., 1998). Phosphorus, unlike other essential elements for life such as carbon, nitrogen, oxygen, and hydrogen, does not circulate in the biosphere due to the absence of a significant gaseous phase. Consequently, phosphorus is a finite resource that takes millions of years to form and accumulate in the Earth's crust before it can be extracted and utilized for human activities (Le et al., 2023). We can address this issue using two approaches: either by increasing the available resources or by reducing their use. Given the challenges associated with traditional P application, green nanofertilizers offer an effective way to supply phosphorus. They can meet the nutrient needs of crops while minimizing environmental risks.

Against this backdrop, this chapter explores the various types of green nanofertilizers, and the mode of action through which these fertilizers interact with plants, promoting growth and enhancing nutrient efficiency. Furthermore, the chapter explores the impact of green nanofertilizers on the environment and ecosystem. Moreover, it provides information on how it mitigates pollution, thus contributing to the broader goal of sustainable agricultural practices. By integrating green nanofertilizers into farming systems, it is possible to not only boost agricultural productivity but also protect the environment, ensuring the long-term viability of agricultural lands.

2. Types of Green Nanofertilizers

Agriculture plays a crucial role in health, nutrition, and economic development. It has undergone significant changes globally due to various factors. Green nanofertilizers can be classified based on their sources and composition (Fig. 1).

Figure 1. Types of green nanofertilizers.

2.1 Biogenic Nanofertilizers

Biogenic nanofertilizers are derived from plant extracts, microbes, and organic materials using biological processes such as microbial synthesis, composting and vermicomposting, fermentation, and enzyme-mediated processes (Thirugnanasambandan, 2018). Biogenic nanofertilizers offer substantial potential for boosting agricultural productivity while reducing environmental pollution. These nanoparticles can penetrate plant tissues, enhancing nutrient uptake. Their high surface area-to-volume ratio improves interactions with roots, leading to increased nutrient absorption. One of the main advantages of biogenic nanofertilizers is their capability for controlled and sustained nutrient release. These nanostructures can effectively store nutrients and gradually release them, ensuring targeted fertilizer delivery. This precision minimizes nutrient leaching and volatilization, optimizing nutrient availability for plants. However, biogenic nanofertilizers encounter significant challenges, particularly in scaling up production. The synthesis of these nanoparticles is complex and resource-intensive, posing considerable hurdles for large-scale implementation. Ongoing efforts aim to develop cost-effective and sustainable production methods for biogenic nanofertilizers. For example, Sharma et al., have utilized green synthesis techniques to produce these nanofertilizers at scale, employing plant extracts, microorganisms, and bio-waste (Sharma et al., 2023). Zhang et al. highlighted the need for thorough evaluations of nanoparticle accumulation and persistence in soil, as well as their effects on soil fertility and biodiversity (Zhang et al., 2022). Biogenic nanofertilizers encounter substantial regulatory challenges, particularly the requirement for standardized environmental and toxicological testing. Effective legislation is crucial to ensure the safe and sustainable production, distribution, and application of these fertilizers. Despite these hurdles, biogenic nanofertilizers hold significant promise for improving agricultural productivity and reducing environmental pollution.

2.2 Bio-based Nanofertilizers

Bio-based nanofertilizers, also known as bio-nanofertilizers, are a new class of fertilizers that use nanotechnology and biotechnology to improve soil health and plant nutrient uptake. Although biogenic nanofertilizers and bio-based nanofertilizers are similar concepts, there is a key difference. Biogenic implies the use of biological processes or organisms to create nanoparticles while bio-based implies the use of biological or renewable raw materials for creating nanoparticles, but not necessarily biological processes. Bionanofertilizers serve as promising eco-friendly alternatives to chemical fertilizers. These biofertilizer preparations comprise nanoscale materials and beneficial microorganisms, which include fungal mycorrhizae, Rhizobium, Blue-Green Algae (BGA), Azotobacter, Azospirillum, and Phosphate-Solubilizing Bacteria (PSB) (Babu et al., 2022). It provides multiple advantages, such as improved seed germination, better soil quality, increased nutrient utilization efficiency, and the breakdown of pesticide residues, which ultimately leads to enhanced crop productivity (Arora et al., 2024). Bio-nanofertilizers are designed for targeted nutrient delivery, controlled release, and reduced environmental pollution, making them a sustainable option for agriculture. This type of fertilizer uses nanoscale polymers for coating through nano-encapsulation in its formulation (Golbashy et al., 2017). The bioorganic components of nano-biofertilizer support nitrogen fixation, phosphate solubilization, and the replenishment of soil nutrients, while the nanoparticle coating of the biofertilizer enables a slow and controlled release of nutrients following the crop's needs (Morales-Díaz et al., 2017). Nano-encapsulation technology is a versatile method for protecting Plant Growth-Promoting Rhizobacteria (PGPR) in biofertilizer components. It enhances their shelf-life and distribution in fertilizer formulations while enabling the controlled release of PGPR (Vejan et al., 2016). The application of nanomaterials like chitosan, zeolite, and polymers in nano-encapsulation techniques enables the gradual and continuous release of organic nutrients to crops. The larger surface area, nanoscale size, and enhanced reactivity of nanoparticle-coated biofertilizers improve nutrient uptake, providing crops with readily available nutrients in a sustainable manner (El-Ghamry et al., 2018). A nano-structured fertilizer that combines neem cake and PGPR has the potential to enhance agricultural yield in certain members of the Fabaceae family by accelerating seedling germination and efficiently providing fixed nutrients to plants (Rahman and Zhang, 2018). The application of nano-biofertilizer on tomatoes enhanced pest resistance (Gatahi, 2017). *Pseudomonas fluorescens, Bacillus subtilis, Paenibacillus elgii*, and *Pseudomonas putida*-based nano-biofertilizers reduce fungal and bacterial infestations in the rhizosphere of Fabaceae family plants (Dikshit et al., 2013). Titanium nanoparticle-coated nano-biofertilizer enhances the adherence of beneficial bacteria to the roots of mustard plants (Mishra and Kumar, 2009).

2.3 Biopolymer-based Green Nanofertilizers

Biopolymers are natural polymers that are abundantly found in nature and can biodegrade with the assistance of microorganisms, thereby promoting ecological sustainability (Sani et al., 2023). Biopolymers can be sourced from a variety of natural origins, including plants, animals, and microorganisms such as bacteria, fungi, and algae. Examples of natural polymers include starch, cellulose, chitosan, and alginate, along with nucleic acids like DNA and RNA. Everyday items like silk, wool, and honey also belong to the category of natural polymers (Panchal et al., 2022). Recent investigations have shown that researchers utilized various chemical compositions of biopolymeric nanoparticles to manage fertilizer release (Vejan et al., 2021) to amplify the uptake of plant fertilizers like urea (Mohammadbagheri et al., 2021). Biopolymers play a crucial role in soil by effectively absorbing water. They can also serve as nanocarriers that gradually release agrochemicals to plants, and can help in managing pathogens and pests. The use of polymer-based hydrogels significantly reduces the release of agrochemicals, minimizing the risk of these chemicals leaching into the soil and aquatic environments. Additionally, these hydrogels can function as absorbent and slow-release agents for agrochemicals, as well as seed-coating materials that protect seeds from damage caused by pathogens and pests, thereby promoting easier germination (Sikder et al., 2021).

2.4 Metal-based-green Nanofertilizers

A key category of nanofertilizers includes nanostructured metal-based materials, which are typically made up of metallic nanoparticles or metal oxides (Kráľová and Jampílek, 2022). These nanofertilizers offer advantages such as controlled nutrient release and enhanced nutrient uptake by plants. Notable examples of metal oxide-based nanoparticles, such as zinc oxide nanoparticles (ZnNPs), iron oxide nanoparticles (Fe_3O_4 NPs), silver nanoparticles (AgNPs), and copper oxide nanoparticles (CuO), can be classified as green nanofertilizers. This is because they can be synthesized by reducing metal ions using plant-derived biomolecules, including polyphenols, alkaloids, flavonoids, and proteins (Bouttier-Figueroa et al., 2023). Moreover, these kinds of metal oxide can be synthesized from a plant extract or other biological resources such as onion, aloe vera, banana, different citrus peel, and so on (Gosavi et al., 2020; Hussein et al., 2019; Medda et al., 2015). This method is environmentally friendly compared to traditional chemical synthesis, as it avoids the use of toxic chemicals and high-energy inputs.

2.5 Bio-nanocomposite Based Nanofertilizers

The bio-nanocomposites are hybrid materials obtained from biopolymers and nanomaterials, and the biodegradable nature of the biopolymer makes the bio-nanocomposite very environmentally friendly (Tanan et al., 2021). Bio-nanocomposite-based fertilizers can be considered green nano fertilizers due to their nature of biodegradability because of using biopolymers instead of using conventional petroleum-based polymers. As a result, bio-nanocomposites produce lower greenhouse gas emissions, which makes them eco-friendly (de Oliveira and Beatrice, 2018). Bio-nanocomposites have been explored as a sustainable alternative to traditional agrochemicals in agricultural applications. Researchers have developed various types of bio-nanocomposites, including slow-release fertilizers, pesticides, herbicides, and soil modifiers (Boonying et al., 2023). These materials have enhanced properties that make them a sustainable and eco-friendly alternative to traditional agrochemicals.

2.6 Silica and Zeolite-based Nanofertilizers

Silica nanoparticles can be considered green nanofertilizers due to their potential to improve plant growth, reduce the need for harmful chemical fertilizers, and enhance soil health. Among the various green nanofertilizers, nano silica has become a crucial component in promoting plant growth and enhancing tolerance to different abiotic and biotic stresses (Hajihashemi and Kazemi, 2022). Nano silica has a higher absorptivity, which leads to an increased uptake of silica in plants, although the exact mechanism is not fully understood. The uptake of nano-silica in the roots and leaves helps reduce the accumulation of Reactive Oxygen Species (ROS) and prevents membrane lipid peroxidation (Mathur and Roy 2020). Nanosilica limits sodium ion and heavy metal absorption in plants. Additionally, its deposition in leaf tissue improves the plants' defense against pathogens.

Nano zeolite fertilizers are a type of nanomaterial-based fertilizer that utilizes zeolites in their nanoscale form. Zeolites are naturally occurring minerals with a unique porous structure, which gives them the ability to retain and exchange nutrients. Nanoporous zeolite is gaining attention in agriculture for its ability to enhance the efficiency of fertilizer use while mitigating the negative effects of chemical fertilizers on agricultural ecosystems. Zeolite can serve as a fertilizer enhancer or a soil amendment, improving the soil's capacity to retain and release nutrients, which in turn promotes plant growth. While zeolite does not provide nutrients directly like traditional fertilizers, when incorporated with macronutrients such as potassium, phosphorus, nitrogen, calcium, and magnesium, it creates a functional nanofertilizer (Khan et al., 2021). This composite fertilizer, based on nanozeolite, enables a gradual release of essential nutrients for plant growth. Thanks to its high surface area, mesoporous structure, and substantial nutrient loading capacity, nanoporous zeolite has been previously noted for its ability to enhance the soil's nutrient retention. This slow release of nutrients through nanocarriers helps improve overall soil quality (Thirunavukkarasu and Subramanian, 2014).

3. Modes of Action for Green Nanofertilizers

Recent laboratory-scale studies have shown that nanofertilizers can improve crop productivity by enhancing seed germination rates, seedling growth, photosynthetic activity, nitrogen metabolism, and the synthesis of carbohydrates and proteins. However, as a developing technology, the ethical and safety concerns surrounding the use of nanoparticles in agriculture are significant and should be carefully evaluated before adopting nanofertilizers in farming practices. These nanostructured fertilizers can boost nutrient use efficiency through mechanisms such as targeted delivery, slow or controlled release, increased bioavailability, improved soil health, reduced chemical usage, and antimicrobial activity (Davarpanah et al., 2016).

3.1 Targeted Nutrient Delivery

Nanoparticles possess a high surface area, significant sorption capacity, and controlled-release kinetics, which make them an effective "smart delivery system." They can release their active ingredients in response to environmental triggers and biological needs. For instance, green nanofertilizers, also known as biogenic nanofertilizers, can store and gradually release nutrients (Jha et al., 2023). Targeted fertilizer delivery improves plant absorption, reduces leaching and combustion, and increases nutrient availability.

3.2 Slow and Controlled Release

Controlled release refers to a highly regulated and precisely engineered release of nutrients from the fertilizer, typically governed by environmental factors (like temperature, moisture, pH) or specific triggers (such as plant demand). This method enhances the steady and predictable release of nutrients over a specific period (Le et al., 2021). While conventional fertilizer applications can effectively target plant areas, many nutrients are lost due to processes such as leaching, runoff, evaporation, hydrolysis from soil moisture, and degradation by soil microorganisms, particularly at low concentrations (Preetha and Balakrishnan, 2017). Ombódi and Saigusa (2000) found that about 40–70% of nitrogen, 80–90% of phosphorus, and 50–90% of potassium fertilizers are lost in the soil. This loss prevents these nutrients from reaching the intended parts of the plants, leading to economic inefficiencies (Ombódi and Saigusa, 2000). As a result, additional fertilizers are often applied to compensate for these losses, disrupting the nutrient balance in the soil, as noted by Baruah and Dutta (Baruah and Dutta, 2009). Slow-release nanofertilizers provide a promising alternative to traditional soluble fertilizers by enabling plants to absorb most nutrients while minimizing loss through leaching (Escribà-Gelonch et al., 2023). Various strategies are used to control nutrient release in green nanofertilizers, including surface coating, encapsulating active ingredients, and utilizing bio-nanocomposites or nano-porous materials (Brady, 1984). Kottegoda et al. reported that nanoparticles of urea-modified hydroxyapatite (HA) were synthesized to slow down the release of nitrogen during crop growth (Kottegoda et al., 2011). Urea-enhanced and modified nanofertilizers release nitrogen more gradually, with effects lasting up to 60 days during plant growth. In contrast, commercial mineral fertilizers are depleted in just 30 days. According to Guo et al., sustained-release nanofertilizers are composed of soluble fertilizers that are coated with nanomaterials to reduce their impact on water-soluble components (Guo et al., 2018). Soluble-coated fertilizers release nutrients into the soil through either simple dispersion or osmosis. Subramanian and Rahale noted that nanofertilizers can release nitrate from urea fertiliser more slowly—over up to 50 days—compared to conventional nitrogen fertilizsers (Subramanian and Rahale, 2009). In another study, Tarafdar et al., examined the application of nanomembranes to coat fertilizer particles, which promoted the slow release of fertilizer additives in the soil (Tarafdar et al., 2012). The examples mentioned above demonstrate that nanofertilizers are effective for controlled release. However, the development and use of controlled-release nanofertilizers come with some drawbacks, including high production costs and potential health concerns. Despite these challenges, the ongoing advancements in controlled-release nanofertilizers and their incorporation into sustainable farming practices represent positive progress in this field.

3.3 Increased Bioavailability

Green nanofertilizers possess exceptional physical and chemical properties, including a small size, a high area-to-volume ratio, increased reactivity, high ionization energy, improved chemical stability, enhanced bioavailability, better pH resistance, and a minimal carbon footprint. They represent a potential solution to the challenges associated with traditional chemical fertilizers by offering a more effective alternative. Additionally, green nanofertilizers have an excellent capacity for nutrient delivery, minimizing nutrient loss and enabling controlled release. This enhances the bioavailability of nutrients to plants (Jha et al., 2023).

3.4 Improved Soil Health

Soil health is essential for plant growth and food production, and it is vital for restoring degraded soils through sustainable technology. In recent years, green nanofertilizers have gained significant attention for their eco-friendly properties. Studies have shown that the excessive use of nitrogen and phosphorus mineral fertilizers has a negative impact on both soil quality and groundwater. This is because crop productivity is diminished when these minerals leach deep into the soil (Yaseen et al., 2020). Green nanofertilizers are environmentally friendly and help restore soil fertility, enhance microbial activity, reduce soil pollution, and improve crop yields (Sharma et al., 2022). Green nanofertilizers release nutrients, reduce soil toxicity, have a target delivery mechanism, and maximize the nutritional efficiency of fertilizers (Cui et al., 2010). Nanofertilizers have several advantages, including a high surface area to volume ratio, specific targeting, greater solubility, increased mobility due to their small size, and low toxicity (Sasson et al., 2007).

3.5 Reduced Chemical use

Green nanofertilizers are an innovative approach to reducing the reliance on traditional chemical fertilizers. They improve the efficiency of nutrient use throughout their production and application. Firstly, green nanofertilizers are designed to release nutrients in a controlled manner, providing a sustained supply over time. This reduces the need for frequent applications of chemical fertilizers, which often experience a significant loss due to leaching, volatilization, or runoff. Secondly, nanoparticles can be specifically engineered to deliver nutrients directly to the plant roots, ensuring efficient absorption. This minimizes waste and reduces the need for excessive fertilizer applications. Moreover, green nanofertilizers break down naturally in the environment, unlike synthetic chemical fertilizers which can persist in the soil and cause long-term harm. Since they utilize natural, eco-friendly materials, they are less likely to contaminate soil and groundwater compared to conventional chemical fertilizers, which can introduce harmful synthetic compounds.

Finally, the ability of green nanofertilizers to target specific plant needs while minimizing the environmental impact makes them an effective tool for decreasing the heavy dependence on chemical fertilizers in agriculture.

3.6 Enhanced Plant Growth

Green nanofertilizers are crucial for enhancing crop physiological and biochemical processes by improving nutrient availability. For example, zinc nanofertilizer significantly boosted plant growth and dry biomass (Vafa et al., 2015). The enhancement of plant quality and quantity may improve physiological processes, including chlorophyll content and antioxidant activity (Rezaei and Abbasi, 2014). Zinc promotes the production of natural auxin and activates essential enzymes involved in biochemical processes, such as carbohydrate and protein metabolism, growth regulator metabolism, pollen formation, and maintaining the integrity of biological membranes (El-Tohamy and El-Greadly, 2007). In addition, the contents of plant growth-promoting hormones are increased when a nano Zn fertilizer is used. Dapkekar et al. observed that nano Zn fertilizer has a 10-fold lower dose requirement than traditional $ZnSO_4$ fertilizer (Dapkekar et al., 2018). Similar results were observed when using a nano-Fe fertilizer on forage corn and *Ocimum basilicum* L. Moreover, green nanofertilizers promote plant growth by increasing total dry matter. They do this by improving nitrogen assimilation, enhancing the photoreduction activities of photosystem II and the electron transport chain, and scavenging reactive oxygen species (Raliya et al., 2015). By promoting these physiological activities green nanofertilizers show the ability to increase crop yield and better quality.

3.7 Antimicrobial Activity

Many green nanofertilizers contain nanoparticles that possess inherent antimicrobial properties. These nanoparticles may be made from naturally occurring substances, such as plant extracts, chitosan (a biopolymer derived from chitin), or metal-based nanoparticles like silver (Ag), copper (Cu), or zinc (Zn) that are known for their antimicrobial effects (Bahrulolum et al., 2021). For example, silver nanoparticles (AgNPs) are widely recognized for their antimicrobial properties (Gosavi et al., 2020). Nanoparticles can disrupt bacterial cell membranes, interfere with metabolic processes, and generate Reactive Oxygen Species (ROS) that damage cellular components, ultimately leading to the death of the microorganisms. Additionally, some nanoparticles derived from plant-based sources, such as plant extracts, exhibit antimicrobial effects. These plant-based compounds contain bioactive molecules that inhibit the growth of harmful microorganisms like bacteria and fungi. This property is beneficial for controlling soilborne pathogens, including bacteria, fungi, and nematodes, which can harm plant roots or crops (Panth et al., 2020). For example, chitosan nanoparticles, which are often used in green nanofertilizers, are known to stimulate plant defense mechanisms and can directly inhibit the growth of certain soil pathogens like Fusarium, Rhizoctonia, or Pythium, which are responsible for diseases in crops (Riseh et al., 2024). Green nanofertilizers possess antimicrobial properties that not only help control harmful microorganisms in the soil but also promote a healthier soil microbiome. This is important because beneficial microorganisms are essential for nutrient cycling and overall plant health. The antimicrobial effects of nanofertilizers are primarily aimed at targeting harmful pathogens without disrupting the balance of soil microbes. By combating these harmful microorganisms, green nanofertilizers can also decrease the incidence of plant diseases. For instance, fungal infections and bacterial diseases are common challenges in crop production, and the antimicrobial properties of nanoparticles can help prevent or lessen the severity of these issues (Li et al., 2023). This leads to better crop yields and reduced reliance on chemical fungicides or bactericides, which are often harmful to the environment.

4. Pollution Impact on the Environment and its Ecosystems

Nanofertilizers are defined as materials engineered at the nanometer scale, typically in the form of nanoparticles, designed to deliver essential macro- and micronutrients to crops in a controlled and efficient manner (Mejías et al., 2021). Nanoparticles (NPs), defined as particles smaller than

100 nanometers, are increasingly utilized in various sectors, including agriculture, industry, and healthcare. While their applications offer numerous benefits, their accumulation in the environment poses significant risks to ecological balance and human health.

4.1 Nanofertilizers in Soil Ecosystems

Nanofertilizers enter soil through agricultural applications, industrial discharges, and sewage sludge, where their accumulation can significantly impact soil microbial communities and disrupt ecological balance. These impacts range from negative to positive effects, depending on factors such as nanoparticle type, concentration, soil properties, and exposure duration. For instance, certain nanoparticles, like silver nanoparticles (AgNPs), can harm soil microbes, altering microbial diversity and reducing activity (Grün et al., 2019). On the other hand, some carbon-based nanoparticles may enhance microbial growth and soil fertility (Zhang et al., 2024). The characteristics of nanoparticles—such as size, surface charge, and functionalization—along with soil texture, pH, and organic matter, influence these interactions. For example, functionalized AgNPs exhibit different levels of toxicity compared to non-functionalized ones (Grün et al., 2019). Soil properties like texture and organic content modulate these effects, as does exposure duration, with long-term exposure posing higher risks (Ren et al., 2022). These disruptions can affect nutrient cycling, soil fertility, and plant health, emphasizing the need for thorough understanding of these interactions to evaluate environmental risks and guide the sustainable use of nanoparticles in agriculture and industry.

4.2 Nanofertilizers in Aquatic Ecosystems

Nanofertilizers enter aquatic ecosystems through pathways such as agricultural runoff, industrial discharges, and wastewater effluents, where they may transform agglomeration, aggregation, and dissolution under the influence of environmental factors such as pH, temperature, salinity, and natural organic matter (Ren et al., 2022). Their presence in water bodies poses significant risks to ecological balance and aquatic organisms. For example, silver nanoparticles (AgNPs) exhibit biotoxicity even at low concentrations, affecting various species and potentially disrupting food chains (Gusev et al., 2021). Moreover, nanoparticles can interact with other pollutants, enhancing their toxicity and facilitating transport, leading to complex exposure scenarios for aquatic life (Gusev et al., 2021). As the use of nanomaterials expands across industries, understanding their behavior, fate, and ecological impact in aquatic environments is crucial. This necessitates comprehensive research to evaluate the risks of nanoparticle pollution and to develop strategies for mitigating their adverse effects on aquatic ecosystems (Mahanta and Kaushik, 2023).

4.3 Nanofertilizers in Atmospheric Ecosystems

Airborne nanoparticles, defined as particles smaller than 100 nanometers, pose significant threats to both human health and ecological systems. Due to their extremely small size, these particles are difficult to detect using traditional methods, as they have low polarizability and are often smaller than the wavelengths of visible light (Rezaei and Johnson, 2021). Once inhaled, nanoparticles can penetrate deep into the respiratory tract, reaching the alveoli, and even translocate into the bloodstream, causing oxidative stress, inflammation, and potential cardiovascular or pulmonary diseases (Rezaei and Johnson, 2021). Ecologically, nanoparticles persist in the atmosphere and deposit into ecosystems, disrupting ecological balances and affecting both flora and fauna (Portugal et al., 2024). Given these risks, improved detection technologies and comprehensive studies are crucial to mitigate the long-term impacts of nanoparticles on human and environmental health.

5. Potential Biological Toxicity

Nanomaterials and nanofertilizers are emerging technologies in agriculture, offering potential benefits for nutrient management and crop productivity. However, their application raises significant concerns regarding soil health, microbial activity, and plant-environment interactions. This review examines the dual-edged nature of these innovations, highlighting their impacts on soil ecosystems and agricultural sustainability.

Nanomaterials, such as silver (Ag), copper (Cu), and zinc (Zn), have been shown to negatively affect the enzymatic activity of beneficial soil microbial communities, essential for nutrient cycling and climate regulation. Exposure to these materials disrupts microbial structure and functionality, impairs critical processes like nitrogen cycling and organic matter decomposition, and induces oxidative stress in microbial cells, reducing enzymatic activities vital for nutrient transformation (Dixit et al., 2024). Moreover, nanomaterial accumulation can shift microbial community composition, decreasing beneficial microbes while promoting pathogenic ones, thereby disrupting soil ecological balance and reducing carbon sequestration capacity.

Nanofertilizers, designed to enhance nutrient delivery and improve nutrient use efficiency, also present risks when improperly applied. Excessive or unbalanced use can lead to phytotoxicity, affecting plant growth, photosynthesis, and nutrient uptake, while inducing oxidative stress, resulting in lipid peroxidation, protein oxidation, and DNA damage (Dixit et al., 2024). Additionally, their prolonged use may alter soil microbiomes, further disrupting nutrient cycling and soil health (Dixit et al., 2024). To address these challenges, adherence to application protocols and further research into their long-term environmental effects are essential for balancing their benefits and risks in sustainable agriculture.

6. Risk to Human Health

Due to their extremely small size and high surface area, nanofertilizer can interact with human cells and tissues in unique and potentially harmful ways. These interactions can lead to oxidative stress, apoptosis, and inflammation, posing concerns about their safety in agricultural and industrial applications.

6.1 Oxidative Stress Induced by Nanoparticles

Nanoparticles have a large surface area that increases their chemical reactivity, which often leads to the generation of reactive oxygen species (ROS). Elevated levels of ROS can induce oxidative stress, causing damage to cellular components such as proteins, lipids, and DNA. This process has been observed with various types of nanoparticles, especially those made from metals like silver and platinum (contributors, 2025c). The oxidative damage induced by nanoparticles highlights the importance of assessing their long-term effects on human health.

6.2 Apoptosis and Cellular Damage

Nanofertilizer exposure has been shown to induce apoptosis, or programmed cell death, in human cells. Studies have demonstrated that silver nanoparticles, for example, can reduce cell viability in a concentration- and time-dependent manner. This effect is particularly concerning in contexts where nanoparticles might accumulate in biological tissues, potentially leading to chronic health issues (contributors, 2025c).

6.3 Inflammatory Responses

Nanofertilizers are also associated with triggering inflammatory responses in human tissues. For instance, inhalation of airborne nanoparticles has been linked to pulmonary inflammation and fibrosis (contributors, 2025a). These inflammatory effects underline the need for evaluating the risks of nanoparticle exposure during manufacturing, application, and unintended environmental release.

6.4 Factors Influencing Nanoparticle Toxicity

The toxicological profiles of nanofertilizers are influenced by several factors, including size, shape, chemical composition, surface charge, and their ability to agglomerate. Smaller nanoparticles, due to their enhanced ability to penetrate biological membranes, may pose greater toxicity risks. Additionally, surface modifications and coatings can significantly alter nanoparticle interactions with cells and biological systems. Such variations necessitate a detailed analysis of the physicochemical properties of nanofertilizers to predict their potential health impacts accurately (contributors, 2025a).

6.5 Routes of Exposure and Organ Accumulation

Nanofertilizers can enter the human body through inhalation, ingestion, or dermal penetration. Once inside, they may accumulate in vital organs, including the liver, kidneys, brain, and heart, potentially disrupting normal physiological functions. For example, silver nanoparticles are known to predominantly accumulate in the liver, but their toxic effects extend to other organs, such as the brain, raising concerns about neurotoxicity (contributors, 2025a).

7. Controlled Release and Stability of Nanofertilizers

The effectiveness of nanofertilizers is closely linked to their ability to release nutrients in a controlled manner, which aligns with the growth requirements of plants. This controlled release mechanism enhances nutrient use efficiency and minimizes environmental impacts. However, various environmental factors can affect the stability of nanofertilizers, leading to unintended nutrient release and associated risks (Padhan et al., 2024).

7.1 Environmental Stability and Premature Nutrient Release

Nanofertilizers are engineered to deliver nutrients gradually and precisely. However, environmental factors such as soil pH, temperature changes, and interactions with soil components can disrupt this balance. These disruptions may compromise the structural integrity of nanofertilizers, resulting in premature nutrient release. Such release diminishes nutrient use efficiency, increasing the risk of environmental pollution through processes like leaching and runoff. Nutrient accumulation in non-target areas can further exacerbate environmental concerns, contributing to eutrophication in aquatic ecosystems and disrupting soil microbial communities (Padhan et al., 2024).

7.2 Advances in Controlled Release Mechanisms

To address the challenge of premature nutrient release, significant research has been devoted to developing advanced controlled-release mechanisms for nanofertilizers. By engineering nanoparticles with coatings or encapsulations, it is possible to regulate nutrient release over extended periods. Polymer-coated nanofertilizers, for instance, have demonstrated the ability to achieve sustained nutrient delivery, aligning with plant demands while reducing environmental losses. This approach reduces the frequency of fertilizer application, enhancing efficiency and mitigating contamination risks (Padhan et al., 2024).

7.3 Designing Stability-enhanced Nanofertilizers

Ensuring the stability of nanofertilizers under diverse environmental conditions is critical for their success. Advanced formulations aim to maintain structural integrity and release nutrients in synchrony with plant growth requirements. These designs not only enhance crop productivity but also contribute to sustainable agricultural practices by minimizing the environmental footprint of fertilization (Gade et al., 2023).

This dual focus on controlled release and environmental stability underscores the importance of designing nanofertilizers capable of addressing the challenges posed by variable environmental conditions while optimizing their agronomic and ecological benefits.

8. Consideration and Challenges of Cost-Effectiveness

Nanofertilizers have garnered significant attention for their potential to enhance agricultural productivity and sustainability. However, their widespread adoption is impeded by several challenges, particularly related to cost-effectiveness.

8.1 Economic Barriers

The synthesis of nanofertilizers requires advanced technology and specialized equipment, contributing to significantly higher production costs compared to conventional fertilizers. These manufacturing processes often involve sophisticated techniques, such as nanoparticle fabrication and stabilization, which demand substantial financial investment in both materials and machinery. Consequently, the elevated costs of production translate into higher market prices, limiting the affordability and accessibility of nanofertilizers. This economic constraint is particularly pronounced for small-scale farmers and those in developing regions, where financial resources are often limited (Nano Fertilizers vs Traditional Fertilizers, 2025).

8.2 Knowledge Gaps and Application Challenges

The optimal dosage and application methods for nanofertilizers remain under-researched. Limited understanding of their behavior in diverse agricultural systems can lead to improper use, resulting in nutrient imbalances, reduced crop yields, or environmental contamination. The absence of standardized guidelines exacerbates this issue, as farmers and agricultural practitioners lack clear instructions for effective and safe application. Furthermore, regulatory frameworks governing the use of nanofertilizers are still underdeveloped in many regions, creating uncertainty regarding their long-term efficacy and safety (Abd-Elsalam et al., 2024).

8.3 Environmental and Health Concerns

The long-term environmental and health impacts of nanofertilizers are not fully understood. Nanoparticles may interact with soil and water ecosystems in unpredictable ways, raising concerns about potential ecological disturbances. These interactions could affect microbial activity, soil health, and water quality, with implications for both environmental sustainability and human health. Comprehensive risk assessments and the establishment of safety guidelines are essential to address these uncertainties and build public confidence in the technology (Abd-Elsalam et al., 2024).

8.4 Infrastructure Limitations

The application of nanofertilizers often requires advanced tools and equipment for precise delivery and monitoring. However, such infrastructure is frequently unavailable in rural or resource-limited areas. In regions lacking access to modern agricultural machinery or technical expertise, the effective utilization of nanofertilizers becomes challenging. This limitation underscores the need for investments in agricultural infrastructure and training programs to support the adoption of nanotechnology in farming practices (CropLife, 2025).

9. Regulation and Supervision

The rapid advancement of nanotechnology has led to the growth of nanomaterials across various industries, including agriculture. Nanofertilizers, which use nanoscale materials to improve nutrient delivery, hold considerable promise for increasing crop yields and minimizing environmental pollution. However, the absence of consistent international standards for assessing and monitoring their environmental and health impacts presents considerable challenges. Without standardized protocols, the unregulated use of nanomaterials, including nanofertilizers, may lead to unintended environmental pollution and health risks (contributors, 2025b).

9.1 Current Standardization Efforts

The International Organization for Standardization (ISO) has initiated efforts to establish standards for manufacturing, terminology, instrumentation, testing methodologies, and safety practices related to nanomaterials. Despite these efforts, comprehensive and universally accepted standards remain underdeveloped. The lack of standardized testing and assessment methods hinders effective evaluation of the toxicological and eco-toxicological properties of nanomaterials. This gap increases the potential for environmental release and human exposure to nanomaterials, including nanofertilizers, with consequences that remain insufficiently understood (contributors, 2025b).

9.2 Environmental and Health Implications of Nanofertilizers

9.2.1 Environmental Impact

Nanoparticles exhibit complex behaviors in soil and water ecosystems, which are not yet fully elucidated. Concerns exist regarding their potential to cause environmental pollution. Specifically, the long-term effects of nanofertilizers on soil health, microbial activity, and water quality remain uncertain (Kumah et al., 2023).

9.2.2 Health Implications

The potential toxicity of nanoparticles to humans and other organisms is a major concern. The long-term effects of nanofertilizers on human health are not yet fully understood, highlighting the need for further research to evaluate their safety comprehensively (Abd-Elsalam et al., 2024).

9.2.3 Research Needs

To ensure the safe and effective use of nanofertilizers, it is imperative to address critical research gaps through extensive studies focusing on:

a) **Toxicological Assessments:** *Comprehensive evaluation of the potential toxicity of nanofertilizers to humans, animals, and other organisms.*

b) **Environmental Fate and Transport:** *Investigation of how nanoparticles move through and persist in various environmental media, including soil, water, and air.*

c) **Long-Term Exposure Studies:** *Analysis of the effects of prolonged exposure to nanofertilizers on ecosystems and human health.*

9.3 Recommendations for Regulatory Frameworks

To mitigate the risks associated with nanofertilizers, it is essential to establish and implement consistent international standards governing their production, application, and disposal. These standards should:

- Facilitate comprehensive risk assessments.
- Encourage safe handling and application practices.
- Ensure that the benefits of nanofertilizers are realized without compromising environmental and public health.

10. Conclusions

This chapter overviews nanofertilizers, discussing their types, modes of action, and potential impact on the environment and ecosystem, potential biological toxicity, potential risks to human health, control release and stability, and cost-effectiveness to reduce pollution in sustainable agriculture and their ecosystems. Nanofertilizers especially green nanofertilizers, derived from biological materials such as plants and microorganisms, are considered eco-friendly and can be absorbed through plant roots or leaves. They enhance plants' resilience to environmental stresses, reduce production costs, and diminish ecological impact. Despite potential health risks associated with the release of nanofertilizers into the environment, which may hamper the ecosystem, they provide a more effective nutrient source compared to traditional fertilizers. But we need to find out the proper composition for producing them at lower costs.

Acknowledgment

All authors acknowledge the South Australian Government's Industry Doctoral Training Centre and Peats Soil & Garden Supplies for providing the necessary funding for this work. Volker Hessel acknowledges financial support from the European Research Council (ERC) under the European Union's Horizon 2020 research and innovation programme (grant agreement No 810182 – SCOPE ERC Synergy project). Mohammed Boshir Ahmed and Volker Hessel acknowledge funding from the ARC Centre of Excellence "Plants for Space" with the grant number CE230100015.

References

Abd-Elsalam, K. A., Seth, C. S., and Alghuthaymi, M. A. (2024). Nanofertilizers: challenges and future trends. pp. 561–570. *In:* Abd-Elsalam, K. A., and Alghuthaymi, M. A. (eds.). *Nanofertilizers for Sustainable Agroecosystems: Recent Advances and Future Trends*. Springer Nature, Switzerland.

Arora, P. K., Tripathi, S., Omar, R. A., Chauhan, P., Sinhal, V. K., Singh, A. et al. (2024). Next-generation fertilizers: the impact of bionanofertilizers on sustainable agriculture. *Microb. Cell Fact.*, 23(1): 254.

Babu, S., Singh, R., Yadav, D., Rathore, S. S., Raj, R., Avasthe, R. et al. (2022). Nanofertilizers for agricultural and environmental sustainability. *Chemosphere*, 292: 133451.

Bahrulolum, H., Nooraei, S., Javanshir, N., Tarrahimofrad, H., Mirbagheri, V. S., Easton, A. J. et al. (2021). Green synthesis of metal nanoparticles using microorganisms and their application in the agrifood sector. *J. Nanobiotechnology*, 19: 1–26.

Baruah, S., and Dutta, J. (2009). Nanotechnology applications in pollution sensing and degradation in agriculture: a review. *Environ. Chem. Lett.*, 7: 191–204.

Batool, A., Nazir, M., and Zargar, S. M. (2024). Nano-pesticides and nano-fertilizers from natural (plant/animal) wastes. *Biocatal. Agric. Biotechnol.*, 60: 103265.

Boonying, P., Boonpavanitchakul, K., Amnuaypanich, S., and Kangwansupamonkon, W. (2023). Natural rubber-lignin composites modified with natural rubber-graft-polyacrylamide as an effective coating for slow-release fertilizers. Ind. Crop Prod., 191: 116018.

Bouttier-Figueroa, D. C., Cortez-Valadez, M., Flores-Acosta, M., and Robles-Zepeda, R. E. (2023). Synthesis of Metallic Nanoparticles Using Plant's Natural Extracts: Synthesis Mechanisms and Applications. *Biotecnia*, 25(3): 125–139.

Brady, N. (1984). The nature and properties of soils. MacMillan Publishing Company, New York, USA.

contributors, W. (2025a). *Nanotoxicology*. Wikimedia Foundation. Retrieved 1 January from https://en.wikipedia.org/wiki/Nanotoxicolog)y.

contributors, W. (2025b. *Pollution from nanomaterials*. Wikimedia Foundation. Retrieved 1 January from https://en.wikipedia.org/wiki/Pollution_from_nanomaterials.

contributors, W. (2025c). *Silver nanoparticle*. Wikimedia Foundation. Retrieved 1 January from https://en.wikipedia.org/wiki/Silver_nanoparticle.

CropLife. (2025). *Revolutionizing Agriculture: Nano Fertilizers Market Set to Surpass USD 12.4 Billion by 2034.* Retrieved 1 January from https://www.croplife.com/crop-inputs/fertilizer/revolutionizing-agriculture-nano-fertilizers-market-set-to-surpass-usd-12-4-billion-by-2034/.

Cui, H., Sun, C., Liu, Q., Jiang, J., and Gu, W. (2010). Applications of nanotechnology in agrochemical formulation, perspectives, challenges and strategies. international conference on Nanoagri, Sao pedro, Brazil.

Daniel, T., Sharpley, A., and Lemunyon, J. L. (1998). Agricultural phosphorus and eutrophication: A symposium overview. *J. Environ. Qual.*, 27(2): 251–257.

Dapkekar, A., Deshpande, P., Oak, M. D., Paknikar, K. M., and Rajwade, J. M. (2018). Zinc use efficiency is enhanced in wheat through nanofertilization. *Sci. Rep.*, 8(1): 6832.

Davarpanah, S., Tehranifar, A., Davarynejad, G., Abadía, J., and Khorasani, R. J. S. h. (2016). Effects of foliar applications of zinc and boron nano-fertilizers on pomegranate (*Punica granatum* cv. Ardestani) fruit yield and quality. *Sci. Hortic.*, 210: 57–64.

de Oliveira, A. D., and Beatrice, C. A. G. (2018). Polymer nanocomposites with different types of nanofiller. Sivasankaran S. *Nanocomposites-Recent Evolutions*. 103–104.

Dikshit, A., Shukla, S. K., and Mishra, R. K. (2013). *Exploring nanomaterials with PGPR in current agricultural scenario: PGPR with special reference to nanomaterials*. Lap Lambert Academic Publishing.

Dixit, V., Goyal, U., Kamal, W. J., Chaubey, K. K., Manjunath, B. T., Xavier, J. et al. (2024). Assessing the Impact of Silver and Zinc on Soil Microbial Structure and Functionality. pp. 179–199. *In:* Bachheti, R. K., Bachheti, A., and Husen, A. (eds.). *Metal and Metal-Oxide Based Nanomaterials: Synthesis, Agricultural, Biomedical and Environmental Interventions*. Springer Nature, Singapore.

El-Ghamry, A., Mosa, A. A., Alshaal, T., and El-Ramady, H. (2018). Nanofertilizers vs. biofertilizers: new insights. *Env. Biodivers. Soil Secur.*, 2(2018): 51–72.

El-Tohamy, W., and El-Greadly, N. (2007). Physiological responses, growth, yield and quality of snap beans in response to foliar application of yeast, vitamin E and zinc under sandy soil conditions. *AJBAS*. 1(3): 294–299

Escribà-Gelonch, M., Butler, G. D., Goswami, A., Tran, N. N., and Hessel, V. (2023). Definition of agronomic circular economy metrics and use for assessment for a nanofertilizer case study. *PPB*. 196: 917–924.

Gade, A., Ingle, P., Nimbalkar, U., Rai, M., Raut, R., Vedpathak, M. et al. (2023). Nanofertilizers: The Next Generation of Agrochemicals for Long-Term Impact on Sustainability in Farming Systems. *Agrochemicals*, 2(2): 257–278.

Gatahi, D. (2017). Use of bio-nanocomposites in enhancing bacterial wilt plant resistance, and water conservation in greenhouse farming tomato production. *JKUAT*.

Golbashy, M., Sabahi, H., Allahdadi, I., Nazokdast, H., and Hosseini, M. (2017). Synthesis of highly intercalated urea-clay nanocomposite via domestic montmorillonite as eco-friendly slow-release fertilizer. Arch. *Acker Pflanzenbau Bodenkd.* 63(1): 84–95.

Gosavi, V. C., Daspute, A. A., Patil, A., Gangurde, A., Wagh, S. G., Sherkhane, A. et al. (2020). Synthesis of green nanobiofertilizer using silver nanoparticles of Allium cepa extract Short title: Green nanofertilizer from Allium cepa. *Int. J. Chem. Stud.*, 8(4): 1690–1694.

Grün, A.-L., Manz, W., Kohl, Y. L., Meier, F., Straskraba, S., Jost, C. et al. (2019). Impact of silver nanoparticles (AgNP) on soil microbial community depending on functionalization, concentration, exposure time, and soil texture. *Environ. Sci. Eur.*, 31(1): 15.

Guo, H., White, J. C., Wang, Z., and Xing, B. (2018). Nano-enabled fertilizers to control the release and use efficiency of nutrients. Curr. Opin. Env. Sci. Hl., 6: 77–83.

Gusev, A. A., Zakharova, O. V., Vasyukova, I. A., Evtushenko, N. A., Vasilieva, S. G., Lukyanov, A. A. et al. (2021). Nanoparticles in the Aquatic Environment: The Risks Associated with Them and the Possibilities of Their Mitigation with Microalgae. *Mosc. Univ. Biol. Bull.*, 76(4): 165–174.

Hajihashemi, S., and Kazemi, S. (2022). The potential of foliar application of nano-chitosan-encapsulated nano-silicon donor in amelioration the adverse effect of salinity in the wheat plant. *BMC Plant Biol.*, 22(1): 148.

Hussein, H., Shaarawy, H., Hussien, N. H., and Hawash, S. (2019). Preparation of nano-fertilizer blend from banana peels. BNRC., 43: 1–9.

Jha, A., Pathania, D., Sonu, Damathia, B., Raizada, P., Rustagi, S. et al. (2023). Panorama of biogenic nano-fertilizers: A road to sustainable agriculture. *Environ. Res.*, 235: 116456.

Khan, A. A., Jilani, G., Akhtar, M. S., Naqvi, S. M. S., and Rasheed, M. (2009). Phosphorus solubilizing bacteria: occurrence, mechanisms and their role in crop production. *J. Agric. Biol. Sci.*, 1(1): 48–58.

Khan, M., Islam, M., Nahar, N., Al-Mamun, M., Khan, M., and Matin, M. (2021). Synthesis and characterization of nanozeolite based composite fertilizer for sustainable release and use efficiency of nutrients. *Heliyon.*, 7(1).

Kottegoda, N., Munaweera, I., Madusanka, N., and Karunaratne, V. (2011). A green slow-release fertilizer composition based on urea-modified hydroxyapatite nanoparticles encapsulated wood. *Curr. Sci.*, 73–78.

Kráľová, K., and Jampílek, J. (2022). Metal-and metalloid-based nanofertilizers and nanopesticides for advanced agriculture. pp. 295–361. *In:* Fernandes Fraceto, L., Pereira de Carvalho, H. W., de Lima, R., Ghoshal, S., Santaella, C. (eds.). *Inorganic Nanopesticides and Nanofertilizers.* Springer, Cham.

Kumah, E. A., Fopa, R. D., Harati, S., Boadu, P., Zohoori, F. V., and Pak, T. (2023). Human and environmental impacts of nanoparticles: a scoping review of the current literature. *BMC Public Health*, 23(1): 1059.

Le, T. N. Q., Robertson, K., Escribà-Gelonch, M., Marschner, P., Tran, N. N., Williams, P. M. et al. (2023). Microflow synthesis of a formulation of phosphorus fertiliser to enhance the P content in soil and P uptake in wheat. Green Chem. 25(22): 9422–9437.

Le, T. N. Q., Tran, N. N., Escribà-Gelonch, M., Serra, C. A., Fisk, I., McClements, D. J. et al. (2021). Microfluidic encapsulation for controlled release and its potential for nanofertilisers. *Chem. Soc. Rev.*, 50(21): 11979–12012.

Li, Y., Zhang, P., Li, M., Shakoor, N., Adeel, M., Zhou, P. et al. (2023). Application and mechanisms of metal-based nanoparticles in the control of bacterial and fungal crop diseases. *Pest Manag. Sci.*, 79(1): 21–36.

Mahanta, M., and Kaushik, K. K. 2023. Nanoparticles in Aquatic Environment: An Overview with Special Reference to Their Ecotoxicity. pp. 385–404. *In:* Malik, J. A., and Sadiq Mohamed, M. J. (eds.). *Modern Nanotechnology: Volume 2: Green Synthesis, Sustainable Energy and Impacts*. Springer Nature Switzerland.

Mathur, P., and Roy, S. (2020). Nanosilica facilitates silica uptake, growth and stress tolerance in plants. *PPB.*, 157: 114–127.

Medda, S., Hajra, A., Dey, U., Bose, P., and Mondal, N. K. (2015). Biosynthesis of silver nanoparticles from Aloe vera leaf extract and antifungal activity against *Rhizopus* sp. and *Aspergillus* sp. Appl. Sci., 5: 875–880.

Mejías, J. H., Salazar, F., Pérez Amaro, L., Hube, S., Rodriguez, M., and Alfaro, M. (2021). Nanofertilizers: A cutting-edge approach to increase nitrogen use efficiency in grasslands. *Front. Environ. Sci.*, 9: 635114.

Mishra, V. K., and Kumar, A. (2009). Impact of metal nanoparticles on the plant growth promoting rhizobacteria. *DJNB.* 4(3): 587–592.

Mohammadbagheri, Z., Rahmati, A., and Hoshyarmanesh, P. (2021). Synthesis of a novel superabsorbent with slow-release urea fertilizer using modified cellulose as a grafting agent and flexible copolymer. *Int. J. Biol. Macromol.*, 182: 1893–1905.

Morales-Díaz, A. B., Ortega-Ortíz, H., Juárez-Maldonado, A., Cadenas-Pliego, G., González-Morales, S., Benavides-Mendoza, A. et al. (2017). Application of nanoelements in plant nutrition and its impact in ecosystems. *ANSN.*, 8(1): 013001.

Nano Fertilizers vs Traditional Fertilizers. (2025). Niche Agriculture. Retrieved 1 January from https://www.nicheagriculture.com/nano-fertilizers-vs-traditional-fertilizers/.

Ogbomo, K. L. (2011). Comparison of growth, yield performance and profitability of tomato (Solanum lycopersicon) under different fertilizer types in humid forest ultisols. *IRJAS.*, 1(8): 332–338.

Ojeniyi, S. (2000). Effect of goat manure on soil nutrients and okra yield in a rain forest area of Nigeria. *ATA.*, 5(1): 7–12.

Ombódi, A., and Saigusa, M. (2000). Reducing nitrate content in lettuce using polyolefin-coated fertilizers. *IJHS.*, 6(4): 35–38.

Padhan, S. R., Kar, I., Mohanty, A., and Panigrahi, K. K. (2024). Nanofertilizers: types, synthesis, methods, and mechanisms. pp. 61–98. *In:* Abd-Elsalam, K. A., and Alghuthaymi, M. A. (eds.). *Nanofertilizers for Sustainable Agroecosystems: Recent Advances and Future Trends*. Springer Nature Switzerland.

Panchal, S. S., and Vasava, D. V. (2022). Fabricating approaches for synthesis of miktoarm star-shaped polymers having tailored biodegradability. *Int. J. Polym. Mater.*, 71(18): 1407–1424.

Panth, M., Hassler, S. C., and Baysal-Gurel, F. (2020). Methods for management of soilborne diseases in crop production. *Agriculture*, 10(1): 16.

Portugal, J., Bedia, C., Amato, F., Juárez-Facio, A. T., Stamatiou, R., Lazou, A. et al. (2024). Toxicity of airborne nanoparticles: Facts and challenges. *Environ. Int.*, 108889.

Preetha, P. S., and Balakrishnan, N. (2017). A review of nano fertilizers and their use and functions in soil. *Int. J. Curr. Microbiol. App. Sci.*, 6(12): 3117–3133.

Rahman, K. A., and Zhang, D. (2018). Effects of fertilizer broadcasting on the excessive use of inorganic fertilizers and environmental sustainability. *Sustain.*, 10(3): 759.

Raliya, R., Nair, R., Chavalmane, S., Wang, W.-N., and Biswas, P. (2015). Mechanistic evaluation of translocation and physiological impact of titanium dioxide and zinc oxide nanoparticles on the tomato (*Solanum lycopersicum* L.) plant. Metallomics., 7(12): 1584–1594.

Ren, W., Liu, H., and Teng, Y. (2022). Impact of nanoparticles on soil ecosystems. pp. 43–81. *In:* Rajput, V. D., Verma, K. K., Sharma, N., and Minkina, T. (eds.). *The Role of Nanoparticles in Plant Nutrition under Soil Pollution: Nanoscience in Nutrient Use Efficiency*. Springer International Publishing.

Rezaei, M., and Abbasi, H. (2014). Foliar application of nanochelate and non-nanochelate of zinc on plant resistance physiological processes in cotton (*Gossipium hirsutum* L.). *Iran. J. Plant Physiol.*, 4(4): 1137–1144.

Rezaei, M., and Johnson, M. S. (2021). Airborne Nanoparticles: Control and Detection. pp. 85–133. *In:* Goodsite, M. E. Johnson, M. S., and Hertel, O. (Eds.), *Air Pollution Sources, Statistics and Health Effects*. Springer, USA.

Riseh, R. S., Vatankhah, M., Hassanisaadi, M., and Varma, R. S. (2024). A review of chitosan nanoparticles: Nature's gift for transforming agriculture through smart and effective delivery mechanisms. *Int. J. Biol. Macromol.*, 129522.

Sani, I. K., Masoudpour-Behabadi, M., Sani, M. A., Motalebinejad, H., Juma, A. S., Asdagh, A. et al. (2023). Value-added utilization of fruit and vegetable processing by-products for the manufacture of biodegradable food packaging films. *Food Chem.*, 405: 134964.

Sasson, Y., Levy-Ruso, G., Toledano, O., and Ishaaya, I. (2007). Nanosuspensions: emerging novel agrochemical formulations. pp. 1–39. *In:* Ishaaya, I., Horowitz, A. R., and Nauen, R. (eds.). Insecticides design using advanced technologies. Springer Berlin, Heidelberg, Germany.

Sharma, B., Tiwari, S., Kumawat, K. C., and Cardinale, M. (2023). Nano-biofertilizers as bio-emerging strategies for sustainable agriculture development: Potentiality and their limitations. *Sci. Total Environ.*, 860: 160476.

Sharma, S., Singh, S., Bahuguna, A., Yadav, B., Barthwal, A., Nandan, R. et al. (2022). Nanotechnology: An efficient tool in plant nutrition management. 165–188. *In:* Jatav, H. S. (ed.). Ecosystem Services. Nova Science Press, New York, USA.

Sikder, A., Pearce, A. K., Parkinson, S. J., Napier, R., and O'Reilly, R. k. (2021). Recent trends in advanced polymer materials in agriculture related applications. *ACS Appl. Polym. Mater.*, 3(3): 1203–1217.

Subramanian, K., and Rahale, C. S. (2009). Synthesis of nanofertiliser formulations for balanced nutrition. *Soil Science-Platinum Jubilee Celebration.* 22–25.

Tanan, W., Panichpakdee, J., Suwanakood, P., and Saengsuwan, S. (2021). Biodegradable hydrogels of cassava starch-g-polyacrylic acid/natural rubber/polyvinyl alcohol as environmentally friendly and highly efficient coating material for slow-release urea fertilizers. *J. Ind. Eng. Chem.*, 101: 237–252.

Tarafdar, J., Xiang, Y., Wang, W., Dong, Q., and Biswas, P. (2012). Nanoparticle synthesis characterization and application to solve some chronic agricultural problems. *Appl. Biol.*, 14: 138–144.

Thirugnanasambandan, T. (2018). Advances and trends in nano-biofertilizers. SSRN. 59.

Thirunavukkarasu, M., Subramanian, K. (2014). Surface modified nano-zeolite used as carrier for slow release of sulphur. *JANS*, 6(1): 19–26.

Tran, N. N., Le, T. N. Q., Pho, H. Q., Tran, T. T., and Hessel, V. (2022). Nanofertilizers and nanopesticides for crop growth. pp. 367–394. *In:* Chen, J. T. (ed.). Plant and Nanoparticles Springer, Singapore.

Vafa, Z. N., Sirousmehr, A. R., Ghanbari, A., Khammari, I., and Falahi, N. (2015). Effects of nano zinc and humic acid on quantitative and qualitative characteristics of savory (*Satureja hortensis* L.). J. BioSci. Biotechnol. 6(3): 124–136.

Vejan, P., Abdullah, R., Khadiran, T., Ismail, S., and Nasrulhaq Boyce, A. J. M. (2016). Role of plant growth promoting rhizobacteria in agricultural sustainability—a review. *Molecules*, 21(5): 573.

Vejan, P., Khadiran, T., Abdullah, R., and Ahmad, N. (2021). Controlled release fertilizer: A review on developments, applications and potential in agriculture. *JCR*, 339: 321–334.

Yaseen, R., IS Ahmed, A., M Omer, A., KM Agha, M., and M Emam, T. J. (2020). Nano-fertilizers: Bio-fabrication, application and biosafety. 4(4): 884–900.

Zhang, H., Zheng, T., Wang, Y., Li, T., and Chi, Q. (2024). Multifaceted impacts of nanoparticles on plant nutrient absorption and soil microbial communities [Review]. *Front. Plant Sci.*, 15.

Zhang, T., Wu, X., Shaheen, S. M., Abdelrahman, H., Ali, E. F., Bolan, et al. (2022). Improving the humification and phosphorus flow during swine manure composting: a trial for enhancing the beneficial applications of hazardous biowastes. *J. Hazard. Mater.*, 425: 127906.

Index

A

abiotic stress 26, 28, 35, 62, 65, 323, 332–334, 336–339, 351, 352, 358, 360
agriculture 205, 206, 207, 211, 212, 217, 225, 226, 288, 289, 290, 294–300

B

biosensors 201
biostimulants 197, 199
Biosynthesis 211
Biotechnology 2, 10
biotic stress 323, 325, 327, 331–334, 336–339, 360, 361

C

changing climate 27, 35
contamination 97–100
crop production 370, 377
crop productivity 28, 33

D

disease management 195, 197, 198
disease resistance 164, 174

E

engineered nanomaterials 259
environment 299
environmental monitoring 145

F

fertilizer 370, 372–377, 380
food production 352
food security 288, 289, 291, 297, 298

G

green nanomaterials 212
green nanotechnology 41, 42, 44, 45, 48–51, 73, 79, 81, 87, 92–94, 236, 237, 252, 253, 288, 290, 292–300
green synthesis 112, 113, 115, 118, 119, 121

H

heat shock proteins 76, 87

I

immobilization 98

N

nano-bioremediation 12
Nanobiosensors 140–150
nanobiotechnology 271
nanocarriers 305
nanofertilizers 3–6, 10, 11, 13, 111, 112, 127, 173, 174, 180, 271, 280, 305, 308
nanoherbicides 3, 8–10
nanomaterials 236, 237, 245, 248, 250, 252, 253, 371, 373–375, 378, 379, 381, 382
nanoparticles 41–50, 57, 59–65, 98–105, 157–170, 172–179, 259–266, 306–309, 311–314, 370, 372–382
nanopesticides 7, 196, 197, 236, 238, 245, 247, 253, 305, 306, 311
nanosensors 180
nanotechnology 26, 27, 34, 35, 111, 112, 121, 122, 127, 140–142, 144, 148, 150, 195–197, 201, 205–207, 214, 225, 226, 236–239, 243–245, 247–253, 322–324, 330, 339, 352, 357, 361, 373, 381
nutrition 288, 289, 296, 298, 300

P

pathogens 157, 162–165, 167, 174, 178, 179
pest 236–253
pesticides 236–240, 242, 243–246, 250–253
PGPR 111, 112, 123, 125–127
phytoviruses 200, 201
plant bacterial disease 207, 211, 212, 214, 220, 224
plant pathogens 205, 212, 213, 224, 225
plant stress tolerance 48
pollution 97
precision agriculture 93, 259, 266

R

RNA interference 195

S

Salinity stress 41–51
soil management 61
stress tolerance 259, 260, 264–266, 337, 352, 359–361
sustainable agriculture 41, 42, 46, 48–50, 60, 79, 81, 91, 94, 111, 112, 121, 124

T

temperature stress resilience 73, 81, 92

W

wireless Nanosensors 148

About the Editor

Jen-Tsung Chen is a Professor of Cell Biology at the Department of Life Sciences, National University of Kaohsiung, Taiwan, where he teaches courses on cell biology, genomics, proteomics, plant physiology, and plant biotechnology. His research interests include bioactive compounds, chromatography techniques, plant molecular biology, plant biotechnology, bioinformatics, and systems pharmacology. In 2023 and 2024, Elsevier and Stanford University recognized Dr. Chen as one of the "World's Top 2% Scientists". In 2025, Dr. Chen received the "Springer Nature Editorial Contribution Award" for his contributions to Plant Methods.

Taylor & Francis
Taylor & Francis Group
http://taylorandfrancis.com

For Product Safety Concerns and Information please contact our
EU representative GPSR@taylorandfrancis.com Taylor & Francis
Verlag GmbH, Kaufingerstraße 24, 80331 München, Germany